Quantum Mechanics in Chemistry

TOPICS IN PHYSICAL CHEMISTRY
A Series of Advanced Textbooks and Monographs

Series Editor, Donald G. Truhlar

F. Iachello and R. D. Levine, *Algebraic Theory of Molecules*

P. Bernath, *Spectra of Atoms and Molecules*

J. Cioslowski, *Electronic Structure Calculations on Fullerenes and Their Derivatives*

E. R. Bernstein, *Chemical Reactions in Clusters*

J. Simons and J. Nichols, *Quantum Mechanics in Chemistry*

Quantum Mechanics in Chemistry

Jack Simons
University of Utah

Jeff Nichols
Pacific Northwest National Laboratory

New York Oxford
OXFORD UNIVERSITY PRESS
1997

Oxford University Press

Oxford New York
Athens Auckland Bangkok Bogota Bombay Buenos Aires
Calcutta Cape Town Dar es Salaam Delhi Florence Hong Kong
Istanbul Karachi Kuala Lumpur Madras Madrid Melbourne
Mexico City Nairobi Paris Singapore Taipei Tokyo Toronto

and associated companies in

Berlin Ibadan

Published by Oxford University Press, Inc.
198 Madison Avenue, New York, New York 10016

Library of Congress Cataloging-in-Publication Data
Simons, Jack
Quantum mechanics in chemistry / Jack Simons, Jeff Nichols.
p. cm. — (Topics in physical chemistry)
Includes index.
ISBN 0-19-508200-1
1. Quantum chemistry. I. Nichols, Jeffrey Allen, 1955–
II. Title. III. Series: Topics in physical chemistry series.
QD42.S53 1997
541.2'8—dc20 96-34013
CIP

9 8 7 6 5 4 3
Printed in the United States of America
on acid-free paper.

Dedication

Much of our knowledge about quantum mechanics and how it applies to chemistry derives from two of our mentors, Professor Henry Eyring and his Ph.D. student, Professor Joseph O. Hirschfelder. We count ourselves as extremely fortunate to have had the opportunity to learn from these outstanding scholars and fine human beings. Henry's intuitive feel for using theory to model nature in the simplest reasonable way, combined with Joe's more quantitative approach, have, we hope, been well blended in our own views of theoretical chemistry. We wish to honor these two friends by dedicating this book in their names and memories.

Contents

Preface xix

SECTION 1 THE BASIC TOOLS OF QUANTUM MECHANICS 1

CHAPTER 1 Quantum mechanics describes matter in terms of wavefunctions and energy levels. Physical measurements are described in terms of operators acting on wavefunctions. 3

I. Operators, Wavefunctions, and the Schrödinger Equation 3
- A. Operators 4
- B. Wavefunctions 5
- C. The Schrödinger Equation 5
  - 1. The Time-Dependent Equation 5
  - 2. The Time-Independent Equation 7

II. Examples of Solving the Schrödinger Equation 8
- A. Free-Particle Motion in Two Dimensions 8
  - 1. The Schrödinger Equation 8
  - 2. Boundary Conditions 9
  - 3. Energies and Wavefunctions for Bound States 10
  - 4. Quantized Action Can also be Used to Derive Energy Levels 10
- B. Other Model Problems 12
  - 1. Particles in Boxes 12
  - 2. One Electron Moving About a Nucleus 14

3. Rotational Motion for a Rigid Diatomic Molecule 21
4. Harmonic Vibrational Motion 22

III. The Physical Relevance of Wavefunctions, Operators and Eigenvalues 24
A. The Basic Rules and Relation to Experimental Measurement 24
B. An Example Illustrating Several of the Fundamental Rules 34

CHAPTER 2 Approximation methods can be used when exact solutions to the Schrödinger equation cannot be found. **37**

I. The Variational Method 37

II. Perturbation Theory 39

CHAPTER 3 The application of the Schrödinger equation to the motions of electrons and nuclei in a molecule lead to the chemists' picture of electronic energy surfaces on which vibration and rotation occurs and among which transitions take place. **43**

I. The Born-Oppenheimer Separation of Electronic and Nuclear Motions 43
A. Time Scale Separation 45
B. Vibration/Rotation States for Each Electronic Surface 45

II. Rotation and Vibration of Diatomic Molecules 45
A. Separation of Vibration and Rotation 46
B. The Rigid Rotor and Harmonic Oscillator 47
C. The Morse Oscillator 47

III. Rotation of Polyatomic Molecules 48
A. Linear Molecules 48
B. Non-Linear Molecules 48

IV. Summary 50
Section Summary 50

Section 1 Exercises, Problems, and Solutions **51**

SECTION 2 SIMPLE MOLECULAR ORBITAL THEORY **123**

CHAPTER 4 Valence atomic orbitals on neighboring atoms combine to form bonding, non-bonding and antibonding molecular orbitals. **125**

I. Atomic Orbitals 125
A. Shapes 125
B. Directions 126
C. Sizes and Energies 126

II. Molecular Orbitals 127
A. Core Orbitals 128
B. Valence Orbitals 129
C. Rydberg Orbitals 131
D. Multicenter Orbitals 132
E. Hybrid Orbitals 133

CHAPTER 5 Molecular orbitals possess specific topology, symmetry, and energy-level patterns. **135**

I. Orbital Interaction Topology 135
II. Orbital Symmetry 138
A. Non-Linear Polyatomic Molecules 140
B. Linear Molecules 144
C. Atoms 146

CHAPTER 6 Along "reaction paths," orbitals can be connected one-to-one according to their symmetries and energies. This is the origin of the Woodward-Hoffmann rules. **149**

I. Reduction in Symmetry 149
II. Orbital Correlation Diagrams 151

CHAPTER 7 The most elementary molecular orbital models contain symmetry, nodal pattern, and approximate energy information. **155**

I. The LCAO-MO Expansion and the Orbital-level Schrödinger Equation 155
II. Determining the Effective Potential V 156
A. The Hückel Parameterization of V 156
B. The Extended Hückel Method 158

Section 2 Exercises, Problems, and Solutions **160**

SECTION 3 ELECTRONIC CONFIGURATIONS, TERM SYMBOLS, AND STATES **187**

Introductory Remarks—The Orbital, Configuration, and State Pictures of Electronic Structure 187

CHAPTER 8 Electrons are placed into orbitals to form configurations, each of which can be labeled by its symmetry. The configurations may "interact" strongly if they have similar energies. **189**

I. Orbitals Do not Provide the Complete Picture; Their Occupancy by the *N* Electrons Must Be Specified 189

II. Even *N*-Electron Configurations Are not Mother Nature's True Energy States 189

III. Mean-Field Models 190

IV. Configuration Interaction (CI) Describes the Correct Electronic States 192

V. Summary 194

CHAPTER 9 Electronic wavefunctions must be constructed to have permutational antisymmetry because the *N* electrons are indistinguishable Fermions. **197**

I. Electronic Configurations 197

II. Antisymmetric Wavefunctions 198

A. General Concepts 198

B. Physical Consequences of Antisymmetry 200

CHAPTER 10 Electronic wavefunctions must also possess proper symmetry. These include angular momentum and point group symmetries. **203**

I. Angular Momentum Symmetry and Strategies for Angular Momentum Coupling 203

A. Electron Spin Angular Momentum 204

B. Vector Coupling of Angular Momenta 206

C. Non-Vector Coupling of Angular Momenta 207

D. Direct Products for Non-Linear Molecules 207

II. Atomic Term Symbols and Wavefunctions 207

A. Non-Equivalent Orbital Term Symbols 207

B. Equivalent Orbital Term Symbols 208

C. Atomic Configuration Wavefunctions 210

D. Inversion Symmetry 211

E. Review of Atomic Cases 212

III. Linear Molecule Term Symbols and Wavefunctions 213

A. Non-Equivalent Orbital Term Symbols 213

B. Equivalent-Orbital Term Symbols 213

C. Linear-Molecule Configuration Wavefunctions 214

D. Inversion Symmetry and σ_v Reflection Symmetry 215

E. Review of Linear Molecule Cases 216

IV. Non-Linear Molecule Term Symbols and Wavefunctions 217

A. Term Symbols for Non-Degenerate Point Group Symmetries 217

B. Wavefunctions for Non-Degenerate, Non-Linear Point Molecules 218

C. Extension to Degenerate Representations for Non-Linear Molecules 219

V. Summary 223

CHAPTER 11 One must be able to evaluate the matrix elements among properly symmetry adapted N-electron configuration functions for any operator, the electronic Hamiltonian in particular. The Slater-Condon rules provide this capability. **225**

I. CSFs Are Used to Express the Full N-Electron Wavefunction 225

II. The Slater-Condon Rules Give Expressions for the Operator Matrix Elements among the CSFs 226

III. Examples of Applying the Slater-Condon Rules 229

IV. Summary 234

CHAPTER 12 Along "reaction paths," configurations can be connected one-to-one according to their symmetries and energies. This is another part of the Woodward-Hoffmann rules. **235**

I. Concepts of Configuration and State Energies 235

A. Plots of CSF Energies Give Configuration Correlation Diagrams 235

B. CSFs Interact and Couple to Produce States and State Correlation Diagrams 235

C. CSFs that Differ by Two Spin-Orbitals Interact Less Strongly than CSFs that Differ by One Spin-Orbital 236

D. State Correlation Diagrams 236

II. Mixing of Covalent and Ionic Configurations 238

A. The H_2 Case in Which Homolytic Bond Cleavage is Favored 239

B. Cases in Which Heterolytic Bond Cleavage Is Favored 239

C. Analysis of Two-Electron, Two-Orbital, Single-Bond Formation 240

1. Orbitals, Configurations, and States 240
2. Orbital, CSF, and State Correlation Diagrams 241
3. Summary 249

III. Various Types of Configuration Mixing 249

A. Essential CI 249

B. Dynamical CI 250

Section 3 Exercises, Problems, and Solutions **252**

SECTION 4 MOLECULAR ROTATION AND VIBRATION 279

CHAPTER 13 Treating the full internal nuclear-motion dynamics of a polyatomic molecule is complicated. It is conventional to examine the rotational movement of a hypothetical "rigid" molecule as well as the vibrational motion of a non-rotating molecule, and to then treat the rotation-vibration couplings using perturbation theory. **281**

I. Rotational Motions of Rigid Molecules 281

A. Linear Molecules 281

1. The Rotational Kinetic Energy Operator 281
2. The Eigenfunctions and Eigenvalues 282

B. Non-Linear Molecules 283

1. The Rotational Kinetic Energy Operator 283
2. The Eigenfunctions and Eigenvalues for Special Cases 284

II. Vibrational Motion within the Harmonic Approximation 286

A. The Newton Equations of Motion for Vibration 286

1. The Kinetic and Potential Energy Matrices 286
2. The Harmonic Vibrational Energies and Normal Mode Eigenvectors 287

B. The Use of Symmetry 288

1. Symmetry Adapted Modes 288
2. Point Group Symmetry of the Harmonic Potential 289

III. Anharmonicity 291

A. The Expansion of $E(v)$ in Powers of $(v + 1/2)$ 291

B. The Birge-Sponer Extrapolation 292

Section Summary 293

Section 4 Exercises, Problems, and Solutions **294**

SECTION 5 TIME-DEPENDENT PROCESSES 309

CHAPTER 14 The interaction of a molecular species with electromagnetic fields can cause transitions to occur among the available molecular energy levels (electronic, vibrational, rotational, and nuclear spin). Collisions among molecular species likewise can cause transitions to occur. Time-dependent perturbation theory and the methods of molecular dynamics can be employed to treat such transitions. **311**

I. The Perturbation Describing Interactions with Electromagnetic Radiation 311
A. The Time-Dependent Vector $A(r,t)$ Potential 311
B. The Electric $E(r,t)$ and Magnetic $H(r,t)$ Fields 312
C. The Resulting Hamiltonian 312
II. Time-Dependent Perturbation Theory 313
A. The Time-Dependent Schrödinger Equation 313
B. Perturbative Solution 313
C. Application to Electromagnetic Perturbations 314
1. First-Order Fermi-Wentzel "Golden Rule" 314
2. Higher Order Results 317
D. The "Long-Wavelength" Approximation 318
1. Electric Dipole Transitions 318
2. Magnetic Dipole and Electric Quadrupole Transitions 319
III. The Kinetics of Photon Absorption and Emission 321
A. The Phenomenological Rate Laws 321
B. Spontaneous and Stimulated Emission 322
C. Saturated Transitions and Transparency 323
D. Equilibrium and Relations between A and B Coefficients 323
E. Summary 324

CHAPTER 15 The tools of time-dependent perturbation theory can be applied to transitions among electronic, vibrational, and rotational states of molecules. **325**

I. Rotational Transitions 325
A. Linear Molecules 327
B. Non-Linear Molecules 330
II. Vibration-Rotation Transitions 330
A. The Dipole Moment Derivatives 331
B. Selection Rules on the Vibrational Quantum Number in the Harmonic Approximation 331
C. Rotational Selection Rules for Vibrational Transitions 331
1. Symmetric Tops 333
2. Linear Molecules 333
III. Electronic-Vibration-Rotation Transitions 335
A. The Electronic Transition Dipole and Use of Point Group Symmetry 335
B. The Franck-Condon Factors 336
C. Vibronic Effects 338
D. Rotational Selection Rules for Electronic Transitions 340

IV. Time Correlation Function Expressions for Transition Rates 341
A. State-to-State Rate of Energy Absorption or Emission 341
B. Averaging over Equilibrium Boltzmann Population of Initial States 341
C. Photon Emission and Absorption 342
D. The Line Shape and Time Correlation Functions 343
E. Rotational, Translational, and Vibrational Contributions to the Correlation Function 343
F. Line Broadening Mechanisms 347
1. Doppler Broadening 349
2. Pressure Broadening 350
3. Rotational Diffusion Broadening 352
4. Lifetime or Heisenberg Homogeneous Broadening 353
5. Site Inhomogeneous Broadening 354

CHAPTER 16 Collisions among molecules can also be viewed as a problem in time-dependent quantum mechanics. The perturbation is the "interaction potential," and the time dependence arises from the movement of the nuclear positions. **357**

I. One-Dimensional Scattering 359
A. Bound States 360
B. Scattering States 362
C. Shape Resonance States 363

II. Multichannel Problems 367
A. The Coupled Channel Equations 369
B. Perturbative Treatment 370
C. Chemical Relevance 371

III. Classical Treatment of Nuclear Motion 374
A. Classical Trajectories 375
B. Initial Conditions 377
C. Analyzing Final Conditions 378

IV. Wavepackets 379

Section 5 Exercises, Problems, and Solutions **382**

SECTION 6 MORE QUANTITATIVE ASPECTS OF ELECTRONIC STRUCTURE CALCULATIONS **399**

CHAPTER 17 Electrons interact via pairwise Coulomb forces; within the "orbital picture" these interactions are modelled by less difficult **401**

to treat "averaged" potentials. The difference between the true Coulombic interactions and the averaged potential is not small, so to achieve reasonable (ca. 1 kcal/mol) chemical accuracy, high-order corrections to the orbital picture are needed.

I. Electron Correlation Requires Moving beyond a Mean-Field Model 402

II. Moving from Qualitative to Quantitative Models 403

III. Atomic Units 403

CHAPTER 18 The single Slater determinant wavefunction (properly spin and symmetry adapted) is the starting point of the most common mean-field potential. It is also the origin of the molecular orbital concept. **405**

I. Optimization of the Energy for a Multiconfiguration Wavefunction 405
- A. The Energy Expression 405
- B. Application of the Variational Method 406
- C. The Fock and Secular Equations 406
- D. One- and Two-Electron Density Matrices 407

II. The Single-Determinant Wavefunction 408

III. The Unrestricted Hartree-Fock Spin Impurity Problem 409

IV. The LCAO-MO Expansion 410

V. Atomic Orbital Basis Sets 411
- A. STOs and GTOs 411
  - 1. Slater-type orbitals 411
  - 2. Cartesian Gaussian-type orbitals 411
- B. Basis Set Libraries 412
- C. The Fundamental Core and Valence Basis 413
- D. Polarization Functions 416
- E. Diffuse Functions 416

VI. The Roothaan Matrix SCF Process 417

VII. Observations on Orbitals and Orbital Energies 418
- A. The Meaning of Orbital Energies 418
- B. Koopmans' Theorem 419
- C. Orbital Energies and the Total Energy 420
- D. The Brillouin Theorem 420

CHAPTER 19 Corrections to the mean-field model are needed to describe the instantaneous Coulombic interactions among the electrons. This is achieved by including more than one Slater determinant in the wavefunction. **423**

I. Different Methods 424

A. Integral Transformations 426

B. Configuration List Choices 427

II. Strengths and Weaknesses of Various Methods 427

A. Variational Methods such as MCSCF, SCF, and CI Produce Energies That are Upper Bounds, but These Energies Are not Size-Extensive 427

B. Non-Variational Methods Such as MPPT/MBPT and CC Do not Produce Upper Bounds, but Yield Size-Extensive Energies 429

C. Which Method Is Best? 430

III. Further Details on Implementing Multiconfigurational Methods 431

A. The MCSCF Method 431

B. The Configuration-Interaction Method 431

C. The MPPT/MBPT Method 433

D. The Coupled-Cluster Method 434

E. Density Functional or X-alpha (X_α) Methods 435

CHAPTER 20 Many physical properties of a molecule can be calculated as expectation values of a corresponding quantum mechanical operator. The evaluation of other properties can be formulated in terms of the "response" (i.e., derivative) of the electronic energy with respect to the application of an external field perturbation. **439**

I. Calculations of Properties Other Than the Energy 439

A. Formulation of Property Calculations as Responses 440

B. The MCSCF Response Case 441

1. The Dipole Moment 441
2. The Geometrical Force 442

C. Responses for Other Types of Wavefunctions 444

D. The Use of Geometrical Energy Derivatives 444

1. Gradients as Newtonian Forces 444
2. Transition State Rate Coefficients 445
3. Harmonic Vibrational Frequencies 446
4. Reaction Path Following 447

II. *Ab Initio*, Semi-Empirical and Empirical Force Field Methods 448

A. *Ab Initio* Methods 448

B. Semi-Empirical and Fully Empirical Methods 448

C. Strengths and Weaknesses 448

Section 6 Exercises, Problems, and Solutions **450**

SECTION 7 APPENDICES 475

APPENDIX A Mathematics Review 477

APPENDIX B The Hydrogen Atom Orbitals 513

APPENDIX C Quantum Mechanical Operators and Commutation 517

APPENDIX D Time-Independent Perturbation Theory 527

APPENDIX E Point Group Symmetry 535

APPENDIX F Qualitative Orbital Picture and Semi-Empirical Methods 557

APPENDIX G Angular Momentum Operator Identities 569

APPENDIX H QMIC Programs 593

APPENDIX I Useful Information and Data 599

Index 605

Preface

WHAT THIS BOOK DOES AND DOES NOT CONTAIN

This is a text dealing with the basics of quantum mechanics and electronic structure theory. It provides an introduction to molecular spectroscopy (although most classes on this subject will require additional material) and to the subject of molecular dynamics (whose classes again will require additional material). This text is intended for use by beginning graduate students and advanced upper division undergraduate students in all areas of chemistry.

It provides:

1. An introduction to the fundamentals of quantum mechanics as they apply to chemistry,
2. Material that provides brief introductions to the subjects of molecular spectroscopy and chemical dynamics,
3. An introduction to computational chemistry applied to the treatment of electronic structures of atoms, molecules, radicals, and ions,
4. A large number of exercises, problems, and detailed solutions.

It does not provide much historical perspective on the development of quantum mechanics. Subjects such as the photoelectric effect, black-body radiation, the dual nature of electrons and photons, and the Davisson and Germer experiments are not even discussed.

To provide a text that students can use to gain introductory level knowledge of quantum mechanics as applied to chemistry problems, such a non-historical approach had to be followed. This text immediately exposes the reader to the machinery of quantum mechanics.

Sections 1 and 2 (i.e., chapters 1–7), together with appendices A, B, C and E, could constitute a one-semester course for most first-year PhD programs in the United States. Section 3 (chapters 8–12) and selected material from other appendices or selections from section 6 would be appropriate for a second-quarter or second-semester course. Chapters 13–15 of sections 4 and 5 would be of use for providing a link to a one-quarter or one-semester class covering molecular spectroscopy. Chapter 16 of section 5 provides a brief introduction to chemical dynamics that could be used at the beginning of a class on this subject.

There are many quantum chemistry and quantum mechanics textbooks that cover material similar to that contained in sections 1 and 2; in fact, our treatment of this material is generally briefer and less detailed than one finds in, for example, H. Eyring, J. Walter, and G. E. Kimball, *Quantum Chemistry*, John Wiley & Sons, New York, (1947); D. A. McQuarrie, *Quantum Chemistry*, University Science Books, Mill Valley, Calif. (1983); P. W. Atkins, *Molecular Quantum Mechanics*, Oxford University Press, Oxford, England (1983); or I. N. Levine, *Quantum Chemistry*, Prentice Hall, Englewood Cliffs, N.J. (1991). Depending on the backgrounds of the students, our coverage may have to be supplemented in these first two sections.

By covering this introductory material in less detail, we are able, within the confines of a text that can be used for a one-year or a two-quarter course, to introduce the student to the more modern subjects treated in sections 3, 5, and 6. Our coverage of modern quantum chemistry methodology is not as detailed as that found in A. Szabo and N. S. Ostlund, *Modern Quantum Chemistry*, McGraw-Hill, New York (1989), which contains little or none of the introductory material of our sections 1 and 2.

By combining both introductory and modern up-to-date quantum chemistry material in a single book designed to serve as a text for one-quarter, one-semester, two-quarter, or one-year classes for first-year graduate students, we offer a unique product.

It is anticipated that a course dealing with atomic and molecular spectroscopy will follow the student's mastery of the material covered in sections 1–4. For this reason, beyond these introductory sections, this text's emphasis is placed on electronic structure applications rather than on vibrational and rotational energy levels, which are traditionally covered in considerable detail in spectroscopy courses.

In brief summary, this book includes the following material:

1. Section 1, **The Basic Tools of Quantum Mechanics,** treats the fundamental postulates of quantum mechanics and several applications to exactly soluble model problems. These problems include the conventional particle-in-a-box (in one and more dimensions), rigid-rotor, harmonic oscillator, and one-electron hydrogenic atomic orbitals. The concept of the Born-Oppenheimer separation of electronic and vibration-rotation motions is introduced here. Moreover, the vibrational and rotational energies, states, and wavefunctions of diatomic, linear polyatomic and non-linear polyatomic molecules are discussed here at an introductory level. This section also introduces the variational method and perturbation theory as tools that are used to deal with problems that can not be solved exactly.

2. Section 2, **Simple Molecular Orbital Theory,** deals with atomic and molecular orbitals in a qualitative manner, including their symmetries, shapes, sizes, and energies. It introduces bonding, non-bonding, and antibonding orbitals, delocalized, hybrid, and Rydberg orbitals, and introduces Hückel-level models for the calculation of molecular orbitals as linear combinations of atomic orbitals (a more extensive treatment of several semi-empirical methods is provided in appendix F). This section also develops the Orbital Correlation Diagram concept that plays a central role in using Woodward-Hoffmann rules to predict whether chemical reactions encounter symmetry-imposed barriers.

3. Section 3, **Electronic Configurations, Term Symbols, and States**, treats the spatial, angular momentum, and spin symmetries of the many-electron wavefunctions that are formed as antisymmetrized products of atomic or molecular orbitals. Proper coupling of angular momenta (orbital and spin) is covered here, and atomic and molecular term symbols are treated. The need to include Configuration Interaction to achieve qualitatively correct descriptions of certain species' electronic structures is treated here. The role of the resultant Configuration Correlation Diagrams in the Woodward-Hoffmann theory of chemical reactivity is also developed.

4. Section 4, **Molecular Rotation and Vibration,** provides an introduction to how vibrational and rotational energy levels and wavefunctions are expressed for diatomic, linear polyatomic, and non-linear polyatomic molecules whose electronic energies are described by a single potential energy surface. Rotations of "rigid" molecules and harmonic vibrations of uncoupled normal modes constitute the starting point of such treatments.

5. Section 5, **Time-Dependent Processes,** uses time-dependent perturbation theory, combined with the classical electric and magnetic fields that arise due to the interaction of photons with the nuclei and electrons of a molecule, to derive expressions for the rates of transitions among atomic or molecular electronic, vibrational, and rotational states induced by photon absorption or emission. Sources of line-broadening and time-correlation function treatments of absorption line shapes are briefly introduced. Finally, transitions induced by collisions rather than by electromagnetic fields are briefly treated to provide an introduction to the subject of theoretical chemical dynamics.

6. Section 6, **More Quantitive Aspects of Electronic Structure Calculations**, introduces many of the computational chemistry methods that are used to quantitatively evaluate molecular orbital and configuration mixing amplitudes. The Hartree-Fock self-consistent field (SCF), configuration interaction (CI), multiconfigurational SCF (MCSCF), many-body and Müller-Plesset perturbation theories, coupled-cluster (CC), and density functional or X_α-like methods are included. The strengths and weaknesses of each of these techniques are discussed in some detail. Having mastered this section, the reader should be familiar with how potential energy hypersurfaces, molecular properties, forces on the individual atomic centers, and responses to externally applied fields or perturbations are evaluated on high speed computers.

HOW TO USE THIS BOOK: OTHER SOURCES OF INFORMATION AND BUILDING NECESSARY BACKGROUND

Other sources of information may be needed to build background in the areas of mathematics and physics. These additional subjects are treated briefly in the associated appendices whose readings are recommended at selected places within the text in the following format:

[Suggested extra reading—appendix A: Mathematics Review].

In most classroom settings, the group of students learning quantum mechanics as it applies to chemistry have quite diverse backgrounds. In particular, the level of preparation in mathematics is likely to vary considerably from student to student, as will the exposure to symmetry and group theory. This text is organized in a manner that allows students to skip material that is already familiar while providing access to most if not all necessary background material. This is accomplished by dividing the material into sections, chapters, and appendices (which fill in the background, provide methodological tools, and provide additional details).

Appendix A, Mathematics Review, and appendix E, Point Group Symmetry, are especially important to master. Neither of these two appendices provides a first-principles treatment of their subject matter. The students are assumed to have fulfilled normal American Chemical Society mathematics requirements for a degree in chemistry, so only a review of the material especially relevant to quantum chemistry is given in the Mathematics Review appendix. Likewise, the student is assumed to have learned or to be simultaneously learning about symmetry and group theory as applied to chemistry, so this subject is treated in a review and practical-application manner in appendix E. If group theory is to be included as an integral part of the class, then this text should be supplemented (e.g., by using the text by F. A. Cotton, *Chemical Applications of Group Theory*, Interscience, New York, (1963)).

The progression of sections leads the reader from the principles of quantum mechanics and several model problems that illustrate these principles and relate to chemical phenomena, through atomic and molecular orbitals, *N*-electron configurations, states, and term symbols, vibrational and rotational energy levels, photon-induced transitions among various levels, and eventually to computational techniques for treating chemical bonding and reactivity.

At the end of each section, a set of review exercises and fully worked out answers are given. Attempting to work these exercises should allow the student to determine whether he or she needs to pursue additional background building via the appendices.

In addition to the review exercises, sets of exercises and problems and their solutions are given at the end of each section. The exercises are brief and highly focused on learning a particular skill. They allow the student to practice the mathematical steps and other material introduced in the section. The problems are more extensive and require that numerous steps be executed. They illustrate application of the material contained in the chapter to chemical phenomena, and they help teach the relevance of this material to experimental chemistry. In many cases, new material is introduced in the problems, so all readers are encouraged to become actively involved in solving all problems.

To further assist the learning process, readers may find it useful to consult other textbooks or literature references. Several particular texts are recommended for additional reading, further details, or simply an alternative point of view. They include the following (in each case, the abbreviated name used in this text is given following the proper reference):

P. W. Atkins, *Molecular Quantum Mechanics*, Oxford University Press, Oxford, England (1983). [Atkins]

E. U. Condon and G. H. Shortley, *The Theory of Atomic Spectra*, Cambridge University Press, Cambridge, England (1963). [Condon and Shortley]

F. A. Cotton, *Chemical Applications of Group Theory*, Interscience, New York, (1963). [Cotton]

P. A. M. Dirac, *The Principles of Quantum Mechanics*, Oxford University Press, Oxford, England (1947). [Dirac]

H. Eyring, J. Walter, and G. E. Kimball, *Quantum Chemistry*, John Wiley & Sons, New York, (1947). [EWK]

E. C. Kemble, *The Fundamental Principles of Quantum Mechanics*, McGraw-Hill, New York, (1937). [Kemble]

I. N. Levine, *Quantum Chemistry*, Prentice Hall, Englewood Cliffs, N.J. (1991). [Levine]

D. A. McQuarrie, *Quantum Chemistry*, University Science Books, Mill Valley, Calif. (1983). [McQuarrie]

L. Pauling and E. B. Wilson, *Introduction to Quantum Mechanics*, Dover, New York, (1963). [Pauling and Wilson]

J. Simons, *Energetic Principles of Chemical Reactions*, Jones and Bartlett, Portola Valley, Calif. (1983).

A. Szabo and N. S. Ostlund, *Modern Quantum Chemistry*, McGraw-Hill, New York (1989). [Szabo and Ostlund]

E. B. Wilson, J. C. Decius, and P. C. Cross, *Molecular Vibrations*, Dover, New York, (1955). [WDC]

R. N. Zare, *Angular Momentum*, John Wiley & Sons, New York, (1988). [Zare]

QMIC COMPUTER PROGRAMS

Included with this text are a set of Quantum Mechanics in Chemistry (QMIC) computer programs. They can be obtained via the World Wide Web, according to instructions given in appendix H. Appendix H also gives more information on what they contain and how to use them.

ACKNOWLEDGMENTS

Many people have contributed to the development and "testing" of this book over several years. As a result of teaching graduate quantum mechanics, spectroscopy, and dynamics classes at the University of Utah, Professors Michael Morse and Randall B. Shirts were able to provide material that we used to generate many of the problems and solutions as well as the appendixes on mathematics and group theory. Dr. Martin Feyereisen, of Cray Research, Inc., contributed significantly to the computer programs associated with this text. Professor Kenneth D. Jordan gave us much feedback after using an early draft of the text in a graduate-level class taught at the University of Pittsburgh. Ms. Michele Pasker provided much assistance in bringing the text to publication quality. Finally, hundreds of graduate students at the University of Utah were used as proverbial guinea pigs for the numerous drafts of this book. To all of these people, we are eternally grateful.

Quantum Mechanics in Chemistry

SECTION

1

THE BASIC TOOLS OF QUANTUM MECHANICS

CHAPTER

1

Quantum mechanics describes matter in terms of wavefunctions and energy levels. Physical measurements are described in terms of operators acting on wavefunctions.

I. OPERATORS, WAVEFUNCTIONS, AND THE SCHRÖDINGER EQUATION

The trends in chemical and physical properties of the elements described beautifully in the periodic table and the ability of early spectroscopists to fit atomic line spectra by simple mathematical formulas and to interpret atomic electronic states in terms of empirical quantum numbers provide compelling evidence that *some* relatively simple framework must exist for understanding the electronic structures of all atoms. The great predictive power of the concept of atomic valence further suggests that molecular electronic structure should be understandable in terms of those of the constituent atoms.

Much of quantum chemistry attempts to make more quantitative these aspects of chemists' view of the periodic table and of atomic valence and structure. By starting from "first principles" and treating atomic and molecular states as solutions of a so-called Schrödinger equation, quantum chemistry seeks to determine *what underlies* the empirical quantum numbers, orbitals, the *aufbau* principle and the concept of valence used by spectroscopists and chemists, in some cases, even prior to the advent of quantum mechanics.

Quantum mechanics is cast in a language that is not familiar to most students of chemistry who are examining the subject for the first time. Its mathematical content and how it relates to experimental measurements both require a great deal of effort to master. With these thoughts in mind, the authors have organized this introductory section in a manner that *first* provides the student with a brief introduction to the two primary constructs of quantum mechanics, operators and wavefunctions that obey a Schrödinger equation, *then* demonstrates the application of these constructs to several chemically relevant model problems, and *finally* returns to examine in more detail the conceptual structure of quantum mechanics.

By learning the solutions of the Schrödinger equation for a few model systems, the student can better appreciate the treatment of the fundamental postulates of quantum mechanics as well as their relation to experimental measurement because the wavefunctions of the known model problems can be used to illustrate.

A. Operators

Each physically measurable quantity has a corresponding operator. The eigenvalues of the operator tell the values of the corresponding physical property that can be observed.

In quantum mechanics, any experimentally measurable physical quantity F (e.g., energy, dipole moment, orbital angular momentum, spin angular momentum, linear momentum, kinetic energy) whose classical mechanical expression can be written in terms of the *cartesian* positions $\{q_i\}$ and momenta $\{p_i\}$ of the particles that comprise the system of interest is assigned a corresponding quantum mechanical operator $\mathbf{F}$. Given F in terms of the $\{q_i\}$ and $\{p_i\}$, $\mathbf{F}$ is formed by replacing p_j by $-i\hbar\partial/\partial q_j$ and leaving q_j untouched.

For example, if

$$F = \sum_{l=1,N} (p_l^2/2m_l + 1/2k(q_l - q_l^0)^2 + L(q_l - q_l^0)),$$

then

$$\mathbf{F} = \sum_{l=1,N} (-\hbar^2/2m_l\partial^2/\partial q_l^2 + 1/2k(q_l - q_l^0)^2 + L(q_l - q_l^0))$$

is the corresponding quantum mechanical operator. Such an operator would occur when, for example, one describes the sum of the kinetic energies of a collection of particles (the $\Sigma_{l=1,N}(p_l^2/2m_l)$ term, plus the sum of "Hookes's Law" parabolic potentials (the $1/2\,\Sigma_{l=1,N}$ $k(q_l - q_l^0)^2$), and (the last term in F) the interactions of the particles with an externally applied field whose potential energy varies linearly as the particles move away from their equilibrium positions $\{q_l^0\}$.

The sum of the z-components of angular momenta of a collection of N particles has

$$F = \sum_{j=1,N} (x_j p_{yj} - y_j p_{xj}),$$

and the corresponding operator is

$$\mathbf{F} = -i\hbar \sum_{j=1,N} (x_j\partial/\partial y_j - y_j\partial/\partial x_j).$$

The x-component of the dipole moment for a collection of N particles has

$$F = \sum_{j=1,N} Z_j e x_j, \text{ and}$$

$$\mathbf{F} = \sum_{j=1,N} Z_j e x_j,$$

where Z_je is the charge on the jth particle.

The mapping from F to $\boldsymbol{F}$ is straightforward only in terms of cartesian coordinates. To map a classical function F, given in terms of curvilinear coordinates (even if they are orthogonal), into its quantum operator is not at all straightforward. Interested readers are referred to Kemble's text on quantum mechanics which deals with this matter in detail. The mapping can always be done in terms of cartesian coordinates after which a transformation of the resulting coordinates and differential operators to a curvilinear system can be performed. The corresponding transformation of the kinetic energy operator to spherical coordinates is treated in detail in appendix A. The text by EWK also covers this topic in considerable detail.

The relationship of these quantum mechanical operators to experimental measurement will be made clear later in this chapter. For now, suffice it to say that these operators define equations whose solutions determine the values of the corresponding physical property that can be observed when a measurement is carried out; *only* the values so determined can be observed. This should suggest the origins of quantum mechanics' prediction that some measurements will produce **discrete** or **quantized** values of certain variables (e.g., energy, angular momentum, etc.).

B. Wavefunctions

The eigenfunctions of a quantum mechanical operator depend on the coordinates upon which the operator acts; these functions are called wavefunctions.

In addition to operators corresponding to each physically measurable quantity, quantum mechanics describes the state of the system in terms of a wavefunction Ψ that is a function of the coordinates $\{q_j\}$ and of time t. The function $|\Psi(q_j,t)|^2 = \Psi^*\Psi$ gives the probability density for observing the coordinates at the values q_j at time t. For a many-particle system such as the H_2O molecule, the wavefunction depends on many coordinates. For the H_2O example, it depends on the x, y, and z (or r, θ, and ϕ) coordinates of the ten electrons and the x, y, and z (or r, θ, and ϕ) coordinates of the oxygen nucleus and of the two protons; a total of thirty-nine coordinates appear in Ψ.

In classical mechanics, the coordinates q_j and their corresponding momenta p_j are functions of time. The state of the system is then described by specifying $q_j(t)$ and $p_j(t)$. In quantum mechanics, the concept that q_j is known as a function of time is replaced by the concept of the probability density for finding q_j at a particular value at a particular time t: $|\Psi(q_j,t)|^2$. Knowledge of the corresponding momenta as functions of time is also relinquished in quantum mechanics; again, only knowledge of the probability density for finding p_j with any particular value at a particular time t remains.

C. The Schrödinger Equation

This equation is an eigenvalue equation for the energy or Hamiltonian operator; its eigenvalues provide the energy levels of the system.

1. The Time-Dependent Equation

If the Hamiltonian operator contains the time variable explicitly, one must solve the time-dependent Schrödinger equation.

How to extract from $\Psi(q_j,t)$ knowledge about momenta is treated below in section III.A, where the structure of quantum mechanics, the use of operators and wavefunctions to make predictions and interpretations about experimental measurements, and the origin of "uncertainty relations" such as the well known Heisenberg uncertainty condition dealing with measurements of coordinates and momenta are also treated.

Before moving deeper into understanding what quantum mechanics "means," it is useful to learn how the wavefunctions Ψ are found by applying the basic equation of quantum mechanics, the *Schrödinger equation*, to a few exactly soluble model problems. Knowing the solutions to these "easy" yet chemically very relevant models will then facilitate learning more of the details about the structure of quantum mechanics because these model cases can be used as "concrete examples."

The Schrödinger equation is a differential equation depending on time and on all of the spatial coordinates necessary to describe the system at hand (thirty-nine for the H_2O example cited above). It is usually written

$$\boldsymbol{H}\Psi = i\hbar\partial\Psi/\partial t,$$

where $\Psi(q_j,t)$ is the unknown wavefunction and $\boldsymbol{H}$ is the operator corresponding to the total energy physical property of the system. This operator is called the Hamiltonian and is formed, as stated above, by first writing down the classical mechanical expression for the total energy (kinetic plus potential) in cartesian coordinates and momenta and then replacing all classical momenta p_j by their quantum mechanical operators $\boldsymbol{p_j} = -i\hbar\partial/\partial q_j$.

For the H_2O example used above, the classical mechanical energy of all 13 particles is

$$E = \sum_i \left\{ p_i^2/2m_e + 1/2\sum_j e^2/r_{i,j} - \sum_a Z_a e^2/r_{i,a} \right\} + \sum_a \left\{ p_a^2/2m_a + 1/2\sum_b Z_a Z_b e^2/r_{a,b} \right\},$$

where the indices i and j are used to label the ten electrons whose 30 cartesian coordinates are $\{q_i\}$ and a and b label the three nuclei whose charges are denoted $\{Z_a\}$, and whose nine cartesian coordinates are $\{q_a\}$. The electron and nuclear masses are denoted m_e and $\{m_a\}$, respectively.

The corresponding Hamiltonian operator is

$$\boldsymbol{H} = \sum_i \left\{ -(\hbar^2/2m_e)\partial^2/\partial q_i^2 + 1/2\sum_j e^2/r_{i,j} - \sum_a Z_a e^2/r_{i,a} \right\}$$

$$+ \sum_a \left\{ -(\hbar^2/2m_a)\partial^2/\partial q_a^2 + 1/2\sum_b Z_a Z_b e^2/r_{a,b} \right\}.$$

Notice that $\boldsymbol{H}$ is a second order differential operator in the space of the 39 cartesian coordinates that describe the positions of the ten electrons and three nuclei. It is a second order operator because the momenta appear in the kinetic energy as p_j^2 and p_a^2, and the quantum mechanical operator for each momentum $\boldsymbol{p} = -i\hbar\partial/\partial q$ is of first order.

The Schrödinger equation for the H_2O example at hand then reads

$$\sum_i \left\{ -(\hbar^2/2m_e)\partial^2/\partial q_i^2 + 1/2\sum_j e^2/r_{i,j} - \sum_a Z_a e^2/r_{i,a} \right\}\Psi$$

$$+ \sum_a \left\{ -(\hbar^2/2m_a)\partial^2/\partial q_a^2 + 1/2\sum_b Z_a Z_b e^2/r_{a,b} \right\}\Psi$$

$$= i\hbar\partial\Psi/\partial t.$$

2. The Time-Independent Equation

If the Hamiltonian operator does not contain the time variable explicitly, one can solve the time-independent Schrödinger equation.

In cases where the classical energy, and hence the quantum Hamiltonian, do *not* contain terms that are explicitly time dependent (e.g., interactions with time varying external electric or magnetic fields would add to the above classical energy expression time dependent terms discussed later in this text), the separations of variables techniques can be used to reduce the Schrödinger equation to a time-independent equation.

In such cases, $\boldsymbol{H}$ is not explicitly time dependent, so one can assume that $\Psi(q_j,t)$ is of the form

$$\Psi(q_j,t) = \Psi(q_j)F(t).$$

Substituting this "ansatz" into the time-dependent Schrödinger equation gives

$$\Psi(q_j)i\hbar\partial F/\partial t = \boldsymbol{H}\Psi(q_j)F(t).$$

Dividing by $\Psi(q_j)F(t)$ then gives

$$F^{-1}(i\hbar\partial F/\partial t) = \Psi^{-1}(\boldsymbol{H}\Psi(q_j)).$$

Since $F(t)$ is only a function of time t, and $\Psi(q_j)$ is only a function of the spatial coordinates $\{q_j\}$, and because the left-hand and right-hand sides must be equal for all values of t and of $\{q_j\}$, both the left- and right-hand sides must equal a constant. If this constant is called E, the *two* equations that are embodied in this separated Schrödinger equation read as follows:

$$\boldsymbol{H}\Psi(q_j) = E\Psi(q_j),$$
$$i\hbar\partial F(t)/\partial t = i\hbar dF(t)/dt = EF(t).$$

The first of these equations is called the time-independent Schrödinger equation; it is a so-called eigenvalue equation in which one is asked to find functions that yield a constant multiple of themselves when acted on by the Hamiltonian operator. Such functions are called eigenfunctions of $\boldsymbol{H}$ and the corresponding constants are called eigenvalues of $\boldsymbol{H}$. For example, if $\boldsymbol{H}$ were of the form $-\hbar^2/2M\partial^2/\partial\phi^2 = \boldsymbol{H}$, then functions of the form $\exp(im\phi)$ would be eigenfunctions because

$$\{-\hbar^2/2M\partial^2/\partial\phi^2\}\exp(im\phi) = \{m^2\hbar^2/2M\}\exp(im\phi).$$

In this case, $\{m^2\hbar^2/2M\}$ is the eigenvalue.

When the Schrödinger equation can be separated to generate a time-independent equation describing the spatial coordinate dependence of the wavefunction, the eigenvalue E must be returned to the equation determining $F(t)$ to find the time dependent part of the wavefunction. By solving

$$i\hbar dF(t)/dt = EF(t)$$

once E is known, one obtains

$$F(t) = \exp(-iEt/\hbar),$$

and the full wavefunction can be written as

$$\Psi(q_j,t) = \Psi(q_j)\exp(-iEt/\hbar).$$

For the above example, the time dependence is expressed by

$$F(t) = \exp(-it\{m^2\hbar^2/2M\}/\hbar).$$

Having been introduced to the concepts of operators, wavefunctions, the Hamiltonian and its Schrödinger equation, it is important to now consider several examples of the applications of these concepts. The examples treated below were chosen to provide the learner with valuable experience in solving the Schrödinger equation; they were also chosen because the models they embody form the most elementary chemical models of electronic motions in conjugated molecules and in atoms, rotations of linear molecules, and vibrations of chemical bonds.

II. EXAMPLES OF SOLVING THE SCHRÖDINGER EQUATION

A. Free-Particle Motion in Two Dimensions

The number of dimensions depends on the number of particles and the number of spatial (and other) dimensions needed to characterize the position and motion of each particle.

1. The Schrödinger Equation

Consider an electron of mass m and charge e moving on a two-dimensional surface that defines the x,y plane (perhaps the electron is constrained to the surface of a solid by a potential that binds it tightly to a narrow region in the z-direction), and assume that the electron experiences a constant potential V_0 at all points in this plane (on any real atomic or molecular surface, the electron would experience a potential that varies with position in a manner that reflects the periodic structure of the surface). The pertinent time independent Schrödinger equation is:

$$-\hbar^2/2m(\partial^2/\partial x^2 + \partial^2/\partial y^2)\Psi(x,y) + V_0\Psi(x,y) = E\Psi(x,y).$$

Because there are no terms in this equation that *couple* motion in the x and y directions (e.g., no terms of the form $x^a y^b$ or $\partial/\partial x\ \partial/\partial y$ or $x\partial/\partial y$), separation of variables can be used to write Ψ as a product $\Psi(x,y) = A(x)B(y)$. Substitution of this form into the Schrödinger equation, followed by collecting together all x-dependent and all y-dependent terms, gives

$$-\hbar^2/2mA^{-1}\partial^2 A/\partial x^2 - \hbar^2/2mB^{-1}\partial^2 B/\partial y^2 = E - V_0.$$

Since the first term contains no y-dependence and the second contains no x-dependence, both must actually be constant (these two constants are denoted E_x and E_y, respectively), which allows two separate Schrödinger equations to be written:

$$-\hbar^2/2mA^{-1}\partial^2 A/\partial x^2 = E_x, \text{ and } -\hbar^2/2mB^{-1}\partial^2 B/\partial y^2 = E_y.$$

The total energy E can then be expressed in terms of these separate energies E_x and E_y as $E_x + E_y = E - V_0$. Solutions to the x- and y-Schrödinger equations are easily seen to be:

$$A(x) = \exp(ix(2mE_x/\hbar^2)^{1/2}) \text{ and } \exp(-ix(2mE_x/\hbar^2)^{1/2}),$$
$$B(y) = \exp(iy(2mE_y/\hbar^2)^{1/2}) \text{ and } \exp(-iy(2mE_y/\hbar^2)^{1/2}).$$

Two independent solutions are obtained for each equation because the x- and y-space Schrödinger equations are both second-order differential equations.

2. Boundary Conditions

The boundary conditions, not the Schrödinger equation, determine whether the eigenvalues will be discrete or continuous.

If the electron is entirely unconstrained within the x,y plane, the energies E_x and E_y can assume any value; this means that the experimenter can "inject" the electron onto the x,y plane with any total energy E and any components E_x and E_y along the two axes as long as $E_x + E_y = E$. In such a situation, one speaks of the energies along both coordinates as being "in the continuum" or "not quantized."

In contrast, if the electron is constrained to remain within a fixed area in the x,y plane (e.g., a rectangular or circular region), then the situation is qualitatively different. Constraining the electron to any such specified area gives rise to so-called boundary conditions that impose additional requirements on the above A and B functions. These constraints can arise, for example, if the potential $V_0(x,y)$ becomes very large for x,y values outside the region, in which case, the probability of finding the electron outside the region is very small. Such a case might represent, for example, a situation in which the molecular structure of the solid surface changes outside the enclosed region in a way that is highly repulsive to the electron.

For example, if motion is constrained to take place within a rectangular region defined by $0 \le x \le L_x$; $0 \le y \le L_y$, then the continuity property that all wavefunctions must obey (because of their interpretation as probability densities, which must be continuous) causes $A(x)$ to vanish at 0 and at L_x. Likewise, $B(y)$ must vanish at 0 and at L_y. To implement these constraints for $A(x)$, one must linearly combine the above two solutions $\exp(ix(2mE_x/\hbar^2)^{1/2})$ and $\exp(-ix(2mE_x/\hbar^2)^{1/2})$ to achieve a function that vanishes at $x = 0$:

$$A(x) = \exp(ix(2mE_x/\hbar^2)^{1/2}) - \exp(-ix(2mE_x/h^2)^{1/2}).$$

One is allowed to linearly combine solutions of the Schrödinger equation that have the same energy (i.e., are degenerate) because Schrödinger equations are linear differential equations. An analogous process must be applied to $B(y)$ to achieve a function that vanishes at $y = 0$:

$$B(y) = \exp(iy(2mE_y/\hbar^2)^{1/2}) - \exp(-iy(2mE_y/\hbar^2)^{1/2}).$$

Further requiring $A(x)$ and $B(y)$ to vanish, respectively, at $x = L_x$ and $y = L_y$, gives equations that can be obeyed only if E_x and E_y assume particular values:

$$\exp(iL_x(2mE_x/\hbar^2)^{1/2}) - \exp(-iL_x(2mE_x/\hbar^2)^{1/2}) = 0, \text{ and}$$
$$\exp(iL_y(2mE_y/\hbar^2)^{1/2}) - \exp(-iL_y(2mE_y/h^2)^{1/2}) = 0.$$

These equations are equivalent to

$$\sin(L_x(2mE_x/\hbar^2)^{1/2}) = \sin(L_y(2mE_y/\hbar^2)^{1/2}) = 0.$$

Knowing that $\sin(\theta)$ vanishes at $\theta = n\pi$, for $n = 1,2,3,\ldots$, (although the $\sin(n\pi)$ function vanishes for $n = 0$, this function vanishes for all x or y, and is therefore unacceptable because it represents zero probability density at all points in space) one concludes that the energies E_x and E_y can assume only values that obey:

$$L_x(2mE_x/\hbar^2)^{1/2} = n_x\pi,\ L_y(2mE_y/\hbar^2)^{1/2} = n_y\pi, \text{ or}$$
$$E_x = n_x^2\pi^2\hbar^2/(2mL_x^2), \text{ and } E_y = n_y^2\pi^2\hbar^2/(2mL_y^2), \text{ with } n_x \text{ and } n_y = 1,2,3,\ldots$$

It is important to stress that it is the imposition of boundary conditions, expressing the fact that the electron is spatially constrained, that gives rise to quantized energies. In the absence of spatial confinement, or with confinement only at $x = 0$ or L_x or only at $y = 0$ or L_y, quantized energies would *not* be realized.

In this example, confinement of the electron to a finite interval along both the x and y coordinates yields energies that are quantized along both axes. If the electron were confined along one coordinate (e.g., between $0 \leq x \leq L_x$) but not along the other (i.e., $B(y)$ is either restricted to vanish at $y = 0$ or at $y = L_y$ or at neither point), then the total energy E lies in the continuum; its E_x component is quantized but E_y is not. Such cases arise, for example, when a linear triatomic molecule has more than enough energy in one of its bonds to rupture it but not much energy in the other bond; the first bond's energy lies in the continuum, but the second bond's energy is quantized.

Perhaps more interesting is the case in which the bond with the higher dissociation energy is excited to a level that is not enough to break it but that is in excess of the dissociation energy of the weaker bond. In this case, one has two degenerate states: (i) the strong bond having high internal energy and the weak bond having low energy (ψ_1), and (ii) the strong bond having little energy and the weak bond having more than enough energy to rupture it (ψ_2). Although an experiment may prepare the molecule in a state that contains only the former component (i.e., $\Psi = C_1\psi_1 + C_2\psi_2$ with $C_1 >> C_2$), coupling between the two degenerate functions (induced by terms in the Hamiltonian $\boldsymbol{H}$ that have been ignored in defining ψ_1 and ψ_2) usually causes the true wavefunction $\Psi = \exp(-it\boldsymbol{H}/\hbar)\Psi$ to acquire a component of the second function as time evolves. In such a case, one speaks of internal vibrational energy flow giving rise to unimolecular decomposition of the molecule.

3. Energies and Wavefunctions for Bound States

For discrete energy levels, the energies are specified functions the depend on quantum numbers, one for each degree of freedom that is quantized.

Returning to the situation in which motion is constrained along both axes, the resultant total energies and wavefunctions (obtained by inserting the quantum energy levels into the expressions for $A(x)$ $B(y)$ are as follows:

$$E_x = n_x^2\pi^2\hbar^2/(2mL_x^2), \text{ and } E_y = n_y^2\pi^2\hbar^2/(2mL_y^2),$$
$$E = E_x + E_y,$$
$$\Psi(x,y) = (1/2L_x)^{1/2}(1/2L_y)^{1/2}[\exp(in_x\pi x/L_x) - \exp(-in_x\pi x/L_x)]$$
$$[\exp(in_y\pi y/L_y) - \exp(-in_y\pi y/L_y)], \text{ with } n_x \text{ and } n_y = 1,2,3,\ldots.$$

The two $(1/2L)^{1/2}$ factors are included to guarantee that ψ is normalized:

$$\int|\Psi(x,y)|^2 dxdy = 1.$$

Normalization allows $|\Psi(x,y)|^2$ to be properly identified as a probability density for finding the electron at a point x, y.

4. Quantized Action Can also be Used to Derive Energy Levels

There is another approach that can be used to find energy levels and is especially straightforward to use for systems whose Schrödinger equations are separable. The so-called classical **action** (denoted S) of a particle moving with momentum $\boldsymbol{p}$ along a path leading from initial coordinate $\boldsymbol{q}_i$ at initial time t_i to a final coordinate $\boldsymbol{q}_f$ at time t_f is defined by:

$$S = \int_{q_i\,;\,t_i}^{q_f\,;\,t_f} \boldsymbol{p} \bullet \boldsymbol{dq}.$$

Here, the momentum vector $\boldsymbol{p}$ contains the momenta along all coordinates of the system, and the coordinate vector $\boldsymbol{q}$ likewise contains the coordinates along all such degrees of freedom. For example, in the two-dimensional particle in a box problem considered above, $\boldsymbol{q} = (x,y)$ has two components as does $\boldsymbol{p} = (p_x,p_y)$, and the action integral is:

$$S = \int_{x_i\,;\,y_i\,;\,t_i}^{x_f\,;\,y_f\,;\,t_f} (p_x\, dx + p_y\, dy)\ .$$

In computing such actions, it is essential to keep in mind the sign of the momentum as the particle moves from its initial to its final positions. An example will help clarify these matters.

For systems such as the above particle in a box example for which the Hamiltonian is separable, the action integral decomposed into a sum of such integrals, one for each degree of freedom. In this two-dimensional example, the additivity of H:

$$\begin{aligned} H &= H_x + H_y = p_x^2/2m + p_y^2/2m + V(x) + V(y) \\ &= -\hbar^2/2m\partial^2/\partial x^2 + V(x) - \hbar^2/2m\partial^2/\partial y^2 + V(y) \end{aligned}$$

means that p_x and p_y can be independently solved for in terms of the potentials $V(x)$ and $V(y)$ as well as the energies E_x and E_y associated with each separate degree of freedom:

$$\begin{aligned} p_x &= \pm\sqrt{2m(E_x - V(x))} \\ p_y &= \pm\sqrt{2m(E_y - V(y))}; \end{aligned}$$

the signs on p_x and p_y must be chosen to properly reflect the motion that the particle is actually undergoing. Substituting these expressions into the action integral yields:

$$\begin{aligned} S &= S_x + S_y \\ &= \int_{x_i\,;\,t_i}^{x_f\,;\,t_f} \pm\sqrt{2m(E_x - V(x))}dx \ + \int_{y_i\,;\,t_i}^{y_f\,;\,t_f} \pm\sqrt{2m(E_y - V(y))}dy. \end{aligned}$$

The relationship between these classical action integrals and existence of quantized energy levels has been show to involve equating the classical action for motion on a *closed path* (i.e., a path that starts and ends at the same place after undergoing motion away from the starting point but eventually returning to the starting coordinate at a later time) to an integral multiple of Planck's constant:

$$S_{closed} = \int_{q_i\,;\,t_i}^{q_f = q_i\,;\,t_f} \boldsymbol{p} \bullet \boldsymbol{dq} = nh. \qquad (n = 1,2,3,4, \ldots)$$

Applied to each of the independent coordinates of the two-dimensional particle in a box problem, this expression reads:

$$n_x h = \int_{x=0}^{x=L_x} \sqrt{2m(E_x - V(x))}\,dx + \int_{x=L_x}^{x=0} -\sqrt{2m(E_x - V(x))}\,dx$$

$$n_y h = \int_{y=0}^{y=L_y} \sqrt{2m(E_y - V(y))}\,dy + \int_{y=L_y}^{y=0} -\sqrt{2m(E_y - V(y))}\,dy.$$

Notice that the sign of the momenta are positive in each of the first integrals appearing above (because the particle is moving from $x = 0$ to $x = L_x$, and analogously for y-motion, and thus has positive momentum) and negative in each of the second integrals (because the motion is from $x = L_x$ to $x = 0$ (and analogously for y-motion) and thus with negative momentum). Within the region bounded by $0 \le x \le L_x$; $0 \le y \le L_y$, the potential vanishes, so $V(x) = V(y) = 0$. Using this fact, and reversing the upper and lower limits, and thus the sign, in the second integrals above, one obtains:

$$n_x h = 2\int_{x=0}^{x=L_x} \sqrt{2mE_x}\,dx = 2\sqrt{2mE_x}\,L_x$$

$$n_y h = 2\int_{y=0}^{y=L_y} \sqrt{2mE_y}\,dy = 2\sqrt{2mE_y}\,L_y.$$

Solving for E_x and E_y, one finds:

$$E_x = \frac{(n_x h)^2}{8mL_x^2}$$

$$E_y = \frac{(n_y h)^2}{8mL_y^2}.$$

These are the same quantized energy levels that arose when the wavefunction boundary conditions were matched at $x = 0$, $x = L_x$ and $y = 0$, $y = L_y$. In this case, one says that the Bohr-Sommerfeld quantization condition:

$$nh = \int_{q_i\,;\,t_i}^{q_f = q_i\,;\,t_f} \mathbf{p} \bullet \mathbf{dq}$$

has been used to obtain the result.

B. Other Model Problems

1. Particles in Boxes

The particle-in-a-box problem provides an important model for several relevant chemical situations.

The above "particle in a box" model for motion in two dimensions can obviously be extended to three dimensions or to one. For two and three dimensions, it provides a crude but useful picture for electronic states on surfaces or in crystals, respectively. Free motion within a spherical volume gives rise to eigenfunctions that are used in nuclear physics to describe the motions of neutrons and protons in nuclei. In the so-called shell model of nuclei, the neutrons and protons fill separate *s, p, d,* etc. orbitals with each type of nucleon forced to obey the Pauli principle. These orbitals are not the same in their radial "shapes" as the *s, p, d,* etc. orbitals of atoms because, in atoms, there is an additional radial potential $V(r) = -Ze^2/r$ present. However, their angular shapes are the same as in atomic structure because, in both cases, the potential is independent of θ and ϕ. This same spherical box model has been used to describe the orbitals of valence electrons in clusters of mono-valent metal atoms such as Cs_n, Cu_n, Na_n and their positive and negative ions. Because of the metallic nature of these species, their valence electrons are sufficiently delocalized to render this simple model rather effective (see T. P. Martin, T. Bergmann, H. Göhlich, and T. Lange, *J. Phys. Chem.* 95, 6421 (1991)).

One-dimensional free particle motion provides a qualitatively correct picture for π-electron motion along the p_π orbitals of a delocalized polyene. The one cartesian dimension then corresponds to motion along the delocalized chain. In such a model, the box length L is related to the carbon-carbon bond length R and the number N of carbon centers involved in the delocalized network $L = (N-1)R$. Below, such a conjugated network involving nine centers is depicted. In this example, the box length would be eight times the C-C bond length.

Conjugated π Network with 9 Centers Involved

The eigenstates $\psi_n(x)$ and their energies E_n represent orbitals into which electrons are placed. In the example case, if nine π electrons are present (e.g., as in the 1,3,5,7-nonatetraene radical), the ground electronic state would be represented by a total wavefunction consisting of a *product* in which the lowest four Ψ's are doubly occupied and the fifth Ψ is singly occupied:

$$\Psi = \psi_1\alpha\psi_1\beta\psi_2\alpha\psi_2\beta\psi_3\alpha\psi_3\beta\psi_4\alpha\psi_4\beta\psi_5\alpha.$$

A product wavefunction is appropriate because the total Hamiltonian involves the kinetic plus potential energies of nine electrons. To the extent that this total energy can be represented as the sum of nine separate energies, one for each electron, the Hamiltonian allows a separation of variables

$$H \cong \sum_j H(j)$$

in which each $H(j)$ describes the kinetic and potential energy of an individual electron. This (approximate) additivity of H implies that solutions of $H\Psi = E\Psi$ are products of solutions to $H(j)\Psi(\mathbf{r}_j) = E_j\Psi(\mathbf{r}_j)$.

The two lowest π-excited states would correspond to states of the form

$$\Psi^* = \psi_1\alpha\psi_1\beta\psi_2\alpha\psi_2\beta\psi_3\alpha\psi_3\beta\psi_4\alpha\psi_5\beta\psi_5\alpha, \text{ and}$$
$$\Psi'^* = \psi_1\alpha\psi_1\beta\psi_2\alpha\psi_2\beta\psi_3\alpha\psi_3\beta\psi_4\alpha\psi_4\beta\psi_6\alpha,$$

where the spin-orbitals (orbitals multiplied by α or β) appearing in the above products depend on the coordinates of the various electrons. For example,

$$\psi_1\alpha\psi_1\beta\psi_2\alpha\psi_2\beta\psi_3\alpha\psi_3\beta\psi_4\alpha\psi_5\beta\psi_5\alpha$$

denotes

$$\psi_1\alpha(\mathbf{r}_1)\psi_1\beta(\mathbf{r}_2)\psi_2\alpha(\mathbf{r}_3)\psi_2\beta(\mathbf{r}_4)\psi_3\alpha(\mathbf{r}_5)\psi_3\beta(\mathbf{r}_6)\psi_4\alpha(\mathbf{r}_7)\psi_5\beta(\mathbf{r}_8)\psi_5\alpha(\mathbf{r}_9).$$

The electronic excitation energies within this model would be

$$\Delta E^* = \pi^2\hbar^2/2m[5^2/L^2 - 4^2/L^2] \text{ and } \Delta E'^* = \pi^2\hbar^2/2m[6^2/L^2 - 5^2/L^2],$$

for the two excited-state functions described above. It turns out that this simple model of π – *electron energies* provides a qualitatively correct picture of such excitation energies.

This simple particle-in-a-box model does not yield orbital energies that relate to ionization energies unless the potential "inside the box" is specified. Choosing the value of this potential V_0 such that $V_0 + \pi^2\hbar^2/2m[5^2/L^2]$ is equal to minus the lowest ionization energy of the 1,3,5,7-nonatetraene radical, gives energy levels (as $E = V_0 + \pi^2\hbar^2/2m[n^2/L^2]$) which then are approximations to ionization energies.

The individual π-molecular orbitals

$$\psi_n = (2/L)^{1/2}\sin(n\pi x/L)$$

are depicted in the figure below for a model of the 1,3,5 hexatriene π-orbital system for which the "box length" L is five times the distance R_{CC} between neighboring pairs of Carbon atoms.

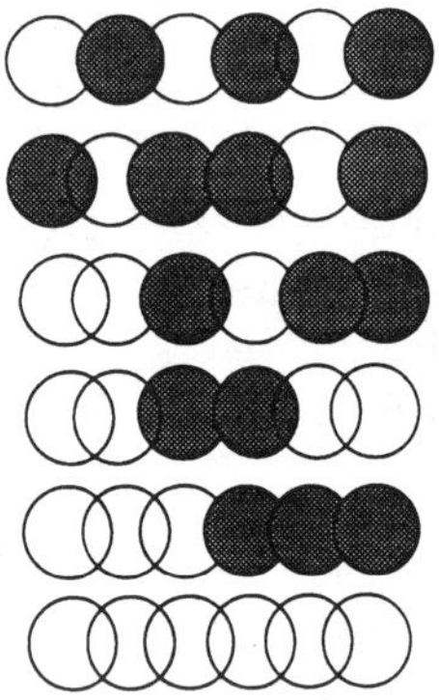

$(2/L)^{1/2}$ sin(nπx/L); L = 5 x R_{CC}

In this figure, positive amplitude is denoted by the clear spheres and negative amplitude is shown by the darkened spheres; the magnitude of the kth C-atom centered atomic orbital in the nth π-molecular orbital is given by $(2/L)^{1/2}\sin(n\pi kR_{CC}/L)$.

This simple model allows one to estimate spin densities at each carbon center and provides insight into which centers should be most amenable to electrophilic or nucleophilic attack. For example, radical attack at the C_5 carbon of the nine-atom system described earlier would be more facile for the ground state Ψ than for either Ψ^* or Ψ'^*. In the former, the unpaired spin density resides in ψ_5, which has non-zero amplitude at the C_5 site $x = L/2$; in Ψ^* and Ψ'^*, the unpaired density is in ψ_4 and ψ_6, respectively, both of which have zero density at C_5. These densities reflect the values $(2/L)^{1/2}\sin(n\pi kR_{CC}/L)$ of the amplitudes for this case in which $L = 8 \times R_{CC}$ for $n = 5, 4,$ and 6, respectively.

2. One Electron Moving About a Nucleus[1]

The Hydrogenic atom problem forms the basis of much of our thinking about atomic structure. To solve the corresponding Schrödinger equation requires separation of the r,θ, and ϕ variables.

---

[1] Suggested extra reading: appendix B: The Hydrogen Atom Orbitals

The Schrödinger equation for a single particle of mass μ moving in a central potential (one that depends only on the radial coordinate r) can be written as

$$-\frac{\hbar^2}{2\mu}\left(\frac{\partial^2}{\partial x^2}+\frac{\partial^2}{\partial y^2}+\frac{\partial^2}{\partial z^2}\right)\psi + V\left(\sqrt{x^2+y^2+z^2}\right)\psi = E\psi.$$

This equation is not separable in cartesian coordinates (x,y,z) because of the way $x,y,$ and z appear together in the square root. However, it is separable in spherical coordinates

$$-\frac{\hbar^2}{2\mu r^2}\left(\frac{\partial}{\partial r}\left(r^2\frac{\partial\psi}{\partial r}\right)+\frac{1}{\sin\theta}\frac{\partial}{\partial\theta}\left(\sin\theta\frac{\partial\psi}{\partial\theta}\right)+\frac{1}{\sin^2\theta}\frac{\partial^2\psi}{\partial\phi^2}\right)+V(r)\psi = E\psi.$$

Subtracting $V(r)\psi$ from both sides of the equation and multiplying by $-2\mu r^2/\hbar^2$ then moving the derivatives with respect to r to the right-hand side, one obtains

$$\frac{1}{\sin\theta}\frac{\partial}{\partial\theta}\left(\sin\theta\frac{\partial\psi}{\partial\theta}\right)+\frac{1}{\sin^2\theta}\frac{\partial^2\psi}{\partial\phi^2} = -\frac{2\mu r^2}{\hbar^2}(E-V(r))\psi - \frac{\partial}{\partial r}\left(r^2\frac{\partial\psi}{\partial r}\right).$$

Notice that the right-hand side of this equation is a function of r only; it contains no θ or ϕ derivatives. Let's call the entire right hand side $\boldsymbol{F}(r)\psi$ to emphasize this fact.

To further separate the θ and ϕ dependence, we multiply by $\sin^2\theta$ and subtract the θ derivative terms from both sides to obtain

$$\frac{\partial^2\psi}{\partial\phi 2} = \boldsymbol{F}(r)\psi\sin^2\theta - \sin\theta\frac{\partial}{\partial\theta}\left(\sin\theta\frac{\partial\psi}{\partial\theta}\right).$$

Now we have separated the ϕ derivatives from the θ and r derivatives. If we now substitute $\psi = \Phi(\phi)Q(r,\theta)$ and divide by ΦQ, we obtain

$$\frac{1}{\Phi}\frac{\partial^2\Phi}{\partial\phi 2} = \frac{1}{Q}\left(\boldsymbol{F}(r)\sin^2\theta Q - \sin\theta\frac{\partial}{\partial\theta}\left(\sin\theta\frac{\partial Q}{\partial\theta}\right)\right).$$

Now all of the ϕ dependence is isolated on the left-hand side; the right-hand side contains only r and θ dependence.

Whenever one has isolated the entire dependence on one variable as we have done above for the ϕ dependence, one can easily see that the left- and right-hand sides of the equation must equal a constant. For the above example, the left-hand side contains no r or θ dependence and the right-hand side contains no ϕ dependence. Because the two sides are equal, they both must actually contain no r, θ, or ϕ dependence; that is, they are constant.

For the above example, we therefore can set both sides equal to a so-called separation constant that we call $-m^2$. It will become clear shortly why we have chosen to express the constant in this form.

a. The Φ Equation The resulting Φ equation reads

$$\Phi'' + m^2\Phi = 0,$$

which has as its most general solution

$$\Phi = Ae^{im\phi} + Be^{-im\phi}.$$

We must require the function Φ to be single-valued, which means that

$$\Phi(\phi) = \Phi(2\pi + \phi)$$

or,

$$Ae^{im\phi}\left(1 - e^{2im\pi}\right) + Be^{-im\phi}\left(1 - e^{-2im\pi}\right) = 0.$$

This is satisfied only when the separation constant is equal to an integer $m = 0, \pm 1, \pm 2, \ldots$ and provides another example of the rule that quantization comes from the boundary conditions on the wavefunction. Here m is restricted to certain discrete values because the wavefunction must be such that when you rotate through 2π about the z-axis, you must get back what you started with.

b. The Θ Equation Now returning to the equation in which the ϕ dependence was isolated from the r and θ dependence and rearranging the θ terms to the left-hand side, we have

$$\frac{1}{\sin\theta}\frac{\partial}{\partial\theta}\left(\sin\theta\frac{\partial Q}{\partial\theta}\right) - \frac{m^2 Q}{\sin^2\theta} = \boldsymbol{F}(r)Q.$$

In this equation we have separated θ and r variations so we can further decompose the wavefunction by introducing $Q = \Theta(\theta)R(r)$, which yields

$$\frac{1}{\Theta}\frac{1}{\sin\theta}\frac{\partial}{\partial\theta}\left(\sin\theta\frac{\partial\Theta}{\partial\theta}\right) - \frac{m^2}{\sin^2\theta} = \frac{\boldsymbol{F}(r)R}{R} = -\lambda,$$

where a second separation constant, $-\lambda$, has been introduced once the r and θ dependent terms have been separated onto the right- and left-hand sides, respectively.

We now can write the θ equation as

$$\frac{1}{\sin\theta}\frac{\partial}{\partial\theta}\left(\sin\theta\frac{\partial\Theta}{\partial\theta}\right) - \frac{m^2\Theta}{\sin^2\theta} = -\lambda\Theta,$$

where m is the integer introduced earlier. To solve this equation for Θ, we make the substitutions $z = \cos\theta$ and $P(z) = \Theta(\theta)$, so $\sqrt{1-z^2} = \sin\theta$, and

$$\frac{\partial}{\partial\theta} = \frac{\partial z}{\partial\theta}\frac{\partial}{\partial z} = -\sin\theta\frac{\partial}{\partial z}.$$

The range of values for θ was $0 \le \theta < \pi$, so the range for z is $-1 < z < 1$. The equation for Θ, when expressed in terms of P and z, becomes

$$\frac{d}{dz}\left((1 - z^2)\frac{dP}{dz}\right) - \frac{m^2 P}{1 - z^2} + \lambda P = 0.$$

Now we can look for polynomial solutions for P, because z is restricted to be less than unity in magnitude. If $m = 0$, we first let

$$P = \sum_{k=0}^{\infty} a_k z^k,$$

and substitute into the differential equation to obtain

$$\sum_{k=0}^{\infty} (k + 2)(k + 1)a_{k+2}z^k - \sum_{k=0}^{\infty} (k + 1)ka_k z^k + \lambda \sum_{k=0}^{\infty} a_k z^k = 0.$$

Equating like powers of z gives

$$a_{k+2} = \frac{a_k(k(k+1)-\lambda)}{(k+2)(k+1)}.$$

Note that for large values of k

$$\frac{a_{k+2}}{a_k} \rightarrow \frac{k^2\left(1+\frac{1}{k}\right)}{k^2\left(1+\frac{2}{k}\right)\left(1+\frac{1}{k}\right)} = 1.$$

Since the coefficients do not decrease with k for large k, this series will diverge for $z = \pm 1$ *unless* it truncates at finite order. This truncation only happens if the separation constant λ obeys $\lambda = l(l+1)$, where l is an integer. So, once again, we see that a boundary condition (i.e., that the wavefunction be normalizable in this case) give rise to quantization. In this case, the values of λ are restricted to $l(l+1)$; before, we saw that m is restricted to $0, \pm 1, \pm 2, \ldots$.

Since this recursion relation links every other coefficient, we can choose to solve for the even and odd functions separately. Choosing a_0 and then determining all of the even a_k in terms of this a_0, followed by rescaling all of these a_k to make the function normalized generates an even solution. Choosing a_1 and determining all of the odd a_k in like manner, generates an odd solution.

For $l = 0$, the series truncates after one term and results in $P_o(z) = 1$. For $l = 1$ the same thing applies and $P_1(z) = z$. For $l = 2$, $a_2 = -6a_o/2 = -3a_o$, so one obtains $P_2 = 3z^2 - 1$, and so on. These polynomials are called Legendre polynomials.

For the more general case where $m \neq 0$, one can proceed as above to generate a polynomial solution for the Θ function. Doing so, results in the following solutions:

$$P_l^m(z) = (1-z^2)^{\frac{|m|}{2}} \frac{d^{|m|} P_l(z)}{dz^{|m|}}.$$

These functions are called Associated Legendre polynomials, and they constitute the solutions to the Θ problem for non-zero m values.

The above P and $e^{im\phi}$ functions, when re-expressed in terms of θ and ϕ, yield the full angular part of the wavefunction for any centrosymmetric potential. These solutions are usually written as

$$Y_{l,m}(\theta,\phi) = P_l^m(\cos\theta)(2\pi)^{-\frac{1}{2}}\exp(im\phi),$$

and are called spherical harmonics. They provide the angular solution of the r, θ, ϕ Schrödinger equation for *any* problem in which the potential depends only on the radial coordinate. Such situations include all one-electron atoms and ions (e.g., H, He^+, Li^{++}, etc.), the rotational motion of a diatomic molecule (where the potential depends only on bond length r), the motion of a nucleon in a spherically symmetrical "box" (as occurs in the shell model of nuclei), and the scattering of two atoms (where the potential depends only on interatomic distance).

c. The R Equation Let us now turn our attention to the radial equation, which is the only place that the explicit form of the potential appears. Using our derived results and specifying $V(r)$ to be the coulomb potential appropriate for an electron in the field of a nucleus of charge $+Ze$, yields:

$$\frac{1}{r^2}\frac{d}{dr}\left(r^2\frac{dR}{dr}\right) + \left(\frac{2\mu}{\hbar^2}\left(E + \frac{Ze^2}{r}\right) - \frac{l(l+1)}{r^2}\right)R = 0.$$

We can simplify things considerably if we choose rescaled length and energy units because doing so removes the factors that depend on μ, $\hbar$, and e. We introduce a new radial coordinate ρ and a quantity σ as follows:

$$\rho = \left(\frac{-8\mu E}{\hbar^2}\right)^{\frac{1}{2}} r, \text{ and}$$

$$\sigma^2 = -\frac{\mu Z^2 e^4}{2E\hbar^2}.$$

Notice that if E is negative, as it will be for bound states (i.e., those states with energy below that of a free electron infinitely far from the nucleus and with zero kinetic energy), ρ is real. On the other hand, if E is positive, as it will be for states that lie in the continuum, ρ will be imaginary. These two cases will give rise to qualitatively different behavior in the solutions of the radial equation developed below.

We now define a function S such that $S(\rho) = R(r)$ and substitute S for R to obtain:

$$\frac{1}{\rho^2}\frac{d}{d\rho}\left(\rho^2\frac{dS}{d\rho}\right) + \left(-\frac{1}{4} - \frac{l(l+1)}{\rho^2} + \frac{\sigma}{\rho}\right)S = 0.$$

The differential operator terms can be recast in several ways using

$$\frac{1}{\rho^2}\frac{d}{d\rho}\left(\rho^2\frac{dS}{d\rho}\right) = \frac{d^2S}{d\rho^2} + \frac{2}{\rho}\frac{dS}{d\rho} = \frac{1}{\rho}\frac{d^2}{d\rho^2}(\rho S).$$

It is useful to keep in mind these three embodiments of the derivatives that enter into the radial kinetic energy; in various contexts it will be useful to employ various of these.

The strategy that we now follow is characteristic of solving second-order differential equations. We will examine the equation for S at large and small ρ values. Having found solutions at these limits, we will use a power series in ρ to "interpolate" between these two limits.

Let us begin by examining the solution of the above equation at small values of ρ to see how the radial functions behave at small r. As $\rho \to 0$, the second term in the brackets will dominate. Neglecting the other two terms in the brackets, we find that, for small values of ρ (or r), the solution should behave like ρ^L and because the function must be normalizable, we must have $L \geq 0$. Since L can be any non-negative integer, this suggests the following more general form for $S(\rho)$:

$$S(\rho) \approx \rho^L e^{-a\rho}.$$

This form will insure that the function is normalizable since $S(\rho) \to 0$ as $r \to \infty$ for all L, as long as ρ is a real quantity. If ρ is imaginary, such a form may not be normalized (see below for further consequences).

Turning now to the behavior of S for large ρ, we make the substitution of $S(\rho)$ into the above equation and keep only the terms with the largest power of ρ (e.g., first term in brackets). Upon so doing, we obtain the equation

$$a^2\rho^L e^{-a\rho} = \frac{1}{4}\rho^L e^{-a\rho},$$

which leads us to conclude that the exponent in the large-ρ behavior of S is $a = 1/2$.

Having found the small- and large-ρ behaviors of $S(\rho)$, we can take S to have the following form to interpolate between large and small ρ-values:

$$S(\rho) = \rho^L e^{-\frac{\rho}{2}} P(\rho),$$

where the function P is expanded in an infinite power series in ρ as

$$P(\rho) = \sum a_k \rho^k.$$

Again, substituting this expression for S into the above equation we obtain

$$P''\rho + P'(2L + 2 - \rho) + P(\sigma - L - l) = 0,$$

and then substituting the power series expansion of P and solving for the a_k 's we arrive at:

$$a_{k+1} = \frac{(k - \sigma + L + l)a_k}{(k+1)(k + 2L + 2)}.$$

For large k, the ratio of expansion coefficients reaches the limit

$$\frac{a_{k+1}}{a_k} = \frac{1}{k},$$

which has the same behavior as the power series expansion of e^ρ. Because the power series expansion of P describes a function that behaves like e^ρ for large ρ, the resulting $S(\rho)$ function would not be normalizable because the

$$e^{-\frac{\rho}{2}}$$

factor would be overwhelmed by this e^ρ dependence. Hence, the series expansion of P must truncate in order to achieve a normalizable S function. Notice that if ρ is imaginary, as it will be if E is in the continuum, the argument that the series must truncate to avoid an exponentially diverging function no longer applies. Thus, we see a key difference between bound (with ρ real) and continuum (with ρ imaginary) states. In the former case, the boundary condition of non-divergence arises; in the latter, it does not.

To truncate at a polynomial of order n', we must have $n' - \sigma + L + l = 0$. This implies that the quantity σ introduced previously is restricted to $\sigma = n' + L + l$, which is certainly an integer; let us call this integer n. If we label states in order of increasing $n = 1,2,3, \ldots$, we see that doing so is consistent with specifying a maximum order (n') in the $P(\rho)$ polynomial $n' = 0,1,2, \ldots$ after which the l-value can run from $l = 0$, in steps of unity up to $L = n - 1$.

Substituting the integer n for σ, we find that the energy levels are quantized because σ is quantized (equal to n):

$$E = -\frac{\mu Z^2 e^4}{2\hbar^2 n^2} \text{ and } \rho = \frac{Zr}{a_o n}.$$

Here, the length a_o is the so called Bohr radius

$$\left(a_o = \frac{\hbar^2}{\mu e^2}\right);$$

it appears once the above E-expression is substituted into the equation for ρ. Using the recursion equation to solve for the polynomial's coefficients a_k for any choice of n and l quantum numbers generates a so-called Laguerre polynomial; $P_{n-L-1}(\rho)$. They contain powers of ρ from zero through $n-l-1$.

This energy quantization does not arise for states lying in the continuum because the condition that the expansion of $P(\rho)$ terminate does not arise. The solutions of the radial equation appropriate to these scattering states (which relate to the scattering motion of an electron in the field of a nucleus of charge Z) are treated on page 90 of EWK.

In summary, separation of variables has been used to solve the full r,θ,ϕ Schrödinger equation for one electron moving about a nucleus of charge Z. The θ and φ solutions are the spherical harmonics $Y_{L,m}(\theta,\phi)$. The bound-state radial solutions

$$R_{n,L}(r) = S(\rho) = \rho^L e^{-\frac{\rho}{2}} P_{n-L-1}(\rho)$$

depend on the n and l quantum numbers and are given in terms of the Laguerre polynomials (see EWK for tabulations of these polynomials).

d. Summary To summarize, the quantum numbers l and m arise through boundary conditions requiring that $\psi(\theta)$ be normalizable (i.e., not diverge) and $\psi(\phi) = \psi(\phi + 2\pi)$. In the texts by Atkins, EWK, and McQuarrie the differential equations obeyed by the θ and φ components of $Y_{l,m}$ are solved in more detail and properties of the solutions are discussed. This differential equation involves the three-dimensional Schrödinger equation's angular kinetic energy operator. That is, the angular part of the above Hamiltonian is equal to $\hbar^2 L^2/2mr^2$, where L^2 is the square of the total angular momentum for the electron.

The radial equation, which is the only place the potential energy enters, is found to possess both bound-states (i.e., states whose energies lie below the asymptote at which the potential vanishes and the kinetic energy is zero) and continuum states lying energetically above this asymptote. The resulting hydrogenic wavefunctions (angular and radial) and energies are summarized in appendix B for principal quantum numbers n ranging from 1 to 3 and in Pauling and Wilson for n up to 5.

There are both bound and continuum solutions to the radial Schrödinger equation for the attractive coulomb potential because, at energies below the asymptote the potential confines the particle between $r = 0$ and an outer turning point, whereas at energies above the asymptote, the particle is no longer confined by an outer turning point (see the figure below).

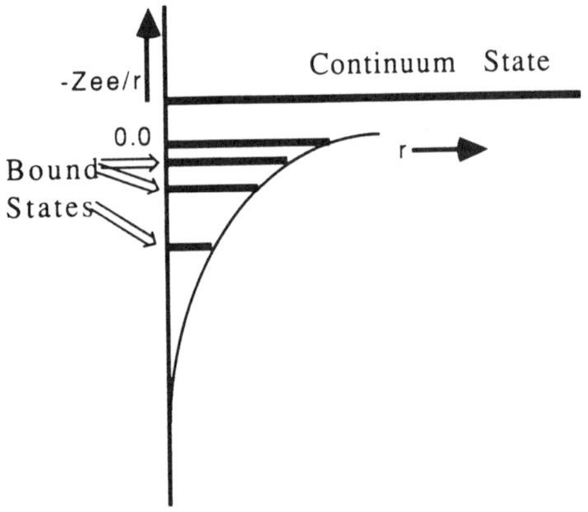

The solutions of this one-electron problem form the qualitative basis for much of atomic and molecular orbital theory. For this reason, the reader is encouraged to use appendix B to gain a firmer understanding of the nature of the radial and angular parts of these wavefunctions. The orbitals that result are labeled by n, l, and m quantum numbers for the bound states and by l and m quantum numbers and the energy E for the continuum states. Much as the particle-in-a-box orbitals are used to qualitatively describe π-electrons in conjugated polyenes, these so-called hydrogen-like orbitals provide qualitative descriptions of orbitals of atoms with more than a single electron. By introducing the concept of screening as a way to represent the repulsive interactions among the electrons of an atom, an effective nuclear charge Z_{eff} can be used in place of Z in the $\psi_{n,l,m}$ and $E_{n,l}$ to generate approximate atomic orbitals to be filled by electrons in a many-electron atom. For example, in the crudest approximation of a carbon atom, the two $1s$ electrons experience the full nuclear attraction so $Z_{eff} = 6$ for them, whereas the $2s$ and $2p$ electrons are screened by the two $1s$ electrons, so $Z_{eff} = 4$ for them. Within this approximation, one then occupies two $1s$ orbitals with $Z = 6$, two $2s$ orbitals with $Z = 4$ and two $2p$ orbitals with $Z = 4$ in forming the full six-electron wavefunction of the lowest-energy state of carbon.

3. Rotational Motion for a Rigid Diatomic Molecule

This Schrödinger equation relates to the rotation of diatomic and linear polyatomic molecules. It also arises when treating the angular motions of electrons in any spherically symmetric potential.

A diatomic molecule with fixed bond length R rotating in the absence of any external potential is described by the following Schrödinger equation:

$$-\hbar^2/2\mu\{(R^2\sin\theta)^{-1}\partial/\partial\theta(\sin\theta\partial/\partial\theta) + (R^2\sin^2\theta)^{-1}\partial^2/\partial\phi^2\}\psi = E\psi$$

or

$$L^2\psi/2\mu R^2 = E\psi.$$

The angles θ and ϕ describe the orientation of the diatomic molecule's axis relative to a laboratory-fixed coordinate system, and μ is the reduced mass of the diatomic molecule $\mu = m_1m_2/(m_1+m_2)$. The differential operators can be seen to be exactly the same as those that arose in the hydrogen-like-atom case, and, as discussed above, these θ and ϕ differential operators are identical to the L^2 angular momentum operator whose general properties are analyzed in appendix G. Therefore, the same spherical harmonics that served as the angular parts of the wavefunction in the earlier case now serve as the entire wavefunction for the so-called rigid rotor: $\psi = Y_{J,M}(\theta,\phi)$. As detailed later in this text, the eigenvalues corresponding to each such eigenfunction are given as

$$E_J = \hbar^2 J(J+1)/(2\mu R^2) = BJ(J+1)$$

and are independent of M. Thus each energy level is labeled by J and is $2J+1$-fold degenerate (because M ranges from $-J$ to J). The so-called rotational constant B (defined as $\hbar^2/2\mu R^2$) depends on the molecule's bond length and reduced mass. Spacings between successive rotational levels (which are of spectroscopic relevance because angular momentum selection rules often restrict ΔJ to 1,0, and -1) are given by

$$\Delta E = B(J+1)(J+2) - BJ(J+1) = 2B(J+1).$$

These energy spacings are of relevance to microwave spectroscopy which probes the rotational energy levels of molecules.

The rigid rotor provides the most commonly employed approximation to the rotational energies and wavefunctions of linear molecules. As presented above, the model restricts the bond length to be fixed. Vibrational motion of the molecule gives rise to changes in R which are then reflected in changes in the rotational energy levels. The coupling between rotational and vibrational motion gives rise to rotational B constants that depend on vibrational state as well as dynamical couplings, called centrifugal distortions, that cause the total ro-vibrational energy of the molecule to depend on rotational and vibrational quantum numbers in a non-separable manner.

4. Harmonic Vibrational Motion

This Schrödinger equation forms the basis for our thinking about bond stretching and angle bending vibrations as well as collective phonon motions in solids.

The radial motion of a diatomic molecule in its lowest ($J = 0$) rotational level can be described by the following Schrödinger equation:

$$-\hbar^2/2\mu r^{-2}\partial/\partial r(r^2\partial/\partial r)\psi + V(r)\psi = E\psi,$$

where μ is the reduced mass $\mu = m_1m_2/(m_1+m_2)$ of the two atoms. By substituting $\psi = F(r)/r$ into this equation, one obtains an equation for $F(r)$ in which the differential operators appear to be less complicated:

$$-\hbar^2/2\mu d^2F/dr^2 + V(r)F = EF.$$

This equation is exactly the same as the equation seen above for the radial motion of the electron in the hydrogen-like atoms except that the reduced mass μ replaces the electron mass m and the potential V(r) is not the coulomb potential.

If the potential is approximated as a quadratic function of the bond displacement $x = r - r_e$ expanded about the point at which V is minimum:

$$V = 1/2k(r-r_e)^2,$$

the resulting harmonic-oscillator equation can be solved exactly. Because the potential V grows without bound as x approaches ∞ or $-\infty$, only bound-state solutions exist for this model problem; that is, the motion is confined by the nature of the potential, so no continuum states exist.

In solving the radial differential equation for this potential (see chapter 5 of McQuarrie), the large-r behavior is first examined. For large-r, the equation reads:

$$d^2F/dx^2 = 1/2kx^2(2\mu/\hbar^2)F,$$

where $x = r - r_e$ is the bond displacement away from equilibrium. Defining $\xi = (\mu k/\hbar^2)^{1/4}x$ as a new scaled radial coordinate allows the solution of the large-r equation to be written as:

$$F_{large-r} = \exp(-\xi^2/2).$$

The general solution to the radial equation is then taken to be of the form:

$$F = \exp(-\xi^2/2)\sum_{n=0}^{\infty}\xi^n C_n,$$

where the C_n are coefficients to be determined. Substituting this expression into the full radial equation generates a set of recursion equations for the C_n amplitudes. As in the solution of the hydrogen-like radial equation, the series described by these coefficients is divergent unless the energy E happens to equal specific values. It is this requirement that the wavefunction not diverge so it can be normalized that yields energy quantization. The energies of the states that arise are given by:

$$E_n = \hbar\,(k/\mu)^{1/2}(n + 1/2),$$

and the eigenfunctions are given in terms of the so-called Hermite polynomials $H_n(y)$ as follows:

$$\psi_n(x) = (n!2^n)^{-1/2}(\alpha/\pi)^{1/4}\exp(-\alpha x^2/2)H_n(\alpha^{1/2}x),$$

where $\alpha = (k\mu/\hbar^2)^{1/2}$. Within this harmonic approximation to the potential, the vibrational energy levels are evenly spaced:

$$\Delta E = E_{n+1} - E_n = \hbar\,(k/\mu)^{1/2}.$$

In experimental data such evenly spaced energy level patterns are seldom seen; most commonly, one finds spacings $E_{n+1} - E_n$ that decrease as the quantum number n increases. In such cases, one says that the progression of vibrational levels displays anharmonicity.

Because the H_n are odd or even functions of x (depending on whether n is odd or even), the wavefunctions $\psi_n(x)$ are odd or even. This splitting of the solutions into two distinct classes is an example of the effect of symmetry; in this case, the symmetry is caused by the symmetry of the harmonic potential with respect to reflection through the origin along the x-axis. Throughout this text, many symmetries will arise; in each case, symmetry properties of the potential will cause the solutions of the Schrödinger equation to be decomposed into various symmetry groupings. Such symmetry decompositions are of great use because they provide additional quantum numbers (i.e., symmetry labels) by which the wavefunctions and energies can be labeled.

The harmonic oscillator energies and wavefunctions comprise the simplest reasonable model for vibrational motion. Vibrations of a polyatomic molecule are often characterized in terms of individual bond-stretching and angle-bending motions each of which is, in turn, approximated harmonically. This results in a total vibrational wavefunction that is written as a product of functions one for each of the vibrational coordinates.

Two of the most severe limitations of the harmonic oscillator model, the lack of anharmonicity (i.e., non-uniform energy level spacings) and lack of bond dissociation, result from the quadratic nature of its potential. By introducing model potentials that allow for proper bond dissociation (i.e., that do not increase without bound as $x \to \infty$), the major shortcomings of the harmonic oscillator picture can be overcome. The so-called Morse potential (see the figure below)

$$V(r) = D_e(1 - exp(-a(r - r_e)))^2,$$

is often used in this regard.

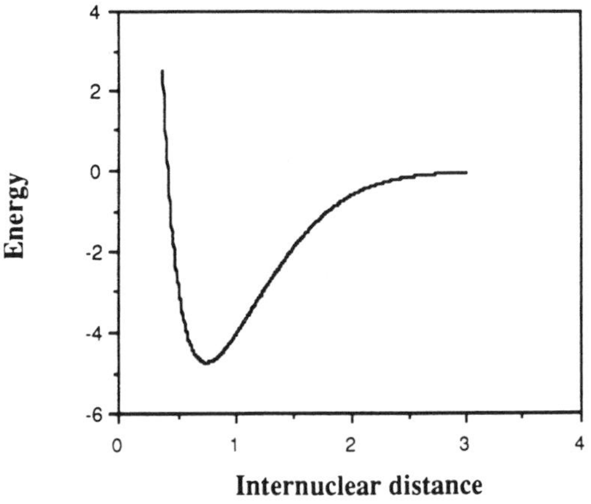

Here, D_e is the bond dissociation energy, r_e is the equilibrium bond length, and a is a constant that characterizes the "steepness" of the potential and determines the vibrational frequencies. The advantage of using the Morse potential to improve upon harmonic-oscillator-level predictions is that its energy levels and wavefunctions are also known exactly. The energies are given in terms of the parameters of the potential as follows:

$$E_n = \hbar\,(k/\mu)^{1/2}\left\{(n+1/2)-(n+1/2)^2\hbar\,(k/\mu)^{1/2}/4D_e\right\},$$

where the force constant k is $k = 2D_e a^2$. The Morse potential supports both bound states (those lying below the dissociation threshold for which vibration is confined by an outer turning point) and continuum states lying above the dissociation threshold. Its degree of anharmonicity is governed by the ratio of the harmonic energy $\hbar\,(k/\mu)^{1/2}$ to the dissociation energy D_e.

III. THE PHYSICAL RELEVANCE OF WAVEFUNCTIONS, OPERATORS AND EIGENVALUES

Having gained experience on the application of the Schrödinger equation to several of the more important model problems of chemistry, it is time to return to the issue of how the wavefunctions, operators, and energies relate to experimental reality. In mastering the sections that follow the reader should keep in mind that:

1. *It is the molecular system that possesses a set of characteristic wavefunctions and energy levels, but*
2. *It is the experimental measurement that determines the nature by which these energy levels and wavefunctions are probed.*

This separation between the "system" with its intrinsic set of energy levels and "observation" or "experiment" with its characteristic interaction with the system forms an important point of view used by quantum mechanics. It gives rise to a point of view in which the measurement itself can "prepare" the system in a wavefunction Ψ that need not be any single eigenstate but can still be represented as a combination of the complete set of eigenstates. For the beginning student of quantum mechanics, these aspects of quantum mechanics are among the more confusing. If it helps, one should rest assured that all of the mathematical and "rule" structure of this subject was created to permit the predictions of quantum mechanics to replicate what has been observed in laboratory experiments.

Note to the Reader: Before moving on to the next section, it would be very useful to work some of the exercises and problems. In particular, exercises 3, 5, and 12 as well as problems 6, 8, and 11 provide insight that would help when the material of the next section is studied. The solution to problem 11 is used throughout this section to help illustrate the concepts introduced here.

A. The Basic Rules and Relation to Experimental Measurement[2]

Quantum mechanics has a set of "ules" that link operators, wavefunctions, and eigenvalues to physically measurable properties. These rules have been formulated not in some arbitrary

[2] Suggested extra reading—appendix C: Quantum Mechanical Operators and Commutation

manner nor by derivation from some higher subject. Rather, the rules were designed to allow quantum mechanics to mimic the experimentally observed *facts as revealed in mother nature's data. The extent to which these rules seem difficult to understand usually reflects the presence of experimental observations that do not fit in with our common experience base.*

The structure of quantum mechanics (QM) relates the wavefunction Ψ and operators $\boldsymbol{F}$ to the "real world" in which experimental measurements are performed through a set of rules (Dirac's text is an excellent source of reading concerning the historical development of these fundamentals). Some of these rules have already been introduced above. Here, they are presented in total as follows:

1. The time evolution of the wavefunction Ψ is determined by solving the time-dependent Schrödinger equation (see pages 23–25 of EWK for a rationalization of how the Schrödinger equation arises from the classical equation governing waves, Einstein's $E = h\nu$, and deBroglie's postulate that $\lambda = h/p$)

$$i\hbar\partial\Psi/\partial t = \boldsymbol{H}\Psi,$$

where $\boldsymbol{H}$ is the Hamiltonian operator corresponding to the total (kinetic plus potential) energy of the system. For an isolated system (e.g., an atom or molecule not in contact with any external fields), $\boldsymbol{H}$ consists of the kinetic and potential energies of the particles comprising the system. To describe interactions with an external field (e.g., an electromagnetic field, a static electric field, or the "crystal field" caused by surrounding ligands), additional terms are added to $\boldsymbol{H}$ to properly account for the system-field interactions.

If $\boldsymbol{H}$ contains no explicit time dependence, then separation of space and time variables can be performed on the above Schrödinger equation $\Psi = \psi\exp(-itE/\hbar)$ to give

$$\boldsymbol{H}\psi = E\psi.$$

In such a case, the time dependence of the state is carried in the phase factor $\exp(-itE/\hbar)$; the spatial dependence appears in $\psi(q_j)$.

The so called time-independent Schrödinger equation $\boldsymbol{H}\psi = E\psi$ must be solved to determine the physically measurable energies E_k and wavefunctions ψ_k of the system. The most general solution to the full Schrödinger equation $i\hbar\partial\Psi/\partial t = \boldsymbol{H}\Psi$ is then given by applying $\exp(-i\boldsymbol{H}t/\hbar)$ to the wavefunction at some initial time

$$(t = 0)\Psi = \sum_k C_k\psi_k$$

to obtain

$$\Psi(t) = \sum_k C_k\psi_k\exp(-itE_k/\hbar).$$

The relative amplitudes C_k are determined by knowledge of the state at the initial time; this depends on how the system has been prepared in an earlier experiment. Just as Newton's laws of motion do not fully determine the time evolution of a classical system (i.e., the coordinates and momenta must be known at some initial time), the Schrödinger equation must be accompanied by initial conditions to fully determine $\Psi(q_j,t)$.

EXAMPLE

Using the results of problem 11 of this chapter to illustrate, the sudden ionization of N_2 in its $v = 0$ vibrational state to generate N_2^+ produces a vibrational wavefunction

$$\Psi_0 = \left(\frac{\alpha}{\pi}\right)^{1/4} e^{-\alpha x^{2/2}} = 3.53333\mathring{A}^{-\frac{1}{2}} e^{-(244.83\mathring{A}^{-2})\,(r-1.09769\mathring{A})^2}$$

that was created by the fast ionization of N_2. Subsequent to ionization, this N_2 function is not an eigenfunction of the new vibrational Schrödinger equation appropriate to N_2^+. As a result, this function will time evolve under the influence of the N_2^+ Hamiltonian. The time evolved wavefunction, according to this first rule, can be expressed in terms of the vibrational functions $\{\Psi v\}$ and energies $\{E_v\}$ of the N_2^+ ion as

$$\Psi(t) = \sum_v C_v \Psi_v \exp(-iE_v t/\hbar).$$

The amplitudes C_v, which reflect the manner in which the wavefunction is prepared (at $t = 0$), are determined by determining the component of each Ψ_v in the function Ψ at $t = 0$. To do this, one uses

$$\int \Psi_{v'}^* \Psi(t=0) d\tau = C_{v'},$$

which is easily obtained by multiplying the above summation by $\Psi_{v'}^*$, integrating, and using the orthonormality of the $\{\Psi_v\}$ functions.

For the case at hand, this results shows that by forming integrals involving products of the N_2 $v = 0$ function $\Psi(t = 0)$

$$\boldsymbol{\Psi_0} = \left(\frac{\alpha}{\pi}\right)^{1/4} e^{-\alpha x^2/2} = 3.53333\mathring{A}^{-\frac{1}{2}} e^{-(244.83\mathring{A}^{-2})\,(r-1.09769\mathring{A})^2}$$

and various N_2^+ vibrational functions Ψ_v, one can determine how Ψ will evolve in time and the amplitudes of all $\{\Psi_v\}$ that it will contain. For example, the N_2 $v = 0$ function, upon ionization, contains the following amount of the N_2^+ $v = 0$ function:

$$\boldsymbol{C_0} = \int \Psi_0^*(N_2^+)\Psi_0(N_2)d\tau$$

$$= \int_{-\infty}^{\infty} 3.47522 e^{-229.113(r-1.11642)^2}\, 3.53333 e^{-244.83(r-1.09769)^2} dr$$

As demonstrated in problem 11, this integral reduces to 0.959. This means that the N_2 $v = 0$ state, subsequent to sudden ionization, can be represented as containing $|0.959|^2 = 0.92$ fraction of the $v = 0$ state of the N_2^+ ion.

This example relates to the well known Franck-Condon principal of spectroscopy in which squares of "overlaps" between the initial electronic state's vibrational wavefunction and the final electronic state's vibrational wavefunctions allow one to estimate the probabilities of populating various final-state vibrational levels.

END EXAMPLE

In addition to initial conditions, solutions to the Schrödinger equation must obey certain other constraints in form. They must be continuous functions of all of their spatial coordinates and must be single valued; these properties allow $\Psi^*\Psi$ to be interpreted as a probability density (i.e., the probability of finding a particle at some position can not be multivalued nor can it be "jerky" or discontinuous). The derivative of the wavefunction must also be continuous except at points where the potential function undergoes an infinite jump (e.g., at the wall of an infinitely high and steep potential barrier). This condition relates to the fact that the momentum must be continuous except at infinitely "steep" potential barriers where the momentum undergoes a "sudden" reversal.

2. An experimental measurement of any quantity (whose corresponding operator is $\boldsymbol{F}$) must result in one of the eigenvalues f_j of the operator $\boldsymbol{F}$. These eigenvalues are obtained by solving

$$\boldsymbol{F}\phi_j = f_j\phi_j,$$

where the ϕ_j are the eigenfunctions of $\boldsymbol{F}$. Once the measurement of $\boldsymbol{F}$ is made, for that sub-population of the experimental sample found to have the particular eigenvalue f_j, the wavefunction becomes ϕ_j.

The equation $\boldsymbol{H}\psi_k = E_k\psi_k$ is but a special case; it is an especially important case because much of the machinery of modern experimental chemistry is directed at placing the system in a particular energy quantum state by detecting its energy (e.g., by spectroscopic means).

The reader is strongly urged to also study appendix C to gain a more detailed and illustrated treatment of this and subsequent rules of quantum mechanics.

3. The operators $\boldsymbol{F}$ corresponding to *all* physically measurable quantities are Hermitian; this means that their matrix representations obey (see appendix C for a description of the "bra" |> and "ket" <| notation used below):

$$<\chi_j|\boldsymbol{F}|\chi_k> = <\chi_k|\boldsymbol{F}|\chi_j>^* = <\boldsymbol{F}\chi_j|\chi_k>$$

in any basis $\{\chi_j\}$ of functions appropriate for the action of $\boldsymbol{F}$ (i.e., functions of the variables on which $\boldsymbol{F}$ operates). As expressed through equality of the first and third elements above, Hermitian operators are often said to "obey the turn-over rule." This means that $\boldsymbol{F}$ can be allowed to operate on the function to its right or on the function to its left if $\boldsymbol{F}$ is Hermitian.

Hermiticity assures that the eigenvalues $\{f_j\}$ are all real, that eigenfunctions $\{\chi_j\}$ having different eigenvalues are orthogonal and can be normalized $<\chi_j|\chi_k> = \delta_{j,k}$, and that eigenfunctions having the same eigenvalues can be made orthonormal (these statements are proven in appendix C).

4. Once a particular value f_j is observed in a measurement of F, this same value will be observed in all subsequent measurements of F as long as the system remains undisturbed by measurements of other properties or by interactions with external fields. In fact, once f_i has been observed, the state of the system becomes an eigenstate of $\boldsymbol{F}$ (if it already was, it remains unchanged):

$$\boldsymbol{F}\Psi = f_i\Psi.$$

This means that the measurement process itself may interfere with the state of the system and even determines what that state will be once the measurement has been made.

EXAMPLE

Again consider the $v = 0$ N_2 ionization treated in problem 11 of this chapter. If, subsequent to ionization, the N_2^+ ions produced were probed to determine their internal vibrational state, a fraction of the sample equal to $|\langle\Psi(N_2; v = 0)|\Psi(N_2^+; v = 0)\rangle|^2 = 0.92$ would be detected in the $v = 0$ state of the N_2^+ ion. For this sub-sample, the vibrational wavefunction becomes, and remains from then on,

$$\Psi(t) = \Psi(N_2^+; v = 0)\exp(-itE^+_{v=0}/\hbar),$$

where $E^+_{v=0}$ is the energy of the N_2^+ ion in its $v = 0$ state. If, at some later time, this sub-sample is again probed, *all* species will be found to be in the $v = 0$ state.

END EXAMPLE

5. The probability P_k of observing a particular value f_k when F is measured, given that the system wavefunction is Ψ prior to the measurement, is given by expanding Ψ in terms of the complete set of normalized eigenstates of $\boldsymbol{F}$

$$\Psi = \sum_j |\phi_j\rangle\langle\phi_j|\Psi\rangle$$

and then computing $P_k = |\langle\phi_K|\Psi\rangle|^2$. For the special case in which Ψ is already one of the eigenstates of $\boldsymbol{F}$ (i.e., $\Psi = \phi_k$), the probability of observing f_j reduces to $P_j = \delta_{j,k}$. The set of numbers $C_j = \langle\phi_j|\Psi\rangle$ are called the expansion coefficients of Ψ in the basis of the $\{\phi_j\}$. These coefficients, when collected together in all possible products as $D_{j,i} = C_i^* C_j$ form the so-called density matrix $D_{j,i}$ of the wavefunction Ψ within the $\{\phi_j\}$ basis.

EXAMPLE

If $\boldsymbol{F}$ is the operator for momentum in the x-direction and $\Psi(x,t)$ is the wave function for x as a function of time t, then the above expansion corresponds to a Fourier transform of Ψ

$$\Psi(x,t) = 1/2\pi\int\exp(ikx)\int\exp(-ikx')\Psi(x',t)dx'dk.$$

Here $(1/2\pi)^{1/2}\exp(ikx)$ is the *normalized* eigenfunction of $F = -i\hbar\partial/\partial x$ corresponding to momentum eigenvalue $\hbar k$. These momentum eigenfunctions are orthonormal:

$$1/2\pi\int\exp(-ikx)\exp(ik'x)dx = \delta(k - k'),$$

and they form a complete set of functions in x-space

$$1/2\pi\int\exp(-ikx)\exp(ikx')dk = \delta(x - x')$$

because F is a Hermitian operator. The function $\int\exp(-ikx')\Psi(x',t)dx'$ is called the momentum-space transform of $\Psi(x,t)$ and is denoted $\Psi(k,t)$; it gives, when used as $\Psi^*(k,t)\Psi(k,t)$, the probability density for observing momentum values $\hbar k$ at time t.

END EXAMPLE

ANOTHER EXAMPLE

Take the initial Ψ to be a superposition state of the form

$$\psi = a(2p_0 + 2p_{-1} - 2p_1) + b(3p_0 - 3p_{-1}),$$

where the a and b are amplitudes that describe the admixture of $2p$ and $3p$ functions in this wavefunction. Then:

a. If L^2 were measured, the value $2\hbar^2$ would be observed with probability $3|a|^2 + 2|b|^2 = 1$, since all of the functions in ψ are p-type orbitals. After said measurement, the wavefunction would still be this same ψ because this entire ψ is an eigenfunction of L^2.

b. If L_z were measured for this

$$\psi = a(2p_0 + 2p - 1 - 2p_1) + b(3p_0 - 3p_{-1}),$$

the values $0\hbar$, $1\hbar$, and $-1\hbar$ would be observed (because these are the only functions with non-zero C_m coefficients for the L_z operator) with respective probabilities $|a|^2 + |b|^2$, $|-a|^2$, and $|a|^2 + |-b|^2$.

c. *After* L_z were measured, if the sub-population for which $-1\hbar$ had been detected were subjected to measurement of L^2 the value $2\hbar^2$ would certainly be found because the *new* wavefunction

$$\psi' = \{a2p_{-1} - b3p_{-1}\}(|a|^2 + |b|^2)^{-1/2}$$

is still an eigenfunction of L^2 with this eigenvalue.

d. Again after L_z were measured, if the sub-population for which $-1\hbar$ had been observed and for which the wavefunction is now

$$\Psi' = \{a2p_{-1} - b3p_{-1}\}(|a|^2 + |b|^2)^{-1/2}$$

were subjected to measurement of the energy (through the Hamiltonian operator), two values would be found. With probability $|a|^2(|a|^2 + |b|^2)^{-1}$ the energy of the $2p_{-1}$ orbital would be observed; with probability $|-b|^2(|a|^2 + |b|^2)^{-1}$, the energy of the $3p_{-1}$ orbital would be observed.

END EXAMPLE

If Ψ is a function of several variables (e.g., when Ψ describes more than one particle in a composite system), and if F is a property that depends on a subset of these variables (e.g., when F is a property of one of the particles in the composite system), then the expansion

$$\Psi = \sum_j |\phi_j\rangle\langle\phi_j|\Psi\rangle$$

is viewed as relating only to Ψ's dependence on the subset of variables related to F. In this case, the integrals $\langle\phi_k|\Psi\rangle$ are carried out over only these variables; thus the probabilities $P_k = |\langle\phi_k|\Psi\rangle|^2$ depend parametrically on the remaining variables.

EXAMPLE

Suppose that $\Psi(r,\theta)$ describes the radial (r) and angular (θ) motion of a diatomic molecule constrained to move on a planar surface. If an experiment were performed to measure the component of the rotational angular momentum of the diatomic molecule perpendicular to the surface ($\boldsymbol{L}_z = -i\hbar\partial/\partial\theta$), only values equal to $m\hbar(m = 0,1,-1,2,-2,3,-3,\ldots)$ could be observed, because these are the eigenvalues of $\boldsymbol{L}_z$:

$$\boldsymbol{L}_z\phi_m = -i\hbar\partial/\partial\theta\phi_m = m\hbar\phi_m, \text{ where}$$
$$\phi_m = (1/2\pi)^{1/2}\exp(im\theta).$$

The quantization of $\boldsymbol{L}_z$ arises because the eigenfunctions $\phi_m(\theta)$ must be periodic in θ:

$$\phi(\theta + 2\pi) = \phi(\theta).$$

Such quantization (i.e., constraints on the values that physical properties can realize) will be seen to occur whenever the pertinent wavefunction is constrained to obey a so-called boundary condition (in this case, the boundary condition is $\phi(\theta + 2\pi) = \phi(\theta)$).

Expanding the θ-dependence of Ψ in terms of the ϕ_μ

$$\Psi = \sum_m <\phi_m|\Psi>\phi_m(\theta)$$

allows one to write the probability that $m\hbar$ is observed if the angular momentum $\boldsymbol{L}_z$ is measured as follows:

$$P_m = |<\phi_m|\Psi>|^2 = |\int\phi_m^*(\theta)\Psi(r,\theta)d\theta|^2.$$

If one is interested in the probability that mh be observed when $\boldsymbol{L}_z$ is measured *regardless* of what bond length r is involved, then it is appropriate to integrate this expression over the r-variable about which one does not care. This, in effect, sums contributions from all r-values to obtain a result that is independent of the r variable. As a result, the probability reduces to:

$$P_m = \int\phi^*(\theta')\left\{\int\Psi^*(r,\theta')\Psi(r,\theta)rdr\right\}\phi(\theta)d\theta'd\theta,$$

which is simply the above result integrated over r with a volume element $r\,dr$ for the two-dimensional motion treated here. If, on the other hand, one were able to measure $\boldsymbol{L}_z$ values when r is equal to some specified bond length (this is only a hypothetical example; there is no known way to perform such a measurement), then the probability would equal:

$$P_m rdr = rdr\int\phi_m^*(\theta')\Psi^*(r,\theta')\Psi(r,\theta)\phi_m(\theta)d\theta'd\theta = |<\phi_m|\Psi>|^2 rdr.$$

END EXAMPLE

6. Two or more properties F, G, J whose corresponding Hermitian operators $\boldsymbol{F}, \boldsymbol{G}, \boldsymbol{J}$ commute

$$\boldsymbol{FG}-\boldsymbol{GF} = \boldsymbol{FJ}-\boldsymbol{JF} = \boldsymbol{GJ}-\boldsymbol{JG} = 0$$

have *complete sets* of simultaneous eigenfunctions (the proof of this is treated in appendix C). This means that the set of functions that are eigenfunctions of one of the operators can be formed into a set of functions that are also eigenfunctions of the others:

$$\boldsymbol{F}\phi_j = f_j\phi_j \rightarrow \boldsymbol{G}\phi_j = g_j\phi_j \rightarrow \boldsymbol{J}\phi_j = j_j\phi_j.$$

EXAMPLE

The p_x, p_y and p_z orbitals are eigenfunctions of the $\boldsymbol{L}^2$ angular momentum operator with eigenvalues equal to $L(L+1)\hbar^2 = 2\hbar^2$. Since $\boldsymbol{L}^2$ and $\boldsymbol{L}_z$ commute and act on the same (angle) coordinates, they possess a complete set of simultaneous eigenfunctions.

Although the p_x, p_y and p_z orbitals are *not* eigenfunctions of $\boldsymbol{L}_z$, they can be combined to form three new orbitals: $p_0 = p_z, p_1 = 2^{-1/2}[p_x + ip_y]$, and $p_{-1} = 2^{-1/2}[p_x - ip_y]$ that are still eigenfunctions of $\boldsymbol{L}^2$ but are now eigenfunctions of $\boldsymbol{L}_z$ also (with eigenvalues $0\hbar$, $1\hbar$, and $-1\hbar$, respectively).

END EXAMPLE

It should be mentioned that if two operators do not commute, they may still have *some* eigenfunctions in common, but they will not have a complete set of simultaneous eigenfunctions. For example, the L_z and L_x components of the angular momentum operator do not commute; however, a wavefunction with $L = 0$ (i.e., an S-state) is an eigenfunction of both operators.

The fact that two operators commute is of great importance. It means that once a measurement of one of the properties is carried out, subsequent measurement of that property or of any of the other properties corresponding to *mutually commuting* operators can be made without altering the system's value of the properties measured earlier. Only subsequent measurement of another property whose operator does not commute with $\boldsymbol{F}$, $\boldsymbol{G}$, or $\boldsymbol{J}$ will destroy precise knowledge of the values of the properties measured earlier.

EXAMPLE

Assume that an experiment has been carried out on an atom to measure its total angular momentum $\boldsymbol{L}^2$. According to quantum mechanics, only values equal to $L(L+1)\hbar^2$ will be observed. Further assume, for the particular experimental sample subjected to observation, that values of $\boldsymbol{L}^2$ equal to $2\hbar^2$ and $0\hbar^2$ were detected in relative amounts of 64% and 36% , respectively. This means that the atom's original wavefunction Ψ could be represented as:

$$\Psi = 0.8P + 0.6S,$$

where P and S represent the P-state and S-state components of Ψ. The squares of the amplitudes 0.8 and 0.6 give the 64% and 36% probabilities mentioned above.

Now assume that a subsequent measurement of the component of angular momentum along the lab-fixed z-axis is to be measured for that sub-population of the original sample found to be in the P-state. For that population, the wavefunction is now a pure P-function:

$$\Psi' = P.$$

However, at this stage we have no information about how much of this Ψ' is of $m = 1, 0$, or -1, nor do we know how much $2p$, $3p$, $4p$, . . . np components this state contains.

Because the property corresponding to the operator $\boldsymbol{L}_z$ is about to be measured, we express the above Ψ' in terms of the eigenfunctions of $\boldsymbol{L}_z$:

$$\Psi' = P = \Sigma_{m=1,0,-1} C'_m P_m.$$

When the measurement of $\boldsymbol{L}_z$ is made, the values $1\hbar$, $0\hbar$, and $-1\hbar$ will be observed with probabilities given by $|C'_1|^2$, $|C'_0|^2$, and$|C'_{-1}|^2$, respectively. For that sub-population found to have, for example, $\boldsymbol{L}_z$ equal to $-1\hbar$, the wavefunction then becomes

$$\Psi'' = P_{-1}.$$

At this stage, we do not know how much of $2p_{-1}$, $3p_{-1}$, $4p_{-1}$, . . . np_{-1} this wavefunction contains. To probe this question another subsequent measurement of the energy (corresponding to the H operator) could be made. Doing so would allow the amplitudes in the expansion of the above $\Psi'' = P_{-1}$

$$\Psi'' = P_{-1} = \sum_n C''_n nP_{-1}$$

to be found.

The kind of experiment outlined above allows one to find the content of each particular component of an initial sample's wavefunction. For example, the original wavefunction has $0.64|C''_n|^2|C'_m|^2$ fractional content of the various nP_m functions. It is analogous to the other examples considered above because all of the operators whose properties are measured commute.

END EXAMPLE

ANOTHER EXAMPLE

Let us consider an experiment in which we begin with a sample (with wavefunction Ψ) that is first subjected to measurement of $\boldsymbol{L}_z$ and then subjected to measurement of L^2 and then of the energy. In this order, one would first find specific values (integer multiples of $\hbar$) of $\boldsymbol{L}_z$ and one would express Ψ as

$$\Psi = \sum_m D_m \Psi_m.$$

At this stage, the nature of each Ψ_m is unknown (e.g., the Ψ_1 function can contain np_1, $n'd_1$, $n''f_1$, etc. components); all that is known is that Ψ_m has $m\hbar$ as its $\boldsymbol{L}_z$ value.

Taking that sub-population ($|D_m|^2$ fraction) with a particular $m\hbar$ value for $\boldsymbol{L}_z$ and subjecting it to subsequent measurement of L^2 requires the current wavefunction Ψ_m to be expressed as

$$\Psi_m = \sum_L D_{L,m} \Psi_{L,m}.$$

When L^2 is measured the value $L(L+1)\hbar^2$ will be observed with probability $|D_{L,m}|^2$, and the wavefunction for that particular sub-population will become

$$\Psi'' = \Psi_{L,m}.$$

At this stage, we know the value of L and of m, but we do not know the energy of the state. For example, we may know that the present sub-population has $L = 1$, $m = -1$, but we have no knowledge (yet) of how much $2p_{-1}$, $3p_{-1}$, . . . np_{-1} the system contains.

To further probe the sample, the above sub-population with $L = 1$ and $m = -1$ can be subjected to measurement of the energy. In this case, the function $\Psi_{1,-1}$ must be expressed as

$$\Psi_{1,-1} = \sum_n D''_n \, nP_{-1}.$$

When the energy measurement is made, the state nP_{-1} will be found $|D''_n|^2$ fraction of the time.

END EXAMPLE

The fact that $\boldsymbol{L_z}$, $\boldsymbol{L^2}$, and $\boldsymbol{H}$ all commute with one another (i.e., are *mutually commutative*) makes the series of measurements described in the above examples more straightforward than if these operators did not commute.

In the first experiment, the fact that they are mutually commutative allowed us to expand the 64% probable $\boldsymbol{L^2}$ eigenstate with $L = 1$ in terms of functions that were eigenfunctions of the operator for which measurement was *about* to be made without destroying our knowledge of the value of L^2. That is, because $\boldsymbol{L^2}$ and $\boldsymbol{L_z}$ *can have simultaneous eigenfunctions*, the $L = 1$ function can be expanded in terms of functions that are eigenfunctions of *both* $\boldsymbol{L^2}$ and $\boldsymbol{L_z}$. This in turn, allowed us to find experimentally the sub-population that had, for example $-1\hbar$ as its value of L_z while retaining knowledge that the state *remains* an eigenstate of $\boldsymbol{L^2}$ (the state at this time had $L = 1$ *and* $m = -1$ and was denoted P_{-1}). Then, when this P_{-1} state was subjected to energy measurement, knowledge of the energy of the sub-population could be gained *without* giving up knowledge of the L^2 and L_z information; upon carrying out said measurement, the state became nP_{-1}.

We therefore conclude that the act of carrying out an experimental measurement disturbs the system in that it causes the system's wavefunction to become an eigenfunction of the operator whose property is measured. If two properties whose corresponding operators commute are measured, the measurement of the second property does *not* destroy knowledge of the first property's value gained in the first measurement.

On the other hand, as detailed further in appendix C, if the two properties ($\boldsymbol{F}$ and $\boldsymbol{G}$) do not commute, the second measurement destroys knowledge of the first property's value. After the first measurement, Ψ is an eigenfunction of $\boldsymbol{F}$; after the second measurement, it becomes an eigenfunction of $\boldsymbol{G}$. If the two non-commuting operators' properties are measured in the opposite order, the wavefunction first is an eigenfunction of $\boldsymbol{G}$, and subsequently becomes an eigenfunction of $\boldsymbol{F}$.

It is thus often said that "measurements for operators that do not commute interfere with one another." The simultaneous measurement of the position and momentum along the same axis provides an example of two measurements that are incompatible. The fact that $\boldsymbol{x} = x$ and $\boldsymbol{p_x} = -i\hbar\partial/\partial x$ do not commute is straightforward to demonstrate:

$$\{x(-i\hbar\partial/\partial x)\,\chi - (-i\hbar\partial/\partial x)x\chi\} = i\hbar\chi.$$

Operators that commute with the Hamiltonian and with one another form a particularly important class because each such operator permits each of the energy eigenstates of the system to be labelled with a corresponding quantum number. These operators are called **symmetry operators.** As will be seen later, they include angular momenta (e.g., L^2, L_z, S^2, S_z, for atoms) and point group symmetries (e.g., planes and rotations about axes). Every operator that qualifies as a symmetry operator provides a quantum number with which the energy levels of the system can be labeled.

7. If a property F is measured for a large number of systems all described by the same Ψ, the average value $\langle F\rangle$ of F for such a set of measurements can be computed as

$$\langle F\rangle = \langle\Psi|\boldsymbol{F}|\Psi\rangle.$$

Expanding Ψ in terms of the complete set of eigenstates of $\boldsymbol{F}$ allows $\langle F\rangle$ to be rewritten as follows:

$$\langle F\rangle = \sum_j f_j|\langle\phi_j|\Psi\rangle|^2,$$

which clearly expresses $<F>$ as the product of the probability P_j of obtaining the particular value f_j when the property F is measured and the value f_j of the property in such a measurement. This same result can be expressed in terms of the density matrix $D_{i,j}$ of the state Ψ defined above as:

$$<F> = \sum_{i,j} <\Psi|\phi_i><\phi_i|\boldsymbol{F}|\phi_j><\phi_j|\Psi> = \sum_{i,j} C_i^* <\phi_i|\boldsymbol{F}|\phi C_j$$

$$= \sum_{i,j} D_{j,i} <\phi_i|\boldsymbol{F}|\phi_j> = Tr(DF).$$

Here, DF represents the matrix product of the density matrix $D_{j,i}$ and the matrix representation $F_{i,j} = <\phi_i|\boldsymbol{F}|\phi_j>$ of the $\boldsymbol{F}$ operator, both taken in the $\{\phi_j\}$ basis, and Tr represents the matrix trace operation.

As mentioned at the beginning of this section, this set of rules and their relationships to experimental measurements can be quite perplexing. The structure of quantum mechanics embodied in the above rules was developed in light of new scientific observations (e.g., the photoelectric effect, diffraction of electrons) that could not be interpreted within the conventional pictures of classical mechanics. Throughout its development, these and other experimental observations placed severe constraints on the structure of the equations of the new quantum mechanics as well as on their interpretations. For example, the observation of discrete lines in the emission spectra of atoms gave rise to the idea that the atom's electrons could exist with only certain discrete energies and that light of specific frequencies would be given off as transitions among these quantized energy states took place.

Even with the assurance that quantum mechanics has firm underpinnings in experimental observations, students learning this subject for the first time often encounter difficulty. Therefore, it is useful to again examine some of the model problems for which the Schrödinger equation can be exactly solved and to learn how the above rules apply to such concrete examples.

The examples examined earlier in this chapter and those given in the exercises and problems serve as useful models for chemically important phenomena: electronic motion in polyenes, in solids, and in atoms as well as vibrational and rotational motions. Their study thus far has served two purposes; it allowed the reader to gain some familiarity with applications of quantum mechanics and it introduced models that play central roles in much of chemistry. Their study now is designed to illustrate how the above seven rules of quantum mechanics relate to experimental reality.

B. An Example Illustrating Several of the Fundamental Rules

The physical significance of the time-independent wavefunctions and energies treated in section II as well as the meaning of the seven fundamental points given above can be further illustrated by again considering the simple two-dimensional electronic motion model.

If the electron were prepared in the eigenstate corresponding to $n_x = 1$, $n_y = 2$, its total energy would be

$$E = \pi^2\hbar^2/2m[1^2/L_x^2 + 2^2/L_y^2] \; .$$

If the energy were experimentally measured, this and only this value would be observed, and this same result would hold for all time as long as the electron is undisturbed.

If an experiment were carried out to measure the momentum of the electron along the y-axis, according to the second postulate above, only values equal to the eigenvalues of $-i\hbar\partial/\partial y$ could be observed. The p_y eigenfunctions (i.e., functions that obey $p_yF = -i\hbar\partial/\partial yF = cF$) are of the form

$$(1/L_y)^{1/2}\exp(ik_y y),$$

where the momentum $\hbar k_y$ can achieve any value; the $(1/L_y)^{1/2}$ factor is used to normalize the eigenfunctions over the range $0 \le y \le L_y$. It is useful to note that the y-dependence of Ψ as expressed above $[\exp(i2\pi y/L_y) - \exp(-i2\pi y/L_y)]$ is already written in terms of two such eigenstates of $-i\hbar\partial/\partial y$:

$$-i\hbar\partial/\partial y \exp(i2\pi y/L_y) = 2h/L_y \exp(i2\pi y/L_y), \text{ and}$$
$$-i\hbar\partial/\partial y \exp(-i2\pi y/L_y) = -2h/L_y \exp(-i2\pi y/L_y).$$

Thus, the expansion of Ψ in terms of eigenstates of the property being measured dictated by the fifth postulate above is already accomplished. The only two terms in this expansion correspond to momenta along the y-axis of $2h/L_y$ and $-2h/L_y$; the probabilities of observing these two momenta are given by the squares of the expansion coefficients of Ψ in terms of the normalized eigenfunctions of $-i\hbar\partial/\partial y$. The functions $(1/L_y)^{1/2}\exp(i2\pi y/L_y)$ and $(1/L_y)^{1/2}\exp(-i2\pi y/L_y)$ are such normalized eigenfunctions; the expansion coefficients of these functions in Ψ are $2^{-1/2}$ and $-2^{-1/2}$, respectively. Thus the momentum $2h/L_y$ will be observed with probability $(2^{-1/2})^2 = 1/2$ and $-2h/L_y$ will be observed with probability $(-2^{-1/2})^2 = 1/2$. If the momentum along the x-axis were experimentally measured, again only two values $1h/L_x$ and $-1h/L_x$ would be found, each with a probability of $1/2$.

The average value of the momentum along the x-axis can be computed either as the sum of the probabilities multiplied by the momentum values:

$$\langle p_x \rangle = 1/2[1h/L_x - 1h/L_x] = 0,$$

or as the so-called *expectation value* integral shown in the seventh postulate:

$$\langle p_x \rangle = \iint \psi^*(-i\hbar\partial\psi/\partial x)dxdy.$$

Inserting the full expression for $\Psi(x,y)$ and integrating over x and y from 0 to L_x and L_y, respectively, this integral is seen to vanish. This means that the result of a large number of measurements of p_x on electrons each described by the same Ψ will yield zero net momentum along the x-axis; half of the measurements will yield positive momenta and half will yield negative momenta of the same magnitude.

The time evolution of the full wavefunction given above for the $n_x = 1$, $n_y = 2$ state is easy to express because this Ψ is an energy eigenstate:

$$\Psi(x,y,t) = \psi(x,y)\exp(-iEt/\hbar).$$

If, on the other hand, the electron had been prepared in a state $\psi(x,y)$ that is not a pure eigenstate (i.e., cannot be expressed as a single energy eigenfunction), then the time evolution is more complicated. For example, if at $t = 0$ ψ were of the form

$$\psi = (2/L_x)^{1/2}(2/L_y)^{1/2}[a\sin(2\pi x/L_x)\sin(1\pi y/L_y) + b\sin(1\pi x/L_x)\sin(2\pi y/L_y)],$$

with a and b both real numbers whose squares give the probabilities of finding the system in the respective states, then the time evolution operator $\exp(-i\boldsymbol{H}t/\hbar)$ applied to ψ would yield the following time dependent function:

$$\Psi = (2/L_x)^{1/2}(2/L_y)^{1/2}[a\exp(-iE_{2,1}t/\hbar)\sin(2\pi x/L_x)$$
$$\sin(1\pi y/L_y) + b\exp(-iE_{1,2}t/\hbar)\sin(1\pi x/L_x)\sin(2\pi y/L_y)],$$

where

$$E_{2,1} = \pi^2\hbar^2/2m[2^2/L_x^2 + 1^2/L_y^2], \text{ and } E_{1,2} = \pi^2\hbar^2/2m[1^2/L_x^2 + 2^2/L_y^2]$$

The probability of finding $E_{2,1}$ if an experiment were carried out to measure energy would be $|a\exp(-iE_{2,1}t/\hbar)|^2 = |a|^2$; the probability for finding $E_{1,2}$ would be $|b|^2$. The spatial probability distribution for finding the electron at points x,y will, in this case, be given by:

$$|\Psi|^2 = |a|^2|\psi_{2,1}|^2 + 1|b|^2|\psi_{1,2}|^2 + 2ab\psi_{2,1}\psi_{1,2}\cos(\Delta Et/\hbar),$$

where ΔE is $E_{2,1} - E_{1,2}$,

$$\psi_{2,1} = (2/L_x)^{1/2}(2/L_y)^{1/2}\sin(2\pi x/L_x)\sin(1\pi y/L_y),$$

and

$$\psi_{1,2} = (2/L_x)^{1/2}(2/L_y)^{1/2}\sin(1\pi x/L_x)\sin(2\pi y/L_y).$$

This spatial distribution is not stationary but evolves in time. So in this case, one has a wavefunction that is not a pure eigenstate of the Hamiltonian (one says that Ψ is a superposition state or a non-stationary state) whose average energy remains constant ($E = E_{2,1}|a|^2 + E_{1,2}|b|^2$) but whose spatial distribution changes with time.

Although it might seem that most spectroscopic measurements would be designed to prepare the system in an eigenstate (e.g., by focusing on the sample light whose frequency matches that of a particular transition), such need not be the case. For example, if very short laser pulses are employed, the Heisenberg uncertainty broadening ($\Delta E\Delta t \geq \hbar$) causes the light impinging on the sample to be very non-monochromatic (e.g., a pulse time of 1×10^{-12} sec corresponds to a frequency spread of approximately 5cm^{-1}). This, in turn, removes any possibility of preparing the system in a particular quantum state with a resolution of better than 30cm^{-1} because the system experiences time oscillating electromagnetic fields whose frequencies range over at least 5cm^{-1}).

Essentially all of the model problems that have been introduced in this chapter to illustrate the application of quantum mechanics constitute widely used, highly successful "starting-point" models for important chemical phenomena. As such, it is important that students retain working knowledge of the energy levels, wavefunctions, and symmetries that pertain to these models.

Thus far, exactly soluble model problems that represent one or more aspects of an atom or molecule's quantum-state structure have been introduced and solved. For example, electronic motion in polyenes was modeled by a particle-in-a-box. The harmonic oscillator and rigid rotor were introduced to model vibrational and rotational motion of a diatomic molecule.

As chemists, we are used to thinking of electronic, vibrational, rotational, and translational energy levels as being (at least approximately) separable. On the other hand, we are aware that situations exist in which energy can flow from one such degree of freedom to another (e.g., electronic-to-vibrational energy flow occurs in radiationless relaxation and vibration-rotation couplings are important in molecular spectroscopy). It is important to understand how the simplifications that allow us to focus on electronic or vibrational or rotational motion arise, how they can be obtained from a first-principles derivation, and what their limitations and range of accuracy are.

CHAPTER

2 Approximation methods can be used when exact solutions to the Schrödinger equation cannot be found.

In applying quantum mechanics to "real" chemical problems, one is usually faced with a Schrödinger differential equation for which, to date, no one has found an analytical solution. This is equally true for electronic and nuclear-motion problems. It has therefore proven essential to develop and efficiently implement mathematical methods which can provide approximate solutions to such eigenvalue equations. Two methods are widely used in this context—the variational method and perturbation theory. These tools, whose use permeates virtually all areas of theoretical chemistry, are briefly outlined here, and the details of perturbation theory are amplified in appendix D.

I. THE VARIATIONAL METHOD

For the kind of potentials that arise in atomic and molecular structure, the Hamiltonian H is a Hermitian operator that is bounded from below (i.e., it has a lowest eigenvalue). Because it is Hermitian, it possesses a complete set of orthonormal eigenfunctions $\{\psi_j\}$. *Any* function Φ that depends on the same spatial and spin variables on which H operates and obeys the same boundary conditions that the $\{\psi_j\}$ obey can be expanded in this complete set

$$\Phi = \sum_j C_j \psi_j.$$

The expectation value of the Hamiltonian for any such function can be expressed in terms of its C_j coefficients and the *exact* energy levels E_j of H as follows:

$$<\Phi|H|\Phi> = \sum_{ij} C_i^* C_j <\psi_i|H|\psi_j> = \sum_j |C_j|^2 E_j.$$

If the function Φ is normalized, the sum $\Sigma_j|C_j|^2$ is equal to unity. Because H is bounded from below, all of the E_j must be greater than or equal to the lowest energy E_0. Combining the latter two observations allows the energy expectation value of Φ to be used to produce a very important inequality:

$$<\Phi|H|\Phi> \geq E_0.$$

The equality can hold only if Φ is equal to ψ_0; if Φ contains components along any of the other ψ_j, the energy of Φ will exceed E_0.

This upper-bound property forms the basis of the so-called *variational method* in which "trial wavefunctions" Φ are constructed:

1. To guarantee that Φ obeys all of the boundary conditions that the exact ψ_j do and that Φ is of the proper spin and space symmetry and is a function of the same spatial and spin coordinates as the ψ_j;
2. With parameters embedded in Φ whose "optimal" values are to be determined by making $<\Phi|H|\Phi>$ *a minimum.*

It is perfectly acceptable to vary any parameters in Φ to attain the lowest possible value for $<\Phi|H|\Phi>$ because the proof outlined above constrains this expectation value to be above the true lowest eigenstate's energy E_0 for *any* Φ. The philosophy then is that the Φ that gives the lowest $<\Phi|H|\Phi>$ is the best because its expectation value is closes to the exact energy.

Quite often a *trial wavefunction* is expanded as a linear combination of other functions

$$\Phi = \sum_J C_J \Phi_J.$$

In these cases, one says that a "linear variational" calculation is being performed. The set of functions $\{\Phi_J\}$ are usually constructed to obey all of the boundary conditions that the exact state Ψ obeys, to be functions of the the same coordinates as Ψ, and to be of the same spatial and spin symmetry as Ψ. Beyond these conditions, the $\{\Phi_J\}$ are nothing more than members of a set of functions that are convenient to deal with (e.g., convenient to evaluate Hamiltonian matrix elements $<\Phi_I|H|\Phi_J>$) and that can, in principle, be made complete if more and more such functions are included.

For such a trial wavefunction, the energy depends quadratically on the "linear variational" C_J coefficients:

$$<\Phi|H|\Phi> = \sum_{IJ} C_I^* C_J <\Phi_I|H|\Phi_J> .$$

Minimization of this energy with the constraint that Φ remain normalized ($<\Phi|\Phi> = 1 = \sum_{IJ} C_I^* C_J <\Phi_I|\Phi_J>$) gives rise to a so-called *secular* or eigenvalue-eigenvector problem:

$$\sum_J [<\Phi_I|H|\Phi_J> - E<\Phi_I|\Phi_J>]C_J = \sum_J [H_{IJ} - ES_{IJ}]C_J = 0.$$

If the functions $\{\Phi_J\}$ are orthonormal, then the overlap matrix S reduces to the unit matrix and the above generalized eigenvalue problem reduces to the more familiar form:

$$\sum_J H_{IJ} C_J = E C_I.$$

The secular problem, in either form, has as many eigenvalues E_i and eigenvectors $\{C_{iJ}\}$ as the dimension of the H_{IJ} matrix. It can also be shown that between successive pairs of the eigenvalues obtained by solving the secular problem at least one exact eigenvalue must occur (i.e., $E_{i+1} > E_{exact} > E_i$, for all i). This observation is referred to as "the bracketing theorem."

Variational methods, in particular the linear variational method, are the most widely used approximation techniques in quantum chemistry. To implement such a method one needs to know the Hamiltonian H whose energy levels are sought and one needs to construct a trial wavefunction in which some "flexibility" exists (e.g., as in the linear variational method where the C_J coefficients can be varied). In section 6 this tool will be used to develop several of the most commonly used and powerful molecular orbital methods in chemistry.

II. PERTURBATION THEORY[1]

Perturbation theory is the second most widely used approximation method in quantum chemistry. It allows one to estimate the splittings and shifts in energy levels and changes in wavefunctions that occur when an external field (e.g., an electric or magnetic field or a field that is due to a surrounding set of "ligands"—a crystal field) or a field arising when a previously ignored term in the Hamiltonian is applied to a species whose "unperturbed" states are known. These "perturbations" in energies and wavefunctions are expressed in terms of the (complete) set of unperturbed eigenstates.

Assuming that *all* of the wavefunctions Φ_k and energies E_k^0 belonging to the unperturbed Hamiltonian H^0 are known

$$H^0\Phi_k = E_k^0\Phi_k,$$

and given that one wishes to find eigenstates (ψ_k and E_k) of the perturbed Hamiltonian

$$H = H^0 + \lambda V,$$

perturbation theory expresses ψ_k and E_k as power series in the perturbation strength λ:

$$\psi_k = \sum_{n=0}^{\infty} \lambda^n \psi_k^{(n)}$$

$$E_k = \sum_{n=0}^{\infty} \lambda^n E_k^{(n)}.$$

The systematic development of the equations needed to determine the $E_k^{(n)}$ and the $\psi_k^{(n)}$ is presented in appendix D. Here, we simply quote the few lowest-order results.

---

[1] Suggested extra reading: appendix D; Time Independent Perturbation Theory

The zeroth-order wavefunctions and energies are given in terms of the solutions of the unperturbed problem as follows:

$$\psi_k^{(0)} = \Phi_k \text{ and } E_k^{(0)} = E_k^0.$$

This simply means that one must be willing to identify one of the unperturbed states as the "best" approximation to the state being sought. This, of course, implies that one must therefore strive to find an unperturbed model problem, characterized by H^0 that represents the true system as accurately as possible, so that one of the Φ_k will be as close as possible to ψ_k.

The first-order energy correction is given in terms of the zeroth-order (i.e., unperturbed) wavefunction as:

$$E_k^{(1)} = \langle\Phi_k|V|\Phi_k\rangle,$$

which is identified as the average value of the perturbation taken with respect to the unperturbed function Φ_k. The so-called *first-order wavefunction* $\psi_k^{(1)}$ expressed in terms of the complete set of unperturbed functions $\{\Phi_J\}$ is:

$$\psi_k^{(1)} = \sum_{j \neq k} \langle\Phi_j|V|\Phi_k\rangle/[E_k^0 - E_j^0]|\Phi_j\rangle.$$

The *second-order energy* correction is expressed as follows:

$$E_k^{(2)} = \sum_{j \neq k} |\langle\Phi_j|V|\Phi_k\rangle|^2/[E_k^0 - E_j^0],$$

and the second-order correction to the wavefunction is expressed as

$$\psi_k^{(2)} = \sum_{j \neq k} [E_k^0 - E_j^0]^{-1} \sum_{l \neq k} \{\langle\Phi_j|V|\Phi_l\rangle - \delta_{j,l} E_k^{(1)}\} \langle\Phi_l|V|\Phi_k\rangle \, [E_k^0 - E_l^0]^{-1} |\Phi_j\rangle.$$

An essential point about perturbation theory is that the energy corrections $E_k^{(n)}$ and wavefunction corrections $\psi_k^{(n)}$ are expressed in terms of integrals over the unperturbed wavefunctions Φ_k involving the perturbation (i.e., $\langle\Phi_j|V|\Phi_l\rangle$) and the unperturbed energies E_j^0. Perturbation theory is most useful when one has, in hand, the solutions to an unperturbed Schrödinger equation that is reasonably "close" to the full Schrödinger equation whose solutions are being sought. In such a case, it is likely that low-order corrections will be adequate to describe the energies and wavefunctions of the full problem.

It is important to stress that although the solutions to the full "perturbed" Schrödinger equation are expressed, as above, in terms of sums over all states of the unperturbed Schrödinger equation, it is improper to speak of the perturbation as creating excited-state species. For example, the polarization of the $1s$ orbital of the Hydrogen atom caused by the application of a static external electric field of strength E along the z-axis is described, in first-order perturbation theory, through the sum

$$\sum_{n=2,\infty} \phi_{np_0} \langle\phi_{np_0}|\boldsymbol{E}\, e\, r \cos\theta|1s\rangle \, [E_{1s} - E_{np_0}]^{-1}$$

over all $p_z = p_0$ orbitals labeled by principal quantum number n. The coefficient multiplying each p_0 orbital depends on the energy gap corresponding to the $1s$-to-np "excitation"'as well as the electric dipole integral $\langle\phi_{np_0}|\boldsymbol{E}\, er\cos\theta|1s\rangle$ between the $1s$ orbital and the np_0 orbital.

This sum describes the polarization of the $1s$ orbital in terms of functions that have p_0 symmetry; by combining an s orbital and p_0 orbitals, one can form a "hybrid-like" orbital that is nothing but a distorted $1s$ orbital. The appearance of the excited np_0 orbitals has nothing to do with forming excited states; these np_0 orbitals simply provide a set of functions that can describe the response of the $1s$ orbital to the applied electric field.

The relative strengths and weaknesses of perturbation theory and the variational method, as applied to studies of the electronic structure of atoms and molecules, are discussed in section 6.

CHAPTER

3 The application of the Schrödinger equation to the motions of electrons and nuclei in a molecule lead to the chemists' picture of electronic energy surfaces on which vibration and rotation occurs and among which transitions take place.

I. THE BORN-OPPENHEIMER SEPARATION OF ELECTRONIC AND NUCLEAR MOTIONS

Many elements of chemists' pictures of molecular structure hinge on the point of view that separates the electronic motions from the vibrational/rotational motions and treats couplings between these (approximately) separated motions as "perturbations." It is essential to understand the origins and limitations of this separated-motions picture.

To develop a framework in terms of which to understand when such separability is valid, one thinks of an atom or molecule as consisting of a collection of N electrons and M nuclei each of which possesses kinetic energy and among which coulombic potential energies of interaction arise. To properly describe the motions of all these particles, one needs to consider the *full* Schrödinger equation $H\Psi = E\Psi$, in which the Hamiltonian H contains the sum (denoted H_e) of the kinetic energies of all N electrons and the coulomb potential energies among the N electrons and the M nuclei as well as the kinetic energy T of the M nuclei

$$T = \sum_{a=1,M} (-\hbar^2/2m_a)\nabla_a^2,$$

$$H = H_e + T$$

$$H_e = \sum_j \left\{(-\hbar^2/2m_e)\nabla_j^2 - \sum_a Z_a e^2/r_{j,a}\right\} + \sum_{j<k} e^2/r_{j,k} + \sum_{a<b} Z_a Z_b e^2/R_{a,b}.$$

Here, m_a is the mass of the nucleus a, $Z_a e^2$ is its charge, and ∇_a^2 is the Laplacian with respect to the three cartesian coordinates of this nucleus (this operator ∇_a^2 is given in spherical polar coordinates in appendix A); $r_{j,a}$ is the distance between the jth electron and the ath nucleus, $r_{j,k}$ is the distance between the jth and k^{th} electrons, m_e is the electron's mass, and $R_{a,b}$ is the distance from nucleus a to nucleus b.

The full Hamiltonian H thus contains differential operators over the $3N$ electronic coordinates (denoted r as a shorthand) and the $3M$ nuclear coordinates (denoted R as a shorthand). In contrast, the electronic Hamiltonian H_e is a Hermitian differential operator in r-space but *not* in R-space. Although H_e is indeed a function of the R-variables, it is not a differential operator involving them.

Because H_e is a Hermitian operator in r-space, its eigenfunctions $\Psi_i(r|R)$ obey

$$H_e\Psi_i(r|R) = E_i(R)\Psi_i(r|R)$$

for any values of the R-variables, and form a *complete set* of functions of r for any values of R. These eigenfunctions and their eigenvalues $E_i(R)$ depend on R only because the potentials appearing in H_e depend on R. The Ψ_i and E_i are the *electronic wavefunctions* and *electronic energies* whose evaluations are treated in the next three chapters.

The fact that the set of $\{\Psi_i\}$ is, in principle, complete in r-space allows the full (electronic and nuclear) wavefunction Ψ to have its r-dependence expanded in terms of the Ψ_i:

$$\Psi(r,R) = \sum_i \Psi_i(r|R)\Xi_i(R).$$

The $\Xi_i(R)$ functions carry the remaining R-dependence of Ψ and are determined by insisting that Ψ as expressed here obey the full Schrödinger equation:

$$(H_e + T - E)\sum_i \Psi_i(r|R)\Xi_i(R) = 0.$$

Projecting this equation against $<\Psi_j(r|R)|$ (integrating only over the electronic coordinates because the Ψ_j are orthonormal only when so integrated) gives:

$$[(E_j(R) - E)\Xi_j(R) + T\Xi_j(R)] =$$

$$-\sum_i \left\{<\Psi_j|T|\Psi_i>(R)\Xi_i(R) + \sum_{a=1,M} (-\hbar^2/m_a)<\Psi_j|\nabla_a|\Psi_i>(R)\cdot\nabla_a\Xi_i(R)\right\},$$

where the (R) notation in $<\Psi_j|T|\Psi_i>(R)$ and $<\Psi_j|\nabla_a|\Psi_i>(R)$ has been used to remind one that the integrals <. . .> are carried out only over the r coordinates and, as a result, still depend on the R coordinates.

In the **Born-Oppenheimer** (BO) approximation, one neglects the so-called non-adiabatic or non-BO couplings on the right-hand side of the above equation. Doing so yields the following equations for the $\Xi_i(R)$ functions:

$$[(E_j(R) - E)\Xi_j^0(R) + T\Xi_j^0(R)] = 0,$$

where the superscript in $\Xi_i^0(R)$ is used to indicate that these functions are solutions within the BO approximation only.

These BO equations can be recognized as the equations for the *translational, rotational, and vibrational* motion of the nuclei on the "potential energy surface" $E_j(R)$. That is, within the BO picture, the electronic energies $E_j(R)$, considered as functions of the nuclear positions R, provide the potentials on which the nuclei move. The electronic and nuclear-motion aspects of the Schrödinger equation are thereby separated.

A. Time Scale Separation

The physical parameters that determine under what circumstances the BO approximation is accurate relate to the motional time scales of the electronic and vibrational/rotational coordinates.

The range of accuracy of this separation can be understood by considering the differences in time scales that relate to electronic motions and nuclear motions under ordinary circumstances. In most atoms and molecules, the electrons orbit the nuclei at speeds much in excess of even the fastest nuclear motions (the vibrations). As a result, the electrons can adjust "quickly" to the slow motions of the nuclei. This means it should be possible to develop a model in which the electrons "follow" smoothly as the nuclei vibrate and rotate.

This picture is that described by the BO approximation. Of course, one should expect large corrections to such a model for electronic states in which "loosely held" electrons exist. For example, in molecular Rydberg states and in anions, where the outer valence electrons are bound by a fraction of an electron volt, the natural orbit frequencies of these electrons are not much faster (if at all) than vibrational frequencies. In such cases, significant breakdown of the BO picture is to be expected.

B. Vibration/Rotation States for Each Electronic Surface

The BO picture is what gives rise to the concept of a manifold of potential energy surfaces on which vibrational/rotational motions occur.

Even within the BO approximation, motion of the nuclei on the various electronic energy surfaces is different because the nature of the chemical bonding differs from surface to surface. That is, the vibrational/rotational motion on the ground-state surface is certainly not the same as on one of the excited-state surfaces. However, there are a complete set of wavefunctions $\Xi^0_{j,m}(R)$ and energy levels $E^0_{j,m}$ for *each* surface $E_j(R)$ because $T + E_j(R)$ is a Hermitian operator in R-space for *each* surface (labelled j):

$$[T + E_j(R)]\ \Xi^0_{j,m}(R) = E^0_{j,m}\ \Xi^0_{j,m}.$$

The eigenvalues $E^0_{j,m}$ must be labelled by the electronic surface (j) on which the motion occurs as well as to denote the particular state (m) on that surface.

II. ROTATION AND VIBRATION OF DIATOMIC MOLECULES

For a diatomic species, the vibration-rotation (V/R) kinetic energy operator can be expressed as follows in terms of the bond length R and the angles θ and ϕ that describe the orientation of the bond axis relative to a laboratory-fixed coordinate system:

$$T_{V/R} = -\hbar^2/2\mu\{R^{-2}\partial/\partial R(R^2\partial/\partial R) - R^{-2}\hbar^{-2}L^2\},$$

where the square of the rotational angular momentum of the diatomic species is

$$L^2 = -\hbar^2\{(\sin\theta)^{-1}\partial/\partial\theta((\sin\theta)\partial/\partial\theta) + (\sin\theta)^{-2}\partial^2/\partial\phi 2\}.$$

Because the potential $E_j(R)$ depends on R but not on θ or ϕ, the V/R function $\Xi^0_{j,m}$ can be written as a product of an angular part and an R-dependent part; moreover, because L^2 contains the full angle-dependence of $T_{V/R}$, $\Xi^0_{j,n}$ can be written as

$$\Xi^0_{j,n} = Y_{J,M}(\theta,\phi)F_{j,J,v}(R).$$

The general subscript n, which had represented the state in the full set of $3M$-3 R-space coordinates, is replaced by the three quantum numbers $J,M,$ and v (i.e., once one focuses on the three specific coordinates R,θ, and ϕ, a total of three quantum numbers arise in place of the symbol n).

Substituting this product form for $\Xi^0_{j,n}$ into the V/R equation gives:

$$-\hbar^2/2\mu\{R^{-2}\partial/\partial R(R^2\partial/\partial R) - R^{-2}J(J+1)\}F_{j,J,v}(R) + E_j(R)F_{j,J,v}(R) = E^0_{j,J,v}F_{j,J,v}$$

as the equation for the vibrational (i.e., R-dependent) wavefunction within electronic state j and with the species rotating with $J(J+1)\hbar^2$ as the square of the total angular momentum and a projection along the laboratory-fixed Z-axis of $M\hbar$. The fact that the $F_{j,J,v}$ functions do not depend on the M quantum number derives from the fact that the $T_{V/R}$ kinetic energy operator does not explicitly contain J_Z; only J^2 appears in $T_{V/R}$.

The solutions for which $J=0$ correspond to vibrational states in which the species has no rotational energy; they obey

$$-\hbar^2/2\mu\{R^{-2}\partial/\partial R(R^2\partial/\partial R)\}F_{j,0,v}(R) + E_j(R)F_{j,0,v}(R) = E^0_{j,0,v}F_{j,0,v}.$$

The differential-operator parts of this equation can be simplified somewhat by substituting $F = R^{-1}\chi$ and thus obtaining the following equation for the new function χ:

$$-\hbar^2/2\mu\partial/\partial R\partial/\partial R\chi_{j,0,v}(R) + E_j(R)\chi_{j,0,v}(R) = E^0_{j,0,v}\chi_{j,0,v}.$$

Solutions for which $J \neq 0$ require the vibrational wavefunction and energy to respond to the presence of the "centrifugal potential" given by $\hbar^2J(J+1)/(2\mu R^2)$; these solutions obey the full coupled V/R equations given above.

A. Separation of Vibration and Rotation

It is common, in developing the working equations of diatomic-molecule rotational/vibrational spectroscopy, to treat the coupling between the two degrees of freedom using perturbation theory as developed later in this chapter. In particular, one can expand the centrifugal coupling $\hbar^2J(J+1)/(2\mu R^2)$ around the equilibrium geometry R_e (which depends, of course, on j):

$$\hbar^2J(J+1)/(2\mu R^2) = \hbar^2J(J+1)/(2\mu[R_e^2(1+\Delta R/R_e)^2]) = \hbar^2J(J+1)/(2\mu R_e^2)[1 - 2\Delta R/R_e + \ldots],$$

and treat the terms containing powers of the bond length displacement ΔR^k as perturbations. The zeroth-order equations read:

$$-\hbar^2/2\mu\{R^{-2}\partial/\partial R(R^2\partial/\partial R)\}F^0_{j,J,v}(R) + E_j(R)F^0_{j,J,v}(R) + \hbar^2J(J+1)/(2\mu R_e^2)F^0_{j,J,v} = E^0_{j,J,v}F^0_{j,J,v},$$

and have solutions whose energies separate

$$E^0_{j,J,v} = \hbar^2J(J+1)/(2\mu R_e^2) + E_{j,v}$$

and whose wavefunctions are independent of J (because the coupling is not R-dependent in zeroth order)

$$F^0_{j,J,v}(R) = F_{j,v}(R).$$

Perturbation theory is then used to express the corrections to these zeroth-order solutions as indicated in appendix D.

B. The Rigid Rotor and Harmonic Oscillator

Treatment of the rotational motion at the zeroth-order level described above introduces the so-called "rigid rotor" energy levels and wavefunctions: $E_J = \hbar^2 J(J+1)/(2\mu R_e^2)$ and $Y_{J,M}(\theta,\phi)$; these same quantities arise when the diatomic molecule is treated as a rigid rod of length R_e. The spacings between successive rotational levels within this approximation are

$$\Delta E_{J+1,J} = 2hcB(J+1),$$

the so-called rotational constant B is given in cm^{-1} as

$$B = h/(8\pi^2 c\mu R_e^2).$$

The rotational level J is $(2J+1)$-fold degenerate because the energy E_J is independent of the M quantum number of which there are $(2J+1)$ values for each J: $M = -J, -J+1, -J+2, \ldots J-2, J-1, J$.

The explicit form of the zeroth-order vibrational wavefunctions and energy levels, $F^0_{j,v}$ and $E^0_{j,v}$, depends on the description used for the electronic potential energy surface $E_j(R)$. In the crudest useful approximation, $E_j(R)$ is taken to be a so-called harmonic potential

$$E_j(R) \approx 1/2k_j(R - R_e)^2;$$

as a consequence, the wavefunctions and energy levels reduce to

$$E^0_{j,v} = E_j(R_e) + \hbar\sqrt{k/\mu}\,(v + 1/2), \text{ and}$$
$$F^0_{j,v}(R) = [2^v v!]^{-1/2}(\alpha/\pi)^{1/4}\exp(-\alpha(R - R_e)^2/2)H_v(\alpha^{1/2}(R - R_e)),$$

where $\alpha = (k_j\mu)^{1/2}/\hbar$ and $H_v(y)$ denotes the Hermite polynomial defined by:

$$H_v(y) = (-1)^v \exp(y^2) d^v/dy^v \exp(-y^2).$$

The solution of the vibrational differential equation

$$-\hbar^2/2\mu\left\{R^{-2}\partial/\partial R(R^2\partial/\partial R)\right\}F_{j,v}(R) + E_j(R)F_{j,v}(R) = E_{j,v}F_{j,v}$$

is treated in EWK, Atkins, and McQuarrie.

These harmonic-oscillator solutions predict evenly spaced energy levels (i.e., no anharmonicity) that persist for all v. It is, of course, known that molecular vibrations display anharmonicity (i.e., the energy levels move closer together as one moves to higher v) and that quantized vibrational motion ceases once the bond dissociation energy is reached.

C. The Morse Oscillator

The Morse oscillator model is often used to go beyond the harmonic oscillator approximation. In this model, the potential $E_j(R)$ is expressed in terms of the bond dissociation energy D_e and a parameter a related to the second derivative k of $E_j(R)$ at R_e, $k = (d^2E_j/dR^2) = 2a^2D_e$ as follows:

$$E_j(R) - E_j(R_e) = D_e\left\{1 - \exp(-a(R - R_e))\right\}^2.$$

The Morse oscillator energy levels are given by

$$E^0_{j,v} = E_j(R_e) + \hbar\sqrt{k/\mu}\,(v + 1/2) - \hbar^2/4(k/\mu D_e)(v + 1/2)^2;$$

the corresponding eigenfunctions are also known analytically in terms of hypergeometric functions (see, for example, *Handbook of Mathematical Functions*, M. Abramowitz and I. A. Stegun, Dover, Inc. New York, N.Y. (1964)). Clearly, the Morse solutions display anharmonicity as reflected in the negative term proportional to $(v + 1/2)^2$.

III. ROTATION OF POLYATOMIC MOLECULES

To describe the orientations of a diatomic or linear polyatomic molecule requires only two angles (usually termed θ and ϕ). For any non-linear molecule, three angles (usually α, β, and γ) are needed. Hence the rotational Schrödinger equation for a non-linear molecule is a differential equation in three dimensions.

There are $3M - 6$ vibrations of a non-linear molecule containing M atoms; a linear molecule has $3M - 5$ vibrations. The linear molecule requires two angular coordinates to describe its orientation with respect to a laboratory-fixed axis system; a non-linear molecule requires three angles.

A. Linear Molecules

The rotational motion of a linear polyatomic molecule can be treated as an extension of the diatomic molecule case. One obtains the $Y_{J,M}(\theta,\phi)$ as rotational wavefunctions and, within the approximation in which the centrifugal potential is approximated at the equilibrium geometry of the molecule (R_e), the energy levels are

$$E^0_J = J(J + 1)h^2/(2I).$$

Here the total moment of inertia I of the molecule takes the place of μR_e^2 in the diatomic molecule case

$$I = \sum_a m_a(R_a - R_{CofM})^2;$$

m_a is the mass of atom a whose distance from the center of mass of the molecule is $(R_a - R_{CofM})$. The rotational level with quantum number J is $(2J + 1)$-fold degenerate again because there are $(2J + 1)$ M-values.

B. Non-Linear Molecules

For a non-linear polyatomic molecule, again with the centrifugal couplings to the vibrations evaluated at the equilibrium geometry, the following terms form the rotational part of the nuclear-motion kinetic energy:

$$T_{rot} = \sum_{i=a,b,c} (J_i^2/2I_i).$$

Here, I_i is the eigenvalue of the moment of inertia tensor:

$$I_{x,x} = \sum_a m_a[(R_a - R_{CofM})^2 - (x_a - x_{CofM})^2]$$

$$I_{x,y} = \sum_a m_a[(x_a - x_{CofM})(y_a - y_{CofM})]$$

expressed originally in terms of the cartesian coordinates of the nuclei (a) and of the center of mass in an arbitrary molecule-fixed coordinate system (and similarly for $I_{z,z}$, $I_{y,y}$, $I_{x,z}$ and $I_{y,z}$). The operator J_i corresponds to the component of the total rotational angular momentum $\boldsymbol{J}$ along the direction belonging to the ith eigenvector of the moment of inertia tensor.

Molecules for which all three principal moments of inertia (the I_i's) are equal are called "spherical tops." For these species, the rotational Hamiltonian can be expressed in terms of the square of the total rotational angular momentum J^2:

$$T_{rot} = J^2/2I,$$

as a consequence of which the rotational energies once again become

$$E_J = \hbar^2 J(J+1)/2I.$$

However, the $Y_{J,M}$ are not the corresponding eigenfunctions because the operator J^2 now contains contributions from rotations about three (no longer two) axes (i.e., the three principal axes). The proper rotational eigenfunctions are the $D^J_{M,K}(\alpha,\beta,\gamma)$ functions known as "rotation matrices" (see sections 3.5 and 3.6 of Zare's book on angular momentum) these functions depend on three angles (the three Euler angles needed to describe the orientation of the molecule in space) and three quantum numbers—J, M, and K. The quantum number M labels the projection of the total angular momentum (as $M\hbar$) along the laboratory-fixed z-axis; $K\hbar$ is the projection along one of the internal principal axes (in a spherical top molecule, all three axes are equivalent, so it does not matter which axis is chosen).

The energy levels of spherical top molecules are $(2J+1)^2$-fold degenerate. Both the M and K quantum numbers run from $-J$, in steps of unity, to J; because the energy is independent of M and of K, the degeneracy is $(2J+1)^2$.

Molecules for which two of the three principal moments of inertia are equal are called symmetric top molecules. Prolate symmetric tops have $I_a < I_b = I_c$; oblate symmetric tops have $I_a = I_b < I_c$ (it is convention to order the moments of inertia as $I_a \le I_b \le I_c$). The rotational Hamiltonian can now be written in terms of J^2 and the component of $\boldsymbol{J}$ along the unique moment of inertia's axis as:

$$T_{rot} = J_a^2(1/2I_a - 1/2I_b) + J^2/2I_b$$

for prolate tops, and

$$T_{rot} = J_c^2(1/2I_c - 1/2I_b) + J^2/2I_b$$

for oblate tops. Again, the $D^J_{M,K}(\alpha,\beta,\gamma)$ are the eigenfunctions, where the quantum number K describes the component of the rotational angular momentum $\boldsymbol{J}$ along the unique molecule-fixed axis (i.e., the axis of the unique moment of inertia). The energy levels are now given in terms of J and K as follows:

$$E_{J,K} = \hbar^2 J(J+1)/2I_b + \hbar^2 K^2(1/2I_a - 1/2I_b)$$

for prolate tops, and

$E_{J,K} = \hbar^2 J(J+1)/2I_b + \hbar^2 K^2(1/2I_c - 1/2I_b)$ for oblate tops.

Because the rotational energies now depend on K (as well as on J), the degeneracies are lower than for spherical tops. In particular, because the energies do not depend on M and depend on the square of K, the degeneracies are $(2J+1)$ for states with $K = 0$ and $2(2J+1)$ for states with $|K| > 0$; the extra factor of 2 arises for $|K| > 0$ states because pairs of states with $K = |K|$ and $K = |-K|$ are degenerate.

IV. SUMMARY

This chapter has shown how the solution of the Schrödinger equation governing the motions and interparticle potential energies of the nuclei and electrons of an atom or molecule (or ion) can be decomposed into two distinct problems: (i) solution of the *electronic* Schrödinger equation for the electronic wavefunctions and energies, both of which depend on the nuclear geometry and (ii) solution of the *vibration/rotation* Schrödinger equation for the motion of the nuclei on any one of the electronic energy surfaces. This decomposition into approximately separable electronic and nuclear-motion problems remains an important point of view in chemistry. It forms the basis of many of our models of molecular structure and our interpretation of molecular spectroscopy. It also establishes how we approach the computational simulation of the energy levels of atoms and molecules; we first compute electronic energy levels at a "grid" of different positions of the nuclei, and we then solve for the motion of the nuclei on a particular energy surface using this grid of data.

The treatment of electronic motion is treated in detail in sections II, III, and VI where molecular orbitals and configurations and their computer evaluation is covered. The vibration/rotation motion of molecules on BO surfaces is introduced above, but should be treated in more detail in a subsequent course in molecular spectroscopy.

SECTION SUMMARY

This introductory section was intended to provide the reader with an overview of the structure of quantum mechanics and to illustrate its application to several exactly solvable model problems. The model problems analyzed play especially important roles in chemistry because they form the basis upon which more sophisticated descriptions of the electronic structure and rotational-vibrational motions of molecules are built. The variational method and perturbation theory constitute the tools needed to make use of solutions of simpler model problems as starting points in the treatment of Schrödinger equations that are impossible to solve analytically.

In sections 2, 3, and 6 of this text, the electronic structures of polyatomic molecules, linear molecules, and atoms are examined in some detail. Symmetry, angular momentum methods, wavefunction antisymmetry, and other tools are introduced as needed throughout the text. The application of modern computational chemistry methods to the treatment of molecular electronic structure is included. Given knowledge of the electronic energy surfaces as functions of the internal geometrical coordinates of the molecule, it is possible to treat vibrational-rotational motion on these surfaces.

Exercises, problems, and solutions are provided for each chapter. Readers are *strongly* encouraged to work these exercises and problems because new material that is used in other chapters is often developed within this context.

Exercises, problems, and solutions

REVIEW EXERCISES

1. Transform (using the coordinate system provided below) the following functions accordingly:

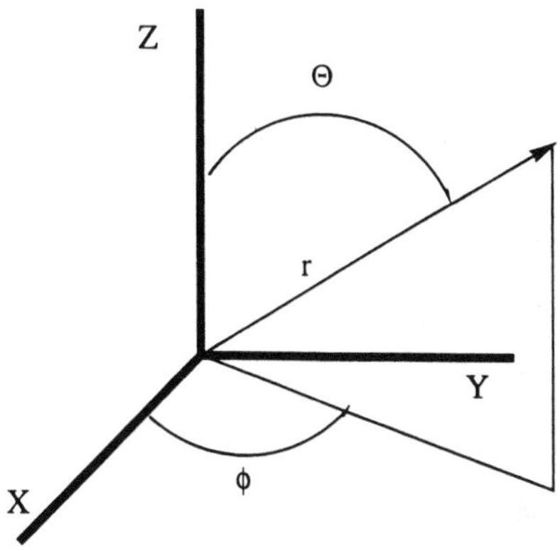

a. From cartesian to spherical polar coordinates

$3x + y - 4z = 12$

b. From cartesian to cylindrical coordinates

$$y^2 + z^2 = 9$$

c. From spherical polar to cartesian coordinates

$$r = 2\sin\theta\cos\phi$$

2. Perform a separation of variables and indicate the general solution for the following expressions:

a. $9x + 16y\dfrac{\partial y}{\partial x} = 0$

b. $2y + \dfrac{\partial y}{\partial x} + 6 = 0$

3. Find the eigenvalues and corresponding eigenvectors of the following matrices:

a. $\begin{bmatrix} -1 & 2 \\ 2 & 2 \end{bmatrix}$

b. $\begin{bmatrix} -2 & 0 & 0 \\ 0 & -1 & 2 \\ 0 & 2 & 2 \end{bmatrix}$

4. For the Hermitian matrix in review exercise 3a show that the eigenfunctions can be normalized and that they are orthogonal.

5. For the Hermitian matrix in review exercise 3b show that the pair of degenerate eigenvalues can be made to have orthonormal eigenfunctions.

6. Solve the following second-order linear differential equation subject to the specified "boundary conditions":

$$\frac{d^2x}{dt^2} + k^2x(t) = 0, \text{ where } x(t = 0) = L, \text{ and } \frac{dx(t = 0)}{dt} = 0.$$

EXERCISES

1. Replace the following classical mechanical expressions with their corresponding quantum mechanical operators.

a. K.E. $= \dfrac{mv^2}{2}$ in three-dimensional space.

b. $\mathbf{p} = m\mathbf{v}$, a three-dimensional cartesian vector.

c. y-component of angular momentum: $L_y = zp_x - xp_z$.

2. Transform the following operators into the specified coordinates:

a. $\mathbf{L}_x = \dfrac{\hbar}{i}\left\{y\dfrac{\partial}{\partial z} - z\dfrac{\partial}{\partial y}\right\}$ from cartesian to spherical polar coordinates.

b. $\mathbf{L}_z = \dfrac{\hbar}{i}\dfrac{\partial}{\partial \phi}$ from spherical polar to cartesian coordinates.

3. Match the eigenfunctions in column B to their operators in column A. What is the eigenvalue for each eigenfunction?

	Column A	*Column B*
a.	$(1-x^2)\dfrac{d^2}{dx^2} - x\dfrac{d}{dx}$	$4x^4 - 12x^2 + 3$
b.	$\dfrac{d^2}{dx^2}$	$5x^4$
c.	$x\dfrac{d}{dx}$	$e^{3x} + e^{-3x}$
d.	$\dfrac{d^2}{dx^2} - 2x\dfrac{d}{dx}$	$x^2 - 4x + 2$
e.	$x\dfrac{d^2}{dx^2} + (1-x)\dfrac{d}{dx}$	$4x^3 - 3x$

4. Show that the following operators are Hermitian.

a. $\boldsymbol{P}_x$

b. $\boldsymbol{L}_x$

5. For the following basis of functions ($\Psi_{2p_{-1}}$, Ψ_{2p_0}, and $\Psi_{2p_{+1}}$), construct the matrix representation of the $\boldsymbol{L_x}$ operator (use the ladder operator representation of $\boldsymbol{L}_x$). Verify that the matrix is Hermitian. Find the eigenvalues and corresponding eigenvectors. Normalize the eigenfunctions and verify that they are orthogonal.

$$\Psi_{2_{p-1}} = \frac{1}{8\pi^{1/2}}\left(\frac{Z}{a}\right)^{5/2} re^{-zr/2a}\sin\theta e^{-i\phi}$$

$$\Psi_{2p_o} = \frac{1}{\pi^{1/2}}\left(\frac{Z}{2a}\right)^{5/2} re^{-zr/2a}\cos\theta$$

$$\Psi_{2p_1} = \frac{1}{8\pi^{1/2}}\left(\frac{Z}{a}\right)^{5/2} re^{-zr/2a}\sin\theta e^{i\phi}$$

6. Using the set of eigenstates (with corresponding eigenvalues) from the preceding problem, determine the probability for observing a z-component of angular momentum equal to $1\hbar$ *if* the state is given by the $\boldsymbol{L_x}$ eigenstate with $0\hbar\boldsymbol{L_x}$ eigenvalue.

7. Use the following definitions of the angular momentum operators:

$$\boldsymbol{L}_x = \frac{\hbar}{i}\left\{y\frac{\partial}{\partial z} - z\frac{\partial}{\partial y}\right\},\ \boldsymbol{L}_y = \frac{\hbar}{i}\left\{z\frac{\partial}{\partial x} - x\frac{\partial}{\partial z}\right\},$$

$$\boldsymbol{L}_z = \frac{\hbar}{i}\left\{x\frac{\partial}{\partial y} - y\frac{\partial}{\partial x}\right\},\ \text{and}\ \boldsymbol{L}^2 = \boldsymbol{L}_x^2 + \boldsymbol{L}_y^2 + \boldsymbol{L}_z^2,$$

and the relationships:

$$[\boldsymbol{x},\boldsymbol{p}_x] = i\hbar, [\boldsymbol{y},\boldsymbol{p}_y] = i\hbar,\ \text{and}\ [\boldsymbol{z},\boldsymbol{p}_z] = i\hbar,$$

to demonstrate the following operator identities:

a. $[L_x,L_y] = i\hbar L_z$,

b. $[L_y,L_z] = i\hbar L_x$,

c. $[L_z,L_x] = i\hbar L_y$,

d. $[L_x,L^2] = 0$,

e. $[L_y,L^2] = 0$,

f. $[L_z,L^2] = 0$.

8. In exercise 7 above you determined whether or not many of the angular momentum operators commute. Now, examine the operators below along with an appropriate given function. Determine if the given function is simultaneously an eigenfunction of *both* operators. Is this what you expected?

a. L_z, L^2, with function: $Y_0^0(\theta,\phi) = \frac{1}{\sqrt{4\pi}}$.

b. L_x, L_z, with function: $Y_0^0(\theta,\phi) = \frac{1}{\sqrt{4\pi}}$.

c. L_z, L^2, with function: $Y_1^0(\theta,\phi) = \sqrt{\frac{3}{4\pi}}\cos\theta$.

d. L_x, L_z, with function: $Y_1^0(\theta,\phi) = \sqrt{\frac{3}{4\pi}}\cos\theta$.

9. For a "particle in a box" constrained along two axes, the wavefunction $\Psi(x,y)$ as given in the text was:

$$\Psi(x,y) = \left(\frac{1}{2L_x}\right)^{\frac{1}{2}}\left(\frac{1}{2L_y}\right)^{\frac{1}{2}}\left[e^{\frac{in_x\pi x}{L_x}} - e^{\frac{-in_x\pi x}{L_x}}\right]\left[e^{\frac{in_y\pi y}{L_y}} - e^{\frac{-in_y\pi y}{L_y}}\right],$$

with n_x and $n_y = 1,2,3,\ldots$. Show that this wavefunction is normalized.

10. Using the same wavefunction, $\Psi(x,y)$, given in exercise 9 show that the expectation value of $\boldsymbol{p}_x$ vanishes.

11. Calculate the expectation value of the $\boldsymbol{x}^2$ operator for the first two states of the harmonic oscillator. Use the $v = 0$ and $v = 1$ harmonic oscillator wavefunctions given below which are normalized such that

$$\int_{-\infty}^{+\infty} \Psi(x)^2 dx = 1.$$

Remember that

$$\Psi_0 = \left(\frac{\alpha}{\pi}\right)^{1/4} e^{-\alpha x2/2} \text{ and } \Psi_1 = \left(\frac{4\alpha^3}{\pi}\right)^{1/4} x e^{-\alpha x2/2}.$$

12. For each of the one-dimensional potential energy graphs shown below, determine:

a. whether you expect symmetry to lead to a separation into odd and even solutions,

b. whether you expect the energy will be quantized, continuous, or both, and

c. the boundary conditions that apply at each boundary (merely stating that Ψ and/or $\frac{\partial\Psi}{\partial x}$ is continuous is all that is necessary).

i.

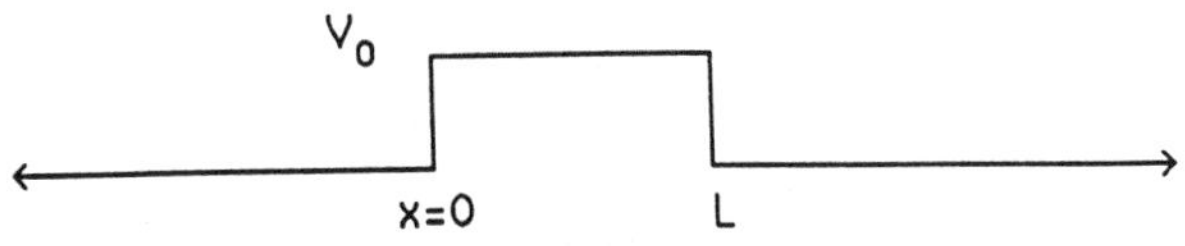

ii.

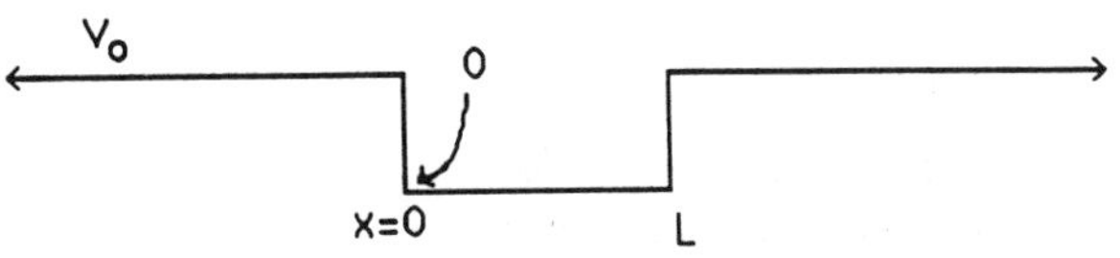

iii.

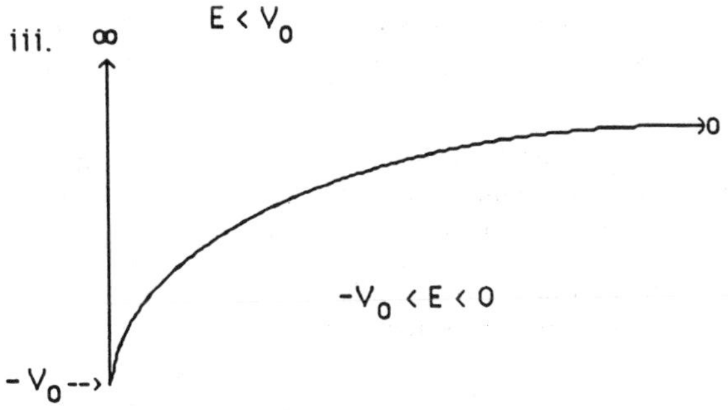

13. Consider a particle of mass m moving in the potential:

$V(x) = \infty$	for	$x < 0$	Region I
$V(x) = 0$	for	$0 \le x \le L$	Region II
$V(x) = V (V > 0)$	for	$x > L$	Region III

a. Write the general solution to the Schrödinger equation for the regions I, II, III, assuming a solution with energy $E < V$ (i.e. a bound state).

b. Write down the wavefunction matching conditions at the interface between regions I and II and between II and III.

c. Write down the boundary conditions on Ψ for $x \to \pm\infty$.

d. Use your answers to a–c to obtain an algebraic equation which must be satisfied for the bound state energies, E.

e. Demonstrate that in the limit $V \to \infty$, the equation you obtained for the bound state energies in d gives the energies of a particle in an infinite box;

$$E_n = \frac{n^2\hbar^2\pi^2}{2mL^2};\ n = 1,2,3,\ldots$$

PROBLEMS

1. A particle of mass m moves in a one-dimensional box of length L, with boundaries at $x = 0$ and $x = L$. Thus, $V(x) = 0$ for $0 \le x \le L$, and $V(x) = \infty$ elsewhere. The normalized eigenfunctions of the Hamiltonian for this system are given by

$$\Psi_n(x) = \left(\frac{2}{L}\right)^{1/2} \sin\frac{n\pi x}{L}, \text{ with } E_n = \frac{n^2\pi^2\hbar^2}{2mL^2},$$

where the quantum number n can take on the values $n = 1,2,3, \ldots.$

a. Assuming that the particle is in an eigenstate, $\Psi_n(x)$, calculate the probability that the particle is found somewhere in the region $0 \le x \le L/4$. Show how this probability depends on n.

b. For what value of n is there the largest probability of finding the particle in $0 \le x \le L/4$?

c. Now assume that Ψ is a superposition of two eigenstates, $\Psi = a\Psi_n + b\Psi_m$, at time $t = 0$. What is Ψ at time t? What energy expectation value does Ψ have at time t and how does this relate to its value at $t = 0$?

d. For an experimental measurement which is capable of distinguishing systems in state Ψ_n from those in Ψ_m, what fraction of a large number of systems each described by Ψ will be observed to be in Ψ_n? What energies will these experimental measurements find and with what probabilities?

e. For those systems originally in $\Psi = a\Psi_n + b\Psi_m$ which were observed to be in Ψ_n at time t, what state (Ψ_n,Ψ_m, or whatever) will they be found in if a second experimental measurement is made at a time t' later than t?

f. Suppose by some method (which need not concern us at this time) the system has been prepared in a nonstationary state (that is, it is not an eigenfunction of $\boldsymbol{H}$). At the time of a measurement of the particle's energy, this state is specified by the normalized wavefunction

$$\Psi = \left(\frac{30}{L^5}\right)^{1/2} x(L - x) \text{ for } 0 \le x \le L, \text{ and } \Psi = 0 \text{ elsewhere.}$$

What is the probability that a measurement of the energy of the particle will give the value

$$E_n = \frac{n^2\pi^2\hbar^2}{2mL^2}$$

for any given value of n?

g. What is the expectation value of $\boldsymbol{H}$, i.e. the average energy of the system, for the wavefunction Ψ given in part f?

2. Show that for a system in a non-stationary state,

$$\Psi = \sum_j C_j\Psi_j e^{-iE_jt/\hbar},$$

the average value of the energy does *not* vary with time but the expectation values of other properties *do* vary with time.

3. A particle is confined to a one-dimensional box of length L having infinitely high walls and is in its lowest quantum state. Calculate: $\langle x\rangle$,$\langle x^2\rangle$,$\langle p\rangle$, and $\langle p^2\rangle$. Using the definition $\Delta A = (\langle A^2\rangle - \langle A\rangle^2)^{1/2}$, to define the uncertainty, ΔA, calculate Δ x and Δp. Verify the Heisenberg uncertainty principle that $\Delta x\Delta p \ge \hbar/2$.

4. It has been claimed that as the quantum number n increases, the motion of a particle in a box becomes more classical. In this problem you will have an oportunity to convince yourself of this fact.

a. For a particle of mass m moving in a one-dimensional box of length L, with ends of the box located at $x = 0$ and $x = L$, the classical probability density can be shown to be independent of x and given by

$$P(x)dx = \frac{dx}{L}$$

regardless of the energy of the particle. Using this probability density, evaluate the probability that the particle will be found within the interval from $x = 0$ to $x = L/4$.

b. Now consider the quantum mechanical particle-in-a-box system. Evaluate the probability of finding the particle in the interval from $x = 0$ to $x = L/4$ for the system in its nth quantum state.

c. Take the limit of the result you obtained in part b as $n \to \infty$. How does your result compare to the classical result you obtained in part a?

5. According to the rules of quantum mechanics as we have developed them, if Ψ is the state function, and ϕ_n are the eigenfunctions of a linear, Hermitian operator, $\boldsymbol{A}$, with eigenvalues a_n, $\boldsymbol{A}\phi_n = a_n\phi_n$, then we can expand Ψ in terms of the complete set of eigenfunctions of $\boldsymbol{A}$ according to

$$\Psi = \sum_n c_n\phi_n, \text{ where } c_n = \int \phi_n^*\Psi d\tau.$$

Furthermore, the probability of making a measurement of the property corresponding to $\boldsymbol{A}$ and obtaining a value a_n is given by $|c_n|^2$, provided both Ψ and ϕ_n are properly normalized. Thus, $P(a_n) = |c_n|^2$. These rules are perfectly valid for operators which take on a discrete set of eigenvalues, but must be generalized for operators which can have a continuum of eigenvalues. An example of this latter type of operator is the momentum operator, $\boldsymbol{p}_x$, which has eigenfunctions given by $\phi_p(x) = Ae^{ipx/\hbar}$ where p is the eigenvalue of the $\boldsymbol{p_x}$ operator and A is a normalization constant. Here p can take on any value, so we have a continuous spectrum of eigenvalues of $\boldsymbol{p}_x$. The obvious generalization to the equation for Ψ is to convert the sum over discrete states to an integral over the continuous spectrum of states:

$$\Psi(x) = \int_{-\infty}^{+\infty} C(p)\phi_p(x)dp = \int_{-\infty}^{+\infty} C(p)Ae^{ipx/\hbar}dp$$

The interpretation of $C(p)$ is now the desired generalization of the equation for the probability $P(p)dp = |C(p)|^2dp$. This equation states that the probability of measuring the momentum and finding it in the range from p to $p + dp$ is given by $|C(p)|^2dp$. Accordingly, the probability of measuring p and finding it in the range from p_1 to p_2 is given by

$$\int_{p_1}^{p_2} P(p)dp = \int_{p_1}^{p_2} C(p)^*C(p)dp.$$

$C(p)$ is thus the probability amplitude for finding the particle with momentum between p and $p + dp$. This is the *momentum representation* of the wavefunction. Clearly we must require $C(p)$ to be normalized, so that

$$\int_{-\infty}^{+\infty} C(p)^*C(p)dp = 1.$$

With this restriction we can derive the normalization constant

$$A = \frac{1}{\sqrt{2\pi\hbar}},$$

giving a direct relationship between the wavefunction in coordinate space, $\Psi(x)$, and the wavefunction in momentum space, $C(p)$:

$$\Psi(x) = \frac{1}{\sqrt{2\pi\hbar}} \int_{-\infty}^{+\infty} C(p) e^{ipx/\hbar} dp, \text{ and by the Fourier integral theorem:}$$

$$C(p) = \frac{1}{\sqrt{2\pi\hbar}} \int_{-\infty}^{+\infty} \Psi(x) e^{ipx/\hbar} dx.$$

Let's use these ideas to solve some problems focusing our attention on the harmonic oscillator; a particle of mass m moving in a one-dimensional potential described by

$$V(x) = \frac{kx^2}{2}.$$

a. Write down the Schrödinger equation in the coordinate representation.

b. Now lets proceed by attempting to write the Schrödinger equation in the momentum representation. Identifying the kinetic energy operator $\boldsymbol{T}$, in the momentum representation is quite straightforward

$$\boldsymbol{T} = \frac{\boldsymbol{p}^2}{2m} = -\frac{\hbar^2}{2m}\frac{\partial^2}{\partial x^2}$$

(this can be seen either by recalling how the kinetic energy is related to linear momentum or by writing the kinetic energy operator in the coordinate representation, as above in part a, and using the definition:

$$\boldsymbol{p} = \frac{\hbar}{i}\frac{\partial}{\partial x}) .$$

Writing the potential, $V(x)$, in the momentum representation is not quite as straightforward. The relationship between position and momentum is realized in their commutation relation $[\boldsymbol{x},\boldsymbol{p}] = i\hbar$, or $(\boldsymbol{xp} - \boldsymbol{px}) = i\hbar$ This commutation relation is easily verified in the coordinate representation leaving $\boldsymbol{x}$ untouched ($\boldsymbol{x} = x\,\cdot$) and using the above definition for $\boldsymbol{p}$. In the momentum representation we want to leave $\boldsymbol{p}$ untouched ($\boldsymbol{p} = p\,\cdot$) and define the operator $\boldsymbol{x}$ in such a manner that the commutation relation is still satisfied. Write the operator $\boldsymbol{x}$ in the momentum representation. Write the full Hamiltonian in the momentum representation and hence the Schrödinger equation in the momentum representation.

c. Verify that Ψ as given below is an eigenfunction of the Hamiltonian in the coordinate representation. What is the energy of the system when it is in this state? Determine the normalization constant C, and write down the normalized ground state wavefunction in coordinate space.

$$\Psi(x) = C\exp\left(-\sqrt{mk}\frac{x^2}{2\hbar}\right).$$

d. Now consider Ψ in the momentum representation. Assuming that an eigenfunction of the Hamiltonian may be found of the form $\Psi(p) = C\exp(-\alpha p^2)$, substitute this form of Ψ into the Schrödinger equation in the momentum representation to find the value of α which makes this an eigenfunction of $\boldsymbol{H}$ having the same energy as $\Psi(x)$ had. Show that this $\Psi(p)$ is the proper Fourier transform of $\Psi(x)$. The following integral may be useful:

$$\int_{-\infty}^{+\infty} e^{-\beta x^2}\cos bx\,dx = \sqrt{\frac{\pi}{\beta}}\,e^{-b^2/4\beta}.$$

Since this Hamiltonian has no degenerate states, you may conclude that $\Psi(x)$ and $\Psi(p)$ represent the same state of the system if they have the same energy.

6. The energy states and wavefunctions for a particle in a three-dimensional box whose lengths are L_1, L_2, and L_3 are given by

$$E(n_1,n_2,n_3) = \frac{h^2}{8m}\left[\left(\frac{n_1}{L_1}\right)^2 + \left(\frac{n_2}{L_2}\right)^2 + \left(\frac{n_3}{L_3}\right)^2\right] \text{ and}$$

$$\Psi(n_1,n_2,n_3) = \left(\frac{2}{L_1}\right)^{\frac{1}{2}}\left(\frac{2}{L_2}\right)^{\frac{1}{2}}\left(\frac{2}{L_3}\right)^{\frac{1}{2}}\sin\left(\frac{n_1\pi x}{L_1}\right)\sin\left(\frac{n_2\pi y}{L_2}\right)\sin\left(\frac{n_3\pi z}{L_3}\right).$$

These wavefunctions and energy levels are sometimes used to model the motion of electrons in a central metal atom (or ion) which is surrounded by six ligands.

a. Show that the lowest energy *level* is nondegenerate and the second energy *level* is triply degenerate if $L_1 = L_2 = L_3$. What values of n_1, n_2, and n_3 characterize the *states* belonging to the triply degenerate level?

b. For a box of volume $V = L_1L_2L_3$, show that for three electrons in the box (two in the nondegenerate lowest "orbital," and one in the next), a lower *total* energy will result if the box undergoes a rectangular distortion ($L_1 = L_2 \neq L_3$). *which preserves the total volume* than if the box remains undistorted (hint: if V is fixed and $L_1 = L_2$, then $L_3 = V/L_1^2$ and L_1 is the only "variable").

c. Show that the degree of distortion (ratio of L_3 to L_1) which will minimize the total energy is $L_3 = \sqrt{2}L_1$. How does this problem relate to Jahn-Teller distortions? Why (in terms of the property of the central atom or ion) do we do the calculation with fixed volume?

d. By how much (in eV) will distortion lower the energy (from its value for a cube, $L_1 = L_2 = L_3$) if $V = 8\text{Å}^3$ and

$$\frac{h^2}{8m} = 6.01 \times 10^{-27}\text{erg cm}^2.$$

$$1eV = 1.6 \times 10^{-12}\text{erg}$$

7. The wavefunction $\Psi = Ae^{-a|x|}$ is an exact eigenfunction of some one-dimensional Schrödinger equation in which x varies from $-\infty$ to $+\infty$. The value of a is: $a = (2\text{Å})^{-1}$. For now, the potential $V(x)$ in the Hamiltonian

$$\boldsymbol{H} = -\frac{\hbar}{2m}\frac{d^2}{dx^2} + V(x)$$

for which $\Psi(x)$ is an eigenfunction is unknown.

a. Find a value of A which makes $\Psi(x)$ normalized. Is this value unique? What units does $\Psi(x)$ have?

b. Sketch the wavefunction for positive and negative values of x, being careful to show the behavior of its slope near $x = 0$. Recall that $|x|$ is defined as:

$$|x| = \begin{matrix} x \text{ if } x > 0 \\ -x \text{ if } x < 0 \end{matrix}$$

c. Show that the derivative of $\Psi(x)$ undergoes a *discontinuity* of magnitude $2(a)^{3/2}$ as x goes through $x = 0$. What does this fact tell you about the potential $V(x)$?

d. Calculate the expectation value of $|x|$ for the above normalized wavefunction (obtain a numerical value and give its units). What does this expectation value give a measure of?

e. The potential $V(x)$ appearing in the Schrödinger equation for which $\Psi = Ae^{-a|x|}$ is an exact solution is given by

$$V(x) = \frac{\hbar^2 a}{m}\delta(x).$$

Using this potential, compute the expectation value of the Hamiltonian

$$H = -\frac{\hbar}{2m}\frac{d^2}{dx^2} + V(x)$$

for your normalized wavefunction. Is $V(x)$ an attractive or repulsive potential? Does your wavefunction correspond to a bound state? Is $<H>$ negative or positive? What does the sign of $<H>$ tell you? To obtain a numerical value for $<H>$ use

$$\frac{\hbar^2}{2m} = 6.06 \times 10^{-28}\text{erg } cm^2 \text{ and } 1eV = 1.6 \prod 10^{-12}\text{erg}.$$

f. Transform the wavefunction, $\Psi = Ae^{-a|x|}$, from coordinate space to momentum space.

g. What is the ratio of the probability of observing a momentum equal to $2a\hbar$ to the probability of observing a momentum equal to $-a\hbar$?

8. The π-orbitals of benzene, C_6H_6, may be modeled very crudely using the wavefunctions and energies of a particle on a ring. Lets first treat the particle on a ring problem and then extend it to the benzene system.

a. Suppose that a particle of mass m is constrained to move on a circle (of radius r) in the xy plane. Further assume that the particle's potential energy is constant (zero is a good choice). Write down the Schrödinger equation in the normal cartesian coordinate representation. Transform this Schrödinger equation to cylindrical coordinates where $x = r\cos\phi$, $y = r\sin\phi$, and $z = z$ ($z = 0$ in this case). Taking r to be held constant, write down the general solution, $\Phi(\phi)$, to this Schrödinger equation. The "boundary" conditions for this problem require that $\Phi(\phi) = \Phi(\phi + 2\pi)$. Apply this boundary condition to the general solution. This results in the quantization of the energy levels of this system. Write down the final expression for the *normalized* wavefunction and quantized energies. What is the physical significance of these quantum numbers which can have both positive and negative values? Draw an energy diagram representing the first five energy levels.

b. Treat the six π-electrons of benzene as particles free to move on a ring of radius 1.40Å, and calculate the energy of the lowest electronic transition. Make sure the Pauli principle is satisfied! What wavelength does this transition correspond to? Suggest some reasons why this differs from the wavelength of the lowest observed transition in benzene, which is 2600Å.

9. A diatomic molecule constrained to rotate on a flat surface can be modeled as a planar rigid rotor (with eigenfunctions, $\Phi(\phi)$, analogous to those of the particle on a ring) with fixed bond length r. At $t = 0$, the rotational (orientational) probability distribution is observed to be described by a wavefunction

$$\psi(\phi,0) = \sqrt{\frac{4}{3\pi}}\,\text{Cos}^2\phi\,.$$

What values, and with what probabilities, of the rotational angular momentum,

$$-i\hbar\frac{\partial}{\partial\phi},$$

could be observed in this system? Explain whether these probabilities would be time dependent as $\Psi(\phi,0)$ evolves into $\Psi(\phi,t)$.

10. A particle of mass m moves in a potential given by

$$V(x,y,z) = \frac{k}{2}(x^2 + y^2 + z^2) = \frac{kr^2}{2}.$$

a. Write down the time-independent Schrödinger equation for this system.

b. Make the substitution $\Psi(x,y,z) = X(x)Y(y)Z(z)$ and separate the variables for this system.

c. What are the solutions to the resulting equations for $X(x)$, $Y(y)$, and $Z(z)$?

d. What is the general expression for the quantized energy levels of this system, in terms of the quantum numbers n_x, n_y, and n_z, which correspond to $X(x)$, $Y(y)$, and $Z(z)$?

e. What is the degree of degeneracy of a state of energy

$$E = 5.5\hbar\sqrt{\frac{k}{m}}$$

for this system?

f. An alternative solution may be found by making the substitution $\Psi(r,\theta,\phi) = F(r)G(\theta,\phi)$. In this substitution, what are the solutions for $G(\theta,\phi)$?

g. Write down the differential equation for $F(r)$ which is obtained when the substitution $\Psi(r,\theta,\phi) = F(r)G(\theta,\phi)$ is made. Do not solve this equation.

11. Consider an N_2 molecule, in the ground vibrational level of the ground electronic state, which is bombarded by 100 eV electrons. This leads to ionization of the N_2 molecule to form N_2^+. In this problem we will attempt to calculate the vibrational distribution of the newly formed N_2^+ ions, using a somewhat simplified approach.

a. Calculate (according to classical mechanics) the velocity (in cm/sec) of a 100eV electron, ignoring any relativistic effects. Also calculate the amount of time required for a 100eV electron to pass an N_2 molecule, which you may estimate as having a length of 2Å.

b. The radial Schrödinger equation for a diatomic molecule treating vibration as a harmonic oscillator can be written as:

$$-\frac{\hbar^2}{2\mu r^2}\left(\frac{\partial}{\partial r}\left(r^2\frac{\partial\Psi}{\partial r}\right)\right) + \frac{k}{2}(r - r_e)^2\Psi = E\Psi,$$

Substituting $\Psi(r) = \frac{F(r)}{r}$, this equation can be rewritten as:

$$-\frac{\hbar^2}{2\mu}\frac{\partial^2}{\partial r^2}F(r) + \frac{k}{2}(r - r_e)^2F(r) = EF(r).$$

The vibrational Hamiltonian for the ground electronic state of the N_2 molecule within this approximation is given by:

$$\boldsymbol{H}(N_2) = -\frac{\hbar^2}{2\mu}\frac{d^2}{dr^2} + \frac{k_{N_2}}{2}(r - r_{N_2})^2,$$

where r_{N_2} and k_{N_2} have been measured experimentally to be:

$$r_{N_2} = 1.09769 \text{ Å}; k_{N_2} = 2.294 \times 10^6 \frac{g}{\text{sec}^2}.$$

The vibrational Hamiltonian for the N_2^+ however, is given by:

$$\boldsymbol{H}(N_2) = -\frac{\hbar^2}{2\mu}\frac{d^2}{dr^2} + \frac{k_{N_2^+}}{2}(r - r_{N_2^+})^2,$$

where $r_{N_2^+}$ and $k_{N_2^+}$ have been measured experimentally to be:

$$r_{N_2^+} = 1.11642 \text{ Å}; \; k_{N_2^+} = 2.009 \times 10^6 \frac{g}{\text{sec}^2}.$$

In both systems the reduced mass is $\mu = 1.1624 \times 10^{-23}$g. Use the above information to write out the ground state vibrational wavefunctions of the N_2 and N_2^+ molecules, giving explicit values for any constants which appear in them. **Note:** For this problem use the "normal" expression for the ground state wavefunction of a harmonic oscillator. You need not solve the differential equation for this system.

c. During the time scale of the ionization event (which you calculated in part a), the vibrational wavefunction of the N_2 molecule has effectively no time to change. As a result, the newly formed N_2^+ ion finds itself in a vibrational state which is *not* an eigenfunction of the *new* vibrational Hamiltonian, $\boldsymbol{H}(N_2^+)$. Assuming that the N_2 molecule was originally in its $v = 0$ vibrational state, calculate the probability that the N_2^+ ion will be produced in its $v = 0$ vibrational state.

12. The force constant, k, of the C-O bond in carbon monoxide is 1.87×10^6g/sec^2. Assume that the vibrational motion of CO is purely harmonic and use the reduced mass $\mu = 6.857$amu.

a. Calculate the spacing between vibrational energy levels in this molecule, in units of ergs and cm^{-1}.

b. Calculate the uncertainty in the internuclear distance in this molecule, assuming it is in its ground vibrational level. Use the ground state vibrational wavefunction ($\Psi_{v=0}$), and calculate $<x>$, $<x^2>$, and $\Delta x = (<x^2> - <x>^2)^{1/2}$.

c. Under what circumstances (i.e., large or small values of k; large or small values of μ) is the uncertainty in internuclear distance large? Can you think of any relationship between this observation and the fact that helium remains a liquid down to absolute zero?

13. Suppose you are given a trial wavefunction of the form:

$$\phi = \frac{Z_e^3}{\pi a_0^3} \exp\left(\frac{-Z_e r_1}{a_0}\right) \exp\left(\frac{-Z_e r_2}{a_0}\right)$$

to represent the electronic structure of a two-electron ion of nuclear charge Z and suppose that you were also lucky enough to be *given* the variational integral, W, (instead of asking you to derive it!):

$$W = \left(Z_e^2 - 2ZZ_e + \frac{5}{8} Z_e\right)\frac{e^2}{a_0}.$$

a. Find the optimum value of the variational parameter Z_e for an arbitrary nuclear charge Z by setting

$$\frac{dW}{dZ_e} = 0.$$

Find both the optimal value of Z_e and the resulting value of W.

b. The total energies of some two-electron atoms and ions have been experimentally determined to be:

$Z = 1$	H^-	– 14.35eV
$Z = 2$	He	– 78.98eV
$Z = 3$	Li^+	– 198.02eV
$Z = 4$	Be^{+2}	– 371.5eV
$Z = 5$	B^{+3}	– 599.3eV
$Z = 6$	C^{+4}	– 881.6eV
$Z = 7$	N^{+5}	– 1218.3eV
$Z = 8$	O^{+6}	– 1609.5eV

Using your optimized expression for W, calculate the estimated total energy of each of these atoms and ions. Also calculate the percent error in your estimate for each ion. What physical reason explains the decrease in percentage error as Z increases?

c. In 1928, when quantum mechanics was quite young, it was not known whether the isolated, gas-phase hydride ion, H^-, was stable with respect to dissociation into a hydrogen atom and an electron. Compare your estimated total energy for H^- to the ground state energy of a hydrogen atom and an isolated electron (system energy $= -13.60$eV), and show that this simple variational calculation erroneously predicts H^- to be unstable. (More complicated variational treatments give a ground state energy of H^- of –14.35eV, in agreement with experiment.)

14. A particle of mass m moves in a one-dimensional potential given by

$$\boldsymbol{H} = -\frac{\hbar^2}{2m}\frac{d^2}{dx^2} + a|x|,$$

where the absolute value function is defined by $|x| = x$ if $x \geq 0$ and $|x| = -x$ if $x \leq 0$.

a. Use the normalized trial wavefunction

$$\phi = \left(\frac{2b}{\pi}\right)^{\frac{1}{4}} e^{-bx^2}$$

to estimate the energy of the ground state of this system, using the variational principle to evaluate $W(b)$.

b. Optimize b to obtain the best approximation to the ground state energy of this system, using a trial function of the form of ϕ, as given above. The numerically calculated exact ground state energy is

$$0.808616\hbar^{\frac{2}{3}} m^{-\frac{1}{3}} a^{-\frac{2}{3}}.$$

What is the percent error in your value?

15. The harmonic oscillator is specified by the Hamiltonian:

$$\boldsymbol{H} = -\frac{\hbar^2}{2m}\frac{d^2}{dx^2} + \frac{1}{2}kx^2.$$

Suppose the ground state solution to this problem were unknown, and that you wish to approximate it using the variational theorem. Choose as your trial wavefunction,

$$\phi = \sqrt{\frac{15}{16}}\, a^{-\frac{5}{2}} (a^2 - x^2) \qquad \text{for } -a < x < a$$

$$\phi = 0 \qquad \text{for } |x| \geq a$$

where a is an arbitrary parameter which specifies the range of the wavefunction. Note that ϕ is properly normalized as given.

a. Calculate

$$\int_{-\infty}^{+\infty} \phi^* \mathbf{H} \phi dx$$

and show it to be given by:

$$\int_{-\infty}^{+\infty} \phi^* \mathbf{H} \phi dx = \frac{5}{4} \frac{\hbar^2}{ma^2} + \frac{ka^2}{14}.$$

b. Calculate

$$\int_{-\infty}^{+\infty} \phi^* \mathbf{H} \phi dx \text{ for } a = b \left(\frac{\hbar^2}{km} \right)^{\frac{1}{4}} \text{ with } b = 0.2, 0.4, 0.6, 0.8, 1.0, 1.5, 2.0, 2.5, 3.0, 4.0, \text{ and } 5.0,$$

and plot the result.

c. To find the best approximation to the true wavefunction and its energy, find the minimum of

$$\int_{-\infty}^{+\infty} \phi^* \mathbf{H} \phi dx$$

by setting

$$\frac{d}{da} \int_{-\infty}^{+\infty} \phi^* \mathbf{H} \phi dx = 0$$

and solving for a. Substitute this value into the expression for

$$\int_{-\infty}^{+\infty} \phi^* \mathbf{H} \phi dx$$

given in part a to obtain the best approximation for the energy of the ground state of the harmonic oscillator.

d. What is the percent error in your calculated energy of part c?

16. Einstein told us that the (relativistic) expression for the energy of a particle having rest mass m and momentum p is $E^2 = m^2c^4 + p^2c^2$.

a. Derive an expression for the relativistic kinetic energy operator which contains terms correct through one higher order than the "ordinary"

$$E = mc^2 + \frac{p^2}{2m}$$

b. Using the first-order correction as a perturbation, compute the first-order perturbation theory estimate of the energy for the $1s$ level of a hydrogen-like atom (general Z). Show the Z dependence of the result.

Note: $\Psi(r)_{1s} = \left(\frac{Z}{a}\right)^{\frac{3}{2}}\left(\frac{1}{\pi}\right)^{\frac{1}{2}} e^{-\frac{Zr}{a}}$ and $E_{1s} = -\frac{Z^2 me^4}{2\hbar^2}$

c. For what value of Z does this first-order relativistic correction amount to 10% of the unperturbed (non-relativistic) $1s$ energy?

17. Consider an electron constrained to move on the surface of a sphere of radius r. The Hamiltonian for such motion consists of a kinetic energy term only

$$\boldsymbol{H}_0 = \frac{\boldsymbol{L}^2}{2m_e r_0^2},$$

where $\boldsymbol{L}$ is the orbital angular momentum operator involving derivatives with respect to the spherical polar coordinates (θ,ϕ). $\boldsymbol{H}_0$ has the complete set of eigenfunctions $\psi_{lm}^{(0)} = Y_{l,m}(\theta,\phi)$.

a. Compute the zeroth-order energy levels of this system.

b. A uniform electric field is applied along the z-axis, introducing a perturbation $V = -e\varepsilon z = -e\varepsilon r_0 \cos\theta$, where ε is the strength of the field. Evaluate the correction to the energy of the lowest level through second-order in perturbation theory, using the identity

$$\cos\theta Y_{l,m}(\theta,\phi) = \sqrt{\frac{(l+m+1)(l-m+1)}{(2l+1)(2l+3)}}\, Y_{l+1,m}(\theta,\phi) + \sqrt{\frac{(l+m)(l-m)}{(2l+1)(2l-1)}}\, Y_{l-1,m}(\theta,\phi).$$

Note that this identity enables you to utilize the orthonormality of the spherical harmonics.

c. The electric polarizability α gives the response of a molecule to an externally applied electric field, and is defined by

$$\alpha = -\left.\frac{\partial^2 E}{\partial^2 \varepsilon}\right|_{\varepsilon = 0}$$

where E is the energy in the presence of the field and ε is the strength of the field. Calculate α for this system.

d. Use this problem as a model to estimate the polarizability of a hydrogen atom, where $r_0 = a_0 = 0.529$Å, and a cesium atom, which has a single 6s electron with $r_0 \approx 2.60$Å. The corresponding experimental values are $\alpha_H = 0.6668$Å^3 and $\alpha_{Cs} = 59.6$Å^3.

18. An electron moving in a conjugated bond framework can be viewed as a particle in a box. An externally applied electric field of strength ε interacts with the electron in a fashion described by the perturbation

$$V = e\varepsilon\left(x - \frac{L}{2}\right),$$

where x is the position of the electron in the box, e is the electron's charge, and L is the length of the box.

a. Compute the first-order correction to the energy of the $n = 1$ state and the first-order wavefunction for the $n = 1$ state. In the wavefunction calculation, you need only compute the contribution to $\Psi_1^{(1)}$ made by $\Psi_2^{(0)}$. Make a rough (no calculation needed) sketch of $\Psi_1^{(0)} + \Psi_1^{(1)}$ as a function of x and physically interpret the graph.

b. Using your answer to part a compute the induced dipole moment caused by the polarization of the electron density due to the electric field effect

$$\mu_{induced} = -e\int \Psi^*\left(x - \frac{L}{2}\right)\Psi dx.$$

You may neglect the term proportional to ε^2; merely obtain the term linear in ε.

c. Compute the polarizability, α, of the electron in the $n = 1$ state of the box, and explain physically why α should depend as it does upon the length of the box L. Remember that

$$\alpha = \left.\frac{\partial\mu}{\partial\varepsilon}\right|_{\varepsilon=0}.$$

SOLUTIONS

Review Exercises

1. The general relationships are as follows:

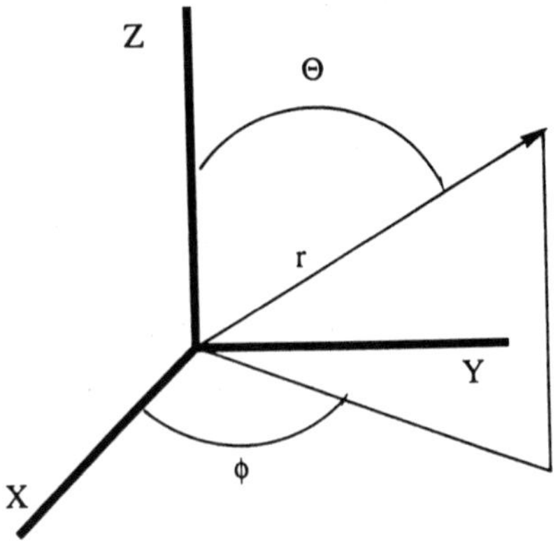

$x = r\sin\theta\cos\phi \qquad r^2 = x^2 + y^2 + z^2$

$y = r\sin\theta\sin\phi \qquad \sin\theta = \dfrac{\sqrt{x^2 + y^2}}{\sqrt{x^2 + y^2 + z^2}}$

$z = r\cos\theta \qquad \cos\theta = \dfrac{z}{\sqrt{x^2 + y^2 + z^2}}$

$\tan\phi = \dfrac{y}{x}$

a. $3x + y - 4z = 12$

$3(r\sin\theta\cos\phi) + r\sin\theta\sin\phi - 4(r\cos\theta) = 12$

$r(3\sin\theta\cos\phi + \sin\theta\sin\phi - 4\cos\theta) = 12$

b. $x = r\cos\phi \qquad r^2 = x^2 + y^2$

$y = r\sin\phi \qquad \tan\phi = \dfrac{y}{x}$

$z = z$

$y^2 + z^2 = 9$

$r^2\sin^2\phi + z^2 = 9$

c. $r = 2\sin\theta\cos\phi$

$$r = 2\left(\frac{x}{r}\right)$$

$$r^2 = 2x$$
$$x^2 + y^2 + z^2 = 2x$$
$$x^2 - 2x + y^2 + z^2 = 0$$
$$x^2 - 2x + 1 + y^2 + z^2 = 1$$
$$(x-1)^2 + y^2 + z^2 = 1$$

2. **a.** $9x + 16y\dfrac{\partial y}{\partial x} = 0$

$$16ydy = -9xdx$$

$$\frac{16}{2}y^2 = -\frac{9}{2}x^2 + c$$

$$16y^2 = -9x^2 + c'$$

$$\frac{y^2}{9} + \frac{x^2}{16} = c'' \text{ (general equation for an ellipse)}$$

b. $2y + \dfrac{\partial y}{\partial x} + 6 = 0$

$$2y + 6 = -\frac{dy}{dx}$$

$$y + 3 = -\frac{dy}{2dx}$$

$$-2dx = \frac{dy}{y+3}$$

$$-2x = ln(y+3) + c$$

$$c'e^{-2x} = y + 3$$

$$y = c'e^{-2x} - 3$$

3. **a.** First determine the eigenvalues:

$$\det\begin{bmatrix} -1-\lambda & 2 \\ 2 & 2-\lambda \end{bmatrix} = 0$$

$$(-1-\lambda)(2-\lambda) - 2^2 = 0$$
$$-2 + \lambda - 2\lambda + \lambda^2 - 4 = 0$$
$$\lambda^2 - \lambda - 6 = 0$$
$$(\lambda - 3)(\lambda + 2) = 0$$
$$\lambda = 3 \text{ or } \lambda = -2.$$

Next, determine the eigenvectors. First, the eigenvector associated with eigenvalue –2:

$$\begin{bmatrix} -1 & 2 \\ 2 & 2 \end{bmatrix}\begin{bmatrix} C_{11} \\ C_{21} \end{bmatrix} = -2\begin{bmatrix} C_{11} \\ C_{21} \end{bmatrix}$$

$$-C_{11} + 2C_{21} = -2C_{11}$$
$$C_{11} = -2C_{21}$$

(Note: The second row offers no new information, e.g., $2C_{11}+2C_{21}=-2C_{21}$)

$C_{11}^2 + C_{21}^2 = 1$ (from normalization)
$(-2C_{21})^2 + C_{21}^2 = 1$
$4C_{21}^2 + C_{21}^2 = 1$
$5C_{21}^2 = 1$
$C_{21}^2 = 0.2$
$C_{21} = \sqrt{0.2}$, and therefore $C_{11} = -2\sqrt{0.2}$.

For the eigenvector associated with eigenvalue 3:

$$\begin{bmatrix} -1 & 2 \\ 2 & 2 \end{bmatrix} \begin{bmatrix} C_{12} \\ C_{22} \end{bmatrix} = 3 \begin{bmatrix} C_{12} \\ C_{22} \end{bmatrix}$$

$-C_{12} + 2C_{22} = 3C_{12}$
$-4C_{12} = -2C_{22}$
$C_{12} = 0.5C_{22}$ (again the second row offers no new information)
$C_{12}^2 + C_{22}^2 = 1$ (from normalization)
$(0.5C_{22})^2 + C_{22}^2 = 1$
$0.25C_{22}^2 + C_{22}^2 = 1$
$1.25C_{22}^2 = 1$
$C_{22}^2 = 0.8$
$C_{22} = \sqrt{0.8} = 2\sqrt{0.2}$, and therefore $C_{12} = \sqrt{0.2}$.

Therefore the eigenvector matrix becomes:

$$\begin{bmatrix} -2\sqrt{0.2} & \sqrt{0.2} \\ \sqrt{0.2} & 2\sqrt{0.2} \end{bmatrix}$$

b. First determine the eigenvalues:

$$\det \begin{bmatrix} -2-\lambda & 0 & 0 \\ 0 & -1-\lambda & 2 \\ 0 & 2 & 2-\lambda \end{bmatrix} = 0$$

$$\det[-2-\lambda]\det \begin{bmatrix} -1-\lambda & 2 \\ 2 & 2-\lambda \end{bmatrix} = 0$$

From 3a, the solutions then become –2, –2, and 3. Next, determine the eigenvectors. First the eigenvector associated with eigenvalue 3 (the third root):

$$\begin{bmatrix} -2 & 0 & 0 \\ 0 & -1 & 2 \\ 0 & 2 & 2 \end{bmatrix} \begin{bmatrix} C_{11} \\ C_{21} \\ C_{31} \end{bmatrix} = 3 \begin{bmatrix} C_{11} \\ C_{21} \\ C_{31} \end{bmatrix}$$

$-2C_{13} = 3C_{13}$ (row one)
$C_{13} = 0$
$-C_{23} + 2C_{33} = 3C_{23}$ (row two)
$2C_{33} = 4C_{23}$
$C_{33} = 2C_{23}$ (again the third row offers no new information)
$C_{13}^2 + C_{23}^2 + C_{33}^2 = 1$ (from normalization)
$0 + C_{23}^2 + (2C_{23})^2 = 1$
$5C_{23}^2 = 1$
$C_{23} = \sqrt{0.2}$, and therefore $C_{33} = 2\sqrt{0.2}$.

Next, find the pair of eigenvectors associated with the degenerate eigenvalue of –2. First, root one eigenvector one:

$-2C_{11} = -2C_{11}$ (no new information from row one)
$-C_{21} + 2C_{31} = -2C_{21}$ (row two)
$C_{21} = -2C_{31}$ (again the third row offers no new information)
$C_{11}^2 + C_{21}^2 + C_{31}^2 = 1$ (from normalization)
$C_{11}^2 + (-2C_{31})^2 + C_{31}^2 = 1$
$C_{11}^2 + 5C_{31}^2 = 1$
$C_{11} = \sqrt{1-5C_{31}^2}$ (**Note:** There are now two equations with three unknowns.)

Second, root two eigenvector two:

$-2C_{12} = -2C_{12}$ (no new information from row one)
$-C_{22} + 2C_{32} = -2C_{22}$ (row two)
$C_{22} = -2C_{32}$ (again the third row offers no new information)
$C_{12}^2 + C_{22}^2 + C_{32}^2 = 1$ (from normalization)
$C_{12}^2 + (-2C_{32})^2 + C_{32}^2 = 1$
$C_{12}^2 + 5C_{32}^2 = 1$
$C_{12} = \sqrt{1 - 5C_{32}^2}$ (**Note**: Again there are now two equations with three unknowns)
$C_{11}C_{12} + C_{21}C_{22} + C_{31}C_{32} = 0$ (from orthogonalization)

Now there are five equations with six unknowns.
Arbitrarily choose $C_{11} = 0$
$C_{11} = 0 = \sqrt{1 - 5C_{31}^2}$
$5C_{31}^2 = 1$
$C_{31} = \sqrt{0.2}$
$C_{21} = -2\sqrt{0.2}$
$C_{11}C_{12} + C_{21}C_{22} + C_{31}C_{32} = 0$ (from orthogonalization)
$0 + -2\sqrt{0.2}(-2C_{32}) + \sqrt{0.2}\,C_{32} = 0$
$5C_{32} = 0$
$C_{32} = 0$, $C_{22} = 0$, and $C_{12} = 1$

Therefore the eigenvector matrix becomes:

$$\begin{bmatrix} 0 & 1 & 0 \\ -2\sqrt{0.2} & 0 & \sqrt{0.2} \\ \sqrt{0.2} & 0 & 2\sqrt{0.2} \end{bmatrix}$$

4. Show: $\langle\phi_1|\phi_1\rangle = 1$, $\langle\phi_2|\phi_2\rangle = 1$, and $\langle\phi_1|\phi_2\rangle = 0$

$\langle\phi_1|\phi_1\rangle \overset{?}{=} 1$

$(-2\sqrt{0.2})^2 + (\sqrt{0.2})^2 \overset{?}{=} 1$

$4(0.2) + 0.2 \overset{?}{=} 1$

$0.8 + 0.2 \overset{?}{=} 1$
$1 = 1$

$\langle\phi_2|\phi_2\rangle \overset{?}{=} 1$

$(\sqrt{0.2})^2 + (2\sqrt{0.2})^2 \overset{?}{=} 1$

$$0.2 + 4(0.2) \stackrel{?}{=} 1$$

$$0.2 + 0.8 \stackrel{?}{=} 1$$

$$1 = 1$$

$$<\phi_1|\phi_2> = <\phi_2|\phi_1> \stackrel{?}{=} 0$$

$$-2\sqrt{0.2}\sqrt{0.2} + \sqrt{0.2}\ 2\sqrt{0.2} \stackrel{?}{=} 0$$

$$-2(0.2) + 2(0.2) \stackrel{?}{=} 0$$

$$-0.4 + 0.4 \stackrel{?}{=} 0$$

$$0 = 0$$

5. Show (for the degenerate eigenvalue; $\lambda = -2$): $<\phi_1|\phi_1> = 1$, $<\phi_2|\phi_2> = 1$, and $<\phi_1|\phi_2> = 0$

$$<\phi_1|\phi_1> \stackrel{?}{=} 1$$

$$0 + (-2\sqrt{0.2})^2 + (\sqrt{0.2})^2 \stackrel{?}{=} 1$$

$$4(0.2) + 0.2 \stackrel{?}{=} 1$$

$$0.8 + 0.2 \stackrel{?}{=} 1$$

$$1 = 1$$

$$<\phi_2|\phi_2> \stackrel{?}{=} 1$$

$$1^2 + 0 + 0 \stackrel{?}{=} 1$$

$$1 = 1$$

$$<\phi_1|\phi_2> = <\phi_2|\phi_1> \stackrel{?}{=} 0$$

$$(0)(1) + (-2\sqrt{0.2})(0) + (\sqrt{0.2})(0) \stackrel{?}{=} 0$$

$$0 = 0$$

6. Suppose the solution is of the form $x(t) = e^{\alpha t}$, with α unknown. Inserting this trial solution into the differential equation results in the following:

$$\frac{d^2}{dt^2}e^{\alpha t} + k^2 e^{\alpha t} = 0$$
$$\alpha^2 e^{\alpha t} + k^2 e^{\alpha t} = 0$$
$$(\alpha^2 + k^2)x(t) = 0$$
$$(\alpha^2 + k^2) = 0$$
$$\alpha^2 = -k^2$$
$$\alpha = \sqrt{-k^2}$$
$$\alpha = \pm ik$$

$\therefore$ Solutions are of the form e^{ikt}, e^{-ikt}, or a combination of both: $x(t) = C_1 e^{ikt} + C_2 e^{-ikt}$. Euler's formula also states that: $e^{\pm i\theta} = \cos\theta \pm i\sin\theta$, so the previous equation for $x(t)$ can also be written as:

$$x(t) = C_1\{\cos(kt) + i\sin(kt)\} + C_2\{\cos(kt) - i\sin(kt)\}$$
$$x(t) = (C_1 + C_2)\cos(kt) + (C_1 + C_2)\, i\sin(kt), \text{ or alternatively}$$
$$x(t) = C_3\cos(kt) + C_4\sin(kt).$$

We can determine these coefficients by making use of the "boundary conditions."

at $t = 0$, $x(0) = L$

$$x(0) = C_3\cos(0) + C_4\sin(0) = L$$
$$C_3 = L$$

at $t = 0$, $\frac{dx(0)}{dt} = 0$

$$\frac{d}{dt}x(t) = \frac{d}{dt}(C_3\cos(kt) + C_4\sin(kt))$$
$$\frac{d}{dt}x(t) = -C_3k\sin(kt) + C_4k\cos(kt)$$
$$\frac{d}{dt}x(0) = 0 = -C_3k\sin(0) + C_4k\cos(0)$$
$$C_4k = 0$$
$$C_4 = 0$$

$\therefore$ The solution is of the form: $x(t) = L\cos(kt)$

Exercises

1. a. $K.E. = \frac{mv^2}{2} = \left(\frac{m}{m}\right)\frac{mv^2}{2} = \frac{(mv)^2}{2m} = \frac{p^2}{2m}$

$$K.E. = \frac{1}{2m}(p_x^2 + p_y^2 + p_z^2)$$

$$K.E. = \frac{1}{2m}\left\{\left(\frac{\hbar}{i}\frac{\partial}{\partial x}\right)^2 + \left(\frac{\hbar}{i}\frac{\partial}{\partial y}\right)^2 + \left(\frac{\hbar}{i}\frac{\partial}{\partial z}\right)^2\right\}$$

$$K.E. = \frac{-\hbar^2}{2m}\left\{\frac{\partial^2}{\partial x^2} + \frac{\partial^2}{\partial y^2} + \frac{\partial^2}{\partial z^2}\right\}$$

b. $\boldsymbol{p} = m\boldsymbol{v} = \boldsymbol{i}p_x + \boldsymbol{j}p_y + \boldsymbol{k}p_z$

$$p = \left\{\boldsymbol{i}\left(\frac{\hbar}{i}\frac{\partial}{\partial x}\right) + \boldsymbol{j}\left(\frac{\hbar}{i}\frac{\partial}{\partial y}\right) + \boldsymbol{k}\left(\frac{\hbar}{i}\frac{\partial}{\partial z}\right)\right\}$$

where $\boldsymbol{i}, \boldsymbol{j}$, and $\boldsymbol{k}$ are unit vectors along the x, y, and z axes.

c. $L_y = zp_x - xp_z$

$$L_y = z\left(\frac{\hbar}{i}\frac{\partial}{\partial x}\right) - x\left(\frac{\hbar}{i}\frac{\partial}{\partial z}\right)$$

2. First derive the general formulas for

$$\frac{\partial}{\partial x}, \frac{\partial}{\partial y}, \frac{\partial}{\partial z}$$

in terms of r, θ, and ϕ, and

$$\frac{\partial}{\partial r}, \frac{\partial}{\partial \theta}, \text{ and } \frac{\partial}{\partial \phi}$$

in terms of x, y, and z. The general relationships are as follows:

$$x = r\sin\theta\cos\phi \qquad r^2 = x^2 + y^2 + z^2$$

$$y = r\sin\theta\sin\phi \qquad \sin\theta = \frac{\sqrt{x^2+y^2}}{\sqrt{x^2+y^2+z^2}}$$

$$z = r\cos\theta \qquad \cos\theta = \frac{z}{\sqrt{x^2+y^2+z^2}}$$

$$\tan\phi = \frac{y}{x}$$

First $\frac{\partial}{\partial x}$, $\frac{\partial}{\partial y}$, and $\frac{\partial}{\partial z}$ from the chain rule:

$$\frac{\partial}{\partial x} = \left(\frac{\partial r}{\partial x}\right)_{y,z}\frac{\partial}{\partial r} + \left(\frac{\partial \theta}{\partial x}\right)_{y,z}\frac{\partial}{\partial \theta} + \left(\frac{\partial \phi}{\partial x}\right)_{y,z}\frac{\partial}{\partial \phi},$$

$$\frac{\partial}{\partial y} = \left(\frac{\partial r}{\partial y}\right)_{x,z}\frac{\partial}{\partial r} + \left(\frac{\partial \theta}{\partial y}\right)_{x,z}\frac{\partial}{\partial \theta} + \left(\frac{\partial \phi}{\partial y}\right)_{x,z}\frac{\partial}{\partial \phi},$$

$$\frac{\partial}{\partial z} = \left(\frac{\partial r}{\partial z}\right)_{x,y}\frac{\partial}{\partial r} + \left(\frac{\partial \theta}{\partial z}\right)_{x,y}\frac{\partial}{\partial \theta} + \left(\frac{\partial \phi}{\partial z}\right)_{x,y}\frac{\partial}{\partial \phi}.$$

Evaluation of the many "coefficients" gives the following:

$$\left(\frac{\partial r}{\partial x}\right)_{y,z} = \sin\theta\cos\phi, \quad \left(\frac{\partial \theta}{\partial x}\right)_{y,z} = \frac{\cos\theta\cos\phi}{r}, \quad \left(\frac{\partial \phi}{\partial x}\right)_{y,z} = -\frac{\sin\phi}{r\sin\theta},$$

$$\left(\frac{\partial r}{\partial y}\right)_{x,z} = \sin\theta\sin\phi, \quad \left(\frac{\partial \theta}{\partial y}\right)_{x,z} = \frac{\cos\theta\sin\phi}{r}, \quad \left(\frac{\partial \phi}{\partial y}\right)_{x,z} = \frac{\cos\phi}{r\sin\theta},$$

$$\left(\frac{\partial r}{\partial z}\right)_{x,y} = \cos\theta, \quad \left(\frac{\partial \theta}{\partial z}\right)_{x,y} = -\frac{\sin\theta}{r}, \text{ and } \left(\frac{\partial \phi}{\partial z}\right)_{x,y} = 0.$$

Upon substitution of these "coefficients":

$$\frac{\partial}{\partial x} = \sin\theta\cos\phi\frac{\partial}{\partial r} + \frac{\cos\theta\cos\phi}{r}\frac{\partial}{\partial \theta} - \frac{\sin\phi}{r\sin\theta}\frac{\partial}{\partial \phi},$$

$$\frac{\partial}{\partial y} = \sin\theta\sin\phi\frac{\partial}{\partial r} + \frac{\cos\theta\sin\phi}{r}\frac{\partial}{\partial \theta} + \frac{\cos\phi}{r\sin\theta}\frac{\partial}{\partial \phi}, \text{ and}$$

$$\frac{\partial}{\partial z} = \cos\theta\frac{\partial}{\partial r} - \frac{\sin\theta}{r}\frac{\partial}{\partial \theta} + 0\frac{\partial}{\partial \phi}.$$

Next $\frac{\partial}{\partial r}$, $\frac{\partial}{\partial \theta}$, and $\frac{\partial}{\partial \phi}$ from the chain rule:

$$\frac{\partial}{\partial r} = \left(\frac{\partial x}{\partial r}\right)_{\theta,\phi}\frac{\partial}{\partial x} + \left(\frac{\partial y}{\partial r}\right)_{\theta,\phi}\frac{\partial}{\partial y} + \left(\frac{\partial z}{\partial r}\right)_{\theta,\phi}\frac{\partial}{\partial z},$$

$$\frac{\partial}{\partial \theta} = \left(\frac{\partial x}{\partial \theta}\right)_{r,\phi}\frac{\partial}{\partial x} + \left(\frac{\partial y}{\partial \theta}\right)_{r,\phi}\frac{\partial}{\partial y} + \left(\frac{\partial z}{\partial \theta}\right)_{r,\phi}\frac{\partial}{\partial z}, \text{ and}$$

$$\frac{\partial}{\partial \phi} = \left(\frac{\partial x}{\partial \phi}\right)_{r,\theta}\frac{\partial}{\partial x} + \left(\frac{\partial y}{\partial \phi}\right)_{r,\theta}\frac{\partial}{\partial y} + \left(\frac{\partial z}{\partial \phi}\right)_{r,\theta}\frac{\partial}{\partial z}.$$

Again evaluation of the many "coefficients" results in:

$$\left(\frac{\partial x}{\partial r}\right)_{\theta,\phi} = \frac{x}{\sqrt{x^2+y^2+z^2}},\ \left(\frac{\partial y}{\partial r}\right)_{\theta,\phi} = \frac{y}{\sqrt{x^2+y^2+z^2}},$$

$$\left(\frac{\partial z}{\partial r}\right)_{\theta,\phi} = \frac{z}{\sqrt{x^2+y^2+z^2}},\ \left(\frac{\partial x}{\partial \theta}\right)_{r,\phi} = \frac{xz}{\sqrt{x^2+y^2}},\ \left(\frac{\partial y}{\partial \theta}\right)_{r,\phi} = \frac{yz}{\sqrt{x^2+y^2}},$$

$$\left(\frac{\partial z}{\partial \theta}\right)_{r,\phi} = -\sqrt{x^2+y^2},\ \left(\frac{\partial x}{\partial \phi}\right)_{r,\theta} = -y,\ \left(\frac{\partial y}{\partial \phi}\right)_{r,\theta} = x, \text{ and } \left(\frac{\partial z}{\partial \phi}\right)_{r,\theta} = 0$$

Upon substitution of these "coefficients":

$$\frac{\partial}{\partial r} = \frac{x}{\sqrt{x^2+y^2+z^2}}\frac{\partial}{\partial x} + \frac{y}{\sqrt{x^2+y^2+z^2}}\frac{\partial}{\partial y} + \frac{z}{\sqrt{x^2+y^2+z^2}}\frac{\partial}{\partial z}$$

$$\frac{\partial}{\partial \theta} = \frac{xz}{\sqrt{x^2+y^2}}\frac{\partial}{\partial x} + \frac{yz}{\sqrt{x^2+y^2}}\frac{\partial}{\partial y} - \sqrt{x^2+y^2}\frac{\partial}{\partial z}$$

$$\frac{\partial}{\partial \phi} = -y\frac{\partial}{\partial x} + x\frac{\partial}{\partial y} + 0\frac{\partial}{\partial z}.$$

Note, these many "coefficients" are the elements which make up the Jacobian matrix used whenever one wishes to transform a function from one coordinate representation to another. One very familiar result should be in transforming the volume element $dxdydz$ to $r^2 \sin\theta\, dr d\theta\, d\phi$. For example:

$$\int f(x,y,z)dxdydz =$$

$$\int f(x(r,\theta,\phi),y(r,\theta,\phi),z(r,\theta,\phi)) \begin{vmatrix} \left(\frac{\partial x}{\partial r}\right)_{\theta\phi} & \left(\frac{\partial x}{\partial \theta}\right)_{r\phi} & \left(\frac{\partial x}{\partial \phi}\right)_{r\theta} \\ \left(\frac{\partial y}{\partial r}\right)_{\theta\phi} & \left(\frac{\partial y}{\partial \theta}\right)_{r\phi} & \left(\frac{\partial y}{\partial \theta}\right)_{r\theta} \\ \left(\frac{\partial z}{\partial r}\right)_{\theta\phi} & \left(\frac{\partial z}{\partial \theta}\right)_{r\phi} & \left(\frac{\partial z}{\partial \theta}\right)_{r\theta} \end{vmatrix} dr d\theta d\phi$$

a. $$L_x = \frac{\hbar}{i}\left\{y\frac{\partial}{\partial z} - z\frac{\partial}{\partial y}\right\}$$

$$L_x = \frac{\hbar}{i}\left(r\sin\theta\sin\phi\left(\cos\theta\frac{\partial}{\partial r} - \frac{\sin\theta}{r}\frac{\partial}{\partial \theta}\right)\right)$$

$$-\frac{\hbar}{i}\left(r\cos\theta\left(\sin\theta\sin\phi\frac{\partial}{\partial r} + \frac{\cos\theta\sin\phi}{r}\frac{\partial}{\partial \theta} + \frac{\cos\phi}{r\sin\theta}\frac{\partial}{\partial \phi}\right)\right)$$

$$L_x = -\frac{\hbar}{i}\left(\sin\phi\frac{\partial}{\partial \theta} + \cos\theta\cos\phi\frac{\partial}{\partial \phi}\right)$$

b. $$L_z = \frac{\hbar}{i}\frac{\partial}{\partial \phi} = -i\hbar\frac{\partial}{\partial \phi}$$

$$L_z = \frac{\hbar}{i}\left(-y\frac{\partial}{\partial x} + x\frac{\partial}{\partial y}\right)$$

3.

	B	*B′*	*B″*
a.	$4x^4 - 12x^2 + 3$	$16x^3 - 24x$	$48x^2 - 24$
b.	$5x^4$	$20x^3$	$60x^2$
c.	$e^{3x} + e^{-3x}$	$3(e^{3x} - e^{-3x})$	$9(e^{3x} + e^{-3x})$
d.	$x^2 - 4x+2$	$2x - 4$	2
e.	$4x^3 - 3x$	$12x^2 - 3$	$24x$

B(e) is an eigenfunction of A(a):

$$(1 - x^2)\frac{d^2}{dx^2} - x\frac{d}{dx}\text{B(e)} = (1 - x^2)(24x) - x(12x^2 - 3)$$
$$= 24x - 24x^3 - 12x^3 + 3x$$
$$= -36x^3 + 27x$$
$$= -9(4x^3 - 3x) \text{ (eigenvalue is } -9)$$

B(c) is an eigenfunction of A(b):

$$\frac{d^2}{dx^2}\text{B(c)} = 9(e^{3x} + e^{-3x}) \text{ (eigenvalue is 9)}$$

B(b) is an eigenfunction of A(c.):

$$x\frac{d}{dx}\text{B(b)} = x(20x^3)$$
$$= 20x^4$$
$$= 4(5x^4) \text{ (eigenvalue is 4)}$$

B(a) is an eigenfunction of A(d):

$$\frac{d^2}{dx^2} - 2x\frac{d}{dx}\text{B(a)} = (48x^2 - 24) - 2x(16x^3 - 24x)$$
$$= 48x^2 - 24 - 32x^4 + 48x^2$$
$$= -32x^4 + 96x^2 - 24$$
$$= -8(4x^4 - 12x^2 + 3) \text{ (eigenvalue is } -8)$$

B(d) is an eigenfunction of A(e):

$$x\frac{d^2}{dx^2} + (1 - x)\frac{d}{dx}\text{B(d)} = x(2) + (1 - x)(2x - 4)$$
$$= 2x + 2x - 4 - 2x^2 + 4x$$
$$= -2x^2 + 8x - 4$$
$$= -2(x^2 - 4x + 2) \text{ (eigenvalue } is \text{ } -2)$$

4. Show that:

$$\int f^* \mathbf{A} g d\tau = \int g(\mathbf{A}f)^* d\tau$$

a. Suppose f and g are functions of x and evaluate the integral on the left-hand side by "integration by parts":

$$\int f(x)^*(-i\hbar\frac{\partial}{\partial x})g(x)dx$$

let $dv = \frac{\partial}{\partial x}g(x)dx$ and $u = -i\hbar f(x)^*$

$v = g(x)$ $du = -i\hbar\frac{\partial}{\partial x}f(x)^*dx$

Now, $\int udv = uv - \int vdu$,

so:

$$\int f(x)^*(-i\hbar\frac{\partial}{\partial x})g(x)dx = -i\hbar f(x)^*g(x) + i\hbar\int g(x)\frac{\partial}{\partial x}f(x)^*dx.$$

Note that in, principle, it is impossible to prove hermiticity *unless* you are given knowledge of the type of function on which the operator is acting. Hermiticity requires (as can be seen in this example) that the term $-i\hbar f(x)^*g(x)$ vanish when evaluated at the integral limits. This, in general, will occur for the "well behaved" functions (e.g., in *bound state* quantum chemistry, the wavefunctions will vanish as the distances among particles approaches infinity). So, in proving the Hermiticity of an operator, one must be careful to specify the behavior of the functions on which the operator is considered to act. This means that an operator may be Hermitian for one class of functions and non-Hermitian for another class of functions. If we assume that f and g vanish at the boundaries, then we have

$$\int f(x)^*(-i\hbar\frac{\partial}{\partial x})g(x)dx = \int g(x)\left(-i\hbar\frac{\partial}{\partial x}f(x)\right)^*dx$$

b. Suppose f and g are functions of y and z and evaluate the integral on the left-hand side by "integration by parts" as in the previous exercise:

$$\int f(y,z)^*\left\{-i\hbar\left(y\frac{\partial}{\partial z} - z\frac{\partial}{\partial y}\right)\right\}g(y,z)dydz$$

$$= \int f(y,z)^*\left(-i\hbar\left(y\frac{\partial}{\partial z}\right)\right)g(y,z)dydz - \int f(y,z)^*\left(-i\hbar\left(z\frac{\partial}{\partial y}\right)\right)g(y,z)dydz$$

For the first integral,

$$\int f(z)^*\left(-i\hbar y\frac{\partial}{\partial z}\right)g(z)dz,$$

let $dv = \frac{\partial}{\partial z}g(z)dz$ $u = -i\hbar y f(z)^*$

$v = g(z)$ $du = -i\hbar y\frac{\partial}{\partial z}f(z)^*dz$

so:

$$\int f(z)^*(-i\hbar y\frac{\partial}{\partial z})g(z)dz = -i\hbar y f(z)^* g(z) + i\hbar y\int g(z)\frac{\partial}{\partial z}f(z)^* dz$$

$$= \int g(z)\left(-i\hbar y\frac{\partial}{\partial z}f(z)\right)^* dz.$$

For the second integral,

$$\int f(y)^*\left(-i\hbar z\frac{\partial}{\partial y}\right)g(y)dy,$$

$$\text{let } dv = \frac{\partial}{\partial y}g(y)dy \qquad u = -i\hbar z f(y)^*$$

$$v = g(y) \qquad du = -i\hbar z\frac{\partial}{\partial y}f(y)^* dy$$

so:

$$\int f(y)^*(-i\hbar z\frac{\partial}{\partial y})g(y)dy = -i\hbar z f(y)^* g(y) + i\hbar z\int g(y)\frac{\partial}{\partial y}f(y)^* dy$$

$$= \int g(y)\left(-i\hbar z\frac{\partial}{\partial y}f(y)\right)^* dy$$

$$\int f(y,z)^*\left\{-i\hbar\left(y\frac{\partial}{\partial z} - z\frac{\partial}{\partial y}\right)\right\}g(y,z)dydz$$

$$= \int g(z)\left(-i\hbar y\frac{\partial}{\partial z}f(z)\right)^* dz - \int g(y)\left(-i\hbar z\frac{\partial}{\partial y}f(y)\right)^* dy$$

$$= \int g(y,z)\left(-i\hbar\left(y\frac{\partial}{\partial z} - z\frac{\partial}{\partial y}\right)f(y,z)\right)^* dydz.$$

Again we have had to assume that the functions f and g vanish at the boundary.

5. $\boldsymbol{L}_+ = \boldsymbol{L}_x + i\boldsymbol{L}_y$

$\boldsymbol{L}_- = \boldsymbol{L}_x - i\boldsymbol{L}_y$, so

$\boldsymbol{L}_+ + \boldsymbol{L}_- = 2\boldsymbol{L}_x$, or $\boldsymbol{L}_x = \frac{1}{2}(\boldsymbol{L}_+ + \boldsymbol{L}_-)$

$\boldsymbol{L}_+ Y_{l,m} = \sqrt{l(l+1) - m(m+1)}\,\hbar Y_{l,m+1}$

$\boldsymbol{L}_- Y_{l,m} = \sqrt{l(l+1) - m(m-1)}\,\hbar Y_{l,m-1}$

Using these relationships:

$\boldsymbol{L}_-\Psi_{2p_{-1}} = 0$, $\boldsymbol{L}_-\Psi_{2p_0} = \sqrt{2}\hbar\Psi_{2p_{-1}}$, $\boldsymbol{L}_-\Psi_{2p_{+1}} = \sqrt{2}\hbar\Psi_{2p_0}$
$\boldsymbol{L}_+\Psi_{2p_{-1}} = \sqrt{2}\hbar\Psi_{2p_0}$, $\boldsymbol{L}_+\Psi_{2p_0} = \sqrt{2}\hbar\Psi_{2p_{+1}}$, $\boldsymbol{L}_+\Psi_{2p_{+1}} = 0$, and

the following L_x matrix elements can be evaluated:

$$L_x(1,1) = \left\langle\Psi_{2p_{-1}}\left|\frac{1}{2}(\boldsymbol{L}_+ + \boldsymbol{L}_-)\right|\Psi_{2p_{-1}}\right\rangle = 0$$

$$L_x(1,2) = <\Psi_{2p_{-1}}\left|\frac{1}{2}(\boldsymbol{L}_+ + \boldsymbol{L}_-)\right|\Psi_{2p_0}> = \frac{\sqrt{2}}{2}\hbar$$

$$L_x(1,3) = <\Psi_{2p_{-1}}\left|\frac{1}{2}(\boldsymbol{L}_+ + \boldsymbol{L}_-)\right|\Psi_{2p_{+1}}> = 0$$

$$L_x(2,1) = <\Psi_{2p_0}\left|\frac{1}{2}(\boldsymbol{L}_+ + \boldsymbol{L}_-)\right|\Psi_{2p_{-1}}> = \frac{\sqrt{2}}{2}\hbar$$

$$L_x(2,2) = <\Psi_{2p_0}\left|\frac{1}{2}(\boldsymbol{L}_+ + \boldsymbol{L}_-)\right|\Psi_{2p_0}> = 0$$

$$L_x(2,3) = <\Psi_{2p_0}\left|\frac{1}{2}(\boldsymbol{L}_+ + \boldsymbol{L}_-)\right|\Psi_{2p_{+1}}> = \frac{\sqrt{2}}{2}\hbar$$

$$L_x(3,1) = <\Psi_{2p_{+1}}\left|\frac{1}{2}(\boldsymbol{L}_+ + \boldsymbol{L}_-)\right|\Psi_{2p_{-1}}> = 0$$

$$L_x(3,2) = <\Psi_{2p_{+1}}\left|\frac{1}{2}(\boldsymbol{L}_+ + \boldsymbol{L}_-)\right|\Psi_{2p_0}> = \frac{\sqrt{2}}{2}\hbar$$

$$L_x(3,3) = 0$$

This matrix:

$$\begin{bmatrix} 0 & \frac{\sqrt{2}}{2}\hbar & 0 \\ \frac{\sqrt{2}}{2}\hbar & 0 & \frac{\sqrt{2}}{2}\hbar \\ 0 & \frac{\sqrt{2}}{2}\hbar & 0 \end{bmatrix},$$

can now be diagonalized:

$$\begin{vmatrix} 0-\lambda & \frac{\sqrt{2}}{2}\hbar & 0 \\ \frac{\sqrt{2}}{2}\hbar & 0-\lambda & \frac{\sqrt{2}}{2}\hbar \\ 0 & \frac{\sqrt{2}}{2}\hbar & 0-\lambda \end{vmatrix} = 0$$

$$\begin{vmatrix} 0-\lambda & \frac{\sqrt{2}}{2}\hbar \\ \frac{\sqrt{2}}{2}\hbar & 0-\lambda \end{vmatrix}(-\lambda) - \begin{vmatrix} \frac{\sqrt{2}}{2}\hbar & \frac{\sqrt{2}}{2}\hbar \\ 0 & 0-\lambda \end{vmatrix}\left(\frac{\sqrt{2}}{2}\hbar\right) = 0$$

Expanding these determinants yields:

$$(\lambda^2 - \frac{\hbar^2}{2})(-\lambda) - \frac{\sqrt{2}\hbar}{2}(-\lambda)\left(\frac{\sqrt{2}\hbar}{2}\right) = 0$$

$$-\lambda(\lambda^2 - \hbar^2) = 0$$

$$-\lambda(\lambda - \hbar)(\lambda + \hbar) = 0$$

with roots: 0, $\hbar$, and $-\hbar$

Next, determine the corresponding eigenvectors:

For $\lambda = 0$:

$$\begin{bmatrix} 0 & \frac{\sqrt{2}}{2}\hbar & 0 \\ \frac{\sqrt{2}}{2}\hbar & 0 & \frac{\sqrt{2}}{2}\hbar \\ 0 & \frac{\sqrt{2}}{2}\hbar & 0 \end{bmatrix} \begin{bmatrix} C_{11} \\ C_{21} \\ C_{31} \end{bmatrix} = 0 \begin{bmatrix} C_{11} \\ C_{21} \\ C_{31} \end{bmatrix}$$

$$\frac{\sqrt{2}}{2}\hbar C_{21} = 0 \text{ (row one)}$$
$$C_{21} = 0$$
$$\frac{\sqrt{2}}{2}\hbar C_{11} + \frac{\sqrt{2}}{2}\hbar C_{31} = 0 \text{ (row two)}$$
$$C_{11} + C_{31} = 0$$
$$C_{11} = -C_{31}$$
$$C_{11}^2 + C_{21}^2 + C_{31}^2 = 1 \text{ (normalization)}$$
$$C_{11}^2 + (-C_{11})^2 = 1$$
$$2C_{11}^2 = 1$$
$$C_{11} = \frac{1}{\sqrt{2}},\ C_{21} = 0, \text{ and } C_{31} = -\frac{1}{\sqrt{2}}$$

For $\lambda = 1\hbar$:

$$\begin{bmatrix} 0 & \frac{\sqrt{2}}{2}\hbar & 0 \\ \frac{\sqrt{2}}{2}\hbar & 0 & \frac{\sqrt{2}}{2}\hbar \\ 0 & \frac{\sqrt{2}}{2}\hbar & 0 \end{bmatrix} \begin{bmatrix} C_{12} \\ C_{22} \\ C_{32} \end{bmatrix} = 1\hbar \begin{bmatrix} C_{12} \\ C_{22} \\ C_{32} \end{bmatrix}$$

$$\frac{\sqrt{2}}{2}\hbar C_{22} = \hbar C_{12} \text{ (row one)}$$
$$C_{12} = \frac{\sqrt{2}}{2}C_{22}$$
$$\frac{\sqrt{2}}{2}\hbar C_{12} + \frac{\sqrt{2}}{2}\hbar C_{32} = \hbar C_{22} \text{ (row two)}$$
$$\frac{\sqrt{2}}{2}\frac{\sqrt{2}}{2}C_{22} + \frac{\sqrt{2}}{2}C_{32} = C_{22}$$
$$\frac{1}{2}C_{22} + \frac{\sqrt{2}}{2}C_{32} = C_{22}$$
$$\frac{\sqrt{2}}{2}C_{32} = \frac{1}{2}C_{22}$$
$$C_{32} = \frac{\sqrt{2}}{2}C_{22}$$
$$C_{12}^2 + C_{22}^2 + C_{32}^2 = 1 \text{ (normalization)}$$

$$\left(\frac{\sqrt{2}}{2}C_{22}\right)^2 + C_{22}^2 + \left(\frac{\sqrt{2}}{2}C_{22}\right)^2 = 1$$

$$\frac{1}{2}C_{22}^2 + C_{22}^2 + \frac{1}{2}C_{22}^2 = 1$$

$$2C_{22}^2 = 1$$

$$C_{22} = \frac{\sqrt{2}}{2}$$

$$C_{12} = \frac{1}{2},\ C_{22} = \frac{\sqrt{2}}{2} \text{ and } C_{32} = \frac{1}{2}$$

For $\lambda = -1\hbar$:

$$\begin{bmatrix} 0 & \frac{\sqrt{2}}{2}\hbar & 0 \\ \frac{\sqrt{2}}{2}\hbar & 0 & \frac{\sqrt{2}}{2}\hbar \\ 0 & \frac{\sqrt{2}}{2}\hbar & 0 \end{bmatrix} \begin{bmatrix} C_{13} \\ C_{23} \\ C_{33} \end{bmatrix} = -1\hbar \begin{bmatrix} C_{13} \\ C_{23} \\ C_{33} \end{bmatrix}$$

$$\frac{\sqrt{2}}{2}\hbar C_{23} = -\hbar C_{13} \text{ (row one)}$$

$$C_{13} = -\frac{\sqrt{2}}{2}C_{23}$$

$$\frac{\sqrt{2}}{2}\hbar C_{13} + \frac{\sqrt{2}}{2}\hbar C_{33} = -\hbar C_{23} \text{ (row two)}$$

$$\frac{\sqrt{2}}{2}\left(-\frac{\sqrt{2}}{2}C_{23}\right) + \frac{\sqrt{2}}{2}C_{33} = -C_{23}$$

$$-\frac{1}{2}C_{23} + \frac{\sqrt{2}}{2}C_{33} = -C_{23}$$

$$\frac{\sqrt{2}}{2}C_{33} = -\frac{1}{2}C_{23}$$

$$C_{33} = -\frac{\sqrt{2}}{2}C_{23}$$

$$C_{13}^2 + C_{23}^2 + C_{33}^2 = 1 \text{ (normalization)}$$

$$\left(-\frac{\sqrt{2}}{2}C_{23}\right)^2 + C_{23}^2 + \left(-\frac{\sqrt{2}}{2}C_{23}\right)^2 = 1$$

$$\frac{1}{2}C_{23}^2 + C_{23}^2 + \frac{1}{2}C_{23}^2 = 1$$

$$2C_{23}^2 = 1$$

$$C_{23} = \frac{\sqrt{2}}{2}$$

$$C_{13} = -\frac{1}{2},\ C_{23} = \frac{\sqrt{2}}{2},\text{ and } C_{33} = -\frac{1}{2}$$

Show: $\langle\phi_1|\phi_1\rangle = 1$, $\langle\phi_2|\phi_2\rangle = 1$, $\langle\phi_3|\phi_3\rangle = 1$, $\langle\phi_1|\phi_2\rangle = 0$, $\langle\phi_1|\phi_3\rangle = 0$, and $\langle\phi_2|\phi_3\rangle = 0$.

$$\langle\phi_1|\phi_1\rangle \stackrel{?}{=} 1$$

$$\left(\frac{\sqrt{2}}{2}\right)^2 + 0 + \left(\frac{-\sqrt{2}}{2}\right)^2 \stackrel{?}{=} 1$$

$$\frac{1}{2} + \frac{1}{2} \stackrel{?}{=} 1$$

$$1 = 1$$

$$\langle\phi_2|\phi_2\rangle \stackrel{?}{=} 1$$

$$\left(\frac{1}{2}\right)^2 + \left(\frac{\sqrt{2}}{2}\right)^2 + \left(\frac{1}{2}\right)^2 \stackrel{?}{=} 1$$

$$\frac{1}{4} + \frac{1}{2} + \frac{1}{4} \stackrel{?}{=} 1$$

$$1 = 1$$

$$\langle\phi_3|\phi_3\rangle \stackrel{?}{=} 1$$

$$\left(-\frac{1}{2}\right)^2 + \left(\frac{\sqrt{2}}{2}\right)^2 + \left(-\frac{1}{2}\right)^2 \stackrel{?}{=} 1$$

$$\frac{1}{4} + \frac{1}{2} + \frac{1}{4} \stackrel{?}{=} 1$$

$$1 = 1$$

$$\langle\phi_1|\phi_2\rangle = \langle\phi_2|\phi_1\rangle \stackrel{?}{=} 0$$

$$\left(\frac{\sqrt{2}}{2}\right)\left(\frac{1}{2}\right) + (0)\left(\frac{\sqrt{2}}{2}\right) + \left(\frac{-\sqrt{2}}{2}\right)\left(\frac{1}{2}\right) \stackrel{?}{=} 0$$

$$\left(\frac{\sqrt{2}}{4}\right) - \left(\frac{\sqrt{2}}{4}\right) \stackrel{?}{=} 0$$

$$0 = 0$$

$$\langle\phi_1|\phi_3\rangle = \langle\phi_3|\phi_1\rangle \stackrel{?}{=} 0$$

$$\left(\frac{\sqrt{2}}{2}\right)\left(-\frac{1}{2}\right) + (0)\left(\frac{\sqrt{2}}{2}\right) + \left(\frac{-\sqrt{2}}{2}\right)\left(-\frac{1}{2}\right) \stackrel{?}{=} 0$$

$$\left(-\frac{\sqrt{2}}{4}\right) + \left(\frac{\sqrt{2}}{4}\right) \stackrel{?}{=} 0$$

$$0 = 0$$

$$\langle\phi_2|\phi_3\rangle = \langle\phi_3|\phi_2\rangle \stackrel{?}{=} 0$$

$$\left(\frac{1}{2}\right)\left(-\frac{1}{2}\right) + \left(\frac{\sqrt{2}}{2}\right)\left(\frac{\sqrt{2}}{2}\right) + \left(\frac{1}{2}\right)\left(-\frac{1}{2}\right) \stackrel{?}{=} 0$$

$$\left(-\frac{1}{4}\right)+\left(\frac{1}{2}\right)+\left(-\frac{1}{4}\right)\stackrel{?}{=}0$$

$$0=0$$

6. $P_{2p_{+1}} = \left|<\phi_{2p_{+1}}, \Psi_{L_x}^{0\hbar}\right|^2$

$$\Psi_{L_x}^{0\hbar} = \frac{1}{\sqrt{2}}\phi_{2p_{-1}} - \frac{1}{\sqrt{2}}\phi_{2p_{+1}}$$

$$P_{2p_{+1}} = \left|-\frac{1}{\sqrt{2}}<\phi_{2p_{+1}}\phi_{2p_{+1}}>\right|^2 = \frac{1}{2} \text{ (or 50\%)}$$

7. It is useful here to use some of the general commutator relations found in appendix C.V.

a. $[L_x, L_y] = [yp_z - zp_y, zp_x - xp_z]$

$$\begin{aligned}
&= [yp_z,zp_x] - [yp_z,xp_z] - [zp_y, zp_x] + [zp_y,xp_z]\\
&= [y,z]p_xp_z + z[y,p_x]p_z + y[p_z,z]p_x + yz[p_z,p_x]\\
&- [y,x]p_zp_z - x[y,p_z]p_z - y[p_z,x]p_z - yx[p_z,p_z]\\
&- [z,z]p_xp_y - z[z,p_x]p_y - z[p_y,z]p_x - zz[p_y,p_x]\\
&+ [z,x]p_zp_y + x[z,p_z]p_y + z[p_y,x]p_z + zx[p_y,p_z]\\
&- [x,x]p_yp_z - x[x,p_y]p_z - x[p_z,x]p_y - xx[p_z,p_y]\\
&+ [x,y]p_xp_z + y[x,p_x]p_z + x[p_z,y]p_x + xy[p_z,p_x]
\end{aligned}$$

Again, as can be easily ascertained, the only non-zero terms are:

$$\begin{aligned}
[L_y, L_z] &= z[p_x,x]p_y + y[x,p_x]p_z\\
&= z(-i\hbar)p_y + y(i\hbar)p_z\\
&= i\hbar(-zp_y + yp_z)\\
&= i\hbar L_x
\end{aligned}$$

c. $[L_z, L_x] = [xp_y - yp_x, yp_z - zp_y]$

$$\begin{aligned}
&= [xp_y,yp_z] - [xp_y,zp_y] - [yp_x,yp_z] + [yp_x,zp_y]\\
&= [x,y]p_zp_y + y[x,p_z]p_y + x[p_y,y]p_z + xy[p_y,p_z]\\
&- [x,z]p_yp_y - z[x,p_y]p_y - x[p_y,z]p_y - xz[p_y,p_y]\\
&- [y,y]p_zp_x - y[y,p_z]p_x - y[p_x,y]p_z - yy[p_x,p_z]\\
&+ [y,z]p_yp_x + z[y,p_y]p_x + y[p_x,z]p_y + yz[p_x,p_y]
\end{aligned}$$

Again, as can be easily ascertained, the only non-zero terms are:

$$\begin{aligned}
[L_z, L_x] &= x[p_y,y]p_z + z[y,p_y]p_x\\
&= x(-i\hbar)p_z + z(i\hbar)p_x\\
&= i\hbar(-xp_z + zp_x)\\
&= i\hbar L_y
\end{aligned}$$

d. $[L_x, L^2] = [L_x,L_x^2 + L_y^2 + L_z^2]$

$$\begin{aligned}
&= [L_x,L_x^2] + [L_x,L_y^2] + [L_x,L_z^2]\\
&= [L_x,L_y^2] + [L_x,L_z^2]\\
&= [L_x,L_y]L_y + L_y[L_x,L_y] + [L_x,L_z]L_z + L_z[L_x,L_z]\\
&= (i\hbar L_z)L_y + L_y(i\hbar L_z) + (-i\hbar L_y)L_z + L_z(-i\hbar L_y)\\
&= (i\hbar)(L_zL_y + L_yL_z - L_yL_z - L_zL_y)\\
&= (i\hbar)([L_z,L_y] + [L_y,L_z]) = 0
\end{aligned}$$

e. $[L_y,L^2] = [L_y,L_x{}^2 + L_y{}^2 + L_{[z}{}^2]$
$= [L_y,L_x{}^2] + [L_y,L_y{}^2] + [L_y,L_z{}^2]$
$= [L_y,L_x{}^2] + [L_y,L_z{}^2]$
$= [L_y,L_x]L_x + L_x[L_y,L_x] + [L_y,L_z]L_z + L_z[L_y,L_z]$
$= (-i\hbar L_z)L_x + L_x(-i\hbar L_z) + (i\hbar L_x)L_z + L_z(i\hbar L_x)$
$= (i\hbar)(-L_zL_x - L_xL_z + L_xL_z + L_zL_x)$
$= (i\hbar)([L_x,L_z] + [L_z,L_x]) = 0$

f. $[L_z,L^2] = [L_z,L_x{}^2 + L_y{}^2 + L_z{}^2]$
$= [L_z,L_x{}^2] + [L_z,L_y{}^2] + [L_z,L_z{}^2]$
$= [L_z,L_x{}^2] + [L_z,L_y{}^2]$
$= [L_z,L_x]L_x + L_x[L_z,L_x] + [L_z,L_y]L_y + L_y[L_z,L_y]$
$= (i\hbar L_y)L_x + L_x(i\hbar L_y) + (-i\hbar L_x)L_y + L_y(-i\hbar L_x)$
$= (i\hbar)(L_yL_x + L_xL_y - L_xL_y - L_yL_x)$
$= (i\hbar)([L_y,L_x] + [L_x,L_y]) = 0$

8. Use the general angular momentum relationships:

$$J^2|j,m> = \hbar^2(j(j+1))|j,m>$$
$$J_z|j,m> = \hbar m|j,m>,$$

and the information used in exercise 5, namely that:

$$L_x = \frac{1}{2}(L_+ + L_-)$$
$$L_+Y_{l,m} = \sqrt{l(l+1) - m(m+1)}\hbar Y_{l,m+1}$$
$$L_-Y_{l,m} = \sqrt{l(l+1) - m(m-1)}\hbar Y_{l,m-1}$$

Given that:

$$Y_{0,0}(\theta,\phi) = \frac{1}{\sqrt{4\pi}} = |0,0>$$
$$Y_{1,0}(\theta,\phi) = \sqrt{\frac{3}{4\pi}}\cos\theta = |1,0>.$$

a. $L_z|0,0> = 0$

$L^2|0,0> = 0$

Since L^2 and L_z commute you would expect $|0,0>$ to be simultaneous eigenfunctions of both.

b. $L_x|0,0> = 0$

$L_z|0,0> = 0$

L_x and L_z *do not* commute. It is unexpected to find a simultaneous eigenfunction ($|0,0>$) of both . . . for sure these operators do not have the same full set of eigenfunctions.

c. $L_z|1,0> = 0$

$L^2|1,0> = 2\hbar^2|1,0>$

Again since L^2 and L_z commute you would expect $|1,0>$ to be simultaneous eigenfunctions of both.

d. $L_x|1,0> = \frac{\sqrt{2}}{2}\hbar|1,-1> + \frac{\sqrt{2}}{2}\hbar|1,1>$
$L_z|1,0> = 0$

Again, L_x and L_z *do not* commute. Therefore it is expected to find differing sets of eigenfunctions for both.

9. For:

$$\Psi(x,y) = \left(\frac{1}{2L_x}\right)^{\frac{1}{2}}\left(\frac{1}{2L_y}\right)^{\frac{1}{2}}\left[e^{in_x\pi x/L_x} - e^{-in_x\pi x/L_x}\right]\left[e^{in_y\pi y/L_y} - e^{-in_y\pi y/L_y}\right]$$

$$<\Psi(x,y)|\Psi(x,y)> \overset{?}{=} 1$$

Let: $a_x = \frac{n_x\pi}{L_x}$, and $a_y = \frac{n_y\pi}{L_y}$ and using Euler's formula, expand the exponentials into sin and cos terms.

$$\Psi(x,y) = \left(\frac{1}{2L_x}\right)^{\frac{1}{2}}\left(\frac{1}{2L_y}\right)^{\frac{1}{2}}[\cos(a_x x) + i\sin(a_x x) - \cos(a_x x) +$$

$$i\sin(a_x x)][\cos(a_y y) + i\sin(a_y y) - \cos(a_y y) + i\sin(a_y y)]$$

$$\Psi(x,y) = \left(\frac{1}{2L_x}\right)^{\frac{1}{2}}\left(\frac{1}{2L_y}\right)^{\frac{1}{2}} 2i\sin(a_x x)\, 2i\sin(a_y y)$$

$$\Psi(x,y) = -\left(\frac{2}{L_x}\right)^{\frac{1}{2}}\left(\frac{2}{L_y}\right)^{\frac{1}{2}} \sin(a_x x)\sin(a_y y)$$

$$<\Psi(x,y)|\Psi(x,y)> = \int\left(-\left(\frac{2}{L_x}\right)^{\frac{1}{2}}\left(\frac{2}{L_y}\right)^{\frac{1}{2}} \sin(a_x x)\sin(a_y y)\right)^2 dxdy$$

$$= \left(\frac{2}{L_x}\right)\left(\frac{2}{L_y}\right)\int \sin^2(a_x x)\sin^2(a_y y)dxdy$$

Using the integral:

$$\int_0^L \sin^2\frac{n\pi x}{L}dx = \frac{L}{2},$$

$$<\Psi(x,y)|\Psi(x,y)> = \left(\frac{2}{L_x}\right)\left(\frac{2}{L_y}\right)\left(\frac{L_x}{2}\right)\left(\frac{L_y}{2}\right) = 1$$

10. $$<\Psi(x,y)|p_x|\Psi(x,y)> = \left(\frac{2}{L_y}\right)\int_0^{L_y}\sin^2(a_y y)dy\left(\frac{2}{L_x}\right)\int_0^{L_x}\sin(a_x x)(-i\hbar\frac{\partial}{\partial x})\sin(a_x x)dx$$

$$= \left(\frac{-i\hbar 2a_x}{L_x}\right)\int_0^{L_x}\sin(a_x x)\cos(a_x x)dx$$

But the integral:

$$\int_0^{L_x} \cos(a_x x)\sin(a_x x)dx = 0,$$

$$\therefore <\Psi(x,y)|p_x|\Psi(x,y)> = 0$$

11. $$<\Psi_0|\boldsymbol{x}^2|\Psi_0> = \left(\frac{\alpha}{\pi}\right)^{\frac{1}{2}} \int_{-\infty}^{+\infty} \left(e^{-\alpha x^2/2}\right)\left(\boldsymbol{x}^2\right)\left(e^{-\alpha x^2/2}\right)dx$$

$$= \left(\frac{\alpha}{\pi}\right)^{\frac{1}{2}} 2 \int_0^{+\infty} x^2 e^{-\alpha x 2} dx$$

Using the integral:

$$\int_0^{+\infty} x^{2n} e^{-\beta x^2} dx = \frac{1 \cdot 3 \cdots (2n-1)}{2^{n+1}} \left(\frac{\pi}{\beta^{2n+1}}\right)^{\frac{1}{2}},$$

$$<\Psi_0|\boldsymbol{x}^2|\Psi_0> = \left(\frac{\alpha}{\pi}\right)^{\frac{1}{2}} 2 \left(\frac{1}{2^2}\right)\left(\frac{\pi}{\alpha^3}\right)^{\frac{1}{2}}$$

$$<\Psi_0|\boldsymbol{x}^2|\Psi_0> = \left(\frac{1}{2\alpha}\right)$$

$$<\Psi_1|\boldsymbol{x}^2|\Psi_1> = \left(\frac{4\alpha^3}{\pi}\right)^{\frac{1}{2}} \int_{-\infty}^{+\infty} \left(xe^{-\alpha x^2/2}\right)\left(\boldsymbol{x}^2\right)\left(xe^{-\alpha x^2/2}\right)dx$$

$$= \left(\frac{4\alpha^3}{\pi}\right)^{\frac{1}{2}} 2 \int_0^{+\infty} x^4 e^{-\alpha x^2/2} dx$$

Using the previously defined integral:

$$<\Psi_1|\boldsymbol{x}^2|\Psi_1> = \left(\frac{4\alpha^3}{\pi}\right)^{\frac{1}{2}} 2 \left(\frac{3}{2^3}\right)\left(\frac{\pi}{\alpha^5}\right)^{\frac{1}{2}}$$

$$<\Psi_1|\boldsymbol{x}^2|\Psi_1> = \left(\frac{3}{2\alpha}\right)$$

12. a.

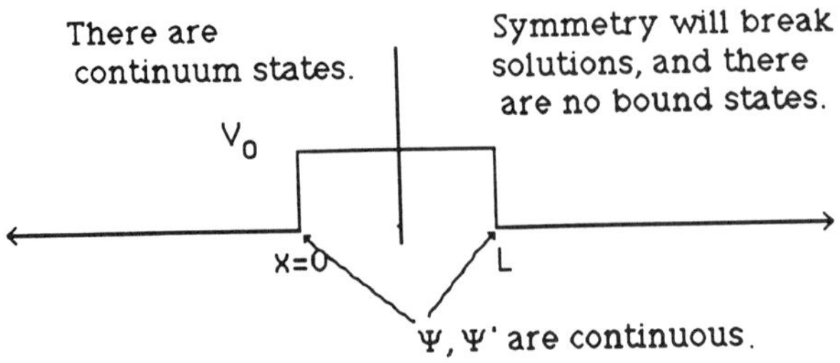
There are continuum states.
Symmetry will break solutions, and there are no bound states.
V0
x=0
L
Ψ, Ψ' are continuous.

b.

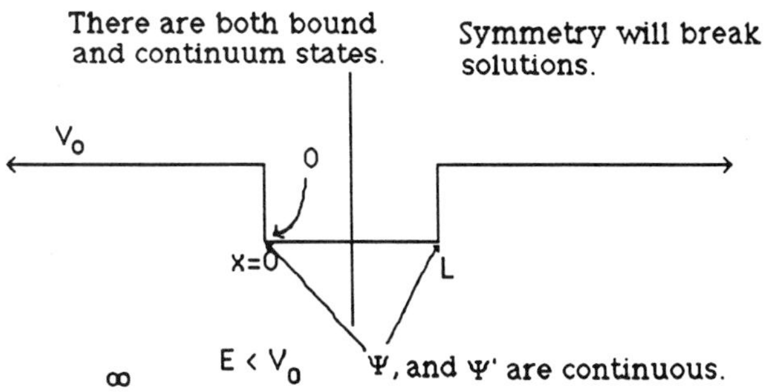

c.

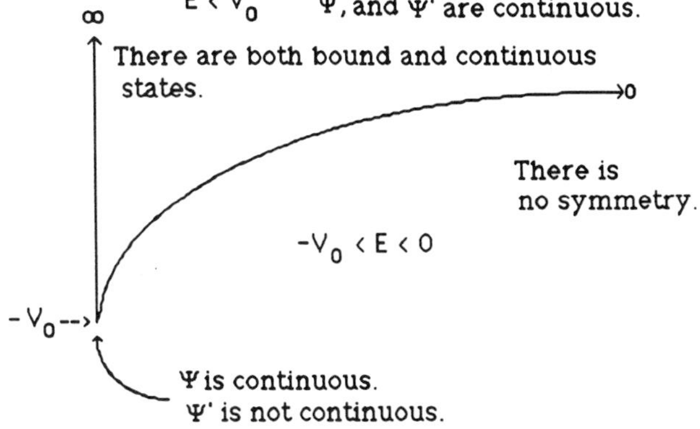

13.

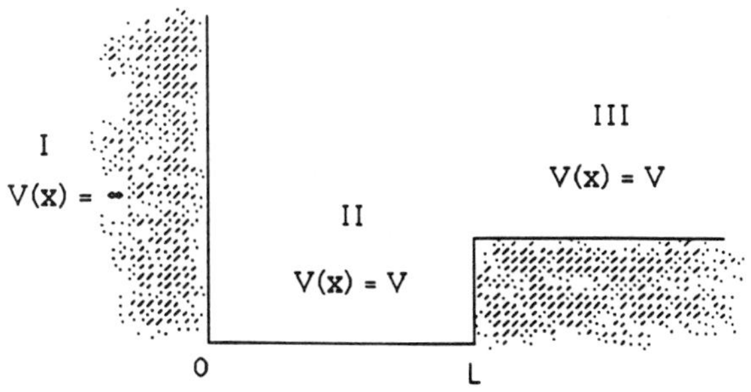

a. $\Psi_I(x) = 0$

$\Psi_{II}(x) = Ae^{i\sqrt{2mE/\hbar^2}\,x} + Be^{-i\sqrt{2mE/\hbar^2}\,x}$

$\Psi_{III}(x) = A'e^{i\sqrt{2m(V-E)/\hbar^2}\,x} + B'e^{-i\sqrt{2m(V-E)/\hbar^2}\,x}$

b. $I \leftrightarrow II$

$\Psi_I(0) = \Psi_{II}(0)$

$\Psi_I(0) = 0 = \Psi_{II}(0) = Ae^{i\sqrt{2mE/\hbar^2}(0)} + Be^{-i\sqrt{2mE/\hbar^2}(0)}$

$0 = A + B$

$B = -A$

$\Psi'_I(0) = \Psi'_{II}(0)$ (this gives no useful information since $\Psi'_I(x)$ does not exist at $x = 0$)

$II \leftrightarrow III$

$\Psi_{II}(L) = \Psi_{III}(L)$

$Ae^{i\sqrt{2mE/\hbar^2}L} + Be^{-i\sqrt{2mE/\hbar^2}L} = A'e^{i\sqrt{2m(V-E)/\hbar^2}L} + B'e^{-i\sqrt{2m(V-E)/\hbar^2}L}$

$\Psi'_{II}(L) = \Psi'_{III}(L)$

$A(i\sqrt{2mE/\hbar^2})e^{i\sqrt{2mE/\hbar^2}L} - B(i\sqrt{2mE/\hbar^2})e^{-i\sqrt{2mE/\hbar^2}L}$

$= A'(i\sqrt{2m(V-E)/\hbar^2})e^{i\sqrt{2m(V-E)/\hbar^2}L} - B'(i\sqrt{2m(V-E)/\hbar^2})e^{-i\sqrt{2m(V-E)/\hbar^2}L}$

c. as $x \to -\infty$, $\Psi_I(x) = 0$

as $x \to +\infty$, $\Psi_{III}(x) = 0 \therefore A' = 0$

d. Rewrite the equations for $\Psi_I(0) = \Psi_{II}(0), \Psi_{II}(L) = \Psi_{III}(L)$, and $\Psi'_{II}(L) = \Psi'_{III}(L)$ using the information in 13c:

$B = -A$ (eqn. 1)

$Ae^{i\sqrt{2mE/\hbar^2}L} + Be^{-i\sqrt{2mE/\hbar^2}L} = B'e^{-i\sqrt{2m(V-E)/\hbar^2}L}$ (eqn. 2)

$A(i\sqrt{2mE/\hbar^2})e^{i\sqrt{2mE/\hbar^2}L} - B(i\sqrt{2mE/\hbar^2})e^{-i\sqrt{2mE/\hbar^2}L} = -B'(i\sqrt{2m(V-E)/\hbar^2})e^{-i\sqrt{2m(V-E)/\hbar^2}L}$ (eqn. 3)

substituting (eqn. 1) into (eqn. 2):

$Ae^{i\sqrt{2mE/\hbar^2}L} - Ae^{-i\sqrt{2mE/\hbar^2}L} = B'e^{-i\sqrt{2m(V-E)/\hbar^2}L}$

$A(\cos(\sqrt{2mE/\hbar^2}L) + i\sin(\sqrt{2mE/\hbar^2}L)) - A(\cos(\sqrt{2mE/\hbar^2}L) - i\sin(\sqrt{2mE/\hbar^2}L)) = B'e^{-i\sqrt{2m(V-E)/\hbar^2}L}$

$2Ai\sin(\sqrt{2mE/\hbar^2}L) = B'e^{-i\sqrt{2m(V-E)/\hbar^2}L}$

$\sin(\sqrt{2mE/\hbar^2}L) = \frac{B'}{2Ai}e^{-i\sqrt{2m(V-E)/\hbar^2}L}$ (eqn. 4)

substituting (eqn. 1) into (eqn. 3):

$A(i\sqrt{2mE/\hbar^2})e^{i\sqrt{2mE/\hbar^2}L} + A(i\sqrt{2mE/\hbar^2})e^{-i\sqrt{2mE/\hbar^2}L} = -B'(i\sqrt{2m(V-E)/\hbar^2})e^{-i\sqrt{2m(V-E)/\hbar^2}L}$

$A(i\sqrt{2mE/\hbar^2})(\cos(\sqrt{2mE/\hbar^2}L) + i\sin(\sqrt{2mE/\hbar^2}L)) \pm A(i\sqrt{2mE/\hbar^2})(\cos(\sqrt{2mE/\hbar^2}L)$

$\quad - i\sin(\sqrt{2mE/\hbar^2}L))$

$+A(i\sqrt{2mE/\hbar^2})(\cos(\sqrt{2mE/\hbar^2}L) - i\sin(\sqrt{2mE/\hbar^2}L))$

$= -B'(i\sqrt{2m(V-E)/\hbar^2})e^{-i\sqrt{2m(V-E)/\hbar^2}L}$

$2Ai\sqrt{2mE/\hbar^2}\cos(\sqrt{2mE/\hbar^2}L) = -B'i\sqrt{2m(V-E)/\hbar^2}e^{-i\sqrt{2m(V-E)/\hbar^2}L}$

$\cos(\sqrt{2mE/\hbar^2}L) = -\frac{B'i\sqrt{2m(V-E)/\hbar^2}}{2Ai\sqrt{2mE/\hbar^2}}e^{-i\sqrt{2m(V-E)/\hbar^2}L}$

$\cos(\sqrt{2mE/\hbar^2}L) = -\frac{B'\sqrt{V-E}}{2A\sqrt{E}}e^{-i\sqrt{2m(V-E)/\hbar^2}L}$ (eqn. 5)

Dividing (eqn. 4) by (eqn. 5):

$$\frac{\sin(\sqrt{2mE/\hbar^2}L)}{\cos(\sqrt{2mE/\hbar^2}L)} = \frac{B'}{2Ai}\frac{-2A\sqrt{E}}{B'\sqrt{V-E}}\frac{e^{-i\sqrt{2m(V-E)/\hbar^2}L}}{e^{-i\sqrt{2m(V-E)/\hbar^2}L}}$$

$$tan(\sqrt{2mE/\hbar^2}L) = -\left(\frac{E}{V-E}\right)^{1/2}$$

e. As $V \to +\infty$, $tan(\sqrt{2mE/\hbar^2}L) \to 0$

So,

$$\sqrt{2mE/\hbar^2}L = n\pi$$

$$E_n = \frac{n^2\pi^2\hbar^2}{2mL^2}$$

Problems

1. a. $\Psi_n(x) = \left(\frac{2}{L}\right)^{\frac{1}{2}} \sin\frac{n\pi x}{L}$

$$P_n(x)dx = \left|\Psi_n\right|^2(x)dx$$

The probability that the particle lies in the interval $0 \le x \le \frac{L}{4}$ is given by:

$$P_n = \int_0^{\frac{L}{4}} P_n(x)dx = \left(\frac{2}{L}\right)\int_0^{\frac{L}{4}} \sin^2\left(\frac{n\pi x}{L}\right)dx$$

This integral can be integrated to give (using integral equation 10 with $\theta = \frac{n\pi x}{L}$):

$$P_n = \left(\frac{L}{n\pi}\right)\left(\frac{2}{L}\right)\int_0^{\frac{n\pi}{4}} \sin^2\left(\frac{n\pi x}{L}\right)d\left(\frac{n\pi x}{L}\right)$$

$$P_n = \left(\frac{L}{n\pi}\right)\left(\frac{2}{L}\right)\int_0^{\frac{n\pi}{4}} \sin^2\theta d\theta$$

$$P_n = \frac{2}{n\pi}\left(-\frac{1}{4}\sin 2\theta + \frac{\theta}{2}\Big|_0^{\frac{n\pi}{4}}\right)$$

$$= \frac{2}{n\pi}\left(-\frac{1}{4}\sin\frac{2n\pi}{4} + \frac{n\pi}{(2)(4)}\right)$$

$$= \frac{1}{4} - \frac{1}{2\pi n}\sin\left(\frac{n\pi}{2}\right)$$

b. If n is even, $\sin\left(\frac{n\pi}{2}\right) = 0$ and $P_n = \frac{1}{4}$.

If n is odd and $n = 1,5,9,13,\ldots$ $\sin\left(\frac{n\pi}{2}\right) = 1$

and $P_n = \frac{1}{4} - \frac{1}{2\pi n}$

If n is odd and $n = 3,7,11,15,\ldots$ $\sin\left(\frac{n\pi}{2}\right) = -1$

and $P_n = \frac{1}{4} + \frac{1}{2\pi n}$

The higher P_n is when $n = 3$. Then $P_n = \frac{1}{4} + \frac{1}{2\pi 3}$

$$P_n = \frac{1}{4} + \frac{1}{6\pi} = 0.303$$

c. $\Psi(t) = e^{\frac{-iHt}{\hbar}}\left[a\Psi_n + b\Psi_m\right] = a\Psi_n e^{\frac{-iE_nt}{\hbar}} + b\Psi_m e^{\frac{-iE_mt}{\hbar}}$

$$H\Psi = a\Psi_n E_n e^{\frac{-iE_nt}{\hbar}} + b\Psi_m E_m e^{\frac{iE_mt}{\hbar}}$$

$$\left\langle \Psi | H | \Psi \right\rangle = |a|^2 E_n + |b|^2 E_m + a^* b e^{\frac{i(E_n - E_m)t}{\hbar}} \left\langle \Psi_n | H | \Psi_m \right\rangle + b^* a e^{\frac{-i(E_m - E_n)t}{\hbar}} \left\langle \Psi_m | H | \Psi_n \right\rangle$$

Since $\left\langle \Psi_n | H | \Psi_m \right\rangle$ and $\left\langle \Psi_m | H | \Psi_n \right\rangle$ are zero,

$$\left\langle \Psi | H | \Psi \right\rangle = |a|^2 E_n + |b|^2 E_m \quad \text{(note the time independence)}$$

d. The fraction of systems observed in Ψ_n is $|a|^2$. The possible energies measured are E_n and E_m. The probabilities of measuring each of these energies is $|a|^2$ and $|b|^2$.

e. Once the system is observed in Ψ_n, it stays in Ψ_n.

f. $P(E_n) = \left|\left\langle \Psi_n | \Psi \right\rangle\right|^2 = |c_n|^2$

$$c_n = \int_0^L \sqrt{\frac{2}{L}} \sin\left(\frac{n\pi x}{L}\right) \sqrt{\frac{30}{L^5}}\, x(L - x)dx$$

$$= \sqrt{\frac{60}{L^6}} \int_0^L x(L - x) \sin\left(\frac{n\pi x}{L}\right) dx$$

$$= \sqrt{\frac{60}{L^6}} \left[L \int_0^L x \sin\left(\frac{n\pi x}{L}\right) dx - \int_0^L x^2 \sin\left(\frac{n\pi x}{L}\right) dx \right]$$

These integrals can be evaluated from integral equations 14 and 16 to give:

$$c_n = \sqrt{\frac{60}{L^6}} \left[L \left(\frac{L^2}{n^2\pi^2} \sin\left(\frac{n\pi x}{L}\right) - \frac{Lx}{n\pi} \cos\left(\frac{n\pi x}{L}\right)\right)\Bigg|_0^L \right]$$

$$-\sqrt{\frac{60}{L^6}} \left[\left(\frac{2xL^2}{n^2\pi^2} \sin\left(\frac{n\pi x}{L}\right) - \left(\frac{n^2\pi^2 x^2}{L^2} - 2\right) \frac{L^3}{n^3\pi^3} \cos\left(\frac{n\pi x}{L}\right)\right)\Bigg|_0^L \right]$$

$$c_n = \sqrt{\frac{60}{L^6}} \left\{ \frac{L^3}{n^2\pi^2}(\sin(n\pi) - \sin(0)) - \frac{L^2}{n\pi}(L\cos(n\pi) - 0\cos 0)) - \left(\frac{2L^2}{n^2\pi^2}(L\sin(n\pi) - 0\sin(0))\right)\right.$$

$$\left. - \left(n^2\pi^2 - 2\right)\frac{L^3}{n^3\pi^3}\cos(n\pi) + \left(\frac{n^2\pi^2(0)}{L^2} - 2\right)\frac{L^3}{n^3\pi^3}\cos(0)) \right.$$

$$c_n = L^{-3}\sqrt{(60)} \left\{ -\frac{L^3}{n\pi}\cos(n\pi) + \left(n^2\pi^2 - 2\right)\frac{L^3}{n^3\pi^3}\cos(n\pi) + \frac{2L^3}{n^3\pi^3} \right\}$$

$$c_n = \sqrt{60}\left(-\frac{1}{n\pi}(-1)^n + \left(n^2\pi^2 - 2\right)\frac{1}{n^3\pi^3}(-1)^n + \frac{2}{n^3\pi^3}\right)$$

$$c_n = \sqrt{60}\left(\left(\frac{-1}{n\pi} + \frac{1}{n\pi} - \frac{2}{n^3\pi^3}\right)(-1)^n + \frac{2}{n^3\pi^3}\right)$$

$$c_n = \frac{2\sqrt{60}}{n^3\pi^3})\left(-(-1)^n+1\right)$$

$$|c_n|^2 = \frac{4(60)}{n^6\pi^6})\left(-(-1)^n + 1\right)^2$$

If n is even then $c_n^2 = 0$

If n is odd then $c_n^2 = \frac{(4)(60)(4)}{n^6\pi^6} = \frac{960}{n^6\pi^6}$

The probability of making a measurement of the energy and obtaining one of the eigenvalues, given by:

$E_n = \frac{n^2\pi^2\hbar^2}{2mL^2}$ is:

$P(E_n) = 0$ if n is even

$P(E_n) = \frac{960}{n^6\pi^6}$ if n is odd

g. $$\left\langle \Psi|H|\Psi \right\rangle = \int_0^L \left(\frac{30}{L^5}\right)^{\frac{1}{2}} x(L-x)\left(\frac{-\hbar^2}{2m}\frac{d^2}{dx^2}\right)\left(\frac{30}{L^5}\right)^{\frac{1}{2}} x(L-x)dx$$

$$= \left(\frac{30}{L^5}\right)\left(\frac{-\hbar^2}{2m}\right)\int_0^L x(L-x)\left(\frac{d^2}{dx^2}\right)\left(xL - x^2\right)dx$$

$$= \left(\frac{-15\hbar^2}{mL^5}\right)\int_0^L x(L-x)(-2)dx$$

$$= \left(\frac{30\hbar^2}{mL^5}\right)\int_0^L xL - x^2 dx$$

$$= \left(\frac{30\hbar^2}{mL^5}\right)\left(L\frac{x^2}{2} - \frac{x^3}{3}\right)\Bigg|_0^L$$

$$= \left(\frac{30\hbar^2}{mL^5}\right)\left(\frac{L^3}{2} - \frac{L^3}{3}\right)$$

$$= \left(\frac{30\hbar^2}{mL^2}\right)\left(\frac{1}{2} - \frac{1}{3}\right)$$

$$= \frac{30\hbar^2}{6mL^2} = \frac{5\hbar^2}{mL^2}$$

2. $$\left\langle \Psi|H|\Psi \right\rangle = \sum_{ij} C_i^* e^{\frac{iE_it}{\hbar}} \left\langle \Psi_i|H|\Psi_j \right\rangle e^{\frac{-iE_jt}{\hbar}} C_j$$

Since $\left\langle \Psi_i|H|\Psi_j \right\rangle = E_j\delta_{ij}$

$$\left\langle \Psi|H|\Psi \right\rangle = \sum_j C_j^* C_j E_j e^{i(E_j - E_j)\frac{t}{\hbar}}$$

$$\left\langle \Psi|H|\Psi \right\rangle = \sum_j C_j^* C_j E_j \text{ (not time dependent)}$$

For other properties:

$$\left\langle \Psi|A|\Psi \right\rangle = \sum_{ij} C_i^* e^{\frac{iE_it}{\hbar}} \left\langle \Psi_i|A|\Psi_j \right\rangle e^{\frac{-iE_jt}{\hbar}} C_j$$

but, $\left\langle \Psi_i|A|\Psi_j \right\rangle$

does not necessarily $= a_j\delta_{ij}$.
This is only true if $[A,H] = 0$.

$$\left\langle \Psi|A|\Psi \right\rangle = \sum_{ij} C_i^* C_j e^{\frac{i(E_i - E_j)t}{\hbar}} \left\langle \Psi_i|A|\Psi_j \right\rangle$$

Therefore, in general, other properties are time dependent.

3. For a particle in a box in its lowest quantum state:

$$\Psi = \sqrt{\frac{2}{L}} \sin\left(\frac{\pi x}{L}\right)$$

$$\left\langle x \right\rangle = \int_0^L \Psi^* x \Psi dx$$

$$= \frac{2}{L} \int_0^L x \sin^2\left(\frac{\pi x}{L}\right) dx$$

Using integral equation 18:

$$= \frac{2}{L}\left(\frac{x^2}{4} - \frac{xL}{4\pi} \sin\left(\frac{2\pi x}{L}\right) - \frac{L^2}{8\pi^2} \cos\left(\frac{2\pi x}{L}\right)\right)\Bigg|_0^L$$

$$= \frac{2}{L}\left(\frac{L^2}{4} - \frac{L^2}{8\pi^2} (\cos(2\pi) - \cos(0))\right)$$

$$= \frac{2}{L}\left(\frac{L^2}{4}\right)$$

$$= \frac{L}{2}$$

$$\left\langle x^2 \right\rangle = \int_0^L \Psi^* x^2 \Psi dx$$

$$= \frac{2}{L} \int_0^L x^2 \sin^2\left(\frac{\pi x}{L}\right) dx$$

Using integral equation 19:

$$= \frac{2}{L}\left(\frac{x^3}{6} - \left(\frac{x^2 L}{4\pi} - \frac{L^3}{8\pi^3}\right)\sin\left(\frac{2\pi x}{L}\right) - \frac{xL^2}{4\pi^2}\cos\left(\frac{2\pi x}{L}\right)\right)\Bigg|_0^L$$

$$= \frac{2}{L}\left(\frac{L^3}{6} - \frac{L^2}{4\pi^2}(L\cos(2\pi) - (0)\cos(0))\right)$$

$$= \frac{2}{L}\left(\frac{L^3}{6} - \frac{L^3}{4\pi^2}\right)$$

$$= \frac{L^2}{3} - \frac{L^2}{2\pi^2}$$

$$\left\langle p \right\rangle = \int_0^L \Psi^* p \Psi dx$$

$$= \frac{2}{L} \int_0^L \sin\left(\frac{\pi x}{L}\right)\left(\frac{\hbar}{i}\frac{d}{dx}\right)\sin\left(\frac{\pi x}{L}\right) dx$$

$$= \frac{2\hbar\pi}{L^2 i} \int_0^L \sin\left(\frac{\pi x}{L}\right)\cos\left(\frac{\pi x}{L}\right) dx$$

$$= \frac{2\hbar}{Li} \int_0^L \sin\left(\frac{\pi x}{L}\right)\cos\left(\frac{\pi x}{L}\right) d\left(\frac{\pi x}{L}\right)$$

integral equation 15 (with $\theta = \frac{\pi x}{L}$):

$$= \frac{2\hbar}{Li}\left(-\frac{1}{2}\cos^2(\theta)\Bigg|_0^\pi\right) = 0$$

$$\left\langle p^2 \right\rangle = \int_0^L \Psi^* p^2 \Psi dx$$

$$= \frac{2}{L} \int_0^L \sin\left(\frac{\pi x}{L}\right)\left(-\hbar^2 \frac{d^2}{dx^2}\right)\sin\left(\frac{\pi x}{L}\right) dx$$

$$= \frac{2\pi^2\hbar^2}{L^3} \int_0^L \sin^2\left(\frac{\pi x}{L}\right) dx$$

$$= \frac{2\pi\hbar^2}{L^2} \int_0^L \sin^2\left(\frac{\pi x}{L}\right) d\left(\frac{\pi x}{L}\right)$$

Using integral equation 10 (with $\theta = \frac{\pi x}{L}$):

$$= \frac{2\pi\hbar^2}{L^2}\left(-\frac{1}{4}\sin(2\theta) + \frac{\theta}{2}\right)\Bigg|_0^{\pi}$$

$$= \frac{2\pi\hbar^2}{L^2}\frac{\pi}{2} = \frac{\pi^2\hbar^2}{L^2}$$

$$\Delta x = \left(\left\langle x^2\right\rangle - \left\langle x\right\rangle^2\right)^{\frac{1}{2}}$$

$$= \left(\frac{L^2}{3} - \frac{L^2}{2\pi^2} - \frac{L^2}{4}\right)^{\frac{1}{2}}$$

$$= L\left(\frac{1}{12} - \frac{1}{2\pi^2}\right)^{\frac{1}{2}}$$

$$\Delta p = \left(\left\langle p^2\right\rangle - \left\langle p\right\rangle^2\right)^{\frac{1}{2}}$$

$$= \left(\frac{\pi^2\hbar^2}{L^2} - 0\right)^{\frac{1}{2}} = \frac{\pi\hbar}{L}$$

$$\Delta x\Delta p = \pi\hbar\left(\frac{1}{12} - \frac{1}{2\pi^2}\right)^{\frac{1}{2}}$$

$$= \frac{\hbar}{2}\left(\frac{4\pi^2}{12} - \frac{4}{2}\right)^{\frac{1}{2}}$$

$$= \frac{\hbar}{2}\left(\frac{\pi^2}{3} - 2\right)^{\frac{1}{2}}$$

Finally, $\frac{\hbar}{2}\left(\frac{\pi^2}{3} - 2\right)^{\frac{1}{2}} > \frac{\hbar}{2}\left(\frac{(3)^2}{3} - 2\right)^{\frac{1}{2}} = \frac{\hbar}{2}$

$\therefore\ \Delta x\Delta p > \frac{\hbar}{2}$

4. a. $\int_0^{L/4} P(x)dx = \int_0^{L/4} \frac{1}{L}dx = \frac{1}{L}x\Big|_0^{L/4}$

$$= \frac{1}{L}\frac{L}{4} = \frac{1}{4} = 25\%$$

$$P_{classical} = \frac{1}{4} \text{ (for interval } 0 - L/4)$$

b. This was accomplished in problem 1a to give:

$$P_n = \frac{1}{4} - \frac{1}{2\pi n}\sin\left(\frac{n\pi}{2}\right)$$

(for interval $0 - L/4$)

c. $$\underset{n \to \infty}{\text{Limit}}\, P_{\text{quantum}} = \underset{n \to \infty}{\text{Limit}}\left(\frac{1}{4} - \frac{1}{2\pi n}\sin\left(\frac{n\pi}{2}\right)\right)$$

$$\underset{n \to \infty}{\text{Limit}}\, P_{\text{quantum}} = \frac{1}{4}$$

Therefore as n becomes large the classical limit is approached.

5. a. The Schrödinger equation for a Harmonic Oscillator in one-dimensional coordinate representation, $\boldsymbol{H}\,\Psi(x) = E_x\Psi(x)$, with the Hamiltonian defined as:

$$\boldsymbol{H} = \frac{-\hbar^2}{2m}\frac{d^2}{dx^2} + \frac{1}{2}kx^2,$$

becomes:

$$\left(\frac{-\hbar^2}{2m}\frac{d^2}{dx^2} + \frac{1}{2}kx^2\right)\Psi(x) = E_x\Psi(x).$$

b. The transformation of the kinetic energy term to the momentum representation is trivial:

$$\boldsymbol{T} = \frac{\boldsymbol{p}_x{}^2}{2m}.$$

In order to maintain the commutation relation $[\boldsymbol{x},\boldsymbol{p}_x] = i\hbar$ and keep the $\boldsymbol{p}$ operator unchanged the coordinate operator must become

$$\boldsymbol{x} = i\hbar\frac{d}{dp_x}.$$

The Schrödinger equation for a Harmonic Oscillator in one-dimensional momentum representation, $\boldsymbol{H}\,\Psi(p_x) = Ep_x\Psi(p_x)$, with the Hamiltonian defined as:

$$\boldsymbol{H} = \frac{1}{2m}p_x{}^2 - \frac{k\hbar^2}{2}\frac{d^2}{dp_x{}^2},$$

becomes:

$$\left(\frac{1}{2m}p_x{}^2 - \frac{k\hbar^2}{2}\frac{d^2}{dp_x{}^2}\right)\Psi(p_x) = Ep_x\Psi(p_x).$$

c. For the wavefunction

$$\Psi(x) = C\exp\left(-\sqrt{mk}\,\frac{x^2}{2\hbar}\right),$$

$$\text{let } a = \frac{\sqrt{mk}}{2\hbar},$$

and hence $\Psi(x) = C\exp(-ax^2)$. Evaluating the derivatives of this expression gives:

$$\frac{d}{dx}\Psi(x) = \frac{d}{dx}C\exp(-ax^2) = -2axC\exp(-ax^2)$$

$$\frac{d^2}{dx^2}\Psi(x) = \frac{d^2}{dx^2}C\exp(-ax^2) = \frac{d}{dx} - 2axC\exp(-ax^2)$$

$$= (-2axC)(-2ax\exp(-ax^2)) + (-2aC)(exp(-ax^2))$$

$$= (4a^2x^2 - 2a)C\exp(-ax^2).$$

$\boldsymbol{H}\,\Psi(x) = E_x\Psi(x)$ then becomes:

$$\boldsymbol{H}\,\Psi(x) = \left(\frac{-\hbar^2}{2m}(4a^2x^2 - 2a) + \frac{1}{2}kx^2\right)\Psi(x).$$

Clearly the energy (eigenvalue) expression must be independent of x and the two terms containing x^2 terms must cancel upon insertion of a:

$$E_x = \frac{-\hbar^2}{2m}\left(4\left(\frac{\sqrt{mk}}{2\hbar}\right)^2 x^2 - 2\frac{\sqrt{mk}}{2\hbar}\right) + \frac{1}{2}kx^2$$

$$= \frac{-\hbar^2}{2m}\left(\frac{4mkx^2}{4\hbar^2}\right) + \frac{\hbar^2}{2m}\frac{2\sqrt{mk}}{2\hbar} + \frac{1}{2}kx^2$$

$$= -\frac{1}{2}kx^2 + \frac{\hbar\sqrt{mk}}{2m} + \frac{1}{2}kx^2$$

$$= \frac{\hbar\sqrt{mk}}{2m}.$$

Normalization of $\Psi(x)$ to determine the constant C yields the equation:

$$C^2\int_{-\infty}^{+\infty}\exp\left(-\sqrt{mk}\,\frac{x^2}{\hbar}\right)dx = 1.$$

Using integral equation (1) gives:

$$C^2 2\left(\frac{1}{2}\sqrt{\pi}\left(\frac{\sqrt{mk}}{\hbar}\right)^{-\frac{1}{2}}\right) = 1$$

$$C^2\left(\frac{\pi\hbar}{\sqrt{mk}}\right)^{\frac{1}{2}} = 1$$

$$C^2 = \left(\frac{\sqrt{mk}}{\pi\hbar}\right)^{\frac{1}{2}}$$

$$C = \left(\frac{\sqrt{mk}}{\pi\hbar}\right)^{\frac{1}{4}}$$

Therefore,

$$\Psi(x) = \left(\frac{\sqrt{mk}}{\pi\hbar}\right)^{\frac{1}{4}} \exp\left(-\sqrt{mk}\,\frac{x^2}{2\hbar}\right).$$

d. Proceeding analogous to part c, for a wavefunction in momentum space

$$\Psi(p) = C\exp(-\alpha p^2),$$

evaluating the derivatives of this expression gives:

$$\frac{d}{dp}\Psi(p) = \frac{d}{dp}C\exp(-\alpha p^2) = -2\alpha pC\exp(-\alpha p^2)$$

$$\frac{d^2}{dp^2}\Psi(p) = \frac{d^2}{dp^2}C\exp(-\alpha p^2) = \frac{d}{dp} - 2\alpha pC\exp(-\alpha p^2)$$

$$= (-2\alpha pC)(-2\alpha p\exp(-\alpha p^2)) + (-2\alpha C)(\exp(-\alpha p^2))$$

$$= (4\alpha^2p^2 - 2\alpha)C exp(-\alpha p^2).$$

$\boldsymbol{H}\,\Psi(p) = E_p\Psi(p)$ then becomes:

$$\boldsymbol{H}\,\Psi(p) = \frac{1}{2m}p^2 - \frac{k\hbar^2}{2}(4\alpha^2p^2 - 2\alpha)\Psi(p)$$

Once again the energy (eigenvalue) expression corresponding to E_p must be independent of p and the two terms containing p^2 terms must cancel with the appropriate choice of α. We also desire our choice of α to give us the same energy we found in part c (in coordinate space).

$$E_p = \frac{1}{2m}p^2 - \frac{k\hbar^2}{2}(4\alpha^2p^2 - 2\alpha)$$

Therefore we can find α either of two ways:

1. $\frac{1}{2m}p^2 = \frac{k\hbar^2}{2}4\alpha^2p^2$, or

2. $\frac{k\hbar^2}{2}2\alpha = \frac{\hbar\sqrt{mk}}{2m}$.

Both equations yield $\alpha = (2\hbar\sqrt{mk})^{-1}$.

Normalization of $\Psi(p)$ to determine the constant C yields the equation:

$$C^2\int_{-\infty}^{+\infty}\exp(-2\alpha p^2)dp = 1.$$

Using integral equation (1) gives:

$$C^2 2\left(\frac{1}{2}\sqrt{\pi}(2\alpha)^{-\frac{1}{2}}\right) = 1$$

$$C^2\sqrt{\pi}\left(2(2\hbar\sqrt{mk})^{-1}\right)^{-\frac{1}{2}} = 1$$

$$C^2\sqrt{\pi}\,(\hbar\sqrt{mk})^{\frac{1}{2}} = 1$$

$$C^2(\pi\hbar\sqrt{mk})^{\frac{1}{2}} = 1$$

$$C^2 = (\pi\hbar\sqrt{mk})^{-\frac{1}{2}}$$

$$C = (\pi\hbar\sqrt{mk})^{-\frac{1}{4}}$$

Therefore, $\Psi(p) = (\pi\hbar\sqrt{mk})^{-\frac{1}{4}}\exp(-p^2/(2\hbar\sqrt{mk})$.

Showing that $\Psi(p)$ is the proper fourier transform of $\Psi(x)$ suggests that the Fourier integral theorem should hold for the two wavefunctions $\Psi(x)$ and $\Psi(p)$ we have obtained, e.g.,

$$\Psi(p) = \frac{1}{\sqrt{2\pi\hbar}}\int_{-\infty}^{+\infty}\Psi(x)e^{ipx/\hbar}dx, \text{ for}$$

$$\Psi(x) = \left(\frac{\sqrt{mk}}{\pi\hbar}\right)^{\frac{1}{4}}\exp\left(-\sqrt{mk}\,\frac{x^2}{2\hbar}\right), \text{ and}$$

$$\Psi(p) = (\pi\hbar\sqrt{mk})^{-\frac{1}{4}}\exp(-p^2/(2\hbar\sqrt{mk}).$$

So, verify that:

$$(\pi\hbar\sqrt{mk})^{-\frac{1}{4}}\exp(-p^2/(2\hbar\sqrt{mk} = \frac{1}{\sqrt{2\pi\hbar}}\int_{-\infty}^{+\infty}\left(\frac{\sqrt{mk}}{\pi\hbar}\right)^{\frac{1}{4}}\exp\left(-\sqrt{mk}\,\frac{x^2}{2\hbar}\right)e^{ipx/\hbar}dx.)$$

Working with the right-hand side of the equation:

$$= \frac{1}{\sqrt{2\pi\hbar}}\left(\frac{\sqrt{mk}}{\pi\hbar}\right)^{\frac{1}{4}}\int_{-\infty}^{+\infty}\exp\left(-\sqrt{mk}\,\frac{x^2}{2\hbar}\right)\left(\cos\left(\frac{px}{\hbar}\right) + i\sin\left(\frac{px}{\hbar}\right)\right)dx,$$

the sin term is odd and the integral will therefore vanish. The remaining integral can be evaluated using the given expression:

$$\int_{-\infty}^{+\infty} e^{-\beta x^2}\cos bx\,dx = \sqrt{\frac{\pi}{\beta}}\,e^{-b^2/4\beta}$$

$$= \frac{1}{\sqrt{2\pi\hbar}}\left(\frac{\sqrt{mk}}{\pi\hbar}\right)^{\frac{1}{4}}\int_{-\infty}^{+\infty}\exp\left(-\frac{\sqrt{mk}}{2\hbar}x^2\right)\cos\left(\frac{p}{\hbar}x\right)dx$$

$$= \frac{1}{\sqrt{2\pi\hbar}}\left(\frac{\sqrt{mk}}{\pi\hbar}\right)^{\frac{1}{4}}\left(\frac{2\pi\hbar}{\sqrt{mk}}\right)^{\frac{1}{2}}\int_{-\infty}^{+\infty}\exp\left(-\left(\frac{p^2}{\hbar^2}\right)\frac{2\hbar}{4\sqrt{mk}}\right)$$

$$= \left(\frac{\sqrt{mk}}{\pi\hbar}\right)^{\frac{1}{4}}\left(\frac{1}{\sqrt{mk}}\right)^{\frac{1}{2}}\int_{-\infty}^{+\infty}\exp\left(-\frac{p^2}{2\hbar\sqrt{mk}}\right)$$

$$= \left(\frac{\sqrt{mk}}{mk\pi\hbar}\right)^{\frac{1}{4}} \int_{-\infty}^{+\infty} \exp\left(-\frac{p^2}{2\hbar\sqrt{mk}}\right)$$

$$= (\hbar\pi\sqrt{mk})^{-\frac{1}{4}} \int_{-\infty}^{+\infty} \exp\left(-\frac{p^2}{2\hbar\sqrt{mk}}\right) = \Psi(p) \qquad Q.E.D.$$

6. a. The lowest energy level for a particle in a 3-dimensional box is when $n_1 = 1$, $n_2 = 1$, and $n_3 = 1$. The total energy (with $L_1 = L_2 = L_3$) will be:

$$E_{total} = \frac{h^2}{8mL^2}\left(n_1^2 + n_2^2 + n_3^2\right) = \frac{3h^2}{8mL^2}$$

Note that $n = 0$ is not possible. The next lowest energy level is when one of the three quantum numbers equals 2 and the other two equal 1:

$n_1 = 1, n_2 = 1, n_3 = 2$
$n_1 = 1, n_2 = 2, n_3 = 1$
$n_1 = 2, n_2 = 1, n_3 = 1.$

Each of these three states have the same energy:

$$E_{total} = \frac{h^2}{8mL^2}\left(n_1^2 + n_2^2 + n_3^2\right) = \frac{6h^2}{8mL^2}$$

Note that these three states are only degenerate if

$L_1 = L_2 = L_3.$

b.

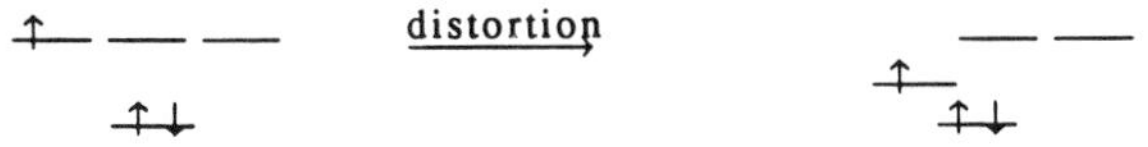

$L_1 = L_2 = L_3$ $\qquad$ $L_3 \neq L_1 = L_2$

For $L_1 = L_2 = L_3$, $V = L_1L_2L_3 = L_1^3$,
$E_{total}(L_1) = 2\varepsilon_1 + \varepsilon_2$

$$= \frac{2h^2}{8m}\left(\frac{1^2}{L_1^2} + \frac{1^2}{L_2^2} + \frac{1^2}{L_3^2}\right) + \frac{1h^2}{8m}\left(\frac{1^2}{L_1^2} + \frac{1^2}{L_2^2} + \frac{2^2}{L_3^2}\right)$$

$$= \frac{2h^2}{8m}\left(\frac{3}{L_1^2}\right) + \frac{1h^2}{8m}\left(\frac{6}{L_1^2}\right) = \frac{h^2}{8m}\left(\frac{12}{L_1^2}\right)$$

For $L_3 \neq L_1 = L_2$, $V = L_1L_2L_3 = L_1^2L_3$, $L_3 = V/L_1^2$

$E_{total}(L_1) = 2\varepsilon_1 + \varepsilon_2$

$$= \frac{2h^2}{8m}\left(\frac{1^2}{L_1^2} + \frac{1^2}{L_2^2} + \frac{1^2}{L_3^2}\right) + \frac{1h^2}{8m}\left(\frac{1^2}{L_1^2} + \frac{1^2}{L_2^2} + \frac{2^2}{L_3^2}\right)$$

$$= \frac{2h^2}{8m}\left(\frac{2}{L_1^2} + \frac{1}{L_3^2}\right) + \frac{1h^2}{8m}\left(\frac{2}{L_1^2} + \frac{4}{L_3^2}\right)$$

$$= \frac{2h^2}{8m}\left(\frac{2}{L_1^2} + \frac{1}{L_3^2} + \frac{1}{L_1^2} + \frac{2}{L_3^2}\right)$$

$$= \frac{2h^2}{8m}\left(\frac{3}{L_1^2} + \frac{3}{L_3^2}\right) = \frac{h^2}{8m}\left(\frac{6}{L_1^2} + \frac{6}{L_3^2}\right)$$

In comparing the total energy *at constant volume* of the undistorted box ($L_1 = L_2 = L_3$) versus the distorted box ($L_3 \neq L_1 = L_2$) it can be seen that:

$$\frac{h^2}{8m}\left(\frac{6}{L_1^2} + \frac{6}{L_3^2}\right) \leq \frac{h^2}{8m}\left(\frac{12}{L_1^2}\right)$$

as long as $L_3 \geq L_1$.

c. In order to minimize the total energy expression, take the derivative of the energy with respect to L_1 and set it equal to zero.

$$\frac{\partial E_{total}}{\partial L_1} = 0$$

$$\frac{\partial}{\partial L_1}\left(\frac{h^2}{8m}\left(\frac{6}{L_1^2} + \frac{6}{L_3^2}\right)\right) = 0$$

But since $V = L_1 L_2 L_3 = L_1^2 L_3$, then $L_3 = V/L_1^2$. This substitution gives:

$$\frac{\partial}{\partial L_1}\left(\frac{h^2}{8m}\left(\frac{6}{L_1^2} + \frac{6L_1^4}{V^2}\right)\right) = 0$$

$$\left(\frac{h^2}{8m}\left(\frac{(-2)6}{L_1^3} + (4)6\frac{L_1^3}{V^2}\right)\right) = 0$$

$$\left(\frac{-12}{L_1^3} + \frac{24L_1^3}{V^2}\right) = 0$$

$$\left(\frac{24L_1^3}{V^2}\right) = \left(\frac{12}{L_1^3}\right)$$

$$24L_1^6 = 12V^2$$

$$L_1^6 = \frac{1}{2}V^2 = \frac{1}{2}\left(L_1^2 L_3\right)^2 = \frac{1}{2}L_1^4 L_3^2$$

$$L_1^2 = \frac{1}{2}L_3^2$$

$$L_3 = \sqrt{2}L_1$$

d. Calculate energy upon distortion:

cube: $V = L_1^3,\ L_1 = L_2 = L_3 = (V)^{\frac{1}{3}}$

distorted: $V = L_1^2 L_3 = L_1^2\sqrt{2}L_1 = \sqrt{2}L_1^3$

$$L_3 = \sqrt{2}\left(\frac{V}{\sqrt{2}}\right)^{\frac{1}{3}} \neq L_1 = L_2 = \left(\frac{V}{\sqrt{2}}\right)^{\frac{1}{3}}$$

$$\Delta E = E_{total}(L_1 = L_2 = L_3) - E_{total}(L_3 \neq L_1 = L_2)$$

$$= \frac{h^2}{8m}\left(\frac{12}{L_1^2}\right) - \frac{h^2}{8m}\left(\frac{6}{L_1^2} + \frac{6}{L_3^2}\right)$$

$$= \frac{h^2}{8m}\left(\frac{12}{V^{2/3}} - \frac{6(2)^{1/3}}{V^{2/3}} - \frac{6(2)^{1/3}}{2V^{2/3}}\right)$$

$$= \frac{h^2}{8m}\left(\frac{12 - 9(2)^{1/3}}{V^{2/3}}\right)$$

Since $V = 8\text{Å}^3, V^{2/3} = 4\text{Å}^2 = 4 \times 10^{-16} cm^2$, and $\frac{h^2}{8m} = 6.01 \times 10^{-27}\text{erg cm}^2$:

$$\Delta E = 6.01 \times 10^{-27}\text{erg cm}^2\left(\frac{12 - 9(2)^{1/3}}{4 \times 10^{-16}\text{cm}^2}\right)$$

$$\Delta E = 6.01 \times 10^{-27}\text{erg cm}^2\left(\frac{0.66}{4 \times 10^{-16}\text{cm}^2}\right)$$

$$\Delta E = 0.99 \times 10^{-11}\text{erg}$$

$$\Delta E = 0.99 \times 10^{-11}\text{erg}\left(\frac{1\text{eV}}{1.6 \times 10^{-12}\text{erg}}\right)$$

$$\Delta E = 6.19\text{eV}$$

7. a. $\int_{-\infty}^{+\infty} \Psi^*(x)\Psi(x)dx = 1.$

$$A^2 \int_{-\infty}^{+\infty} e^{-2a|x|}dx = 1.$$

$$A^2 \int_{-\infty}^{0} e^{2ax}dx + A^2 \int_{0}^{+\infty} e^{-2ax}dx = 1$$

Making use of integral equation (4) this becomes:

$$A^2\left(\frac{1}{2a} + \frac{1}{2a}\right) = \frac{2A^2}{2a} = 1$$

$A^2 = a$

$A = \pm\sqrt{a}$, therefore A is not unique.

$\Psi(x) = Ae^{-a|x|} = \pm\sqrt{a}\,e^{-a|x|}$

Since a has units of $Å^{-1}$, $\Psi(x)$ must have units of $Å^{-\frac{1}{2}}$.

b. $|x| = \begin{Bmatrix} x \text{ if } x \geq 0 \\ -x \text{ if } x \leq 0 \end{Bmatrix}$

$$\Psi(x) = \sqrt{a}\begin{Bmatrix} e^{-ax} \text{ if } x \geq 0 \\ e^{ax} \text{ if } x \leq 0 \end{Bmatrix}$$

Sketching this wavefunction with respect to x (keeping constant a fixed; $a = 1$) gives:

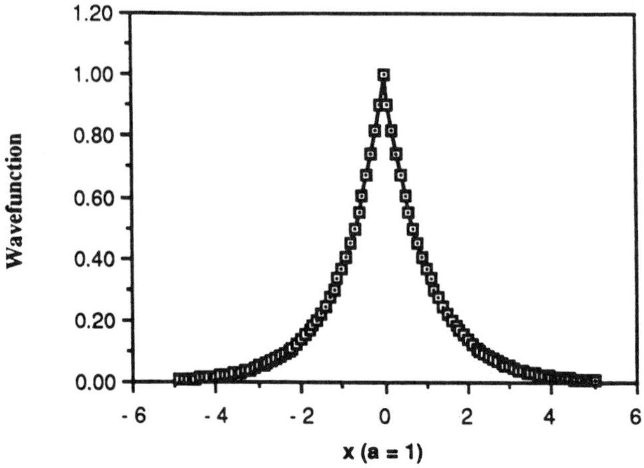

c. $\dfrac{d\Psi(x)}{dx} = \sqrt{a}\begin{Bmatrix} -ae^{-ax} \text{ if } x \geq 0 \\ ae^{ax} \text{ if } x \leq 0 \end{Bmatrix}$

$$\frac{d\Psi(x)}{dx}\Big|_{0+\varepsilon} = -a\sqrt{a}$$

$$\frac{d\Psi(x)}{dx}\Big|_{0-\varepsilon} = a\sqrt{a}$$

The magnitude of discontinuity is $a\sqrt{a} + a\sqrt{a} = 2a\sqrt{a}$ as x goes through $x = 0$. This also indicates that the potential V undergoes a discontinuity of ∞ magnitude at $x = 0$.

d. $\left\langle |x| \right\rangle = \displaystyle\int_{-\infty}^{+\infty} \Psi^*(x)|x|\Psi(x)dx$

$$= (\sqrt{a})^2 \int_{-\infty}^{0} e^{2ax}(-x)dx + (\sqrt{a})^2 \int_{0}^{+\infty} e^{-2ax}(x)dx$$

$$= 2a \int_{0}^{\infty} e^{-2ax}(x)dx$$

Making use of integral equation (4) again this becomes:

$$= 2a\frac{1}{(2a)^2} = \frac{1}{2a} = \frac{1}{2(2Å)^{-1}}$$

$$\left\langle |x| \right\rangle = 1Å$$

This expectation value is a measure of the average distance ($|x|$) from the origin.

e. $\Psi(x) = \sqrt{a}\begin{Bmatrix} e^{-ax} \text{ if } x \geq 0 \\ e^{ax} \text{ if } x \leq 0 \end{Bmatrix}$

$$\frac{d\,\Psi(x)}{dx} = \sqrt{a}\begin{Bmatrix} -ae^{-ax} \text{ if } x \geq 0 \\ ae^{ax} \text{ if } x \leq 0 \end{Bmatrix}$$

$$\frac{d^2\Psi(x)}{dx^2} = \sqrt{a}\begin{Bmatrix} a^2e^{-ax} \text{ if } x \geq 0 \\ a^2e^{ax} \text{ if } x \leq 0 \end{Bmatrix} = a^2\Psi(x)$$

$$\left\langle H \right\rangle = \left\langle \left(-\frac{\hbar}{2m}\frac{d^2}{dx^2} - \frac{\hbar^2 a}{m}\delta(x) \right) \right\rangle$$

$$\left\langle H \right\rangle = \int_{-\infty}^{+\infty} \Psi^*(x)\left(-\frac{\hbar}{2m}\frac{d^2}{dx^2} \right)\Psi(x)dx - \int_{-\infty}^{+\infty} \Psi^*(x)\left(\frac{\hbar^2 a}{m}\delta(x) \right)\Psi(x)dx$$

$$= -\frac{\hbar a^2}{2m}\int_{-\infty}^{+\infty} \Psi^*(x)\Psi(x)dx - \frac{\hbar^2 a}{m}\int_{-\infty}^{+\infty} \Psi^*(x)(\delta(x))\Psi(x)dx$$

Using the integral equation:

$$\int_a^b f(x)\delta(x - x_0)dx = \begin{Bmatrix} f(x_0) \text{ if } a < x_0 < b \\ 0 \text{ otherwise} \end{Bmatrix}$$

$$\left\langle H \right\rangle = -\frac{\hbar a^2}{2m}(1) - \frac{\hbar^2 a}{m}(\sqrt{a})^2 = -\frac{3\hbar a^2}{2m}$$

$$= -3(6.06 \times 10^{-28}\text{erg cm}^2)(2 \times 10^{-8}\text{cm})^{-2}$$
$$= -4.55 \times 10^{-12}\text{erg}$$
$$= -2.84\text{eV}.$$

f. In problem 5 the relationship between $\Psi(p)$ and $\Psi(x)$ was derived:

$$\Psi(p) = \frac{1}{\sqrt{2\pi\hbar}}\int_{-\infty}^{+\infty} \Psi(x)e^{-ipx/\hbar}dx.$$

$$\Psi(p) = \frac{1}{\sqrt{2\pi\hbar}}\int_{-\infty}^{+\infty} \sqrt{a}\,e^{-a|x|}e^{-ipx/\hbar}dx.$$

$$\Psi(p) = \frac{1}{\sqrt{2\pi\hbar}}\int_{-\infty}^{0} \sqrt{a}\,e^{ax}e^{-ipx/\hbar}dx + \frac{1}{\sqrt{2\pi\hbar}}\int_{0}^{+\infty} \sqrt{a}\,e^{-ax}e^{-ipx/\hbar}dx.$$

$$= \sqrt{\frac{a}{2\pi\hbar}}\left(\frac{1}{a - ip/\hbar} + \frac{1}{a + ip/\hbar} \right)$$

$$= \sqrt{\frac{a}{2\pi\hbar}}\left(\frac{2a}{a^2 + p^2/\hbar^2} \right)$$

g. $$\frac{|\Psi(p=2a\hbar)|^2}{|\Psi(p=-a\hbar)|^2} = \left|\frac{1/(a^2+(2a\hbar)^2/\hbar^2)}{1/(a^2+(-a\hbar)^2/\hbar^2)}\right|^2$$

$$= \left|\frac{1/(a^2+4a^2)}{1/(a^2+a^2)}\right|^2$$

$$= \left|\frac{1/(5a^2)}{1/(2a^2)}\right|^2$$

$$= \left|\frac{2}{5}\right|^2 = 0.16 = 16\%$$

8. a. $$\boldsymbol{H} = \frac{-\hbar^2}{2m}\left\{\frac{\partial^2}{\partial x^2} + \frac{\partial^2}{\partial y^2}\right\} \text{ (cartesian coordinates)}$$

Finding $\frac{\partial}{\partial x}$ and $\frac{\partial}{\partial y}$ from the chain rule gives:

$$\frac{\partial}{\partial x} = \left(\frac{\partial r}{\partial x}\right)_y \frac{\partial}{\partial r} + \left(\frac{\partial \phi}{\partial x}\right)_y \frac{\partial}{\partial \phi}, \quad \frac{\partial}{\partial y} = \left(\frac{\partial r}{\partial y}\right)_x \frac{\partial}{\partial r} + \left(\frac{\partial \phi}{\partial y}\right)_x \frac{\partial}{\partial \phi},$$

Evaluation of the "coefficients" gives the following:

$$\left(\frac{\partial r}{\partial x}\right)_y = \cos\phi, \left(\frac{\partial \phi}{\partial x}\right)_y = -\frac{\sin\phi}{r},$$

$$\left(\frac{\partial r}{\partial y}\right)_x = \sin\phi, \text{ and } \left(\frac{\partial \phi}{\partial y}\right)_x = \frac{\cos\phi}{r},$$

Upon substitution of these "coefficients":

$$\frac{\partial}{\partial x} = \cos\phi\frac{\partial}{\partial r} - \frac{\sin\phi}{r}\frac{\partial}{\partial \phi} = -\frac{\sin\phi}{r}\frac{\partial}{\partial \phi}\text{; at fixed } r.$$

$$\frac{\partial}{\partial y} = \sin\phi\frac{\partial}{\partial r} + \frac{\cos\phi}{r}\frac{\partial}{\partial \phi} = \frac{\cos\phi}{r}\frac{\partial}{\partial \phi}\text{; at fixed } r.$$

$$\frac{\partial^2}{\partial x^2} = \left(-\frac{\sin\phi}{r}\frac{\partial}{\partial \phi}\right)\left(-\frac{\sin\phi}{r}\frac{\partial}{\partial \phi}\right) = \frac{\sin^2\phi}{r^2}\frac{\partial^2}{\partial \phi^2} + \frac{\sin\phi\cos\phi}{r^2}\frac{\partial}{\partial \phi}\text{; at fixed } r.$$

$$\frac{\partial^2}{\partial y^2} = \left(\frac{\cos\phi}{r}\frac{\partial}{\partial \phi}\right)\left(\frac{\cos\phi}{r}\frac{\partial}{\partial \phi}\right) = \frac{\cos^2\phi}{r^2}\frac{\partial^2}{\partial \phi 2} - \frac{\cos\phi\sin\phi}{r^2}\frac{\partial}{\partial \phi}\text{; at fixed } r.$$

$$\frac{\partial^2}{\partial x^2} + \frac{\partial^2}{\partial y^2} = \frac{\sin^2\phi}{r^2}\frac{\partial^2}{\partial \phi^2} + \frac{\sin\phi\cos\phi}{r^2}\frac{\partial}{\partial \phi} + \frac{\cos^2\phi}{r^2}\frac{\partial^2}{\partial \phi^2} - \frac{\cos\phi\sin\phi}{r^2}\frac{\partial}{\partial \phi} = \frac{1}{r^2}\frac{\partial^2}{\partial \phi^2}\text{; at fixed } r.$$

So, $$\boldsymbol{H} = \frac{-\hbar^2}{2mr^2}\frac{\partial^2}{\partial \phi^2} \text{ (cylindrical coordinates, fixed } r)$$

$$= \frac{-\hbar^2}{2I}\frac{\partial^2}{\partial \phi^2}$$

The Schrödinger equation for a particle on a ring then becomes:

$$H\Psi = E\Psi$$

$$\frac{-\hbar^2}{2I}\frac{\partial^2\Phi}{\partial\phi^2} = E\Phi$$

$$\frac{\partial^2\Phi}{\partial\phi^2} = \left(\frac{-2IE}{\hbar^2}\right)\Phi$$

The general solution to this equation is the now familiar expression:

$$\Phi(\phi) = C_1e^{-im\phi} + C_2e^{im\phi}, \text{ where } m = \left(\frac{2IE}{\hbar^2}\right)^{\frac{1}{2}}$$

Application of the cyclic boundary condition, $\Phi(\phi) = \Phi(\phi + 2\pi)$, results in the quantization of the energy expression:

$$E = \frac{m^2\hbar^2}{2I} \text{ where } m = 0, \pm 1, \pm 2, \pm 3, \ldots$$

It can be seen that the $\pm m$ values correspond to angular momentum of the same magnitude but opposite directions. Normalization of the wavefunction (over the region 0 to 2π) corresponding to + or $-m$ will result in a value of

$$\left(\frac{1}{2\pi}\right)^{\frac{1}{2}}$$

for the normalization constant.

$$\therefore\ \Phi(\phi) = \left(\frac{1}{2\pi}\right)^{\frac{1}{2}} e^{im\phi}$$

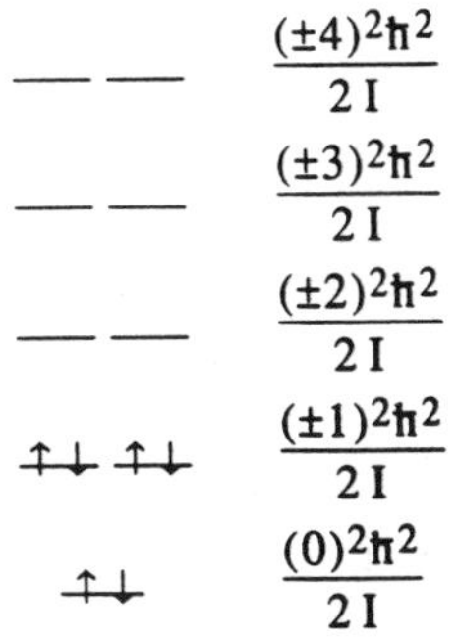

b. $\dfrac{\hbar^2}{2m} = 6.06 \times 10^{-28}\text{erg cm}^2$

$$\frac{\hbar^2}{2mr^2} = \frac{6.06 \times 10^{-28}\text{erg cm}^2}{(1.4 \times 10^{-8}\text{cm})^2} = 3.09 \times 10^{-12}\text{ erg}$$

$$\Delta E = (2^2 - 1^2)3.09 \times 10^{-12}\text{ erg} = 9.27 \times 10^{-12}\text{ erg}$$

but $\Delta E = h\nu = hc/\lambda$

so $\lambda = hc/\Delta E$

$$\lambda = \frac{(6.63 \times 10^{-27} \text{ erg sec})(3.00 \times 10^{10} \text{ cm sec}^{-1})}{9.27 \times 10^{-12} \text{ erg}} = 2.14 \times 10^{-5} \text{ cm} = 2.14 \times 10^{3} \text{Å}$$

Sources of error in this calculation include:

i. The attractive force of the carbon nuclei is not included in the Hamiltonian.

ii. The repulsive force of the other π-electrons is not included in the Hamiltonian.

iii. Benzene is not a ring.

iv. Electrons move in three dimensions not one.

v. Etc.

9. $\Psi(\phi,0) = \sqrt{\frac{4}{3\pi}}\cos^2\phi.$

This wavefunction needs to be expanded in terms of the eigenfunctions of the angular momentum operator,

$$\left(-i\hbar\frac{\partial}{\partial\phi}\right).$$

This is most easily accomplished by an exponential expansion of the cos function.

$$\Psi(\phi,0) = \sqrt{\frac{4}{3\pi}}\left(\frac{e^{i\phi}+e^{-i\phi}}{2}\right)\left(\frac{e^{i\phi}+e^{-i\phi}}{2}\right)$$

$$= \left(\frac{1}{4}\right)\sqrt{\frac{4}{3\pi}}\left(e^{2i\phi}+e^{-2i\phi}+2e^{(0)i\phi}\right)$$

The wavefunction is now written in terms of the eigenfunctions of the angular momentum operator,

$$\left(-i\hbar\frac{\partial}{\partial\phi}\right),$$

but they need to include their normalization constant, $\frac{1}{\sqrt{2\pi}}$.

$$\Psi(\phi,0) = \left(\frac{1}{4}\right)\sqrt{\frac{4}{3\pi}}\sqrt{2\pi}\left(\frac{1}{\sqrt{2\pi}}e^{2i\phi}+\frac{1}{\sqrt{2\pi}}e^{-2i\phi}+2\frac{1}{\sqrt{2\pi}}e^{(0)i\phi}\right)$$

$$= \left(\sqrt{\frac{1}{6}}\right)\left(\frac{1}{\sqrt{2\pi}}e^{2i\phi}+\frac{1}{\sqrt{2\pi}}e^{-2i\phi}+2\frac{1}{\sqrt{2\pi}}e^{(0)i\phi}\right)$$

Once the wavefunction is written in this form (in terms of the normalized eigenfunctions of the angular momentum operator having $m\hbar$ as eigenvalues) the probabilities for observing angular momentums of $0\hbar$, $2\hbar$, and $-2\hbar$ can be easily identified as the square of the coefficients of the corresponding eigenfunctions.

$$P_{2\hbar} = \left(\sqrt{\frac{1}{6}}\right)^2 = \frac{1}{6}$$

$$P_{-2\hbar} = \left(\sqrt{\frac{1}{6}}\right)^2 = \frac{1}{6}$$

$$P_{0\hbar} = \left(2\sqrt{\frac{1}{6}}\right)^2 = \frac{4}{6}$$

10. a. $\left(\frac{-\hbar^2}{2m}\right)\left(\frac{\partial^2}{\partial x^2}+\frac{\partial^2}{\partial y^2}+\frac{\partial^2}{\partial z^2}\right)\Psi(x,y,z)+\frac{1}{2}k(x^2+y^2+z^2)\Psi(x,y,z)=E\,\Psi(x,y,z).$

b. Let $\Psi(x,y,z)=X(x)Y(y)Z(z)$

$$\left(\frac{-\hbar^2}{2m}\right)\left(\frac{\partial^2}{\partial x^2}+\frac{\partial^2}{\partial y^2}+\frac{\partial^2}{\partial z^2}\right)X(x)Y(y)Z(z)+\frac{1}{2}k(x^2+y^2+z^2)X(x)Y(y)Z(z)=EX(x)Y(y)Z(z).$$

$$\left(\frac{-\hbar^2}{2m}\right)Y(y)Z(z)\frac{\partial^2X(x)}{\partial x^2}+\left(\frac{-\hbar^2}{2m}\right)X(x)Z(z)\frac{\partial^2Y(y)}{\partial y^2}+\left(\frac{-\hbar^2}{2m}\right)X(x)Y(y)\frac{\partial^2Z(z)}{\partial z^2}+$$

$$\frac{1}{2}kx^2X(x)Y(y)Z(z)+\frac{1}{2}ky^2X(x)Y(y)Z(z)+\frac{1}{2}kz^2X(x)Y(y)Z(z)=EX(x)Y(y)Z(z).$$

Dividing by $X(x)Y(y)Z(z)$ you obtain:

$$\left(\frac{-\hbar^2}{2m}\right)\left(\frac{1}{X(x)}\right)\frac{\partial^2X(x)}{\partial x^2}+\frac{1}{2}kx^2+\left(\frac{-\hbar^2}{2m}\right)\left(\frac{1}{Y(y)}\right)\frac{\partial^2Y(y)}{\partial y^2}+\frac{1}{2}ky^2+\left(\frac{-\hbar^2}{2m}\right)\left(\frac{1}{Z(z)}\right)\frac{\partial^2Z(z)}{\partial z^2}+\frac{1}{2}kz^2=E.$$

Now you have each variable isolated:

$F(x)+G(y)+H(z)=\text{constant}$

So,

$$\left(\frac{-\hbar^2}{2m}\right)\left(\frac{1}{X(x)}\right)\frac{\partial^2X(x)}{\partial x^2}+\frac{1}{2}kx^2=E_x\Rightarrow\left(\frac{-\hbar^2}{2m}\right)\frac{\partial^2X(x)}{\partial x^2}+\frac{1}{2}kx^2X(x)=E_xX(x),$$

$$\left(\frac{-\hbar^2}{2m}\right)\left(\frac{1}{Y(y)}\right)\frac{\partial^2Y(y)}{\partial y^2}+\frac{1}{2}ky^2=E_y\Rightarrow\left(\frac{-\hbar^2}{2m}\right)\frac{\partial^2Y(y)}{\partial y^2}+\frac{1}{2}ky^2Y(y)=E_yY(y),$$

$$\left(\frac{-\hbar^2}{2m}\right)\left(\frac{1}{Z(z)}\right)\frac{\partial^2Z(z)}{\partial z^2}+\frac{1}{2}kz^2=E_z\Rightarrow\left(\frac{-\hbar^2}{2m}\right)\frac{\partial^2Z(z)}{\partial z^2}+\frac{1}{2}kz^2Z(z)=E_zZ(z),$$

and $E=E_x+E_y+E_z$.

c. All three of these equations are one-dimensional harmonic oscillator equations and thus each have one-dimensional harmonic oscillator solutions which taken from the text are:

$$X_n(x)=\left(\frac{1}{n!2^n}\right)^{\frac{1}{2}}\left(\frac{\alpha}{\pi}\right)^{\frac{1}{4}}e^{\left(\frac{-\alpha x^2}{2}\right)}H_n(\alpha^{\frac{1}{2}}x),$$

$$Y_n(y)=\left(\frac{1}{n!2^n}\right)^{\frac{1}{2}}\left(\frac{\alpha}{\pi}\right)^{\frac{1}{4}}e^{\left(\frac{-\alpha y^2}{2}\right)}H_n(\alpha^{\frac{1}{2}}y),\text{ and}$$

$$Z_n(z)=\left(\frac{1}{n!2^n}\right)^{\frac{1}{2}}\left(\frac{\alpha}{\pi}\right)^{\frac{1}{4}}e^{\left(\frac{-\alpha z^2}{2}\right)}H_n(\alpha^{\frac{1}{2}}z),$$

where $\alpha=\left(\frac{k\mu}{\hbar^2}\right)^{\frac{1}{2}}$.

d. $E_{n_x,n_y,n_z} = E_{n_x} + E_{n_y} + E_{n_z}$

$$= \left(\frac{\hbar^2 k}{\mu}\right)^{\frac{1}{2}} \left\{\left(n_x + \frac{1}{2}\right) + \left(n_y + \frac{1}{2}\right) + \left(n_z + \frac{1}{2}\right)\right\}$$

e. Suppose $E = 5.5\left(\frac{\hbar^2 k}{\mu}\right)^{\frac{1}{2}}$

$$= \left(\frac{\hbar^2 k}{\mu}\right)^{\frac{1}{2}} \left\{\left(n_x + n_y + n_z + \frac{3}{2}\right)\right\}$$

$$5.5 = \left(n_x + n_y + n_z + \frac{3}{2}\right)$$

So, $n_x + n_y + n_z = 4$. This gives rise to a degeneracy of 15. They are:

States 1–3			States 4–6			States 7–9		
n_x	n_y	n_z	n_x	n_y	n_z	n_x	n_y	n_z
4	0	0	3	1	0	0	3	1
0	4	0	3	0	1	1	0	3
0	0	4	1	3	0	0	1	3

States 10–12			States 13–15		
n_x	n_y	n_z	n_x	n_y	n_z
2	2	0	2	1	1
2	0	2	1	2	1
0	2	2	1	1	2

f. Suppose $V = \frac{1}{2}kr^2$(independent of θ and ϕ)

The solutions $G(\theta,\phi)$ are the spherical harmonics $Y_{l,m}(\theta,\phi)$.

g. $$-\frac{\hbar^2}{2\mu r^2}\left(\frac{\partial}{\partial r}\left(r^2\frac{\partial\Psi}{\partial r}\right)\right) + \frac{1}{r^2\sin\theta}\frac{\partial}{\partial\theta}\left(\sin\theta\frac{\partial\Psi}{\partial\theta}\right) + \frac{1}{r^2\sin^2\theta}\frac{\partial^2\Psi}{\partial\phi^2} + \frac{k}{2}(r-r_e)^2\Psi = E\Psi,$$

$$+\frac{1}{r^2\sin^2\theta}\frac{\partial^2\Psi}{\partial\phi 2} + \frac{k}{2}(r-r_e)^2\Psi = E\Psi,$$

If $\Psi(r,\theta,\phi)$is replaced by $F(r)G(\theta,\phi)$:

$$-\frac{\hbar^2}{2\mu r^2}\left(\frac{\partial}{\partial r}\left(r^2\frac{\partial F(r)G(\theta,\phi)}{\partial r}\right)\right) + \frac{F(r)}{r^2\sin\theta}\frac{\partial}{\partial\theta}\left(\sin\theta\frac{\partial G(\theta,\phi)}{\partial\theta}\right)$$

$$+\frac{F(r)}{r^2\sin^2\theta}\frac{\partial^2 G(\theta,\phi)}{\partial\phi 2} + \frac{k}{2}(r-r_e)^2F(r)G(\theta,\phi) = EF(r)G(\theta,\phi),$$

and the angle dependence is recognized as the L^2 angular momentum operator. Division by $G(\theta,\phi)$ further reduces the equation to:

$$\frac{\hbar^2}{2\mu r^2}\left(\frac{\partial}{\partial r}\left(r^2\frac{\partial F(r)}{\partial r}\right)\right)+\frac{J(J+1)\hbar^2}{2\mu r_e^2 F(r)}+\frac{k}{2}(r-r_e)^2F(r)=EF(r)\,.$$

11. a. $$\frac{1}{2}mv^2 = 100\text{eV}\left(\frac{1.602\times10^{-12}\text{erg}}{1\text{eV}}\right)$$

$$v^2 = \left(\frac{(2)1.602\times10^{-10}\text{erg}}{9.109\times10^{-28}\text{g}}\right)$$

$$v = 0.593\times10^9\text{cm/sec}$$

The length of the N_2 molecule is $2\,Å = 2\times10^{-8}$cm.

$$v=\frac{d}{t}$$

$$t=\frac{d}{v}=\frac{2\times10^{-8}\text{cm}}{0.593\times10^9\text{cm/sec}}=3.37\times10^{-17}\text{sec}$$

b. The normalized ground state harmonic oscillator can be written (from both in the text and in exercise 11) as:

$$\Psi_0=\left(\frac{\alpha}{\pi}\right)^{1/4}e^{-\alpha x^2/2},\qquad \text{where } \alpha=\left(\frac{k\mu}{\hbar^2}\right)^{\frac{1}{2}} \text{ and } x=r-r_e$$

Calculating constants;

$$\alpha_{N_2}=\left(\frac{(2.294\times10^6\text{g sec}^{-2})(1.1624\times10^{-23}\text{g})}{(1.0546\times10^{-27}\text{erg sec})^2}\right)^{\frac{1}{2}}$$

$$=0.48966\times10^{19}\text{cm}^{-2}=489.66Å^{-2}$$

For N_2: $$\Psi_0(r)=3.53333Å^{-\frac{1}{2}}e^{-(244.83Å^{-2})(r-1.09769Å)^2}$$

$$\alpha_{N_2^+}=\left(\frac{(2.009\times10^6\text{g sec}^{-2})(1.1624\times10^{-23}\text{g})}{(1.0546\times10^{-27}\text{erg sec})^2}\right)^{\frac{1}{2}}$$

$$=0.45823\times10^{19}\text{cm}^{-2}=458.23Å^{-2}$$

For N_2^+: $$\Psi_0(r)=3.47522Å^{-\frac{1}{2}}e^{-(229.113Å^{-2})(r-1.11642Å)^2}$$

c. $$P(v=0)=\left|\left\langle\Psi_{v=0}(N_2^+)\middle|\Psi_{v=0}(N_2)\right\rangle\right|^2$$

Let $P(v=0)=I^2$ where I = integral:

$$I=\int_{-\infty}^{+\infty}(3.47522Å^{-\frac{1}{2}}e^{-(229.113Å^{-2})(r-1.11642Å)^2})\cdot(3.53333Å^{-\frac{1}{2}}e^{-(244.830Å^{-2})(r-1.09769Å)^2})dr$$

Let $C_1=3.47522Å^{-\frac{1}{2}}$, $\quad C_2=3.53333Å^{-\frac{1}{2}}$,

$A_1=229.113Å^{-2}$, $\quad A_2=244.830Å^{-2}$,

$r_1=1.11642Å$, $\quad r_2=1.09769Å$,

$$I = C_1C_2 \int_{-\infty}^{+\infty} e^{-A_1(r-r_1)^2} e^{-A_2(r-r_2)^2} dr.$$

Focusing on the exponential:

$$-A_1(r-r_1)^2 - A_2(r-r_2)^2 = -A_1(r^2 - 2r_1r + r_1^2) - A_2(r^2 - 2r_2r + r_2^2)$$

$$= -(A_1 + A_2)r^2 + (2A_1r_1 + 2A_2r_2)r - A_1r_1^2 - A_2r_2^2$$

Let $A = A_1 + A_2$,
$B = 2A_1r_1 + 2A_2r_2$,
$C = C_1C_2$, and
$D = A_1r_1^2 + A_2r_2^2$.

$$I = C \int_{-\infty}^{+\infty} e^{-Ar^2+Br-D} dr$$

$$= C \int_{-\infty}^{+\infty} e^{-A(r-r_0)^2+D'} dr$$

where $-A(r-r_0)^2 + D' = -Ar^2 + Br - D$

$-A(r^2 - 2rr_0 + r_0^2) + D' = -Ar^2 + Br - D$

such that, $2Ar_0 = B$

$-Ar_0^2 + D' = -D$

and, $r_0 = \dfrac{B}{2A}$

$$D' = Ar_0^2 - D = A\frac{B^2}{4A^2} - D = \frac{B^2}{4A} - D.$$

$$I = C \int_{-\infty}^{+\infty} e^{-A(r-r_0)^2 + D'} dr$$

$$= Ce^{D'} \int_{-\infty}^{+\infty} e^{-Ay^2} dy$$

$$= Ce^{D'} \sqrt{\frac{\pi}{A}}$$

Now back substituting all of these constants:

$$I = C_1C_2\sqrt{\left(\frac{\pi}{A_1 + A_2}\right)} \exp\left(\frac{(2A_1r_1+2A_2r_2)^2}{4(A_1 + A_2)} - A_1r_1^2 - A_2r_2^2\right)$$

$$I = (3.47522)(3.53333)\sqrt{\frac{\pi}{(229.113) + (244.830)}}$$

$$\cdot \exp\left(\frac{(2(229.113)(1.11642) + 2(244.830)(1.09769))^2}{4((229.113) + (244.830))}\right)$$

$$\cdot \exp\left(-(229.113)(1.11642)^2 - (244.830)(1.09769)^2\right)$$

$$I = 0.959$$

$$P(v = 0) = I^2 = 0.92$$

12. a. $E_\nu = \left(\frac{\hbar^2 k}{\mu}\right)^{\frac{1}{2}}\left(\nu + \frac{1}{2}\right)$

$$\Delta E = E_{\nu+1} - E_\nu$$

$$= \left(\frac{\hbar^2 k}{\mu}\right)^{\frac{1}{2}}\left\{\nu + 1 + \frac{1}{2} - \nu - \frac{1}{2}\right\} = \left(\frac{\hbar^2 k}{\mu}\right)$$

$$= \left(\frac{(1.0546 \times 10^{-27}\text{erg sec})^2(1.87 \times 10^6\text{g sec}^{-2})}{6.857\text{g}/6.02 \times 10^{23}}\right)^{\frac{1}{2}}$$

$$= 4.27 \times 10^{-13}\text{erg}$$

$$\Delta E = \frac{hc}{\lambda}$$

$$\lambda = \frac{hc}{\Delta E} = \frac{(6.626 \times 10^{-27}\text{erg sec})(3.00 \times 10^{10}\text{cm sec}^{-1})}{4.27 \times 10^{-13}\text{erg}}$$

$$= 4.66 \times 10^{-4}\text{cm}$$

$$\frac{1}{\lambda} = 2150\text{cm}^{-1}$$

b. $\Psi_0 = \left(\frac{\alpha}{\pi}\right)^{1/4} e^{-\alpha x^2/2}$

$$\left\langle x \right\rangle = \left\langle \Psi_{v=0} | x | \Psi_{v=0} \right\rangle$$

$$= \int_{-\infty}^{+\infty} \Psi_0^* x \Psi_0 dx$$

$$= \int_{-\infty}^{+\infty} \left(\frac{\alpha}{\pi}\right)^{1/2} x e^{-\alpha x^2} dx$$

$$= \int_{-\infty}^{+\infty} \left(\frac{\alpha}{-\alpha^2 \pi}\right)^{1/2} e^{-\alpha x^2} d(-\alpha x^2)$$

$$= \left(\frac{-1}{\alpha \pi}\right)^{1/2} e^{-\alpha x^2}\Big|_{-\infty}^{+\infty} = 0$$

$$\left\langle x^2 \right\rangle = \left\langle \Psi_{v=0} | x^2 | \Psi_{v=0} \right\rangle$$

$$= \int_{-\infty}^{+\infty} \Psi_0^* x^2 \Psi_0 dx$$

$$= \int_{-\infty}^{+\infty} \left(\frac{\alpha}{\pi}\right)^{1/2} x^2 e^{-\alpha x^2} dx$$

$$= 2\left(\frac{\alpha}{\pi}\right)^{1/2} \int_{0}^{+\infty} x^2 e^{-\alpha x^2} dx)$$

Using integral equation (4) this becomes:

$$= 2\left(\frac{\alpha}{\pi}\right)^{1/2}\left(\frac{1}{2^{1+1}\alpha}\right)\left(\frac{\pi}{\alpha}\right)^{1/2}$$

$$= \left(\frac{1}{2\alpha}\right)$$

$$\Delta x = (< x^2 > - < x > ^2)^{1/2} = \left(\frac{1}{2\alpha}\right)$$

$$= \left(\frac{\hbar}{2\sqrt{k\mu}}\right)^{\frac{1}{2}}$$

$$= \left(\frac{(1.0546 \times 10^{-27}\text{erg sec})^2}{4(1.87 \times 10^6\text{g sec}^{-2})(6.857\text{g}/6.02 \times 10^{23})}\right)^{\frac{1}{4}}$$

$$= 3.38 \times 10^{-10}\text{cm} = 0.0338\text{Å}$$

c. $\Delta x = \left(\frac{\hbar}{2\sqrt{k\mu}}\right)^{\frac{1}{2}}$

The smaller k and μ become, the larger the uncertainty in the internuclear distance becomes. Helium has a small μ and small force between atoms. This results in a very large Δx. This implies that it is extremely difficult for He atoms to "vibrate" with small displacement as a solid even as absolute zero is approached.

13. a. $W = \left(Z_e^2 - 2ZZ_e + \frac{5}{8}Z_e\right)\frac{e^2}{a_0}$

$$\frac{dW}{dZ_e} = \left(2Z_e - 2Z + \frac{5}{8}\right)\frac{e^2}{a_0} = 0$$

$$2Z_e - 2Z + \frac{5}{8} = 0$$

$$2Z_e = 2Z - \frac{5}{8}$$

$$Z_e = Z - \frac{5}{16} = Z - 0.3125$$

(Note this is the shielding factor of one $1s$ electron to the other).

$$W = Z_e\left(Z_e - 2Z + \frac{5}{8}\right)\frac{e^2}{a_0}$$

$$W = \left(Z - \frac{5}{16}\right)\left(\left(Z - \frac{5}{16}\right) - 2Z + \frac{5}{8}\right)\frac{e^2}{a_0}$$

$$W = \left(Z - \frac{5}{16}\right)\left(-Z + \frac{5}{16}\right)\frac{e^2}{a_0}$$

$$W = -\left(Z - \frac{5}{16}\right)\left(Z - \frac{5}{16}\right)\frac{e^2}{a_0} = -\left(Z - \frac{5}{16}\right)^2\frac{e^2}{a_0}$$

$$= -(Z - 0.3125)^2(27.21)\text{eV}$$

b. Using the above result for W and the percent error as calculated below we obtain the following:

$$\%\text{error} = \frac{(\text{Experimental} - \text{Theoretical})}{\text{Experimental}} * 100$$

Z	Atom	Experimental	Calculated	% Error
$Z = 1$	H^-	−14.35eV	−12.86eV	10.38%
$Z = 2$	He	−78.98eV	−77.46eV	1.92%
$Z = 3$	Li^+	−198.02eV	−196.46eV	0.79%
$Z = 4$	Be^{+2}	−371.5eV	−369.86eV	0.44%
$Z = 5$	B^{+3}	−599.3eV	−597.66eV	0.27%
$Z = 6$	C^{+4}	−881.6eV	−879.86eV	0.19%
$Z = 7$	N^{+5}	−1218.3eV	−1216.48eV	0.15%
$Z = 8$	O^{+6}	−1609.5eV	−1607.46eV	0.13%

The ignored electron correlation effects are essentially constant over the range of Z, but this correlation effect is a larger percentage error at small Z. At large Z the dominant interaction is electron attraction to the nucleus completely overwhelming the ignored electron correlation and hence reducing the overall percent error.

c. Since −12.86eV(H^-) is greater than −13.6 eV (H + e) this simple variational calculation erroneously predicts H^- to be unstable.

14. a. $W = \int_{-\infty}^{\infty} \phi^* H\phi dx$

$$W = \left(\frac{2b}{\pi}\right)^{\frac{1}{2}} \int_{-\infty}^{\infty} e^{-bx^2}\left(-\frac{\hbar^2}{2m}\frac{d^2}{dx^2} + a|x|\right)e^{-bx^2}dx$$

$$\frac{d^2}{dx^2}e^{-bx^2} = \frac{d}{dx}\left(-2bxe^{-bx^2}\right)$$

$$= (-2bx)\left(-2bxe^{-bx^2}\right) + \left(e^{-bx^2}\right)(-2b)$$

$$= \left(4b^2x^2e^{-bx^2}\right) + \left(-2be^{-bx^2}\right)$$

Making this substitution results in the following three integrals:

$$W = \left(\frac{2b}{\pi}\right)^{\frac{1}{2}}\left(-\frac{\hbar^2}{2m}\right)\int_{-\infty}^{\infty} e^{-bx^2}4b^2x^2e^{-bx^2}dx + \left(\frac{2b}{\pi}\right)^{\frac{1}{2}}\left(-\frac{\hbar^2}{2m}\right)\int_{-\infty}^{\infty} e^{-bx^2}-2be^{-bx^2}dx +$$

$$\left(\frac{2b}{\pi}\right)^{\frac{1}{2}}\int_{-\infty}^{\infty} e^{-bx^2}a|x|e^{-bx^2}dx$$

$$= \left(\frac{2b}{\pi}\right)^{\frac{1}{2}}\left(-\frac{2b^2\hbar^2}{m}\right)\int_{-\infty}^{\infty} x^2e^{-2bx^2}dx + \left(\frac{2b}{\pi}\right)^{\frac{1}{2}}\left(\frac{b\hbar^2}{m}\right)\int_{-\infty}^{\infty} e^{-2bx^2}dx + \left(\frac{2b}{\pi}\right)^{\frac{1}{2}}a\int_{-\infty}^{\infty} |x|e^{-2bx^2}dx$$

Using integral equations (1), (2), and (3) this becomes:

$$= \left(\frac{2b}{\pi}\right)^{\frac{1}{2}}\left(-\frac{2b^2\hbar^2}{m}\right)2\left(\frac{1}{2^2 2b}\right)\sqrt{\frac{\pi}{2b}} + \left(\frac{2b}{\pi}\right)^{\frac{1}{2}}\left(\frac{b\hbar^2}{m}\right)2\left(\frac{1}{2}\right)\sqrt{\frac{\pi}{2b}} + \left(\frac{2b}{\pi}\right)^{\frac{1}{2}}a\left(\frac{0!}{2b}\right)$$

$$= \left(-\frac{b\hbar^2}{m}\right)\left(\frac{1}{2}\right) + \left(\frac{b\hbar^2}{m}\right) + \left(\frac{2b}{\pi}\right)^{\frac{1}{2}}\left(\frac{a}{2b}\right)$$

$$W = \left(\frac{b\hbar^2}{2m}\right) + a\left(\frac{1}{2b\pi}\right)^{\frac{1}{2}}$$

b. Optimize b by evaluating $\frac{dW}{db} = 0$

$$\frac{dW}{db} = \frac{d}{db}\left(\left(\frac{b\hbar^2}{2m}\right) + a\left(\frac{1}{2b\pi}\right)^{\frac{1}{2}}\right)$$

$$= \left(\frac{\hbar^2}{2m}\right) - \frac{a}{2}\left(\frac{1}{2\pi}\right)^{\frac{1}{2}} b^{-\frac{3}{2}}$$

So, $\frac{a}{2}\left(\frac{1}{2\pi}\right)^{\frac{1}{2}} b^{-\frac{3}{2}} = \left(\frac{\hbar^2}{2m}\right)$

or, $b^{-\frac{3}{2}} = \left(\frac{\hbar^2}{2m}\right)\frac{2}{a}\left(\frac{1}{2\pi}\right)^{-\frac{1}{2}} = \left(\frac{\hbar^2}{ma}\right)\sqrt{2\pi}$, and, $b = \left(\frac{ma}{\sqrt{2\pi}\hbar^2}\right)^{\frac{2}{3}}$.

Substituting this value of b into the expression for W gives:

$$W = \left(\frac{\hbar^2}{2m}\right)\left(\frac{ma}{\sqrt{2\pi}\hbar^2}\right)^{\frac{2}{3}} + a\left(\frac{1}{2\pi}\right)^{\frac{1}{2}}\left(\frac{ma}{\sqrt{2\pi}\hbar^2}\right)^{-\frac{1}{3}}$$

$$= \left(\frac{\hbar^2}{2m}\right)\left(\frac{ma}{\sqrt{2\pi}\hbar^2}\right)^{\frac{2}{3}} + a\left(\frac{1}{2\pi}\right)^{\frac{1}{2}}\left(\frac{ma}{\sqrt{2\pi}\hbar^2}\right)^{-\frac{1}{3}}$$

$$= 2^{-\frac{4}{3}}\pi^{-\frac{1}{3}}\hbar^{\frac{2}{3}}a^{\frac{2}{3}}m^{-\frac{1}{3}} + 2^{-\frac{1}{3}}\pi^{-\frac{1}{3}}\hbar^{\frac{2}{3}}a^{\frac{2}{3}}m^{-\frac{1}{3}}$$

$$= \left(2^{-\frac{4}{3}}\pi^{-\frac{1}{3}} + 2^{-\frac{1}{3}}\pi^{-\frac{1}{3}}\right)\hbar^{\frac{2}{3}}a^{\frac{2}{3}}m^{-\frac{1}{3}} = \frac{3}{2}(2\pi)^{-\frac{1}{3}}\hbar^{\frac{2}{3}}a^{\frac{2}{3}}m^{-\frac{1}{3}}$$

$$= 0.812889106\hbar^{\frac{2}{3}}a^{\frac{2}{3}}m^{-\frac{1}{3}} \text{ in error} = 0.5284\%!!!!!$$

15. a. $H = -\frac{\hbar^2}{2m}\frac{d^2}{dx^2} + \frac{1}{2}kx^2$

$\phi = \sqrt{\frac{15}{16}}a^{-\frac{5}{2}}(a^2 - x^2)$ for $-a < x < a$

$\phi = 0$ for $|x| \geq a$

$$\int_{-\infty}^{+\infty}\phi^*H\phi dx = \int_{-a}^{+a}\sqrt{\frac{15}{16}}a^{-\frac{5}{2}}(a^2-x^2)\left(-\frac{\hbar^2}{2m}\frac{d^2}{dx^2} + \frac{1}{2}kx^2\right)\sqrt{\frac{15}{16}}a^{-\frac{5}{2}}(a^2-x^2)dx$$

$$= \left(\frac{15}{16}\right)a^{-5}\int_{-a}^{+a}(a^2-x^2)\left(-\frac{\hbar^2}{2m}\frac{d^2}{dx^2} + \frac{1}{2}kx^2\right)(a^2-x^2)dx$$

$$= \left(\frac{15}{16}\right)a^{-5}\int_{-a}^{+a}(a^2-x^2)\left(-\frac{\hbar^2}{2m}\right)\frac{d^2}{dx^2}(a^2-x^2)dx + \left(\frac{15}{16}\right)a^{-5}\int_{-a}^{+a}(a^2-x^2)\frac{1}{2}kx^2(a^2-x^2)dx$$

$$= \left(\frac{15}{16}\right)a^{-5}\int_{-a}^{+a}(a^2-x^2)\left(-\frac{\hbar^2}{2m}\right)(-2)dx + \left(\frac{15}{32}\right)a^{-5}\int_{-a}^{+a}(kx^2)(a^4 - 2a^2x^2 + x^4)dx$$

$$= \left(\frac{15\hbar^2}{16m}\right)a^{-5}\int_{-a}^{+a}(a^2-x^2)\,dx + \left(\frac{15}{32}\right)a^{-5}\int_{-a}^{+a}a^4kx^2 - 2a^2kx^4 + kx^6dx$$

$$= \left(\frac{15\hbar^2}{16m}\right)a^{-5}\left(a^2x\Big|_{-a}^{a} - \frac{1}{3}x^3\Big|_{-a}^{a}\right) + \left(\frac{15}{32}\right)a^{-5}\left(\frac{a^4k}{3}x^3\Big|_{-a}^{a} - \frac{2a^2k}{5}x^5\Big|_{-a}^{a} + \frac{k}{7}x^7\Big|_{-a}^{a}\right)$$

$$= \left(\frac{15\hbar^2}{16m}\right)a^{-5}\left(2a^3 - \frac{2}{3}a^3\right) + \left(\frac{15}{32}\right)a^{-5}\left(\frac{2a^7k}{3} - \frac{4a^7k}{5} + \frac{2k}{7}a^7\right)$$

$$= \left(\frac{15}{16}\right)a^{-5}\left(\frac{4\hbar^2}{3m}a^3 + \frac{a^7k}{3} - \frac{2a^7k}{5} + \frac{k}{7}a^7\right)$$

$$= \left(\frac{15}{16}\right)a^{-5}\left(\frac{4\hbar^2}{3m}a^3 + \left(\frac{k}{3} - \frac{2k}{5} + \frac{k}{7}\right)a^7\right)$$

$$= \left(\frac{15}{16}\right)a^{-5}\left(\frac{4\hbar^2}{3m}a^3 + \left(\frac{35k}{105} - \frac{42k}{105} + \frac{15k}{105}\right)a^7\right)$$

$$= \left(\frac{15}{16}\right)a^{-5}\left(\frac{4\hbar^2}{3m}a^3 + \left(\frac{8k}{105}\right)a^7\right) = \frac{5\hbar^2}{4ma^2} + \frac{ka^2}{14}$$

b. Substituting $a = b\left(\frac{\hbar^2}{km}\right)^{\frac{1}{4}}$ into the above expression for E we obtain:

$$E = \frac{5\hbar^2}{4b^2m}\left(\frac{km}{\hbar^2}\right)^{\frac{1}{2}} + \frac{kb^2}{14}\left(\frac{\hbar^2}{km}\right)^{\frac{1}{2}}$$

$$= \hbar k^{\frac{1}{2}} m^{-\frac{1}{2}}\left(\frac{5}{4}b^{-2} + \frac{1}{14}b^2\right)$$

Plotting this expression for the energy with respect to b having values of 0.2, 0.4, 0.6, 0.8, 1.0, 1.5, 2.0, 2.5, 3.0, 4.0, and 5.0 gives:

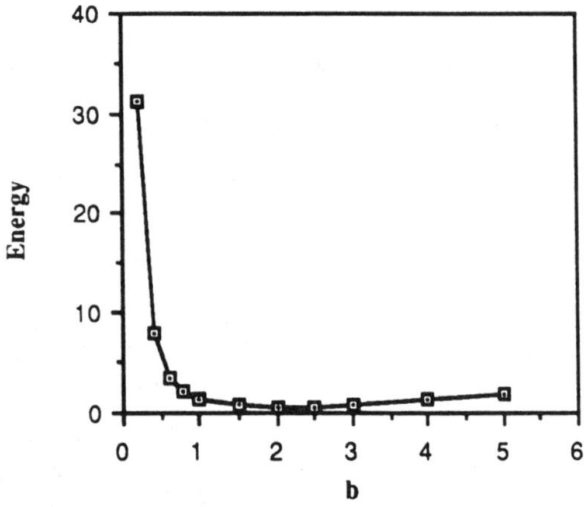

c. $E = \frac{5\hbar^2}{4ma^2} + \frac{ka^2}{14}$

$$\frac{dE}{da} = -\frac{10\hbar^2}{4ma^3} + \frac{2ka}{14} = -\frac{5\hbar^2}{2ma^3} + \frac{ka}{7} = 0$$

$$\frac{5\hbar^2}{2ma^3} = \frac{ka}{7} \text{ and } 35\hbar^2 = 2mka^4$$

So,

$$a^4 = \frac{35\hbar^2}{2mk}, \text{ or } a = \left(\frac{35\hbar^2}{2mk}\right)^{\frac{1}{4}}$$

Therefore

$$\phi_{best} = \sqrt{\frac{15}{16}}\left(\frac{35\hbar^2}{2mk}\right)^{-\frac{5}{8}}\left(\left(\frac{35\hbar^2}{2mk}\right)^{\frac{1}{2}} - x^2\right)$$

and

$$E_{best} = \frac{5\hbar^2}{4m}\left(\frac{2mk}{35\hbar^2}\right)^{\frac{1}{2}} + \frac{k}{14}\left(\frac{35\hbar^2}{2mk}\right)^{\frac{1}{2}} = \hbar k^{\frac{1}{2}} m^{-\frac{1}{2}}\left(\frac{5}{14}\right)^{\frac{1}{2}}.$$

d. $$\frac{E_{best}-E_{true}}{E_{true}} = \frac{\hbar k^{\frac{1}{2}} m^{-\frac{1}{2}}\left(\left(\frac{5}{14}\right)^{\frac{1}{2}} -0.5\right)}{\hbar k^{\frac{1}{2}} m^{-\frac{1}{2}} 0.5} = \frac{\left(\frac{5}{14}\right)^{\frac{1}{2}}-0.5}{0.5} = \frac{0.0976}{0.5} = 0.1952 = 19.52\%$$

16. a $E^2 = m^2c^4 + p^2c^2.$

$$= m^2c^4\left(1+\frac{p^2}{m^2c^2}\right)$$

$$E = mc^2\sqrt{\left(1+\frac{p^2}{m^2c^2}\right)}$$

$$\approx mc^2\left(1+\frac{p^2}{2m^2c^2}-\frac{p^4}{8m^4c^4}+\ldots\right)$$

$$= mc^2 + \frac{p^2}{2m} - \frac{p^4}{8m^3c^2} + \ldots)$$

Let

$$V = -\frac{p^4}{8m^3c^2}$$

b. $E_{1s}^{(1)} = \left\langle \Psi(r)_{1s}^{*} V \Psi(r)_{1s} \right\rangle$

$$E_{1s}^{(1)} = \int \left(\frac{Z}{a}\right)^{\frac{3}{2}}\left(\frac{1}{\pi}\right)^{\frac{1}{2}} e^{-\frac{Zr}{a}}\left(-\frac{p^4}{8m^3c^2}\right)\left(\frac{Z}{a}\right)^{\frac{3}{2}}\left(\frac{1}{\pi}\right)^{\frac{1}{2}} e^{-\frac{Zr}{a}} d\tau$$

Substituting $p = -i\hbar\nabla$, $d\tau = r^2 dr \sin\theta\, d\theta d\phi$, and pulling out constants gives:

$$E_{1s}^{(1)} = \left(-\frac{\hbar^4}{8m^3c^2}\right)\left(\frac{Z}{a}\right)^3\left(\frac{1}{\pi}\right)\int_0^\infty e^{-\frac{Zr}{a}} \nabla^2\nabla^2 e^{-\frac{Zr}{a}} r^2 dr \int_0^\pi \sin\theta\, d\theta \int_0^{2\pi} d\phi.$$

The integrals over the angles are easy,

$$\int_0^{2\pi} d\phi = 2\pi \text{ and } \int_0^\pi \sin\theta d\,\theta = 2.$$

The work remaining is in evaluating the integral over r. Substituting

$$\nabla^2 = \frac{1}{r^2}\frac{\partial}{\partial r} r^2 \frac{\partial}{\partial r}$$

we obtain:

$$\nabla^2 e^{-\frac{Zr}{a}} = \frac{1}{r^2}\frac{\partial}{\partial r} r^2 \frac{\partial}{\partial r} e^{-\frac{Zr}{a}} = \frac{1}{r^2}\frac{\partial}{\partial r} r^2 \frac{-Z}{a} e^{-\frac{Zr}{a}} = \frac{-Z}{a}\frac{1}{r^2}\frac{\partial}{\partial r} r^2 e^{-\frac{Zr}{a}}$$

$$= \frac{-Z}{a}\frac{1}{r^2}\left(\frac{\partial}{\partial r} r^2 e^{-\frac{Zr}{a}}\right) = \frac{-Z}{a}\frac{1}{r^2}\left(r^2\frac{-Z}{a} e^{-\frac{Zr}{a}} + e^{-\frac{Zr}{a}} 2r\right)$$

$$= \frac{-Z}{a}\left(\frac{-Z}{a}+\frac{2}{r}\right)e^{-\frac{Zr}{a}} = \left(\left(\frac{Z}{a}\right)^2 - \frac{2Z}{ar}\right)e^{-\frac{Zr}{a}}.$$

The integral over r then becomes:

$$\int_0^\infty e^{-\frac{Zr}{a}}\nabla^2\nabla^2 e^{-\frac{Zr}{a}}r^2dr = \int_0^\infty \left(\left(\frac{Z}{a}\right)^2 - \frac{2Z}{ar}\right)^2 e^{-\frac{2Zr}{a}}r^2dr$$

$$= \int_0^\infty \left(\left(\frac{Z}{a}\right)^4 - \frac{4}{r}\left(\frac{Z}{a}\right)^3 + \frac{4}{r^2}\left(\frac{Z}{a}\right)^2\right)r^2 e^{-\frac{2Zr}{a}}dr$$

$$= \int_0^\infty \left(\left(\frac{Z}{a}\right)^4 r^2 - 4\left(\frac{Z}{a}\right)^3 r + 4\left(\frac{Z}{a}\right)^2\right)e^{-\frac{2Zr}{a}}dr$$

Using integral equation (4) these integrals can easily be evaluated:

$$= 2\left(\frac{Z}{a}\right)^4\left(\frac{a}{2Z}\right)^3 - 4\left(\frac{Z}{a}\right)^3\left(\frac{a}{2Z}\right)^2 + 4\left(\frac{Z}{a}\right)^2\left(\frac{a}{2Z}\right)$$

$$= \left(\frac{Z}{4a}\right) - \left(\frac{Z}{a}\right) + 2\left(\frac{Z}{a}\right) = \left(\frac{5Z}{4a}\right)$$

So, $E_{1s}^{(1)} = \left(-\frac{\hbar^4}{8m^3c^2}\right)\left(\frac{Z}{a}\right)^3\left(\frac{1}{\pi}\right)\left(\frac{5Z}{4a}\right)4\pi = -\frac{5\hbar^4 Z^4}{8m^3c^2a^4}$

Substituting $a_0 = \frac{\hbar^2}{m_e e^2}$ gives:

$$E_{1s}^{(1)} = -\frac{5\hbar^4 Z^4 m^4 e^8}{8m^3c^2\hbar^8} = -\frac{5Z^4 me^8}{8c^2\hbar^4}$$

Notice that $E_{1s} = -\frac{Z^2me^4}{2\hbar^2}$, so, $E_{1s}{}^2 = \frac{Z^4m^2e^8}{4\hbar^4}$ and that $E_{1s}^{(1)} = -\frac{5m}{2}E_{1s}{}^2$

c. $\frac{E_{1s}^{(1)}}{E_{1s}} = \left(-\frac{5Z^4me^8}{8c^2\hbar^4}\right)\left(-\frac{2\hbar^2}{Z^2me^4}\right) = 10\% = 0.1$

$\frac{5Z^2e^4}{4c^2\hbar^2} = 0.1$, so, $Z^2 = \frac{(0.1)4c^2\hbar^2}{5e^4}$

$$Z^2 = \frac{(0.1)(4)(3.00\times 10^{10})^2(1.05\times 10^{-27})^2}{(5)(4.8\times 10^{-10})^4}$$

$Z^2 = 1.50\times 10^3$

$Z = 39$

17. a. $\boldsymbol{H}_0\Psi_{lm}^{(0)} = \frac{\boldsymbol{L}^2}{2m_er_0{}^2}\Psi_{lm}^{(0)} = \frac{\boldsymbol{L}^2}{2m_er_0{}^2}Y_{l,m}(\theta,\phi) = \frac{1}{2m_er_0{}^2}\hbar^2 l(l+1)Y_{l,m}(\theta,\phi)$

$$= \frac{1}{2m_e r_0^2} \hbar^2 l(l+1) Y_{l,m}(\theta,\phi)$$

$$E_{lm}^{(0)} = \frac{\hbar^2}{2m_e r_0^2} l(l+1)$$

b. $V = -e\varepsilon z = -e\varepsilon r_0 \cos\theta$

$$E_{00}^{(1)} = \left\langle Y_{00}|V|Y_{00}\right\rangle = \left\langle Y_{00}| -e\varepsilon r_0 \cos\theta|Y_{00}\right\rangle$$

$$= -e\varepsilon r_0 \left\langle Y_{00}|\cos\theta|Y_{00}\right\rangle$$

Using the given identity this becomes:

$$E_{00}^{(1)} = -e\varepsilon r_0 \left\langle Y_{00}|Y_{10}\right\rangle \sqrt{\frac{(0+0+1)(0-0+1)}{(2(0)+1)(2(0)+3)}} + -e\varepsilon r_0 \left\langle Y_{00}|Y_{-10}\right\rangle \sqrt{\frac{(0+0)(0-0)}{(2(0)+1)(2(0)-1)}}$$

$$\left\langle Y_{lm}|V|Y_{00}\right\rangle = -\frac{e\varepsilon r_0}{\sqrt{3}} \left\langle Y_{lm}|Y_{10}\right\rangle$$

The spherical harmonics are orthonormal, thus

$\left\langle Y_{00}|Y_{10}\right\rangle = \left\langle Y_{00}|Y_{-10}\right\rangle = 0$, and $E_{00}^{(1)} = 0$.

$$E_{00}^{(2)} = \sum_{lm \neq 00} \frac{\left|\left\langle Y_{lm}|V|Y_{00}\right\rangle\right|^2}{E_{00}^{(0)} - E_{lm}^{(0)}}$$

$$\left\langle Y_{lm}|V|Y_{00}\right\rangle = -e\varepsilon r_0 \left\langle Y_{lm}|\cos\theta|Y_{00}\right\rangle$$

Using the given identity this becomes:

$$\left\langle Y_{lm}|V|Y_{00}\right\rangle = -e\varepsilon r_0 \left\langle Y_{lm}|Y_{10}\right\rangle \sqrt{\frac{(0+0+1)(0-0+1)}{(2(0)+1)(2(0)+3)}} +$$

$$-e\varepsilon r_0 \left\langle Y_{lm}|Y_{-10}\right\rangle \sqrt{\frac{(0+0)(0-0)}{(2(0)+1)(2(0)-1)}}$$

$$\left\langle Y_{lm}|V|Y_{00}\right\rangle = -\frac{e\varepsilon r_0}{\sqrt{3}} \left\langle Y_{lm}|Y_{10}\right\rangle$$

This indicates that the only term contributing to the sum in the expression for $E_{00}^{(2)}$ is when $lm = 10(l = 1$, and $m = 0)$, otherwise

$$\left\langle Y_{lm}|V|Y_{00}\right\rangle$$

vanishes (from orthonormality). In quantum chemistry when using orthonormal functions it is typical to write the term

$$\left\langle Y_{lm}|Y_{10}\right\rangle$$

as a delta function, for example $\delta_{lm,10}$, which only has values of 1 or 0; $\delta_{ij} = 1$ when $i = j$ and 0 when $i \neq j$.This delta function when inserted into the sum then eliminates the sum by "picking out" the non-zero component. For example,

$$\left\langle Y_{lm}|V|Y_{00}\right\rangle = -\frac{e\varepsilon r_0}{\sqrt{3}}\delta_{lm,10}, \text{ so}$$

$$E_{00}^{(2)} = \sum_{lm\neq 00}\frac{e^2\varepsilon^2 r_0^2}{3}\frac{\delta_{lm'10}{}^2}{E_{00}^{(0)} - E_{lm}^{(0)}} = \frac{e^2\varepsilon^2 r_0^2}{3}\frac{1}{E_{00}^{(0)} - E_{10}^{(0)}}$$

$$E_{00}^{(0)} = \frac{\hbar^2}{2m_e r_0^2}0(0+1) = 0 \text{ and } E_{10}^{(0)} = \frac{\hbar^2}{2m_e r_0^2}1(1+1) = \frac{\hbar^2}{m_e r_0^2}$$

Inserting these energy expressions above yields:

$$E_{00}^{(2)} = -\frac{e^2\varepsilon^2 r_0^2}{3}\frac{m_e r_0{}^2}{\hbar^2} = -\frac{m_e e^2\varepsilon^2 r_0^4}{3\hbar^2}$$

c. $E_{00} = E_{00}^{(0)} + E_{00}^{(1)} + E_{00}^{(2)} + \ldots$

$$= 0 + 0 - \frac{m_e e^2\varepsilon^2 r_0^4}{3\hbar^2}$$

$$= -\frac{m_e e^2\varepsilon^2 r_0^4}{3\hbar^2}$$

$$\alpha = -\frac{\partial^2 E}{\partial^2\varepsilon} = \frac{\partial^2}{\partial^2\varepsilon}\left(\frac{m_e e^2\varepsilon^2 r_0^4}{3\hbar^2}\right)$$

$$= \frac{2m_e e^2 r_0{}^4}{3\hbar^2}$$

d. $\alpha = \dfrac{2(9.1095\times 10^{-28}g)(4.80324\times 10^{-10}g^{\frac{1}{2}}cm^{\frac{3}{2}}s^{-1})^2 r_0^4}{3(1.05459\times 10^{-27}g\ cm^2 s^{-1})^2}$

$\alpha = r_0^4/12598\times 10^6\text{cm}^{-1} = r_0^4 1.2598\text{Å}^{-1}$

$\alpha_H = 0.0987\text{Å}^3$

$\alpha_{Cs} = 57.57\text{Å}^3$

18. a. $V = e\varepsilon\left(x - \frac{L}{2}\right), \Psi_n^{(0)} = \left(\frac{2}{L}\right)^{\frac{1}{2}}\sin\left(\frac{n\pi x}{L}\right)$, and

$$E_n^{(0)} = \frac{\hbar^2\pi^2 n^2}{2mL^2}.$$

$$E_{n=1}^{(1)} = \left\langle\Psi_{n=1}^{(0)}|V|\Psi_{n=1}^{(0)}\right\rangle = \left\langle\Psi_{n=1}^{(0)}\,|\,e\varepsilon\left(x - \frac{L}{2}\right)|\,\Psi_{n=1}^{(0)}\right\rangle$$

$$= \left(\frac{2}{L}\right)\int_0^L \sin^2\left(\frac{\pi x}{L}\right)e\varepsilon\left(x - \frac{L}{2}\right)dx$$

$$= \left(\frac{2e\varepsilon}{L}\right)\int_0^L \sin^2\left(\frac{\pi x}{L}\right)x dx - \left(\frac{2e\varepsilon}{L}\right)\frac{L}{2}\int_0^L \sin^2\left(\frac{\pi x}{L}\right)dx$$

The first integral can be evaluated using integral equation (18) with $a = \frac{\pi}{L}$:

$$\int_0^L \sin^2(ax)x dx = \frac{x^2}{4} - \frac{x\sin(2ax)}{4a} - \frac{\cos(2ax)}{8a^2}\Bigg|_0^L = \frac{L^2}{4}$$

The second integral can be evaluated using integral equation (10) with $\theta = \frac{\pi x}{L}$ and $d\theta = \frac{\pi}{L}dx$:

$$\int_0^L \sin^2\left(\frac{\pi x}{L}\right)dx = \frac{L}{\pi}\int_0^\pi \sin^2\theta d\theta$$

$$\int_0^\pi \sin^2\theta d\theta = -\frac{1}{4}\sin(2\theta) + \frac{\theta}{2}\Bigg|_0^\pi = \frac{\pi}{2}$$

Making all of these appropriate substitutions we obtain:

$$E_{n=1}^{(1)} = \left(\frac{2e\varepsilon}{L}\right)\left(\frac{L^2}{4} - \frac{L}{2}\frac{L}{\pi}\frac{\pi}{2}\right) = 0$$

$$\Psi_{n=1}^{(1)} = \frac{\left\langle \Psi_{n=2}^{(0)}|e\varepsilon\left(x - \frac{L}{2}\right)|\Psi_{n=1}^{(0)}\right\rangle \Psi_{n=2}^{(0)}}{E_{n=1}^{(0)} - E_{n=2}^{(0)}}$$

$$\Psi_{n=1}^{(1)} = \frac{\left(\frac{2}{L}\right)\int_0^L \sin\left(\frac{2\pi x}{L}\right)e\varepsilon\left(x - \frac{L}{2}\right)\sin\left(\frac{\pi x}{L}\right)dx}{\frac{\hbar^2\pi^2}{2mL^2}\left(1^2 - 2^2\right)}\left(\frac{2}{L}\right)^{\frac{1}{2}}\sin\left(\frac{2\pi x}{L}\right)$$

The two integrals in the numerator need to be evaluated:

$$\int_0^L x\sin\left(\frac{2\pi x}{L}\right)\sin\left(\frac{\pi x}{L}\right)dx, \text{ and } \int_0^L \sin\left(\frac{2\pi x}{L}\right)\sin\left(\frac{\pi x}{L}\right)dx.$$

Using trigonometric identity (20), the integral

$\int x\cos(ax)dx = \frac{1}{a^2}\cos(ax) + \frac{x}{a}\sin(ax)$, and the integral $\int \cos(ax)dx = \frac{1}{a}\sin(ax)$,

we obtain the following:

$$\int_0^L \sin\left(\frac{2\pi x}{L}\right)\sin\left(\frac{\pi x}{L}\right)dx = \frac{1}{2}\left[\int_0^L \cos\left(\frac{\pi x}{L}\right)dx - \int_0^L \cos\left(\frac{3\pi x}{L}\right)dx\right]$$

$$= \frac{1}{2}\left[\frac{L}{\pi}\sin\left(\frac{\pi x}{L}\right)\Bigg|_0^L - \frac{L}{3\pi}\sin\left(\frac{3\pi x}{L}\right)\Bigg|_0^L\right] = 0$$

$$\int_0^L x\sin\left(\frac{2\pi x}{L}\right)\sin\left(\frac{\pi x}{L}\right)dx = \frac{1}{2}\left[\int_0^L x\cos\left(\frac{\pi x}{L}\right)dx - \int_0^L x\cos\left(\frac{3\pi x}{L}\right)dx\right]$$

$$= \frac{1}{2}\left[\left(\frac{L^2}{\pi^2}\cos\left(\frac{\pi x}{L}\right) + \frac{Lx}{\pi}\sin\left(\frac{\pi x}{L}\right)\right)\Big|_0^L - \left(\frac{L^2}{9\pi^2}\cos\left(\frac{3\pi x}{L}\right) + \frac{Lx}{3\pi}\sin\left(\frac{3\pi x}{L}\right)\right)\Big|_0^L\right]$$

$$= \frac{L^2}{2\pi^2}(\cos(\pi) - \cos(0)) + \frac{L^2}{2\pi}\sin(\pi) - 0 - \frac{L^2}{18\pi^2}(\cos(3\pi) - \cos(0)) - \frac{L^2}{6\pi}\sin(3\pi) + 0$$

$$= \frac{-2L^2}{2\pi^2} - \frac{-2L^2}{18\pi^2} = \frac{L^2}{9\pi^2} - \frac{L^2}{\pi^2} = -\frac{8L^2}{9\pi^2}$$

Making all of these appropriate substitutions we obtain:

$$\Psi^{(1)}_{n=1} = \frac{\left(\frac{2}{L}\right)(e\varepsilon)\left(-\frac{8L^2}{9\pi^2} - \frac{L}{2}(0)\right)}{\frac{-3\hbar^2\pi^2}{2mL^2}}\left(\frac{2}{L}\right)^{\frac{1}{2}}\sin\left(\frac{2\pi x}{L}\right)$$

$$\Psi^{(1)}_{n=1} = \frac{32mL^3e\varepsilon}{27\hbar^2\pi^4}\left(\frac{2}{L}\right)^{\frac{1}{2}}\sin\left(\frac{2\pi x}{L}\right)$$

Crudely sketching $\Psi^{(0)}_{n=1} + \Psi^{(1)}_{n=1}$ gives:

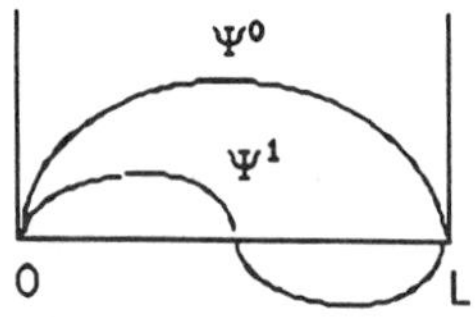

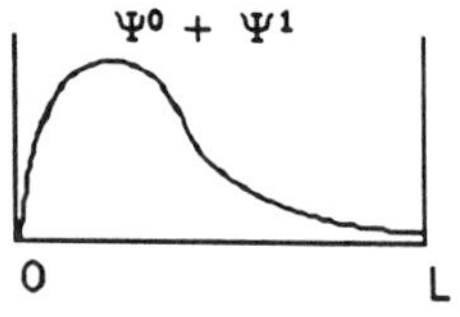

Note that the electron density has been pulled to the left side of the box by the external field!

b. $\mu_{\text{induced}} = -e\int \Psi^*\left(x - \frac{L}{2}\right)\Psi dx$, where, $\Psi = \left(\Psi^{(0)}_1 + \Psi^{(1)}_1\right)$.

$$\mu_{induced} = -e\int_0^L \left(\Psi^{(0)}_1 + \Psi^{(1)}_1\right)^*\left(x - \frac{L}{2}\right)\left(\Psi^{(0)}_1 + \Psi^{(1)}_1\right)dx$$

$$= -e\int_0^L \Psi^{(0)*}_1\left(x - \frac{L}{2}\right)\Psi^{(0)}_1 dx - e\int_0^L \Psi^{(0)*}_1\left(x - \frac{L}{2}\right)\Psi^{(1)}_1 dx$$

$$-e\int_0^L \Psi^{(1)*}_1\left(x - \frac{L}{2}\right)\Psi^{(0)}_1 dx - e\int_0^L \Psi^{(1)*}_1\left(x - \frac{L}{2}\right)\Psi^{(1)}_1 dx$$

The first integral is zero (see the evaluation of this integral for $E^{(1)}_1$ above in part a). The fourth integral is neglected since it is proportional to ε^2. The second and third integrals are the same and are combined:

$$\mu_{\text{induced}} = -2e \int_0^L \Psi_1^{(0)*}\left(x - \frac{L}{2}\right)\Psi_1^{(1)}dx$$

Substituting

$$\Psi_1^{(0)} = \left(\frac{2}{L}\right)^{\frac{1}{2}} \sin\left(\frac{\pi x}{L}\right) \text{and } \Psi_1^{(1)} = \frac{32mL^3e\varepsilon}{27\hbar^2\pi^4}\left(\frac{2}{L}\right)^{\frac{1}{2}} \sin\left(\frac{2\pi x}{L}\right),$$

we obtain:

$$\mu_{\text{induced}} = -2e\,\frac{32mL^3e\varepsilon}{27\hbar^2\pi^4}\left(\frac{2}{L}\right)\int_0^L \sin\left(\frac{\pi x}{L}\right)\left(x - \frac{L}{2}\right)\sin\left(\frac{2\pi x}{L}\right)dx$$

These integrals are familiar from part a:

$$\mu_{\text{induced}} = -2e\,\frac{32mL^3e\varepsilon}{27\hbar^2\pi^4}\left(\frac{2}{L}\right)\left(-\frac{8L^2}{9\pi^2}\right)$$

$$\mu_{\text{induced}} = \frac{mL^4e^2\varepsilon}{\hbar^2\pi^6}\,\frac{2^{10}}{3^5}$$

c. $\alpha = \left(\frac{\partial\mu}{\partial\varepsilon}\right)_{\varepsilon=0} = \frac{mL^4e^2}{\hbar^2\pi^6}\,\frac{2^{10}}{3^5}$

The larger the box (molecule), the more polarizable the electron density.

SECTION

2

SIMPLE MOLECULAR ORBITAL THEORY

In this section, the conceptual framework of molecular orbital theory is developed. Applications are presented and problems are given and solved within qualitative and semi-empirical models of electronic structure. *Ab Initio* approaches to these same matters, whose solutions require the use of digital computers, are treated later in section 6. Semi-empirical methods, most of which also require access to a computer, are treated in this section and in appendix F.

Unlike most texts on molecular orbital theory and quantum mechanics, this text treats polyatomic molecules before linear molecules before atoms. The finite point-group symmetry (appendix E provides an introduction to the use of point group symmetry) that characterizes the orbitals and electronic states of non-linear polyatomics is more straightforward to deal with because fewer degeneracies arise. In turn, linear molecules, which belong to an axial rotation group, possess fewer degeneracies (e.g., π orbitals or states are no more degenerate than δ, ϕ, or γ orbitals or states; all are doubly degenerate) than atomic orbitals and states (e.g., p orbitals or states are 3-fold degenerate, d's are 5-fold, etc.). Increased orbital degeneracy, in turn, gives rise to more states that can arise from a given orbital occupancy (e.g., the $2p^2$ configuration of the C atom yields fifteen states, the π^2 configuration of the NH molecule yields six, and the $\pi\pi^*$ configuration of ethylene gives four states). For these reasons, it is more straightforward to treat low-symmetry cases (i.e., non-linear polyatomic molecules) first and atoms last.

It is recommended that the reader become familiar with the point-group symmetry tools developed in appendix E before proceeding with this section. In particular, it is important to know how to label atomic orbitals as well as the various hybrids that can be formed from them according to the irreducible representations of the molecule's point group and how to construct symmetry adapted combinations of atomic, hybrid, and molecular orbitals using projection operator methods. If additional material on group theory is needed, Cotton's book on this subject is very good and provides many excellent chemical applications.

CHAPTER

4 Valence atomic orbitals on neighboring atoms combine to form bonding, non-bonding and antibonding molecular orbitals.

I. ATOMIC ORBITALS

In section 1 the Schrödinger equation for the motion of a single electron moving about a nucleus of charge Z was explicitly solved. The energies of these orbitals relative to an electron infinitely far from the nucleus with zero kinetic energy were found to depend strongly on Z and on the principal quantum number n, as were the radial "sizes" of these hydrogenic orbitals. Closed analytical expressions for the r, θ, and ϕ dependence of these orbitals are given in appendix B. The reader is advised to also review this material before undertaking study of this section.

A. Shapes

Shapes of atomic orbitals play central roles in governing the types of directional bonds an atom can form.

All atoms have sets of bound and continuum *s*, *p*, *d*, *f*, *g*, etc. orbitals. Some of these orbitals may be unoccupied in the atom's low energy states, but they are still present and able to accept electron density if some physical process (e.g., photon absorption, electron attachment, or Lewis-base donation) causes such to occur. For example, the Hydrogen atom has 1*s*, 2*s*, 2*p*, 3*s*, 3*p*, 3*d*, etc. orbitals. Its negative ion H^- has states that involve $1s2s$, $2p^2$, $3s^2$, $3p^2$, etc. orbital occupancy. Moreover, when an H atom is placed in an external electronic field, its charge density polarizes in the direction of the field. This polarization can be described in terms of the orbitals of the isolated atom being combined to yield distorted orbitals (e.g., the 1*s* and 2*p* orbitals can "mix" or combine to yield *sp* hybrid orbitals, one directed toward increasing field and the other directed in the opposite direction). Thus in many situations it is important to keep in mind that each atom has a full set of orbitals available to it even if some of these orbitals are not occupied in the lowest-energy state of the atom.

B. Directions

Atomic orbital directions also determine what directional bonds an atom will form.

Each set of p orbitals has three distinct directions or three different angular momentum m-quantum numbers as discussed in appendix G. Each set of d orbitals has five distinct directions or m-quantum numbers, etc; s orbitals are unidirectional in that they are spherically symmetric, and have only $m = 0$. Note that the degeneracy of an orbital $(2l + 1)$, which is the number of distinct spatial orientations or the number of m-values, grows with the angular momentum quantum number l of the orbital without bound.

It is because of the *energy degeneracy* within a set of orbitals, that these distinct directional orbitals (e.g., x, y, z for p orbitals) may be combined to give new orbitals which no longer possess specific spatial directions but which have specified angular momentum characteristics. The act of combining these degenerate orbitals does not change their energies. For example, the $2^{-1/2}(p_x + ip_y)$ and $2^{-1/2}(p_x - ip_y)$ combinations no longer point along the x and y axes, but instead correspond to specific angular momenta $(+1\hbar$ and $-1\hbar)$ about the z axis. The fact that they are angular momentum eigenfunctions can be seen by noting that the x and y orbitals contain ϕ dependences of $\cos(\phi)$ and $\sin(\phi)$, respectively. Thus the above combinations contain $\exp(i\phi)$ and $exp(-i\phi)$, respectively. The sizes, shapes, and directions of a few s, p, and d orbitals are illustrated below (the light and dark areas represent positive and negative values, respectively).

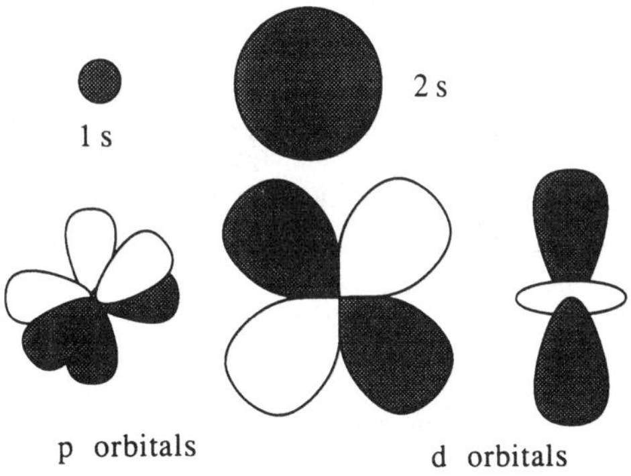

C. Sizes and Energies

Orbital energies and sizes go hand-in-hand; small "tight" orbitals have large electron binding energies (i.e., low energies relative to a detached electron). For orbitals on neighboring atoms to have large (and hence favorable to bond formation) overlap, the two orbitals should be of comparable size and hence of similar electron binding energy.

The size (e.g., average value or expectation value of the distance from the atomic nucleus to the electron) of an atomic orbital is determined primarily by its principal quantum number n and by the strength of the potential attracting an electron in this orbital to the atomic center (which has some l-dependence too). The energy (with negative energies corresponding to bound states in which the electron is attached to the atom with positive binding energy and positive energies corresponding to unbound scattering states) is also determined by n and by the electrostatic potential produced by the nucleus and by the other electrons. Each atom has an infinite set of orbitals of each l quantum number ranging from those with low energy and small size to those with higher energy and larger size.

Atomic orbitals are solutions to an orbital-level Schrödinger equation in which an electron moves in a potential energy field provided by the nucleus and all the other electrons. Such one-electron Schrödinger equations are discussed, as they pertain to qualitative and semi-empirical models of electronic structure in appendix F. The spherical symmetry of the one-electron potential appropriate to atoms and atomic ions is what makes sets of the atomic orbitals degenerate. Such degeneracies arise in molecules too, but the extent of degeneracy is lower because the molecule's nuclear coulomb and electrostatic potential energy has lower symmetry than in the atomic case. As will be seen, it is the symmetry of the potential experienced by an electron moving in the orbital that determines the kind and degree of orbital degeneracy which arises.

Symmetry operators leave the electronic Hamiltonian $\boldsymbol{H}$ invariant because the potential and kinetic energies are not changed if one applies such an operator R to the coordinates and momenta of *all* the electrons in the system. Because symmetry operations involve reflections through planes, rotations about axes, or inversions through points, the application of such an operation to a product such as $\boldsymbol{H}\Psi$ gives the product of the operation applied to each term in the original product. Hence, one can write:

$$R(\boldsymbol{H}\Psi) = (R\boldsymbol{H})(R\Psi).$$

Now using the fact that $\boldsymbol{H}$ is invariant to R, which means that $(R\boldsymbol{H}) = \boldsymbol{H}$, this result reduces to:

$$R(\boldsymbol{H}\Psi) = \boldsymbol{H}(R\Psi),$$

which says that R commutes with $\boldsymbol{H}$:

$$[R,\boldsymbol{H}] = 0.$$

Because symmetry operators commute with the electronic Hamiltonian, the wavefunctions that are eigenstates of $\boldsymbol{H}$ can be labeled by the symmetry of the point group of the molecule (i.e., those operators that leave $\boldsymbol{H}$ invariant). It is for this reason that one constructs symmetry-adapted atomic basis orbitals to use in forming molecular orbitals.

II. MOLECULAR ORBITALS

Molecular orbitals (mos) are formed by combining atomic orbitals (aos) of the constituent atoms. This is one of the most important and widely used ideas in quantum chemistry. Much of chemists' understanding of chemical bonding, structure, and reactivity is founded on this point of view.

When aos are combined to form mos, core, bonding, nonbonding, antibonding, and Rydberg molecular orbitals can result. The mos ϕ_i are usually expressed in terms of the constituent atomic orbitals χ_a in the linear-combination-of-atomic-orbital-molecular-orbital (LCAO-MO) manner:

$$\phi_i = \sum_a C_{ia}\chi_a.$$

The orbitals on one atom are orthogonal to one another because they are eigenfunctions of a hermitian operator (the atomic one-electron Hamiltonian) having different eigenvalues. However, those on one atom are not orthogonal to those on another atom because they are eigenfunctions of different operators (the one-electron Hamiltonia of the different atoms). Therefore, in practice, the primitive atomic orbitals must be orthogonalized to preserve maximum identity of each primitive orbital in the resultant orthonormalized orbitals before they can be used in the LCAO-MO process. This is both computationally expedient and conceptually useful. Throughout this book, the atomic orbitals (aos) will be assumed to consist of such orthonormalized primitive orbitals once the nuclei are brought into regions where the "bare" aos interact.

Sets of orbitals that are not orthonormal can be combined to form new orthonormal functions in many ways. One technique that is especially attractive when the original functions are orthonormal in the absence of "interactions" (e.g., at large interatomic distances in the case of atomic basis orbitals) is the so-called symmetric orthonormalization (SO) method. In this method, one first forms the so-called overlap matrix

$$S_{\mu\nu} = \langle\chi_\mu|\chi_\nu\rangle$$

for all functions χ_μ to be orthonormalized. In the atomic-orbital case, these functions include those on the first atom, those on the second, etc.

Since the orbitals belonging to the individual atoms are themselves orthonormal, the overlap matrix will contain, along its diagonal, blocks of unit matrices, one for each set of individual atomic orbitals. For example, when a carbon and oxygen atom, with their core $1s$ and valence $2s$ and $2p$ orbitals are combined to form CO, the 10×10 $S_{\mu,\nu}$ matrix will have two 5×5 unit matrices along its diagonal (representing the overlaps among the carbon and among the oxygen atomic orbitals) and a 5×5 block in its upper right and lower left quadrants. The latter block represents the overlaps $\langle\chi^{C_\mu}|\chi^{O_\nu}\rangle$ among carbon and oxygen atomic orbitals.

After forming the overlap matrix, the new orthonormal functions χ'_μ are defined as follows:

$$\chi'_\mu = \sum_\nu (S^{-1/2})_{\mu\nu}\chi_\nu.$$

As shown in appendix A, the matrix $S^{-1/2}$ is formed by finding the eigenvalues $\{\lambda_i\}$ and eigenvectors $\{V_{i\mu}\}$ of the S matrix and then constructing:

$$(S^{-1/2})_{\mu\nu} = \sum_i V_{i\mu}V_{i\nu}(\lambda_i)^{-1/2}.$$

The new functions $\{\chi'_\mu\}$ have the characteristic that they evolve into the original functions as the "coupling," as represented in the $S_{\mu,\nu}$ matrix's off-diagonal blocks, disappears.

It is important to keep in mind that the orbitals that one employs in formulating the fundamental bonding, non-bonding, and antibonding interactions in molecules must be made orthonormal. If they are constructed from the atomic orbitals of the constituent atoms, such a procedure must be applied to them before they can be used in most of the equations developed in this text (because these equations assume that the orbitals being used are orthonormal). Once the constituent atomic orbitals are in hand, the formation of various classes of molecular orbitals can begin.

A. Core Orbitals

Core orbitals describe the inner shells that are not intimately involved in the chemical bonding.

Core orbitals do not enter into significant bonding-antibonding, etc. interactions (except when the two atomic centers are compressed to very short internuclear distances) because they do not directly overlap with one another; only the outer (valence) orbitals interact directly. The energies of the core orbitals become affected secondarily through changes in the valence orbitals caused by chemical bonding. Because each orbital with electrons in it contributes to the potential energy experienced by the other orbitals (see below and appendix F), if the nature of the valence orbitals changes, the electrostatic potential experienced by the core (and other) orbitals is altered, as a result of which the core orbitals themselves can change. Such changes in core-orbital energies due to "environmental" effects (i.e., effects that take place within the valence orbitals) can be observed experimentally as *shifts* in photoelectron spectroscopy ionization energies compared to the "isolated-atom" value.

B. Valence Orbitals

Valence orbitals describe the aos and resultant mos that are intimately involved in the chemical bonding.

Valence orbitals on neighboring atoms are coupled by changes in the electrostatic potential due to the other atoms (coulomb attraction to the other nuclei and repulsions from electrons on the other atoms). These coupling potentials vanish when the atoms are far apart and become significant only when the valence orbitals overlap one another. In the most qualitative picture, such interactions are described in terms of off-diagonal Hamiltonian matrix elements (h_{ab}; see below and in appendix F) between pairs of atomic orbitals which interact (the diagonal elements h_{aa} represent the energies of the various orbitals and are related via Koopmans' theorem (see section 6, chapter 18.VII.B) to the ionization energy of the orbital). Such a matrix embodiment of the molecular orbital problem arises, as developed below and in appendix F, by using the above LCAO-MO expansion in a variational treatment of the one-electron Schrödinger equation appropriate to the mos $\{\phi_i\}$.

In the simplest two-center, two-valence-orbital case (which could relate, for example, to the Li_2 molecule's two $2s$ orbitals), this gives rise to a 2×2 matrix eigenvalue problem (h_{11}, h_{12}, h_{22}) with a low-energy mo ($E = (h_{11} + h_{22})/2 - 1/2[(h_{11} - h_{22})^2 + 4h_{12}^2]^{1/2}$) and a higher energy mo ($E = (h_{11} + h_{22})/2 + 1/2[(h_{11} - h_{22})^2 + 4h_{12}^2]^{1/2}$) corresponding to bonding and antibonding orbitals (because their energies lie below and above the lowest and highest interacting atomic orbital energies, respectively). The mos themselves are expressed $\phi_i = \Sigma C_{ia}\chi_a$ where the LCAO-MO coefficients C_{ia} are obtained from the normalized eigenvectors of the h_{ab} matrix. Note that the bonding-antibonding orbital energy splitting depends on h_{ab}^2 and on the energy difference $(h_{aa} - h_{bb})$; the best bonding (and worst antibonding) occur when two orbitals couple strongly (have large h_{ab}) and are similar in energy ($h_{aa} \cong h_{bb}$).

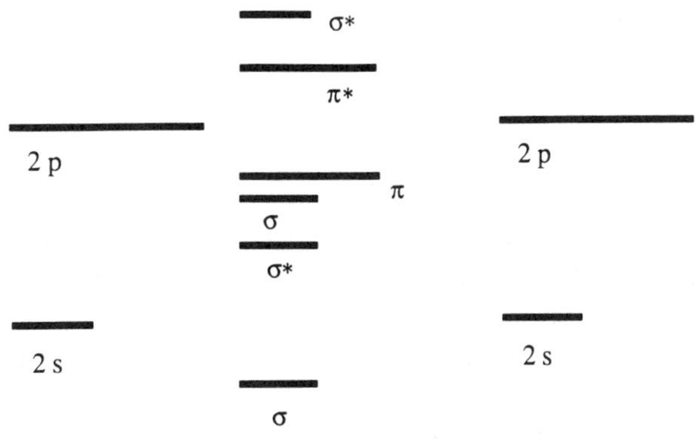

Homonuclear Bonding With 2s and 2p Orbitals

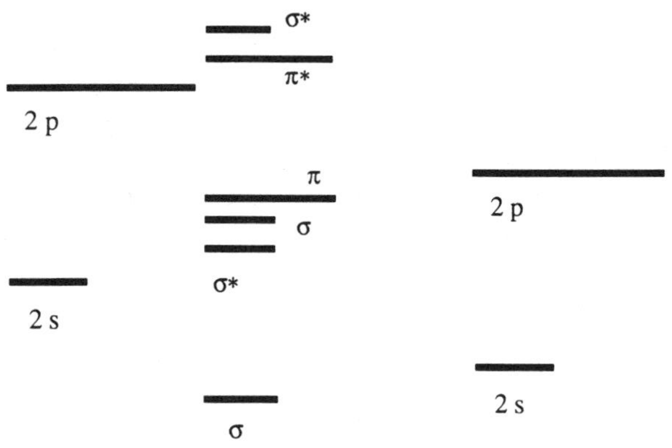

Heteronuclear Bonding With 2s and 2p Orbitals

In both the homonuclear and heteronuclear cases depicted above, the energy ordering of the resultant mos depends upon the energy ordering of the constituent aos as well as the strength of the bonding-antibonding interactions among the aos. For example, if the $2s$-$2p$ atomic orbital energy splitting is large compared with the interaction matrix elements coupling orbitals on neighboring atoms $h_{2s,2s}$ and $h_{2p,2p}$, then the ordering shown above will result. On the other hand, if the $2s$-$2p$ splitting is small, the two $2s$ and two $2p$ orbitals can all participate in the formation of the four σ mos. In this case, it is useful to think of the atomic $2s$ and $2p$ orbitals forming *sp* hybrid orbitals with each atom having one hybrid directed toward the other atom and one hybrid directed away from the other atom. The resultant pattern of four σ mos will involve one bonding orbital (i.e., an in-phase combination of two *sp* hybrids), two non-bonding orbitals (those directed away from the other atom) and one antibonding orbital (an out-of-phase combination of two *sp* hybrids). Their energies will be ordered as shown in the figure below.

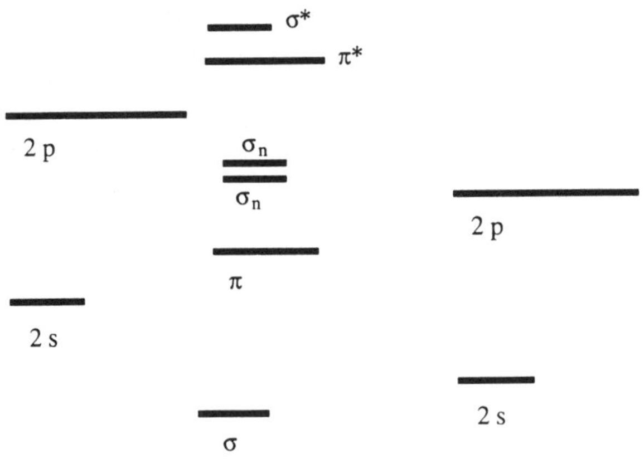

Here σ_n is used to denote the non-bonding σ-type orbitals and σ, σ^*, π, and π^* are used to denote bonding and antibonding σ- and π-type orbitals.

Notice that the total number of σ orbitals arising from the interaction of the $2s$ and $2p$ orbitals is equal to the number of aos that take part in their formation. Notice also that this is true regardless of whether one thinks of the interactions involving bare 2s and 2p atomic orbitals or hybridized orbitals. The only advantage that the hybrids provide is that they permit one to foresee the fact that two of the four mos must be non-bonding because two of the four hybrids are directed away from all other valence orbitals and hence can not form bonds. In all such qualitative mo analyses, the final results (i.e., how many mos there are of any given symmetry) will *not* depend on whether one thinks of the interactions involving atomic or hybrid orbitals. However, it is often easier to "guess" the bonding, non-bonding, and antibonding nature of the resultant mos when thought of as formed from hybrids because of the directional properties of the hybrid orbitals.

C. Rydberg Orbitals

It is essential to keep in mind that all atoms possess "excited'"orbitals that may become involved in bond formation if one or more electrons occupies these orbitals. Whenever aos with principal quantum number one or more unit higher than that of the conventional aos becomes involved in bond formation, Rydberg mos are formed.

Rydberg orbitals (i.e., very diffuse orbitals having principal quantum numbers higher than the atoms' valence orbitals) can arise in molecules just as they do in atoms. They do not usually give rise to bonding and antibonding orbitals because the valence-orbital interactions bring the atomic centers so close together that the Rydberg orbitals of each atom subsume both atoms. Therefore as the atoms are brought together, the atomic Rydberg orbitals usually pass through the internuclear distance region where they experience (weak) bonding-antibonding interactions all the way to much shorter distances at which they have essentially reached their united-atom limits. As a result, molecular Rydberg orbitals are molecule-centered and display little, if any, bonding or antibonding character. They are usually labeled with principal quantum numbers beginning one higher than the highest n value of the constituent atomic valence orbitals, although they are sometimes labeled by the n quantum number to which they correlate in the united-atom limit.

An example of the interaction of $3s$ Rydberg orbitals of a molecule whose $2s$ and $2p$ orbitals are the valence orbitals and of the evolution of these orbitals into united-atom orbitals is given below.

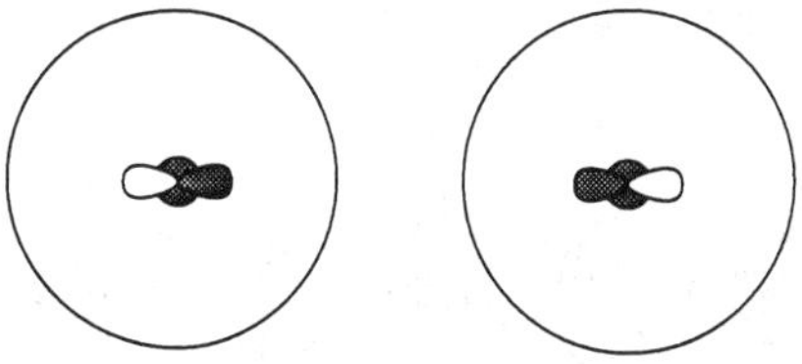

2s and 2p Valence Orbitals and 3s Rydberg Orbitals For Large R Values

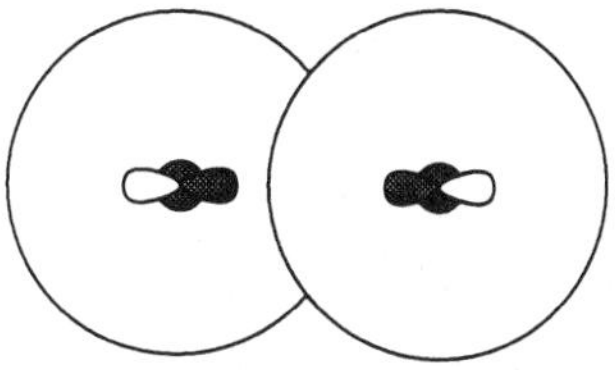

Overlap of the Rydberg Orbitals Begins

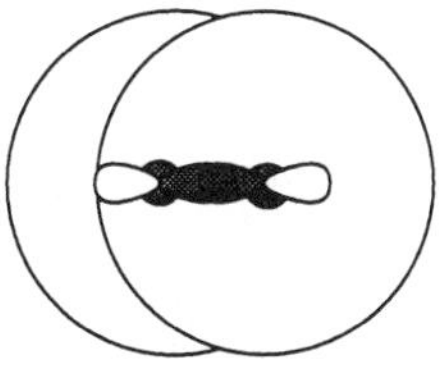

Rydberg Overlap is Strong and Bond Formation Occurs

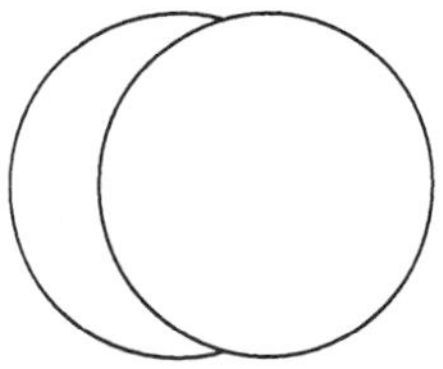

The In-Phase (3s + 3s) Combination of Rydberg Orbitals Correlates to an s-type Orbital of the United Atom

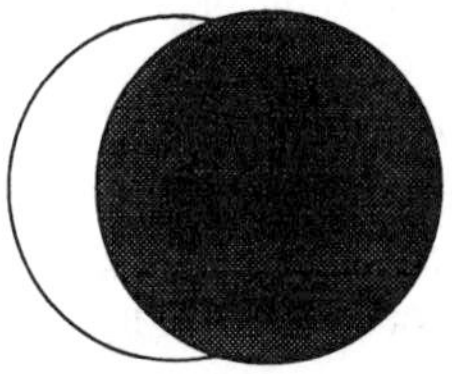

The Out-of-Phase Combination of Rydberg Orbitals (3s - 3s) Correlates to a p-type United-Atom Orbital

D. Multicenter Orbitals

If aos on one atom overlap aos on more than one neighboring atom, mos that involve amplitudes on three or more atomic centers can be formed. Such mos are termed delocalized or multicenter mos.

Situations in which more than a pair of orbitals interact can, of course, occur. Three-center bonding occurs in Boron hydrides and in carbonyl bridge bonding in transition metal complexes as well as in delocalized conjugated π orbitals common in unsaturated organic hydrocarbons. The three p_π orbitals on the allyl radical (considered in the absence of the underlying σ orbitals) can be described qualitatively in terms of three p_π aos on the three carbon atoms. The couplings h_{12} and h_{23} are equal (because the two CC bond lengths are the same) and h_{13} is approximated as 0 because orbitals 1 and 3 are too far away to interact. The result is a 3×3 secular matrix (see below and in appendix F):

$$\begin{matrix} h_{11} & h_{12} & 0 \\ h_{21} & h_{22} & h_{23} \\ 0 & h_{32} & h_{33} \end{matrix}$$

whose eigenvalues give the molecular orbital energies and whose eigenvectors give the LCAO-MO coefficients C_{ia}.

This 3×3 matrix gives rise to a bonding, a non-bonding and an antibonding orbital (see the figure below). Since all of the h_{aa} are equal and $h_{12} = h_{23}$, the resultant orbital energies are: $h_{11} + \sqrt{2}h_{12}$, h_{11}, and $h_{11} - \sqrt{2}h_{12}$, and the respective LCAO-MO coefficients C_{ia} are (0.50, 0.707, 0.50), (0.707, 0.00, –0.707), and (0.50, –0.707, 0.50). Notice that the sign (i.e., phase) relations of the bonding orbital are such that overlapping orbitals interact constructively, whereas for the antibonding orbital they interact out of phase. For the nonbonding orbital, there are no interactions because the central C orbital has zero amplitude in this orbital and only h_{12} and h_{23} are non-zero.

Allyl System π Orbitals

E. Hybrid Orbitals

It is sometimes convenient to combine aos to form hybrid orbitals that have well defined directional character and to then form mos by combining these hybrid orbitals. This recombination of aos to form hybrids is never *necessary and never provides any information that could be achieved in its absence. However, forming hybrids often allows one to focus on those interactions among directed orbitals on neighboring atoms that are most important.*

When atoms combine to form molecules, the molecular orbitals can be thought of as being constructed as linear combinations of the constituent atomic orbitals. This clearly is the only reasonable picture when each atom contributes only one orbital to the particular interactions being considered (e.g., as each Li atom does in Li_2 and as each C atom does in the π orbital aspect of the allyl system). However, when an atom uses more than one of its valence orbitals within particular bonding, non-bonding, or antibonding interactions, it is sometimes useful to combine the constituent atomic orbitals into hybrids and to then use the hybrid orbitals to describe the interactions. As stated above, the directional nature of hybrid orbitals often makes it more straightforward to "guess" the bonding, non-bonding, and antibonding nature of the resultant mos. It should be stressed, however, that exactly the same quantitative results are obtained if one forms mos from primitive aos or from hybrid orbitals; the hybrids span exactly the same space as the original aos and can therefore contain no additional information. This point is illustrated in chapter 5 when the H_2O and N_2 molecules are treated in both the primitive ao and hybrid orbital bases.

CHAPTER

5 Molecular orbitals possess specific topology, symmetry, and energy-level patterns.

In this chapter the symmetry properties of atomic, hybrid, and molecular orbitals are treated. It is important to keep in mind that *both symmetry and characteristics of orbital energetics and bonding "topology,"* as embodied in the orbital energies themselves and the interactions (i.e., $h_{j,k}$ values) among the orbitals, are involved in determining the pattern of molecular orbitals that arise in a particular molecule.

I. ORBITAL INTERACTION TOPOLOGY

The pattern of mo energies can often be "guessed" by using qualitative information about the energies, overlaps, directions, and shapes of the aos that comprise the mos.

The orbital interactions determine how many and which mos will have low (bonding), intermediate (non-bonding), and higher (antibonding) energies, with all energies viewed relative to those of the constituent atomic orbitals. The **general patterns** that are observed in most compounds can be summarized as follows:

1. If the energy splittings among a given atom's aos with the same principal quantum number are small, hybridization can easily occur to produce hybrid orbitals that are directed toward (and perhaps away from) the other atoms in the molecule. In the first-row elements (Li, Be, B, C, N, O, and F), the $2s$-$2p$ splitting is small, so hybridization is common. In contrast, for Ca, Ga, Ge, As, and Br it is less common, because the $4s$-$4p$ splitting is larger. Orbitals directed toward other atoms can form bonding and antibonding mos; those directed toward no other atoms will form nonbonding mos.

2. In attempting to gain a qualitative picture of the electronic structure of any given molecule, it is advantageous to begin by hybridizing the aos of those atoms which contain more than one ao in their valence shell. Only those aos that are not involved in π-orbital interactions should be so hybridized.

3. Atomic or hybrid orbitals that are not directed in a σ-interaction manner toward other aos or hybrids on neighboring atoms can be involved in π-interactions or in nonbonding interactions.

4. Pairs of aos or hybrid orbitals on neighboring atoms directed toward one another interact to produce bonding and antibonding orbitals. The more the bonding orbital lies below the lower-energy ao or hybrid orbital involved in its formation, the higher the antibonding orbital lies above the higher-energy ao or hybrid orbital.

For example, in formaldehyde, H_2CO, one forms sp^2 hybrids on the C atom; on the O atom, either *sp* hybrids (with one *p* orbital "reserved" for use in forming the π and π^* orbitals and another *p* orbital to be used as a non-bonding orbital lying in the plane of the molecule) or sp^2 hybrids (with the remaining *p* orbital reserved for the π and π^* orbitals) can be used. The H atoms use their 1*s* orbitals since hybridization is not feasible for them. The C atom clearly uses its sp^2 hybrids to form two CH and one CO σ bonding-antibonding orbital pairs.

The O atom uses one of its *sp* or sp^2 hybrids to form the CO σ bond and antibond. When *sp* hybrids are used in conceptualizing the bonding, the other *sp* hybrid forms a lone pair orbital directed away from the CO bond axis; one of the atomic *p* orbitals is involved in the CO π and π^* orbitals, while the other forms an in-plane non-bonding orbital. Alternatively, when sp^2 hybrids are used, the two sp^2 hybrids that do not interact with the C-atom sp^2 orbital form the two non-bonding orbitals. Hence, the final picture of bonding, non-bonding, and antibonding orbitals does not depend on which hybrids one uses as intermediates.

As another example, the 2*s* and 2*p* orbitals on the two N atoms of N_2 can be formed into pairs of *sp* hybrids on each N atom plus a pair of p_π atomic orbitals on each N atom. The *sp* hybrids directed toward the other N atom give rise to bonding σ and antibonding σ^* orbitals, and the *sp* hybrids directed away from the other N atom yield nonbonding σ orbitals. The p_π orbitals, which consist of 2*p* orbitals on the N atoms directed perpendicular to the N-N bond axis, produce bonding π and antibonding π^* orbitals.

5. In general, σ interactions for a given pair of atoms interacting are stronger than π interactions (which, in turn, are stronger than δ interactions, etc.) for any given sets (i.e., principal quantum number) of aos that interact. Hence, σ bonding orbitals (originating from a given set of aos) lie below π bonding orbitals, and σ^* orbitals lie above π^* orbitals that arise from the same sets of aos. In the N_2 example, the σ bonding orbital formed from the two *sp* hybrids lies below the π bonding orbital, but the π^* orbital lies below the σ^* orbital. In the H_2CO example, the two CH and the one CO bonding orbitals have low energy; the CO π bonding orbital has the next lowest energy; the two O-atom non-bonding orbitals have intermediate energy; the CO π^* orbital has somewhat higher energy; and the two CH and one CO antibonding orbitals have the highest energies.

6. If a given ao or hybrid orbital interacts with or is coupled to orbitals on more than a single neighboring atom, multicenter bonding can occur. For example, in the allyl radical the central carbon atom's p_π orbital is coupled to the p_π orbitals on both neighboring atoms; in linear Li_3, the central Li atom's $2s$ orbital interacts with the $2s$ orbitals on both terminal Li atoms; in triangular Cu_3, the $4s$ orbitals on each Cu atom couple to each of the other two atoms' $4s$ orbitals.

7. Multicenter bonding that involves "linear" chains containing N atoms (e.g., as in conjugated polyenes or in chains of Cu or Na atoms for which the valence orbitals on one atom interact with those of its neighbors on both sides) gives rise to mo energy patterns in which there are $N/2$ (if N is even) or $N/2 - 1$ non-degenerate bonding orbitals and the same number of antibonding orbitals (if N is odd, there is also a single non-bonding orbital).

8. Multicenter bonding that involves "cyclic" chains of N atoms (e.g., as in cyclic conjugated polyenes or in rings of Cu or Na atoms for which the valence orbitals on one atom interact with those of its neighbors on both sides and the entire net forms a closed cycle) gives rise to mo energy patterns in which there is a lowest non-degenerate orbital and then a progression of doubly degenerate orbitals. If N is odd, this progression includes $(N-1)/2$ levels; if N is even, there are $(N-2)/2$ doubly degenerate levels and a final non-degenerate highest orbital. These patterns and those that appear in linear multicenter bonding are summarized in the figures shown below.

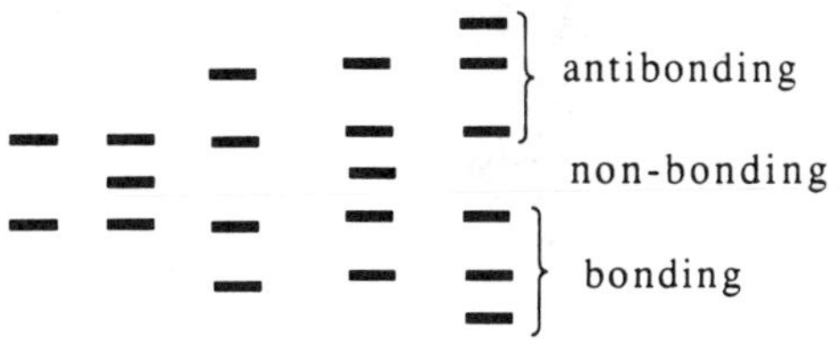

Pattern for Linear Multicenter
Bonding Situation: N=2, 3, ..6

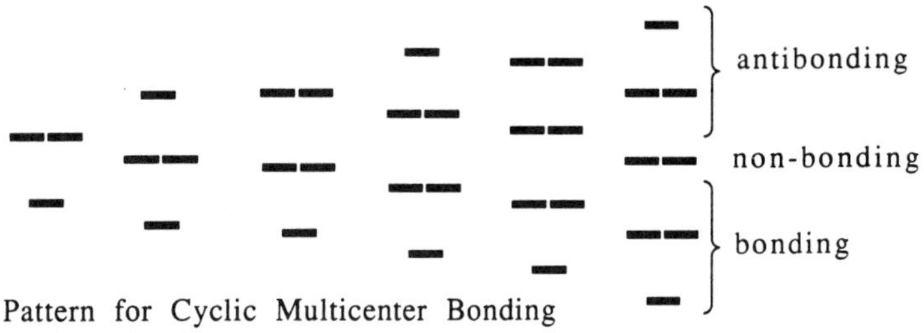

Pattern for Cyclic Multicenter Bonding
N= 3, 4, 5, ...8

9. In extended systems such as solids, atom-based orbitals combine as above to form so-called "bands" of molecular orbitals. These bands are continuous rather than discrete as in the above cases involving small polyenes. The energy 'spread' within a band depends on the overlap among the atom-based orbitals that form the band; large overlap gives rise to a large band width, while small overlap produces a narrow band. As one moves from the bottom (i.e., the lower energy part) of a band to the top, the number of nodes in the corresponding band orbital increases, as a result of which its bonding nature decreases. In the figure shown below, the bands of a metal such as Ni (with $3d$, $4s$, and $4p$ orbitals) is illustrated. The d-orbital band is narrow because the $3d$ orbitals are small and hence do not overlap appreciably; the 4*s* and 4*p* bands are wider because the larger 4*s* and 4*p* orbitals overlap to a greater extent. The *d*-band is split into σ, π, and δ components corresponding to the nature of the overlap interactions among the constituent atomic *d* orbitals. Likewise, the *p*-band is split into σ and π components. The widths of the σ components of each band are larger than those of the π components because the corresponding σ overlap interactions are stronger. The intensities of the bands at energy *E* measure the densities of states at that *E*. The total integrated intensity under a given band is a measure of the total number of atomic orbitals that form the band.

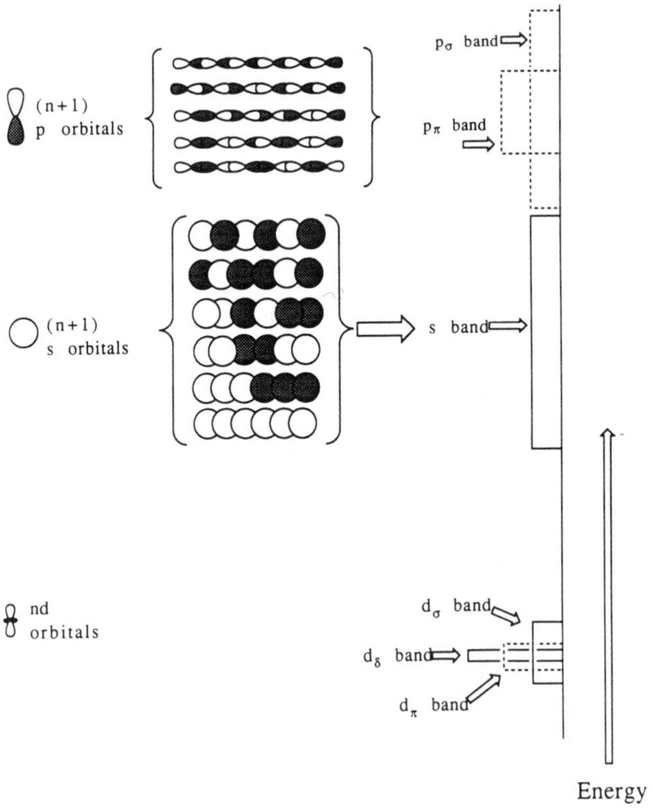

II. ORBITAL SYMMETRY

Symmetry provides additional quantum numbers or labels to use in describing the mos. Each such quantum number further sub-divides the collection of all mos into sets that have vanishing Hamiltonian matrix elements among members belonging to different sets.

Orbital interaction "*topology*" as discussed above plays a most- important role in determining the orbital energy level patterns of a molecule. *Symmetry* also comes into play but in a different manner. Symmetry can be used to characterize the core, bonding, non-bonding, and antibonding molecular orbitals. Much of this chapter is devoted to how this can be carried out in a systematic manner. Once the various mos have been labeled according to symmetry, it may be possible to recognize additional degeneracies that may not have been apparent on the basis of orbital-interaction considerations alone. Thus, topology provides the basic energy ordering pattern and then symmetry enters to identify additional degeneracies.

For example, the three NH bonding and three NH antibonding orbitals in NH_3, when symmetry adapted within the C_{3v} point group, cluster into a_1 and e mos as shown in the figure below. The N-atom localized non-bonding lone pair orbital and the N-atom 1s core orbital also belong to a_1 symmetry.

In a second example, the three CH bonds, three CH antibonds, CO bond and antibond, and three O-atom non-bonding orbitals of the methoxy radical H_3C-O also cluster into a_1 and e orbitals as shown below. In these cases, point group symmetry allows one to identify degeneracies that may not have been apparent from the structure of the orbital interactions alone.

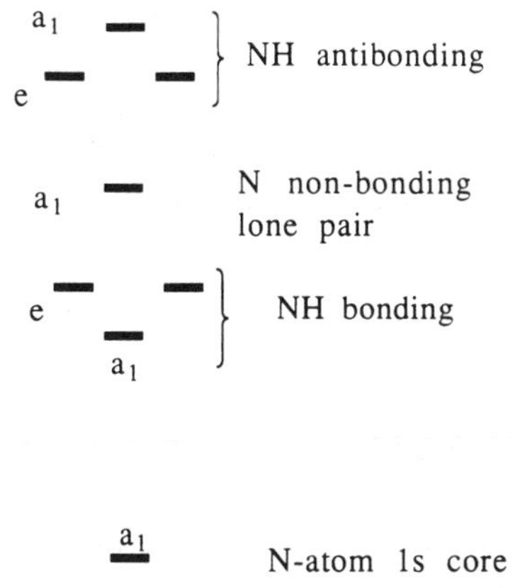

Orbital Character and Symmetry in NH_3

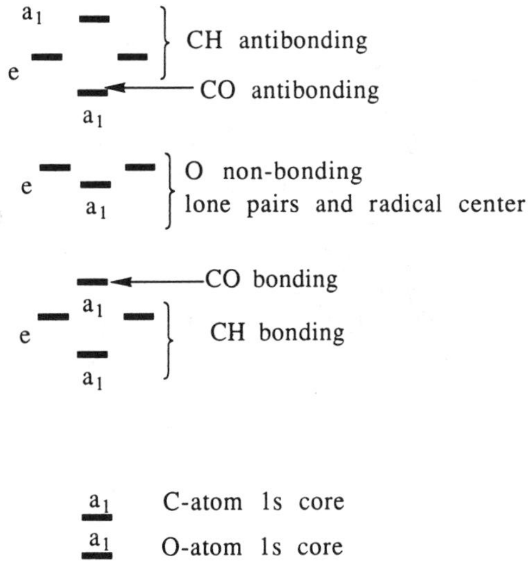

It should be noted that the splitting of the three O-atom non-bonding orbitals into a_1 and e symmetries results no matter how one conceptualizes the formation of these orbitals. Thinking of sp^3 orbitals on the O atom, one uses one of these hybrids to form the CO σ bond and the remaining three sp^3 hybrids can be combined, just as the three CH bonds and antibonds are, to form a_1 and e orbitals. Alternatively, if one uses sp hybrids on the O atom, the sp hybrid that is directed away from the CO bond axis is an a_1 non-bonding orbital, while the two O-atom p orbitals perpendicular to the CO bond axis form the e non-bonding orbitals. Finally, if one uses sp^2 hybrids on the O atom, the two hybrids that are not involved in the CO bonding can be symmetry adapted. Doing so yields one a_1 orbital ((the $sp^2 + sp^2$) combination of the two hybrids) and another combination ($sp^2 - sp^2$) that, together with the remaining p orbital, form non-bonding e orbitals.

Having introduced the concepts of orbital interaction topology and symmetry as tools for analyzing molecular orbital bonding, non-bonding, and antibonding characteristics, it is time to proceed to develop in a systematic manner the tools needed to carry out symmetry analyses of atomic, hybrid, and molecular orbitals. If the material contained in appendix E has not yet been mastered, it is strongly suggested that the reader do so before proceeding in this chapter. The tools provided by group theory are very powerful; every student of quantum chemistry should be fully skilled in their use.

A. Non-Linear Polyatomic Molecules

The symmetry of nonlinear molecules is easier to handle than those of linear molecules or of atoms. The more symmetry a species has, the more quantum numbers arise to label the states, but also the more complicated is the symmetry analysis.

The symmetry of the electrostatic potential caused by the nuclei and the other electrons gives rise to symmetry in the resultant molecular orbitals. For polyatomic molecules, this symmetry is cast in terms of conventional point group symmetry; for linear molecules, the symmetry of the axial rotation group is relevant; and for atoms, for which full spherical symmetry is present, the full rotation group must be used.

To understand why the electronic Hamiltonian of a molecule is invariant under the symmetry operations (R) of that molecule (i.e., why $[\boldsymbol{H},R] = 0$), it may be helpful to consider, as an example, the electronic Hamiltonian of the CO molecule. This Hamiltonian consists of a kinetic energy part T and a potential energy part V: $H = T + V$. Each electron (see the figure below) has a momentum vector $\boldsymbol{p}_i$ which contributes to T in the form $p_i^2/2m_e$. The position $\boldsymbol{r}_i$ of the electron enters into V which consists of the sum of the electrostatic attractions of this electron to all of the nuclei in the molecule plus its (repulsive) electrostatic interactions with the other electrons. As the figure below illustrates, the three operations (reflection of the electron's coordinates and momentum through a σ_v plane, inversion of the electron's coordinates and momentum through the center of the molecule, and rotation of the electron by any amount (γ) about the CO bond axis) alter the direction of *each* electron's momentum vector $\boldsymbol{p}_i$ but not the magnitude of p_i and hence not the kinetic energy of the electrons. Of these three operations, σ_v and C_γ also leave the elecrostatic potential experienced by the electrons unchanged because they do not alter the distances (which is what V depends on) among the electrons and between the nuclei and the electrons. In contrast, the inversion operation i changes the $e^- - C$ and $e^- - O$ distances and, therefore, does not preserve V. For these reasons, σ_v and C_γ commute with H and are valid symmetry operations for CO; i does not commute with H and is not a valid symmetry operation (see appendix E) for CO.

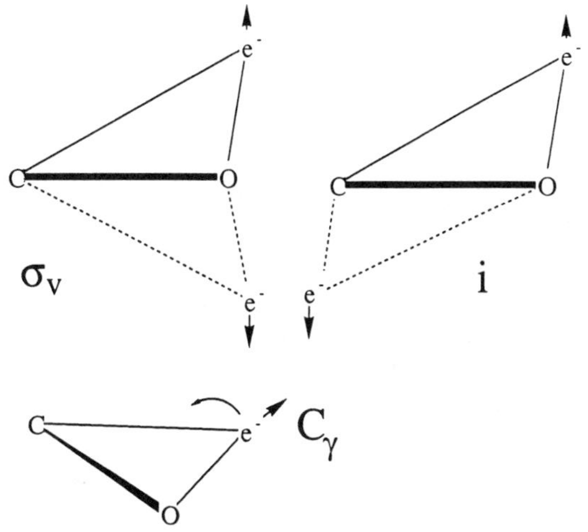

Labeling orbitals by their symmetry is useful primarily because the Hamiltonian can only couple orbitals of the same symmetry (because the Hamiltonian is invariant to the symmetry operations of the molecular or atomic system; that is, it commutes with all symmetry operations $[\boldsymbol{H},R] = 0$). As a result, forming symmetry-adapted orbitals (by combining the original orbitals of the molecule or atom) can sometimes be used to break apart a complex set of atomic orbital interactions into smaller subsets, one for each symmetry type.

In the case of a single bond between two identical atoms involving one orbital on each atom, symmetry actually gives the mos directly since the LCAO-MO coefficients C_{ia} are the symmetry adaptation coefficients (0.707, 0.707) and (0.707, –0.707) that relate the symmetry orbitals to the atomic basis orbitals. The bonding and antibonding orbital energies can then be found in terms of 1×1 subblocks of the secular matrix as

$$E_S = \langle \phi_S | h | \phi_S \rangle = 0.707^2 h_{11} + 0.707^2 h_{22} + 2(0.707^2) h_{12},$$

and

$$E_A = \langle \phi_A | h | \phi_A \rangle = 0.707^2 h_{11} + 0.707^2 h_{22} - 2(0.707^2) h_{12},$$

where ϕ_S and ϕ_A represent the symmetric and antisymmetric symmetry-adapted orbitals, respectively (recall that h_{12} is negative if the two atomic orbitals have positive overlap, so $E_S < E_A$).

In the allyl radical case, symmetry can be used to reduce the 3×3 secular problem. In particular, by forming combinations of the three atomic orbitals that are either even or odd under the σ_v plane perpendicular to the plane containing the three Carbon atoms, one arrives at three symmetry adapted orbitals (ϕ_s) which are related to the original three aos χ_a through the transformation vectors: C_{sa} = (0.707, 0.0, 0.707), (0.0, 1.00, 0.0), and (0.707, 0.0, –0.707). The first two symmetry-adapted orbitals are even under σ_v; the third is odd.

Note that it does no good to use the σ_v plane in which the three Carbon atoms lie to further symmetry decompose the three π molecular orbitals because all three of the constituent atomic p_π orbitals are odd under this plane; only when the constituent orbitals can be combined to produce symmetry-adapted orbitals which are of different symmetries under a given symmetry operation will that symmetry operation be useful in breaking the secular problem into smaller subblocks.

In terms of the three new symmetry-adapted orbitals described above, the 3×3 secular matrix reduces to

$(h_{11}+h_{33})/2$	$0.707(h_{12}+h_{23})$	0.0
$0.707(h_{12}+h_{23})$	h_{22}	0.0
0.0	0.0	$(h_{11}+h_{33})/2$

The two symmetric orbitals, whose C_{sa} coefficients are (0.707, 0.0, 0.707) and (0.0, 1.00, 0.0) lead to the 2×2 block and the one antisymmetric orbital whose C_{sa} coefficients are (0.707, 0.0, –0.707) leads to the 1×1 block.

The three resultant molecular orbital energies are, of course, identical to those obtained without symmetry above. The three LCAO-MO coefficients, now expressing the mos in terms of the symmetry adapted orbitals are C_{is} = (0.707, 0.707, 0.0) for the bonding orbital, (0.0, 0.0, 1.00) for the nonbonding orbital, and (0.707, –0.707, 0.0) for the antibonding orbital. These coefficients, when combined with the symmetry adaptation coefficients C_{sa} given earlier, express the three mos in terms of the three aos as

$$\phi_i = \sum_{sa} C_{is} C_{sa} \chi_a;$$

the sum

$$\sum_s C_{is} C_{sa}$$

gives the LCAO-MO coefficients C_{ia} which, for example, for the bonding orbital, are (0.707^2, 0.707, 0.707^2), in agreement with what was found earlier without using symmetry.

The low energy orbitals of the H_2O molecule can be used to illustrate the use of symmetry within the primitive ao basis as well as in terms of hybrid orbitals. The $1s$ orbital on the Oxygen atom is clearly a nonbonding core orbital. The Oxygen $2s$ orbital and its three $2p$ orbitals are of valence type, as are the two Hydrogen $1s$ orbitals. In the absence of symmetry, these six valence orbitals would give rise to a 6×6 secular problem. By combining the two Hydrogen $1s$ orbitals into $0.707(1s_L + 1s_R)$ and $0.707(1s_L - 1s_R)$ symmetry adapted orbitals (labeled a_1 and b_2 within the C_{2v} point group; see the figure below), and recognizing that the Oxygen $2s$ and $2p_z$ orbitals belong to a_1 symmetry (the z axis is taken as the C_2 rotation axis and the x axis is taken to be perpendicular to the plane in which the three nuclei lie) while the $2p_x$ orbital is b_1 and the $2p_y$ orbital is b_2, allows the 6×6 problem to be decomposed into a 3×3 (a_1) secular problem, a 2×2 (b_2) secular problem and a 1×1 (b_1) problem. These decompositions allow one to conclude that there is one nonbonding b_1 orbital (the Oxygen $2p_x$ orbital), bonding and antibonding b_2 orbitals (the O-H bond and antibond formed by the Oxygen $2p_y$ orbital interacting with $0.707(1s_L - 1s_R)$), and, finally, a set of bonding, nonbonding, and antibonding a_1 orbitals (the O-H bond and antibond formed by the Oxygen $2s$ and $2p_z$ orbitals interacting with $0.707(1s_L + 1s_R)$ and the nonbonding orbital formed by the Oxygen $2s$ and $2p_z$ orbitals combining to form the "lone pair" orbital directed along the z-axis away from the two Hydrogen atoms).

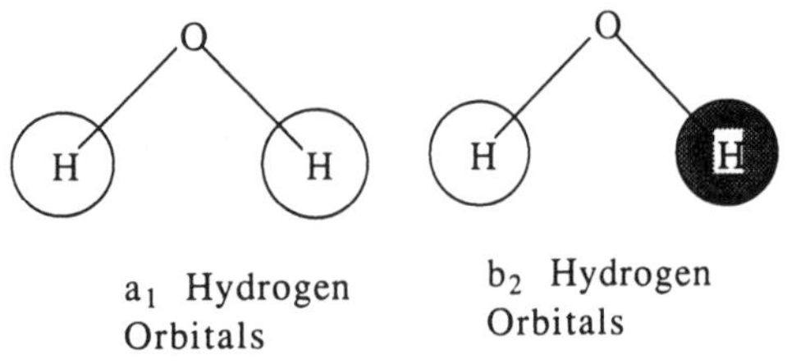

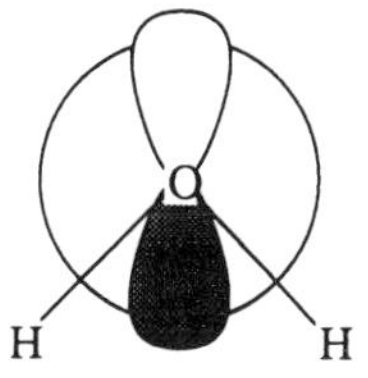

Oxygen a_1 Orbitals

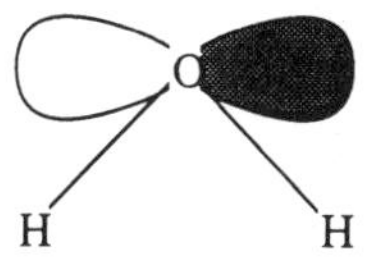

Oxygen b_2 Orbital

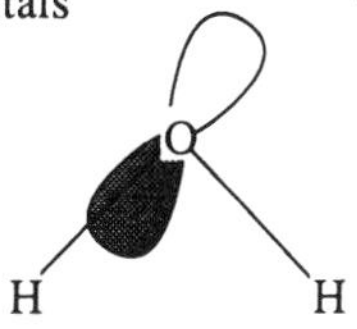

Oxygen b_1 Orbital

Alternatively, to analyze the H_2O molecule in terms of hybrid orbitals, one first combines the Oxygen $2s$, $2p_z$, $2p_x$ and $2p_y$ orbitals to form four sp^3 hybrid orbitals. The valence-shell electron-pair repulsion (VSEPR) model of chemical bonding (see R. J. Gillespie and R. S. Nyholm, *Quart. Rev. 11*, 339 (1957) and R. J. Gillespie, *J. Chem. Educ. 40*, 295 (1963)) directs one to involve all of the Oxygen valence orbitals in the hybridization because four σ-bond or nonbonding electron pairs need to be accommodated about the Oxygen center; no π orbital interactions are involved, of course.

Having formed the four sp^3 hybrid orbitals, one proceeds as with the primitive aos; one forms symmetry adapted orbitals. In this case, the two Hydrogen $1s$ orbitals are combined exactly as above to form $0.707(1s_L + 1s_R)$ and $0.707(1s_L - 1s_R)$. The two sp^3 hybrids which lie in the plane of the H and O nuclei (label them L and R) are combined to give symmetry adapted hybrids: $0.707(L+R)$ and $0.707(L-R)$, which are of a_1 and b_2 symmetry, respectively (see the figure below). The two sp^3 hybrids that lie above and below the plane of the three nuclei (label them T and B) are also symmetry adapted to form $0.707(T+B)$ and $0.707(T-B)$, which are of a_1 and b_1 symmetry, respectively. Once again, one has broken the 6×6 secular problem into a 3×3 a_1 block, a 2×2 b_2 block and a 1×1 b_1 block. Although the resulting bonding, nonbonding and antibonding a_1 orbitals, the bonding and antibonding b_2 orbitals and the nonbonding b_1 orbital are now viewed as formed from symmetry adapted Hydrogen orbitals and four Oxygen sp^3 orbitals, they are, of course, *exactly the same* molecular orbitals as were obtained earlier in terms of the symmetry adapted primitive aos. The formation of hybrid orbitals was an intermediate step which could not alter the final outcome.

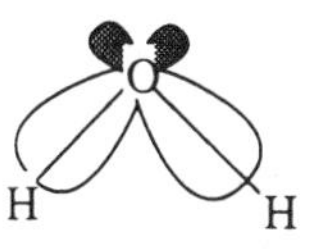

L + R a_1 Hybrid
Symmetry Orbital

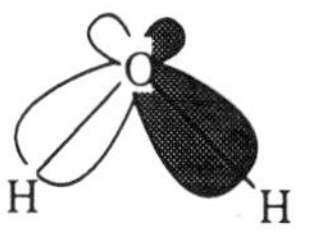

L - R b_2 Hybrid
Symmetry Orbital

T + B Hybrid
Symmetry Orbital
Seen From the Side

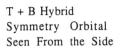

T - B Hybrid
Symmetry Orbital
Seen From the Side

That no degenerate molecular orbitals arose in the above examples is a result of the fact that the C_{2v} point group to which H_2O and the allyl system belong (and certainly the C_s subgroup which was used above in the allyl case) has no degenerate representations. Molecules with higher symmetry such as NH_3, CH_4, and benzene have energetically degenerate orbitals because their molecular point groups have degenerate representations.

B. Linear Molecules

Linear molecules belong to the axial rotation group. Their symmetry is intermediate in complexity between nonlinear molecules and atoms.

For linear molecules, the symmetry of the electrostatic potential provided by the nuclei and the other electrons is described by either the $C_{\infty v}$ or $D_{\infty h}$ group. The essential difference between these symmetry groups and the finite point groups which characterize the non-linear molecules lies in the fact that the electrostatic potential which an electron feels is invariant to rotations of *any* amount about the molecular axis (i.e., $V(\gamma + \delta\gamma) = V(\gamma)$, for any angle increment $\delta\gamma$). This means that the operator $C_{\delta\gamma}$ which generates a rotation of the electron's azimuthal angle γ by an amount $\delta\gamma$ about the molecular axis commutes with the Hamiltonian $[h, C_{\delta\gamma}] = 0$. $C\delta\gamma$ can be written in terms of the quantum mechanical operator $L_z = -i\hbar\partial/\partial\gamma$ describing the orbital angular momentum of the electron about the molecular (z) axis:

$$C_{\delta\gamma} = \exp(i\delta\gamma L_z/\hbar).$$

Because $C_{\delta\gamma}$ commutes with the Hamiltonian and $C_{\delta\gamma}$ can be written in terms of L_z, L_z must commute with the Hamiltonian. As a result, the molecular orbitals ϕ of a linear molecule must be eigenfunctions of the z-component of angular momentum L_z:

$$-i\hbar\partial/\partial\gamma\phi = m\hbar\phi.$$

The electrostatic potential is not invariant under rotations of the electron about the x or y axes (those perpendicular to the molecular axis), so L_x and L_y do *not* commute with the Hamiltonian. Therefore, only L_z provides a "good quantum number" in the sense that the operator L_z commutes with the Hamiltonian.

In summary, the molecular orbitals of a linear molecule can be labeled by their m quantum number, which plays the same role as the point group labels did for non-linear polyatomic molecules, and which gives the eigenvalue of the angular momentum of the orbital about the molecule's symmetry axis. Because the kinetic energy part of the Hamiltonian contains $(\hbar^2/2m_e r^2)\partial^2/\partial\gamma^2$, whereas the potential energy part is independent of γ, the energies of the molecular orbitals depend on the *square* of the m quantum number. Thus, pairs of orbitals with $m = \pm 1$ are energetically degenerate; pairs with $m = \pm 2$ are degenerate, and so on. The absolute value of

m, which is what the energy depends on, is called the λ quantum number. Molecular orbitals with $\lambda = 0$ are called σ orbitals; those with $\lambda = 1$ are π orbitals; and those with $\lambda = 2$ are δ orbitals.

Just as in the non-linear polyatomic-molecule case, the atomic orbitals which constitute a given molecular orbital must have the same symmetry as that of the molecular orbital. This means that σ, π, and δ molecular orbitals are formed, via LCAO-MO, from $m = 0$, $m = \pm 1$, and $m = \pm 2$ atomic orbitals, respectively. In the diatomic N_2 molecule, for example, the core orbitals are of σ symmetry as are the molecular orbitals formed from the $2s$ and $2p_z$ atomic orbitals (or their hybrids) on each Nitrogen atom. The molecular orbitals formed from the atomic $2p_{-1} = (2p_x - i2p_y)$ and the $2p_{+1} = (2p_x + i2p_y)$ orbitals are of π symmetry and have $m = -1$ and $+1$.

For homonuclear diatomic molecules and other linear molecules which have a center of symmetry, the inversion operation (in which an electron's coordinates are inverted through the center of symmetry of the molecule) is also a symmetry operation. Each resultant molecular orbital can then also be labeled by a quantum number denoting its parity with respect to inversion. The symbols g (for gerade or even) and u (for ungerade or odd) are used for this label. Again for N_2, the core orbitals are of σ_g and σ_u symmetry, and the bonding and antibonding σ orbitals formed from the $2s$ and $2p_\sigma$ orbitals on the two Nitrogen atoms are of σ_g and σ_u symmetry.

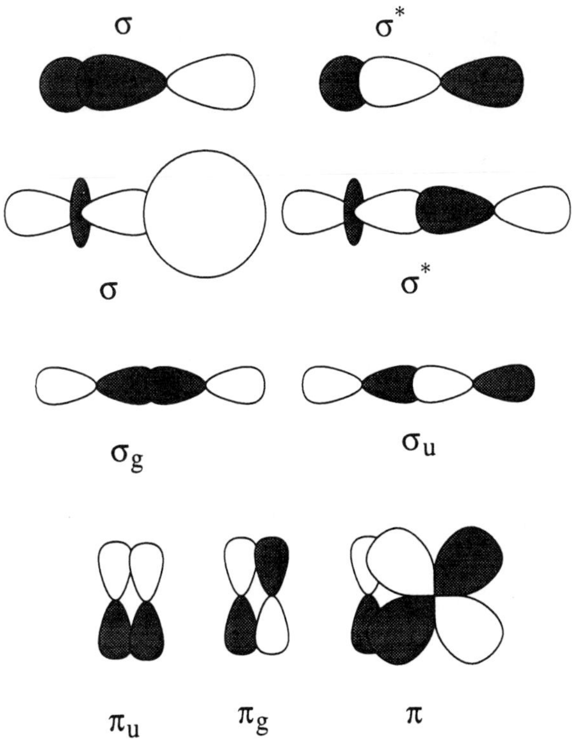

The bonding π molecular orbital pair (with $m = +1$ and -1) is of π_u symmetry whereas the corresponding antibonding orbital is of π_g symmetry. Examples of such molecular orbital symmetries are shown above.

The use of hybrid orbitals can be illustrated in the linear-molecule case by considering the N_2 molecule. Because two π bonding and antibonding molecular orbital pairs are involved in N_2 (one with $m = +1$, one with $m = -1$), VSEPR theory guides one to form sp hybrid orbitals from each of the Nitrogen atom's $2s$ and $2p_z$ (which is also the $2p$ orbital with $m = 0$) orbitals. Ignoring the core orbitals, which are of σ_g and σ_u symmetry as noted above, one then symmetry adapts the four sp hybrids (two from each atom) to build one σ_g orbital involving a bonding interaction between two sp hybrids pointed toward one another, an antibonding σ_u orbital involving the same pair of sp orbitals but coupled with opposite signs, a nonbonding σ_g orbital composed of two sp hybrids pointed away from the interatomic region combined with like sign, and a nonbonding σ_u orbital made of the latter two sp hybrids combined with opposite signs. The two $2p_m$ orbitals ($m = +1$ and -1) on each Nitrogen atom are then symmetry adapted to produce a pair of bonding π_u orbitals (with $m = +1$ and -1) and a pair of antibonding π_g orbitals (with $m = +1$ and -1). This hybridization and symmetry adaptation thereby reduces the 8×8 secular problem (which would be 10×10 if the core orbitals were included) into a 2×2 σ_g problem (one bonding and one nonbonding), a 2×2 σ_u problem (one bonding and one nonbonding), an identical pair of 1×1 π_u problems (bonding), and an identical pair of 1×1 π_g problems (antibonding).

Another example of the equivalence among various hybrid and atomic orbital points of view is provided by the CO molecule. Using, for example, sp hybrid orbitals on C and O, one obtains a picture in which there are: two core σ orbitals corresponding to the O-atom $1s$ and C-atom $1s$ orbitals; one CO bonding, two non-bonding, and one CO antibonding orbitals arising from the four sp hybrids; a pair of bonding and a pair of antibonding π orbitals formed from the two p orbitals on O and the two p orbitals on C. Alternatively, using sp^2 hybrids on both C and O, one obtains: the two core σ orbitals as above; a CO bonding and antibonding orbital pair formed from the sp^2 hybrids that are directed along the CO bond; and a single π bonding and antibonding π^* orbital set. The remaining two sp^2 orbitals on C and the two on O can then be symmetry adapted by forming $\pm$ combinations within each pair to yield: an a_1 non-bonding orbital (from the + combination) on each of C and O directed away from the CO bond axis; and a p_π orbital on each of C and O that can subsequently overlap to form the second π bonding and π^* antibonding orbital pair.

It should be clear from the above examples, that no matter what particular hybrid orbitals one chooses to utilize in conceptualizing a molecule's orbital interactions, symmetry ultimately returns to force one to form proper symmetry adapted combinations which, in turn, renders the various points of view equivalent. In the above examples and in several earlier examples, symmetry adaptation of, for example, sp^2 orbital pairs (e.g., $sp_L^2 \pm sp_R^2$) generated orbitals of pure spatial symmetry. In fact, symmetry combining hybrid orbitals in this manner amounts to forming other hybrid orbitals. For example, the above $\pm$ combinations of sp^2 hybrids directed to the left (L) and right (R) of some bond axis generate a new sp hybrid directed along the bond axis but opposite to the sp^2 hybrid used to form the bond and a non-hybridized p orbital directed along the L-to-R direction. In the CO example, these combinations of sp^2 hybrids on O and C produce sp hybrids on O and C and p_π orbitals on O and C.

C. Atoms

Atoms belong to the full rotation symmetry group; this makes their symmetry analysis the most complex to treat.

In moving from linear molecules to atoms, additional symmetry elements arise. In particular, the potential field experienced by an electron in an orbital becomes invariant to rotations of arbitrary amounts about the x, y, and z axes; in the linear-molecule case, it is invariant only to rotations of the electron's position about the molecule's symmetry axis (the z axis). These invariances are, of course, caused by the spherical symmetry of the potential of any atom. This additional symmetry of the potential causes the Hamiltonian to commute with all three components of the electron's angular momentum: $[L_x,H] = 0$, $[L_y,H] = 0$, and $[L_z,H] = 0$. It is straightforward to show that H also commutes with the operator $L^2 = L_x^2 + L_y^2 + L_z^2$, defined as the sum of the squares of the three individual components of the angular momentum. Because L_x, L_y, and L_z do not commute with one another, orbitals which are eigenfunctions of H cannot be simultaneous eigenfunctions of all three angular momentum operators. Because L_x, L_y, and L_z do commute with L^2, orbitals can be found which are eigenfunctions of H, of L^2 and of any one component of $\mathbf{L}$; it is convention to select L_z as the operator which, along with H and L^2, form a mutually commutative operator set of which the orbitals are simultaneous eigenfunctions.

So, for any atom, the orbitals can be labeled by both l and m quantum numbers, which play the role that point group labels did for non-linear molecules and λ did for linear molecules. Because (i) the kinetic energy operator in the electronic Hamiltonian explicitly contains $L^2/2m_e r^2$, (ii) the Hamiltonian does not contain additional L_z, L_x, or L_y factors, and (iii) the potential energy part of the Hamiltonian is spherically symmetric (and commutes with L^2 and L_z), the energies of atomic orbitals depend upon the l quantum number and are independent of the m quantum number. This is the source of the $2l + 1$-fold degeneracy of atomic orbitals.

The angular part of the atomic orbitals is described in terms of the spherical harmonics $Y_{l,\,m}$; that is, each atomic orbital ϕ can be expressed as

$$\phi_{n,l,m} = Y_{l,m}(\theta,\varphi)R_{n,l}(r).$$

The explicit solutions for the $Y_{l,\,m}$ and for the radial wavefunctions $R_{n,l}$ are given in appendix B. The variables r,θ,φ give the position of the electron in the orbital in spherical coordinates. These angular functions are, as discussed earlier, related to the cartesian (i.e., spatially oriented) orbitals by simple transformations; for example, the orbitals with $l = 2$ and $m = 2, 1, 0, -1, -2$ can be expressed in terms of the d_{xy}, d_{xz}, d_{yz}, d_{xx-yy}, and d_{zz} orbitals. Either set of orbitals is acceptable in the sense that each orbital is an eigenfunction of H; transformations within a degenerate set of orbitals do not destroy the Hamiltonian-eigenfunction feature. The orbital set labeled with l and m quantum numbers is most useful when one is dealing with isolated atoms (which have spherical symmetry), because m is then a valid symmetry label, or with an atom in a local environment which is axially symmetric (e.g., in a linear molecule) where the m quantum number remains a useful symmetry label. The cartesian orbitals are preferred for describing an atom in a local environment which displays lower than axial symmetry (e.g., an atom interacting with a diatomic molecule in C_{2v} symmetry).

The radial part of the orbital $R_{n,l}(r)$ as well as the orbital energy $\varepsilon_{n,l}$ depend on l because the Hamiltonian itself contains $l\,(l + 1)\hbar^2/2m_e r^2$; they are independent of m because the Hamiltonian has no m-dependence. For bound orbitals, $R_{n,l}(r)$ decays exponentially for large r (as $\exp(-2r\sqrt{2\varepsilon_{n,l}})$), and for unbound (scattering) orbitals, it is oscillatory at large r with an oscillation period related to the deBroglie wavelength of the electron. In $R_{n,l}(r)$ there are (n-l-1) radial nodes lying between $r = 0$ and $r = \infty$. These nodes provide differential stabilization of low-l orbitals over high-l orbitals of the same principal quantum number n. That is, penetration of outer shells is greater for low-l orbitals because they have more radial nodes; as a result, they have larger amplitude near the atomic nucleus and thus experience enhanced attraction to the positive nuclear charge. The average size (e.g., average value of r; $\langle r \rangle = \int R_{n,l}^2 r r^2 dr$) of an orbital depends strongly on n, weakly on l and is independent of m; it also depends strongly on the nuclear charge and on the potential produced by the other electrons. This potential is often characterized qualitatively in terms of an effective nuclear charge Z_{eff} which is the true nuclear charge of the atom Z minus a screening component Z_{sc} which describes the repulsive effect of the electron density lying radially inside the electron under study. Because, for a given n, low-l orbitals penetrate closer to the nucleus than do high-l orbitals, they have higher Z_{eff} values (i.e., smaller Z_{sc} values) and correspondingly smaller average sizes and larger binding energies.

CHAPTER

6

Along "reaction paths," orbitals can be connected one-to-one according to their symmetries and energies. This is the origin of the Woodward-Hoffmann rules.

I. REDUCTION IN SYMMETRY

As fragments are brought together to form a larger molecule, the symmetry of the nuclear framework (recall the symmetry of the coulombic potential experienced by electrons depends on the locations of the nuclei) changes. However, in some cases, certain symmetry elements persist throughout the path connecting the fragments and the product molecule. These preserved symmetry elements can be used to label the orbitals throughout the "reaction."

The point-group, axial- and full-rotation group symmetries which arise in non-linear molecules, linear molecules, and atoms, respectively, are seen to provide quantum numbers or symmetry labels which can be used to characterize the orbitals appropriate for each such species. In a physical event such as interaction with an external electric or magnetic field or a chemical process such as collision or reaction with another species, the atom or molecule can experience a change in environment which causes the electrostatic potential which its orbitals experience to be of lower symmetry than that of the isolated atom or molecule. For example, when an atom interacts with another atom to form a diatomic molecule or simply to exchange energy during a collision, each atom's environment changes from being spherically symmetric to being axially symmetric. When the formaldehyde molecule undergoes unimolecular decomposition to produce $CO + H_2$ along a path that preserves C_{2v} symmetry, the orbitals of the CO moiety evolve from C_{2v} symmetry to axial symmetry.

It is important, therefore to be able to label the orbitals of atoms, linear, and non-linear molecules in terms of their full symmetries as well in terms of the groups appropriate to lower-symmetry situations. This can be done by knowing how the representations of a higher symmetry group decompose into representations of a lower group. For example, the $Y_{l,m}$ functions appropriate for spherical symmetry, which belong to a $2l + 1$ fold degenerate set in this higher symmetry, decompose into doubly degenerate pairs of functions $Y_{l,l}, Y_{l,-l}$; $Y_{l,l-1}, Y_{l,-l+1}$; etc., plus a single non-degenerate function $Y_{l,0}$, in axial symmetry. Moreover, because L^2 no longer commutes with the Hamiltonian whereas L_z does, orbitals with different l-values but the same m-values can be coupled. As the N_2 molecule is formed from two N atoms, the $2s$ and $2p_z$ orbitals, both of which belong to the same (σ) symmetry in the axial rotation group but which are of different symmetry in the isolated-atom spherical symmetry, can mix to form the σ_g bonding orbital, the σ_u antibonding, as well as the σ_g and σ_u nonbonding lone-pair orbitals. The fact that 2s and 2p have different l-values no longer uncouples these orbitals as it did for the isolated atoms, because l is no longer a "good" quantum number.

Another example of reduced symmetry is provided by the changes that occur as H_2O fragments into OH and H. The σ bonding orbitals (a_1 and b_2) and in-plane lone pair (a_1) and the σ^* antibonding (a_1 and b_2) of H_2O become a' orbitals (see the figure below); the out-of-plane b_1 lone pair orbital becomes a'' (in appendix IV of *Electronic Spectra and Electronic Structure of Polyatomic Molecules*, G. Herzberg, Van Nostrand Reinhold Co., New York, N.Y. (1966) tables are given which allow one to determine how particular symmetries of a higher group evolve into symmetries of a lower group).

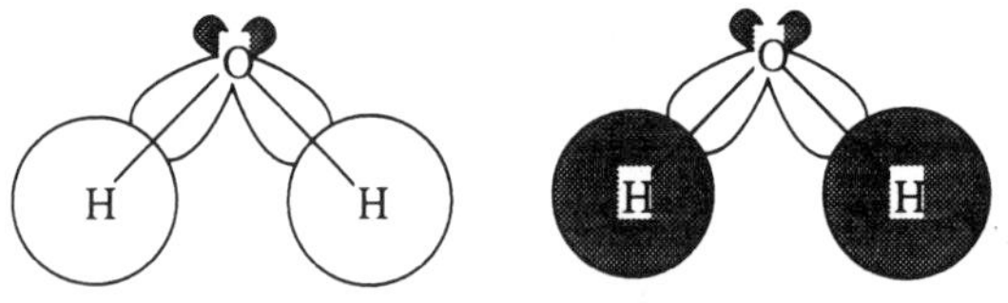

a_1 σ bonding orbital

a_1 σ^* antibonding orbital

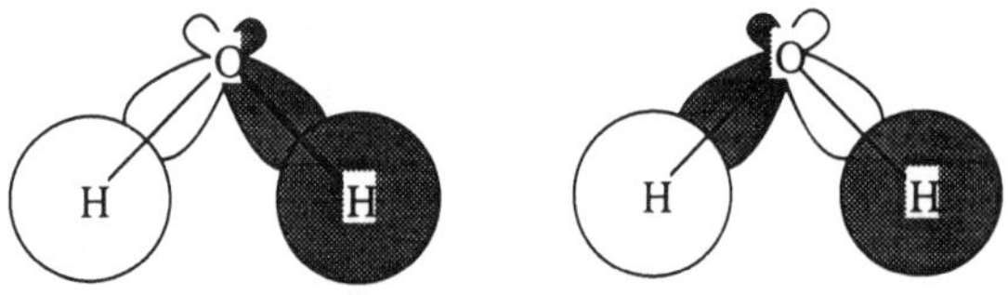

b_2 σ bonding orbital

b_2 σ^* antibonding orbital

To further illustrate these points dealing with orbital symmetry, consider the insertion of CO into H_2 along a path which preserves C_{2v} symmetry. As the insertion occurs, the degenerate π bonding orbitals of CO become b_1 and b_2 orbitals. The antibonding π^* orbitals of CO also become b_1 and b_2. The σ_g bonding orbital of H_2 becomes a_1, and the σ_u antibonding H_2 orbital becomes b_2. The orbitals of the reactant H_2CO are energy-ordered and labeled according to C_{2v} symmetry in the figure shown below as are the orbitals of the product $H_2 + CO$.

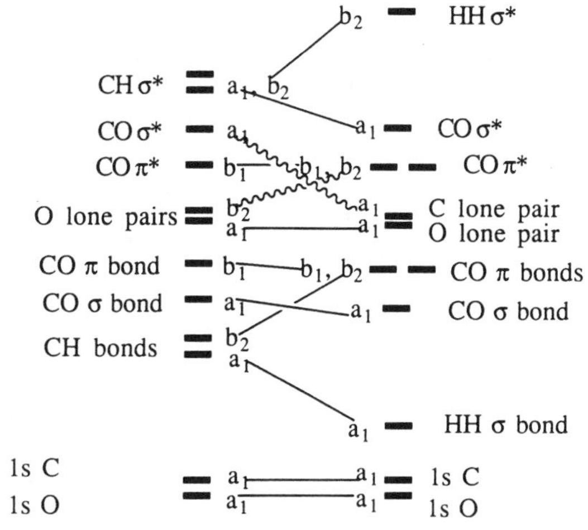

H_2CO ==> H_2 + CO Orbital Correlation Diagram in C_{2v} Symmetry

When these orbitals are connected according to their symmetries as shown above, one reactant orbital to one product orbital starting with the low-energy orbitals and working to increasing energy, an orbital correlation diagram (OCD) is formed. These diagrams play essential roles in analyzing whether reactions will have symmetry-imposed energy barriers on their potential energy surfaces along the reaction path considered in the symmetry analysis. The essence of this analysis, which is covered in detail in chapter 12, can be understood by noticing that the 16 electrons of ground-state H_2CO do *not* occupy their orbitals with the same occupancy pattern, symmetry-by-symmetry, as do the 16 electrons of ground-state $H_2 + CO$. In particular, H_2CO places a pair of electrons in the second b_2 orbital while $H_2 + CO$ does not; on the other hand, $H_2 + CO$ places two electrons in the sixth a_1 orbital while H_2CO does not. The mismatch of the orbitals near the $5a_1$, $6a_1$, and $2b_2$ orbitals is the source of the mismatch in the electronic configurations of the ground-states of H_2CO and $H_2 + CO$. These mismatches give rise, as shown in chapter 12, to symmetry-caused energy barriers on the $H_2CO \Rightarrow H_2 + CO$ reaction potential energy surface.

II. ORBITAL CORRELATION DIAGRAMS

Connecting the energy-ordered orbitals of reactants to those of products according to symmetry elements that are preserved throughout the reaction produces an orbital correlation diagram.

In each of the examples cited above, symmetry reduction occurred as a molecule or atom approached and interacted with another species. The "path" along which this approach was thought to occur was characterized by symmetry in the sense that it preserved certain symmetry elements while destroying others. For example, the collision of two Nitrogen atoms to produce N_2 clearly occurs in a way which destroys spherical symmetry but preserves axial symmetry. In the other example used above, the formaldehyde molecule was postulated to decompose along a path which preserves C_{2v} symmetry while destroying the axial symmetries of CO and H_2. The actual decomposition of formaldehyde may occur along some other path, but *if* it were to occur along the proposed path, then the symmetry analysis presented above would be useful.

The symmetry reduction analysis outlined above allows one to see new orbital interactions that arise (e.g., the 2s and $2p_z$ interactions in the $N + N ==> N_2$ example) as the interaction increases. It also allows one to construct orbital correlation diagrams (OCD's) in which the orbitals of the "reactants" and "products" are energy ordered and labeled by the symmetries which are preserved throughout the "path," and the orbitals are then correlated by drawing lines connecting the orbitals of a given symmetry, one-by-one in increasing energy, from the reactants side of the diagram to the products side. As noted above, such orbital correlation diagrams play a central role in using symmetry to predict whether photochemical and thermal chemical reactions will experience activation barriers along proposed reaction paths (this subject is treated in chapter 12).

To again illustrate the construction of an OCD, consider the π orbitals of 1,3-butadiene as the molecule undergoes disrotatory closing (notice that this is where a particular path is postulated; the actual reaction may or may not occur along such a path) to form cyclobutene. Along this path, the plane of symmetry which bisects and is perpendicular to the C_2-C_3 bond is preserved, so the orbitals of the reactant and product are labeled as being even-*e* or odd-*o* under reflection through this plane. It is *not* proper to label the orbitals with respect to their symmetry under the plane containing the four C atoms; although this plane is indeed a symmetry operation for the reactants and products, it does not remain a valid symmetry throughout the reaction path.

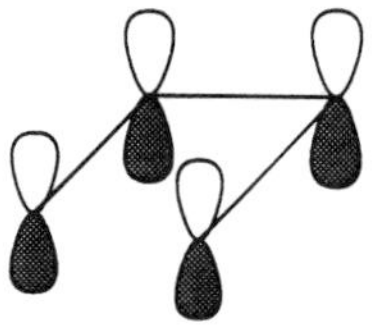

Lowest π orbital of 1,3- butadiene denoted π_1

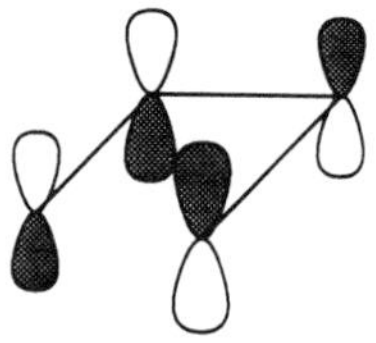

π_2

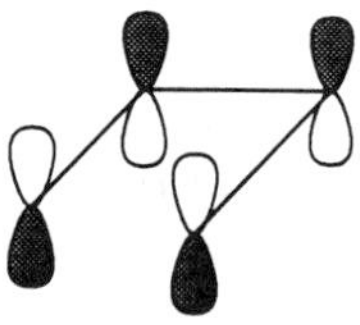

π_3

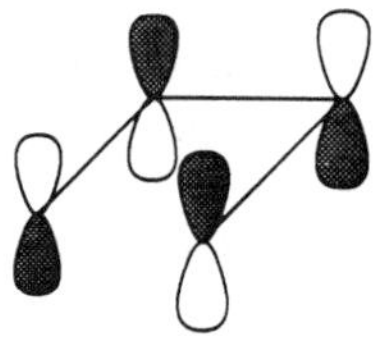

π_4

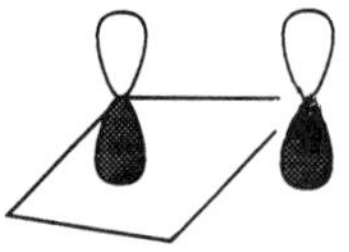

π orbital of cyclobutene

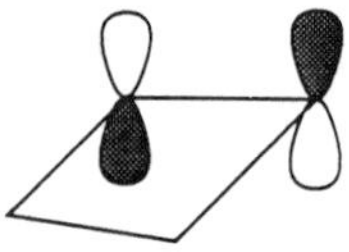

π* orbital of cyclobutene

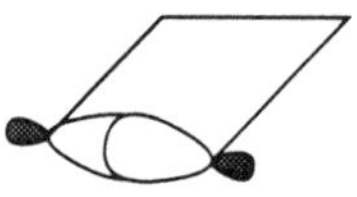

σ orbital of cyclobutene

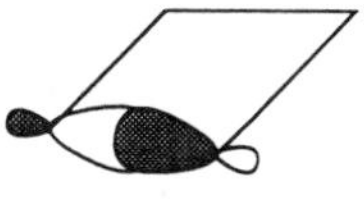

σ* orbital of cyclobutene

The four π orbitals of 1,3-butadiene are of the following symmetries under the preserved plane (see the orbitals in the figure above): $\pi_1 = e$, $\pi_2 = o$, $\pi_3 = e$, $\pi_4 = o$. The π and π^* and σ and σ^* orbitals of cyclobutane which evolve from the four active orbitals of the 1,3-butadiene are of the following symmetry and energy order: $\sigma = e$, $\pi = e$, $\pi^* = o$, $\sigma^* = o$. Connecting these orbitals by symmetry, starting with the lowest energy orbital and going through the highest energy orbital, gives the following OCD:

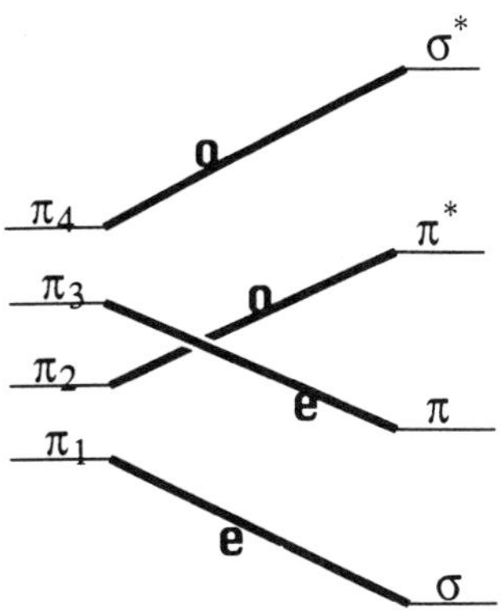

The fact that the lowest two orbitals of the reactants, which are those occupied by the four π electrons of the reactant, do not correlate to the lowest two orbitals of the products, which are the orbitals occupied by the two σ and two π electrons of the products, will be shown later in chapter 12 to be the origin of the activation barrier for the thermal disrotatory rearrangement (in which the four active electrons occupy these lowest two orbitals) of 1,3-butadiene to produce cyclobutene.

If the reactants could be prepared, for example by photolysis, in an excited state having orbital occupancy $\pi_1^2\pi_2^1\pi_3^1$, then reaction along the path considered would not have any symmetry-imposed barrier because this singly excited configuration correlates to a singly-excited configuration $\sigma^2\pi^1\pi*^1$ of the products. The fact that the reactant and product configurations are of equivalent excitation level causes there to be no symmetry constraints on the photochemically induced reaction of 1,3-butadiene to produce cyclobutene. In contrast, the thermal reaction considered first above has a symmetry-imposed barrier because the orbital occupancy is forced to rearrange (by the occupancy of two electrons) from the ground-state wavefunction of the reactant to smoothly evolve into that of the product.

It should be stressed that although these symmetry considerations may allow one to anticipate barriers on reaction potential energy surfaces, they have nothing to do with the thermodynamic energy differences of such reactions. Symmetry says whether there will be symmetry-imposed barriers above and beyond any thermodynamic energy differences. The enthalpies of formation of reactants and products contain the information about the reaction's overall energy balance.

As another example of an OCD, consider the $N + N ==> N_2$ recombination reaction mentioned above. The orbitals of the atoms must first be labeled according to the axial rotation group (including the inversion operation because this is a homonuclear molecule). The core $1s$ orbitals are symmetry adapted to produce $1\sigma_g$ and $1\sigma_u$ orbitals (the number 1 is used to indicate that these are the lowest energy orbitals of their respective symmetries); the $2s$ orbitals generate $2\sigma_g$ and $2\sigma_u$ orbitals; the $2p$ orbitals combine to yield $3\sigma_g$, a pair of $1\pi_u$ orbitals, a pair of $1\pi_g$ orbitals, and the $3\sigma_u$ orbital, whose bonding, nonbonding, and antibonding nature was detailed earlier. In the two separated Nitrogen atoms, the two orbitals derived from the $2s$ atomic orbitals are degenerate, and the six orbitals derived from the Nitrogen atoms' $2p$ orbitals are degenerate. At the equilibrium geometry of the N_2 molecule, these degeneracies are lifted, Only the degeneracies of the $1\pi_u$ and $1\pi_g$ orbitals, which are dictated by the degeneracy of $+m$ and $-m$ orbitals within the axial rotation group, remain.

As one proceeds inward past the equilibrium bond length of N_2, toward the united-atom limit in which the two Nitrogen nuclei are fused to produce a Silicon nucleus, the energy ordering of the orbitals changes. Labeling the orbitals of the Silicon atom according to the axial rotation group, one finds the $1s$ is σ_g, the $2s$ is σ_g, the $2p$ orbitals are σ_u and π_u, the $3s$ orbital is σ_g, the $3p$ orbitals are σ_u and π_u, and the $3d$ orbitals are σ_g, π_g, and δ_g. The following OCD is obtained when one connects the orbitals of the two separated Nitrogen atoms (properly symmetry adapted) to those of the N_2 molecule and eventually to those of the Silicon atom.

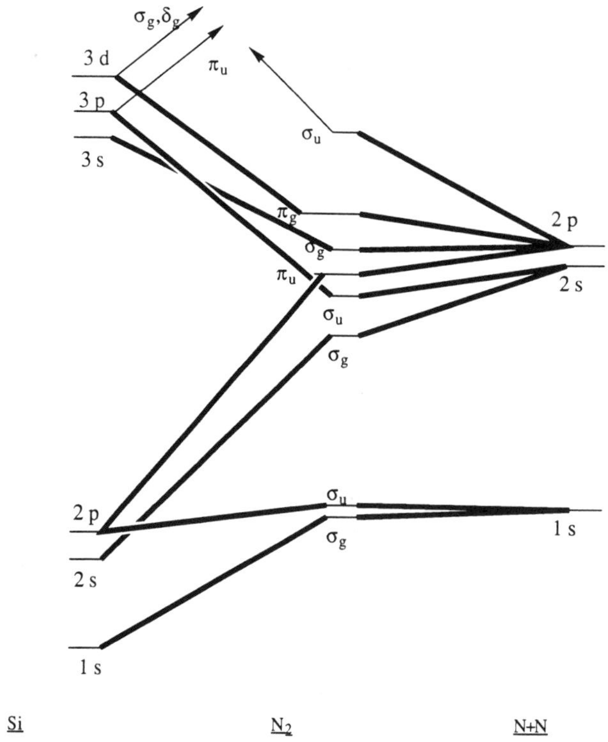

The fact that the separated-atom and united-atom limits involve several crossings in the OCD can be used to explain barriers in the potential energy curves of such diatomic molecules which occur at short internuclear distances. It should be noted that the Silicon atom's $3p$ orbitals of π_u symmetry and its $3d$ orbitals of σ_g and δ_g symmetry correlate with higher energy orbitals of N_2 not with the valence orbitals of this molecule, and that the $3\sigma_u$ antibonding orbital of N_2 correlates with a higher energy orbital of Silicon (in particular, its $4p$ orbital).

CHAPTER

7 The most elementary molecular orbital models contain symmetry, nodal pattern, and approximate energy information.

I. THE LCAO-MO EXPANSION AND THE ORBITAL-LEVEL SCHRÖDINGER EQUATION

In the simplest picture of chemical bonding, the valence molecular orbitals ϕ_i are constructed as linear combinations of valence atomic orbitals χ_μ according to the LCAO-MO formula:

$$\phi_i = \sum_\mu C_{i\mu}\chi_\mu.$$

The core electrons are not explicitly included in such a treatment, although their effects are felt through an electrostatic potential V that has the following properties:

1. V contains contributions from all of the nuclei in the molecule exerting coulombic attractions on the electron, as well as coulombic repulsions and exchange interactions exerted by the other electrons on this electron;
2. As a result of the (assumed) cancellation of attractions from distant nuclei and repulsions from the electron clouds (i.e., the core, lone-pair, and valence orbitals) that surround these distant nuclei, the effect of V on any particular mo ϕ_i depends primarily on the atomic charges and local bond polarities of the atoms over which ϕ_i is delocalized.

As a result of these assumptions, qualitative molecular orbital models can be developed in which one assumes that each mo ϕ_i obeys a one-electron Schrödinger equation

$$h\phi_i = \varepsilon_i\phi_i.$$

Here the orbital-level Hamiltonian h contains the kinetic energy of motion of the electron and the potential V mentioned above:

$$[-\hbar^2/2m_e\nabla^2 + V]\phi_i = \varepsilon_i\phi_i.$$

Expanding the mo ϕ_i in the LCAO-MO manner, substituting this expansion into the above Schrödinger equation, multiplying on the left by χ_ν, and integrating over the coordinates of the electron generates the following orbital-level eigenvalue problem:

$$\Sigma_\mu<\chi_\nu| - \hbar^2/2m_e\nabla^2 + V|\chi_\mu > C_{i\mu} = \varepsilon_i\sum_\mu <\chi_2|\chi_\nu>C_{i\mu}.$$

If the constituent atomic orbitals $\{\chi_\mu\}$ have been orthonormalized as discussed earlier in this chapter, the overlap integrals $<\chi_\nu|\chi_\mu>$ reduce to $\delta_{\mu,\nu}$.

II. DETERMINING THE EFFECTIVE POTENTIAL *V*

In the most elementary models of orbital structure, the quantities that explicitly define the potential V are not computed from first principles as they are in so-called *ab initio* methods (see section 6). Rather, either experimental data or results of *ab initio* calculations are used to determine the parameters in terms of which V is expressed. The resulting empirical or semi-empirical methods discussed below differ in the sophistication used to include electron-electron interactions as well as in the manner experimental data or *ab initio* computational results are used to specify V.

If experimental data is used to parameterize a semi-empirical model, then the model should not be extended beyond the level at which it has been parameterized. For example, experimental bond energies, excitation energies, and ionization energies may be used to determine molecular orbital energies which, in turn, are summed to compute total energies. In such a parameterization it would be incorrect to subsequently use these mos to form a wavefunction, as in sections 3 and 6, that goes beyond the simple "product of orbitals" description. To do so would be inconsistent because the more sophisticated wavefunction would duplicate what using the experimental data (which already contains mother nature's electronic correlations) to determine the parameters had accomplished.

Alternatively, if results of *ab initio* theory at the single-configuration orbital-product wavefunction level are used to define the parameters of a semi-empirical model, it would then be proper to use the semi-empirical orbitals in a subsequent higher-level treatment of electronic structure as done in section 6.

A. The Hückel Parameterization of *V*

In the most simplified embodiment of the above orbital-level model, the following additional approximations are introduced:

1. The diagonal values $<\chi_\mu| - \hbar^2/2m_e\nabla^2 + V|\chi_\mu>$, which are usually denoted α_μ, are taken to be equal to the energy of an electron in the atomic orbital χ_μ and, as such, are evaluated in terms of atomic ionization energies (IP's) and electron affinities (EA's):

$$<\chi_\mu| - \hbar^2/2m_e\nabla^2 + V|\chi_\mu> = -IP_\mu,$$

for atomic orbitals that are occupied in the atom, and

$$<\chi_\mu| - \hbar^2/2m_e\nabla^2 + V|\chi_\mu> = -EA_\mu,$$

for atomic orbitals that are not occupied in the atom.

These approximations assume that contributions in V arising from coulombic attraction to nuclei other than the one on which χ_μ is located, and repulsions from the core, lone-pair, and valence electron clouds surrounding these other nuclei cancel to an extent that $<\chi_\mu|V|\chi_\mu>$ contains only potentials from the atom on which χ_μ sits.

It should be noted that the IP's and EA's of valence-state orbitals are not identical to the experimentally measured IP's and EA's of the corresponding atom, but can be obtained from such information. For example, the $2p$ valence-state IP (VSIP) for a Carbon atom is the energy difference associated with the hypothetical process

$$C(1s^2 2s 2p_x 2p_y 2p_z) \Longrightarrow C^+(1s^2 2s 2p_x 2p_y).$$

If the energy differences for the "promotion" of C

$$C(1s^2 2s^2 2p_x 2p_y) \Longrightarrow C(1s^2 2s 2p_x 2p_y 2p_z); \Delta E_C$$

and for the promotion of C^+

$$C^+(1s^2 2s^2 2p_x) \Longrightarrow C^+(1s^2 2s 2p_x 2p_y); \Delta E_C +$$

are known, the desired VSIP is given by:

$$IP_{2p_z} = IP_C + \Delta E_C + -\Delta E_C.$$

The EA of the $2p$ orbital is obtained from the

$$C(1s^2 2s^2 2p_x 2p_y) \Longrightarrow C^-(1s^2 2s^2 2p_x 2p_y 2p_z)$$

energy gap, which means that $EA_{2p_z} = EA_C$. Some common IP's of valence $2p$ orbitals in eV are as follows: C(11.16), N(14.12), N^+(28.71), O(17.70), O^+(31.42), F^+(37.28).

2. The off-diagonal elements $<\chi_\nu| - \hbar^2/2m_e\nabla^2 + V|\chi_\mu>$ are taken as zero if χ_μ and χ_ν belong to the same atom because the atomic orbitals are assumed to have been constructed to diagonalize the one-electron Hamiltonian appropriate to an electron moving in that atom. They are set equal to a parameter denoted $\beta_{\mu,\nu}$ if χ_μ and χ_ν reside on neighboring atoms that are chemically bonded. If χ_μ and χ_ν reside on atoms that are not bonded neighbors, then the off-diagonal matrix element is set equal to zero.

3. The geometry dependence of the $\beta_{\mu,\nu}$ parameters is often approximated by assuming that $\beta_{\mu,\nu}$ is proportional to the overlap $S_{\mu,\nu}$ between the corresponding atomic orbitals:

$$\beta_{\mu,\nu} = \beta^o_{\mu,\nu} S_{\mu,\nu}.$$

Here $\beta^o_{\mu,\nu}$ is a constant (having energy units) characteristic of the bonding interaction between χ_μ and χ_ν; its value is usually determined by forcing the molecular orbital energies obtained from such a qualitative orbital treatment to yield experimentally correct ionization potentials, bond dissociation energies, or electronic transition energies.

The particular approach described thus far forms the basis of the so-called *Hückel model*. Its implementation requires knowledge of the atomic α_μ and $\beta^0_{\mu,\nu}$ values, which are eventually expressed in terms of experimental data, as well as a means of calculating the geometry dependence of the $\beta_{\mu,\nu}$'s (e.g., some method for computing overlap matrices $S_{\mu,\nu}$) .

B. The Extended Hückel Method

It is well known that bonding and antibonding orbitals are formed when a pair of atomic orbitals from neighboring atoms interact. The energy splitting between the bonding and antibonding orbitals depends on the overlap between the pair of atomic orbitals. Also, the energy of the antibonding orbital lies higher above the arithmetic mean $E_{ave} = E_A + E_B$ of the energies of the constituent atomic orbitals (E_A and E_B) than the bonding orbital lies below E_{ave}. If overlap is ignored, as in conventional Hückel theory (except in parameterizing the geometry dependence of $\beta_{\mu,\nu}$), the differential destabilization of antibonding orbitals compared to stabilization of bonding orbitals can not be accounted for.

By parameterizing the off-diagonal Hamiltonian matrix elements in the following overlap-dependent manner:

$$h_{\nu,\mu} = <\chi_\nu| - \hbar^2/2m_e\nabla^2 + V|\chi_\mu> = 0.5K(h_{\mu,\mu} + h_{\nu,\nu})S_{\mu,\nu},$$

and explicitly treating the overlaps among the constituent atomic orbitals $\{\chi_\mu\}$ in solving the orbital-level Schrödinger equation

$$\sum_\mu <\chi_\nu| - \hbar^2/2m_e\nabla^2 + V|\chi_\mu>C_{i\mu} = \varepsilon_i \sum_\mu <\chi_\nu|\chi_\mu>C_{iu},$$

Hoffmann introduced the so-called extended Hückel method. He found that a value for $K = 1.75$ gave optimal results when using Slater-type orbitals as a basis (and for calculating the $S_{\mu,\nu}$). The diagonal $h_{\mu,\mu}$ elements are given, as in the conventional Hückel method, in terms of valence-state IP's and EA's. Cusachs later proposed a variant of this parameterization of the off-diagonal elements:

$$h_{\nu,\mu} = 0.5K(h_{\mu,\mu} + h_{\nu,\nu})S_{\mu,\nu}(2-|S_{\mu,\nu}|) .$$

For first- and second-row atoms, the $1s$ or ($2s$, $2p$) or ($3s$, $3p$, $3d$) valence-state ionization energies (α_μ's), the number of valence electrons (#Elec.) as well as the orbital exponents (e_s, e_p and e_d) of Slater-type orbitals used to calculate the overlap matrix elements $S_{\mu,\nu}$ corresponding are given on the next page.

In the Hückel or extended Hückel methods no *explicit* reference is made to electron-electron interactions although such contributions are absorbed into the V potential, and hence into the α_μ and $\beta_{\mu,\nu}$ parameters of Hückel theory or the $h_{\mu,\mu}$ and $h_{\mu,\nu}$ parameters of extended Hückel theory. As electron density flows from one atom to another (due to electronegativity differences), the electron-electron repulsions in various atomic orbitals changes. To account for such charge-density-dependent coulombic energies, one must use an approach that includes explicit reference to

Atom	# Elec.	$e_s = e_p$	e_d	$\alpha_s(eV)$	$\alpha_p(eV)$	$\alpha_d(eV)$
H	1	1.3		−13.6		
Li	1	0.650		−5.4	−3.5	
Be	2	0.975		−10.0	−6.0	
B	3	1.300		−15.2	−8.5	
C	4	1.625		−21.4	−11.4	
N	5	1.950		−26.0	−13.4	
O	6	2.275		−32.3	−14.8	
F	7	2.425		−40.0	−18.1	
Na	1	0.733		−5.1	−3.0	
Mg	2	0.950		−9.0	−4.5	
Al	3	1.167		−12.3	−6.5	
Si	4	1.383	1.383	−17.3	−9.2	−6.0
P	5	1.600	1.400	−18.6	−14.0	−7.0
S	6	1.817	1.500	−20.0	−13.3	−8.0
Cl	7	2.033	2.033	−30.0	−15.0	−9.0

inter-orbital coulomb and exchange interactions. There exists a large family of semi-empirical methods that permit explicit treatment of electronic interactions; some of the more commonly used approaches are discussed in appendix F.

Exercises, Problems, and Solutions

REVIEW EXERCISES

1. Draw qualitative shapes of the (1) *s*, (3) *p* and (5) *d* "tangent sphere" atomic orbitals (note that these orbitals represent only the angular portion and *do not* contain the radial portion of the hydrogen like atomic wavefunctions) Indicate with $\pm$ the relative signs of the wavefunctions and the position(s) (if any) of any nodes.

2. Define the symmetry adapted "core" and "valence" orbitals of the following systems:

a. NH_3 in the C_{3v} point group,

b. H_2O in the C_{2v} point group,

c. H_2O_2 (cis) in the C_2 point group,

d. N in $D_{\infty h}$, D_{2h}, C_{2v}, and C_s point groups,

e. N_2 in $D_{\infty h}$, D_{2h}, C_{2v}, and C_s point groups.

3. Plot the radial portions of the 4*s*, 4*p*, 4*d*, and 4*f* hydrogen like atomic wavefunctions.

4. Plot the radial portions of the 1s, 2*s*, 2*p*, 3*s*, and 3*p* hydrogen like atomic wavefunctions for the Si atom using screening concepts for any inner electrons.

EXERCISES

1. In quantum chemistry it is quite common to use combinations of more familiar and easy-to-handle "basis functions" to approximate atomic orbitals. Two common types of basis functions are the Slater-type orbitals (STO's) and gaussian type orbitals (GTO's). STO's have the normalized form:

$$\left(\frac{2\zeta}{a_o}\right)^{n+\frac{1}{2}}\left(\frac{1}{(2n)!}\right)^{\frac{1}{2}} r^{n-1} e^{\left(\frac{-\zeta r}{a_o}\right)} Y_{l,m}(\theta,\phi),$$

whereas GTO's have the form:

$$N r^{l} e^{\left(-\zeta r^{2}\right)} Y_{l,\,m}(\theta,\phi).$$

Orthogonalize (using Löwdin (symmetric) orthogonalization) the following $1s$ (core), $2s$ (valence), and $3s$ (Rydberg) STO's for the Li atom given:

$$\mathrm{Li}_{1s}\zeta = 2.6906$$
$$\mathrm{Li}_{2s}\zeta = 0.6396$$
$$\mathrm{Li}_{3s}\zeta = 0.1503.$$

Express the three resultant orthonormal orbitals as linear combinations of these three normalized STO's.

2. Calculate the expectation value of r for each of the orthogonalized $1s$, $2s$, and $3s$ Li orbitals found in exercise 1.

3. Draw a plot of the radial probability density (e.g., $r^2[R_{nl}(r)]^2$ with R referring to the radial portion of the STO) versus r for each of the orthonormal Li s orbitals found in exercise 1.

PROBLEMS

1. Given the following orbital energies (in hartrees) for the N atom and the coupling elements between two like atoms (these coupling elements are the Fock matrix elements from standard *ab-initio* minimum-basis SCF calculations), calculate the molecular orbital energy levels and 1-electron wavefunctions. Draw the orbital correlation diagram for formation of the N_2 molecule. Indicate the symmetry of each atomic and molecular orbital. Designate each of the molecular orbitals as bonding, non-bonding, or antibonding.

$$\mathrm{N}_{1s} = -15.31^{*}$$
$$\mathrm{N}_{2s} = -0.86^{*}$$
$$\mathrm{N}_{2p} = -0.48^{*}$$
$$\mathrm{N}_2\ \sigma_g \text{ Fock matrix}^{*}$$

$$\begin{bmatrix} -6.52 & & \\ -6.22 & -7.06 & \\ 3.61 & 4.00 & -3.92 \end{bmatrix}$$

$$\mathrm{N}_2\ \pi_g \text{ Fock matrix}^{*}$$

$$[0.28]$$

$N_2\ \sigma_u$ Fock matrix*

$$\begin{bmatrix} 1.02 & & \\ -0.60 & -7.59 & \\ 0.02 & 7.42 & -8.53 \end{bmatrix}$$

$N_2\ \pi_u$ Fock matrix*

[−0.58]

*The Fock matrices (and orbital energies) were generated using standard STO3G minimum basis set SCF calculations. The Fock matrices are in the orthogonal basis formed from these orbitals.

2. Given the following valence orbital energies for the C atom and H_2 molecule draw the orbital correlation diagram for formation of the CH_2 molecule (via a C_{2v} insertion of C into H_2 resulting in bent CH_2). Designate the symmetry of each atomic and molecular orbital in both their highest point group symmetry and in that of the reaction path (C_{2v}).

$C_{1s} = -10.91^*$ $H_2\ \sigma_g = -0.58^*$

$C_{2s} = -0.60^*$ $H_2\ \sigma_u = 0.67^*$

$C_{2p} = -0.33^*$

*The orbital energies were generated using standard STO3G minimum basis set SCF calculations.

3. Using the empirical parameters given below for *C* and *H* (taken from appendix F and *The HMO Model and its Applications* by E. Heilbronner and H. Bock, Wiley-Interscience, N.Y., 1976), apply the Hückel model to ethylene in order to determine the valence electronic structure of this system. Note that you will be obtaining the *1*-electron energies and wavefunctions by solving the *secular equation* (as you *always* will when the energy is dependent upon a set of linear parameters like the MO coefficients in the LCAO-MO approach) using the definitions for the matrix elements found in appendix F.

$C\alpha_{2p\pi} = -11.4\text{eV}$
$C\alpha_{sp^2} = -14.7\text{eV}$
$H\alpha_s = -13.6eV$
C-C $\beta_{2p\pi-2p\pi} = -1.2\text{eV}$
C-C $\beta_{sp^2-sp^2} = -5.0\text{eV}$
C-H $\beta_{sp^2-s} = -4.0\text{eV}$

a. Determine the $C = C(2p\pi)$ *1*-electron molecular orbital energies and wavefunctions. Calculate the $\pi \rightarrow \pi^*$ transition energy for ethylene within this model.

b. Determine the C-C (sp^2) *1*-electron molecular orbital energies and wavefunctions.

c. Determine the C-H ($sp^2 - s$) *1*-electron molecular orbital energies and wavefunctions (note that appropriate choice of symmetry will reduce this 8×8 matrix down to four matrices; that is, you are encouraged to symmetry adapt the atomic orbitals before starting the Hückel calculation). Draw a qualitative orbital energy diagram using the HMO energies you have calculated.

4. Using the empirical parameters given below for *B* and *H* (taken from appendix F and *The HMO Model and its Applications* by E. Heilbronner and H. Bock, Wiley-Interscience, N.Y., 1976), apply the Hückel model to borane (BH_3) in order to determine the valence electronic structure of this system.

$B\alpha_{2p\pi} = -8.5\text{eV}$
$B\alpha_{sp^2} = -10.7\text{eV}$
$H\alpha_s = -13.6\text{eV}$
$B - H\ \beta_{sp^2-s} = -3.5\text{eV}$

Determine the symmetries of the resultant molecular orbitals in the D_{3h} point group. Draw a qualitative orbital energy diagram using the HMO energies you have calculated.

5. Qualitatively analyze the electronic structure (orbital energies and *1*-electron wavefunctions) of PF_5. Analyze only the 3*s* and 3*p* electrons of P and the *one* 2*p* bonding electron of each F. Proceed with a D_{3h} analysis in the following manner:

- **a.** Symmetry adapt the top and bottom F atomic orbitals.
- **b.** Symmetry adapt the three (trigonal) F atomic orbitals.
- **c.** Symmetry adapt the P 3*s* and 3*p* atomic orbitals.
- **d.** Allow these three sets of D_{3h} orbitals to interact and draw the resultant orbital energy diagram. Symmetry label each of these molecular energy levels. Fill this energy diagram with 10 "valence" electrons.

SOLUTIONS

Review Exercises

1.

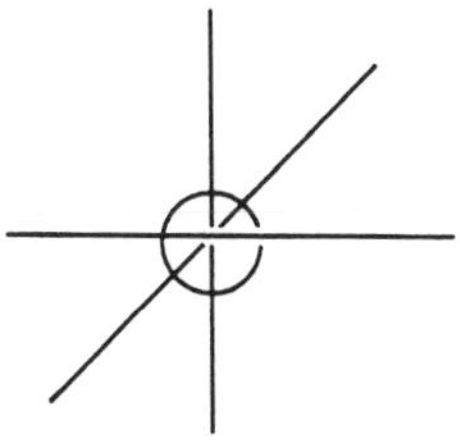

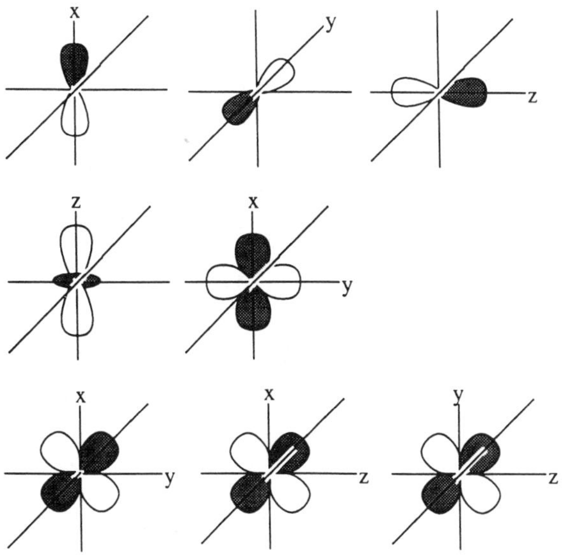

2. **a.** In ammonia the only "core" orbital is the N 1s and this becomes an a_1 orbital in C_{3v} symmetry. The N 2s orbitals and 3 H 1s orbitals become 2 a_1 and an e set of orbitals. The remaining N 2p orbitals also become 1a_1 and a set of e orbitals. The total valence orbitals in C_{3v} symmetry are 3a_1 and 2e orbitals.

b. In water the only core orbital is the O 1s and this becomes an a_1 orbital in C_{2v} symmetry. Placing the molecule in the yz plane allows us to further analyze the remaining valence orbitals as: O $2p_z = a_1$, O $2p_y$ as b_2, and O $2p_x$ as b_1. The H 1s + H 1s combination is an a_1 whereas the H 1s – H 1s combination is a b_2.

c. Placing the oxygens of H_2O_2 in the yz plane (z bisecting the oxygens) and the (cis) hydrogens distorted slightly in +x and –x directions allows us to analyze the orbitals as follows. The core O 1s + O 1s combination is an a orbital whereas the O 1s – O 1s combination is a b orbital. The valence orbitals are:

O $2s$ + O $2s = a$, O $2s$ – O $2s = b$, O $2p_x$ + O $2p_x = b$, O $2p_x$ – O $2p_x = a$, O $2p_y$ + O $2p_y = a$, O $2p_y$ – O $2p_y = b$, O $2p_z$ + O $2p_z = b$, O $2p_z$ – O $2p_z = a$, H 1s + H 1$s = a$,

and finally the H 1s – H 1$s = b$.

d. For the next two problems we will use the convention of choosing the z axis as principal axis for the $D_\infty h$, D_{2h}, and C_{2v} point groups and the xy plane as the horizontal reflection plane in C_s symmetry.

	$D_\infty h$	D_{2h}	C_{2v}	C_s
N 1s	σ_g	a_g	a_1	a'
N2s	σ_g	a_g	a_1	a'
N2p_x	π_{xu}	b_{3u}	b_1	a'
N2p_y	π_{yu}	b_{2u}	b_2	a'
N2p_z	σ_u	b_{1u}	a_1	a''

2. **e.** The Nitrogen molecule is in the yz plane for all point groups except the C_s in which case it is placed in the xy plane.

	$D\int h$	D_{2h}	C_{2v}	C_s
N1s + N1s	σ_g	a_g	a_1	a'
N1s – N1s	σ_u	b_{1u}	b_2	a'
N2s + N2s	σ_g	a_g	a_1	a'
N2s – N2s	σ_u	b_{1u}	b_2	a'
N2p_x + N2p_x	π_{xu}	b_{3u}	b_1	a'
N2p_x – N2p_x	π_{xg}	b_{2g}	a_2	a'
N2p_y + N2p_y	π_{yu}	b_{2u}	a_1	a'
N2p_y – N2p_y	π_{yg}	b_{3g}	b_2	a'
N2p_z + N2p_z	σ_u	b_{1u}	b_2	a''
N2p_z – N2p_z	σ_g	a_g	a_1	a''

3.

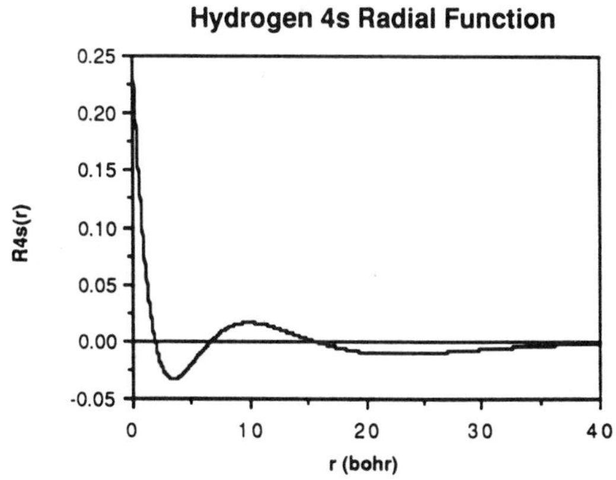
Hydrogen 4s Radial Function
R4s(r)
0.25
0.20
0.15
0.10
0.05
0.00
-0.05
0
10
20
30
40
r (bohr)

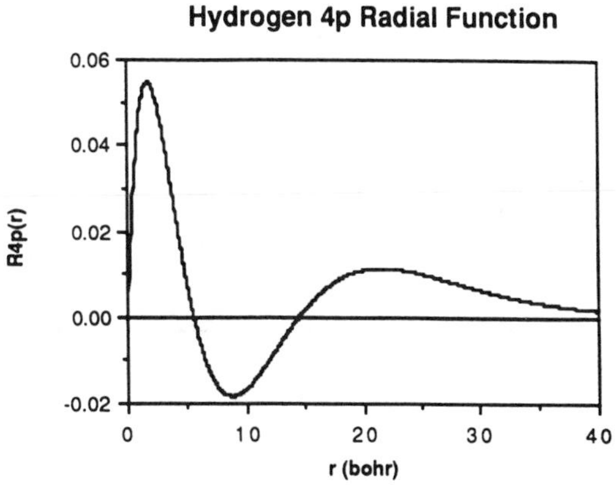
Hydrogen 4p Radial Function
R4p(r)
0.06
0.04
0.02
0.00
-0.02
0
10
20
30
40
r (bohr)

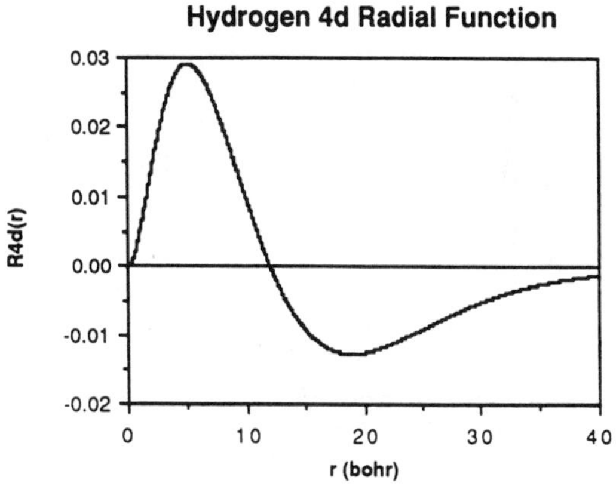
Hydrogen 4d Radial Function
R4d(r)
0.03
0.02
0.01
0.00
-0.01
-0.02
0
10
20
30
40
r (bohr)

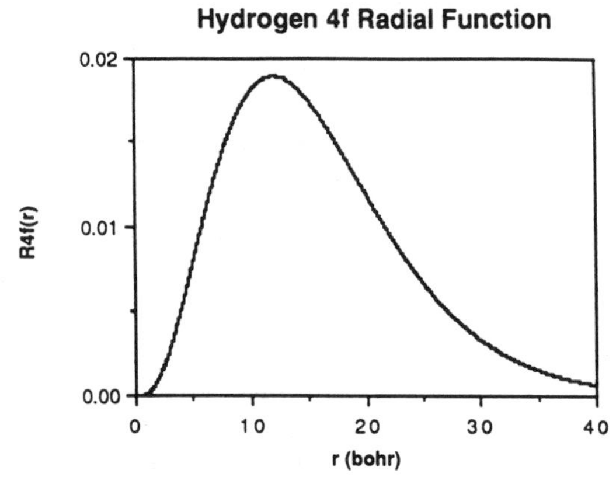
Hydrogen 4f Radial Function
R4f(r)
0.02
0.01
0.00
0
10
20
30
40
r (bohr)

4.

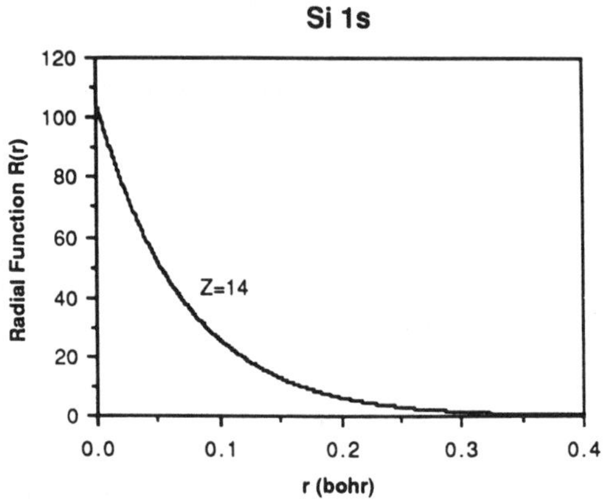
Si 1s
Radial Function R(r)
120
100
80
60
40
20
0
Z=14
0.0
0.1
0.2
0.3
0.4
r (bohr)

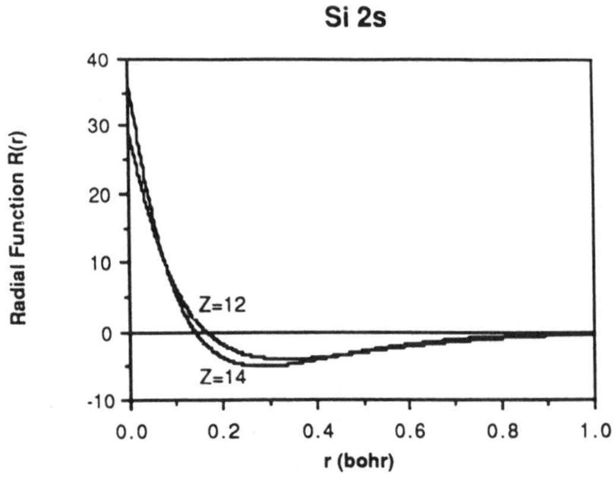
Si 2s
Radial Function R(r)
40
30
20
10
0
-10
Z=12
Z=14
0.0
0.2
0.4
0.6
0.8
1.0
r (bohr)

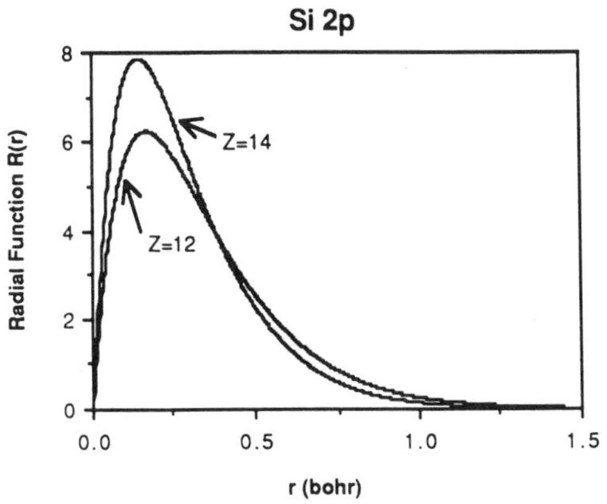

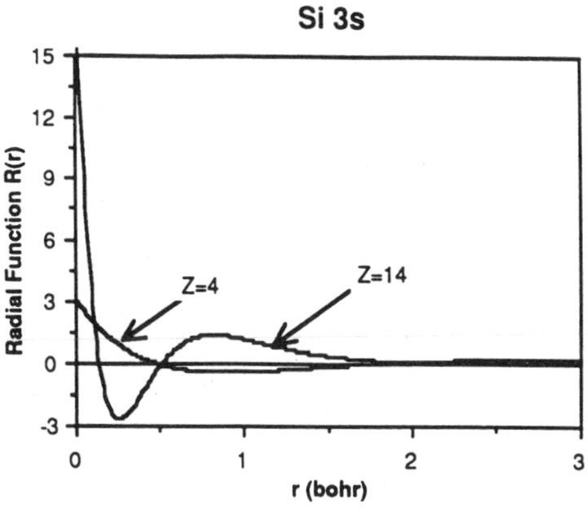

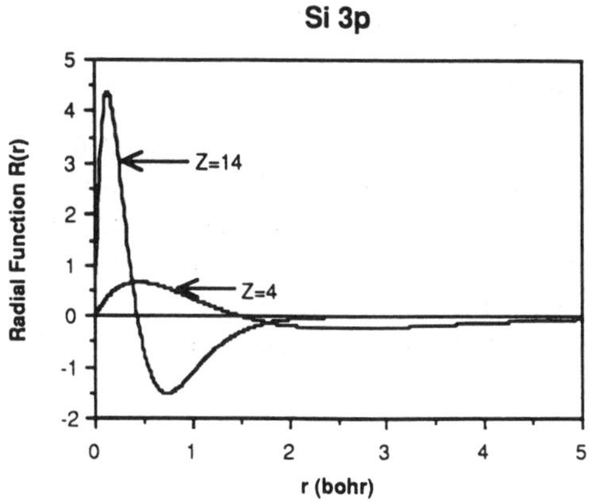

Exercises

1. Two Slater-type orbitals, *i* and *j*, centered on the same point results in the following overlap integrals:

$$S_{ij} = \int_0^{2\pi}\int_0^{\pi}\int_0^{\infty} \left(\frac{2\zeta_i}{a_0}\right)^{n_i+\frac{1}{2}} \left(\frac{1}{(2n_i)!}\right)^{\frac{1}{2}} r^{(n_i-1)} e^{\left(\frac{-\zeta_i r}{a_0}\right)} Y_{l_i,m_i}(\theta,\phi)\,\cdot$$

$$\left(\frac{2\zeta_j}{a_0}\right)^{n_j+\frac{1}{2}} \left(\frac{1}{(2n_j)!}\right)^{\frac{1}{2}} r^{(n_j-1)} e^{\left(\frac{-\zeta_j r}{a_0}\right)} Y_{l_j,m_j}(\theta,\phi)\cdot r^2 \sin\theta\, dr d\theta d\phi \,.$$

For these s orbitals $l = m = 0$ and $Y_{0,0}(\theta,\phi) = \frac{1}{\sqrt{4\pi}}$.

Performing the integrations over θ and ϕ yields 4π which then cancels with these Y terms. The integral then reduces to:

$$S_{ij} = \left(\frac{2\zeta_i}{a_0}\right)^{n_i+\frac{1}{2}} \left(\frac{1}{(2n_i)!}\right)^{\frac{1}{2}} \left(\frac{2\zeta_j}{a_0}\right)^{n_j+\frac{1}{2}} \left(\frac{1}{(2n_j)!}\right)^{\frac{1}{2}} \int_0^{\infty} r^{(n_i-1+n_j-1)} e^{\left(\frac{-(\zeta_i+\zeta_j)r}{a_0}\right)} r^2 dr$$

$$= \left(\frac{2\zeta_i}{a_0}\right)^{n_i+\frac{1}{2}} \left(\frac{1}{(2n_i)!}\right)^{\frac{1}{2}} \left(\frac{2\zeta_j}{a_0}\right)^{n_j+\frac{1}{2}} \left(\frac{1}{(2n_j)!}\right)^{\frac{1}{2}} \int_0^{\infty} r^{(n_i+n_j)} e^{\left(\frac{-(\zeta_i+\zeta_j)r}{a_0}\right)} dr$$

Using integral equation (4) the integral then reduces to:

$$S_{ij} = \left(\frac{2\zeta_i}{a_0}\right)^{n_i+\frac{1}{2}} \left(\frac{1}{(2n_i)!}\right)^{\frac{1}{2}} \left(\frac{2\zeta_j}{a_0}\right)^{n_j+\frac{1}{2}} \left(\frac{1}{(2n_j)!}\right)^{\frac{1}{2}} (n_i+n_j)! \left(\frac{a_0}{\zeta_i+\zeta_j}\right)^{n_i+n_j+1}$$

We then substitute in the values for each of these constants:

for $i = 1$; $n = 1$, $l = m = 0$, and $\zeta = 2.6906$
for $i = 2$; $n = 2$, $l = m = 0$, and $\zeta = 0.6396$
for $i = 3$; $n = 3$, $l = m = 0$, and $\zeta = 0.1503$.

Evaluating each of these matrix elements we obtain:

$S_{11} = (12.482992)(0.707107)(12.482992)(0.707107)(2.000000)(0.006417) = 1.000000$
$S_{21} = S_{12} = (1.850743)(0.204124)(12.482992)(0.707107)(6.000000)(0.008131) = 0.162673$
$S_{22} = (1.850743)(0.204124)(1.850743)(0.204124)(24.000000)(0.291950) = 1.000000$
$S_{31} = S_{13} = (0.014892)(0.037268)(12.482992)(0.707107)(24.000000)(0.005404) = 0.000635$
$S_{32} = S_{23} = (0.014892)(0.037268)(1.850743)(0.204124)(120.000000)(4.116872) = 0.103582$
$S_{33} = (0.014892)(0.037268)(0.014892)(0.037268)(720.000000)(4508.968136) = 1.000000$

$$S = \begin{bmatrix} 1.000000 & & \\ 0.162673 & 1.000000 & \\ 0.000635 & 0.103582 & 1.000000 \end{bmatrix}$$

We now solve the matrix eigenvalue problem $SU = \lambda U$. The eigenvalues, λ, of this overlap matrix are:

[0.807436 0.999424 1.193139],

and the corresponding eigenvectors, U, are:

$$\begin{bmatrix} 0.596540 & -0.537104 & -0.596372 \\ -0.707634 & -0.001394 & -0.706578 \\ 0.378675 & 0.843515 & -0.380905 \end{bmatrix}.$$

The $\lambda^{-\frac{1}{2}}$ matrix becomes:

$$\lambda^{-\frac{1}{2}} = \begin{bmatrix} 1.112874 & 0.000000 & 0.000000 \\ 0.000000 & 1.000288 & 0.000000 \\ 0.000000 & 0.000000 & 0.915492 \end{bmatrix}.$$

Back transforming into the original eigenbasis gives $S^{-\frac{1}{2}}$, e.g.,

$$S^{-\frac{1}{2}} = U\lambda^{-\frac{1}{2}}U^T$$

$$S^{-\frac{1}{2}} = \begin{bmatrix} 1.010194 & & \\ -0.083258 & 1.014330 & \\ 0.006170 & -0.052991 & 1.004129 \end{bmatrix}$$

The old ao matrix can be written as:

$$C = \begin{bmatrix} 1.000000 & 0.000000 & 0.000000 \\ 0.000000 & 1.000000 & 0.000000 \\ 0.000000 & 0.000000 & 1.000000 \end{bmatrix}.$$

The new ao matrix (which now gives each ao as a linear combination of the original aos) then becomes:

$$C' = S^{-\frac{1}{2}}C = \begin{bmatrix} 1.010194 & -0.083258 & 0.006170 \\ -0.083258 & 1.014330 & -0.052991 \\ 0.006170 & -0.052991 & 1.004129 \end{bmatrix}$$

These new aos have been constructed to meet the orthonormalization requirement $C'^TSC' = 1$ since:

$$\left(S^{-\frac{1}{2}}C\right)^T SS^{-\frac{1}{2}}C = C^TS^{-\frac{1}{2}}SS^{-\frac{1}{2}}C = C^TC = 1.$$

But, it is always good to check our result and indeed:

$$C'^TSC' = \begin{bmatrix} 1.000000 & 0.000000 & 0.000000 \\ 0.000000 & 1.000000 & 0.000000 \\ 0.000000 & 0.000000 & 1.000000 \end{bmatrix}$$

2. The least time consuming route here is to evaluate each of the needed integrals first. These are evaluated analogous to exercise 1, letting χ_i denote each of the individual Slater-type orbitals.

$$\int_0^\infty \chi_i r \chi_j r^2 dr = \langle r \rangle_{ij}$$

$$= \left(\frac{2\zeta_i}{a_0}\right)^{n_i+\frac{1}{2}} \left(\frac{1}{(2n_i)!}\right)^{\frac{1}{2}} \left(\frac{2\zeta_j}{a_0}\right)^{n_j+\frac{1}{2}} \left(\frac{1}{(2n_j)!}\right)^{\frac{1}{2}} \int_0^\infty r^{(n_i+n_j+1)} e^{\left(\frac{-(\zeta_i+\zeta_j)r}{a_0}\right)} dr$$

Once again using integral equation (4) the integral reduces to:

$$= \left(\frac{2\zeta_i}{a_0}\right)^{n_i+\frac{1}{2}} \left(\frac{1}{(2n_i)!}\right)^{\frac{1}{2}} \left(\frac{2\zeta_j}{a_0}\right)^{n_j+\frac{1}{2}} \left(\frac{1}{(2n_j)!}\right)^{\frac{1}{2}} (n_i+n_j+1)! \left(\frac{a_0}{\zeta_i+\zeta_j}\right)^{n_i+n_j+2}$$

Again, upon substituting in the values for each of these constants, evaluation of these expectation values yields:

$$<r>_{11} = (12.482992)(0.707107)(12.482992)(0.707107)(6.000000)(0.001193) = 0.557496$$
$$<r>_{21} = <r>_{12} = (1.850743)(0.204124)(12.482992)(0.707107)(24.000000)(0.002441) = 0.195391$$
$$<r>_{22} = (1.850743)(0.204124)(1.850743)(0.204124)(120.000000)(0.228228) = 3.908693$$
$$<r>_{31} = <r>_{13} = (0.014892)(0.037268)(12.482992)(0.707107)(120.000000)(0.001902) = 0.001118$$
$$<r>_{32} = <r>_{23} = (0.014892)(0.037268)(1.850743)(0.204124)(720.000000)(5.211889) = 0.786798$$
$$<r>_{33} = (0.014892)(0.037268)(0.014892)(0.037268)(5040.000000)(14999.893999) = 23.286760$$

$$\int_0^\infty \chi_i r \chi_j r^2 dr = <r>_{ij} = \begin{bmatrix} 0.557496 & & \\ 0.195391 & 3.908693 & \\ 0.001118 & 0.786798 & 23.286760 \end{bmatrix}$$

Using these integrals one then proceeds to evaluate the expectation values of each of the orthogonalized aos, χ'_n, as:

$$\int_0^\infty \chi'_n r \chi'_n r^2 dr = \sum_{i=1}^{3} \sum_{j=1}^{3} C'_{ni} C'_{nj} <r>_{ij}.$$

This results in the following expectation values (in atomic units):

$$\int_0^\infty \chi'_{1s} r \chi'_{1s} r^2 dr = 0.563240 \text{bohr}$$

$$\int_0^\infty \chi'_{2s} r \chi'_{2s} r^2 dr = 3.973199 \text{bohr}$$

$$\int_0^\infty \chi'_{3s} r \chi'_{3s} r^2 dr = 23.406622 \text{bohr}$$

3. The radial density for each orthogonalized orbital, χ'_n, assuming integrations over θ and ϕ have already been performed can be written as:

$$\int_0^\infty \chi'_n \chi'_n r^2 dr = \sum_{i=1}^{3} \sum_{j=1}^{3} C'_{ni} C'_{nj} \int_0^\infty R_i R_j r^2 dr,$$

where R_i and R_j are the radial portions of the individual Slater-type orbitals, e.g.,

$$R_i R_j r^2 = \left(\frac{2\zeta_i}{a_0}\right)^{n_i+\frac{1}{2}} \left(\frac{1}{(2n_i)!}\right)^{\frac{1}{2}} \left(\frac{2\zeta_j}{a_0}\right)^{n_j+\frac{1}{2}} \left(\frac{1}{(2n_j)!}\right)^{\frac{1}{2}} r^{(n_i+n_j)} e^{\left(\frac{-(\zeta_i+\zeta_j)r}{a_0}\right)}$$

Therefore a plot of the radial probability for a given orthogonalized atomic orbital, n, will be:

$$\sum_{i=1}^{3} \sum_{j=1}^{3} C'_{ni} C'_{nj} R_i R_j r^2 \text{ vs } r.$$

Plot of the orthogonalized 1*s* orbital probability density *vs* *r*; note there are no nodes.

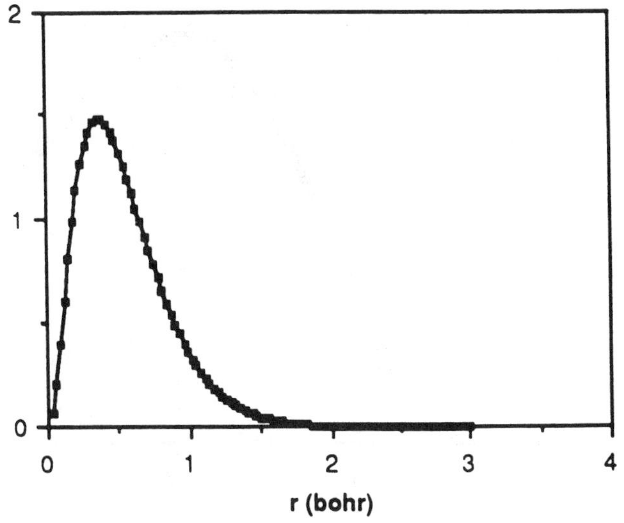

Plot of the orthogonalized 2*s* orbital probability density *vs* *r*; note there is one node.

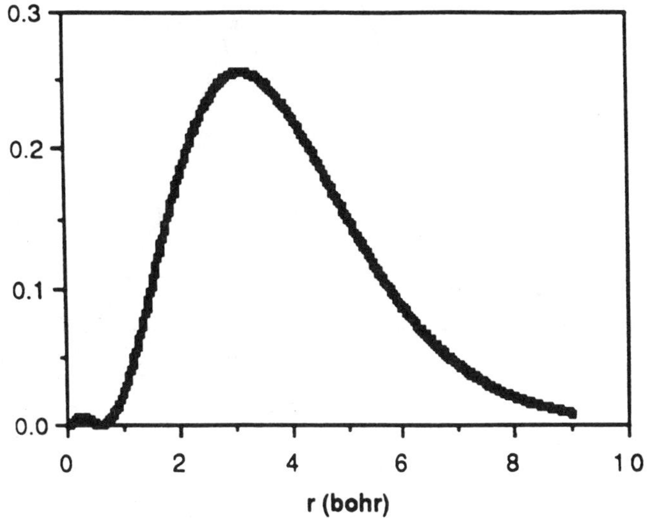

Plot of the orthogonalized 3*s* orbital probability density *vs* *r*; note there are two nodes in the 0–5 bohr region but they are not distinguishable as such. A duplicate plot with this nodal region expanded follows.

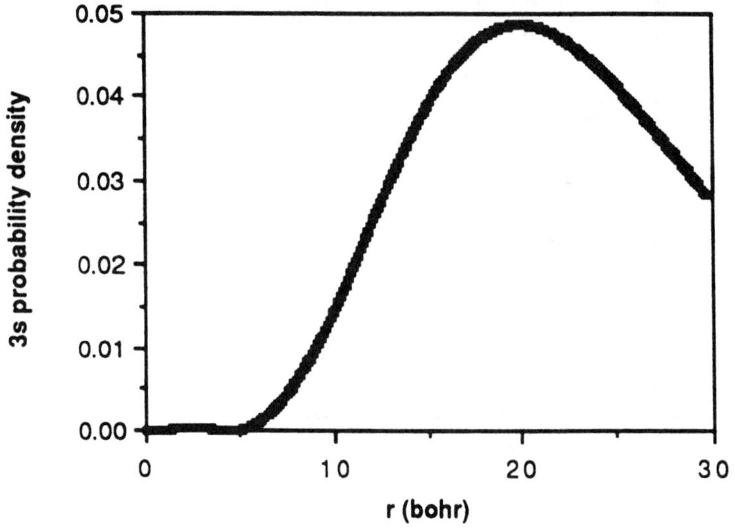

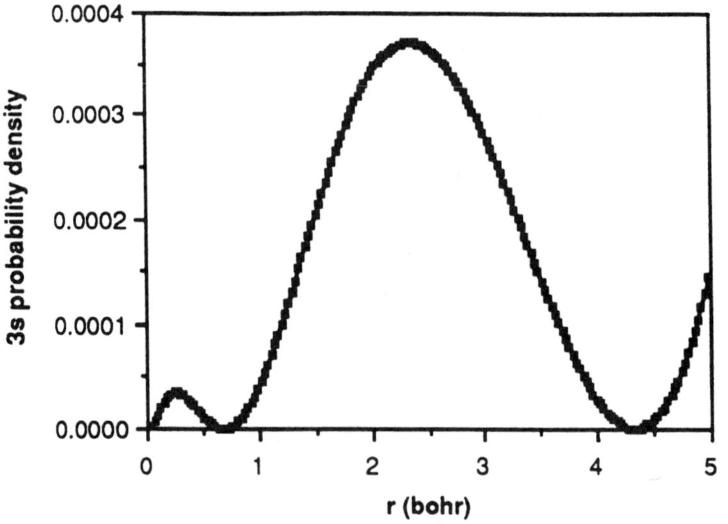

Problems

1.

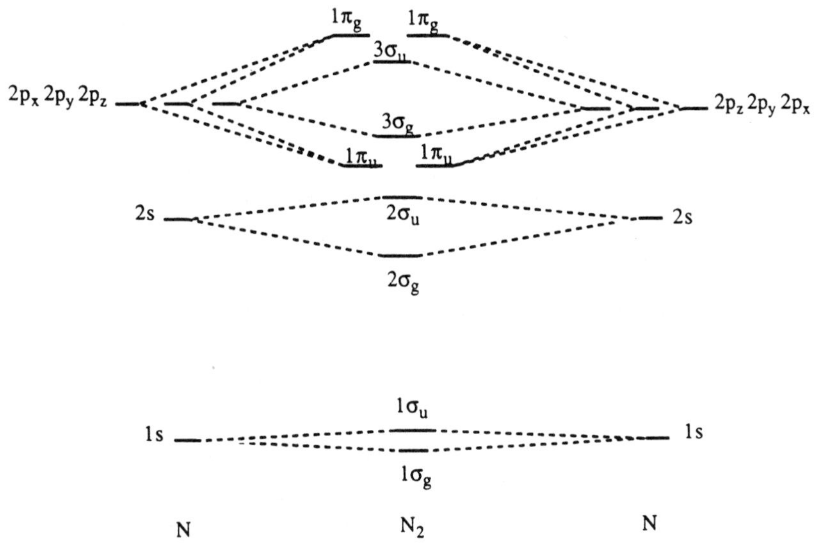

The above diagram indicates how the SALC-AOs are formed from the $1s$, $2s$, and $2p$ N atomic orbitals. It can be seen that there are $3\sigma_g$, $3\sigma_u$, $1\pi_{ux}$, $1\pi_{uy}$, $1\pi_{gx}$, and $1\pi_{gy}$ SALC-AOs. The Hamiltonian matrices (Fock matrices) are given. Each of these can be diagonalized to give the following MO energies:

$3\sigma_g$; –15.52, –1.45, and –0.54 (hartrees)
$3\sigma_u$; –15.52, –0.72, and 1.13
$1\pi_{ux}$; –0.58
$1\pi_{uy}$; –0.58
$1\pi_{gx}$; 0.28
$1\pi_{gy}$; 0.28

It can be seen that the $3\sigma_g$ orbitals are bonding, the $3\sigma_u$ orbitals are antibonding, the $1\pi_{ux}$ and $1\pi_{uy}$ orbitals are bonding, and the $1\pi_{gx}$ and $1\pi_{gy}$ orbitals are antibonding. The eigenvectors one obtains are in the orthogonal basis and therefore pretty meaningless. Back transformation into the original basis will generate the expected results for the $1e^-$ MOs (expected combinations of SALC-AOs).

2. Using these approximate energies we can draw the following MO diagram:

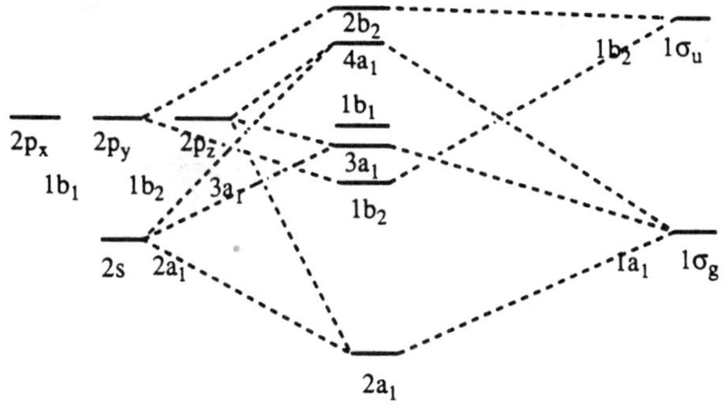

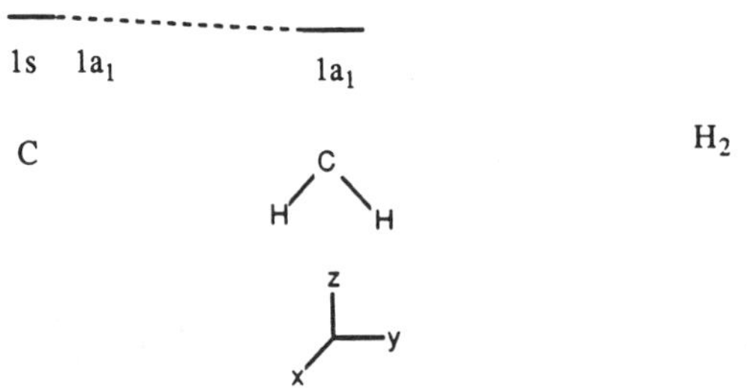

This MO diagram is *not* an orbital correlation diagram but can be used to help generate one. The energy levels on each side (C and H_2) can be "superimposed" to generate the left side of the orbital correlation diagram and the center CH_2 levels can be used to form the right side. Ignoring the core levels this generates the following orbital correlation diagram.

Orbital-correlation diagram for the reaction $C + H_2$ -----> CH_2 (bent)

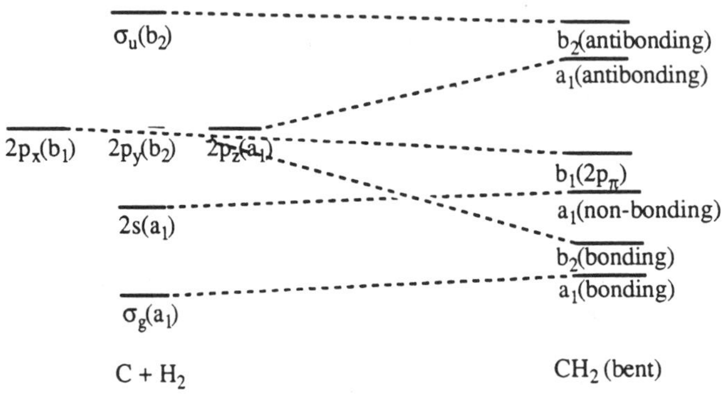

3.

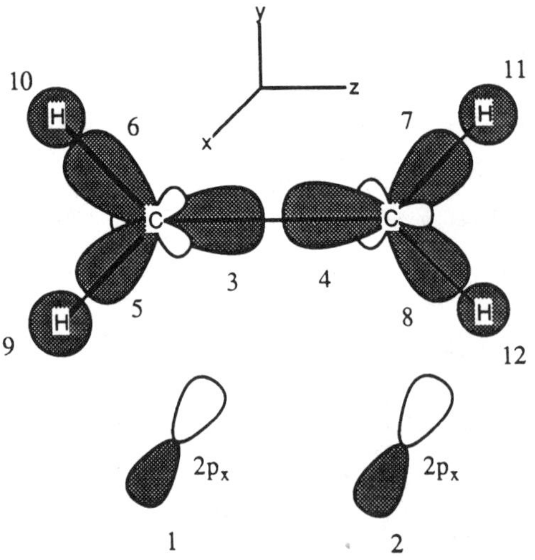

Using D_{2h} symmetry and labeling the orbitals ($f_1 - f_{12}$) as shown above proceed by using the orbitals to define a reducible representation.which may be subsequently reduced to its irreducible components. Use projectors to find the SALC-AOs for these irreps.

a. The $2P_x$ orbitals on each carbon form the following reducible representation:

D_{2h}	E	$C_2(z)$	$C_2(y)$	$C_2(x)$	i	$\sigma(xy)$	$\sigma(xz)$	$\sigma(yz)$
Γ_{2p_x}	2	−2	0	0	0	0	2	−2

The number of irreducible representations may be found by using the following formula:

$$n_{irrep} = \frac{1}{g} \sum_R \chi_{red}(R)\chi_{irrep}(R) ,$$

where g = the order of the point group (8 for D_{2h}).

$$n_{Ag} = \frac{1}{8} \sum_R \Gamma_{2p_x}(R) \cdot A_g(R)$$

$$= \frac{1}{8}\left[(2)(1) + (-2)(1) + (0)(1) + (0)(1) + (0)(1) + (0)(1) + (2)(1) + (-2)(1)\right] = 0$$

Similarly,

$$n_{B_{1g}} = 0$$
$$n_{B_{2g}} = 1$$
$$n_{B_{3g}} = 0$$
$$n_{A_u} = 0$$
$$n_{B_{1u}} = 0$$
$$n_{B_{2u}} = 0$$
$$n_{B_{3u}} = 1$$

Projectors using the formula:

$$P_{irrep} = \sum_R \chi_{irrep}(R)R,$$

may be used to find the SALC-AOs for these irreducible representations.

$$P_{B_{2g}} = \sum_R \chi_{B_{2g}}(R)R ,$$

$$P_{B_{2g}}f_1 = (1)Ef_1 + (-1)C_2(z)f_1 + (1)C_2(y)f_1 + (-1)C_2(x)f_1 + (1)if_1 + (-1)\sigma(xy)f_1 + (1)\sigma(xz)f_1 + (-1)\sigma(yz)f_1$$
$$= (1)f_1 + (-1) - f_1 + (1) - f_2 + (-1)f_2 + (1) - f_2 + (-1)f_2 + (1)f_1 + (-1) - f_1$$
$$= f_1 + f_1 - f_2 - f_2 - f_2 - f_2 + f_1 + f_1$$
$$= 4f_1 - 4f_2$$

Normalization of this SALC-AO (and representing the SALC-AOs with ϕ) yields:

$$\int N(f_1 - f_2)N(f_1 - f_2)d\tau = 1$$

$$N^2\left(\int f_1f_1d\tau - \int f_1f_2d\tau - \int f_2f_1d\tau + \int f_2f_2d\tau\right) = 1$$

$$N^2(1 + 1) = 1$$

$$2N^2 = 1$$
$$N = \frac{1}{\sqrt{2}}$$
$$\phi_{1b_{2g}} = \frac{1}{\sqrt{2}}(f_1 - f_2).$$

The B_{3u} SALC-AO may be found in a similar fashion:

$$\begin{aligned} P_{B3u}f_1 &= (1)f_1 + (-1) - f_1 + (-1) - f_2 + (1)f_2 + (-1) - f_2 + (1)f_2 + (1)f_1 + (-1) - f_1 \\ &= f_1 + f_1 + f_2 + f_2 + f_2 + f_2 + f_1 + f_1 \\ &= 4f_1 + 4f_2 \end{aligned}$$

Normalization of this SALC-AO yields:

$$\phi_{1b_{3u}} = \frac{1}{\sqrt{2}}(f_1 + f_2).$$

Since there are only two SALC-AOs and both are of different symmetry types these SALC-AOs are MOs and the 2 × 2 Hamiltonian matrix reduces to two 1 × 1 matrices.

$$\begin{aligned} H_{1b_{2g},1b_{2g}} &= \int \frac{1}{\sqrt{2}}(f_1 - f_2)H\frac{1}{\sqrt{2}}(f_1 - f_2)d\tau \\ &= \frac{1}{2}\left(\int f_1Hf_1d\tau - 2\int f_1Hf_2d\tau + \int f_2Hf_2d\tau\right) \\ &= \frac{1}{2}\left(\alpha_{2p\pi} - 2\beta_{2p\pi-2p\pi} + \alpha_{2p\pi}\right) \\ &= \alpha_{2p\pi} - \beta_{2p\pi-2p\pi} \\ &= -11.4 - (-1.2) = -10.2 \end{aligned}$$

$$\begin{aligned} H_{1b3u,1b3u} &= \int \frac{1}{\sqrt{2}}(f_1 + f_2)H\frac{1}{\sqrt{2}}(f_1 + f_2)d\tau \\ &= \frac{1}{2}\left(\int f_1Hf_1d\tau + 2\int f_1Hf_2d\tau + \int f_2Hf_2d\tau\right) \\ &= \frac{1}{2}\left(\alpha_{2p\pi} + 2\beta_{2p\pi-2p\pi} + \alpha_{2p\pi}\right) \\ &= \alpha_{2p\pi} + \beta_{2p\pi-2p\pi} \\ &= -11.4 + (-1.2) = -12.6 \end{aligned}$$

This results in a $\pi \rightarrow \pi^*$ splitting of 2.4 eV.

b. The sp^2 orbitals forming the C-C bond generate the following reducible representation:

D_{2h}	E	$C_2(z)$	$C_2(y)$	$C_2(x)$	i	$\sigma(xy)$	$\sigma(xz)$	$\sigma(yz)$
Γ_{Csp2}	2	2	0	0	0	0	2	2

This reducible representation reduces to $1A_g$ and $1\ B_{1u}$ irreducible representations. Projectors are used to find the SALC-AOs for these irreducible representations.

$$\begin{aligned} P_{A_g}f_3 &= (1)Ef_3 + (1)C_2(z)f_3 + (1)C_2(y)f_3 + (1)C_2(x)f_3 + (1)if_3 + (1)\sigma(xy)f_3 + (1)\sigma(xz)f_3 + (1)\sigma(yz)f_3 \\ &= (1)f_3 + (1)f_3 + (1)f_4 + (1)f_4 + (1)f_4 + (1)f_4 + (1)f_3 + (1)f_3 \\ &= 4f_3 + 4f_4 \end{aligned}$$

Normalization of this SALC-AO yields:

$$\phi_{1ag} = \frac{1}{\sqrt{2}}(f_3 + f_4).$$

The B_{1u} SALC-AO may be found in a similar fashion:

$$\begin{aligned} P_{B1u}f_3 &= (1)f_3 + (1)f_3 + (-1)f_4 + (-1)f_4 + (-1)f_4 + (-1)f_4 + (1)f_3 + (1)f_3 \\ &= 4f_3 - 4f_4 \end{aligned}$$

Normalization of this SALC-AO yields:

$$\phi_{1b_{3u}} = \frac{1}{\sqrt{2}}(f_3 - f_4).$$

Again since there are only two SALC-AOs and both are of different symmetry types these SALC-AOs are MOs and the 2×2 Hamiltonian matrix reduces to two 1×1 matrices.

$$\begin{aligned} H_{1_{a_g},1_{a_g}} &= \int \frac{1}{\sqrt{2}}(f_3 + f_4)H\frac{1}{\sqrt{2}}(f_3 + f_4)d\tau \\ &= \frac{1}{2}\left(\int f_3Hf_3d\tau + 2\int f_3Hf_4d\tau + \int f_4Hf_4d\tau\right) \\ &= \frac{1}{2}\left(\alpha_{sp^2} + 2\beta_{sp^2-sp^2} + \alpha_{sp^2}\right) \\ &= \alpha_{sp^2} + \beta_{sp^2-sp^2} \\ &= -14.7 + (-5.0) = -19.7 \end{aligned}$$

$$\begin{aligned} H_{1b_{1u},1b_{1u}} &= \int \frac{1}{\sqrt{2}}(f_3 - f_4)H\frac{1}{\sqrt{2}}(f_3 - f_4)d\tau \\ &= \frac{1}{2}\left(\int f_3Hf_3d\tau - 2\int f_3Hf_4d\tau + \int f_4Hf_4d\tau\right) \\ &= \frac{1}{2}\left(\alpha_{sp^2} - 2\beta_{sp^2-sp^2} + \alpha_{sp^2}\right) \\ &= \alpha_{sp^2} - \beta_{sp^2-sp^2} \\ &= -14.7 - (-5.0) = -9.7 \end{aligned}$$

c. The C sp^2 orbitals and the H s orbitals forming the C-H bonds generate the following reducible representation:

D_{2h}	E	$C_2(z)$	$C_2(y)$	$C_2(x)$	i	$\sigma(xy)$	$\sigma(xz)$	$\sigma(yz)$
Γ_{sp^2-s}	8	0	0	0	0	0	0	8

This reducible representation reduces to $2A_g$, $2B_{3g}$, $2B_{1u}$ and $2B_{2u}$ irreducible representations. Projectors are used to find the SALC-AOs for these irreducible representations.

$$\begin{aligned} P_{A_g}f_6 &= (1)Ef_6 + (1)C_2(z)f_6 + (1)C_2(y)f_6 + (1)C_2(x)f_6 + (1)if_6 + (1)\sigma(xy)f_6 + (1)\sigma(xz)f_6 + (1)\sigma(yz)f_6 \\ &= (1)f_6 + (1)f_5 + (1)f_7 + (1)f_8 + (1)f_8 + (1)f_7 + (1)f_5 + (1)f_6 \\ &= 2f_5 + 2f_6 + 2f_7 + 2f_8 \end{aligned}$$

Normalization yields: $\phi_{2_{a_g}} = \frac{1}{2}(f_5 + f_6 + f_7 + f_8)$.

$$\begin{aligned} P_{A_g}f_{10} &= (1)Ef_{10} + (1)C_2(z)f_{10} + (1)C_2(y)f_{10} + (1)C_2(x)f_{10} + \\ &\quad (1)if_{10} + (1)\sigma(xy)f_{10} + (1)\sigma(xz)f_{10} + (1)\sigma(yz)f_{10} \\ &= (1)f_{10} + (1)f_9 + (1)f_{11} + (1)f_{12} + (1)f_{12} + (1)f_{11} + (1)f_9 + (1)f_{10} \\ &= 2f_9 + 2f_{10} + 2f_{11} + 2f_{12} \end{aligned}$$

Normalization yields: $\phi_{3_{a_g}} = \frac{1}{2}(f_9 + f_{10} + f_{11} + f_{12})$.

$$P_{B_{3g}} f_6 = (1)f_6 + (-1)f_5 + (-1)f_7 + (1)f_8 + (1)f_8 + (-1)f_7 + (-1)f_5 + (1)f_6$$
$$= -2f_5 + 2f_6 - 2f_7 + 2f_8$$

Normalization yields: $\phi_{1b_{3_g}} = \frac{1}{2}(-f_5 + f_6 - f_7 + f_8)$.

$$P_{B_{3g}} f_{10} = (1)f_{10} + (-1)f_9 + (-1)f_{11} + (1)f_{12} + (1)f_{12} + (-1)f_{11} + (-1)f_9 + (1)f_{10}$$
$$= -2f_9 + 2f_{10} - 2f_{11} + 2f_{12}$$

Normalization yields: $\phi_{2b_{3_g}} = \frac{1}{2}(-f_9 + f_{10} - f_{11} + f_{12})$.

$$P_{B1u} f_6 = (1)f_6 + (1)f_5 + (-1)f_7 + (-1)f_8 + (-1)f_8 + (-1)f_7 + (1)f_5 + (1)f_6$$
$$= 2f_5 + 2f_6 - 2f_7 - 2f_8$$

Normalization yields: $\phi_{2b_{1_u}} = \frac{1}{2}(f_5 + f_6 - f_7 - f_8)$.

$$P_{B1u} f_{10} = (1)f_{10} + (1)f_9 + (-1)f_{11} + (-1)f_{12} + (-1)f_{12} + (-1)f_{11} + (1)f_9 + (1)f_{10}$$
$$= 2f_9 + 2f_{10} - 2f_{11} - 2f_{12}$$

Normalization yields: $\phi_{3b_{1_u}} = \frac{1}{2}(f_9 + f_{10} - f_{11} - f_{12})$.

$$P_{B_{2u}} f_6 = (1)f_6 + (-1)f_5 + (1)f_7 + (-1)f_8 + (-1)f_8 + (1)f_7 + (-1)f_5 + (1)f_6$$
$$= -2f_5 + 2f_6 + 2f_7 - 2f_8$$

Normalization yields: $\phi_{1b_{2_u}} = \frac{1}{2}(-f_5 + f_6 + f_7 - f_8)$.

$$P_{B_{2u}} f_{10} = (1)f_{10} + (-1)f_9 + (1)f_{11} + (-1)f_{12} + (-1)f_{12} + (1)f_{11} + (-1)f_9 + (1)f_{10}$$
$$= -2f_9 + 2f_{10} + 2f_{11} - 2f_{12}$$

Normalization yields: $\phi_{2b_{2_u}} = \frac{1}{2}(-f_9 + f_{10} + f_{11} - f_{12})$.

Each of these four 2×2 symmetry blocks generate identical Hamiltonian matrices. This will be demonstrated for the B_{3g} symmetry, the others proceed analogously:

$$H_{1b_{3_g},1b_{3_g}} = \int \frac{1}{2}(-f_5 + f_6 - f_7 + f_8) H \frac{1}{2}(-f_5 + f_6 - f_7 + f_8) d\tau$$
$$= \frac{1}{4}\Big\{ \int f_5 H f_5 d\tau - \int f_5 H f_6 d\tau + \int f_5 H f_7 d\tau - \int f_5 H f_8 d\tau -$$
$$\int f_6 H f_5 d\tau + \int f_6 H f_6 d\tau - \int f_6 H f_7 d\tau + \int f_6 H f_8 d\tau +$$
$$\int f_7 H f_5 d\tau - \int f_7 H f_6 d\tau + \int f_7 H f_7 d\tau - \int f_7 H f_8 d\tau -$$
$$\int f_8 H f_5 d\tau + \int f_8 H f_6 d\tau - \int f_8 H f_7 d\tau + \int f_8 H f_8 d\tau \Big\}$$
$$= \frac{1}{4}\Big\{ \alpha_{sp^2} - 0 + 0 - 0 - 0 + \alpha_{sp^2} - 0 + 0 + 0 - 0 + \alpha_{sp^2} - 0 - 0 + 0 - 0 + \alpha_{sp^2} \Big\} = \alpha_{sp^2}$$

$$H_{1b_{3g},2b_{3g}} = \int \frac{1}{2}(-f_5 + f_6 - f_7 + f_8)H\frac{1}{2}(-f_9 + f_{10} - f_{11} + f_{12})d\tau$$

$$= \frac{1}{4}\left\{\int f_5Hf_9d\tau - \int f_5Hf_{10}d\tau + \int f_5Hf_{11}d\tau - \int f_5Hf_{12}d\tau - \right.$$

$$\int f_6Hf_9d\tau + \int f_6Hf_{10}d\tau - \int f_6Hf_{11}d\tau + \int f_6Hf_{12}d\tau +$$

$$\int f_7Hf_9d\tau - \int f_7Hf_{10}d\tau + \int f_7Hf_{11}d\tau - \int f_7Hf_{12}d\tau -$$

$$\left.\int f_8Hf_9d\tau + \int f_8Hf_{10}d\tau - \int f_8Hf_{11}d\tau + \int f_8Hf_{12}d\tau\right\}$$

$$= \frac{1}{4}\left\{ /\beta_{sp^2-s} - 0 + 0 - 0 - 0 + \beta_{sp^2-s} - 0 + 0 + 0 - 0 + \beta_{sp^2-s} - 0 - 0 + 0 - 0 + \beta_{sp^2-s} \right\} = \beta_{sp^2-s}$$

$$H_{2b_{3g},2b_{3g}} = \int \frac{1}{2}(-f_9 + f_{10} - f_{11} + f_{12})H\frac{1}{2}(-f_9 + f_{10} - f_{11} + f_{12})d\tau$$

$$= \frac{1}{4}\left\{\int f_9Hf_9d\tau - \int f_9Hf_{10}d\tau + \int f_9Hf_{11}d\tau - \int f_9Hf_{12}d\tau - \right.$$

$$\int f_{10}Hf_9d\tau + \int f_{10}Hf_{10}d\tau - \int f_{10}Hf_{11}d\tau + \int f_{10}Hf_{12}d\tau +$$

$$\int f_{11}Hf_9d\tau - \int f_{11}Hf_{10}d\tau + \int f_{11}Hf_{11}d\tau - \int f_{11}Hf_{12}d\tau -$$

$$\left.\int f_{12}Hf_9d\tau + \int f_{12}Hf_{10}d\tau - \int f_{12}Hf_{11}d\tau + \int f_{12}Hf_{12}d\tau\right\}$$

$$= \frac{1}{4}\left\{ \alpha_s - 0 + 0 - 0 - 0 + \alpha_s - 0 + 0 + 0 - 0 + \alpha_s - 0 - 0 + 0 - 0 + \alpha_s \right\} = \alpha_s$$

This matrix eigenvalue problem then becomes:

$$\begin{vmatrix} \alpha_{sp^2} - \varepsilon & \beta_{sp^2-s} \\ \beta_{sp^2-s} & \alpha_s - \varepsilon \end{vmatrix} = 0$$

$$\begin{vmatrix} -14.7 - \varepsilon & -4.0 \\ -4.0 & -13.6 - \varepsilon \end{vmatrix} = 0$$

Solving this yields eigenvalues of:

$$\begin{vmatrix} -18.19 & -10.11 \end{vmatrix}$$

and corresponding eigenvectors:

$$\begin{vmatrix} -0.7537 & -0.6572 \\ -0.6572 & 0.7537 \end{vmatrix}$$

This results in an orbital energy diagram:

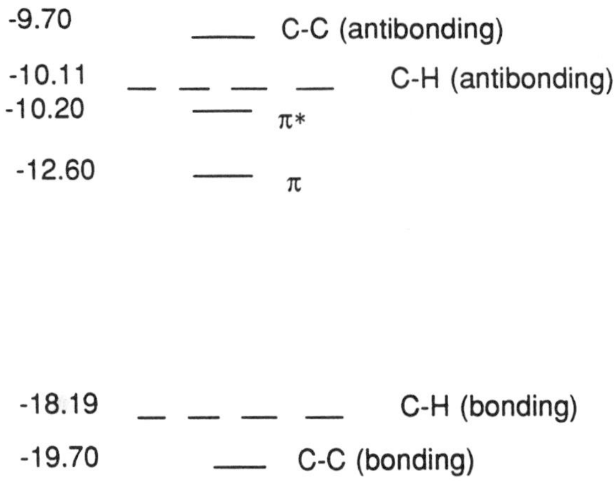

For the ground state of ethylene you would fill the bottom 3 levels (the C-C, C-H, and π bonding orbitals), with 12 electrons.

4.

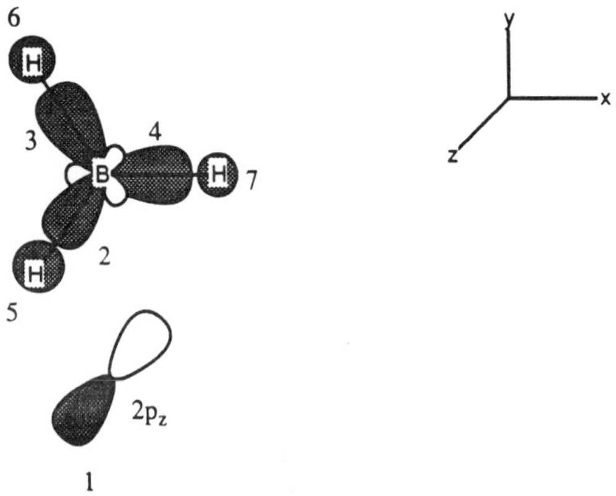

Using the hybrid atomic orbitals as labeled above (functions f_1–f_7) and the D_{3h} point group symmetry it is easiest to construct three sets of reducible representations:

a. the B $2p_z$ orbital (labeled function 1)

b. the 3 B sp^2 hybrids (labeled functions 2–4)

c. the 3 H 1s orbitals (labeled functions 5–7).

a. The B $2p_z$ orbital generates the following irreducible representation:

D_{3h}	E	$2C_3$	$3C_2$	σ_h	$2S_3$	$3\sigma_v$
Γ_{2p_z}	1	1	−1	−1	−1	1

This irreducible representation is A_2'' and is its own SALC-AO.

b. The B sp^2 orbitals generate the following reducible representation:

D_{3h}	E	$2C_3$	$3C_2$	σ_h	$2S_3$	$3\sigma_v$
Γ_{sp2}	3	0	1	3	0	1

This reducible representation reduces to $1A_1'$ and $1E'$ irreducible representations. Projectors are used to find the SALC-AOs for these irreducible representations. Define: C_3 = 120 degree rotation, C_3' = 240 degree rotation, C_2 = rotation around f_4, C_2' = rotation around f_2, and C_2 = rotation around f_3. S_3 and S_3' are defined analogous to C_3 and $C3'$ with accompanying horizontal reflection. σ_v = a reflection plane through f_4, σ_v' = a reflection plane through f_2, and σ_v'' = a reflection plane through f_3.

$$\begin{aligned} P_{A1'}f_2 &= (1)Ef_2 + (1)C_3f_2 + (1)C_3'f_2 + (1)C_2f_2 + (1)C_2'f_2 + (1)C_2''f_2 + (1)\sigma_hf_2 + (1)S_3f_2 + (1)S_3'f_2 + \\ &\quad (1)\sigma_vf_2 + (1)\sigma_v'f_2 + (1)\sigma_v''f_2 \\ &= (1)f_2 + (1)f_3 + (1)f_4 + (1)f_3 + (1)f_2 + (1)f_4 + (1)f_2 + (1)f_3 + (1)f_4 + (1)f_3 + (1)f_2 + (1)\sigma f_4 \\ &= 4f_2 + 4f_3 + 4f_4 \end{aligned}$$

Normalization yields: $\phi_{1_{a_1}'} = \frac{1}{\sqrt{3}}(f_2 + f_3 + f_4)$.

$$\begin{aligned} P_{E'}f_2 &= (2)Ef_2 + (-1)C_3f_2 + (-1)C_3'f_2 + (0)C_2f_2 + (0)C_2'f_2 + (0)C_2''f_2 + (2)\sigma_hf_2 + (-1)S_3f_2 + (-1)S_3'f_2 \\ &\quad (0)\sigma_vf_2 + (0)\sigma_v'f_2 + (0)\sigma_v''f_2 \\ &= (2)f_2 + (-1)f_3 + (-1)f_4 + (2)f_2 + (-1)f_3 + (-1)f_4 + \\ &= 4f_2 - 2f_3 - 2f_4 \end{aligned}$$

Normalization yields: $\phi_{1e'} = \frac{1}{\sqrt{6}}(2f_2 - f_3 - f_4)$.

To find the second e'(orthogonal to the first), projection on f_3 yields $(2f_3 - f_2 - f_4)$ and projection on f_4 yields $(2f_4 - f_2 - f_3)$. Neither of these functions are orthogonal to the first, but a combination of the two $(2f_3 - f_2 - f_4) - (2f_4 - f_2 - f_3)$ yields a function which is orthogonal to the first.

Normalization yields: $\phi_{2e'} = \frac{1}{\sqrt{2}}(f_3 - f_4)$.

c. The H $1s$ orbitals generate the following reducible representation:

D_{3h}	E	$2C_3$	$3C_2$	σ_h	$2S_3$	$3\sigma_v$
Γ_{sp^2}	3	0	1	3	0	1

This reducible representation reduces to $1A_1'$ and $1E'$ irreducible representations exactly like part b. and in addition the projectors used to find the SALC-AOs for these irreducible representations is exactly analogous to part b.

$$\phi_{2a_{1'}} = \frac{1}{\sqrt{3}}(f_5 + f_6 + f_7)$$

$$\phi_{3e'} = \frac{1}{\sqrt{6}}(2f_5 - f_6 - f_7)$$

$$\phi_{4e'} = \frac{1}{\sqrt{2}}(f_6 - f_7)$$

So, there are $1A_2''$,$2A_1'$ and $2E'$ orbitals. Solving the Hamiltonian matrix for each symmetry block yields:

A_2'' Block:

$$\begin{aligned} H_{1a_{2'},1a_{2'}} &= \int f_1 H f_1 d\tau \\ &= \alpha_{2p\pi} = -8.5 \end{aligned}$$

A_1' Block:

$$H_{1a_{1'},1a_{1'}} = \int \frac{1}{\sqrt{3}}(f_2 + f_3 + f_4) H \frac{1}{\sqrt{3}}(f_2 + f_3 + f_4) d\tau$$

$$= \frac{1}{3} \left\{ \int f_2 H f_2 d\tau + \int f_2 H f_3 d\tau + \int f_2 H f_4 d\tau + \right.$$

$$\int f_3 H f_2 d\tau + \int f_3 H f_3 d\tau + \int f_3 H f_4 d\tau +$$

$$\left. \int f_4 H f_2 d\tau + \int f_4 H f_3 d\tau + \int f_4 H f_4 d\tau \right\}$$

$$= \frac{1}{3} \left| \alpha_{sp^2} + 0 + 0 + 0 + \alpha_{sp^2} + 0 + 0 + 0 + \alpha_{sp^2} \right| = \alpha_{sp^2}$$

$$H_{1a_{1'},2a_{1'}} = \int \frac{1}{\sqrt{3}} (f_2 + f_3 + f_4) H \frac{1}{\sqrt{3}} (f_5 + f_6 + f_7) d\tau$$

$$= \frac{1}{3} \left\{ \int f_2 H f_5 d\tau + \int f_2 H f_6 d\tau + \int f_2 H f_7 d\tau + \right.$$

$$\int f_3 H f_5 d\tau + \int f_3 H f_6 d\tau + \int f_3 H f_7 d\tau +$$

$$\left. \int f_4 H f_5 d\tau + \int f_4 H f_6 d\tau + \int f_4 H f_7 d\tau \right\}$$

$$= \frac{1}{3} \left| \beta_{sp^2-s} + 0 + 0 + 0 + \beta_{sp^2-s} + 0 + 0 + 0 + \beta_{sp^2-s} \right| = \beta_{sp^2-s}$$

$$H_{2_{a_{1'}},2_{a_{1'}}} = \int \frac{1}{\sqrt{3}} (f_5 + f_6 + f_7) H \frac{1}{\sqrt{3}} (f_5 + f_6 + f_7) d\tau$$

$$= \frac{1}{3} \left\{ \int f_5 H f_5 d\tau + \int f_5 H f_6 d\tau + \int f_5 H f_7 d\tau + \right.$$

$$\int f_6 H f_5 d\tau + \int f_6 H f_6 d\tau + \int f_6 H f_7 d\tau +$$

$$\left. \int f_7 H f_5 d\tau + \int f_7 H f_6 d\tau + \int f_7 H f_7 d\tau \right\}$$

$$= \frac{1}{3} \left| \alpha_s + 0 + 0 + 0 + \alpha_s + 0 + 0 + 0 + \alpha_s \right| = \alpha_s$$

This matrix eigenvalue problem then becomes:

$$\begin{vmatrix} \alpha_{sp^2} - \varepsilon & \beta_{sp^2-s} \\ \beta_{sp^2-s} & \alpha_s - \varepsilon \end{vmatrix} = 0$$

$$\begin{vmatrix} -10.7 - \varepsilon & -3.5 \\ -3.5 & -13.6 - \varepsilon \end{vmatrix} = 0$$

Solving this yields eigenvalues of:

$$\left| -15.94 \quad -8.36 \right|$$

and corresponding eigenvectors:

$$\begin{vmatrix} -0.5555 & -0.8315 \\ -0.8315 & 0.5555 \end{vmatrix}$$

E' Block:

This 4×4 symmetry block factors to two 2×2 blocks: where one 2×2 block includes the SALC-AOs

$$\phi_{e'} = \frac{1}{\sqrt{6}}(2f_2 - f_3 - f_4)$$

$$\phi_{e'} = \frac{1}{\sqrt{6}}(2f_5 - f_6 - f_7),$$

and the other includes the SALC-AOs

$$\phi_{e'} = \frac{1}{\sqrt{2}}(f_3 - f_4)$$

$$\phi_{e'} = \frac{1}{\sqrt{2}}(f_6 - f_7).$$

Both of these 2×2 matrices are identical to the A_1' 2×2 array and therefore yield identical energies and MO coefficients. This results in an orbital energy diagram:

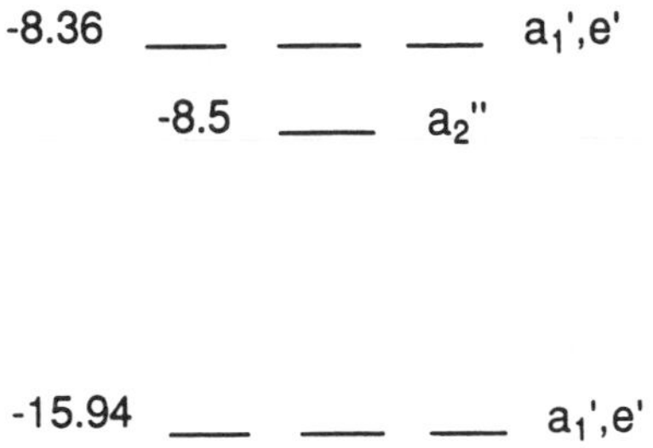

For the ground state of BH_3 you would fill the bottom level (B-H bonding), a_1' and e' orbitals, with six electrons.

5.

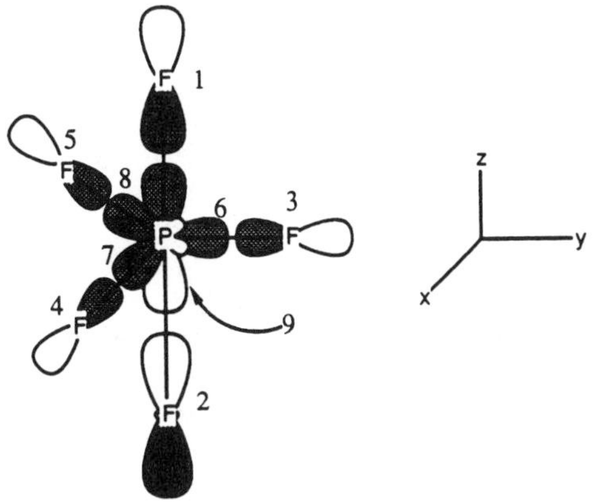

a. The two F p orbitals (top and bottom) generate the following reducible representation:

D_{3h}	E	$2C_3$	$3C_2$	σ_h	$2S_3$	$3\sigma_v$
Γ_p	2	2	0	0	0	2

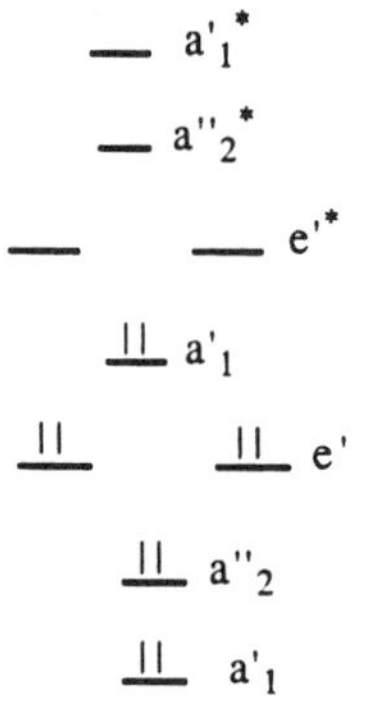

This reducible representation reduces to $1A_1'$ and $1A_2''$ irreducible representations. Projectors may be used to find the SALC-AOs for these irreducible representations.

$$\phi_{a'_1} = \frac{1}{\sqrt{2}}(f_1 - f_2)$$

$$\phi_{a''_2} = \frac{1}{\sqrt{2}}(f_1 + f_2)$$

b. The three trigonal F p orbitals generate the following reducible representation:

D_{3h}	E	$2C_3$	$3C_2$	σ_h	$2S_3$	$3\sigma_v$
Γ_p	3	0	1	3	0	1

This reducible representation reduces to $1A_1'$ and $1E'$ irreducible representations. Projectors may be used to find the SALC-AOs for these irreducible representations (but they are exactly analogous to the previous few problems):

$$\phi_{a'_1} = \frac{1}{\sqrt{3}}(f_3 + f_4 + f_5)$$

$$\phi_{e'} = \frac{1}{\sqrt{6}}(2f_3 - f_4 - f_5)$$

$$\phi_{e'} = \frac{1}{\sqrt{2}}(f_4 - f_5).$$

c. The 3Psp^2 orbitals generate the following reducible representation:

D_{3h}	E	$2C_3$	$3C_2$	σ_h	$2S_3$	$3\sigma_v$
Γ_{sp2}	3	0	1	3	0	1

This reducible representation reduces to $1A_1'$ and $1E'$ irreducible representations. Again, projectors may be used to find the SALC-AOs for these irreducible representations.(but again they are exactly analogous to the previous few problems):

$$\phi_{a'_1} = \frac{1}{\sqrt{3}}(f_6 + f_7 + f_8)$$

$$\phi_{e'} = \frac{1}{\sqrt{6}}(2f_6 - f_7 - f_8)$$

$$\phi_{e'} = \frac{1}{\sqrt{2}}(f_7 - f_8).$$

The leftover P p_z orbital generate the following irreducible representation:

D_{3h}	E	$2C_3$	$3C_2$	σ_h	$2S_3$	$3\sigma_v$
Γ_{p_z}	1	1	−1	−1	−1	1

This irreducible representation is an A_2''

$\phi_{a''_2} = f_9$.

Drawing an energy level diagram using these SALC-AOs would result in the following:

SECTION

3

ELECTRONIC CONFIGURATIONS, TERM SYMBOLS, AND STATES

INTRODUCTORY REMARKS—THE ORBITAL, CONFIGURATION, AND STATE PICTURES OF ELECTRONIC STRUCTURE

One of the goals of quantum chemistry is to allow practicing chemists to use knowledge of the electronic states of fragments (atoms, radicals, ions, or molecules) to predict and understand the behavior (i.e., electronic energy levels, geometries, and reactivities) of larger molecules. In the preceding section, orbital correlation diagrams were introduced to connect the orbitals of the fragments along a "reaction path" leading to the orbitals of the products. In this section, analogous connections are made among the fragment and product electronic states, again labeled by appropriate symmetries. To realize such connections, one must first write down N-electron wavefunctions that possess the appropriate symmetry; this task requires combining symmetries of the occupied orbitals to obtain the symmetries of the resulting states.

CHAPTER

8

Electrons are placed into orbitals to form configurations, each of which can be labeled by its symmetry. The configurations may "interact" strongly if they have similar energies.

I. ORBITALS DO NOT PROVIDE THE COMPLETE PICTURE; THEIR OCCUPANCY BY THE *N* ELECTRONS MUST BE SPECIFIED

Knowing the orbitals of a particular species provides information about the sizes, shapes, directions, symmetries, and energies of those regions of space that are *available* to the electrons (i.e., the complete set of orbitals that are available). This knowledge does *not* determine into which orbitals the electrons are placed. It is by describing the electronic configurations (i.e., orbital occupancies such as $1s^22s^22p^2$ or $1s^22s^22p^13s^1$) appropriate to the energy range under study that one focuses on how the electrons occupy the orbitals. Moreover, a given configuration may give rise to several energy levels whose energies differ by chemically important amounts. For example, the $1s^22s^22p^2$ configuration of the Carbon atom produces nine degenerate ^3P states, five degenerate ^1D states, and a single ^1S state. These three energy levels differ in energy by 1.5eV and 1.2eV, respectively.

II. EVEN *N*-ELECTRON CONFIGURATIONS ARE NOT MOTHER NATURE'S TRUE ENERGY STATES

Moreover, even single-configuration descriptions of atomic and molecular structure (e.g., $1s^22s^22p^4$ for the Oxygen atom) do not provide fully correct or highly accurate representations of the respective electronic wavefunctions. As will be shown in this section and in more detail in section 6, the picture of *N* electrons occupying orbitals to form a configuration is based on a so-called "mean field" description of the coulomb interactions among electrons. In such models, an electron at $\boldsymbol{r}$ is viewed as interacting with an "averaged" charge density arising from the N–1 remaining electrons:

$$V_{\text{mean field}} = \int \rho_{N-1}(\boldsymbol{r}')e^2/|\boldsymbol{r} - \boldsymbol{r}'|d\boldsymbol{r}'.$$

Here $\rho_{N-1}(\boldsymbol{r}')$ represents the probability density for finding electrons at $\boldsymbol{r}'$, and $e^2/|\boldsymbol{r}-\boldsymbol{r}'|$ is the mutual coulomb repulsion between electron density at $\boldsymbol{r}$ and $\boldsymbol{r}'$. Analogous mean-field models arise in many areas of chemistry and physics, including electrolyte theory (e.g., the Debye-Hückel theory), statistical mechanics of dense gases (e.g., where the Mayer-Mayer cluster expansion is used to improve the ideal-gas mean field model), and chemical dynamics (e.g., the vibrationally averaged potential of interaction).

In each case, the mean-field model forms only a starting point from which one attempts to build a fully correct theory by effecting systematic corrections (e.g., using perturbation theory) to the mean-field model. The ultimate value of any particular mean-field model is related to its accuracy in describing experimental phenomena. If predictions of the mean-field model are far from the experimental observations, then higher-order corrections (which are usually difficult to implement) must be employed to improve its predictions. In such a case, one is motivated to search for a better model to use as a starting point so that lower-order perturbative (or other) corrections can be used to achieve chemical accuracy (e.g., ±1kcal/mole).

In electronic structure theory, the single-configuration picture (e.g., the $1s^22s^22p^4$ description of the Oxygen atom) forms the mean-field starting point; the configuration interaction (CI) or perturbation theory techniques are then used to systematically improve this level of description.

The single-configuration mean-field theories of electronic structure neglect *correlations* among the electrons. That is, in expressing the interaction of an electron at $\boldsymbol{r}$ with the N–1 other electrons, they use a probability density $\rho_{N-1}(\boldsymbol{r}')$ that is independent of the fact that another electron resides at $\boldsymbol{r}$. In fact, the so-called conditional probability density for finding one of N–1 electrons at $\boldsymbol{r}'$, given that an electron is at $\boldsymbol{r}$ certainly depends on $\boldsymbol{r}$. As a result, the mean-field coulomb potential felt by a $2p_x$ orbital's electron in the $1s^22s^22p_x2p_y$ single-configuration description of the Carbon atom is:

$$V_{\text{mean field}} = 2\int |1s(\boldsymbol{r}')|^2 e^2/|\boldsymbol{r} - \boldsymbol{r}'|d\boldsymbol{r}'$$

$$+2\int |2s(\boldsymbol{r}')|^2 e^2/|\boldsymbol{r} - \boldsymbol{r}'|d\boldsymbol{r}'$$

$$+\int |2p_y(\boldsymbol{r}')|^2 e^2/|\boldsymbol{r} - \boldsymbol{r}'|d\boldsymbol{r}'.$$

In this example, the density $\rho_{N-1}(\boldsymbol{r}')$ is the sum of the charge densities of the orbitals occupied by the five other electrons $2|1s(\boldsymbol{r}')|^2 + 2|2s(\boldsymbol{r}')|^2 + |2p_y(\boldsymbol{r}')|^2$, and is not dependent on the fact that an electron resides at $\boldsymbol{r}$.

III. MEAN-FIELD MODELS

The mean-field model, which forms the basis of chemists' pictures of electronic structure of molecules, is not very accurate.

The magnitude and "shape" of such a mean-field potential is shown below for the Beryllium atom. In this figure, the nucleus is at the origin, and one electron is placed at a distance from the nucleus equal to the maximum of the $1s$ orbital's radial probability density (near 0.13Å). The radial coordinate of the second is plotted as the abscissa; this second electron is arbitrarily constrained to lie on the line connecting the nucleus and the first electron (along this direction, the inter-electronic interactions are largest). On the ordinate, there are two quantities plotted: (i) the Self-Consistent Field (SCF) mean-field potential

$$\int |1s(\mathbf{r}')|^2 e^2/|\mathbf{r} - \mathbf{r}'| d\mathbf{r}',$$

and (ii) the so-called Fluctuation potential (F), which is the true coulombic $e^2/|\mathbf{r} - \mathbf{r}'|$ interaction potential minus the SCF potential.

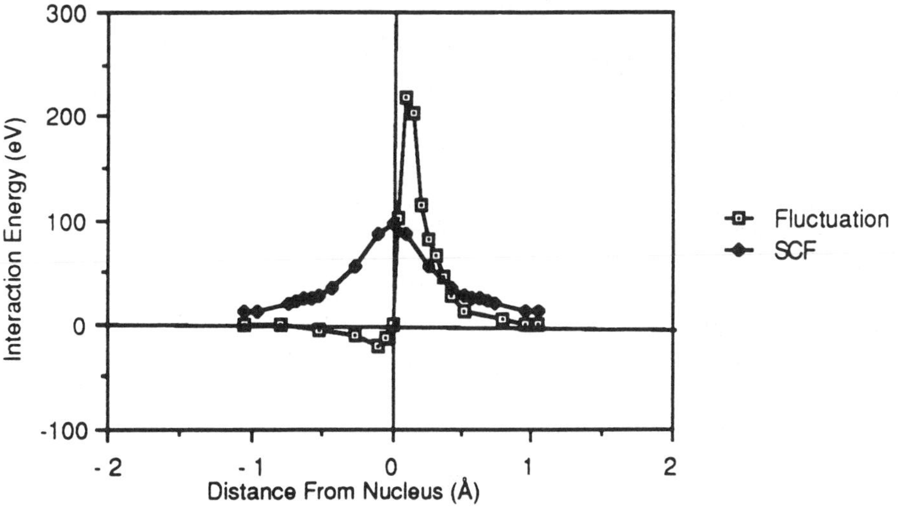

As a function of the inter-electron distance, the fluctuation potential decays to zero more rapidly than does the SCF potential. For this reason, approaches in which F is treated as a perturbation and corrections to the mean-field picture are computed perturbatively might be expected to be rapidly convergent (whenever perturbations describing long-range interactions arise, convergence of perturbation theory is expected to be slow or not successful). However, the magnitude of F is quite large and remains so over an appreciable range of inter-electron distances.

The resultant corrections to the SCF picture are therefore quite large when measured in kcal/mole. For example, the differences ΔE between the true (state-of-the-art quantum chemical calculation) energies of interaction among the four electrons in Be and the SCF mean-field estimates of these interactions are given in the table shown below in eV (recall that 1eV = 23.06 kcal/mole).

Orb. Pair	$1s\alpha 1s\beta$	$1s\alpha 2s\alpha$	$1s\alpha 2s\beta$	$1s\beta 2s\alpha$	$1s\beta 2s\beta$	$2s\alpha 2s\beta$
ΔE in eV	1.126	0.022	0.058	0.058	0.022	1.234

To provide further insight why the SCF mean-field model in electronic structure theory is of limited accuracy, it can be noted that the average value of the kinetic energy plus the attraction to the Be nucleus plus the SCF interaction potential for one of the $2s$ orbitals of Be with the three remaining electrons in the $1s^2 2s^2$ configuration is:

$$<2s|-\hbar^2/2m_e\nabla^2 - 4e^2/r + V_{SCF}|2s> = -15.4\text{eV};$$

the analogous quantity for the $2p$ orbital in the $1s^2 2s2p$ configuration is:

$$<2p|-\hbar^2/2m_e\nabla^2 - 4e^2/r + V'_{SCF}|2p> = -12.28\text{eV};$$

the corresponding value for the $1s$ orbital is (negative and) of even larger magnitude. The SCF average coulomb interaction between the two $2s$ orbitals of $1s^2 2s^2$ Be is:

$$\int |2s(\mathbf{r})|^2 |2s(\mathbf{r}')|^2 e^2/|\mathbf{r} - \mathbf{r}'| d\mathbf{r} d\mathbf{r}' = 5.95\text{eV}.$$

This data clearly shows that corrections to the SCF model (see the above table) represent significant fractions of the inter-electron interaction energies (e.g., 1.234eV compared to 5.95 – 1.234 = 4.72eV for the two $2s$ electrons of Be), and that the inter-electron interaction energies, in turn, constitute significant fractions of the total energy of each orbital (e.g., 5.95 – 1.234eV = 4.72eV out of –15.4 eV for a $2s$ orbital of Be).

The task of describing the electronic states of atoms and molecules from first principles and in a chemically accurate manner (±1kcal/mole) is clearly quite formidable. The orbital picture and its accompanying SCF potential take care of "most" of the interactions among the N electrons (which interact via long-range coulomb forces and whose dynamics requires the application of quantum physics and permutational symmetry). However, the residual fluctuation potential, although of shorter range than the bare coulomb potential, is large enough to cause significant corrections to the mean-field picture. This, in turn, necessitates the use of more sophisticated and computationally taxing techniques (e.g., high order perturbation theory or large variational expansion spaces) to reach the desired chemical accuracy.

Mean-field models are obviously approximations whose accuracy must be determined so scientists can know to what degree they can be "trusted." For electronic structures of atoms and molecules, they require quite substantial corrections to bring them into line with experimental fact. Electrons in atoms and molecules undergo dynamical motions in which their coulomb repulsions cause them to "avoid" one another at every instant of time, not only in the average-repulsion manner that the mean-field models embody. The inclusion of instantaneous spatial correlations among electrons is necessary to achieve a more accurate description of atomic and molecular electronic structure.

IV. CONFIGURATION INTERACTION (CI) DESCRIBES THE CORRECT ELECTRONIC STATES

The most commonly employed tool for introducing such spatial correlations into electronic wavefunctions is called configuration interaction (CI); this approach is described briefly later in this section and in considerable detail in section 6.

Briefly, one employs the (in principle, complete as shown by P. O. Löwdin, *Rev. Mod. Phys. 32*, 328 (1960)) set of N-electron configurations that (i) can be formed by placing the N electrons into orbitals of the atom or molecule under study, and that (ii) possess the spatial, spin, and angular momentum symmetry of the electronic state of interest. This set of functions is then used, in a linear variational function, to achieve, via the CI technique, a more accurate and dynamically correct description of the electronic structure of that state. For example, to describe the ground ^1S state of the Be atom, the $1s^22s^2$ configuration (which yields the mean-field description) is augmented by including other configurations such as $1s^23s^2$, $1s^22p^2$, $1s^23p^2$, $1s^22s3s$, $3s^22s^2$, $2p^22s^2$, etc., all of which have overall ^1S spin and angular momentum symmetry. The excited ^1S states are also combinations of all such configurations. Of course, the ground-state wavefunction is dominated by the $|1s^22s^2|$ and excited states contain dominant contributions from $|1s^22s3s|$, etc. configurations. The resultant CI wavefunctions are formed as shown in section 6 as linear combinations of all such configurations.

To clarify the physical significance of mixing such configurations, it is useful to consider what are found to be the two most important such configurations for the ground ^1S state of the Be atom:

$$\Psi \cong C_1|1s^22s^2| - C_2[|1s^22p_x^2| + |1s^22p_y^2| + |1s^22p_z^2|] .$$

As proven in chapter 17.III, this two-configuration description of Be's electronic structure is equivalent to a description is which two electrons reside in the $1s$ orbital (with opposite, α and β spins) while the other pair reside in $2s$-$2p$ hybrid orbitals (more correctly, polarized orbitals) in a manner that instantaneously correlates their motions:

$$\begin{aligned}\Psi \cong 1/6C_1|1s^2\{ &[(2s - a2p_x)\alpha(2s + a2p_x)\beta - (2s - a2p_x)\beta(2s + a2p_x)\alpha] \\ &+[(2s - a2p_y)\alpha(2s + a2p_y)\beta - (2s - a2p_y)\beta(2s + a2p_y)\alpha] \\ &+[(2s - a2p_z)\alpha(2s + a2p_z)\beta - (2s - a2p_z)\beta(2s + a2p_z)\alpha]\}|,\end{aligned}$$

where $a = \sqrt{3C_2/C_1}$. The so-called polarized orbital pairs ($2s \pm a2p_{x,y,\text{ or } z}$) are formed by mixing into the $2s$ orbital an amount of the $2p_{x,y,\text{ or } z}$ orbital, with the mixing amplitude determined by the ratio of C_2 to C_1. As will be detailed in section 6, this ratio is proportional to the magnitude of the coupling $\langle|1s^22s^2|H|1s^22p^2|\rangle$ between the two configurations and inversely proportional to the energy difference $[\langle|1s^22s^2|H|1s^22s^2|\rangle - \langle|1s^22p^2|H|1s^22p^2|\rangle]$ for these configurations. So, in general, configurations that have similar energies (Hamiltonian expectation values) and couple strongly give rise to strongly mixed polarized orbital pairs. The result of forming such polarized orbital pairs are described pictorially below.

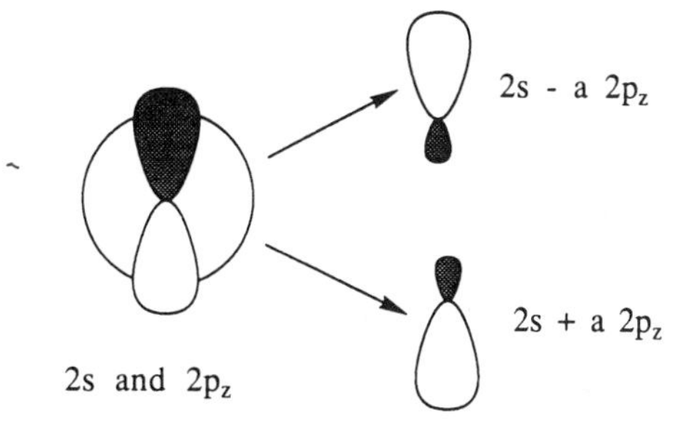

Polarized Orbital 2s and 2p z Pairs

In each of the three equivalent terms in this wavefunction, one of the valence electrons moves in a $2s + a2p$ orbital polarized in one direction while the other valence electron moves in the $2s - a2p$ orbital polarized in the opposite direction. For example, the first term $[(2s - a2p_x)\alpha(2s + a2p_x)\beta - (2s - a2p_x)\beta(2s + a2p_x)\alpha]$ describes one electron occupying a $2s - a2p_x$ polarized orbital while the other electron occupies the $2s + a2p_x$ orbital. In this picture, the electrons reduce their mutual coulomb repulsion by occupying *different* regions of space; in the SCF mean-field picture, both electrons reside in the same $2s$ region of space. In this particular example, the electrons undergo *angular correlation* to "avoid" one another. The fact that equal amounts of x, y, and z orbital polarization appear in Ψ is what preserves the ^1S symmetry of the wavefunction.

The fact that the CI wavefunction

$$\Psi \cong C_1|1s^2 2s^2| - C_2[|1s^2 2p_x^2| + |1s^2 2p_y^2| + |1s^2 2p_z^2|]$$

mixes its two configurations with *opposite sign* is of significance. As will be seen later in section 6, solution of the Schrödinger equation using the CI method in which two configurations (e.g., $|1s^2 2s^2|$ and $|1s^2 2p^2|$) are employed gives rise to two solutions. One approximates the ground state wave function; the other approximates an excited state. The former is the one that mixes the two configurations with opposite sign.

To understand why the latter is of higher energy, it suffices to analyze a function of the form

$$\Psi' \cong C_1|1s^2 2s^2| + C_2[|1s^2 2p_x^2|+|1s^2 2p_y^2| + |1s^2 2p_z^2|]$$

in a manner analogous to above. In this case, it can be shown that

$$\begin{aligned}\Psi' \cong 1/6C_1|1s^2\{&[(2s - ia2p_x)\alpha(2s + ia2p_x)\beta - (2s-ia2p_x)\beta(2s + ia2p_x)\alpha]\\ &+[(2s - ia2p_y)\alpha(2s + ia2p_y)\beta - (2s - ia2p_y)\beta(2s + ia2p_y)\alpha]\\ &+[(2s - ia2p_z)\alpha(2s + ia2p_z)\beta - (2s - ia2p_z)\beta(2s + ia2p_z)\alpha]|\}.\end{aligned}$$

There is a fundamental difference, however, between the polarized orbital pairs introduced earlier $\phi_\pm = (2s \pm a2p_{x,y,\text{ or } z})$ and the corresponding functions $\phi'_\pm = (2s \pm ia2p_{x,y,\text{ or } z})$ appearing here. The probability densities embodied in the former

$$|\phi_\pm|^2 = |2s|^2 + a^2|2p_{x,y,\text{ or } z}|^2 \pm 2a(2s2p_{x,y,\text{ or } z})$$

describe constructive (for the + case) and destructive (for the – case) superposition of the probabilities of the $2s$ and $2p$ orbitals. The probability densities of $\phi'_\pm$ are

$$\begin{aligned}|\phi'_\pm|^2 &= (2s \pm ia2p_{x,y,\text{ or } z})^*(2s \pm ia2p_{x,y,\text{ or } z})\\ &= |2s|^2 + a^2|2p_{x,y,\text{ or } z}|^2.\end{aligned}$$

These densities are identical to one another and do not describe polarized orbital densities. Therefore, the CI wavefunction which mixes the two configurations with like sign, when analyzed in terms of orbital pairs, places the electrons into orbitals $\phi'_\pm = (2s \pm ia2p_{x,y,\text{ or } z})$ whose densities do not permit the electrons to avoid one another. Rather, both orbitals have the same spatial density $|2s|^2 + a^2|2p_{x,y,\text{ or } z}|^2$, which gives rise to higher coulombic interaction energy for this state.

V. SUMMARY

In summary, the dynamical interactions among electrons give rise to instantaneous spatial correlations that must be handled to arrive at an accurate picture of atomic and molecular structure. The simple, single-configuration picture provided by the mean-field model is a useful starting point, but improvements are often needed. In section 6, methods for treating electron correlation will be discussed in greater detail.

For the remainder of this section, the primary focus is placed on forming proper N-electron wavefunctions by occupying the orbitals available to the system in a manner that guarantees that the resultant N-electron function is an eigenfunction of those operators that commute with the N-electron Hamiltonian.

For polyatomic molecules, these operators include point-group symmetry operators (which act on *all* N electrons) and the spin angular momentum (S^2 and S_z) of *all* of the electrons taken as a whole (this is true in the absence of spin-orbit coupling which is treated later as a perturbation). For linear molecules, the point group symmetry operations involve rotations R_z of all N electrons about the principal axis, as a result of which the total angular momentum L_z of the N electrons (taken as a whole) about this axis commutes with the Hamiltonian, H. Rotation of all N electrons about the x and y axes does not leave the total coulombic potential energy unchanged, so L_x and L_y do not commute with H. Hence for a linear molecule, L_z, S^2, and S_z are the operators that commute with H. For atoms, the corresponding operators are L^2, L_z, S^2, and S_z (again, in the absence of spin-orbit coupling) where each operator pertains to the total orbital or spin angular momentum of the N electrons.

To construct N-electron functions that are eigenfunctions of the spatial symmetry or orbital angular momentum operators as well as the spin angular momentum operators, one has to "couple" the symmetry or angular momentum properties of the individual spin-orbitals used to construct the N-electrons functions. This coupling involves forming direct product symmetries in the case of polyatomic molecules that belong to finite point groups, it involves vector coupling orbital and spin angular momenta in the case of atoms, and it involves vector coupling spin angular momenta and axis coupling orbital angular momenta when treating linear molecules. Much of this section is devoted to developing the tools needed to carry out these couplings.

CHAPTER

9

Electronic wavefunctions must be constructed to have permutational antisymmetry because the *N* electrons are indistinguishable Fermions.

I. ELECTRONIC CONFIGURATIONS

Atoms, linear molecules, and non-linear molecules have orbitals which can be labeled either according to the symmetry appropriate for that isolated species or for the species in an environment which produces lower symmetry. These orbitals should be viewed as regions of space in which electrons can move, with, of course, at most two electrons (of opposite spin) in each orbital. Specification of a particular occupancy of the set of orbitals available to the system gives an *electronic configuration.* For example, $1s^2 2s^2 2p^4$ is an electronic configuration for the Oxygen atom (and for the F^{+1} ion and the N^{-1} ion); $1s^2 2s^2 2p^3 3p^1$ is another configuration for O, F^{+1}, or N^{-1}. These configurations represent situations in which the electrons occupy low-energy orbitals of the system and, as such, are likely to contribute strongly to the true ground and low-lying excited states and to the low-energy states of molecules formed from these atoms or ions.

Specification of an electronic configuration does not, however, specify a particular electronic state of the system. In the above $1s^2 2s^2 2p^4$ example, there are many ways (fifteen, to be precise) in which the $2p$ orbitals can be occupied by the four electrons. As a result, there are a total of fifteen states which cluster into three energetically distinct *levels*, lying within this single configuration. The $1s^2 2s^2 2p^3 3p^1$ configuration contains thirty-six states which group into six distinct energy levels (the word *level* is used to denote one or more state with the same energy). Not all states which arise from a given electronic configuration have the same energy because various states occupy the degenerate (e.g., $2p$ and $3p$ in the above examples) orbitals differently. That is, some states have orbital occupancies of the form $2p_1^2 2p_0^1 2p_{-1}^1$ while others have $2p_1^2 2p_0^2 2p_{-1}^0$; as a result, the states can have quite different coulombic repulsions among the electrons (the state with two doubly occupied orbitals would lie higher in energy than that with two singly occupied orbitals). Later in this section and in appendix G techniques for constructing wavefunctions for each state contained within a particular configuration are given in detail. Mastering these tools is an important aspect of learning the material in this text.

In summary, an atom or molecule has many orbitals (core, bonding, non-bonding, Rydberg, and antibonding) available to it; occupancy of these orbitals in a particular manner gives rise to a configuration. If some orbitals are partially occupied in this configuration, more than one state will arise; these states can differ in energy due to differences in how the orbitals are occupied. In particular, if degenerate orbitals are partially occupied, many states can arise and have energies which differ substantially because of differences in electron repulsions arising in these states. Systematic procedures for extracting all states from a given configuration, for labeling the states according to the appropriate symmetry group, for writing the wavefunctions corresponding to each state and for evaluating the energies corresponding to these wavefunctions are needed. Much of chapters 10 and 11 are devoted to developing and illustrating these tools.

II. ANTISYMMETRIC WAVEFUNCTIONS

A. General Concepts

The total electronic Hamiltonian

$$H = \sum_i (-\hbar^2/2m_e\nabla_i^2 - \sum_a Z_a e^2/r_{ia}) + \sum_{i>j} e^2/r_{ij} + \sum_{a>b} Z_a Z_b e^2/r_{ab},$$

where i and j label electrons and a and b label the nuclei (whose charges are denoted Z_a), commutes with the operators P_{ij} which permute the names of the electrons i and j. This, in turn, requires eigenfunctions of H to be eigenfunctions of P_{ij}. In fact, the set of such permutation operators form a group called the symmetric group (a good reference to this subject is contained in chapter 7 of *Group Theory*, M. Hamermesh, Addison-Wesley, Reading, Mass., 1962). In the present text, we will not exploit the full group theoretical nature of these operators; we will focus on the simple fact that all wavefunctions must be eigenfunctions of the P_{ij} (additional material on this subject is contained in chapter XIV of Kemble).

Because P_{ij} obeys $P_{ij} * P_{ij} = 1$, the eigenvalues of the P_{ij} operators must be +1 or −1. Electrons are Fermions (i.e., they have half-integral spin), and they have wavefunctions which are odd under permutation of any pair: $P_{ij}\Psi = -\Psi$. Bosons such as photons or deuterium nuclei (i.e., species with integral spin quantum numbers) have wavefunctions which obey $P_{ij}\Psi = +\Psi$.

These permutational symmetries are not only characteristics of the exact eigenfunctions of H belonging to any atom or molecule containing more than a single electron but they are also conditions which must be placed on any acceptable model or trial wavefunction (e.g., in a variational sense) which one constructs.

In particular, within the orbital model of electronic structure (which is developed more systematically in section 6), one can not construct trial wavefunctions which are simple spin-orbital products (i.e., an orbital multiplied by an α or β spin function for each electron) such as $1s\alpha 1s\beta 2s\alpha 2s\beta 2p_1\alpha 2p_0\alpha$. Such spin-orbital product functions *must* be made permutationally antisymmetric if the N-electron trial function is to be properly antisymmetric. This can be accomplished for any such product wavefunction by applying the following *antisymmetrizer operator*:

$$\boldsymbol{A} = \left(\frac{1}{\sqrt{N!}}\right)\sum_p s_p P,$$

where N is the number of electrons, P runs over all $N!$ permutations, and s_p is +1 or −1 depending on whether the permutation P contains an even or odd number of pairwise permutations (e.g., 231 can be reached from 123 by two pairwise permutations − 123 ==> 213 ==> 231, so 231 would have $s_p = 1$). The permutation operator P in $\mathbf{A}$ acts on a product wavefunction and permutes the ordering of the spin-orbitals. For example,

$$\mathbf{A}\phi_1\phi_2\phi_3 = (1/\sqrt{6})[\phi_1\phi_2\phi_3 - \phi_1\phi_3\phi_2 - \phi_3\phi_2\phi_1 - \phi_2\phi_1\phi_3 + \phi_3\phi_1\phi_2 + \phi_2\phi_3\phi_1],$$

where the convention is that electronic coordinates r_1, r_2, and r_3 correspond to the orbitals as they appear in the product (e.g., the term $\phi_3\phi_2\phi_1$ represents $\phi_3(r_1)\phi_2(r_2)\phi_1(r_3)$).

It turns out that the permutations P can be allowed either to act on the "names" or labels of the electrons, keeping the order of the spin-orbitals fixed, or to act on the spin-orbitals, keeping the order and identity of the electrons' labels fixed. The resultant wavefunction, which contains $N!$ terms, is exactly the same regardless of how one allows the permutations to act. Because we wish to use the above convention in which the order of the electronic labels remains fixed as 1, 2, 3, . . . N, we choose to think of the permutations acting on the names of the spin-orbitals.

It should be noted that the effect of $\mathbf{A}$ on any spin-orbital product is to produce a function that is a sum of $N!$ terms. In each of these terms the same spin-orbitals appear, but the order in which they appear differs from term to term. Thus antisymmetrization does not alter the overall orbital occupancy; it simply "scrambles" any knowledge of which electron is in which spin-orbital.

The antisymmetrized orbital product $\mathbf{A}\phi_1\phi_2\phi_3$ is represented by the short hand $|\phi_1\phi_2\phi_3|$ and is referred to as a *Slater determinant.* The origin of this notation can be made clear by noting that $(1/\sqrt{N!})$ times the determinant of a matrix whose rows are labeled by the index i of the spin-orbital ϕ_i and whose columns are labeled by the index j of the electron at $\mathbf{r}_j$ is equal to the above function: $\mathbf{A}\phi_1\phi_2\phi_3 = (1/\sqrt{3}!)\det(\phi_i(\mathbf{r}_j))$. The general structure of such Slater determinants is illustrated below:

$$(1/N!)^{1/2}\det\{\phi_j(\mathbf{r}_i)\} = (1/N!)^{1/2}\begin{bmatrix} \phi_1(1)\phi_2(1)\phi_3(1)\ldots\phi_k(1)\ldots\ldots\phi_N(1) \\ \phi_1(2)\phi_2(2)\phi_3(2)\ldots\phi_k(2)\ldots\ldots\phi_N(2) \\ \cdot \\ \cdot \\ \cdot \\ \cdot \\ \phi_1(N)\phi_2(N)\phi_3(N)\ldots\phi_k(N)\ldots\ldots\phi_N(N) \end{bmatrix}$$

The antisymmetry of many-electron spin-orbital products places constraints on any acceptable model wavefunction, which give rise to important physical consequences. For example, it is antisymmetry that makes a function of the form $|1s\alpha 1s\alpha|$ vanish (thereby enforcing the Pauli exclusion principle) while $|1s\alpha 2s\alpha|$ does not vanish, except at points $\mathbf{r}_1$ and $\mathbf{r}_2$ where $1s(\mathbf{r}_1) = 2s(\mathbf{r}_2)$, and hence is acceptable. The Pauli principle is embodied in the fact that if any two or more columns (or rows) of a determinant are identical, the determinant vanishes. Antisymmetry also enforces indistinguishability of the electrons in that $|1s\alpha 1s\beta 2s\alpha 2s\beta| = -|1s\alpha 1s\beta 2s\beta 2s\alpha|$. That is, two wavefunctions which differ simply by the ordering of their spin-orbitals are equal to within a sign (±1); such an overall sign difference in a wavefunction has no physical consequence because all physical properties depend on the product $\Psi^*\Psi$, which appears in any expectation value expression.

B. Physical Consequences of Antisymmetry

Once the rules for evaluating energies of determinental wavefunctions and for forming functions which have proper spin and spatial symmetries have been put forth (in chapter 11), it will be clear that antisymmetry and electron spin considerations, in addition to orbital occupancies, play substantial roles in determining energies and that it is precisely these aspects that are responsible for energy splittings among states arising from one configuration. A single example may help illustrate this point. Consider the $\pi^1\pi*^1$ configuration of ethylene (ignore the other orbitals and focus on the properties of these two). As will be shown below when spin angular momentum is treated in full, the triplet spin states of this configuration are:

$$|S = 1, M_S = 1> = |\pi\alpha\pi*\alpha|,$$
$$|S = 1, M_S = -1> = |\pi\beta\pi*\beta|,$$

and

$$|S = 1, M_S = 0> = 2^{-1/2}[|\pi\alpha\pi*\beta| + |\pi\beta\pi*\alpha|].$$

The singlet spin state is:

$$|S = 0, M_S = 0> = 2^{-1/2}[|\pi\alpha\pi*\beta| - |\pi\beta\pi*\alpha|].$$

To understand how the three triplet states have the same energy and why the singlet state has a different energy, and an energy different than the $M_S = 0$ triplet even though these two states are composed of the same two determinants, we proceed as follows:

1. We express the bonding π and antibonding $\pi*$ orbitals in terms of the atomic p-orbitals from which they are formed: $\pi = 2^{-1/2}[L + R]$ and $\pi* = 2^{-1/2}[L - R]$, where R and L denote the p-orbitals on the left and right carbon atoms, respectively.
2. We substitute these expressions into the Slater determinants that form the singlet and triplet states and collect terms and throw out terms for which the determinants vanish.
3. This then gives the singlet and triplet states in terms of atomic-orbital occupancies where it is easier to see the energy equivalences and differences.

Let us begin with the triplet states:

$$|\pi\alpha\pi*\alpha| = 1/2[|L\alpha L\alpha| - |R\alpha R\alpha| + |R\alpha L\alpha| - |L\alpha R\alpha|]$$
$$= |R\alpha L\alpha|;$$
$$2^{-1/2}[|\pi\alpha\pi*\beta| + |\pi\beta\pi*\alpha|] = 2^{-1/2}1/2[|L\alpha L\beta| - |R\alpha R\beta| + |R\alpha L\beta| -$$
$$|L\alpha R\beta| + |L\beta L\alpha| - |R\beta R\alpha| + |R\beta L\alpha| - |L\beta R\alpha|]$$
$$= 2^{-1/2}[|R\alpha L\beta| + |R\beta L\alpha|];$$
$$|\pi\beta\pi*\beta| = 1/2[|L\beta L\beta| - |R\beta R\beta| + |R\beta L\beta| - |L\beta R\beta|]$$
$$= |R\beta L\beta|.$$

The singlet state can be reduced in like fashion:

$$2^{-1/2}[|\pi\alpha\pi*\beta| - |\pi\beta\pi*\alpha|] = 2^{-1/2}1/2[|L\alpha L\beta| - |R\alpha R\beta| + |R\alpha L\beta| -$$
$$|L\alpha R\beta| - |L\beta L\alpha| + |R\beta R\alpha| - |R\beta L\alpha| + |L\beta R\alpha|]$$
$$= 2^{-1/2}[|L\alpha L\beta| - |R\beta R\alpha|].$$

Notice that all three triplet states involve atomic orbital occupancy in which one electron is on one atom while the other is on the second carbon atom. In contrast, the singlet state places both electrons on one carbon (it contains two terms; one with the two electrons on the left carbon and the other with both electrons on the right carbon).

In a "valence bond" analysis of the physical content of the singlet and triplet $\pi^1\pi*^1$ states, it is clear that the energy of the triplet states will lie below that of the singlet because the singlet contains "zwitterion" components that can be denoted C^+C^- and C^-C^+, while the three triplet states are purely "covalent." This case provides an excellent example of how the spin and permutational symmetries of a state "conspire" to qualitatively affect its energy and even electronic character as represented in its atomic orbital occupancies. Understanding this should provide ample motivation for learning how to form proper antisymmetric spin (and orbital) angular momentum eigenfunctions for atoms and molecules.

CHAPTER

10

Electronic wavefunctions must also possess proper symmetry. These include angular momentum and point group symmetries.

I. ANGULAR MOMENTUM SYMMETRY AND STRATEGIES FOR ANGULAR MOMENTUM COUPLING

Because the total Hamiltonian of a many-electron atom or molecule forms a mutually commutative set of operators with S^2, S_z, and

$$A = \left(\frac{1}{\sqrt{N!}}\right) \sum_p s_p P,$$

the exact eigenfunctions of H must be eigenfunctions of these operators. Being an eigenfunction of A forces the eigenstates to be odd under all P_{ij}. Any acceptable model or trial wavefunction should be constrained to also be an eigenfunction of these symmetry operators.

If the atom or molecule has additional symmetries (e.g., full rotation symmetry for atoms, axial rotation symmetry for linear molecules and point group symmetry for non-linear polyatomics), the trial wavefunctions should also conform to these spatial symmetries. This chapter addresses those operators that commute with H, P_{ij}, S^2, and S_z and among one another for atoms, linear, and non-linear molecules.

As treated in detail in appendix G, the full non-relativistic N-electron Hamiltonian of an atom or molecule

$$H = \sum_j (-\hbar^2/2m\nabla_j^2 - \sum_a Z_a e^2/r_{j,a}) + \sum_{j<k} e^2/r_{j,k}$$

commutes with the following operators:

1. The inversion operator i and the three components of the total orbital angular momentum

$$L_z = \sum_j L_z(j),\ L_y,\ L_x,$$

as well as the components of the total spin angular momentum S_z, S_x, and S_y **for atoms** (but not the individual electrons' $L_z(j)$, $S_z(j)$, etc.). Hence, L^2, L_z, S^2, S_z are the operators we need to form eigenfunctions of, and L, M_L, S, and M_S are the "good" quantum numbers.

2. $$L_z = \sum_j L_z(j),$$

as well as the N-electron S_x, S_y, and S_z **for linear molecules** (also i, if the molecule has a center of symmetry). Hence, L_z, S^2, and S_z are the operators we need to form eigenfunctions of, and M_L, S, and M_S are the "good" quantum numbers; L no longer is!

3. S_x, S_y, and S_z as well as all point group operations **for non-linear polyatomic molecules**. Hence S^2, S_z, and the point group operations are used to characterize the functions we need to form. When we include spin-orbit coupling into H (this adds another term to the potential that involves the spin and orbital angular momenta of the electrons), L^2, L_z, S^2, S_z no longer commute with H. However, $J_z = S_z + L_z$ and $J^2 = (\mathbf{L} + \mathbf{S})^2$ now do commute with H.

A. Electron Spin Angular Momentum

Individual electrons possess intrinsic spin characterized by angular momentum quantum numbers s and m_s; for electrons, $s = 1/2$ and $m_s = 1/2$, or $-1/2$. The $m_s = 1/2$ spin state of the electron is represented by the symbol α and the $m_s = -1/2$ state is represented by β. These spin functions obey: $S^2\alpha = 1/2(1/2 + 1)\hbar^2\alpha$, $S_z\alpha = 1/2\hbar\alpha$, $S^2\beta = 1/2(1/2 + 1)\hbar^2\beta$, and $S_z\beta = -1/2\hbar\beta$. The α and β spin functions are connected via lowering S_- and raising S_+ operators, which are defined in terms of the x and y components of $\mathbf{S}$ as follows: $S_+ = S_x + iS_y$, and $S_- = S_x - iS_y$. In particular $S_+\beta = \hbar\alpha$, $S_+\alpha = 0$, $S_-\alpha = \hbar\beta$, and $S_-\beta = 0$. These expressions are examples of the more general relations (these relations are developed in detail in appendix G) which all angular momentum operators and their eigenstates obey:

$$J^2\,|j,m\rangle = j(j+1)\hbar^2|j,m\rangle,$$
$$J_z\,|j,m\rangle = m\hbar|j,m\rangle,$$
$$J_+\,|j,m\rangle = \hbar\{j(j+1) - m(m+1)\}^{1/2}|j,m+1\rangle, \text{ and}$$
$$J_-\,|j,m\rangle = \hbar\{j(j+1) - m(m-1)\}^{1/2}|j,m-1\rangle.$$

In a many-electron system, one must combine the spin functions of the individual electrons to generate eigenfunctions of the total

$$S_z = \sum_i S_z(i)$$

(expressions for

$$S_x = \sum_i S_x(i)$$

and

$$S_y = \sum_i S_y(i)$$

also follow from the fact that the total angular momentum of a collection of particles is the sum of the angular momenta, component-by-component, of the individual angular momenta) and total S^2 operators because only these operators commute with the full Hamiltonian, H, and with the permutation operators P_{ij}. No longer are the individual $S^2(i)$ and $S_z(i)$ good quantum numbers; these operators do not commute with P_{ij}.

Spin states which are eigenfunctions of the total S^2 and S_z can be formed by using angular momentum coupling methods or the explicit construction methods detailed in appendix G. In the latter approach, one forms, consistent with the given electronic configuration, the spin state having maximum S_z eigenvalue (which is easy to identify as shown below and which corresponds to a state with S equal to this maximum S_z eigenvalue) and then generating states of lower S_z values and lower S values using the angular momentum raising and lowering operators

$$\left(S_- = \sum_i S_-(i) \text{ and } S_+ = \sum_i S_+(i)\right).$$

To illustrate, consider a three-electron example with the configuration $1s2s3s$. Starting with the determinant $|1s\alpha 2s\alpha 3s\alpha|$, which has the maximum $M_s = 3/2$ and hence has $S = 3/2$ (this function is denoted $|3/2,3/2>$), apply S_- in the additive form $S_- = \sum_i S_-(i)$ to generate the following combination of three determinants:

$$\hbar[|1s\beta 2s\alpha 3s\alpha| + |1s\alpha 2s\beta 3s\alpha| + |1s\alpha 2s\alpha 3s\beta|],$$

which, according to the above identities, must equal

$$\hbar\sqrt{3/2(3/2+1) - 3/2(3/2-1)}\,|3/2, 1/2>.$$

So the state $|3/2,1/2>$ with $S = 3/2$ and $M_s = 1/2$ can be solved for in terms of the three determinants to give

$$|3/2,1/2> = 1/\sqrt{3}\,[|1s\beta 2s\alpha 3s\alpha| + |1s\alpha 2s\beta 3s\alpha| + |1s\alpha 2s\alpha 3s\beta|].$$

The states with $S = 3/2$ and $M_s = -1/2$ and $-3/2$ can be obtained by further application of S_- to $|3/2,1/2>$ (actually, the $M_s = -3/2$ can be identified as the "spin flipped" image of the state with $M_s = 3/2$ and the one with $M_s = -1/2$ can be formed by interchanging all α's and β's in the $M_s = 1/2$ state).

Of the eight total spin states (each electron can take on either α or β spin and there are three electrons, so the number of states is 2^3), the above process has identified proper combinations which yield the four states with $S = 3/2$. Doing so consumed the determinants with $M_s = 3/2$ and $-3/2$, one combination of the three determinants with $M_S = 1/2$, and one combination of the three determinants with $M_s = -1/2$. There still remain two combinations of the $M_s = 1/2$ and two combinations of the $M_s = -1/2$ determinants to deal with. These functions correspond to two sets of $S = 1/2$ eigenfunctions having $M_s = 1/2$ and $-1/2$. Combinations of the determinants must be used in forming the $S = 1/2$ functions to keep the $S = 1/2$ eigenfunctions orthogonal to the above $S = 3/2$ functions (which is required because S^2 is a hermitian operator whose eigenfunctions belonging to different eigenvalues must be orthogonal). The two independent $S = 1/2$, $M_s = 1/2$ states can be formed by simply constructing combinations of the above three determinants with

$M_s = 1/2$ which are orthogonal to the $S = 3/2$ combination given above and orthogonal to each other. For example,

$$|1/2,1/2> = 1/\sqrt{2}[|1s\beta 2s\alpha 3s\alpha| - |1s\alpha 2s\beta 3s\alpha| + 0 \times |1s\alpha 2s\alpha 3s\beta|],$$
$$|1/2,1/2> = 1/\sqrt{6}[|1s\beta 2s\alpha 3s\alpha| + |1s\alpha 2s\beta 3s\alpha| - 2 \times |1s\alpha 2s\alpha 3s\beta|]$$

are acceptable (as is any combination of these two functions generated by a unitary transformation). A pair of independent orthonormal states with $S = 1/2$ and $M_s = -1/2$ can be generated by applying S_- to each of these two functions (or by constructing a pair of orthonormal functions which are combinations of the three determinants with $M_s = -1/2$ and which are orthogonal to the $S = 3/2, M_s = -1/2$ function obtained as detailed above).

The above treatment of a three-electron case shows how to generate quartet (spin states are named in terms of their spin degeneracies $2S + 1$) and doublet states for a configuration of the form $1s2s3s$. Not all three-electron configurations have both quartet and doublet states; for example, the $1s^2 2s$ configuration only supports one doublet state. The methods used above to generate $S = 3/2$ and $S = 1/2$ states are valid for any three-electron situation; however, some of the determinental functions vanish if doubly occupied orbitals occur as for $1s^2 2s$. In particular, the $|1s\alpha 1s\alpha 2s\alpha|$ and $|1s\beta 1s\beta 2s\beta|$ $M_s = 3/2,-3/2$ and $|1s\alpha 1s\alpha 2s\beta|$ and $|1s\beta 1s\beta 2s\alpha|$ $M_s = 1/2,-1/2$ determinants vanish because they violate the Pauli principle; only $|1s\alpha 1s\beta 2s\alpha|$ and $|1s\alpha 1s\beta 2s\beta|$ do not vanish. These two remaining determinants form the $S = 1/2, M_s = 1/2,-1/2$ doublet spin functions which pertain to the $1s^2 2s$ configuration. It should be noted that all closed-shell components of a configuration (e.g., the $1s^2$ part of $1s^2 2s$ or the $1s^2 2s^2 2p^6$ part of $1s^2 2s^2 2p^6 3s^1 3p^1$) must involve α and β spin functions for each doubly occupied orbital and, as such, can contribute nothing to the total M_s value; only the open-shell components need to be treated with the angular momentum operator tools to arrive at proper total-spin eigenstates.

In summary, proper spin eigenfunctions must be constructed from antisymmetric (i.e., determinental) wavefunctions as demonstrated above because the *total* S^2 and *total* S_z remain valid symmetry operators for many-electron systems. Doing so results in the spin-adapted wavefunctions being expressed as combinations of determinants with coefficients determined via spin angular momentum techniques as demonstrated above. In configurations with closed-shell components, not all spin functions are possible because of the antisymmetry of the wavefunction; in particular, any closed-shell parts must involve $\alpha\beta$ spin pairings for each of the doubly occupied orbitals, and, as such, contribute zero to the total M_s.

B. Vector Coupling of Angular Momenta

Given two angular momenta (of any kind) $\boldsymbol{L}_1$ and $\boldsymbol{L}_2$, when one generates states that are eigenstates of their vector sum $\boldsymbol{L} = \boldsymbol{L}_1 + \boldsymbol{L}_2$, one can obtain L values of $L_1 + L_2, L_1 + L_2 - 1, \ldots |L_1 - L_2|$. This can apply to two electrons for which the total spin S can be 1 or 0 as illustrated in detail above, or to a p and a d orbital for which the total orbital angular momentum L can be 3, 2, or 1. Thus for a $p^1 d^1$ electronic configuration, ^3F, ^1F, ^3D, ^1D, ^3P, and ^1P energy levels (and corresponding wavefunctions) arise. Here the term symbols are specified as the spin degeneracy $(2S + 1)$ and the letter that is associated with the L-value. If spin-orbit coupling is present, the ^3F level further splits into $J = 4$, 3, and 2 levels which are denoted $^3F_4$, $^3F_3$, and $^3F_2$.

This simple "vector coupling" method applies to any angular momenta. However, if the angular momenta are "equivalent" in the sense that they involve indistinguishable particles that occupy the same orbital shell (e.g., $2p^3$ involves three equivalent electrons; $2p^1 3p^1 4p^1$ involves three non-equivalent electrons; $2p^2 3p^1$ involves two equivalent electrons and one non-equivalent electron), the Pauli principle eliminates some of the expected term symbols (i.e., when the corresponding wavefunctions are formed, some vanish because their Slater determinants vanish). Later in this section, techniques for dealing with the equivalent-angular momenta case are introduced. These techniques involve using the above tools to obtain a list of candidate term symbols after which Pauli-violating term symbols are eliminated.

C. Non-Vector Coupling of Angular Momenta

For linear molecules, one does *not* vector couple the orbital angular momenta of the individual electrons (because only L_z not L^2 commutes with H), but one does vector couple the electrons' spin angular momenta. Coupling of the electrons' orbital angular momenta involves simply considering the various L_z eigenvalues that can arise from adding the L_z values of the individual electrons. For example, coupling two π orbitals (each of which can have $m = \pm 1$) can give $M_L = 1 + 1, 1 - 1, -1 + 1$, and $-1 - 1$, or 2, 0, 0, and −2. The level with $M_L = \pm 2$ is called a Δ state (much like an orbital with $m = \pm 2$ is called a δ orbital), and the two states with $M_L = 0$ are called Σ states. States with L_z eigenvalues of M_L and $-M_L$ are degenerate because the total energy is independent of which direction the electrons are moving about the linear molecule's axis (just as π_{+1} and π_{-1} orbitals are degenerate).

Again, if the two electrons are non-equivalent, all possible couplings arise (e.g., a $\pi^1\pi'^1$ configuration yields $^3\Delta$, $^3\Sigma$, $^3\Sigma$, $^1\Delta$, $^1\Sigma$, and $^1\Sigma$ states). In contrast, if the two electrons are equivalent, certain of the term symbols are Pauli forbidden. Again, techniques for dealing with such cases are treated later in this chapter.

D. Direct Products for Non-Linear Molecules

For non-linear polyatomic molecules, one vector couples the electrons' spin angular momenta but their orbital angular momenta are not even considered. Instead, their point group symmetries must be combined, by forming direct products, to determine the symmetries of the resultant spin-orbital product states. For example, the $b_1^1 b_2^1$ configuration in C_{2v} symmetry gives rise to $^3A_2$ and $^1A_2$ term symbols. The $e^1e'^1$ configuration in C_{3v} symmetry gives ^3E, $^3A_2$, $^3A_1$, ^1E, $^1A_2$, and $^1A_1$ term symbols. For two equivalent electrons such as in the e^2 configuration, certain of the ^3E, $^3A_2$, $^3A_1$, ^1E, $^1A_2$, and $^1A_1$ term symbols are Pauli forbidden. Once again, the methods needed to identify which term symbols arise in the equivalent-electron case are treated later.

One needs to learn how to tell which term symbols will be Pauli excluded, and to learn how to write the spin-orbit product wavefunctions corresponding to each term symbol and to evaluate the corresponding term symbols' energies.

II. ATOMIC TERM SYMBOLS AND WAVEFUNCTIONS

A. Non-Equivalent Orbital Term Symbols

When coupling non-equivalent angular momenta (e.g., a spin and an orbital angular momenta or two orbital angular momenta of non-equivalent electrons), one vector couples using the fact that the coupled angular momenta range from the sum of the two individual angular momenta to the absolute value of their difference. For example, when coupling the spins of two electrons, the total spin S can be 1 or 0; when coupling a p and a d orbital, the total orbital angular momentum can be 3, 2, or 1. Thus for a p^1d^1 electronic configuration, ^3F, ^1F, ^3D, ^1D, ^3P, and ^1P energy levels (and corresponding wavefunctions) arise. The energy differences among these levels has to do with the different electron-electron repulsions that occur in these levels; that is, their wavefunctions involve different occupancy of the p and d orbitals and hence different repulsion energies. If spin-orbit coupling is present, the L and S angular momenta are further vector coupled. For example, the ^3F level splits into $J = 4$, 3, and 2 levels which are denoted $^3F_4$, $^3F_3$, and $^3F_2$. The energy differences among these J-levels are caused by spin-orbit interactions.

B. Equivalent Orbital Term Symbols

If equivalent angular momenta are coupled (e.g., to couple the orbital angular momenta of a p^2 or d^3 configuration), one must use the "box" method to determine which of the term symbols, that would be expected to arise if the angular momenta were non-equivalent, violate the Pauli principle. To carry out this step, one forms all possible unique (determinental) product states with non-negative M_L and M_S values and arranges them into groups according to their M_L and M_S values. For example, the boxes appropriate to the p^2 orbital occupancy are shown below:

	M_L	2	1	0
M_S	1		$\|p_1\alpha p_0\alpha\|$	$\|p_1\alpha p_{-1}\alpha\|$
	0	$\|p_1\alpha p_1\beta\|$	$\|p_1\alpha p_0\beta\|$, $\|p_0\alpha p_1\beta\|$	$\|p_1\alpha p_{-1}\beta\|$, $\|p_{-1}\alpha p_1\beta\|$, $\|p_0\alpha p_0\beta\|$

There is no need to form the corresponding states with negative M_L or negative M_S values because they are simply "mirror images" of those listed above. For example, the state with $M_L = -1$ and $M_S = -1$ is $|p_{-1}\beta p_0\beta|$, which can be obtained from the $M_L = 1$, $M_S = 1$ state $|p_1\alpha p_0\alpha|$ by replacing α by β and replacing p_1 by p_{-1}.

Given the box entries, one can identify those term symbols that arise by applying the following procedure over and over until all entries have been accounted for:

1. One identifies the highest M_S value (this gives a value of the total spin quantum number that arises, S) in the box. For the above example, the answer is $S = 1$.
2. For all product states of *this* M_S value, one identifies the highest M_L value (this gives a value of the total orbital angular momentum, L, that can arise *for this* S). For the above example, the highest M_L within the $M_S = 1$ states is $M_L = 1$ (not $M_L = 2$), hence $L = 1$.
3. Knowing an S, L combination, one knows the first term symbol that arises from this configuration. In the p^2 example, this is ^3P.
4. Because the level with this L and S quantum numbers contains $(2L + 1)(2S + 1)$ states with M_L and M_S quantum numbers running from $-L$ to L and from $-S$ to S, respectively, one must remove from the original box this number of product states. To do so, one simply erases from the box one entry with each such M_L and M_S value. Actually, since the box need only show those entries with non-negative M_L and M_S values, only these entries need be explicitly deleted. In the ^3P example, this amounts to deleting nine product states with M_L, M_S values of 1,1; 1,0; 1,–1; 0,1; 0,0; 0,–1; -1,1; –1,0; –1,–1.

5. After deleting these entries, one returns to step 1 and carries out the process again. For the p^2 example, the box after deleting the first nine product states looks as follows (those that appear in italics should be viewed as already cancelled in counting all of the ^3P states):

	M_L	2	1	0
M_S	1		*$\vert p_1\alpha p_0\alpha\vert$*	*$\vert p_1\alpha p_{-1}\alpha\vert$*
	0	$\vert p_1\alpha p_1\beta\vert$	*$\vert p_1\alpha p_0\beta\vert$*, $\vert p_0\alpha p_1\beta\vert$	*$\vert p_1\alpha p_{-1}\beta\vert$*, $\vert p_{-1}\alpha p_1\beta\vert$, $\vert p_0\alpha p_0\beta\vert$

It should be emphasized that the process of deleting or crossing off entries in various M_L, M_S boxes involves only *counting* how many states there are; by no means do we identify the particular L, S, M_L, M_S wavefunctions when we cross out any particular entry in a box. For example, when the $|p_1\alpha p_0\beta|$ product is deleted from the $M_L = 1, M_S = 0$ box in accounting for the states in the ^3P level, we do not claim that $|p_1\alpha p_0\beta|$ itself is a member of the ^3P level; the $|p_0\alpha p_1\beta|$ product state could just as well been eliminated when accounting for the ^3P states. As will be shown later, the ^3P state with $M_L = 1, M_S = 0$ will be a combination of $|p_1\alpha p_0\beta|$ and $|p_0\alpha p_1\beta|$.

Returning to the p^2 example at hand, after the ^3P term symbol's states have been accounted for, the highest M_S value is 0 (hence there is an $S = 0$ state), and within this M_S value, the highest M_L value is 2 (hence there is an $L = 2$ state). This means there is a ^1D level with five states having $M_L = 2, 1, 0, -1, -2$. Deleting five appropriate entries from the above box (again denoting deletions by italics) leaves the following box:

	M_L	2	1	0
M_S	1		*$\vert p_1\alpha p_0\alpha\vert$*	*$\vert p_1\alpha p_{-1}\alpha\vert$*
	0	*$\vert p_1\alpha p_1\beta\vert$*	*$\vert p_1\alpha p_0\beta\vert$*, *$\vert p_0\alpha p_1\beta\vert$*	*$\vert p_1\alpha p_{-1}\beta\vert$*, *$\vert p_{-1}\alpha p_1\beta\vert$*, $\vert p_0\alpha p_0\beta\vert$

The only remaining entry, which thus has the highest M_S and M_L values, has $M_S = 0$ and $M_L = 0$. Thus there is also a ^1S level in the p^2 configuration.

Thus, unlike the non-equivalent $2p^1 3p^1$ case, in which ^3P, ^1P, ^3D, ^1D, ^3S, and ^1S levels arise, only the ^3P, ^1D, and ^1S arise in the p^2 situation. This "box method" is necessary to carry out whenever one is dealing with equivalent angular momenta.

If one has mixed equivalent and non-equivalent angular momenta, one can determine *all* possible couplings of the equivalent angular momenta using this method and then use the simpler vector coupling method to add the non-equivalent angular momenta to *each* of these coupled angular momenta. For example, the p^2d^1 configuration can be handled by vector coupling (using the straightforward non-equivalent procedure) $L = 2$(the d orbital) and $S = 1/2$ (the third electron's spin) to *each* of ^3P, ^1D, and ^1S. The result is ^4F, ^4D, ^4P, ^2F, ^2D, ^2P, ^2G, ^2F, ^2D, ^2P, ^2S, and ^2D .

C. Atomic Configuration Wavefunctions

To express, in terms of Slater determinants, the wavefunctions corresponding to each of the states in each of the levels, one proceeds as follows:

1. For each M_S, M_L combination for which one can write down only one product function (i.e., in the non-equivalent angular momentum situation, for each case where only one product function sits at a given box row and column point), that product function *itself* is one of the desired states. For the p^2 example, the $|p_1\alpha p_0\alpha|$ and $|p_1\alpha p_{-1}\alpha|$ (as well as their four other M_L and M_S "mirror images") are members of the ^3P level (since they have $M_S = \pm 1$) and $|p_1\alpha p_1\beta|$ and its M_L mirror image are members of the ^1D level (since they have $M_L = \pm 2$).
2. After identifying as many such states as possible by inspection, one uses $L_\pm$ and $S_\pm$ to generate states that belong to the same term symbols as those already identified but which have higher or lower M_L and/or M_S values.
3. If, after applying the above process, there are term symbols for which states have not yet been formed, one may have to construct such states by forming linear combinations that are orthogonal to all those states that have thus far been found.

To illustrate the use of raising and lowering operators to find the states that can not be identified by inspection, let us again focus on the p^2 case. Beginning with three of the ^3P states that are easy to recognize, $|p_1\alpha p_0\alpha|$, $|p_1\alpha p_{-1}\alpha|$, and $|p_{-1}\alpha p_0\alpha|$, we apply S_- to obtain the $M_S = 0$ functions:

$$\begin{aligned} S_-{}^3P(M_L = 1, M_S = 1) &= [S_-(1) + S_-(2)]|p_1\alpha p_0\alpha| \\ &= \hbar(1(2) - 1(0))^{1/2}\,{}^3P(M_L = 1, M_S = 0) \\ &= \hbar(1/2(3/2) - 1/2(-1/2))^{1/2}|p_1\beta p_0\alpha| + \hbar(1)^{1/2}|p_1\alpha p_0\beta|, \end{aligned}$$

so,

$$^3P(M_L = 1, M_S = 0) = 2^{-1/2}[|p_1\beta p_0\alpha| + |p_1\alpha p_0\beta|].$$

The same process applied to $|p_1\alpha p_{-1}\alpha|$ and $|p_{-1}\alpha p_0\alpha|$ gives

$$1/\sqrt{2}\,[|p_1\alpha p_{-1}\beta| + |p_1\beta p_{-1}\alpha|] \text{ and } 1/\sqrt{2}\,[|p_{-1}\alpha p_0\beta| + |p_{-1}\beta p_0\alpha|],$$

respectively.

The $^3P(M_L = 1, M_S = 0) = 2^{-1/2}[|p_1\beta p_0\alpha| + |p_1\alpha p_0\beta|$ function can be acted on with L_- to generate $^3P(M_L = 0, M_S = 0)$:

$$\begin{aligned} L_-{}^3P(M_L = 1, M_S = 0) &= [L_-(1) + L_-(2)]2^{-1/2}[|p_1\beta p_0\alpha| + |p_1\alpha p_0\beta|] \\ &= \hbar(1(2) - 1(0))^{1/2}\,{}^3P(M_L = 0, M_S = 0) \\ &= \hbar(1(2) - 1(0))^{1/2}2^{-1/2}[|p_0\beta p_0\alpha| + |p_0\alpha p_0\beta|] \\ &\quad +\hbar(1(2) - 0(-1))^{1/2}2^{-1/2}[|p_1\beta p_{-1}\alpha| + |p_1\alpha p_{-1}\beta|], \end{aligned}$$

so,

$$^3P(M_L = 0, M_S = 0) = 2^{-1/2}[|p_1\beta p_{-1}\alpha| + |p_1\alpha p_{-1}\beta|].$$

The ^1D term symbol is handled in like fashion. Beginning with the $M_L = 2$ state $|p_1\alpha p_1\beta|$, one applies L_- to generate the $M_L = 1$ state:

$$\begin{aligned}L_-{}^1D(M_L=2, M_S=0) &= [L_-(1)+L_-(2)]|p_1\alpha p_1\beta| \\ &= \hbar(2(3)-2(1))^{1/2}\,{}^1D(M_L=1, M_S=0) \\ &= \hbar(1(2)-1(0))^{1/2}[|p_0\alpha p_1\beta| + |p_1\alpha p_0\beta|],\end{aligned}$$

so,

$${}^1D(M_L=1, M_S=0) = 2^{-1/2}[|p_0\alpha p_1\beta| + |p_1\alpha p_0\beta|].$$

Applying L_- once more generates the ${}^1D(M_L=0, M_S=0)$ state:

$$\begin{aligned}L_-{}^1D(M_L=1, M_S=0) &= [L_-(1)+L_-(2)]2^{-1/2}[|p_0\alpha p_1\beta| + |p_1\alpha p_0\beta|] \\ &= \hbar(2(3)-1(0))^{1/2}\,{}^1D(M_L=0, M_S=0) \\ &= \hbar(1(2)-0(-1))^{1/2}2^{-1/2}[|p_{-1}\alpha p_1\beta| + |p_1\alpha p_{-1}\beta|] \\ &\quad +\hbar(1(2)-1(0))^{1/2}2^{-1/2}[|p_0\alpha p_0\beta| + |p_0\alpha p_0\beta|],\end{aligned}$$

so,

$${}^1D(M_L=0, M_S=0) = 6^{-1/2}[2|p_0\alpha p_0\beta| + |p_{-1}\alpha p_1\beta| + |p_1\alpha p_{-1}\beta|]\ .$$

Notice that the $M_L=0$, $M_S=0$ states of ${}^3P$ and of ${}^1D$ are given in terms of the three determinants that appear in the "center" of the p^2 box diagram:

$$\begin{aligned}&{}^1D(M_L=0, M_S=0) = 6^{-1/2}[2|p_0\alpha p_0\beta| + |p_{-1}\alpha p_1\beta| + |p_1\alpha p_{-1}\beta|], \\ &{}^3P(M_L=0, M_S=0) = 2^{-1/2}[|p_1\beta p_{-1}\alpha| + |p_1\alpha p_{-1}\beta|] = 2^{-1/2}[-|p_{-1}\alpha p_1\beta| + |p_1\alpha p_{-1}\beta|].\end{aligned}$$

The only state that has eluded us thus far is the ${}^1S$ state, which also has $M_L=0$ and $M_S=0$. To construct this state, which must also be some combination of the three determinants with $M_L=0$ and $M_S=0$, we use the fact that the ${}^1S$ wavefunction *must* be orthogonal to the ${}^3P$ and ${}^1D$ functions ${}^1S$, ${}^3P$, and ${}^1D$ are eigenfunctions of the hermitian operator L^2 having different eigenvalues. The state that is normalized and is a combination of 1 $p_0\alpha p_0\beta|$, $|p_{-1}\alpha p_1\beta|$, and $|p_1\alpha p_{-1}\beta|$ is given as follows:

$${}^1S = 3^{-1/2}[|p_0\alpha p_0\beta| - |p_{-1}\alpha p_1\beta| - |p_1\alpha p_{-1}\beta|].$$

The procedure used here to form the ${}^1S$ state illustrates point 3 in the above prescription for determining wavefunctions. Additional examples for constructing wavefunctions for atoms are provided later in this chapter and in appendix G.

D. Inversion Symmetry

One more quantum number, that relating to the inversion (i) symmetry operator can be used in atomic cases because the total potential energy V is unchanged when *all* of the electrons have their position vectors subjected to inversion ($i\mathbf{r} = -\mathbf{r}$). This quantum number is straightforward to determine. Because each L, S, M_L, M_S, H state discussed above consist of a few (or, in the case of configuration interaction several) symmetry adapted combinations of Slater determinant functions, the effect of the inversion operator on such a wavefunction Ψ can be determined by:

1. applying i to each orbital occupied in Ψ thereby generating a ± 1 factor for each orbital (+1 for *s*, *d*, *g*, *i*, etc. orbitals; -1 for *p*, *f*, *h*, *j*, etc. orbitals),
2. multiplying these ± 1 factors to produce an overall sign for the character of Ψ under i.

When this overall sign is positive, the function Ψ is termed "even" and its term symbol is appended with an "e" superscript (e.g., the ^3P level of the O atom, which has $1s^22s^22p^4$ occupancy is labeled $^3P^e$); if the sign is negative Ψ is called "odd" and the term symbol is so amended (e.g., the ^3P level of $1s^22s^12p^1$ B$^+$ ion is labeled $^3P^o$).

E. Review of Atomic Cases

The orbitals of an atom are labeled by l and m quantum numbers; the orbitals belonging to a given energy and l value are $2l+1$-fold degenerate. The many-electron Hamiltonian, H, of an atom and the antisymmetrizer operator

$$\boldsymbol{A} = (\sqrt{1/N!})\sum_p s_p P$$

commute with total

$$L_z = \sum_i L_z(i),$$

as in the linear-molecule case. The additional symmetry present in the spherical atom reflects itself in the fact that L_x, and L_y now also commute with H and A. However, since L_z does not commute with L_x or L_y, new quantum numbers can not be introduced as symmetry labels for these other components of $\boldsymbol{L}$. A new symmetry label does arise when $L^2 = L_z^2 + L_x^2 + L_y^2$ is introduced; L^2 commutes with H, A, and L_z, so proper eigenstates (and trial wavefunctions) can be labeled with L, M_L, S, M_s, and H quantum numbers.

To identify the states which arise from a given atomic configuration and to construct properly symmetry-adapted determinental wave functions corresponding to these symmetries, one must employ L and M_L and S and M_S angular momentum tools. One first identifies those determinants with maximum M_S (this then defines the maximum S value that occurs); within *that set* of determinants, one then identifies the determinant(s) with maximum M_L (this identifies the highest L value). This determinant has S and L equal to its M_s and M_L values (this can be verified, for example for L, by acting on this determinant with L^2 in the form

$$L^2 = L_-L_+ + L_z^2 + \hbar L_z$$

and realizing that L_+ acting on the state must vanish); other members of this L, S energy level can be constructed by sequential application of S_- and

$$L_- = \sum_i L_-(i).$$

Having exhausted a set of $(2L+1)(2S+1)$ combinations of the determinants belonging to the given configuration, one proceeds to apply the same procedure to the remaining determinants (or combinations thereof). One identifies the maximum M_s and, within it, the maximum M_L which thereby specifies another S, L label and a new "maximum" state. The determinental functions corresponding to these L, S (and various M_L, $M_{s)}$ values can be constructed by applying S_- and L_- to this "maximum" state. This process is continued until all of the states and their determinental wave functions are obtained.

As illustrated above, any p^2 configuration gives rise to $^3P^e$, $^1D^e$, and $^1S^e$ levels which contain nine, five, and one state respectively. The use of L and S angular momentum algebra tools allows one to identify the wavefunctions corresponding to these states. As shown in detail in appendix G, in the event that *spin-orbit* coupling causes the Hamiltonian, $\boldsymbol{H}$, not to commute with $\boldsymbol{L}$ or with $\boldsymbol{S}$ but only with their vector sum $\boldsymbol{J} = \boldsymbol{L} + \boldsymbol{S}$, then these $L^2S^2L_zS_z$ eigenfunctions must be coupled (i.e., recombined) to generate J^2J_z eigenstates. The steps needed to effect this coupling are developed and illustrated for the above p^2 configuration case in appendix G.

In the case of a pair of *non-equivalent p* orbitals (e.g., in a $2p^13p^1$ configuration), even more states would arise. They can also be found using the tools provided above. Their symmetry labels can be obtained by vector coupling (see appendix G) the spin and orbital angular momenta of the two subsystems. The orbital angular momentum coupling with $l = 1$ and $l = 1$ gives $L = 2$, 1, and 0 or D, P, and S states. The spin angular momentum coupling with $s = 1/2$ and $s = 1/2$ gives $S = 1$ and 0, or triplet and singlet states. So, vector coupling leads to the prediction that $^3D^e$, $^1D^e$, $^3P^e$, $^1P^e$, $^3S^e$, and $^1S^e$ states can be formed from a pair of non-equivalent p orbitals. It is seen that more states arise when non-equivalent orbitals are involved; for equivalent orbitals, some determinants vanish, thereby decreasing the total number of states that arise.

III. LINEAR MOLECULE TERM SYMBOLS AND WAVEFUNCTIONS

A. Non-Equivalent Orbital Term Symbols

Equivalent angular momenta arising in linear molecules also require use of specialized angular momentum coupling. Their spin angular momenta are coupled exactly as in the atomic case because both for atoms and linear molecules, S^2 and S_z commute with H. However, unlike atoms, linear molecules no longer permit L^2 to be used as an operator that commutes with H; L_z still does, but L^2 does not. As a result, when coupling non-equivalent linear molecule angular momenta, one vector couples the electron spins as before. However, in place of vector coupling the individual orbital angular momenta, one *adds* the individual L_z values to obtain the L_z values of the coupled system. For example, the $\pi^1\pi'^1$ configuration gives rise to $S = 1$ and $S = 0$ spin states. The individual m_l values of the two pi-orbitals can be added to give $M_L = 1 + 1, 1 - 1, -1 + 1$, and $-1 - 1$, or 2, 0, 0, and -2. The $M_L = 2$ and -2 cases are degenerate (just as the $m_l = 2$ and -2 δ orbitals are and the $m_l = 1$ and -1 π orbitals are) and are denoted by the term symbol Δ; there are two distinct $M_L = 0$ states that are denoted Σ. Hence, the $\pi^1\pi'^1$ configuration yields $^3\Delta$, $^3\Sigma$, $^3\Sigma$, $^1\Delta$, $^1\Sigma$, and $^1\Sigma$ term symbols.

B. Equivalent-Orbital Term Symbols

To treat the equivalent-orbital case π^2, one forms a box diagram as in the atom case:

	M_L	2	1	0
M_S	1			$\lvert\pi_1\alpha\pi_{-1}\alpha\rvert$
	0	$\lvert\pi_1\alpha\pi_1\beta\rvert$		$\lvert\pi_1\alpha\pi_{-1}\beta\rvert$, $\lvert\pi_{-1}\alpha\pi_1\beta\rvert$

The process is very similar to that used for atoms. One first identifies the highest M_S value (and hence an S value that occurs) and *within* that M_S, the highest M_L. However, the highest M_L

does *not* specify an L-value, because L is no longer a "good quantum number" because L^2 no longer commutes with H. Instead, we simply take the highest M_L value (and minus this value) as specifying a Σ, Π, Δ, Φ, Γ, etc. term symbol. In the above example, the highest M_S value is $M_S = 1$, so there is an $S = 1$ level. Within $M_S = 1$, the highest $M_L = 0$; hence, there is a $^3\Sigma$ level.

After deleting from the box diagram entries corresponding to M_S values ranging from $-S$ to S and M_L values of M_L and $-M_L$, one has (again using italics to denote the deleted entries):

M_L	2	1	0
M_S 1			$/\pi_1\alpha\pi_{-1}\alpha/$
0	$\|\pi_1\alpha\pi_1\beta\|$		$/\pi_1\alpha\pi_{-1}\beta/$, $\|\pi_{-1}\alpha\pi_1\beta\|$

Among the remaining entries, the highest M_S value is $M_S = 0$, and within this M_S the highest M_L is $M_L = 2$. Thus, there is a $^1\Delta$ state. Deleting entries with $M_S = 0$ and $M_L = 2$ and -2, one has left the following box diagram:

M_L	2	1	0
M_S 1			$/\pi_1\alpha\pi_{-1}\alpha/$
0	$/\pi_1\alpha\pi_1\beta/$		$/\pi_1\alpha\pi_{-1}\beta/$, $\|\pi_{-1}\alpha\pi_1\beta\|$

There still remains an entry with $M_S = 0$ and $M_L = 0$; hence, there is also a $^1\Sigma$ level.

Recall that the non-equivalent $\pi^1\pi'^1$ case yielded $^3\Delta$, $^3\Sigma$, $^3\Sigma$, $^1\Delta$, $^1\Sigma$, and $^1\Sigma$ term symbols. The equivalent π^2 case yields only $^3\Sigma$, $^1\Delta$, and $^1\Sigma$ term symbols. Again, whenever one is faced with equivalent angular momenta in a linear-molecule case, one must use the box method to determine the allowed term symbols. If one has a mixture of equivalent and non-equivalent angular momenta, it is possible to treat the equivalent angular momenta using boxes and to then add in the non-equivalent angular momenta using the more straightforward technique. For example, the $\pi^2\delta^1$ configuration can be treated by coupling the π^2 as above to generate $^3\Sigma$, $^1\Delta$, and $^1\Sigma$ and then vector coupling the spin of the third electron and additively coupling the $m_l = 2$ and -2 of the third orbital. The resulting term symbols are $^4\Delta$, $^2\Delta$, $^2\Gamma$, $^2\Sigma$, $^2\Sigma$, and $^2\Delta$ (e.g., for the $^1\Delta$ intermediate state, adding the δ orbital's m_l values gives total M_L values of $M_L = 2 + 2, 2 - 2, -2 + 2$, and $-2 - 2$, or 4, 0, 0, and -4).

C. Linear-Molecule Configuration Wavefunctions

Procedures analogous to those used for atoms can be applied to linear molecules. However, in this case only $S_\pm$ can be used; $L_\pm$ no longer applies because L is no longer a good quantum number. One begins as in the atom case by identifying determinental functions for which M_L and M_S are unique. In the π^2 example considered above, these states include $|\pi_1\alpha\pi_{-1}\alpha|$, $|\pi_1\alpha\pi_1\beta|$, and their mirror images. These states are members of the $^3\Sigma$ and $^1\Delta$ levels, respectively, because the first has $M_S = 1$ and because the latter has $M_L = 2$.

Applying S_- to this $^3\Sigma$ state with $M_S = 1$ produces the $^3\Sigma$ state with $M_S = 0$:

$$\begin{aligned} S_-{}^3\Sigma(M_L = 0, M_S = 1) &= [S_-(1) + S_-(2)]|\pi_1\alpha\pi_{-1}\alpha| \\ &= \hbar(1(2) - 1(0))^{1/2}\,{}^3\Sigma(M_L = 0, M_S = 0) \\ &= \hbar(1)^{1/2}[|\pi_1\beta\pi_{-1}\alpha| + |\pi_1\alpha\pi_{-1}\beta|], \end{aligned}$$

so,

$$^3\Sigma(M_L = 0, M_S = 0) = 2^{-1/2}[|\pi_1\beta\pi_{-1}\alpha| + |\pi_1\alpha\pi_{-1}\beta|].$$

The only other state that can have $M_L = 0$ and $M_S = 0$ is the $^1\Sigma$ state, which must itself be a combination of the two determinants, $|\pi_1\beta\pi_{-1}\alpha|$ and $|\pi_1\alpha\pi_{-1}\beta|$, with $M_L = 0$ and $M_S = 0$. Because the $^1\Sigma$ state has to be orthogonal to the $^3\Sigma$ state, the combination must be

$$^1\Sigma = 2^{-1/2}[|\pi_1\beta\pi_{-1}\alpha| - |\pi_1\alpha\pi_{-1}\beta|]\ .$$

As with the atomic systems, additional examples are provided later in this chapter and in appendix G.

D. Inversion Symmetry and σ_v Reflection Symmetry

For homonuclear molecules (e.g., O_2, N_2, etc.) the inversion operator i (where inversion of all electrons now takes place through the center of mass of the nuclei rather than through an individual nucleus as in the atomic case) is also a valid symmetry, so wavefunctions Ψ may also be labeled as even or odd. The former functions are referred to as ***gerade*** (g) and the latter as ***ungerade*** (u) (derived from the German words for even and odd). The g or u character of a term symbol is straightforward to determine. Again one

1. applies i to each orbital occupied in Ψ thereby generating a ±1 factor for each orbital (+1 for σ, π^*, δ, ϕ^*, etc. orbitals; -1 for σ^*, π, δ^*, ϕ, etc. orbitals),
2. multiplying these ±1 factors to produce an overall sign for the character of Ψ under i.

When this overall sign is positive, the function Ψ is gerade and its term symbol is appended with a "g" subscript (e.g., the $^3\Sigma$ level of the O_2 molecule, which has $\pi_u^4\pi_g{*}^2$ occupancy is labeled $^3\Sigma_g$); if the sign is negative, Ψ is ungerade and the term symbol is so amended (e.g., the $^3\Pi$ level of the $1\sigma_g^2 1\sigma_u^2 2\sigma_g^1 1\pi_u^1$ configuration of the Li_2 molecule is labeled $^3\Pi_u$).

Finally, for linear molecules in Σ states, the wavefunctions can be labeled by one additional quantum number that relates to their symmetry under reflection of *all* electrons through a σ_v plane passing through the molecule's C_∞ axis. If Ψ is even, a + sign is appended as a superscript to the term symbol; if Ψ is odd, a – sign is added.

To determine the σ_v symmetry of Ψ, one first applies σ_v to each orbital in Ψ. Doing so replaces the azimuthal angle ϕ of the electron in that orbital by $2\pi - \phi$; because orbitals of linear molecules depend on ϕ as $\exp(im\phi)$, this changes the orbital into $\exp(im(-\phi))\exp(2\pi im) = \exp(-im\phi)$. In effect, σ_v applied to Ψ changes the signs of all of the m values of the orbitals in Ψ. One then determines whether the resultant $\sigma_v\Psi$ is equal to or opposite in sign from the original Ψ by inspection. For example, the $^3\Sigma_g$ ground state of O_2, which has a Slater determinant function

$$\begin{aligned} |S = 1, M_S = 1\rangle &= |\pi^*{}_1\alpha\pi^*{}_{-1}\alpha| \\ &= 2^{-1/2}[\pi^*{}_1(\boldsymbol{r}_1)\alpha_1\pi^*{}_{-1}(\boldsymbol{r}_2)\alpha_2 - \pi^*{}_1(\boldsymbol{r}_2)\alpha_2\pi^*{}_{-1}(\boldsymbol{r}_1)\alpha_1]. \end{aligned}$$

Recognizing that $\sigma_v\pi^*{}_1 = \pi^*{}_{-1}$ and $\sigma_v\pi^*{}_{-1} = \pi^*{}_1$, then gives

$$\sigma_v|S = 1, M_S = 1\rangle = |\pi^*{}_1\alpha\pi^*{}_{-1}\alpha|$$
$$= 2^{-1/2}[\pi^*{}_{-1}(\boldsymbol{r}_1)\alpha_1\pi^*{}_1(\boldsymbol{r}_2)\alpha_2 - \pi^*{}_{-1}(\boldsymbol{r}_2)\alpha_2\pi^*{}_1(\boldsymbol{r}_1)\alpha_1]$$
$$= (-1)2^{-1/2}[\pi^*{}_1(\boldsymbol{r}_1)\alpha_1\pi^*{}_{-1}(\boldsymbol{r}_2)\alpha_2 - \pi^*{}_1(\boldsymbol{r}_2)\alpha_2\pi^*{}_{-1}(\boldsymbol{r}_1)\alpha_1],$$

so this wavefunction is odd under σ_v which is written as ${}^3\Sigma_g^-$.

E. Review of Linear Molecule Cases

Molecules with axial symmetry have orbitals of σ, π, δ, ϕ, etc. symmetry; these orbitals carry angular momentum about the z-axis in amounts (in units of $\hbar$) 0, +1 and −1, +2 and −2, +3 and −3, etc. The axial point-group symmetries of configurations formed by occupying such orbitals can be obtained by adding, in all possible ways, the angular momenta contributed by each orbital to arrive at a set of possible total angular momenta. The eigenvalue of total

$$L_z = \sum_i L_z(i)$$

is a valid quantum number because total L_z commutes with the Hamiltonian and with P_{ij}; one obtains the eigenvalues of total L_z by adding the individual spin-orbitals' m eigenvalues because of the additive form of the L_z operator. L^2 no longer commutes with the Hamiltonian, so it is no longer appropriate to construct N-electron functions that are eigenfunctions of L^2. Spin symmetry is treated as usual via the spin angular momentum methods described in the preceding sections and in appendix G. For molecules with centers of symmetry (e.g., for homonuclear diatomics or ABA linear triatomics), the many-electron spin-orbital product inversion symmetry, which is equal to the product of the individual spin-orbital inversion symmetries, provides another quantum number with which the states can be labeled. Finally the σ_v symmetry of Σ states can be determined by changing the m values of all orbitals in Ψ and then determining whether the resultant function is equal to Ψ or to $-\Psi$.

If, instead of a π^2 configuration like that treated above, one had a δ^2 configuration, the above analysis would yield ${}^1\Gamma$, ${}^1\Sigma$ and ${}^3\Sigma$ symmetries (because the two δ orbitals' m values could be combined as 2 + 2, 2 − 2 , −2 + 2, and −2 − 2); the wavefunctions would be identical to those given above with the π_1 orbitals replaced by δ_2 orbitals and π_{-1} replaced by δ_{-2}. Likewise, ϕ^2 gives rise to ${}^1I$, ${}^1\Sigma$, and ${}^3\Sigma$ symmetries.

For a $\pi^1\pi'^1$ configuration in which two non-equivalent π orbitals (i.e., orbitals which are of π symmetry but which are not both members of the same degenerate set; an example would be the π and π^* orbitals in the B_2 molecule) are occupied, the above analysis must be expanded by including determinants of the form: $|\pi_1\alpha\pi'_1\alpha|$, $|\pi_{-1}\alpha\pi'_{-1}\alpha|$, $|\pi_1\beta\pi'_1\beta|$, $|\pi_{-1}\beta\pi'_{-1}\beta|$. Such determinants were excluded in the π^2 case because they violated the Pauli principle (i.e., they vanish identically when $\pi' = \pi$). Determinants of the form $|\pi'_1\alpha\pi_{-1}\alpha|$, $|\pi'_1\alpha\pi_1\beta|$, $|\pi'_{-1}\alpha\pi_{-1}\beta|$, $|\pi'_1\beta\pi_{-1}\beta|$, $|\pi'_1\alpha\pi_{-1}\beta|$, and $|\pi'_1\beta\pi_{-1}\alpha|$ are now distinct and must be included as must the determinants $|\pi_1\alpha\pi'_{-1}\alpha|$, $|\pi_1\alpha\pi'_1\beta|$, $|\pi_{-1}\alpha\pi'_{-1}\beta|$, $|\pi_1\beta\pi'_{-1}\beta|$, $|\pi_1\alpha\pi'_{-1}\beta|$, and $|\pi_1\beta\pi'_{-1}\alpha|$, which are analogous to those used above. The result is that there are more possible determinants in the case of non-equivalent orbitals. However, the techniques for identifying space-spin symmetries and creating proper determinental wavefunctions are the same as in the equivalent-orbital case.

For any π^2 configuration, one finds ${}^1\Delta$, ${}^1\Sigma$, and ${}^3\Sigma$ wavefunctions as detailed earlier; for the $\pi^1\pi'^1$ case, one finds ${}^3\Delta$, ${}^1\Delta$, ${}^3\Sigma$, ${}^1\Sigma$, ${}^3\Sigma$, and ${}^1\Sigma$ wavefunctions by starting with the determinants with the maximum M_s value, identifying states by their $|M_L|$ values, and using spin angular momentum algebra and orthogonality to generate states with lower M_s and, subsequently, lower S

values. Because L^2 is not an operator of relevance in such cases, raising and lowering operators relating to $\boldsymbol{L}$ are *not* used to generate states with lower Λ values. States with specific Λ values are formed by occupying the orbitals in all possible manners and simply computing Λ as the absolute value of the sum of the individual orbitals' m-values.

If a center of symmetry is present, all of the states arising from π^2 are gerade; however, the states arising from $\pi^1\pi'^1$ can be gerade if π and π' are both g or both u or ungerade if π and π' are of opposite inversion symmetry.

The state symmetries appropriate to the non-equivalent $\pi^1\pi'^1$ case can, alternatively, be identified by "coupling" the spin and L_z angular momenta of two "independent" subsystems—the π^1 system which gives rise to $^2\Pi$ symmetry (with $M_L = 1$ and -1 and $S = 1/2$) and the π'^1 system which also give $^2\Pi$ symmetry. The coupling gives rise to triplet and singlet spins (whenever two full vector angular momenta $|j,m\rangle$ and $|j',m'\rangle$ are coupled, one can obtain total angular momentum values of $J = j + j', j + j' - 1, j + j' - 2, \ldots |j - j'|$; see appendix G for details) and to M_L values of $1 + 1 = 2, -1 - 1 = -2, 1 - 1 = 0$ and $-1 + 1 = 0$ (i.e., to Δ, Σ, and Σ states). The L_z angular momentum coupling is *not* carried out in the full vector coupling scheme used for the electron spins because, unlike the spin case where one is forming eigenfunctions of total S^2 and S_z, one is only forming L_z eigenstates (i.e., L^2 is not a valid quantum label). In the case of axial angular momentum coupling, the various possible M_L values of each subsystem are added to those of the other subsystem to arrive at the total M_L value. This angular momentum coupling approach gives the same set of symmetry labels ($^3\Delta$, $^1\Delta$, $^3\Sigma$, $^1\Sigma$, $^3\Sigma$, and $^1\Sigma$) as are obtained by considering all of the determinants of the composite system as treated above.

IV. NON-LINEAR MOLECULE TERM SYMBOLS AND WAVEFUNCTIONS

A. Term Symbols for Non-Degenerate Point Group Symmetries

The point group symmetry labels of the individual orbitals which are occupied in any determinental wave function can be used to determine the overall spatial symmetry of the determinant. When a point group symmetry operation is applied to a determinant, it acts on all of the electrons in the determinant; for example, $\sigma_v|\phi_1\phi_2\phi_3| = |\sigma_v\phi_1\sigma_v\phi_2\sigma_v\phi_3|$. If each of the spin-orbitals ϕ_i belong to non-degenerate representations of the point group, $\sigma_v\phi_i$ will yield the character $\chi_i(\sigma_v)$ appropriate to that spin-orbital multiplying ϕ_i. As a result, $\sigma_v|\phi_1\phi_2\phi_3|$ will involve the product of the three characters (one for each spin-orbital) $\Pi_i\chi_i(\sigma_v)\times|\phi_1\phi_2\phi_3|$. This gives an example of how the symmetry of a spin-orbital product (or an antisymmetrized product) is given as the *direct product* of the symmetries of the individual spin-orbitals in the product; the point group symmetry operator, because of its product nature, passes through or commutes with the antisymmetrizer. It should be noted that any closed-shell parts of the determinant (e.g., $1a_1^2 2a_1^2 1b_2^2$ in the configuration $1a_1^2 2a_1^2 1b_2^2 1b_1^1$) contribute unity to the direct product because the squares of the characters of any non-degenerate point group for any group operation equals unity. Therefore, only the open-shell parts need to be considered further in the symmetry analysis. For a brief introduction to point group symmetry and the use of direct products in this context, see appendix E.

An example will help illustrate these ideas. Consider the formaldehyde molecule H_2CO in C_{2v} symmetry. The configuration which dominates the ground-state wavefunction has doubly occupied O and C $1s$ orbitals, two CH bonds, a CO σ bond, a CO π bond, and two O-centered lone pairs; this configuration is described in terms of symmetry adapted orbitals as follows: ($1a_1^2 2a_1^2 3a_1^2 1b_2^2 4a_1^2 1b_1^2 5a_1^2 2b_2^2$) and is of $^1A_1$ symmetry because it is entirely closed-shell (note that lowercase letters are used to denote the symmetries of orbitals and capital letters are used for many-electron functions' symmetries).

The lowest-lying $n \Rightarrow \pi^*$ states correspond to a configuration (only those orbitals whose occupancies differ from those of the ground state are listed) of the form $2b_2^1 2b_1^1$, which gives rise to $^1A_2$ and $^3A_2$ wavefunctions (the direct product of the open-shell spin orbitals is used to obtain the symmetry of the product wavefunction: $A_2 = b_1 \times b_2$). The $\pi \Rightarrow \pi^*$ excited configuration $1b_1^1 2b_1^1$ gives $^1A_1$ and $^3A_1$ states because $b_1 \times b_1 = A_1$.

The only angular momentum coupling that occurs in non-linear molecules involves the electron spin angular momenta, which are treated in a vector coupling manner. For example, in the lowest-energy state of formaldehyde, the orbitals are occupied in the configuration $1a_1^2 2a_1^2 3a_1^2 1b_2^2 4a_1^2 1b_1^2 5a_1^2 2b_2^2$. This configuration has only a single entry in its "box." Its highest M_S value is $M_S = 0$, so there is a singlet $S = 0$ state. The spatial symmetry of this singlet state is totally symmetric A_1 because this is a closed-shell configuration.

The lowest-energy $n\pi*$ excited configuration of formaldehyde has a $1a_1^2 2a_1^2 3a_1^2 1b_2^2 4a_1^2 1b_1^2 5a_1^2$ $2b_2^1 2b_1^1$ configuration, which has a total of four entries in its "box" diagram:

$$
\begin{aligned}
M_S &= 1 & &|2b_2^1\alpha 2b_1^1\alpha|, \\
M_S &= 0 & &|2b_2^1\alpha 2b_1^1\beta|, \\
M_S &= 0 & &|2b_2^1\beta 2b_1^1\alpha|, \\
M_S &= -1 & &|2b_2^1\beta 2b_1^1\beta| .
\end{aligned}
$$

The highest M_S value is $M_S = 1$, so there is an $S = 1$ state. After deleting one entry each with $M_S = 1$, 0, and -1, there is one entry left with $M_S = 0$. Thus, there is an $S = 0$ state also.

As illustrated above, the spatial symmetries of these four $S = 1$ and $S = 0$ states are obtained by forming the direct product of the "open-shell" orbitals that appear in this configuration: $b_2 \times b_1 = A_2$. All four states have this spatial symmetry. In summary, the above configuration yields $^3A_2$ and $^1A_2$ term symbols. The $\pi^1\pi*^1$ configuration $1a_1^2 2a_1^2 3a_1^2 1b_2^2 4a_1^2 1b_1^1 5a_1^2 2b_2^2 2b_1^1$ produces $^3A_1$ and $^1A_1$ term symbols (because $b_1 \times b_1 = A_1$).

B. Wavefunctions for Non-Degenerate, Non-Linear Point Molecules

The techniques used earlier for linear molecules extend easily to non-linear molecules. One begins with those states that can be straightforwardly identified as unique entries within the box diagram. For polyatomic molecules with no degenerate representations, the spatial symmetry of each box entry is identical and is given as the direct product of the open-shell orbitals. For the formaldehyde example considered earlier, the spatial symmetries of the $n\pi*$ and $\pi\pi*$ states were A_2 and A_1, respectively.

After the unique entries of the box have been identified, one uses $S_\pm$ operations to find the other functions. For example, the wavefunctions of the $^3A_2$ and $^1A_2$ states of the $n\pi* 1a_1^2 2a_1^2 3a_1^2 1b_2^2$ $4a_1^2 1b_1^2 5a_1^2 2b_2^1 2b_1^1$ configuration of formaldehyde are formed by first identifying the $M_S = \pm 1$ components of the $S = 1$ state as $|2b_2\alpha 2b_1\alpha|$ and $|2b_2\beta 2b_1\beta|$ (all of the closed-shell components of the determinants are not explicitly given). Then, applying S_- to the $M_S = 1$ state, one obtains the $M_S = 0$ component $(1/2)^{1/2}[|2b_2\beta 2b_1\alpha| + |2b_2\alpha 2b_1\beta|]$. The singlet state is then constructed as the combination of the two determinants appearing in the $S = 1$, $M_S = 0$ state that is orthogonal to this triplet state. The result is $(1/2)^{1/2}[|2b_2\beta 2b_1\alpha| - |2b_2\alpha 2b_1\beta|]$.

The results of applying these rules to the $n\pi^*$ and $\pi\pi^*$ states are as follows:

$$^3A_2(M_s = 1) = |1a_1\alpha 1a_1\beta 2a_1\alpha 2a_1\beta 3a_1\alpha 3a_1\beta 1b_2\alpha 1b_2\beta 4a_1\alpha 4a_1\beta 1b_1\alpha 1b_1\beta$$
$$5a_1\alpha 5a_1\beta 2b_2\alpha 2b_1\alpha|,$$
$$^3A_2(M_s = 0) = 1/\sqrt{2}[|2b_2\alpha 2b_1\beta| + |2b_2\beta 2b_1\alpha|],$$
$$^3A_2(M_S = -1) = |2b_2\beta 2b_1\beta|,$$
$$^1A_2 = 1/\sqrt{2}[|2b_2\alpha 2b_1\beta| - |2b_2\beta 2b_1\alpha|]\ .$$

The lowest $\pi\pi*$ states of triplet and singlet spin involve the following:

$$^3A_1(M_s = 1) = |1b_1\alpha 2b_1\alpha|,$$
$$^1A_1 = 1/\sqrt{2}[|1b_1\alpha 2b_1\beta| - |1b_1\beta 2b_1\alpha|].$$

In summary, forming spatial- and spin-adapted determinental functions for molecules whose point groups have no degenerate representations is straightforward. The direct product of all of the open-shell spin orbitals gives the point-group symmetry of the determinant. The spin symmetry is handled using the spin angular momentum methods introduced and illustrated earlier.

C. Extension to Degenerate Representations for Non-Linear Molecules

Point groups in which degenerate orbital symmetries appear can be treated in like fashion but require more analysis because a symmetry operation R acting on a degenerate orbital generally yields a linear combination of the degenerate orbitals rather than a multiple of the original orbital (i.e., $R\phi_i = \chi_i(R)\phi_i$ is no longer valid). For example, when a pair of degenerate orbitals (denoted e_1 and e_2) are involved, one has

$$Re_i = \sum_j R_{ij}e_j,$$

where R_{ij} is the 2×2 matrix representation of the effect of R on the two orbitals. The effect of R on a product of orbitals can be expressed as:

$$Re_ie_j = \sum_{k,l} R_{ik}R_{jl}e_ke_l.$$

The matrix $R_{ij,kl} = R_{ik}R_{jl}$ represents the effect of R on the orbital products in the same way R_{ik} represents the effect of R on the orbitals. One says that the orbital products also form a basis for a representation of the point group. The character (i.e., the trace) of the representation matrix $R_{ij,kl}$ appropriate to the orbital product basis is seen to equal the product of the characters of the matrix R_{ik} appropriate to the orbital basis: $\chi_e^2(R) = \chi_e(R)\chi_e(R)$, which is, of course, why the term "direct product" is used to describe this relationship.

For point groups which contain no degenerate representations, the direct product of one symmetry with another is equal to a unique symmetry; that is, the characters $\chi(R)$ obtained as $\chi_a(R)\chi_b(R)$ belong to a pure symmetry and can be immediately identified in a point-group character table. However, for point groups in which degenerate representations occur, such is not the case. The direct product characters will, in general, not correspond to the characters of a single representation; they will contain contributions from more than one representation and these contributions will have to be sorted out using the tools provided below.

A concrete example will help clarify these concepts. In C_{3v} symmetry, the π orbitals of the cyclopropenyl anion transform according to a_1 and e symmetries

and can be expressed as LCAO-MO's in terms of the individual p_i orbitals as follows:

$$a_1 = 1/\sqrt{3}\,[p_1 + p_2 + p_3],\ e_1 = 1/\sqrt{2}\,[p_1 - p_3],$$

and

$$e_2 = 1/\sqrt{6}\,[2p_2 - p_1 - p_3]\ .$$

For the anion's lowest energy configuration, the orbital occupancy $a_1^2 e^2$ must be considered, and hence the spatial and spin symmetries arising from the e^2 configuration are of interest. The character table shown below:

C_{3v}	E	$3\sigma_v$	$2C_3$
a_1	1	1	1
a_2	1	–1	1
e	2	0	–1

allows one to compute the characters appropriate to the direct product $(e \times e)$ as $\chi(E) = 2 \times 2 = 4$, $\chi(\sigma_v) = 0 \times 0 = 0$, $\chi(C_3) = (-1) \times (-1) = 1$.

This *reducible* representation (the occupancy of two e orbitals in the anion gives rise to more than one state, so the direct product $e \times e$ contains more than one symmetry component) can be decomposed into pure symmetry components (labels Γ are used to denote the irreducible symmetries) by using the decomposition formula given in appendix E:

$$n(\Gamma) = 1/g \sum_R \chi(R)\chi_\Gamma(R).$$

Here g is the order of the group (the number of symmetry operations in the group—six in this case) and $\chi_\Gamma(R)$ is the character for the particular symmetry Γ whose component in the direct product is being calculated.

For the case given above, one finds $n(a_1) = 1$, $n(a_2) = 1$, and $n(e) = 1$; so within the configuration e^2 there is one A_1 wavefunction, one A_2 wavefunction and a pair of E wavefunctions (where the symmetry labels now refer to the symmetries of the determinental wavefunctions). This analysis tells one how many different wavefunctions of various spatial symmetries are contained in a configuration in which degenerate orbitals are fractionally occupied. Considerations of spin symmetry and the construction of proper determinental wavefunctions, as developed earlier in this section, still need to be applied to each spatial symmetry case.

To generate the proper A_1, A_2, and E wavefunctions of singlet and triplet spin symmetry (thus far, it is not clear which spin can arise for each of the three above spatial symmetries; however, only singlet and triplet spin functions can arise for this two-electron example), one can apply the following (un-normalized) symmetry projection operators (see appendix E where these projectors are introduced) to all determinental wavefunctions arising from the e^2 configuration:

$$P_\Gamma = \sum_R \chi_\Gamma(R) R.$$

Here, $\chi_\Gamma(R)$ is the character belonging to symmetry Γ for the symmetry operation R. Applying this projector to a determinental function of the form $|\phi_i\phi_j|$ generates a sum of determinants with coefficients determined by the matrix representations R_{ik}:

$$P_\Gamma|\phi_i\phi_j| = \sum_R \sum_{kl} \chi_\Gamma(R) R_{ik} R_{jl} |\phi_k\phi_l| \ .$$

For example, in the e^2 case, one can apply the projector to the determinant with the maximum M_s value to obtain

$$P_\Gamma|e_1\alpha e_2\alpha| = \sum_R \chi_\Gamma(R)[R_{11}R_{22}|e_1\alpha e_2\alpha| + R_{12}R_{21}|e_2\alpha e_1\alpha|]$$

$$= \sum_R \chi_\Gamma(R)[R_{11}R_{22} - R_{12}R_{21}]|e_1\alpha e_2\alpha|,$$

or to the other two members of this triplet manifold, thereby obtaining

$$P_\Gamma|e_1\beta e_2\beta| = \sum_R \chi_\Gamma(R)[R_{11}R_{22} - R_{12}R_{21}]|e_1\beta e_2\beta|$$

and

$$P_\Gamma 1/\sqrt{2}[|e_1\alpha e_2\beta| + |e_1\beta e_2\alpha|] = \sum_R \chi_\Gamma(R)[R_{11}R_{22} - R_{12}R_{21}]1/\sqrt{2}[|e_1\alpha e_2\beta| + |e_1\beta e_2\alpha|].$$

The other (singlet) determinants can be symmetry analyzed in like fashion and result in the following:

$$P_\Gamma|e_1\alpha e_1\beta| = \sum_R \chi_\Gamma(R)\{R_{11}R_{11}|e_1\alpha e_1\beta| + R_{12}R_{12}|e_2\alpha e_2\beta| + R_{11}R_{12}[|e_1\alpha e_2\beta| - |e_1\beta e_2\alpha|]\}$$

$$P_\Gamma|e_2\alpha e_2\beta| = \sum_R \chi_\Gamma(R)\{R_{22}R_{22}|e_2\alpha e_2\beta| + R_{21}R_{21}|e_1\alpha e_1\beta| + R_{22}R_{21}\ [|e_2\alpha e_1\beta| - |e_2\beta e_1\alpha|]\}$$

and

$$P_\Gamma 1/\sqrt{2}[|e_1\alpha e_2\beta| - |e_1\beta e_2\alpha|] = \sum_R \chi_\Gamma(R)\{\sqrt{2}R_{11}R_{21}|e_1\alpha e_1\beta|$$
$$+ \sqrt{2}R_{22}R_{12}|e_2\alpha e_2\beta| + (R_{11}R_{22} + R_{12}R_{21})[|e_1\alpha e_2\beta| - |e_1\beta e_2\alpha|]\}$$

To make further progress, one needs to evaluate the R_{ik} matrix elements for the particular orbitals given above and to then use these explicit values in the above equations. The matrix representations for the two e orbitals can easily be formed and are as follows:

$$\begin{pmatrix} 1 & 0 \\ 0 & 1 \end{pmatrix} \quad \begin{pmatrix} -1 & 0 \\ 0 & 1 \end{pmatrix} \quad \begin{pmatrix} -1/2 & \sqrt{3}/2 \\ \sqrt{3}/2 & 1/2 \end{pmatrix}$$

$$E \qquad \sigma_v \qquad \sigma'_v$$

$$\begin{pmatrix} -1/2 & -\sqrt{3}/2 \\ -\sqrt{3}/2 & 1/2 \end{pmatrix} \begin{pmatrix} -1/2 & \sqrt{3}/2 \\ -\sqrt{3}/2 & -1/2 \end{pmatrix} \begin{pmatrix} -1/2 & -\sqrt{3}/2 \\ \sqrt{3}/2 & -1/2 \end{pmatrix}$$

$$\sigma''_v \qquad C_3 \qquad C'_3$$

Turning first to the three triplet functions, one notes that the effect of the symmetry projector acting on each of these three was the following multiple of the respective function:

$$\sum_R \chi_\Gamma(R)[R_{11}R_{22} - R_{12}R_{21}].$$

Evaluating this sum for each of the three symmetries $\Gamma = A_1$, A_2, and E, one obtains values of 0, 2, and 0, respectively. That is, the projection of the each of the original triplet determinants gives zero except for A_2 symmetry. This allows one to conclude that there are no A_1 or E triplet functions in this case; the triplet functions are of pure ${}^3A_2$ symmetry.

Using the explicit values for R_{ik} matrix elements in the expressions given above for the projection of each of the singlet determinental functions, one finds only the following non-vanishing contributions:

1. For A_1 symmetry—$P|e_1\alpha e_1\beta| = 3[|e_1\alpha e_1\beta| + |e_2\alpha e_2\beta|] = P|e_2\alpha e_2\beta|$,
2. For A_2 symmetry—all projections vanish,
3. For E symmetry—$P|e_1\alpha e_1\beta| = 3/2[|e_1\alpha e_1\beta| - |e_2\alpha e_2\beta|] = -P|e_2\alpha e_2\beta|$ and $P1/\sqrt{2}[|e_1\alpha e_2\beta| - |e_1\beta e_2\alpha|] = 3/\sqrt{2}[|e_1\alpha e_2\beta| - |e_1\beta e_2\alpha|]$.

Remembering that the projection process does not lead to a normalized function, although it does generate a function of pure symmetry, one can finally write down the normalized symmetry-adapted singlet functions as:

1. ${}^1A_1 = 1/\sqrt{2}[|e_1\alpha e_1\beta| + |e_2\alpha e_2\beta|]$,
2. ${}^1E = \{1/\sqrt{2}[|e_1\alpha e_1\beta| - |e_2\alpha e_2\beta|]$, and $1/\sqrt{2}[|e_1\alpha e_2\beta| - |e_1\beta e_2\alpha|]\}$.

The triplet functions given above are:

3. ${}^3A_2 = \{|e_1\alpha e_2\alpha|, 1/\sqrt{2}[|e_1\alpha e_2\beta| + |e_1\beta e_2\alpha|]$, and $|e_1\beta e_2\beta|\}$.

In summary, whenever one has partially occupied degenerate orbitals, the characters corresponding to the direct product of the open-shell orbitals (as always, closed-shells contribute nothing to the symmetry analysis and can be ignored, although their presence must, of course, be specified when one finally writes down complete symmetry-adapted wavefunctions) must be reduced to identify the spatial symmetry components of the configuration. Given knowledge of the various spatial symmetries, one must then form determinental wavefunctions of each possible space and spin symmetry. In doing so, one starts with the maximum M_s function and uses spin angular momentum algebra and orthogonality to form proper spin eigenfunctions, and then employs point group projection operators (which require the formation of the R_{ik} representation matrices). Antisymmetry, as embodied in the determinants, causes some space-spin symmetry combinations to vanish (e.g., ${}^3A_1$ and ${}^3E$ and ${}^1A_2$ in the above example) thereby enforcing the

Pauli principle. This procedure, although tedious, is guaranteed to generate all space- and spin-symmetry adapted determinants for any configuration involving degenerate orbitals. The results of certain such combined spin and spatial symmetry analyses have been tabulated. For example, in appendix 11 of Atkins such information is given in the form of tables of direct products for several common point groups.

For cases in which one has a *non-equivalent* set of degenerate orbitals (e.g., for a configuration whose open-shell part is $e^1e'^1$), the procedure is exactly the same as above except that the determination of the possible space-spin symmetries is more straightforward. In this case, singlet and triplet functions exist for all three space symmetries: A_1, A_2, and E, because the Pauli principle does not exclude determinants of the form $|e_1\alpha e'_1\alpha|$ or $|e_2\beta e'_2\beta|$, whereas the equivalent determinants ($|e_1\alpha e_1\alpha|$ or $|e_2\beta e_2\beta|$) vanish when the degenerate orbitals belong to the same set (in which case, one says that the orbitals are equivalent).

For all point, axial rotation, and full rotation group symmetries, this observation holds: if the orbitals are equivalent, certain space-spin symmetry combinations will vanish due to antisymmetry; if the orbitals are not equivalent, all space-spin symmetry combinations consistent with the content of the direct product analysis are possible. In either case, one must proceed through the construction of determinental wavefunctions as outlined above.

V. SUMMARY

The ability to identify all term symbols and to construct all determinental wavefunctions that arise from a given electronic configuration is important. This knowledge allows one to understand and predict the changes (i.e., physical couplings due to external fields or due to collisions with other species and chemical couplings due to interactions with orbitals and electrons of a "ligand" or another species) that each state experiences when the atom or molecule is subjected to some interaction. Such understanding plays central roles in interpreting the results of experiments in spectroscopy and chemical reaction dynamics.

The essence of this analysis involves being able to write each wavefunction as a combination of determinants each of which involves occupancy of particular spin-orbitals. Because different spin-orbitals interact differently with, for example, a colliding molecule, the various determinants will interact differently. These differences thus give rise to different interaction potential energy surfaces.

For example, the Carbon-atom

$$^3P(M_L = 1, M_S = 0) = 2^{-1/2}[|p_1\beta p_0\alpha| + |p_1\alpha p_0\beta|]$$

and

$$^3P(M_L = 0, M_S = 0) = 2^{-1/2}[|p_1\beta p_{-1}\alpha| + |p_1\alpha p_{-1}\beta|]$$

states interact quite differently in a collision with a closed-shell Ne atom. The $M_L = 1$ state's two determinants both have an electron in an orbital directed toward the Ne atom (the $2p_0$ orbital) as well as an electron in an orbital directed perpendicular to the C-Ne internuclear axis (the $2p_1$ orbital); the $M_L = 0$ state's two determinants have both electrons in orbitals directed perpendicular to the C-Ne axis. Because Ne is a closed-shell species, any electron density directed toward it will produce a "repulsive" antibonding interaction. As a result, we expect the $M_L = 1$ state to undergo a more repulsive interaction with the Ne atom than the $M_L = 0$ state.

Although one may be tempted to "guess" how the various $^3P(M_L)$ states interact with a Ne atom by making an analogy between the three M_L states within the ^3P level and the three orbitals that comprise a set of p-orbitals, such analogies are not generally valid. The wavefunctions that correspond to term symbols are N-electron functions; they describe how N spin-orbitals are occupied and, therefore, how N spin-orbitals will be affected by interaction with an approaching "ligand" such as a Ne atom. The net effect of the ligand will depend on the occupancy of all N spin-orbitals.

To illustrate this point, consider how the ^1S state of Carbon would be expected to interact with an approaching Ne atom. This term symbol's wavefunction $^1S = 3^{-1/2}[|p_0\alpha p_0\beta| - |p_{-1}\alpha p_1\beta| - |p_1\alpha p_{-1}\beta|]$ contains three determinants, each with a $1/3$ probability factor. The first, $|p_0\alpha p_0\beta|$, produces a repulsive interaction with the closed-shell Ne; the second and third, $|p_{-1}\alpha p_1\beta|$ and $|p_1\alpha p_{-1}\beta|$, produce attractive interactions because they allow the Carbon's vacant p_0 orbital to serve in a Lewis acid capacity and accept electron density from Ne. The net effect is likely to be an attractive interaction because of the equal weighting of these three determinants in the ^1S wavefunction. This result could not have been "guessed" by making analogy with how an s-orbital interacts with a Ne atom; the ^1S state and an s-orbital are distinctly different in this respect.

CHAPTER

11

One must be able to evaluate the matrix elements among properly symmetry adapted *N*-electron configuration functions for any operator, the electronic Hamiltonian in particular. The Slater-Condon rules provide this capability.

I. CSFs ARE USED TO EXPRESS THE FULL *N*-ELECTRON WAVEFUNCTION

It has been demonstrated that a given electronic configuration can yield several space- and spin-adapted determinental wavefunctions; such functions are referred to as configuration state functions (CSFs). These CSF wavefunctions are *not* the exact eigenfunctions of the many-electron Hamiltonian, *H;* they are simply functions which possess the space, spin, and permutational symmetry of the exact eigenstates. As such, they comprise an acceptable set of functions to use in, for example, a linear variational treatment of the true states.

In such variational treatments of electronic structure, the *N*-electron wavefunction Ψ is expanded as a sum over *all* CSFs that possess the desired spatial and spin symmetry:

$$\Psi = \sum_J C_J \Phi_J.$$

Here, the Φ_J represent the CSFs that are of the correct symmetry, and the C_J are their expansion coefficients to be determined in the variational calculation. If the spin-orbitals used to form the determinants, that in turn form the CSFs $\{\Phi_J\}$, are orthonormal one-electron functions (i.e., $<\phi_k|\phi_j> = \delta_{k,j}$), then the CSFs can be shown to be orthonormal functions of *N* electrons

$$<\Phi_J|\Phi_K> = \delta_{J,K}.$$

In fact, the Slater determinants themselves also are orthonormal functions of *N* electrons whenever orthonormal spin-orbitals are used to form the determinants.

The above expansion of the full N-electron wavefunction is termed a "configuration-interaction" (CI) expansion. It is, in principle, a mathematically rigorous approach to expressing Ψ because the set of *all* determinants that can be formed from a complete set of spin-orbitals can be shown to be complete. In practice, one is limited to the number of orbitals that can be used and in the number of CSFs that can be included in the CI expansion. Nevertheless, the CI expansion method forms the basis of the most commonly used techniques in quantum chemistry.

In general, the optimal variational (or perturbative) wavefunction for any (i.e., the ground or excited) state will include contributions from spin-and space-symmetry adapted determinants derived from all possible configurations. For example, although the determinant with $L = 1$, $S = 1$, $M_L = 1$, $M_s = 1$ arising from the $1s^22s^22p^2$ configuration may contribute strongly to the true ground electronic state of the Carbon atom, there will be contributions from all configurations which can provide these L, S, M_L, and M_s values (e.g., the $1s^22s^22p^13p^1$ and $2s^22p^4$ configurations will also contribute, although the $1s^22s^22p^13s^1$ and $1s^22s^12p^23p^1$ will not because the latter two configurations are odd under inversion symmetry whereas the state under study is even).

The mixing of CSFs from many configurations to produce an optimal description of the true electronic states is referred to as configuration interaction (CI). Strong CI (i.e., mixing of CSFs with large amplitudes appearing for more than one dominant CSF) can occur, for example, when two CSFs from different electronic configurations have nearly the same Hamiltonian expectation value. For example, the $1s^22s^2$ and $1s^22p^2$ ^1S configurations of Be and the analogous ns^2 and np^2 configurations of all alkaline earth atoms are close in energy because the ns-np orbital energy splitting is small for these elements; the π^2 and π^{*2} configurations of ethylene become equal in energy, and thus undergo strong CI mixing, as the CC π bond is twisted by 90° in which case the π and π^* orbitals become degenerate.

Within a variational treatment, the relative contributions of the spin- and space-symmetry adapted CSFs are determined by solving a secular problem for the eigenvalues (E_i) and eigenvectors (C_i) of the matrix representation H of the full many-electron Hamiltonian $\boldsymbol{H}$ within this CSF basis:

$$\sum_L H_{K,L} C_{i,L} = E_i C_{i,K}.$$

The eigenvalue E_i gives the variational estimate for the energy of the ith state, and the entries in the corresponding eigenvector $C_{i,K}$ give the contribution of the Kth CSF to the ith wavefunction Ψ_i in the sense that

$$\Psi_i = \sum_K C_{i,K} \Phi_K,$$

where Φ_K is the Kth CSF.

II. THE SLATER-CONDON RULES GIVE EXPRESSIONS FOR THE OPERATOR MATRIX ELEMENTS AMONG THE CSFs

To form the $H_{K,L}$ matrix, one uses the so-called *Slater-Condon rules* which express all non-vanishing determinental matrix elements involving either one- or two-electron operators (one-electron operators are additive and appear as

$$F = \sum_i f(i);$$

two-electron operators are pairwise additive and appear as

$$G = \sum_{ij} g(i,j)).$$

Because the CSFs are simple linear combinations of determinants with coefficients determined by space and spin symmetry, the $H_{I,J}$ matrix in terms of determinants can be used to generate the $H_{K,L}$ matrix over CSFs.

The Slater-Condon rules give the matrix elements between two determinants

$$|> = |\phi_1\phi_2\phi_3 \ldots \phi_N|$$

and

$$|'> = |\phi'_1\phi'_2\phi'_3 \ldots \phi'_N|$$

for *any* quantum mechanical operator that is a sum of one- and two-electron operators ($F + G$). It expresses these matrix elements in terms of one-and two-electron integrals involving the spin-orbitals that appear in |> and |'> and the operators f and g.

As a first step in applying these rules, one must examine | > and |'> and determine by how many (if any) spin-orbitals | > and |'> differ. In so doing, one may have to reorder the spin-orbitals in one of the determinants to achieve maximal coincidence with those in the other determinant; it is essential to keep track of the number of permutations (N_p) that one makes in achieving maximal coincidence. The results of the Slater-Condon rules given below are then multiplied by $(-1)^{N_p}$ to obtain the matrix elements between the original | > and |'>. The final result does not depend on whether one chooses to permute | > or |'>.

Once maximal coincidence has been achieved, the Slater-Condon (SC) rules provide the following prescriptions for evaluating the matrix elements of any operator $F + G$ containing a one-electron part

$$F = \sum_i f(i)$$

and a two-electron part

$$G = \sum_{ij} g(i,j)$$

(the Hamiltonian is, of course, a specific example of such an operator; the electric dipole operator

$$\sum_i e\mathbf{r}_i$$

and the electronic kinetic energy

$$-\hbar^2/2m_e \sum_i \nabla_i^2$$

are examples of one-electron operators (for which one takes $g = 0$) ; the electron-electron coulomb interaction

$$\sum_{i>j} e^2/r_{ij}$$

is a two-electron operator (for which one takes $f = 0$)):

The Slater-Condon Rules

1. If $| >$ and $|'>$ are identical, then

   $$<|F + G|> = \sum_i <\phi_i|f|\phi_i> + \sum_{i>j} [<\phi_i\phi_j|g|\phi_i\phi_j> - <\phi_i\phi_j|g|\phi_j\phi_i>] ,$$

   where the sums over i and j run over all spin-orbitals in $| >$;
2. If $| >$ and $|'>$ differ by a single spin-orbital mismatch ($\phi_p \neq \phi'_p$),

   $$<|F + G|'> = <\phi_p|f| \phi'_p> + \sum_j [<\phi_p\phi_j|g|\phi'_p\phi_j> - <\phi_p\phi_j|g|\phi_j\phi'_p>],$$

   where the sum over j runs over all spin-orbitals in $| >$ except ϕ_p;
3. If $| >$ and $|'>$ differ by two spin-orbitals ($\phi_p \neq \phi'_p$ and $\phi_q \neq \phi'_q$),

   and $<|F + G|'> = <\phi_p\phi_q|g|\phi'_p\phi'_q> - <\phi_p\phi_q|g|\phi'_q\phi'_p>$

   (note that the F contribution vanishes in this case);
4. If $| >$ and $|'>$ differ by three or more spin orbitals, then $<|F + G|'> = 0$;
5. For the identity operator I, the matrix elements $<|I|'> = 0$ if $| >$ and $|'>$ differ by one or more spin-orbitals (i.e., the Slater determinants are orthonormal if their spin-orbitals are).

Recall that each of these results is subject to multiplication by a factor of $(-1)^{N_p}$ to account for possible ordering differences in the spin-orbitals in $| >$ and $|'>$.

In these expressions,

$$<\phi_i|f|\phi_j>$$

is used to denote the one-electron integral

$$\int \phi_i^*(r)f(r)\phi_j(r)dr$$

and

$<\phi_i\phi_j|g|\phi_k\phi_l>$ (or in short hand notation $<ij|kl>$)

represents the two-electron integral

$$\int \phi_i^*(r)\phi_j^*(r')g(r,r')\phi_k(r)\phi_l(r')drdr'.$$

The notation $<ij|kl>$ introduced above gives the two-electron integrals for the $g(r,r')$ operator in the so-called Dirac notation, in which the i and k indices label the spin-orbitals that refer to the coordinates r and the j and l indices label the spin-orbitals referring to coordinates r'. The r and r' denote r, θ, ϕ, σ and r', θ', ϕ', σ' (with σ and σ' being the α or β spin functions). The fact that r and r' are integrated and hence represent "dummy" variables introduces index permutational symmetry into this list of integrals. For example,

$$<ij|kl> = <ji|lk> = <kl|ij>^* = <lk|ji>^*;$$

the final two equivalences are results of the Hermitian nature of $g(r,r')$.

It is also common to represent these same two-electron integrals in a notation referred to as Mulliken notation in which:

$$\int \phi_i^*(r)\phi_j^*(r')g(r,r')\phi_k(r)\phi_l(r')drdr' = (ik|jl).$$

Here, the indices i and k, which label the spin-orbital having variables r are grouped together, and j and l, which label spin-orbitals referring to the r' variables appear together. The above permutational symmetries, when expressed in terms of the Mulliken integral list read:

$$(ik|jl) = (jl|ik) = (ki|lj)* = (lj|ki)*.$$

If the operators f and g do not contain any electron spin operators, then the spin integrations implicit in these integrals (all of the ϕ_i are spin-orbitals, so each ϕ is accompanied by an α or β spin function and each ϕ^* involves the adjoint of one of the α or β spin functions) can be carried out as $<\alpha|\alpha> = 1$, $<\alpha|\beta> = 0$, $<\beta|\alpha> = 0$, $<\beta|\beta> = 1$, thereby yielding integrals over spatial orbitals. These spin integration results follow immediately from the general properties of angular momentum eigenfunctions detailed in appendix G; in particular, because α and β are eigenfunctions of S_z with different eigenvalues, they must be orthogonal $<\alpha|\beta> = <\beta|\alpha> = 0$.

The essential results of the Slater-Condon rules are:

1. The full $N!$ terms that arise in the N-electron Slater determinants do not have to be treated explicitly, nor do the $N!(N! + 1)/2$ Hamiltonian matrix elements among the $N!$ terms of one Slater determinant and the $N!$ terms of the same or another Slater determinant.
2. All such matrix elements, for *any* one- and/or two-electron operator can be expressed in terms of one- or two-electron integrals over the spin-orbitals that appear in the determinants.
3. The integrals over orbitals are three- or six-dimensional integrals, regardless of how many electrons N there are.
4. These integrals over mo's can, through the LCAO-MO expansion, ultimately be expressed in terms of one- and two-electron integrals over the primitive atomic orbitals. It is only these ao-based integrals that can be evaluated explicitly (on high speed computers for all but the smallest systems).

III. EXAMPLES OF APPLYING THE SLATER-CONDON RULES

It is wise to gain some experience using the SC rules, so let us consider a few illustrative example problems.

1. What is the contribution to the total energy of the ^3P level of Carbon made by the two $2p$ orbitals alone? Of course, the two $1s$ and two $2s$ spin-orbitals contribute to the total energy, but we artificially ignore all such contributions in this example to simplify the problem.

Because all nine of the ^3P states have the same energy, we can calculate the energy of any one of them; it is therefore prudent to choose an "easy" one

$$^3P(M_L = 1, M_S = 1) = |p_1\alpha p_0\alpha| .$$

The energy of this state is $<|p_1\alpha p_0\alpha|H|p_1\alpha p_0\alpha|>$. The SC rules tell us this equals:

$$I_{2p_1} + I_{2p_0} + <2p_1 2p_0|2p_1 2p_0> - <2p_1 2p_0|2p_0 2p_1>,$$

where the short hand notation $I_j = <j|f|j>$ is introduced.

If the contributions from the two $1s$ and two $2s$ spin-orbitals are now taken into account, one obtains a *total* energy that also contains

$$2I_{1s} + 2I_{2s} + \langle 1s1s|1s1s\rangle + 4\langle 1s2s|1s2s\rangle - 2\langle 1s2s|2s1s\rangle + \langle 2s2s|2s2s\rangle + 2\langle 1s2p_1|1s2p_1\rangle$$
$$- \langle 1s2p_1|2p_1 1s\rangle + 2\langle 1s2p_0|1s2p_0\rangle - \langle 1s2p_0|2p_0 1s\rangle + 2\langle 2s2p_1|2s2p_1\rangle - \langle 2s2p_1|2p_1 2s\rangle$$
$$+ 2\langle 2s2p_0|2s2p_0\rangle - \langle 2s2p_0|2p_0 2s\rangle.$$

2. Is the energy of another ^3P state equal to the above state's energy? Of course, but it may prove informative to prove this.

Consider the $M_S = 0$, $M_L = 1$ state whose energy is:

$$2^{-1/2}\langle[|p_1\alpha p_0\beta| + |p_1\beta p_0\alpha|]|H|\langle[|p_1\alpha p_0\beta| + |p_1\beta p_0\alpha|]\rangle 2^{-1/2}$$
$$= 1/2\{I_{2p1} + I_{2p0} + \langle 2p_1 2p_0|2p_1 2p_0\rangle + I_{2p1} + I_{2p0} + \langle 2p_1 2p_0|2p_1 2p_0\rangle\}$$
$$+ 1/2\{-\langle 2p_1 2p_0|2p_0 2p_1\rangle - \langle 2p_1 2p_0|2p_0 2p_1\rangle\}$$
$$= I_{2p_1} + I_{2p_0} + \langle 2p_1 2p_0|2p_1 2p_0\rangle - \langle 2p_1 2p_0|2p_0 2p_1\rangle.$$

Which is, indeed, the same as the other ^3P energy obtained above.

3. What energy would the singlet state $2^{-1/2}\langle[|p_1\alpha p_0\beta| - |p_1\beta p_0\alpha|]$ have?

The ^3P $M_S = 0$ example can be used (changing the sign on the two determinants) to give

$$E = I_{2p_1} + I_{2p_0} + \langle 2p_1 2p_0|2p_1 2p_0\rangle + \langle 2p_1 2p_0|2p_0 2p_1\rangle.$$

Note, this is the $M_L = 1$ component of the ^1D state; it is, of course, not a ^1P state because no such state exists for two equivalent p electrons.

4. What is the CI matrix element coupling $|1s^2 2s^2|$ and $|1s^2 3s^2|$?

These two determinants differ by two spin-orbitals, so

$$\langle|1s\alpha 1s\beta 2s\alpha 2s\beta|H|1s\alpha 1s\beta 3s\alpha 3s\beta|\rangle = \langle 2s2s|3s3s\rangle = \langle 2s3s|3s2s\rangle$$

(note, this is an exchange-type integral).

5. What is the CI matrix element coupling $|\pi\alpha\pi\beta|$ and $|\pi^*\alpha\pi^*\beta|$?

These two determinants differ by two spin-orbitals, so

$$\langle|\pi\alpha\pi\beta|H|\pi^*\alpha\pi^*\beta|\rangle = \langle\pi\pi|\pi^*\pi^*\rangle = \langle\pi\pi^*|\pi^*\pi\rangle$$

(note, again this is an exchange-type integral).

6. What is the Hamiltonian matrix element coupling $|\pi\alpha\pi\beta|$ and $2^{-1/2}[|\pi\alpha\pi^*\beta| - |\pi\beta\pi^*\alpha|]$?

The first determinant differs from the π^2 determinant by one spin-orbital, as does the second (after it is placed into maximal coincidence by making one permutation), so

$$\langle|\pi\alpha\pi\beta|H|2^{-1/2}[|\pi\alpha\pi^*\beta| - |\pi\beta\pi^*\alpha|]\rangle$$
$$= 2^{-1/2}[\langle\pi|f|\pi^*\rangle + \langle\pi\pi|\pi^*\pi\rangle] - (-1)2^{-1/2}[\langle\pi|f|\pi^*\rangle + \langle\pi\pi|\pi^*\pi\rangle]$$
$$= 2^{1/2}[\langle\pi|f|\pi^*\rangle + \langle\pi\pi|\pi^*\pi\rangle].$$

7. What is the element coupling $|\pi\alpha\pi\beta|$ and $2^{-1/2}[|\pi\alpha\pi^*\beta| + |\pi\beta\pi^*\alpha|]$?

$$\langle|\pi\alpha\pi\beta|H|2^{-1/2}[|\pi\alpha\pi^*\beta| + |\pi\beta\pi^*\alpha|]\rangle$$
$$= 2^{-1/2}[\langle\pi|f|\pi^*\rangle + \langle\pi\pi|\pi^*\pi\rangle] + (-1)2^{-1/2}[\langle\pi|f|\pi^*\rangle + \langle\pi\pi|\pi^*\pi\rangle] = 0.$$

This result should not surprise you because $|\pi\alpha\pi\beta|$ is an $S = 0$ singlet state while $2^{-1/2}[|\pi\alpha\pi^*\beta| + |\pi\beta\pi^*\alpha|]$ is the $M_S = 0$ component of the $S = 1$ triplet state.

8. What is the

$$\boldsymbol{r} = \sum_j e\boldsymbol{r}_j$$

electric dipole matrix element between $|p_1\alpha p_1\beta|$ and $2^{-1/2}[|p_1\alpha p_0\beta| + |p_0\alpha p_1\beta|]$? Is the second function a singlet or triplet? It is a singlet in disguise; by interchanging the $p_0\alpha$ and $p_1\beta$ and thus introducing a (–1), this function is clearly identified as $2^{-1/2}[|p_1\alpha p_0\beta| - |p_1\beta p_0\alpha|]$ which is a singlet.

The first determinant differs from the latter two by one spin orbital in each case, so

$$<|p_1\alpha p_1\beta|\boldsymbol{r}|2^{-1/2}[|p_1\alpha p_0\beta| + |p_0\alpha p_1\beta|]> =$$
$$2^{-1/2}[<p_1|\boldsymbol{r}|p_0> + <p_1|\boldsymbol{r}|p_0>] = 2^{1/2}<p_1|\boldsymbol{r}|p_0>.$$

9. What is the electric dipole matrix elements between the $^1\Delta = |\pi_1\alpha\pi_1\beta|$ state and the $^1\Sigma = 2^{-1/2}[|\pi_1\alpha\pi_{-1}\beta| + |\pi_{-1}\alpha\pi_1\beta|]$ state?

$$<2^{-1/2}[|\pi_1\alpha\pi_{-1}\beta| + |\pi_{-1}\alpha\pi_1\beta|]|\boldsymbol{r}|\pi_1\alpha\pi_1\beta|>$$
$$= 2^{-1/2}[<\pi_{-1}|\boldsymbol{r}|\pi_1> + <\pi_{-1}|\boldsymbol{r}|\pi_1>]$$
$$= 2^{1/2}<\pi_{-1}|\boldsymbol{r}|\pi_1>.$$

10. As another example of the use of the SC rules, consider the configuration interaction which occurs between the $1s^22s^2$ and $1s^22p^2\ ^1S$ CSFs in the Be atom.

The CSFs corresponding to these two configurations are as follows:

$$\Phi_1 = |1s\alpha 1s\beta 2s\alpha 2s\beta|$$

and

$$\Phi_2 = 1/\sqrt{3}\,[|1s\alpha 1s\beta 2p_0\alpha 2p_0\beta| - |1s\alpha 1s\beta 2p_1\alpha 2p_{-1}\beta| - |1s\alpha 1s\beta 2p_{-1}\alpha 2p_1\beta|].$$

The determinental Hamiltonian matrix elements needed to evaluate the $2 \times 2\ H_{K,L}$ matrix appropriate to these two CSFs are evaluated via the SC rules. The first such matrix element is:

$$<|1s\alpha 1s\beta 2s\alpha 2s\beta|H|1s\alpha 1s\beta 2s\alpha 2s\beta|> = 2h_{1s} + 2h_{2s} + J_{1s,1s} + 4J_{1s,2s} + J_{2s,2s} - 2K_{1s,2s},$$

where

$$h_i = <\phi_i| - \hbar^2/2m_e\nabla^2 - 4e^2/r|\phi_i>,$$
$$J_{i,j} = <\phi_i\phi_j|e^2/r_{12}|\phi_i\phi_j>, \text{ and}$$
$$K_{ij} = <\phi_i\phi_j|e^2/r_{12}|\phi_j\phi_i>$$

are the orbital-level *one-electron, coulomb,* and *exchange integrals*, respectively.

Coulomb integrals J_{ij} describe the coulombic interaction of one charge density (ϕ_i^2 above) with another charge density (ϕ_j^2 above); exchange integrals K_{ij} describe the interaction of an *overlap* charge density (i.e., a density of the form $\phi_i\phi_j$) with itself ($\phi_i\phi_j$ with $\phi_i\phi_j$ in the above).

The spin functions α and β which accompany each orbital in $|1s\alpha 1s\beta 2s\alpha 2s\beta|$ have been eliminated by carrying out the spin integrations as discussed above. Because H contains no spin operators, this is straightforward and amounts to keeping integrals $<\phi_i|f|\phi_j>$ only if ϕ_i and ϕ_j are of the same spin and integrals $<\phi_i\phi_j|g|\phi_k\phi_l>$ only if ϕ_i and ϕ_k are of the same spin *and* ϕ_j and ϕ_l are of the same spin. The physical content of the above energy (i.e., Hamiltonian expectation value) of the $|1s\alpha 1s\beta 2s\alpha 2s\beta|$ determinant is clear: $2h_{1s} + 2h_{2s}$ is the sum of the expectation values of the one-electron (i.e., kinetic energy and electron-nuclear coulomb interaction) part of the Hamiltonian for the four occupied spin-orbitals; $J_{1s,1s} + 4J_{1s,2s} + J_{2s,2s} - 2K_{1s,2s}$ contains the coulombic repulsions among all pairs of occupied spin-orbitals minus the exchange interactions among pairs of spin-orbitals with like spin.

The determinental matrix elements linking Φ_1 and Φ_2 are as follows:

$$<|1s\alpha 1s\beta 2s\alpha 2s\beta|H|1s\alpha 1s\beta 2p_0\alpha 2p_0\beta|> = <2s2s|2p_0 2p_0>,$$
$$<|1s\alpha 1s\beta 2s\alpha 2s\beta|H|1s\alpha 1s\beta 2p_1\alpha 2p_{-1}\beta|> = <2s2s|2p_1 2p_{-1}>,$$
$$<|1s\alpha 1s\beta 2s\alpha 2s\beta|H|1s\alpha 1s\beta 2p_{-1}\alpha 2p_1\beta|> = <2s2s|2p_{-1} 2p_1>,$$

where the Dirac convention has been introduced as a shorthand notation for the two-electron integrals (e.g., $<2s2s|2p_0 2p_0>$ represents $\int 2s^*(r_1)2s^*(r_2)e^2/r_{12}2p_0(r_1)2p_0(r_2)dr_1 dr_2$).

The three integrals shown above can be seen to be equal and to be of the exchange-integral form by expressing the integrals in terms of integrals over cartesian functions and recognizing identities due to the equivalence of the $2p_x$, $2p_y$, and $2p_z$ orbitals. For example,

$$\begin{aligned} <2s2s|2p_1 2p_{-1}> &= (1\sqrt{2})^2\{<2s2s|[2p_x + i2p_y][2p_x - i2p_y]>\} \\ &= 1/2\{<2s2s|xx> + <2s2s|yy> + i<2s2s|yx> - i<2s2s|xy>\} \\ &= <2s2s|xx> = K_{2s,x} \end{aligned}$$

(here the two imaginary terms cancel and the two remaining real integrals are equal);

$$<2s2s2p_0 2p_0> = <2s2s|zz> = <2s2s|xx> = K_{2s,x}$$

(this is because $K_{2s,z} = K_{2s,x} = K_{2s,y}$);

$$\begin{aligned} &<2s2s|2p_{-1}2p_1> = 1/2\{<2s2s|[2p_x - i2p_y][2p_x + i2p_y]>\} = \\ &<2s2s|xx> = \int 2s^*(r_1)2s^*(r_2)e^2/r_{12}2p_x(r_1)2p_x(r_2)dr_1 dr_2 = K_{2s,x}. \end{aligned}$$

These integrals are clearly of the exchange type because they involve the coulombic interaction of the $2s2p_{x,y,\text{ or } z}$ overlap charge density with itself.

Moving on, the matrix elements among the three determinants in Φ_2 are given as follows:

$$\begin{aligned} &<|1s\alpha 1s\beta 2p_0\alpha 2p_0\beta|H|1s\alpha 1s\beta 2p_0\alpha 2p_0\beta|> \\ &\quad = 2h_{1s} + 2h_{2p} + J_{1s,1s} + J_{2pz,2pz} + 4J_{1s,2p} - 2K_{1s,2p} \end{aligned}$$

($J_{1s,2p}$ and $K_{1s,2p}$ are independent of whether the $2p$ orbital is $2p_x$, $2p_y$, or $2p_z$);

$$\begin{aligned} &<|1s\alpha 1s\beta 2p_1\alpha 2p_{-1}\beta|H|1s\alpha 1s\beta 2p_1\alpha 2p_{-1}\beta|> \\ &= 2h_{1s} + 2h_{2p} + J_{1s,1s} + 4J_{1s,2p} - 2K_{1s,2p} + <2p_1 2p_{-1}|2p_1 2p_{-1}>; \\ &<|1s\alpha 1s\beta 2p_{-1}\alpha 2p_1\beta|H|1s\alpha 1s\beta 2p_{-1}\alpha 2p_1\beta|> \\ &2h_{1s} + 2h_{2p} + J_{1s,1s} + 4J_{1s,2p} - 2K_{1s,2p} + <2p_{-1}2p_1|2p_{-1}2p_1>; \\ &<|1s\alpha 1s\beta 2p_0\alpha 2p_0\beta|H|1s\alpha 1s\beta 2p_1\alpha 2p_{-1}\beta|> = <2p_0 2p_0|2p_1 2p_{-1}> \\ &<|1s\alpha 1s\beta 2p_0\alpha 2p_0\beta|H|1s\alpha 1s\beta 2p_{-1}\alpha 2p_1\beta|> = <2p_0 2p_0|2p_{-1} 2p_1> \\ &<|1s\alpha 1s\beta 2p_1\alpha 2p_{-1}\beta|H|1s\alpha 1s\beta 2p_{-1}\alpha 2p_1\beta|> = <2p_1 2p_{-1}|2p_{-1}2p_1>. \end{aligned}$$

Certain of these integrals can be recast in terms of cartesian integrals for which equivalences are easier to identify as follows:

$$\begin{aligned} <2p_0 2p_0|2p_1 2p_{-1}> &= <2p_0 2p_0|2p_{-1}2p_1> = <zz|xx> = K_{z,x}; \\ <2p_1 2p_{-1}|2p_{-1}2p_1> &= <xx|yy> + 1/2[<xx|xx> - <xy|xy>] \\ &= K_{x,y} + 1/2[J_{x,x} - J_{x,y}]; \\ <2p_1 2p_{-1}|2p_1 2p_{-1}> &= <2p_{-1}2p_1|2p_{-1}2p_1> = 1/2(J_{x,x} + J_{x,y}). \end{aligned}$$

Finally, the 2×2 CI matrix corresponding to the CSFs Φ_1 and Φ_2 can be formed from the above determinental matrix elements; this results in:

$$H_{11} = 2h_{1s} + 2h_{2s} + J_{1s,1s} + 4J_{1s,2s} + J_{2s,2s} - 2K_{1s,2s};$$
$$H_{12} = -K_{2s,x}/\sqrt{3};$$
$$H_{22} = 2h_{1s} + 2h_{2p} + J_{1s,1s} + 4J_{1s,2p} - 2K_{1s,2p} + J_{z,z} - 2/3K_{z,x}.$$

The lowest eigenvalue of this matrix provides this CI calculation's estimate of the ground-state ^1S energy of Be; its eigenvector provides the CI amplitudes for Φ_1 and Φ_2 in this ground-state wavefunction. The other root of the 2×2 secular problem gives an approximation to another ^1S state of higher energy, in particular, a state dominated by the $3^{-1/2}[|1s\alpha 1s\beta 2p_0\alpha 2p_0\beta| - |1s\alpha 1s\beta 2p_1\alpha 2p_{-1}\beta| - |1s\alpha 1s\beta 2p_{-1}\alpha 2p_1\beta|]$ *CSF.*

11. As another example, consider the matrix elements which arise in electric dipole transitions between two singlet electronic states: $\langle\Psi_1|\boldsymbol{E}\cdot\Sigma_i e\boldsymbol{r}_i|\Psi_2\rangle$. Here $\boldsymbol{E}\cdot\Sigma_i e\boldsymbol{r}_i$ is the one-electron operator describing the interaction of an electric field of magnitude and polarization $\boldsymbol{E}$ with the instantaneous dipole moment of the electrons (the contribution to the dipole operator arising from the nuclear charges

$$-\sum_a Z_a e^2 \boldsymbol{R_a}$$

does not contribute because, when placed between Ψ_1 and Ψ_2, this zero-electron operator yields a vanishing integral because Ψ_1 and Ψ_2 are orthogonal).

When the states Ψ_1 and Ψ_2 are described as linear combinations of CSFs as introduced earlier

$$\left(\Psi_i = \sum_K C_{iK}\Phi_K\right),$$

these matrix elements can be expressed in terms of CSF-based matrix elements

$$\left\langle\Phi_K\middle|\sum_i e\boldsymbol{r}_i\middle|\Phi_L\right\rangle.$$

The fact that the electric dipole operator is a one-electron operator, in combination with the SC rules, guarantees that only states for which the dominant determinants differ by at most a single spin-orbital (i.e., those which are "singly excited") can be connected via electric dipole transitions through first order (i.e., in a one-photon transition to which the

$$\left\langle\Psi_1\middle|\sum_i e\boldsymbol{r}_i\middle|\Psi_2\right\rangle$$

matrix elements pertain). It is for this reason that light with energy adequate to ionize or excite deep core electrons in atoms or molecules usually causes such ionization or excitation rather than double ionization or excitation of valence-level electrons; the latter are two-electron events.

In, for example, the $\pi \Rightarrow \pi^*$ excitation of an olefin, the ground and excited states are dominated by CSFs of the form (where all but the "active" π and π^* orbitals are not explicitly written):

$$\Phi_1 = |\ldots \pi\alpha\pi\beta|$$

and

$$\Phi_2 = 1/\sqrt{2}[| \ldots \pi\alpha\pi^*\beta| - | \ldots \pi\beta\pi^*\alpha|].$$

The electric dipole matrix element between these two CSFs can be found, using the SC rules, to be

$$e/\sqrt{2}[<\pi|\mathbf{r}|\pi^*> + <\pi|\mathbf{r}|\pi^*>] = \sqrt{2}e<\pi|\mathbf{r}|\pi^*>.$$

Notice that in evaluating the second determinental integral $<| \ldots \pi\alpha\pi\beta|e\mathbf{r}| \ldots \pi\beta\pi^*\alpha|>$, a sign change occurs when one puts the two determinants into maximum coincidence; this sign change then makes the minus sign in Φ_2 yield a positive sign in the final result.

IV. SUMMARY

In all of the above examples, the SC rules were used to reduce matrix elements of one- or two-electron operators between determinental functions to one- or two-electron integrals over the orbitals which appear in the determinants. In any *ab initio* electronic structure computer program there must exist the capability to form symmetry-adapted CSFs and to evaluate, using these SC rules, the Hamiltonian and other operators' matrix elements among these CSFs in terms of integrals over the mos that appear in the CSFs. The SC rules provide not only the tools to compute quantitative matrix elements; they allow one to understand in qualitative terms the strengths of interactions among CSFs. In the following section, the SC rules are used to explain why chemical reactions in which the reactants and products have dominant CSFs that differ by two spin-orbital occupancies often display activation energies that exceed the reaction endoergicity.

CHAPTER

12

Along "reaction paths," configurations can be connected one-to-one according to their symmetries and energies. This is another part of the Woodward-Hoffmann rules.

I. CONCEPTS OF CONFIGURATION AND STATE ENERGIES

A. Plots of CSF Energies Give Configuration Correlation Diagrams

The energy of a particular electronic state of an atom or molecule has been expressed in terms of Hamiltonian matrix elements, using the SC rules, over the various spin- and spatially-adapted determinants or CSFs which enter into the state wavefunction.

$$E = \sum_{I,J} <\Phi_I|H|\Phi_J> C_I C_J.$$

The diagonal matrix elements of H in the CSF basis multiplied by the appropriate CI amplitudes $<\Phi_I|H|\Phi_I>C_I C_I$ represent the energy of the Ith CSF weighted by the strength (C_I^2) of that CSF in the wavefunction. The off-diagonal elements represent the effects of mixing among the CSFs; mixing is strongest whenever two or more CSFs have nearly the same energy (i.e., $<\Phi_I|H|\Phi_I> \cong <\Phi_J|H|\Phi_J>$) and there is strong coupling (i.e., $<\Phi_I|H|\Phi_J>$ is large). Whenever the CSFs are widely separated in energy, each wavefunction is dominated by a single CSF.

B. CSFs Interact and Couple to Produce States and State Correlation Diagrams

Just as orbital energies connected according to their symmetries and plotted as functions of geometry constitute an orbital correlation diagram, plots of the *diagonal CSF energies*, connected according to symmetry, constitute a *configuration correlation diagram* (CCD). If, near regions where energies of CSFs of the same symmetry cross (according to the direct product rule of group theory discussed in appendix E, only CSFs of the same symmetry mix because only they have non-vanishing $\langle\Phi_I|H|\Phi_J\rangle$ matrix elements), CI mixing is allowed to couple the CSFs to give rise to "avoided crossings," then the CCD is converted into a so-called *state correlation diagram* (SCD).

C. CSFs that Differ by Two Spin-Orbitals Interact Less Strongly than CSFs that Differ by One Spin-Orbital

The strengths of the couplings between pairs of CSFs whose energies cross are evaluated through the SC rules. CSFs that differ by more than two spin-orbital occupancies do not couple; the SC rules give vanishing Hamiltonian matrix elements for such pairs. Pairs that differ by two spin-orbitals (e.g., $|\ldots\phi_a\ldots\phi_b\ldots|$ vs $|\ldots\phi_{a'}\ldots\phi_{b'}\ldots|$) have interaction strengths determined by the two-electron integrals $\langle ab|a'b'\rangle - \langle ab|b'a'\rangle$. Pairs that differ by a single spin-orbital (e.g., $|\ldots\phi_a\ldots\ldots|$ vs $|\ldots\phi_{a'}\ldots\ldots|$) are coupled by the one- and two- electron parts of H:

$$\langle a|f|b\rangle + \sum_j [\langle aj|bj\rangle - \langle aj|jb\rangle].$$

Usually, couplings among CSFs that differ by two spin-orbitals are substantially weaker than those among CSFs that differ by one spin-orbital. In the latter case, the full strength of H is brought to bear, whereas in the former, only the electron-electron coulomb potential is operative.

D. State Correlation Diagrams

In the SCD, the energies are connected by symmetry but the configurational nature as reflected in the C_I coefficients changes as one passes through geometries where crossings in the CCD occur. The SCD is the ultimate product of an orbital and configuration symmetry and energy analysis and gives one the most useful information about whether reactions will or will not encounter barriers on the ground and excited state surfaces.

As an example of the application of CCD's and SCD's, consider the disrotatory closing of 1,3-butadiene to produce cyclobutene. The OCD given earlier for this proposed reaction path is reproduced below.

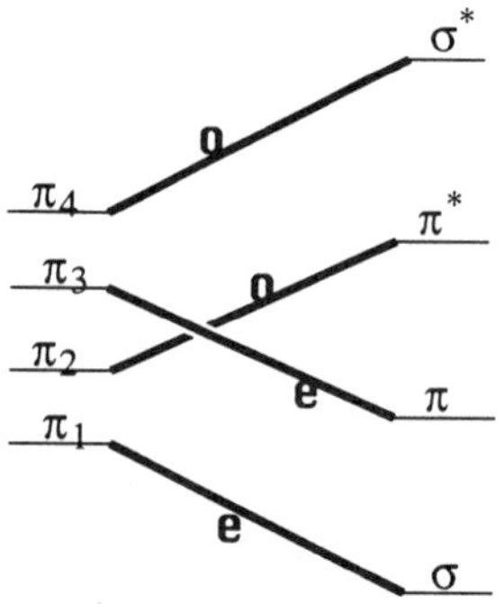

Recall that the symmetry labels *e* and *o* refer to the symmetries of the orbitals under reflection through the one C_v plane that is *preserved* throughout the proposed disrotatory closing. Low-energy configurations (assuming one is interested in the thermal or low-lying photochemically excited-state reactivity of this system) for the reactant molecule and their overall space and spin symmetry are as follows:

1. $\pi_1^2\pi_2^2 = 1e^2 1o^2$, ^{1}Even
2. $\pi_1^2\pi_2^1\pi_3^1 = 1e^2 1o^1 2e^1$, ^{3}Odd and ^{1}Odd.

For the product molecule, on the other hand, the low-lying states are

3. $\sigma^2\pi^2 = 1e^2 2e^2$, ^{1}Even
4. $\sigma^2\pi^1\pi^{*1} = 1e^2 2e^1 1o^1$, ^{3}Odd , ^{1}Odd.

Notice that although the lowest energy configuration at the reactant geometry $\pi_1^2\pi_2^2 = 1e^2 1o^2$ and the lowest energy configuration at the product geometry $\sigma^2\pi^2 = 1e^2 2e^2$ are both of ^{1}Even symmetry, they are *not* the same configurations; they involve occupancy of different symmetry orbitals.

In constructing the CCD, one must trace the energies of all four of the above CSFs (actually there are more because the singlet and triplet excited CSFs must be treated independently) along the proposed reaction path. In doing so, one must realize that the $1e^2 1o^2$ CSF has low energy on the reactant side of the CCD because it corresponds to $\pi_1{}^2\pi_2{}^2$ orbital occupancy, but on the product side, it corresponds to $\sigma^2\pi^{*2}$ orbital occupancy and is thus of very high energy. Likewise, the $1e^2 2e^2$ CSF has low energy on the product side where it is $\sigma^2\pi^2$ but high energy on the reactant side where it corresponds to $\pi_1^2\pi_3^2$. The low-lying singly excited CSFs are $1e^2 2e^1 1o^1$ at both reactant and product geometries; in the former case, they correspond to $\pi_1^2\pi_2^1\pi_3^1$ occupancy and at the latter to $\sigma^2\pi^1\pi^{*1}$ occupancy. Plotting the energies of these CSFs along the disrotatory reaction path results in the CCD shown below.

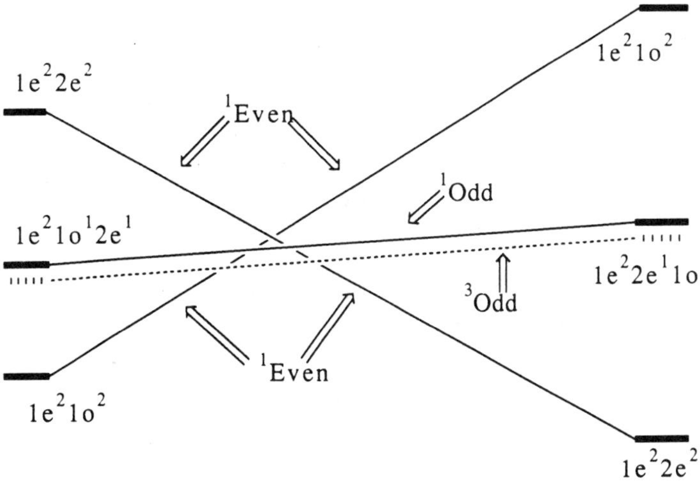

If the two ^{1}Even CSFs which cross are allowed to interact (the SC rules give their interaction strength in terms of the exchange integral $\langle |1e^2 1o^2| H | 1e^2 2e^2| \rangle = \langle 1o1o|2e2e\rangle = K_{1o,2e}$) to produce states which are combinations of the two ^{1}Even CSFs, the following SCD results:

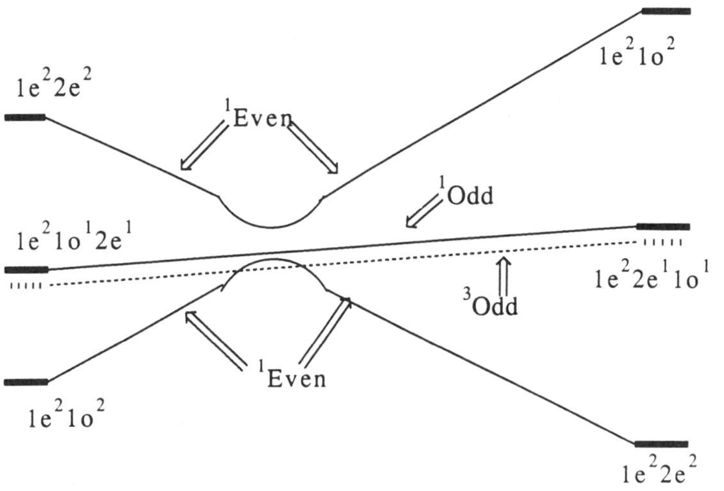

This SCD predicts that the thermal (i.e., on the ground electronic surface) disrotatory rearrangement of 1,3-butadiene to produce cyclobutene will experience a *symmety-imposed barrier* which arises because of the avoided crossing of the two ^{1}Even configurations; this avoidance occurs because the orbital occupancy pattern (i.e., the configuration) which is best for the ground state of the reactant is not identical to that of the product molecule. The SCD also predicts that there should be no symmetry-imposed barrier for the singlet or triplet excited-state rearrangement, although the reaction leading from excited 1,3-butadiene to excited cyclobutene may be endothermic on the grounds of bond strengths alone.

It is also possible to infer from the SCD that excitation of the lowest singlet $\pi\pi^*$ state of 1,3-butadiene would involve a low quantum yield for producing cyclobutene and would, in fact, produce ground-state butadiene. As the reaction proceeds along the singlet $\pi\pi^*$ surface this ^{1}Odd state intersects the ground ^{1}Even surface on the *reactant side* of the diagram; internal conversion (i.e., quenching from the ^{1}Odd to the ^{1}Even surfaces induced by using a vibration of odd symmetry to "digest" the excess energy, much like vibronic borrowing in spectroscopy) can lead to production of ground-state reactant molecules. Some fraction of such events will lead to the system remaining on the ^{1}Odd surface until, further along the reaction path, the ^{1}Odd surface again intersects the ^{1}Even surface on the *product side* at which time quenching to produce ground-state products can occur. Although, in principle, it is possible for some fraction of the events to follow the ^{1}Odd surface beyond this second intersection and to thus lead to ^{1}Odd product molecules that might fluoresce, quenching is known to be rapid in most polyatomic molecules; as a result, reactions which are chemiluminescent are rare. An appropriate introduction to the use of OCD's, CCD's, and SCD's as well as the radiationless processes that can occur in thermal and photochemical reactions is given in the text *Energetic Principles of Chemical Reactions*, J. Simons, Jones, and Bartlett, Boston (1983).

II. MIXING OF COVALENT AND IONIC CONFIGURATIONS

As chemists, much of our intuition concerning chemical bonds is built on simple models introduced in undergraduate chemistry courses. The detailed examination of the H_2 molecule via the valence bond and molecular orbital approaches forms the basis of our thinking about bonding when confronted with new systems. Let us examine this model system in further detail to explore the electronic states that arise by occupying two orbitals (derived from the two 1s orbitals on the two hydrogen atoms) with two electrons.

In total, there exist *six* electronic states for all such two-orbital, two-electron systems. The heterolytic fragments *X*+*Y*: and *X*: + *Y* produce two singlet states; the homolytic fragments *X*• + *Y*• produce one singlet state and a set of three triplet states having $M_S = 1, 0$, and -1. Understanding the relative energies of these six states, their bonding and antibonding characters, and which molecular state dissociates to which asymptote are important.

Before proceeding, it is important to clarify the notation (e.g., *X*•, *Y*•, *X*, *Y*:, etc.), which is designed to be applicable to neutral as well as charged species. In all cases considered here, only two electrons play active roles in the bond formation. These electrons are represented by the dots. The symbols *X*• and *Y*• are used to denote species in which a single electron is attached to the respective fragment. By *X*:, we mean that both electrons are attached to the *X*-fragment; *Y* means that neither electron resides on the *Y*-fragment. Let us now examine the various bonding situations that can occur; these examples will help illustrate and further clarify this notation.

A. The H_2 Case in Which Homolytic Bond Cleavage is Favored

To consider why the two-orbital two-electron single bond formation case can be more complex than often thought, let us consider the H_2 system in more detail. In the molecular orbital description of H_2, both bonding σ_g and antibonding σ_u mos appear. There are two electrons that can both occupy the σ_g mo to yield the ground electronic state

$$H_2\left({}^1\Sigma_g^+, \sigma_g{}^2\right);$$

however, they can also occupy both orbitals to yield

$${}^3\Sigma_u^+\ (\sigma_g{}^1\sigma_u{}^1)$$

and

$${}^1\Sigma_u^+\ (\sigma_g{}^1\sigma_u{}^1),$$

or both can occupy the σ_u mo to give the

$${}^1\Sigma_g^+\ (\sigma_u{}^2)$$

state. As demonstrated explicitly below, these latter two states dissociate heterolytically to $X + Y{:} = \mathrm{H}^+ + \mathrm{H}^-$, and are sufficiently high in energy relative to *X*• + *Y*• = H + H that we ordinarily can ignore them. However, their presence and character are important in the development of a full treatment of the molecular orbital model for H_2 and are *essential* to a proper treatment of cases in which heterolytic bond cleavage is favored.

B. Cases in Which Heterolytic Bond Cleavage Is Favored

For some systems one or both of the heterolytic bond dissociation asymptotes (e.g., *X* + *Y*: or *X*: + *Y*) may be *lower* in energy than the homolytic bond dissociation asymptote. Thus, the states that are analogues of the

$${}^1\Sigma_u^+\ (\sigma_g{}^1\sigma_u{}^1)$$

and

$$^1\Sigma_g^+ (\sigma_u^2)$$

states of H_2 can no longer be ignored in understanding the valence states of the XY molecules. This situation arises quite naturally in systems involving transition metals, where interactions between empty metal or metal ion orbitals and two-electron donor ligands are ubiquitous.

Two classes of systems illustrate cases for which heterolytic bond dissociation lies lower than the homolytic products. The first involves transition metal dimer cations, M_2^+. Especially for metals to the right side of the periodic table, such cations can be considered to have ground-state electron configurations with $\sigma^2 d^n d^{n+1}$ character, where the d electrons are not heavily involved in the bonding and the σ bond is formed primarily from the metal atom s orbitals. If the σ bond is homolytically broken, one forms $X\bullet + Y\bullet = M(s^1d^{n+1}) + M^+(s^1d^n)$. For most metals, this dissociation asymptote lies higher in energy than the heterolytic products $X{:} + Y = M(s^2d^n) + M^+(s^0d^{n+1})$, since the latter electron configurations correspond to the ground states for the neutrals and ions, respectively. A prototypical species which fits this bonding picture is Ni_2^+.

The second type of system in which heterolytic cleavage is favored arises with a metal-ligand complex having an atomic metal ion (with a s^0d^{n+1} configuration) and a two-electron donor, $L{:}$. A prototype is $(AgC_6H_6)^+$ which was observed to photodissociate to form $X\bullet + Y\bullet = Ag(^2S, s^1d^{10}) + C_6H_6^+(^2B_1)$ rather than the lower energy (heterolytically cleaved) dissociation limit $Y + X{:} = Ag^+(^1S, s^0d^{10}) + C_6H_6(^1A_1)$.

C. Analysis of Two-Electron, Two-Orbital, Single-Bond Formation

1. Orbitals, Configurations, and States

The resultant family of six-electronic states can be described in terms of the six-configuration state functions (CSFs) that arise when one occupies the pair of bonding σ and antibonding σ* molecular orbitals with two electrons. The CSFs are combinations of Slater determinants formed to generate proper spin- and spatial-symmetry functions.

The spin- and spatial-symmetry adapted N-electron functions referred to as CSFs can be formed from one or more Slater determinants. For example, to describe the singlet CSF corresponding to the closed-shell σ^2 orbital occupancy, a single Slater determinant

$$^1\Sigma(0) = |\sigma\alpha\sigma\beta| = (2)^{-1/2}\{\sigma\alpha(1)\sigma\beta(2) - \sigma\beta(1)\sigma\alpha(2)\}$$

suffices. An analogous expression for the $(\sigma*)^2$ CSF is given by

$$^1\Sigma^{**}(0) = |\sigma*\alpha\sigma*\beta| = (2)^{-1/2}\{\sigma*\alpha(1)\sigma*\beta(2) - \sigma*\alpha(2)\sigma*\beta(1)\}.$$

Also, the $M_S = 1$ component of the triplet state having σσ* orbital occupancy can be written as a single Slater determinant:

$$^3\Sigma^*(1) = |\sigma\alpha\sigma*\alpha| = (2)^{-1/2}\{\sigma\alpha(1)\sigma*\alpha(2) - \sigma*\alpha(1)\sigma\alpha(2)\},$$

as can the $M_S = -1$ component of the triplet state

$$^3\Sigma^*(-1) = |\sigma\beta\sigma*\beta| = (2)^{-1/2}\{\sigma\beta(1)\sigma*\beta(2) - \sigma*\beta(1)\sigma\beta(2)\}.$$

However, to describe the singlet CSF and $M_S = 0$ triplet CSF belonging to the σσ* occupancy, two Slater determinants are needed:

$$^1\Sigma^*(0) = \frac{1}{\sqrt{2}}[|\sigma\alpha\sigma*\beta| - |\sigma\beta\sigma*\alpha|]$$

is the singlet CSF and

$$^3\Sigma^*(0) = \frac{1}{\sqrt{2}}[|\sigma\alpha\sigma^*\beta| + |\sigma\beta\sigma^*\alpha|]$$

is the triplet CSF. In each case, the spin quantum number S, its z-axis projection M_S, and the Λ quantum number are given in the conventional $^{2S+1}\Lambda(M_S)$ notation.

2. Orbital, CSF, and State Correlation Diagrams

i. Orbital Diagrams The two orbitals of the constituent atoms or functional groups (denoted s_x and s_y for convenience and in anticipation of considering groups X and Y that possess valence s orbitals) combine to form a bonding $\sigma = \sigma_g$ molecular orbital and an antibonding $\sigma^* = \sigma_u$ molecular orbital (mo). As the distance R between the X and Y fragments is changed from near its equilibrium value of R_e and approaches infinity, the energies of the σ and σ^* orbitals vary in a manner well known to chemists as depicted below.

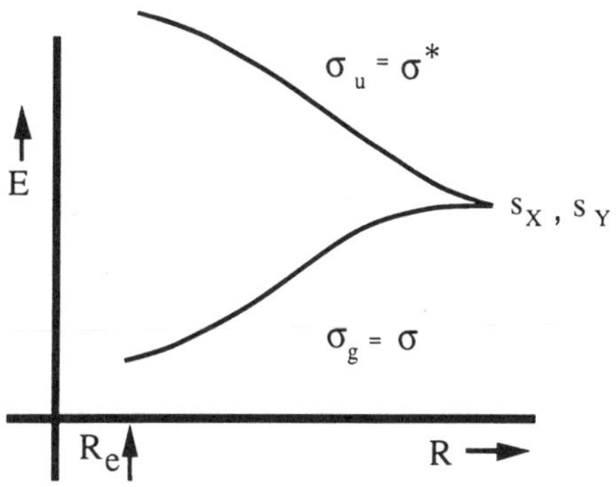

Energies of the bonding σ and antibonding σ^* orbitals as functions of interfragment distance; R_e denotes a distance near the equilibrium bond length for XY.

In the heteronuclear case, the s_x and s_y orbitals still combine to form a bonding σ and an antibonding σ^* orbital, although these orbitals no longer belong to g and u symmetry. The energies of these orbitals, for R values ranging from near R_e to $R \rightarrow \infty$, are depicted below.

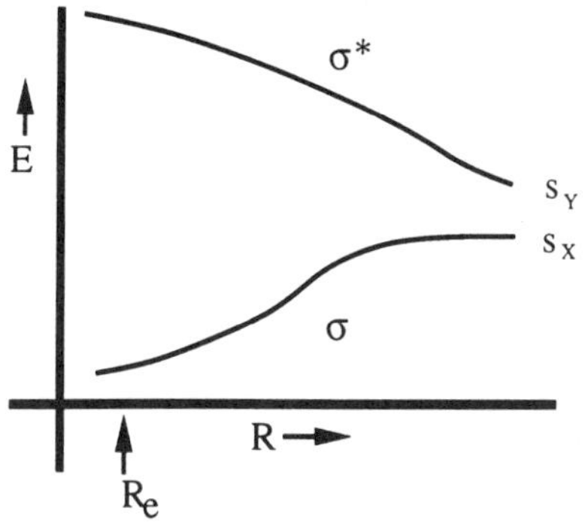

Energies of the bonding σ and antibonding σ^* orbitals as functions of internuclear distance. Here, X is more electronegative than Y.

For the homonuclear case, as R approaches ∞, the energies of the σ_g and σ_u orbitals become degenerate. Moreover, as $R \rightarrow 0$, the orbital energies approach those of the united atom. In the heteronuclear situation, as R approaches ∞, the energy of the σ orbital approaches the energy of the s_x orbital, and the $\sigma*$ orbital converges to the s_y orbital energy. Unlike the homonuclear case, the σ and $\sigma*$ orbitals are *not* degenerate as $R \rightarrow \infty$. The energy "gap" between the σ and $\sigma*$ orbitals at $R = \infty$ depends on the electronegativity difference between the groups X and Y. If this gap is small, it is expected that the behavior of this (slightly) heteronuclear system should approach that of the homonuclear X_2 and Y_2 systems. Such similarities are demonstrated in the next section.

ii. Configuration and State Diagrams The energy variation in these orbital energies gives rise to variations in the energies of the six CSFs and of the six electronic states that arise as combinations of these CSFs. The three singlet ($^1\Sigma(0)$, $^1\Sigma^*(0)$, and $^1\Sigma^{**}(0)$) and three triplet ($^3\Sigma^*(1)$, $^3\Sigma^*(0)$ and $^3\Sigma^*(-1)$) CSFs are, by no means, the true electronic eigenstates of the system; they are simply spin and spatial angular momentum adapted antisymmetric spin-orbital products. In principle, the set of CSFs Φ_I of the same symmetry must be combined to form the proper electronic eigenstates Ψ_K of the system:

$$\Psi_K = \sum_I C_I^K \Phi_I.$$

Within the approximation that the valence electronic states can be described adequately as combinations of the above valence CSFs, the three $^1\Sigma$, $^1\Sigma^*$, and $^1\Sigma^{**}$ CSFs must be combined to form the three lowest energy valence electronic states of $^1\Sigma$ symmetry. For the homonuclear case, the $^1\Sigma^*$ CSF does not couple with the other two because it is of ungerade symmetry, while the other CSFs $^1\Sigma$ and $^1\Sigma^{**}$ have gerade symmetry and do combine.

The relative amplitudes C_I^K of the CSFs Φ_I within each state Ψ_K are determined by solving the configuration-interaction (CI) secular problem:

$$\sum_J \langle\Phi_I|H|\Phi_J\rangle C_J^K = E_K C_I^K$$

for the state energies E_K and state CI coefficient vectors C_I^K. Here, H is the electronic Hamiltonian of the molecule.

To understand the extent to which the $^1\Sigma$ and $^1\Sigma^{**}$ (and $^1\Sigma^*$ for heteronuclear cases) CSFs couple, it is useful to examine the energies $\langle\Phi_I|H|\Phi_I\rangle$ of these CSFs for the range of internuclear distances of interest $R_e < R < \infty$. Near R_e, where the energy of the σ orbital is substantially below that of the $\sigma*$ orbital, the σ^2 $^1\Sigma$ CSF lies significantly below the $\sigma\sigma^*$ $^1\Sigma^*$ CSF which, in turn lies below the σ^{*2} $^1\Sigma^{**}$ CSF; the large energy splittings among these three CSFs simply reflecting the large gap between the σ and σ^* orbitals. The $^3\Sigma^*$ CSF generally lies below the corresponding $^1\Sigma^*$ CSF by an amount related to the exchange energy between the σ and $\sigma*$ orbitals.

As $R \rightarrow \infty$, the CSF energies $\langle\Phi_I|H|\Phi_J\rangle$ are more difficult to "intuit" because the σ and $\sigma*$ orbitals become degenerate (in the homonuclear case) or nearly so. To pursue this point and arrive at an energy ordering for the CSFs that is appropriate to the $R \rightarrow \infty$ region, it is useful to express each of the above CSFs in terms of the atomic orbitals s_x and s_y that comprise σ and $\sigma*$. To do so, the LCAO-MO expressions for σ and $\sigma*$,

$$\sigma = C[s_x + zs_y]$$

and

$$\sigma^* = C^*[zs_x - s_y],$$

are substituted into the Slater determinant definitions of the CSFs. Here C and C* are the normalization constants. The parameter z is 1.0 in the homonuclear case and deviates from 1.0 in relation to the s_x and s_y orbital energy difference (if s_x lies below s_y, then $z < 1.0$; if s_x lies above s_y, $z > 1.0$).

To simplify the analysis of the above CSFs, the familiar homonuclear case in which $z = 1.0$ will be examined first. The process of substituting the above expressions for σ and σ * into the Slater determinants that define the singlet and triplet CSFs can be illustrated as follows:

$$
\begin{aligned}
{}^1\Sigma(0) &= |\sigma\alpha\sigma\beta| = C^2|(s_x + s_y)\alpha(s_x + s_y)\beta| \\
&= C^2[|s_x\alpha s_x\beta| + |s_y\alpha s_y\beta| + |s_x\alpha s_y\beta| + |s_y\alpha s_x\beta|]
\end{aligned}
$$

The first two of these atomic-orbital-based Slater determinants ($|s_x\alpha s_x\beta|$ and $|s_y\alpha s_y\beta|$) are denoted "ionic" because they describe atomic orbital occupancies, which are appropriate to the $R \rightarrow \infty$ region, that correspond to $X{:} + Y$ and $X + Y{:}$ valence bond structures, while $|s_x\alpha s_y\beta|$ and $|s_y\alpha s_x\beta|$ are called "covalent" because they correspond to $X\bullet + Y\bullet$ structures.

In similar fashion, the remaining five CSFs may be expressed in terms of atomic-orbital-based Slater determinants. In so doing, use is made of the antisymmetry of the Slater determinants $|\phi_1\phi_2\phi_3| = -|\phi_1\phi_3\phi_2|$, which implies that any determinant in which two or more spin-orbitals are identical vanishes $|\phi_1\phi_2\phi_2| = -|\phi_1\phi_2\phi_2| = 0$. The result of decomposing the mo-based CSFs into their atomic orbital components is as follows:

$$
\begin{aligned}
{}^1\Sigma^{**}(0) &= |\sigma^*\alpha\sigma^*\beta| = \\
&\qquad C^{*2}[|s_x\alpha s_x\beta| + |s_y\alpha s_y\beta| - |s_x\alpha s_y\beta| - |s_y\alpha s_x\beta|] \\
{}^1\Sigma^*(0) &= \frac{1}{\sqrt{2}}[|\sigma\alpha\sigma^*\beta| - |\sigma\beta\sigma^*\alpha|] \\
&= CC^*\sqrt{2}[|s_x\alpha s_x\beta| - |s_y\alpha s_y\beta|] \\
{}^3\Sigma^*(1) &= |\sigma\alpha\sigma^*\alpha| \\
&= CC^*2|s_y\alpha s_x\alpha| \\
{}^3\Sigma^*(0) &= \frac{1}{\sqrt{2}}[|\sigma\alpha\sigma^*\beta| + |\sigma\beta\sigma^*\alpha|] \\
&= CC^*\sqrt{2}[|s_y\alpha s_x\beta| - |s_x\alpha s_y\beta|] \\
{}^3\Sigma^*(-1) &= |\sigma\alpha\sigma^*\alpha| \\
&= CC^*2|s_y\beta s_x\beta|
\end{aligned}
$$

These decompositions of the six valence CSFs into atomic-orbital or valence bond components allow the $R = \infty$ energies of the CSFs to be specified. For example, the fact that both ${}^1\Sigma$ and ${}^1\Sigma^{**}$ contain 50% ionic and 50% covalent structures implies that, as $R \rightarrow \infty$, both of their energies will approach the average of the covalent and ionic atomic energies

$$1/2[E(X\bullet) + E(Y\bullet) + E(Y) + E(X{:})].$$

The ${}^1\Sigma^*$ CSF energy approaches the purely ionic value $E(Y) + E(X{:})$ as $R \rightarrow \infty$. The energies of ${}^3\Sigma^*(0)$, ${}^3\Sigma^*(1)$ and ${}^3\Sigma^*(-1)$ all approach the purely covalent value $E(X\bullet) + E(Y\bullet)$ as $R \rightarrow \infty$.

The behaviors of the energies of the six valence CSFs as R varies are depicted below for situations in which the homolytic bond cleavage is energetically favored (i.e., for which $E(X\bullet) + E(Y\bullet) < E(Y) + E(X{:})$).

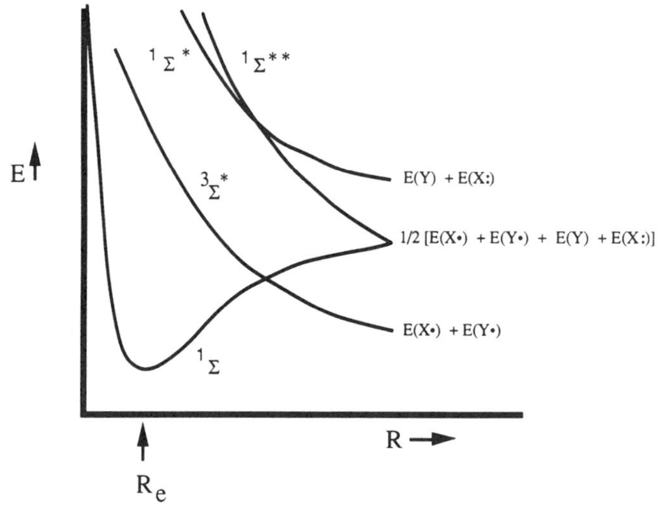

Configuration correlation diagram for homonuclear case in which homolytic bond cleavage is energetically favored.

When heterolytic bond cleavage is favored, the configuration energies as functions of internuclear distance vary as shown below.

Configuration correlation diagram for a homonuclear case in which heterolytic bond cleavage is energetically favored.

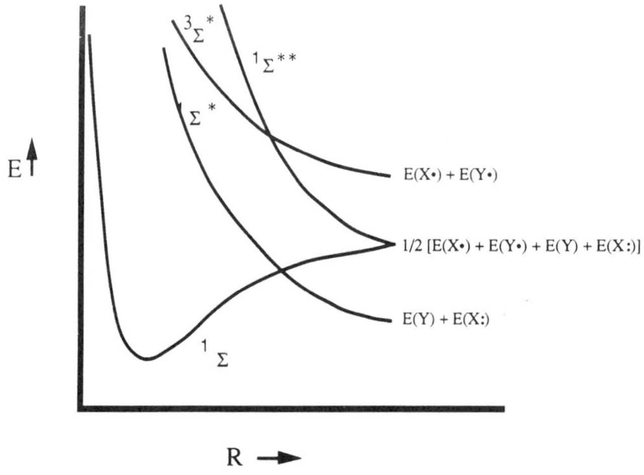

It is essential to realize that the energies $\langle\Phi_I|H|\Phi_I\rangle$ of the CSFs do *not* represent the energies of the true electronic states E_K; the CSFs are simply spin- and spatial- symmetry adapted antisymmetric functions that form a *basis* in terms of which to expand the true electronic states. For R-values at which the CSF energies are separated widely, the true E_K are rather well approximated by individual $\langle\Phi_I|H|\Phi_I\rangle$ values; such is the case near R_e.

For the homonuclear example, the $^1\Sigma$ and $^1\Sigma^{**}$ CSFs undergo CI coupling to form a pair of states of $^1\Sigma$ symmetry (the $^1\Sigma^*$ CSF cannot partake in this CI mixing because it is of ungerade symmetry; the $^3\Sigma*$ states can not mix because they are of triplet spin symmetry). The CI mixing of the $^1\Sigma$ and $^1\Sigma^{**}$ CSFs is described in terms of a 2×2 secular problem

$$\begin{bmatrix} <^1\Sigma|H|^1\Sigma> & <^1\Sigma|H|^1\Sigma^{**}> \\ <^1\Sigma^{**}|H|^1\Sigma> & <^1\Sigma^{**}|H|^1\Sigma^{**}> \end{bmatrix} \begin{bmatrix} A \\ B \end{bmatrix} = E \begin{bmatrix} A \\ B \end{bmatrix}$$

The diagonal entries are the CSF energies depicted in the above two figures. Using the Slater-Condon rules, the off-diagonal coupling can be expressed in terms of an exchange integral between the σ and $\sigma*$ orbitals:

$$<^1\Sigma|H|^1\Sigma^{**}> = <|\sigma\alpha\sigma\beta|H||\sigma*\alpha\sigma*\beta|> = <\sigma\sigma|\frac{1}{r_{12}}|\sigma*\sigma*> = K_{\sigma\sigma*}$$

At $R \to \infty$, where the $^1\Sigma$ and $^1\Sigma^{**}$ CSFs are degenerate, the two solutions to the above CI secular problem are:

$$E_{\mp} = 1/2[E(X\bullet) + E(Y\bullet) + E(Y) + E(X\textbf{:})] \mp <\sigma\sigma|\frac{1}{r_{12}}|\sigma*\sigma*>$$

with respective amplitudes for the $^1\Sigma$ and $^1\Sigma**$ CSFs given by

$$A_{\mp} = \pm\frac{1}{\sqrt{2}}; \quad B_{\mp} = \mp\frac{1}{\sqrt{2}}.$$

The first solution thus has

$$\Psi_{-} = \frac{1}{\sqrt{2}}[|\sigma\alpha\sigma\beta| - |\sigma*\alpha\sigma*\beta|]$$

which, when decomposed into atomic valence bond components, yields

$$\Psi_{-} = \frac{1}{\sqrt{2}}[|s_x\alpha s_y\beta| - |s_x\beta s_y\alpha|].$$

The other root has

$$\Psi_{+} = \frac{1}{\sqrt{2}}[|\sigma\alpha\sigma\beta| + |\sigma*\alpha\sigma*\beta|]$$
$$= \frac{1}{\sqrt{2}}[|s_x\alpha s_x\beta| + |s_y\alpha s_y\beta|].$$

Clearly, $^1\Sigma$ and $^1\Sigma^{**}$, which both contain 50% ionic and 50% covalent parts, combine to produce Ψ_- which is purely covalent and Ψ_+ which is purely ionic.

The above strong CI mixing of $^1\Sigma$ and $^1\Sigma^{**}$ as $R \to \infty$ qualitatively alters the configuration correlation diagrams shown above. Descriptions of the resulting valence singlet and triplet Σ *states* are given below for homonuclear situations in which covalent products lie below and above ionic products, respectively. Note that in both cases, there exists a single attractive curve and five (n.b., the triplet state has three curves superposed) repulsive curves.

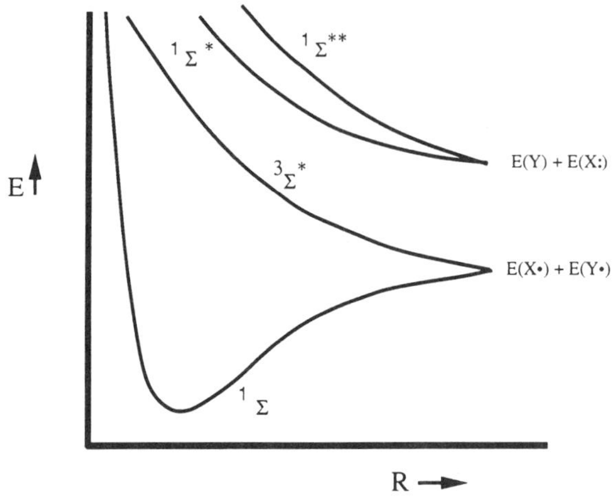

State correlation diagram for homonuclear case in which homolytic bond cleavage is energetically favored.

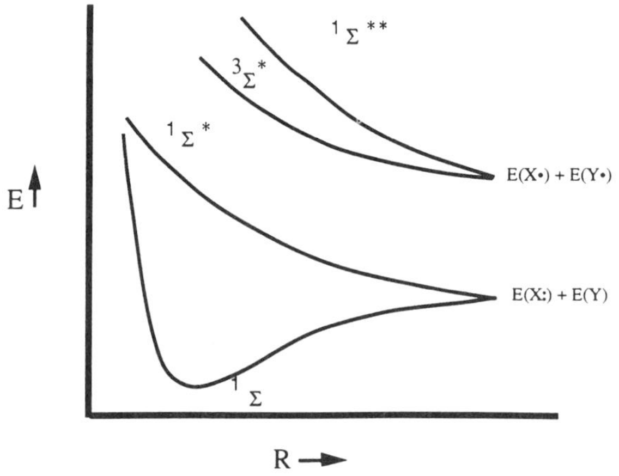

State correlation diagram for homonuclear case in which heterolytic bond cleavage is energetically favored.

If the energies of the s_x and s_y orbitals do not differ significantly (compared to the coulombic interactions between electron pairs), it is expected that the essence of the findings described above for homonuclear species will persist even for heteronuclear systems. A decomposition of the six CSFs listed above, using the *heteronuclear* molecular orbitals introduced earlier yields:

$$^1\Sigma(0) = C^2[|s_x\alpha s_x\beta| + z^2|s_y\alpha s_y\beta| + z|s_x\alpha s_y\beta| + z|s_y\alpha s_x\beta|]$$
$$^1\Sigma^{**}(0) = C^{*2}[z^2|s_x\alpha s_x\beta| + |s_y\alpha s_y\beta| - z|s_x\alpha s_y\beta| - z|s_y\alpha s_x\beta|]$$
$$^1\Sigma^*(0) = \frac{CC^*}{\sqrt{2}}[2z|s_x\alpha s_x\beta| - 2z|s_y\alpha s_y\beta| + (z^2 - 1)|s_y\alpha s_x\beta| + (z^2 - 1)|s_x\alpha s_y\beta|]$$

$$^3\Sigma^*(0) = \frac{CC^*}{\sqrt{2}}(z^2+1)[|s_y\alpha s_x\beta| - |s_x\alpha s_y\beta|]$$
$$^3\Sigma^*(1) = CC^*(z^2+1)|s_y\alpha s_x\alpha|$$
$$^3\Sigma^*(-1) = CC^*(z^2+1)|s_y\beta s_x\beta|$$

Clearly, the three $^3\Sigma^*$ CSFs retain purely covalent $R \to \infty$ character even in the heteronuclear case. The $^1\Sigma$, $^1\Sigma^{**}$, and $^1\Sigma^*$ (all three of which can undergo CI mixing now) possess one covalent and two ionic components of the form $|s_x\alpha s_y\beta| + |s_y\alpha s_x\beta|$, $|s_x\alpha s_x\beta|$, and $|s_y\alpha s_y\beta|$. The three singlet CSFs therefore can be combined to produce a singlet covalent product function $|s_x\alpha s_y\beta| + |s_y\alpha s_x\beta|$ as well as *both* $X + Y\!:$ and $X\!: + Y$ ionic product wavefunctions $|s_y\alpha s_y\beta|$ and $|s_x\alpha s_x\beta|$, respectively. In most situations, the energy ordering of the homolytic and heterolytic dissociation products will be either

$$E(X\bullet) + E(Y\bullet) < E(X\!:) + E(Y) < E(X) + E(Y\!:)$$

or

$$E(X\!:) + E(Y) < E(X\bullet) + E(Y\bullet) < E(X) + E(Y\!:).$$

The extensions of the state correlation diagrams given above to the heteronuclear situations are described below.

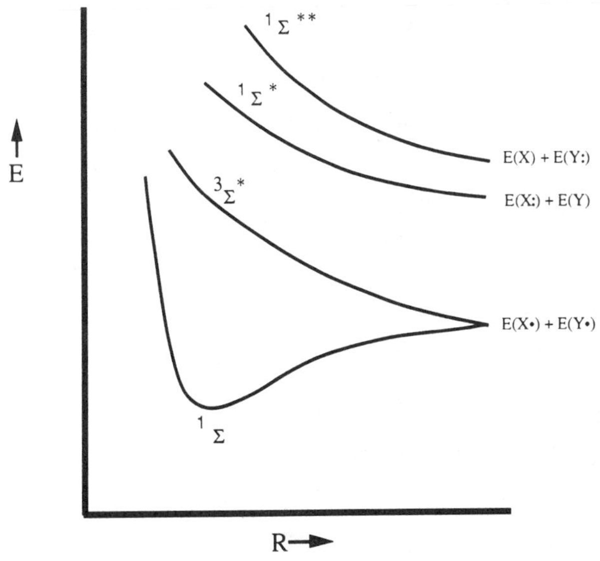

State correlation diagram for heteronuclear case in which homolytic bond cleavage is energetically favored.

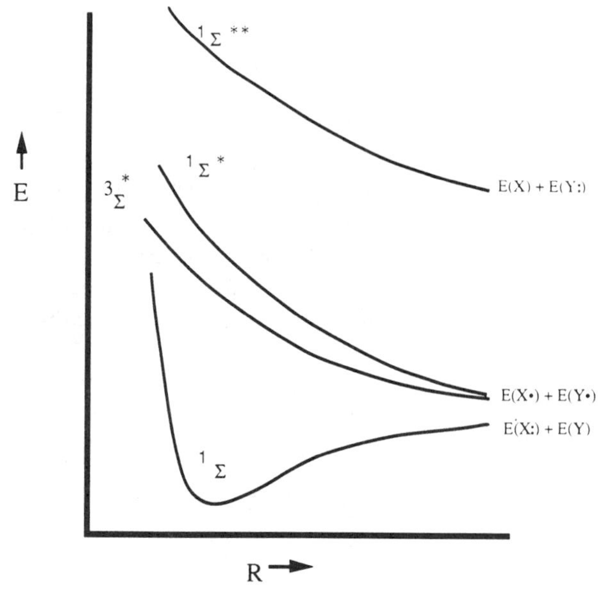

State correlation diagram for heteronuclear case in which heterolytic bond cleavage to one product is energetically favored but homolytic cleavage lies below the second heterolytic asymptote.

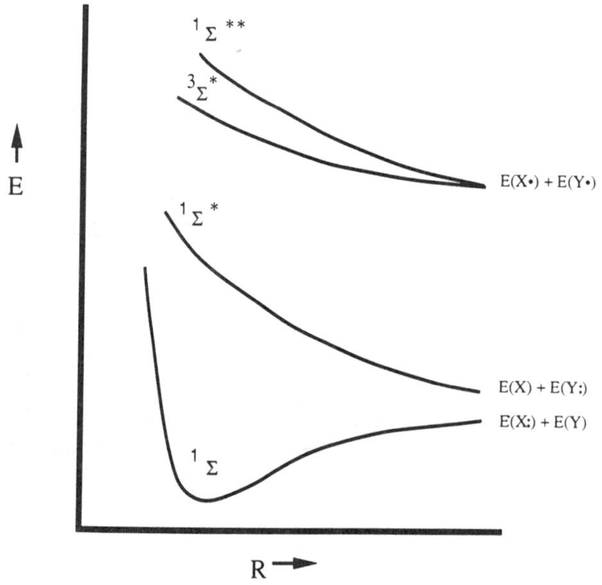

State correlation diagram for heteronuclear case in which both heterolytic bond cleavage products are energetically favored relative to homolytic cleavage.

Again note that only one curve is attractive and five are repulsive in all cases. In these heteronuclear cases, it is the mixing of the $^1\Sigma$, $^1\Sigma*$, and $^1\Sigma**$ CSFs, which varies with R, that determines which molecular state connects to which asymptote. As the energy ordering of the asymptotes varies, so do these correlations.

3. Summary

Even for the relatively simple two-electron, two-orbital single-bond interactions between a pair of atoms or functional groups, the correlations among energy-ordered molecular states and energy-ordered asymptotic states is complex enough to warrant considerations beyond what is taught in most undergraduate and beginning graduate inorganic and physical chemistry classes. In particular, the correlations that arise when one (or both) of the heterolytic bond dissociation aysmptotes lies below the homolytic cleavage products are important to realize and keep in mind.

In all cases treated here, the three singlet states that arise produce one and only one attractive (bonding) potential energy curve; the other two singlet surfaces are repulsive. The three triplet surfaces are also repulsive. Of course, in arriving at these conclusions, we have considered only contributions to the inter-fragment interactions that arise from valence-orbital couplings; no consideration has been made of attractive or repulsive forces that result from one or both of the X- and Y-fragments possessing net charge. In the latter case, one must, of course, add to the qualitative potential surfaces described here any coulombic, charge-dipole, or charge-induced-dipole energies. Such additional factors can lead to attractive long-range interactions in typical ion-molecule complexes.

The necessity of the analysis developed above becomes evident when considering dissociation of diatomic transition metal ions. Most transition metal atoms have ground states with electron configurations of the form s^2d^n (for first-row metals, exceptions include Cr(s^1d^5), Cu(s^1d^{10}), and the s^1d^9 state of Ni is basically isoenergetic with the s^2d^8 ground state). The corresponding positive ions have ground states with s^1d^n(Sc, Ti, Mn, Fe) or s^0d^{n+1}(V, Cr, Co, Ni, Cu) electron configurations. For each of these elements, the alternate electron configuration leads to low-lying excited states.

One can imagine forming a M_2^+ metal dimer ion with a configuration described as $\sigma_g^2 d^{2n+1}$, where the σ_g bonding orbital is formed primarily from the metal s orbitals and the d orbitals are largely nonbonding (as is particularly appropriate towards the right hand side of the periodic table). Cleavage of such a σ bond tends to occur heterolytically since this forms lower energy species, $M(s^2d^n) + M^+(s^0d^{n+1})$, than homolytic cleavage to $M(s^1d^{n+1}) + M^+(s^1d^n)$. For example, Co_2^+ dissociates to Co(s^2d^7) + Co$^+$(s^0d^8) rather than to Co(s^1d^8) + Co$^+$(s^1d^7),[2] which lies 0.85 eV higher in energy.

Qualitative aspects of the above analysis for homonuclear transition metal dimer ions will persist for heteronuclear ions. For example, the ground-state dissociation asymptote for CoNi$^+$ is the heterolytic cleavage products Co(s^2d^7) + Ni$^+$(s^0d^9). The alternative heterolytic cleavage to form Co$^+$(s^0d^8) + Ni(s^2d^8) is 0.23 eV higher in energy, while homolytic cleavage can lead to Co$^+$(s^1d^7) + Ni(s^1d^9),0.45 eV higher, or Co(s^1d^8) + Ni$^+$(s^1d^8), 1.47 eV higher. This is the situation illustrated in the last figure above.

III. VARIOUS TYPES OF CONFIGURATION MIXING

A. Essential CI

The above examples of the use of CCD's show that, as motion takes place along the proposed reaction path, geometries may be encountered at which it is *essential* to describe the electronic wavefunction in terms of a linear combination of more than one CSF:

$$\Psi = \sum_I C_I \Phi_I,$$

where the Φ_I are the CSFs which are undergoing the avoided crossing. Such essential configuration mixing is often referred to as treating "*essential CI*".

B. Dynamical CI

To achieve reasonable chemical accuracy (e.g., ±5 kcal/mole) in electronic structure calculations it is necessary to use a multiconfigurational Ψ even in situations where no obvious strong configuration mixing (e.g., crossings of CSF energies) is present. For example, in describing the π^2 bonding electron pair of an olefin or the ns^2 electron pair in alkaline earth atoms, it is important to mix in doubly excited CSFs of the form $(\pi^*)^2$ and np^2, respectively. The reasons for introducing such a CI-level treatment were treated for an alkaline earth atom earlier in this chapter.

Briefly, the physical importance of such doubly-excited CSFs can be made clear by using the identity:

$$\begin{aligned} &C_1|\ldots\phi\alpha\phi\beta\ldots| - C_2|\ldots\phi'\alpha\phi'\beta\ldots| \\ &\quad = C_1/2\{|\ldots(\phi - x\phi')\alpha(\phi + x\phi')\beta\ldots| - |\ldots(\phi - x\phi')\beta(\phi + x\phi')\alpha\ldots|\}, \end{aligned}$$

where

$$x = (C_2/C_1)^{1/2}.$$

This allows one to interpret the combination of two CSFs which differ from one another by a double excitation from one orbital (ϕ) to another (ϕ') as equivalent to a singlet coupling of two different (non-orthogonal) orbitals ($\phi - x\phi'$) and ($\phi + x\phi'$). This picture is closely related to the so-called generalized valence bond (GVB) model that W. A. Goddard and his co-workers have developed (see, for example, W. A. Goddard and L. B. Harding, *Annu. Rev. Phys. Chem. 29*, 363 (1978)). In the simplest embodiment of the GVB model, each electron pair in the atom or molecule is correlated by mixing in a CSF in which that electron pair is "doubly excited" to a correlating orbital. The direct product of all such pair correlations generates the GVB-type wavefunction. In the GVB approach, these electron correlations are not specified in terms of double excitations involving CSFs formed from orthonormal spin orbitals; instead, explicitly non-orthogonal GVB orbitals are used as described above, but the result is the same as one would obtain using the direct product of doubly excited CSFs.

In the olefin example mentioned above, the two non-orthogonal "polarized orbital pairs" involve mixing the π and π^* orbitals to produce two left-right polarized orbitals as depicted below:

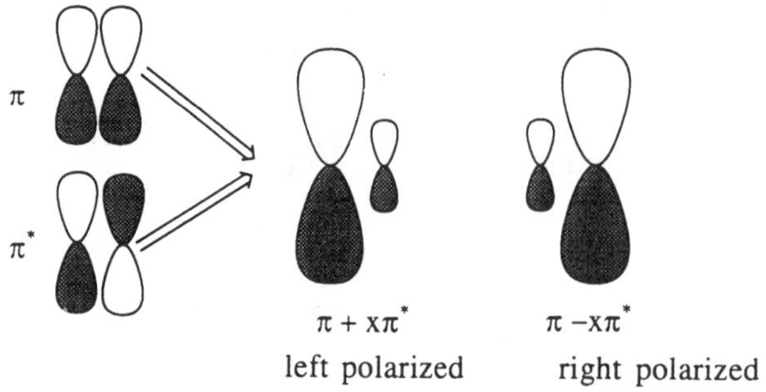

In this case, one says that the π^2 electron pair undergoes left-right correlation when the $(\pi*)^2$ CSF is mixed into the CI wavefunction. In the alkaline earth atom case, the polarized orbital pairs are formed by mixing the ns and np orbitals (actually, one must mix in equal amounts of p_1, p_{-1}, and p_0 orbitals to preserve overall ^1S symmetry in this case), and give rise to angular correlation of the electron pair. Use of an $(n + 1)s^2$ CSF for the alkaline earth calculation would contribute in-out or radial correlation because, in this case, the polarized orbital pair formed from the ns and $(n + 1)s$ orbitals would be radially polarized.

The use of doubly excited CSFs is thus seen as a mechanism by which Ψ can place electron *pairs*, which in the single-configuration picture occupy the same orbital, into different regions of space (i.e., one into a member of the polarized orbital pair) thereby lowering their mutual coulombic repulsions. Such electron correlation effects are referred to as "*dynamical electron correlation*"; they are extremely important to include if one expects to achieve chemically meaningful accuracy (i.e., ±5 kcal/mole).

Exercises, Problems, and Solutions

REVIEW EXERCISES

1. For the given orbital occupations (configurations) of the following systems, determine all possible states (all possible allowed combinations of spin and space states). There is no need to form the determinental wavefunctions simply label each state with its proper *term symbol*. One method commonly used is Harry Grays "box method" found in *Electrons and Chemical Bonding*.

a. CH_2 $1a_1^2 2a_1^2 1b_2^2 3a_1^1 1b_1^1$

b. B_2 $1\sigma_g^2 1\sigma_u^2 2\sigma_g^2 2\sigma_u^2 1\pi_u^1 2\pi_u^1$

c. O_2 $1\sigma_g^2 1\sigma_u^2 2\sigma_g^2 2\sigma_u^2 1\pi_u^4 3\sigma_g^2 1\pi_g^2$

d. Ti $1s^2 2s^2 2p^6 3s^2 3p^6 4s^2 3d^1 4d^1$

e. Ti $1s^2 2s^2 2p^6 3s^2 3p^6 4s^2 3d^2$

EXERCISES

1. Show that the configuration (determinant) corresponding to the Li^+ $1s(\alpha)1s(\alpha)$ state vanishes.

2. Construct the 3 triplet and 1 singlet wavefunctions for the Li^+ $1s^1 2s^1$ configuration. Show that each state is a proper eigenfunction of S^2 and S_z(use raising and lowering operators for S^2).

3. Construct wavefunctions for each of the following states of CH_2:

a. $^1B_1$ $(1a_1^2 2a_1^2 1b_2^2 3a_1^1 1b_1^1)$

b. $^3B_1$ $(1a_1^2 2a_1^2 1b_2^2 3a_1^1 1b_1^1)$

c. $^1A_1$ $(1a_1^2 2a_1^2 1b_2^2 3a_1^2)$

4. Construct wavefunctions for each state of the $1\sigma^2 2\sigma^2 3\sigma^2 1\pi^2$ configuration of NH.

5. Construct wavefunctions for each state of the $1s^1 2s^1 3s^1$ configuration of Li.

6. Determine all term symbols that arise from the $1s^2 2s^2 2p^2 3d^1$ configuration of the excited N atom.

7. Calculate the energy (using Slater-Condon rules) associated with the $2p$ valence electrons for the following states of the C atom.

a. $^3P(M_L = 1, M_S = 1)$,

b. $^3P(M_L = 0, M_S = 0)$,

c. $^1S(M_L = 0, M_S = 0)$, and

d. $^1D(M_L = 0, M_S = 0)$.

8. Calculate the energy (using Slater-Condon rules) associated with the π valence electrons for the following states of the NH molecule.

a. $^1\Delta(M_L = 2, M_S = 0)$,

b. $^1\Sigma(M_L = 0, M_S = 0)$, and

c. $^3\Sigma(M_L = 0, M_S = 0)$.

PROBLEMS

1. Let us investigate the reactions:

$CH_2(^1A_1) \rightarrow H_2 + C$, and

$CH_2(^3B_1) \rightarrow H_2 + C$,

under an assumed C_{2v} reaction pathway utilizing the following information:

C atom: $^3P \xrightarrow{29.2\ \text{kcal/mole}} {^1D} \xrightarrow{32.7\ \text{kcal/mole}} {^1S}$

$C(^3P) + H_2 \rightarrow CH_2(^3B_1)$ $\Delta E = -78.8$ kcal/mole

$C(^1D) + H_2 \rightarrow CH_2(^1A_1)$ $\Delta E = -97.0$ kcal/mole

$IP(H_2) > IP(2s\ carbon)$.

a. Write down (first in terms of $2p_{1,0,-1}$ orbitals and then in terms of $2p_{x,y,z}$ orbitals) the:

i. three Slater determinant (SD) wavefunctions belonging to the ^3P state all of which have $M_S = 1$,

ii. five ^1D SD wavefunctions, and

iii. one ^1S SD wavefunction.

b. Using the coordinate system shown below, label the hydrogen orbitals σ_g, σ_u and the carbon $2s$, $2p_x$, $2p_y$, $2p_z$, orbitals as a_1, $b_1(x)$, $b_2(y)$, or a_2. Do the same for the σ, σ, σ^*, σ^*, n, and p_π orbitals of CH_2.

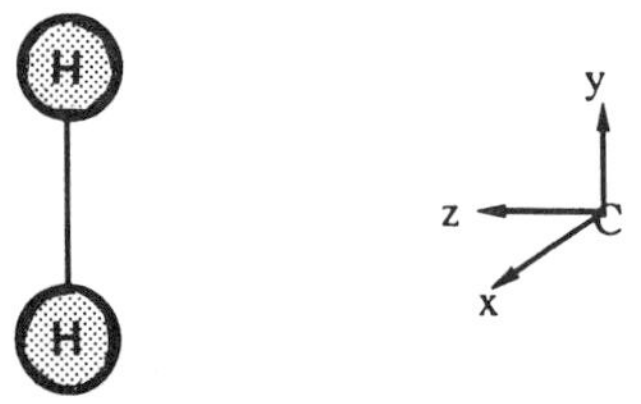

c. Draw an orbital correlation diagram for the $CH_2 \rightarrow H_2 + C$ reactions. Try to represent the relative energy orderings of the orbitals correctly.

d. Draw (on graph paper) a configuration correlation diagram for $CH_2(^3B_1) \rightarrow H_2 + C$ showing *all* configurations which arise from the $C(^3P) + H_2$ products. You can assume that doubly excited configurations lie much (100 kcal/mole) above their parent configurations.

e. Repeat step d. for $CH_2(^1A_1) \rightarrow H_2 + C$ again showing *all* configurations which arise from the $C(^1D) + H_2$ products.

f. Do you expect the reaction $C(^3P) + H_2 \rightarrow CH_2$ to have a large activation barrier? About how large? What state of CH_2 is produced in this reaction? Would distortions away from C_{2v} symmetry be expected to raise of lower the activation barrier? Show how one could estimate where along the reaction path the barrier top occurs.

g. Would $C(^1D) + H_2 \rightarrow CH_2$ be expected to have a larger or smaller barrier than you found for the ^3P C reaction?

2. The decomposition of the ground-state singlet carbene,

to produce acetylene and ^1D carbon is known to occur with an activation energy equal to the reaction endothermicity. However, when triplet carbene decomposes to acetylene and ground-state (triplet) carbon, the activation energy exceeds this reaction's endothermicity.

Construct orbital, configuration, and state correlation diagrams which permit you to explain the above observations. Indicate whether single configuration or configuration interaction wavefunctions would be required to describe the above singlet and triplet decomposition processes.

3. We want to carry out a configuration interaction calculation on H_2 at $R = 1.40$ au. A minimal basis consisting of normalized $1s$ Slater orbitals with $\zeta = 1.0$ gives rise to the following overlap (S), one-electron (h), and two-electron atomic integrals:

$$\left\langle 1s_A | 1s_B \right\rangle = 0.753 \equiv S,$$

$$\left\langle 1s_A | h | 1s_A \right\rangle = -1.110, \quad \left\langle 1s_B | h | 1s_A \right\rangle = -0.968,$$

$$\left\langle 1s_A 1s_A | h | 1s_A 1s_A \right\rangle = 0.625 \equiv \left\langle AA | AA \right\rangle$$

$$\left\langle AA | BB \right\rangle = 0.323, \quad \left\langle AB | AB \right\rangle = 0.504 \text{ and } \left\langle AA | AB \right\rangle = 0.426.$$

a. The normalized and orthogonal molecular orbitals we will use for this minimal basis will be determined purely by symmetry:

$\sigma_g = (2+2S)^{-\frac{1}{2}}\left(1s_A + 1s_B\right)$, and

$\sigma_u = (2+2S)^{-\frac{1}{2}}\left(1s_A - 1s_B\right)$. Show that these orbitals are indeed orthogonal.

b. Evaluate (using the one- and two-electron atomic integrals given above) the unique one- and two-electron integrals over this molecular orbital basis (this is called a transformation from the ao to the mo basis). For example, evaluate

$\left\langle u|h|u\right\rangle, \left\langle uu|uu\right\rangle, \left\langle gu|gu\right\rangle$, etc.

c. Using the

$^1\Sigma_g^+$

configurations σ_g^2, and σ_u^2, show that the elements of the 2×2 configuration interaction Hamiltonian matrix are –1.805, 0.140, and –0.568.

d. Using *this* configuration interaction matrix, find the configuration interaction (CI) approximation to the ground and excited state energies and wavefunctions.

e. Evaluate and make a rough sketch of the polarized orbitals which result from the above ground state σ_g^2 and σ_u^2 CI wavefunction.

SOLUTIONS

Review Exercises

1. a. For non-degenerate point groups one can simply multiply the representations (since only one representation will be obtained):

$a_1 \otimes b_1 = b_1$

Constructing a "box" in this case is unnecessary since it would only contain a single row. Two unpaired electrons will result in a singlet ($S=0, M_S=0$), and three triplets

$(S=1, M_S=1; S=1, M_S=0; S=1, M_S=-1)$.

The states will be:

$^3B_1(M_S=1)$, $^3B_1(M_S=0)$, $^3B_1(M_S=-1)$, and $^1B_1(M_S=0)$.

b. Remember that when coupling non-equivalent linear molecule angular momenta, one simple adds the individual L_z values and vector couples the electron spin. So, in this case $(1\pi_u{}^1 2\pi_u{}^1)$, we have M_L values of $1+1$, $1-1$, $-1+1$, and $-1-1$ (2, 0, 0, and –2). The term symbol Δ is used to denote the spatially doubly degenerate level ($M_L = \pm2$) and there are two distinct spatially non-degenerate levels denoted by the term symbol $\Sigma(M_L=0)$.

Again, two unpaired electrons will result in a singlet ($S=0, M_S=0$), and three triplets

$(S=1, M_S=1; S=1, M_S=0; S=1, M_S=-1)$.

The states generated are then:

$^1\Delta(M_L = 2)$; one state ($M_S = 0$),
$^1\Delta(M_L = -2)$; one state ($M_S = 0$),
$^3\Delta(M_L = 2)$; three states ($M_S = 1$, 0, and -1),
$^3\Delta(M_L = -2)$; three states ($M_S = 1$, 0, and -1),
$^1\Sigma(M_L = 0)$; one state ($M_S = 0$),
$^1\Sigma(M_L = 0)$; one state ($M_S = 0$),
$^3\Sigma(M_L = 0)$; three states ($M_S = 1$, 0, and -1), and
$^3\Sigma(M_L = 0)$; three states ($M_S = 1$, 0, and -1).

c. Constructing the "box" for two equivalent π electrons one obtains:

M_S \ M_L	2	1	0
1			$\lvert\pi_1\alpha\pi_{-1}\alpha\rvert$
0	$\lvert\pi_1\alpha\pi_1\beta\rvert$		$\lvert\pi_1\alpha\pi_{-1}\beta\rvert$, $\lvert\pi_{-1}\alpha\pi_1\beta\rvert$

From this "box" one obtains six states:

$^1\Delta(M_L = 2)$; one state ($M_S = 0$),
$^1\Delta(M_L = -2)$; one state ($M_S = 0$),
$^1\Sigma(M_L = 0)$; one state ($M_S = 0$),
$^3\Sigma(M_L = 0)$; three states ($M_S = 1$, 0, and -1).

d. It is not necessary to construct a "box" when coupling non-equivalent angular momenta since the vector coupling results in a range from the sum of the two individual angular momenta to the absolute value of their difference. In this case, $3d^14d^1$, $L = 4, 3, 2, 1, 0$, and $S = 1, 0$. The term symbols are: ^3G, ^1G, ^3F, ^1F, ^3D, ^1D, ^3P, ^1P, ^3S, and ^1S. The L and S angular momenta can be vector coupled to produce further splitting into levels:

$J = L + S \ldots |L - S|$.

Denoting J as a term symbol subscript one can identify all the levels and subsequent $(2J + 1)$ states:

$^3G_5$ (11 states),
$^3G_4$ (9 states),
$^3G_3$ (7 states),
$^1G_4$ (9 states),
$^3F_4$ (9 states),
$^3F_3$ (7 states),
$^3F_2$ (5 states),
$^1F_3$ (7 states),
$^3D_3$ (7 states),
$^3D_2$ (5 states),
$^3D_1$ (3 states),

$^1D_2$ (5 states),
$^3P_2$ (5 states),
$^3P_1$ (3 states),
$^3P_0$ (1 state),
$^1P_1$ (3 states),
$^3S_1$ (3 states), and
$^1S_0$ (1 state).

e. Construction of a "box" for the two equivalent d electrons generates (note the "box" has been turned side ways for convenience):

M_S / M_L	1	0
4		$\|d_2\alpha d_2\beta\|$
3	$\|d_2\alpha d_1\alpha\|$	$\|d_2\alpha d_1\beta\|$, $\|d_2\beta d_1\alpha\|$
2	$\|d_2\alpha d_0\alpha\|$	$\|d_2\alpha d_0\beta\|$, $\|d_2\beta d_0\alpha\|$, $\|d_1\alpha d_1\beta\|$
1	$\|d_1\alpha d_0\alpha\|$, $\|d_2\alpha d_{-1}\alpha\|$	$\|d_1\alpha d_0\beta\|$, $\|d_1\beta d_0\alpha\|$, $\|d_2\alpha d_{-1}\beta\|$, $\|d_2\beta d_{-1}\alpha\|$
0	$\|d_2\alpha d_{-2}\alpha\|$, $\|d_1\alpha d_{-1}\alpha\|$	$\|d_2\alpha d_{-2}\beta\|$, $\|d_2\beta d_{-2}\alpha\|$, $\|d_1\alpha d_{-1}\beta\|$, $\|d_1\beta d_{-1}\alpha\|$, $\|d_0\alpha d_0\beta\|$

The term symbols are: ^1G, ^3F, ^1D, ^3P, and ^1S. The L and S angular momenta can be vector coupled to produce further splitting into levels:

$^1G_4$ (9 states),
$^3F_4$ (9 states),
$^3F_3$ (7 states),
$^3F_2$ (5 states),
$^1D_2$ (5 states),
$^3P_2$ (5 states),
$^3P_1$ (3 states),
$^3P_0$ (1 state), and
$^1S_0$ (1 state).

Exercises

1. Constructing the Slater determinant corresponding to the "state" $1s(\alpha)1s(\alpha)$ with the rows labeling the orbitals and the columns labeling the electron gives:

$$
\begin{aligned}
|1s\alpha 1s\alpha| &= \frac{1}{\sqrt{2!}}\begin{vmatrix} 1s\alpha(1) & 1s\alpha(2) \\ 1s\alpha(1) & 1s\alpha(2) \end{vmatrix} \\
&= \frac{1}{\sqrt{2}}(1s\alpha(1)1s\alpha(2) - 1s\alpha(1)1s\alpha(2)) \\
&= 0
\end{aligned}
$$

2. Starting with the $M_S = 1\ {}^3S$ state (which in a "box" for this $M_L = 0, M_S = 1$ case would contain only one product function; $|1s\alpha 2s\alpha|$) and applying S_- gives:

$$S_-\ {}^3S(S=1, M_S=1) = \sqrt{1(1+1) - 1(1-1)}\ \hbar\ {}^3S(S=1, M_S=0)$$
$$= \hbar\sqrt{2}\ {}^3S(S=1, M_S=0)$$
$$= \Big(S_-(1) + S_-(2)\Big)|1s\alpha 2s\alpha|$$
$$= S_-(1)|1s\alpha 2s\alpha| + S_-(2)|1s\alpha 2s\alpha|$$
$$= \hbar\sqrt{\frac{1}{2}\left(\frac{1}{2}+1\right) - \frac{1}{2}\left(\frac{1}{2}-1\right)}\ |1s\beta 2s\alpha|$$
$$+ \hbar\sqrt{\frac{1}{2}\left(\frac{1}{2}+1\right) - \frac{1}{2}\left(\frac{1}{2}-1\right)}\ |1s\alpha 2s\beta|$$
$$= \hbar(|1s\beta 2s\alpha| + |1s\alpha 2s\beta|)$$

So,

$$\hbar\sqrt{2}\ {}^3S(S=1, M_S=0) = \hbar(|1s\beta 2s\alpha| + |1s\alpha 2s\beta|)$$
$${}^3S(S=1, M_S=0) = \frac{1}{\sqrt{2}}(|1s\beta 2s\alpha| + |1s\alpha 2s\beta|)$$

The three triplet states are then:

$${}^3S(S=1, M_S=1) = |1s\alpha 2s\alpha|,$$
$${}^3S(S=1, M_S=0) = \frac{1}{\sqrt{2}}(|1s\beta 2s\alpha| + |1s\alpha 2s\beta|),\ \text{and}$$
$${}^3S(S=1, M_S=-1) = |1s\beta 2s\beta|..$$

The singlet state which must be constructed orthogonal to the three singlet states (and in particular to the ${}^3S(S=1, M_S=0)$ state) can be seen to be:

$${}^1S(S=0, M_S=0) = \frac{1}{\sqrt{2}}(|1s\beta 2s\alpha| - |1s\alpha 2s\beta|).$$

Applying S^2 and S_z to each of these states gives:

$$S_z|1s\alpha 2s\alpha| = \Big(S_z(1) + S_z(2)\Big)|1s\alpha 2s\alpha|$$
$$= S_z(1)|1s\alpha 2s\alpha| + S_z(2)|1s\alpha 2s\alpha|$$
$$= \hbar\left(\frac{1}{2}\right)|1s\alpha 2s\alpha| + \hbar\left(\frac{1}{2}\right)|1s\alpha 2s\alpha|$$
$$= \hbar|1s\alpha 2s\alpha|$$
$$S^2|1s\alpha 2s\alpha| = (S_-S_+ + S_z^2 + \hbar S_z)|1s\alpha 2s\alpha|$$
$$= S_-S_+|1s\alpha 2s\alpha| + S_z^2|1s\alpha 2s\alpha| + \hbar S_z|1s\alpha 2s\alpha|$$
$$= 0 + \hbar^2|1s\alpha 2s\alpha| + \hbar^2|1s\alpha 2s\alpha|$$
$$= 2\hbar^2|1s\alpha 2s\alpha|$$

$$
\begin{aligned}
S_z\frac{1}{\sqrt{2}}(|1s\beta 2s\alpha| + |1s\alpha 2s\beta|) &= \left(S_z(1) + S_z(2)\right)\frac{1}{\sqrt{2}}(|1s\beta 2s\alpha| + |1s\alpha 2s\beta|) \\
&= \frac{1}{\sqrt{2}}\left(S_z(1) + S_z(2)\right)|1s\beta 2s\alpha| \\
&\quad + \frac{1}{\sqrt{2}}\left(S_z(1) + S_z(2)\right)|1s\alpha 2s\beta| \\
&= \frac{1}{\sqrt{2}}\left(\hbar\left(-\frac{1}{2}\right) + \hbar\left(\frac{1}{2}\right)\right)|1s\beta 2s\alpha| + \frac{1}{\sqrt{2}}\left(\hbar\left(\frac{1}{2}\right) + \hbar\left(-\frac{1}{2}\right)\right)|1s\alpha 2s\beta| \\
&= 0\hbar\frac{1}{\sqrt{2}}(|1s\beta 2s\alpha| + |1s\alpha 2s\beta|)
\end{aligned}
$$

$$
\begin{aligned}
S^2\frac{1}{\sqrt{2}}(|1s\beta 2s\alpha| + |1s\alpha 2s\beta|) &= (S_-\,S_+ + S_z^2 + \hbar S_z)\frac{1}{\sqrt{2}}(|1s\beta 2s\alpha| + |1s\alpha 2s\beta|) \\
&= S_-\,S_+\frac{1}{\sqrt{2}}(|1s\beta 2s\alpha| + |1s\alpha 2s\beta|) \\
&= \frac{1}{\sqrt{2}}\left(S_-(S_+(1) + S_+(2))|1s\beta 2s\alpha| + S_-(S_+(1) + S_+(2))|1s\alpha 2s\beta|\right) \\
&= \frac{1}{\sqrt{2}}\left(S_-\hbar|1s\alpha 2s\alpha| + S_-\hbar|1s\alpha 2s\alpha|\right) \\
&= 2\hbar\frac{1}{\sqrt{2}}\left((S_-(1) + S_-(2))|1s\alpha 2s\alpha|\right) \\
&= 2\hbar\frac{1}{\sqrt{2}}(\hbar|1s\beta 2s\alpha| + \hbar|1s\alpha 2s\beta|) \\
&= 2\hbar^2\frac{1}{\sqrt{2}}(|1s\beta 2s\alpha| + |1s\alpha 2s\beta|)
\end{aligned}
$$

$$
\begin{aligned}
S_z|1s\beta 2s\beta| &= \left(S_z(1) + S_z(2)\right)|1s\beta 2s\beta| \\
&= S_z(1)|1s\beta 2s\beta| + S_z(2))|1s\beta 2s\beta| \\
&= \hbar\left(-\frac{1}{2}\right)|1s\beta 2s\beta| + \hbar\left(-\frac{1}{2}\right)|1s\beta 2s\beta| \\
&= -\hbar|1s\beta 2s\beta|
\end{aligned}
$$

$$
\begin{aligned}
S^2|1s\beta 2s\beta| &= (S_+\,S_- + S_z^2 - \hbar S_z)|1s\beta 2s\beta| \\
&= S_+\,S_-|1s\beta 2s\beta| + S_z^2|1s\beta 2s\beta| - \hbar S_z|1s\beta 2s\beta| \\
&= 0 + \hbar^2|1s\beta 2s\beta| + \hbar^2|1s\beta 2s\beta| \\
&= 2\hbar^2|1s\beta 2s\beta|
\end{aligned}
$$

$$
\begin{aligned}
S_z\frac{1}{\sqrt{2}}(|1s\beta 2s\alpha| - |1s\alpha 2s\beta|) &= \left(S_z(1) + S_z(2)\right)\frac{1}{\sqrt{2}}(|1s\beta 2s\alpha| - |1s\alpha 2s\beta|) \\
&= \frac{1}{\sqrt{2}}\left(S_z(1) + S_z(2)\right)|1s\beta 2s\alpha| - \frac{1}{\sqrt{2}}\left(S_z(1) + S_z(2)\right)|1s\alpha 2s\beta| \\
&= \frac{1}{\sqrt{2}}\left(\hbar\left(-\frac{1}{2}\right) + \hbar\left(\frac{1}{2}\right)\right)|1s\beta 2s\alpha| - \frac{1}{\sqrt{2}}\left(\hbar\left(\frac{1}{2}\right) + \hbar\left(-\frac{1}{2}\right)\right)|1s\alpha 2s\beta| \\
&= 0\hbar\frac{1}{\sqrt{2}}(|1s\beta 2s\alpha| - |1s\alpha 2s\beta|)
\end{aligned}
$$

$$
\begin{aligned}
S^2\frac{1}{\sqrt{2}}(|1s\beta 2s\alpha| - |1s\alpha 2s\beta|) &= (S_-\,S_+ + S_z^2 + \hbar S_z)\frac{1}{\sqrt{2}}(|1s\beta 2s\alpha| - |1s\alpha 2s\beta|) \\
&= S_-\,S_+\frac{1}{\sqrt{2}}(|1s\beta 2s\alpha| - |1s\alpha 2s\beta|) \\
&= \frac{1}{\sqrt{2}}\left(S_-(S_+(1) + S_+(2))|1s\beta 2s\alpha| - S_-(S_+(1) + S_+(2))|1s\alpha 2s\beta|\right)
\end{aligned}
$$

$$= \frac{1}{\sqrt{2}}\left(S_{-}\hbar|1s\alpha 2s\alpha| - S_{-}\hbar|1s\alpha 2s\alpha|\right)$$

$$= 0\hbar\frac{1}{\sqrt{2}}\left((S_{-}(1) + S_{-}(2))|1s\alpha 2s\alpha|\right)$$

$$= 0\hbar\frac{1}{\sqrt{2}}(\hbar|1s\beta 2s\alpha| - \hbar|1s\alpha 2s\beta|)$$

$$= 0\hbar^2\frac{1}{\sqrt{2}}(|1s\beta 2s\alpha| - |1s\alpha 2s\beta|)$$

3. a. Once the spatial symmetry has been determined by multiplication of the irreducible representations, the spin coupling is identical to exercise 2 and gives the result:

$$\frac{1}{\sqrt{2}}\left(|3a_1\alpha 1b_1\beta| - |3a_1\beta 1b_1\alpha|\right)$$

b. There are three states here (again analogous to exercise 2):

i. $|3a_1\alpha 1b_1\alpha|$,

ii. $\frac{1}{\sqrt{2}}\left(|3a1\alpha 1b_1\beta| + |3a_1\beta 1b_1\alpha|\right)$, and

iii. $|3a_1\beta 1b_1\beta|$

c. $|3a_1\alpha 3a_1\beta|$

4. As shown in review exercise 1c, for two equivalent π electrons one obtains six states:

$^1\Delta(M_L = 2)$; one state ($M_S = 0$),
$^1\Delta(M_L = -2)$; one state ($M_S = 0$),
$^1\Sigma(M_L = 0)$; one state ($M_S = 0$), and
$^3\Sigma(M_L = 0)$; three states ($M_S = 1, 0$, and -1).

By inspecting the "box" in review exercise 1c, it should be fairly straightforward to write down the wavefunctions for each of these:

$^1\Delta(M_L = 2)$; $|\pi_1\alpha\pi_1\beta|$
$^1\Delta(M_L = -2)$; $|\pi_{-1}\alpha\pi_{-1}\beta|$
$^1\Sigma(M_L = 0)$; $\frac{1}{\sqrt{2}}\left(|\pi_1\beta\pi_{-1}\alpha| - |\pi_1\alpha\pi_{-1}\beta|\right)$
$^3\Sigma(M_L = 0, M_S = 1)$; $|\pi_1\alpha\pi_{-1}\alpha|$
$^3\Sigma(M_L = 0, M_S = 0)$; $\frac{1}{\sqrt{2}}\left(|\pi_1\beta\pi_{-1}\alpha| + |\pi_1\alpha\pi_{-1}\beta|\right)$
$^3\Sigma(M_L = 0, M_S = -1)$; $|\pi_1\beta\pi_{-1}\beta|$

5. We can conveniently couple another s electron to the states generated from the $1s^1 2s^1$ configuration in exercise 2:

$^3S(L = 0, S = 1)$ with $3s^1(L = 0, S = 1/2)$ giving:
$L = 0$, $S = \frac{3}{2}, \frac{1}{2}$; ^4S (4 states) and ^2S (2 states).
$^1S(L = 0, S = 0)$ with $3s^1(L = 0, S = 1/2)$ giving:
$L = 0$, $S = \frac{1}{2}$; ^2S (2 states).

Constructing a "box" for this case would yield:

M_S \ M_L	0
$\frac{3}{2}$	$\|1s\alpha 2s\alpha 3s\alpha\|$
$\frac{1}{2}$	$\|1s\alpha 2s\alpha 3s\alpha\|$, $\|1s\beta 2s\beta 3s\beta\|$, $\|1s\alpha 2s\alpha 3s\alpha\|$

One can immediately identify the wavefunctions for two of the quartets (they are single entries):

$$^4S(S=\frac{3}{2}, M_S=\frac{3}{2}): |1s\alpha 2s\alpha 3s\alpha|$$
$$^4S(S=\frac{3}{2}, M_S=-\frac{3}{2}): |1s\beta 2s\beta 3s\beta|$$

Applying

$$S_- \text{ to } {}^4S(S=\frac{3}{2}, M_S=\frac{3}{2}) \text{ yields:}$$
$$S_-\, {}^4S(S=\frac{3}{2}, M_S=\frac{3}{2}) = \hbar\sqrt{\frac{3}{2}(\frac{3}{2}+1)-\frac{3}{2}(\frac{3}{2}-1)}\;\; {}^4S(S=\frac{3}{2}, M_S=\frac{1}{2})$$
$$= \hbar\sqrt{3}\;\; {}^4S(S=\frac{3}{2}, M_S=\frac{1}{2})$$
$$S_-|1s\alpha 2s\alpha 3s\alpha| = \hbar(|1s\beta 2s\alpha 3s\alpha| + |1s\alpha 2s\beta 3s\alpha| + |1s\alpha 2s\alpha 3s\beta|)$$

So,

$$^4S(S=\frac{3}{2}, M_S=\frac{1}{2}) = \frac{1}{\sqrt{3}}(|1s\beta 2s\alpha 3s\alpha| + |1s\alpha 2s\beta 3s\alpha| + |1s\alpha 2s\alpha 3s\beta|)$$

Applying

$$S_+ \text{ to } {}^4S\left(S=\frac{3}{2}, M_S=-\frac{3}{2}\right) \text{ yields:}$$
$$S_+\, {}^4S\left(S=\frac{3}{2}, M_S=-\frac{3}{2}\right) = \hbar\sqrt{\frac{3}{2}\left(\frac{3}{2}+1\right)-\frac{3}{2}\left(-\frac{3}{2}+1\right)}\;\; {}^4S\left(S=\frac{3}{2}, M_S=-\frac{1}{2}\right)$$
$$= \hbar\sqrt{3}\;{}^4S\left(S=\frac{3}{2}, M_S=-\frac{1}{2}\right)$$
$$S_+|1s\beta 2s\beta 3s\beta| = \hbar(|1s\alpha 2s\beta 3s\beta| + |1s\beta 2s\alpha 3s\beta| + |1s\beta 2s\beta 3s\alpha|)$$

So,

$$^4S\left(S=\frac{3}{2}, M_S=-\frac{1}{2}\right) = \frac{1}{\sqrt{3}}(|1s\alpha 2s\beta 3s\beta| + |1s\beta 2s\alpha 3s\beta| + |1s\beta 2s\beta 3s\alpha|)$$

It only remains to construct the doublet states which are orthogonal to these quartet states. Recall that the orthogonal combinations for systems having three equal components (for example when symmetry adapting the $3sp^2$ hybrids in C_{2v} or D_{3h} symmetry) give results of + + +, +2 − −, and 0 + −. Notice that the quartets are the + + + combinations and therefore the doublets can be recognized as:

$$^2S\left(S=\frac{1}{2}, M_S=\frac{1}{2}\right)=\frac{1}{\sqrt{6}}(|1s\beta 2s\alpha 3s\alpha| + |1s\alpha 2s\beta 3s\alpha| - 2|1s\alpha 2s\alpha 3s\beta|)$$

$$^2S\left(S=\frac{1}{2}, M_S=\frac{1}{2}\right)=\frac{1}{\sqrt{2}}(|1s\beta 2s\alpha 3s\alpha| - |1s\alpha 2s\beta 3s\alpha| + 0|1s\alpha 2s\alpha 3s\beta|)$$

$$^2S\left(S=\frac{1}{2}, M_S=-\frac{1}{2}\right)=\frac{1}{\sqrt{6}}(|1s\alpha 2s\beta 3s\beta| + |1s\beta 2s\alpha 3s\beta| - 2|1s\beta 2s\beta 3s\alpha|)$$

$$^2S\left(S=\frac{1}{2}, M_S=-\frac{1}{2}\right)=\frac{1}{\sqrt{3}}(|1s\alpha 2s\beta 3s\beta| - |1s\beta 2s\alpha 3s\beta| + 0|1s\beta 2s\beta 3s\alpha|)$$

6. As illustrated in this chapter a p^2 configuration (two equivalent p electrons) gives rise to the term symbols: ^3P, ^1D, and ^1S. Coupling an additional electron ($3d^1$) to this p^2 configuration will give the desired $1s^22s^22p^23d^1$ term symbols:

$^3P(L=1, S=1)$ with $^2D\left(L=2, S=\frac{1}{2}\right)$ generates;

$L=3,2,1$, and $S=\frac{3}{2},\frac{1}{2}$ with term symbols ^4F, ^2F, ^4D, ^2D, ^4P, and ^2P,

$^1D(L=2, S=0)$ with $^2D\left(L=2, S=\frac{1}{2}\right)$ generates;

$L=4,3,2,1,0$, and $S=\frac{1}{2}$ with term symbols ^2G, ^2F, ^2D, ^2P, and ^2S,

$^1S(L=0, S=0)$ with $^2D\left(L=2, S=\frac{1}{2}\right)$ generates;

$L=2$ and $S=\frac{1}{2}$ with term symbol ^2D.

7. The notation used for the Slater-Condon rules will be the same as used in the text:

zero (spin orbital) difference;

$$\left\langle |F+G| \right\rangle = \sum_i \left\langle \phi_i|f|\phi_i \right\rangle + \sum_{i>j}\left(\left\langle \phi_i\phi_j|g|\phi_i\phi_j \right\rangle - \left\langle \phi_i\phi_j|g|\phi_j\phi_i \right\rangle\right)$$

$$= \sum_i f_{ii} + \sum_{i>j}\left(g_{ijij} - g_{ijji}\right)$$

one (spin orbital) difference ($\phi_p \neq \phi_{p'}$) ;

$$\left\langle |F+G| \right\rangle = \left\langle \phi_p|f|\phi_{p'} \right\rangle + \sum_{j\neq p;p'}\left(\left\langle \phi_p\phi_j|g|\phi_{p'}\phi_j \right\rangle - \left\langle \phi_p\phi_j|g|\phi_j\phi_{p'} \right\rangle\right)$$

$$= f_{pp'} + \sum_{j\neq p;p'}\left(g_{pjp'j} - g_{pjjp'}\right)$$

two (spin orbital) differences ($\phi_p \neq \phi_{p'}$ and $\phi_q \neq \phi_{q'}$) ;

$$\left\langle |F+G| \right\rangle = \left\langle \phi_p\phi_q|g|\phi_{p'}\phi_{q'} \right\rangle - \left\langle \phi_p\phi_q|g|\phi_{q'}\phi_{p'} \right\rangle$$
$$= g_{pqp'q'} - g_{pqq'p'}$$

three or more (spin orbital) differences;

$$\left\langle |F+G| \right\rangle = 0$$

a. $^3P(M_L = 1, M_S = 1) = |p_1\alpha p_0\alpha|$

$$\left\langle |p_1\alpha p_0\alpha|H|p_1\alpha p_0\alpha| \right\rangle = \left\langle |10|H|10| \right\rangle$$

(note that the notation used here has the spin integrated out and that the notation referring to the p orbital has been dropped since all the orbitals of interest are p orbitals). Using the Slater-Condon rule a above (SCa):

$$\left\langle |10|H|10| \right\rangle = f_{11} + f_{00} + g_{1010} - g_{1001}$$

b. $^3P(M_L = 0, M_S = 0) = \frac{1}{\sqrt{2}}\left(|p_1\alpha p_{-1}\beta| + |p_1\beta p_{-1}\alpha|\right)$

$$\left\langle {}^3P(M_L = 0, M_S = 0)|H|{}^3P(M_L = 0, M_S = 0) \right\rangle$$

$$= \frac{1}{2}\left\langle |p_1\alpha p_{-1}\beta|H|p_1\alpha p_{-1}\beta| \right\rangle + \left\langle |p_1\alpha p_{-1}\beta|H|p_1\beta p_{-1}\alpha| \right\rangle$$

$$+ \left\langle |p_1\beta p_{-1}\alpha|H|p_1\alpha p_{-1}\beta| \right\rangle + \left\langle |p_1\beta p_{-1}\alpha|H|p_1\beta p_{-1}\alpha| \right\rangle$$

Evaluating each matrix element gives:

$$\left\langle |p_1\alpha p_{-1}\beta|H|p_1\alpha p_{-1}\beta| \right\rangle = f_{1\alpha 1\alpha} + f_{-1\beta-1\beta} + g_{1\alpha-1\beta 1\alpha-1\beta} - g_{1\alpha-1\beta-1\beta 1\alpha} \text{ (SCa)}$$

$$= f_{11} + f_{-1-1} + g_{1-11-1} - 0$$

$$\left\langle |p_1\alpha p_{-1}\beta|H|p_1\beta p_{-1}\alpha| \right\rangle = g_{1\alpha-1\beta 1\beta-1\alpha} - g_{1\alpha-1\beta-1\alpha 1\beta} \text{ (SCc)}$$

$$= 0 - g_{1-1-11}$$

$$\left\langle |p_1\beta p_{-1}\alpha|H|p_1\alpha p_{-1}\beta| \right\rangle = g_{1\beta-1\alpha 1\alpha-1\beta} - g_{1\beta-1\alpha-1\beta 1\alpha} \text{ (SCc)}$$

$$= 0 - g_{1-1-11}$$

$$\left\langle |p_1\beta p_{-1}\alpha|H|p_1\beta p_{-1}\alpha| \right\rangle = f_{1\beta 1\beta} + f_{-1\alpha-1\alpha} + g_{1\beta-1\alpha 1\beta-1\alpha} - g_{1\beta-1\alpha-1\alpha 1\beta} \text{ (SCa)}$$

$$= f_{11} + f_{-1-1} + g_{1-11-1} - 0$$

Substitution of these expressions give:

$$\left\langle {}^3P(M_L=0,M_S=0)|H|{}^3P(M_L=0,M_S=0)\right\rangle$$
$$=1/2(f_{11}+f_{-1-1}+g_{1-11-1}-g_{1-1-11}-g_{1-1-11}+f_{11}+f_{-1-1}+g_{1-11-1})$$
$$=f_{11}+f_{-1-1}+g_{1-11-1}-g_{1-1-11}$$

c. ${}^1S(M_L=0,M_S=0);\frac{1}{\sqrt{3}}(|p_0\alpha p_0\beta| - |p_1\alpha p_{-1}\beta| - |p_{-1}\alpha p_1\beta|)$

$$\left\langle {}^1S(M_L=0,M_S=0)|H|{}^1S(M_L=0,M_S=0)\right\rangle$$

$$=\frac{1}{3}\Big(\left\langle |p_0\alpha p_0\beta|H|p_0\alpha p_0\beta| \right\rangle - \left\langle |p_0\alpha p_0\beta|H|p_1\alpha p_{-1}\beta| \right\rangle$$

$$-\left\langle |p_0\alpha p_0\beta|H|p_{-1}\alpha p_1\beta| \right\rangle - \left\langle |p_1\alpha p_{-1}\beta|H|p_0\alpha p_0\beta| \right\rangle$$

$$+\left\langle |p_1\alpha p_{-1}\beta|H|p_1\alpha p_{-1}\beta| \right\rangle + \left\langle |p_1\alpha p_{-1}\beta|H|p_{-1}\alpha p_1\beta| \right\rangle$$

$$-\left\langle |p_{-1}\alpha p_1\beta|H|p_0\alpha p_0\beta| \right\rangle + \left\langle |p_{-1}\alpha p_1\beta|H|p_1\alpha p_{-1}\beta| \right\rangle$$

$$+\left\langle |p_{-1}\alpha p_1\beta|H|p_{-1}\alpha p_1\beta| \right\rangle\Big)$$

Evaluating each matrix element gives:

$$\left\langle |p_0\alpha p_0\beta|H|p_0\alpha p_0\beta| \right\rangle = f_{0\alpha0\alpha}+f_{0\beta0\beta}+g_{0\alpha0\beta0\alpha0\beta}-g_{0\alpha0\beta0\beta0\alpha} \text{ (SCa)}$$

$$=f_{00}+f_{00}+g_{0000}-0$$

$$\left\langle |p_0\alpha p_0\beta|H|p_1\alpha p_{-1}\beta| \right\rangle = \left\langle |p_1\alpha p_{-1}\beta|H|p_0\alpha p_0\beta| \right\rangle$$

$$=g_{0\alpha0\beta1\alpha-1\beta}-g_{0\alpha0\beta-1\beta1\alpha} \text{ (SCc)}$$
$$=g_{001-1}-0$$

$$\left\langle |p_0\alpha p_0\beta|H|p_{-1}\alpha p_1\beta| \right\rangle = \left\langle |p_{-1}\alpha p_1\beta|H|p_0\alpha p_0\beta| \right\rangle$$

$$=g_{0\alpha0\beta-1\alpha1\beta}-g_{0\alpha0\beta1\beta-1\alpha} \text{ (SCc)}$$
$$=g_{00-11}-0$$

$$\left\langle |p_1\alpha p_{-1}\beta|H|p_1\alpha p_{-1}\beta| \right\rangle = f_{1\alpha1\alpha}+f_{-1\beta-1\beta}+g_{1\alpha-1\beta1\alpha-1\beta}-g_{1\alpha-1\beta-1\beta1\alpha} \text{ (SCa)}$$

$$=f_{11}+f_{-1-1}+g_{1-11-1}-0$$

$$\left\langle |p_1\alpha p_{-1}\beta|H|p_{-1}\alpha p_1\beta| \right\rangle = \left\langle |p_{-1}\alpha p_1\beta|H|p_1\alpha p_{-1}\beta| \right\rangle$$

$$=g_{1\alpha-1\beta-1\alpha1\beta}-g_{1\alpha-1\beta1\beta-1\alpha} \text{ (SCc)}$$
$$=g_{1-1-11}-0$$

$$\left\langle |p_{-1}\alpha p_1\beta|H|p_{-1}\alpha p_1\beta| \right\rangle = f_{-1\alpha-1\alpha} + f_{1\beta1\beta} + g_{-1\alpha1\beta-1\alpha1\beta} - g_{-1\alpha1\beta1\beta-1\alpha} \text{ (SCa)}$$
$$= f_{-1-1} + f_{11} + g_{-11-11} - 0$$

Substitution of these expressions give:

$$\left\langle {}^1S(M_L = 0, M_S = 0)|H|{}^1S(M_L = 0, M_S = 0) \right\rangle$$
$$= \frac{1}{3}(f_{00} + f_{00} + g_{0000} - g_{001-1} - g_{00-11} - g_{001-1} + f_{11} + f_{-1-1}$$
$$+ g_{1-11-1} + g_{1-1-11} - g_{00-11} + g_{1-1-11} + f_{-1-1} + f_{11} + g_{-11-11})$$
$$= \frac{1}{3}(2f_{00} + 2f_{11} + 2f_{-1-1} + g_{0000} - 4g_{001-1} + 2g_{1-11-1} + 2g_{1-1-11})$$

d. ${}^1D(M_L = 0, M_S = 0) = \frac{1}{\sqrt{6}}\left(2|p_0\alpha p_0\beta| + |p_1\alpha p_{-1}\beta| + |p_{-1}\alpha p_1\beta|\right)$
Evaluating

$$\left\langle {}^1D(M_L = 0, M_S = 0)|H|{}^1D(M_L = 0, M_S = 0) \right\rangle$$

we note that all the Slater-Condon matrix elements generated are the same as those evaluated in part c. (the signs for the wavefunction components and the multiplicative factor of two for one of the components, however, are different).

$$\left\langle {}^1D(M_L = 0, M_S = 0)|H|{}^1D(M_L = 0, M_S = 0) \right\rangle$$
$$= \frac{1}{6}(4f_{00} + 4f_{00} + 4g_{0000} + 2g_{001-1} + 2g_{00-11} + 2g_{001-1} + f_{11}$$
$$+ f_{-1-1} + g_{1-11-1} + g_{1-1-11} + 2g_{00-11} + g_{1-1-11} + f_{-1-1} + f_{11} + g_{-11-11})$$
$$= 1/6(8f_{00} + 2f_{11} + 2f_{-1-1} + 4g_{0000} + 8g_{001-1} + 2g_{1-11-1} + 2g_{1-1-11})$$

8. a. ${}^1\Delta(M_L = 2, M_S = 0) = |\pi_1\alpha\pi_1\beta|$

$$\left\langle {}^1\Delta(M_L = 2, M_S = 0)|H|{}^1\Delta(M_L = 2, M_S = 0) \right\rangle$$

$$= \left\langle |\pi_1\alpha\pi_1\beta|H|\pi_1\alpha\pi_1\beta| \right\rangle$$

$$= f_{1\alpha1\alpha} + f_{1\beta1\beta} + g_{1\alpha1\beta1\alpha1\beta} - g_{1\alpha1\beta1\beta1\alpha} \text{ (SCa)}$$
$$= f_{11} + f_{11} + g_{1111} - 0$$
$$= 2f_{11} + g_{1111}$$

b. ${}^1\Sigma(M_L = 0, M_S = 0) = \frac{1}{\sqrt{2}}\left(|\pi_1\alpha\pi_{-1}\beta| - |\pi_1\beta\pi_{-1}\alpha|\right)$

$$\left\langle {}^3\Sigma(M_L = 0, M_S = 0)|H|{}^3\Sigma(M_L = 0, M_S = 0) \right\rangle$$

$$= \frac{1}{2}\left(\left\langle |\pi_1\alpha\pi_{-1}\beta|H|\pi_1\alpha\pi_{-1}\beta| \right\rangle - \left\langle |\pi_1\alpha\pi_{-1}\beta|H|\pi_1\beta\pi_{-1}\alpha| \right\rangle\right.$$

$$\left. - \left\langle |\pi_1\beta\pi_{-1}\alpha|H|\pi_1\alpha\pi_{-1}\beta| \right\rangle + \left\langle |\pi_1\beta\pi_{-1}\alpha|H|\pi_1\beta\pi_{-1}\alpha| \right\rangle\right)$$

Evaluating each matrix element gives:

$$\left\langle |\pi_1\alpha\pi_{-1}\beta|H|\pi_1\alpha\pi_{-1}\beta| \right\rangle = f_{1\alpha1\alpha} + f_{-1\beta-1\beta} + g_{1\alpha-1\beta1\alpha-1\beta} - g_{1\alpha-1\beta-1\beta1\alpha} \text{ (SCa)}$$

$$= f_{11} + f_{-1-1} + g_{1-11-1} - 0$$

$$\left\langle |\pi_1\alpha\pi_{-1}\beta|H|\pi_1\beta\pi_{-1}\alpha| \right\rangle = g_{1\alpha-1\beta1\beta-1\alpha} - g_{1\alpha-1\beta-1\alpha1\beta} \text{ (SCc)}$$

$$= 0 - g_{1-1-11}$$

$$\left\langle |\pi_1\beta\pi_{-1}\alpha|H|\pi_1\alpha\pi_{-1}\beta| \right\rangle = g_{1\beta-1\alpha1\alpha-1\beta} - g_{1\beta-1\alpha-1\beta1\alpha} \text{ (SCc)}$$

$$= 0 - g_{1-1-11}$$

$$\left\langle |\pi_1\beta\pi_{-1}\alpha|H|\pi_1\beta\pi_{-1}\alpha| \right\rangle = f_{1\beta1\beta} + f_{-1\alpha-1\alpha} + g_{1\beta-1\alpha1\beta-1\alpha} - g_{1\beta-1\alpha-1\alpha1\beta} \text{ (SCa)}$$

$$= f_{11} + f_{-1-1} + g_{1-11-1} - 0$$

Substitution of these expressions gives:

$$\left\langle {}^3\Sigma(M_L = 0, M_S = 0)|H|{}^3\Sigma(M_L = 0, M_S = 0) \right\rangle$$

$$= \frac{1}{2}(f_{11} + f_{-1-1} + g_{1-11-1} + g_{1-1-11} + g_{1-1-11} + f_{11} + f_{-1-1} + g_{1-11-1})$$

$$= f_{11} + f_{-1-1} + g_{1-11-1} + g_{1-1-11}$$

c. $^3\Sigma(M_L = 0, M_S = 0) = \frac{1}{\sqrt{2}}\left(|\pi_1\alpha\pi_{-1}\beta| + |\pi_1\beta\pi_{-1}\alpha|\right)$

$$\left\langle {}^3\Sigma(M_L = 0, M_S = 0)|H|{}^3\Sigma(M_L = 0, M_S = 0) \right\rangle$$

$$= \frac{1}{2}\left(\left\langle |\pi_1\alpha\pi_{-1}\beta|H|\pi_1\alpha\pi_{-1}\beta| \right\rangle + \left\langle |\pi_1\alpha\pi_{-1}\beta|H|\pi_1\beta\pi_{-1}\alpha| \right\rangle \right.$$

$$\left. + \left\langle |\pi_1\beta\pi_{-1}\alpha|H|\pi_1\alpha\pi_{-1}\beta| \right\rangle + \left\langle |\pi_1\beta\pi_{-1}\alpha|H|\pi_1\beta\pi_{-1}\alpha| \right\rangle \right)$$

Evaluating each matrix element gives:

$$\left\langle |\pi_1\alpha\pi_{-1}\beta|H|\pi_1\alpha\pi_{-1}\beta| \right\rangle = f_{1\alpha1\alpha} + f_{-1\beta-1\beta} + g_{1\alpha-1\beta1\alpha-1\beta} - g_{1\alpha-1\beta-1\beta1\alpha} \text{ (SCa)}$$

$$= f_{11} + f_{-1-1} + g_{1-11-1} - 0$$

$$\left\langle |\pi_1\alpha\pi_{-1}\beta|H|\pi_1\beta\pi_{-1}\alpha| \right\rangle = g_{1\alpha-1\beta1\beta-1\alpha} - g_{1\alpha-1\beta-1\alpha1\beta} \text{ (SCc)}$$

$$= 0 - g_{1-1-11}$$

$$\left\langle |\pi_1\beta\pi_{-1}\alpha|H|\pi_1\alpha\pi_{-1}\beta| \right\rangle = g_{1\beta-1\alpha1\alpha-1\beta} - g_{1\beta-1\alpha-1\beta1\alpha} \text{ (SCc)}$$

$$= 0 - g_{1-1-11}$$

$$\left\langle |\pi_1\beta\pi_{-1}\alpha|H|\pi_1\beta\pi_{-1}\alpha| \right\rangle = f_{1\beta1\beta} + f_{-1\alpha-1\alpha} + g_{1\beta-1\alpha1\beta-1\alpha} - g_{1\beta-1\alpha-1\alpha1\beta} \text{ (SCa)}$$

$$= f_{11} + f_{-1-1} + g_{1-11-1} - 0$$

Substitution of these expressions give:

$$\left\langle {}^3\Sigma(M_L = 0, M_S = 0)|H|{}^3\Sigma(M_L = 0, M_S = 0) \right\rangle$$

$$= \frac{1}{2}(f_{11} + f_{-1-1} + g_{1-11-1} - g_{1-1-11} - g_{1-1-11} + f_{11} + f_{-1-1} + g_{1-11-1})$$
$$= f_{11} + f_{-1-1} + g_{1-11-1} - g_{1-1-11}$$

Problems

1. **a.** All the Slater determinants have in common the $|1s\alpha1s\beta2s\alpha2s\beta|$ "core" and hence this component will not be written out explicitly for each case.

$$
\begin{aligned}
{}^3P(M_L = 1, M_S = 1) &= |p_1\alpha p_0\alpha| \\
&= |\frac{1}{\sqrt{2}}(p_x + ip_y)\alpha(p_z)\alpha| \\
&= \frac{1}{\sqrt{2}}\left(|p_x\alpha p_z\alpha| + i|p_y\alpha p_z\alpha|\right) \\
{}^3P(M_L = 0, M_S = 1) &= |p_1\alpha p_{-1}\alpha| \\
&= |\frac{1}{\sqrt{2}}(p_x + ip_y)\alpha\frac{1}{\sqrt{2}}(p_x - ip_y)\alpha| \\
&= \frac{1}{2}\left(|p_x\alpha p_x\alpha| - i|p_x\alpha p_y\alpha| + i|p_y\alpha p_x\alpha| + |p_y\alpha p_y\alpha|\right) \\
&= \frac{1}{2}\left(0 - i|p_x\alpha p_y\alpha| - i|p_x\alpha p_y\alpha| + 0\right) \\
&= \frac{1}{2}\left(-2i|p_x\alpha p_y\alpha|\right) \\
&= -i|p_x\alpha p_y\alpha| \\
{}^3P(M_L = -1, M_S = 1) &= |p_{-1}\alpha p_0\alpha| \\
&= |\frac{1}{\sqrt{2}}(p_x - ip_y)\alpha(p_z)\alpha| \\
&= \frac{1}{\sqrt{2}}\left(|p_x\alpha p_z\alpha| - i|p_y\alpha p_z\alpha|\right)
\end{aligned}
$$

As you can see, the symmetries of each of these states cannot be labeled with a single irreducible representation of the C_{2v} point group. For example, $|p_x\alpha p_z\alpha|$ is $xz(B_1)$ and $|p_y\alpha p_z\alpha|$ is $yz(B_2)$ and hence the ${}^3P(M_L = 1, M_S = 1)$ state is a combination of B_1 and B_2 symmetries. But, the three ${}^3P(M_L, M_S = 1)$ functions are degenerate for the C atom and any combination of these three functions would also be degenerate. Therefore we can choose new combinations which can be labeled with "pure" C_{2v} point group labels.

$$
\begin{aligned}
{}^3P(xz, M_S = 1) &= |p_x\alpha p_z\alpha| \\
&= \frac{1}{\sqrt{2}}\left({}^3P(M_L = 1, M_S = 1) + {}^3P(M_L = -1, M_S = 1)\right) = {}^3B_1
\end{aligned}
$$

$$^{3}P(yx,M_S=1) = |p_y\alpha p_x\alpha|$$
$$= \frac{1}{i}\left(^{3}P(M_L=0,M_S=1)\right) = {}^{3}A_2$$

$$^{3}P(yz,M_S=1) = |p_y\alpha p_z\alpha|$$
$$= \frac{1}{i\sqrt{2}}\left(^{3}P(M_L=1,M_S=1) - {}^{3}P(M_L=-1,M_S=1)\right) = {}^{3}B_2$$

Now we can do likewise for the five degenerate ^{1}D states:

$$^{1}D(M_L=2,M_S=0) = |p_1\alpha p_1\beta|$$
$$= |\frac{1}{\sqrt{2}}(p_x+ip_y)\alpha\frac{1}{\sqrt{2}}(p_x+ip_y)\beta|$$
$$= \frac{1}{2}\left(|p_x\alpha p_x\beta| + i|p_x\alpha p_y\beta| + i|p_y\alpha p_x\beta| - |p_y\alpha p_y\beta|\right)$$

$$^{1}D(M_L=-2,M_S=0) = |p_{-1}\alpha p_{-1}\beta|$$
$$= |\frac{1}{\sqrt{2}}(p_x-ip_y)\alpha\frac{1}{\sqrt{2}}(p_x-ip_y)\beta|$$
$$= \frac{1}{2}\left(|p_x\alpha p_x\beta| - i|p_x\alpha p_y\beta| - i|p_y\alpha p_x\beta| - |p_y\alpha p_y\beta|\right)$$

$$^{1}D(M_L=1,M_S=0) = \frac{1}{\sqrt{2}}\left(|p_0\alpha p_1\beta| - |p_0\beta p_1\alpha|\right)$$
$$= \frac{1}{\sqrt{2}}\left(|(p_z)\alpha\frac{1}{\sqrt{2}}(p_x+ip_y)\beta| - |(p_z)\beta\frac{1}{\sqrt{2}}(p_x+ip_y)\alpha|\right)$$
$$= \frac{1}{2}\left(|p_z\alpha p_x\beta| + i|p_z\alpha p_y\beta| - |p_z\beta p_x\alpha| - i|p_z\beta p_y\alpha|\right)$$

$$^{1}D(M_L=-1,M_S=0) = \frac{1}{\sqrt{2}}\left(|p_0\alpha p_{-1}\beta| - |p_0\beta p_{-1}\alpha|\right)$$
$$= \frac{1}{\sqrt{2}}\left(|(p_z)\alpha\frac{1}{\sqrt{2}}(p_x-ip_y)\beta| - |(p_z)\beta\frac{1}{\sqrt{2}}(p_x-ip_y)\alpha|\right)$$
$$= \frac{1}{2}\left(|p_z\alpha p_x\beta| - i|p_z\alpha p_y\beta| - |p_z\beta p_x\alpha| + i|p_z\beta p_y\alpha|\right)$$

$$^{1}D(M_L=0,M_S=0) = \frac{1}{\sqrt{6}}\left(2|p_0\alpha p_0\beta| + |p_1\alpha p_{-1}\beta| + |p_{-1}\alpha p_1\beta|\right)$$
$$= \frac{1}{\sqrt{6}}(2|p_z\alpha p_z\beta| + |\frac{1}{\sqrt{2}}(p_x+ip_y)\alpha\frac{1}{\sqrt{2}}(p_x-ip_y)\beta|$$
$$+|\frac{1}{\sqrt{2}}(p_x-ip_y)\alpha\frac{1}{\sqrt{2}}(p_x+ip_y)\beta|)$$
$$= \frac{1}{\sqrt{6}}\Big(\; 2|p_z\alpha p_z\beta|$$
$$+\frac{1}{2}\left(|p_x\alpha p_x\beta| - i|p_x\alpha p_y\beta| + i|p_y\alpha p_x\beta| + |p_y\alpha p_y\beta|\right)$$

$$+\frac{1}{2}\left(|p_x\alpha p_x\beta| + i|p_x\alpha p_y\beta| - i|p_y\alpha p_x\beta| + |p_y\alpha p_y\beta|\right)\Big)$$

$$= \frac{1}{\sqrt{6}}\left(2|p_z\alpha p_z\beta| + |p_x\alpha p_x\beta| + |p_y\alpha p_y\beta|\right)$$

Analogous to the three ^3P states we can also choose combinations of the five degenerate ^1D states which can be labeled with "pure" C_{2v} point group labels:

$$^1D(xx - yy, M_S = 0) = |p_x\alpha p_x\beta| - |p_y\alpha p_y\beta|$$
$$= \left({}^1D(M_L = 2, M_S = 0) + {}^1D(M_L = -2, M_S = 0)\right) = {}^1A_1$$
$$^1D(yx, M_S = 0) = |p_x\alpha p_y\beta| + |p_y\alpha p_x\beta|$$
$$= \frac{1}{i}\left({}^1D(M_L = 2, M_S = 0) - {}^1D(M_L = -2, M_S = 0)\right) = {}^1A_2$$
$$^1D(zx, M_S = 0) = |p_z\alpha p_x\beta| - |p_z\beta p_x\alpha|$$
$$= \left({}^1D(M_L = 1, M_S = 0) + {}^1D(M_L = -1, M_S = 0)\right) = {}^1B_1$$
$$^1D(zy, M_S = 0) = |p_z\alpha p_y\beta| - |p_z\beta p_y\alpha|$$
$$= \frac{1}{i}\left({}^1D(M_L = 1, M_S = 0) - {}^1D(M_L = -1, M_S = 0)\right) = {}^1B_2$$
$$^1D(2zz + xx + yy, M_S = 0) = \frac{1}{\sqrt{6}}\left(2|p_z\alpha p_z\beta| + |p_x\alpha p_x\beta| + |p_y\alpha p_y\beta|\right))$$
$$= {}^1D(M_L = 0, M_S = 0) = {}^1A_1$$

The only state left is the ^1S:

$$^1S(M_L = 0, M_S = 0) = \frac{1}{\sqrt{3}}\left(|p_0\alpha p_0\beta| - |p_1\alpha p_{-1}\beta| - |p_{-1}\alpha p_1\beta|\right)$$
$$= \frac{1}{\sqrt{3}}(|p_z\alpha p_z\beta| - |\frac{1}{\sqrt{2}}(p_x + ip_y)\alpha\frac{1}{\sqrt{2}}(p_x - ip_y)\beta|$$
$$- |\frac{1}{\sqrt{2}}(p_x - ip_y)\alpha\frac{1}{\sqrt{2}}(p_x + ip_y)\beta|)$$
$$= \frac{1}{\sqrt{3}}\Big(|p_z\alpha p_z\beta|$$
$$- \frac{1}{2}\left(|p_x\alpha p_x\beta| - i|p_x\alpha p_y\beta| + i|p_y\alpha p_x\beta| + |p_y\alpha p_y\beta|\right)$$
$$- \frac{1}{2}\left(|p_x\alpha p_x\beta| + i|p_x\alpha p_y\beta| - i|p_y\alpha p_x\beta| + |p_y\alpha p_y\beta|\right)\Big)$$
$$= \frac{1}{\sqrt{3}}\left(|p_z\alpha p_z\beta| - |p_x\alpha p_x\beta| - |p_y\alpha p_y\beta|\right)$$

Each of the components of this state are A_1 and hence this state has A_1 symmetry.

b. Forming SALC-AOs from the C and H atomic orbitals would generate the following:

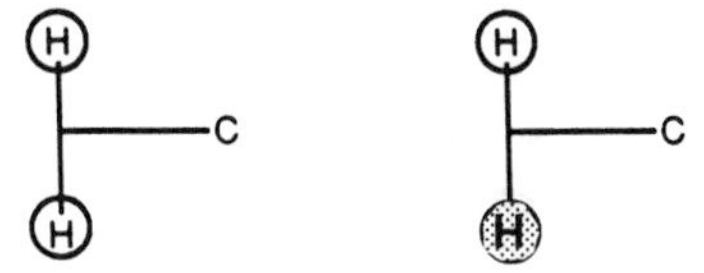

$H_{1s} + H_{1s} = \sigma_g = a_1$ $\quad$ $H_{1s} - H_{1s} = \sigma_u = b_2$

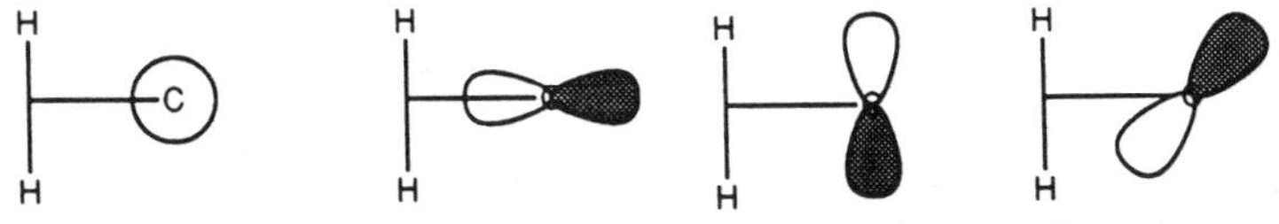

$C_{2s} = a_1$ $C_{2p_z} = a_1$ $C_{2p_y} = b_2$ $C_{2p_x} = b_1$

The bonding, nonbonding, and antibonding orbitals of CH_2 can be illustrated in the following manner:

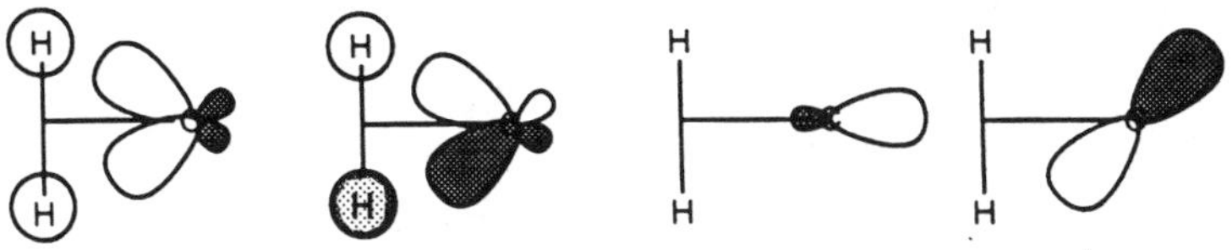

$\sigma = a_1$ $\sigma = b_2$ $n = a_1$ $p_\pi = b_1$

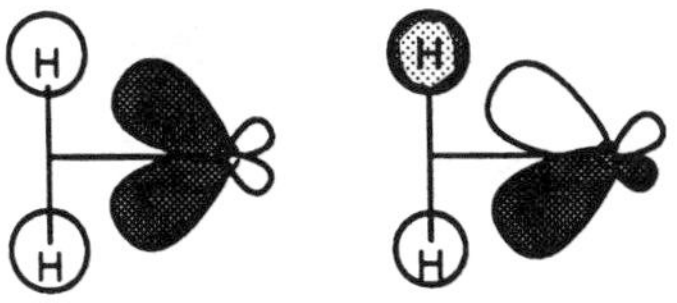

$\sigma^* = a_1$ $\sigma^* = b_2$

c.

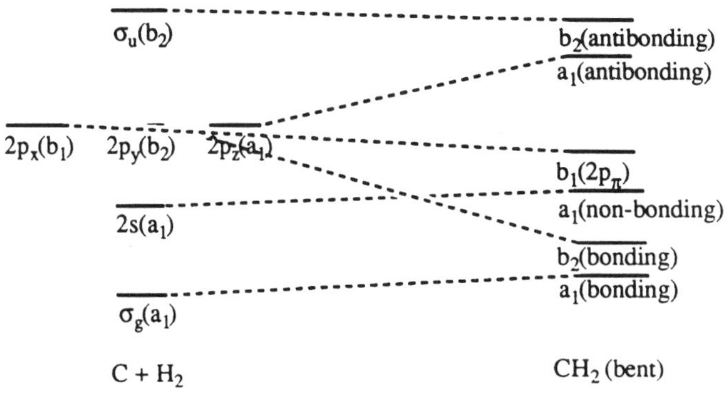

d.–e. It is necessary to determine how the wavefunctions found in part a correlate with states of the CH_2 molecule:

$$^3P(xz, M_S = 1); {}^3B_1 = \sigma_g^2 s^2 p_x p_z \longrightarrow \sigma^2 n^2 p_\pi \sigma^*$$
$$^3P(yx, M_S = 1); {}^3A_2 = \sigma_g^2 s^2 p_x p_y \longrightarrow \sigma^2 n^2 p_\pi \sigma$$
$$^3P(yz, M_S = 1); {}^3B_2 = \sigma_g^2 s^2 p_y p_z \longrightarrow \sigma^2 n^2 \sigma\sigma^*$$
$$^1D(xx - yy, M_S = 0); {}^1A_1 \longrightarrow \sigma^2 n^2 p_\pi^2 - \sigma^2 n^2 \sigma^2$$

$^1D(yx, M_S = 0); ^1A_2 \longrightarrow \sigma^2 n^2 \sigma p_\pi$
$^1D(zx, M_S = 0); ^1B_1 \longrightarrow \sigma^2 n^2 \sigma{*} p_\pi$
$^1D(zy, M_S = 0); ^1B_2 \longrightarrow \sigma^2 n^2 \sigma{*}\sigma$
$^1D(2zz + xx + yy, M_S = 0); ^1A_1 \longrightarrow 2\sigma^2 n^2 \sigma{*}^2 + \sigma^2 n^2 p_\pi^2 + \sigma^2 n^2 \sigma^2$

Note, the C + H_2 state to which the lowest $^1A_1(\sigma^2 n^2 \sigma^2)CH_2$ state decomposes would be $\sigma_g^2 s^2 p_y^2$. This state ($\sigma_g^2 s^2 p_y^2$) cannot be obtained by a simple combination of the ^1D states. In order to obtain pure $\sigma_g^2 s^2 p_y^2$ it is necessary to combine ^1S with ^1D. For example,

$$\sigma_g^2 s^2 p_y^2 = \frac{1}{6}\left(\sqrt{6}\,{}^1D(0,0) - 2\sqrt{3}\,{}^1S(0,0)\right) - \frac{1}{2}\left({}^1D(2,0) + {}^1D(-2,0)\right).$$

This indicates that a CCD must be drawn with a barrier near the ^1D asymptote to represent the fact that $^1A_1$ CH_2 correlates with a mixture of ^1D and ^1S carbon plus hydrogen. The C + H_2 state to which the lowest $^3B_1(\sigma^2 n \sigma^2 p_\pi)CH_2$ state decomposes would be $\sigma_g^2 s p_y^2 p_x$.

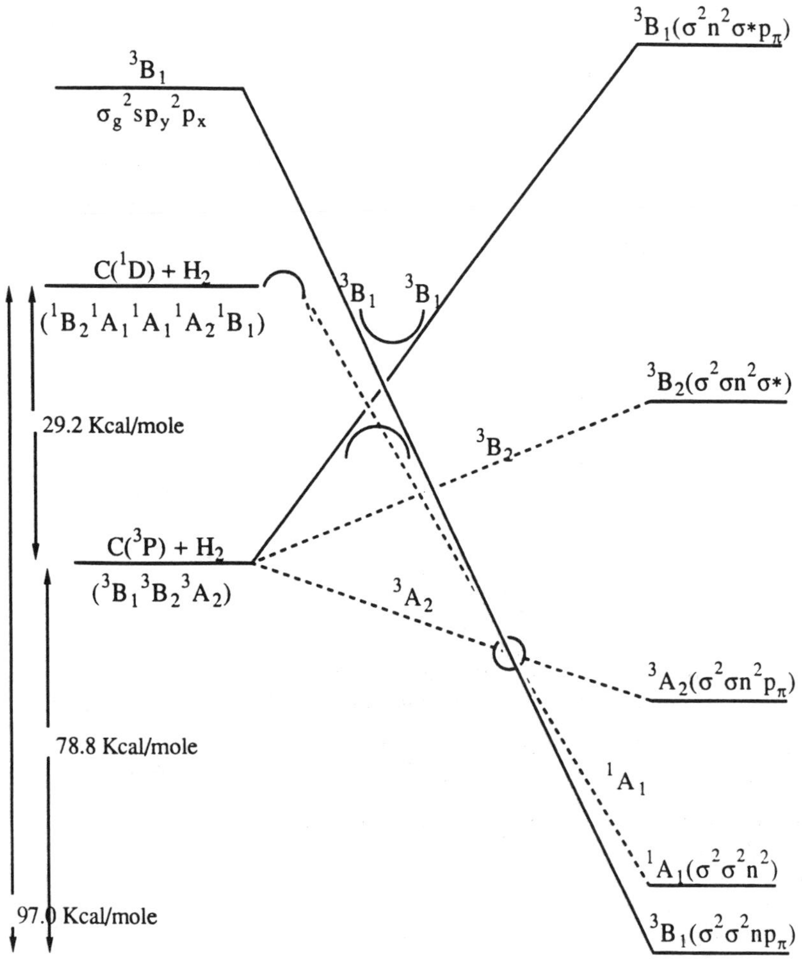

f. If you follow the $^3B_1$ component of the $C(^3P) + H_2$ (since it leads to the ground-state products) to $^3B_1$ CH_2 you must go over an approximately 20Kcal/mole barrier. Of course this path produces $^3B_1$ CH_2 product. Distortions away from C_{2v} symmetry, for example to C_s symmetry, would make the a_1 and b_2 orbitals identical in symmetry (a'). The b_1 orbitals would maintain their identity going to a'' symmetry. Thus $^3B_1$ and $^3A_2$ (both $^3A''$ in C_s symmetry and *odd* under reflection through the molecular plane) can mix. The system could thus follow the $^3A_2$ component of the $C(^3P) + H_2$ surface to the place (marked with a circle on the CCD) where it crosses the $^3B_1$ surface upon which it then moves and continues to products. As a result, the barrier would be lowered.

You can estimate when the barrier occurs (late or early) using thermodynamic information for the reaction (i.e. slopes and asymptotic energies). For example, an early barrier would be obtained for a reaction with the characteristics:

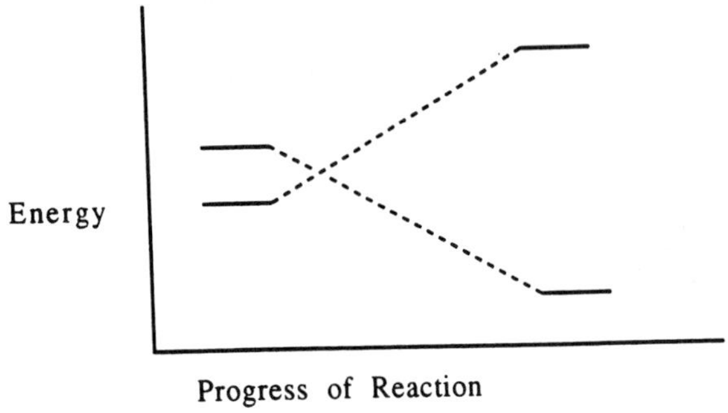

and a late barrier would be obtained for a reaction with the characteristics:

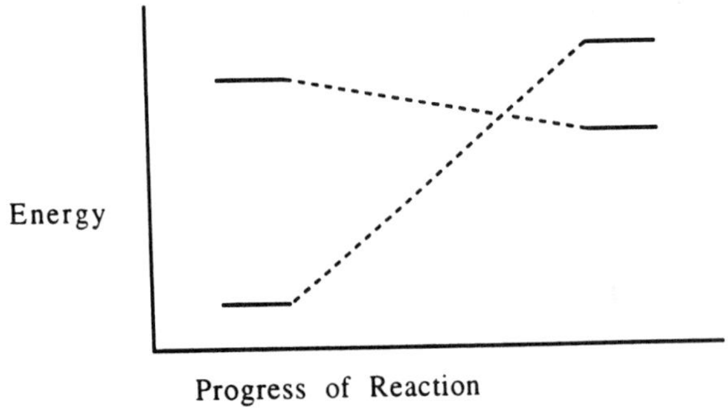

This relation between reaction endothermicity or exothermicity is known as the Hammond postulate. Note that the $C(^3P_1) + H_2 \longrightarrow CH_2$ reaction of interest here (see the CCD) has an early barrier.

g. The reaction $C(^1D) + H_2 \longrightarrow CH_2(^1A_1)$ should have no symmetry barrier (this can be recognized by following the $^1A_1(C(^1D) + H_2)$ reactants down to the $^1A_1(CH_2)$ products on the CCD).

2. This problem in many respects is analogous to problem 1. The $^3B_1$ surface certainly requires a two configuration CI wavefunction; the $\sigma^2\sigma^2np_x(\pi^2p_y^2sp_x)$ and the $\sigma^2n^2p_x\sigma*(\pi^2s^2p_xp_z)$. The $^1A_1$ surface could use the $\sigma^2\sigma^2n^2(\pi^2s^2p_y^2)$ only but once again there is no combination of ^1D determinants which gives purely this configuration $(\pi^2s^2p_y^2)$. Thus mixing of both ^1D and ^1S determinants are necessary to yield the required $\pi^2s^2p_y^2$ configuration. Hence even the $^1A_1$ surface would require a multiconfigurational wavefunction for adequate description.

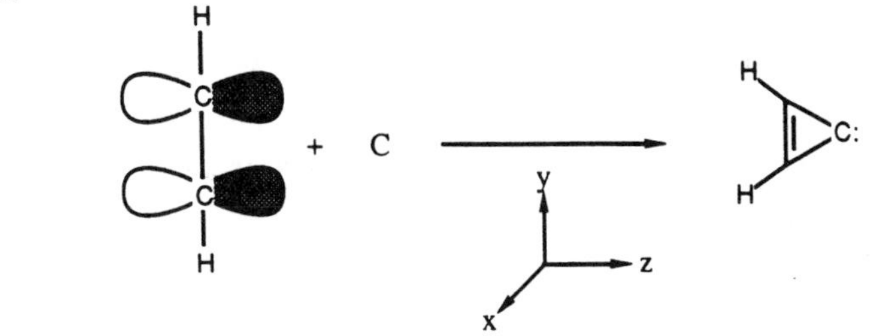

Orbital-correlation diagram for the reaction C_2H_2 + C -----> C_3H_2

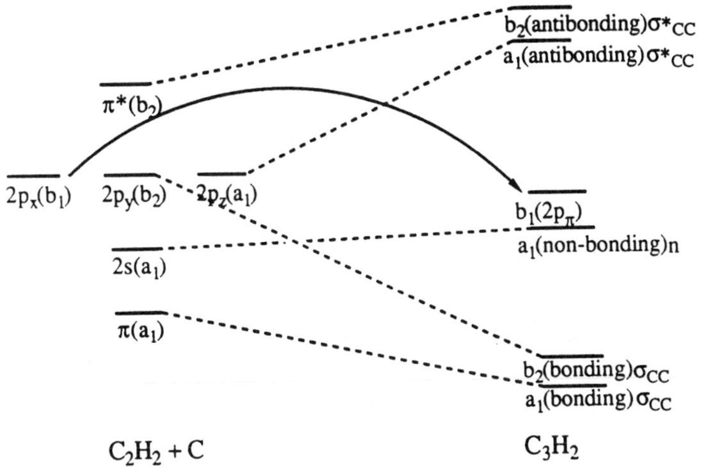

Configuration correlation diagram for the reaction C_2H_2 + C ——> C_3H_2.

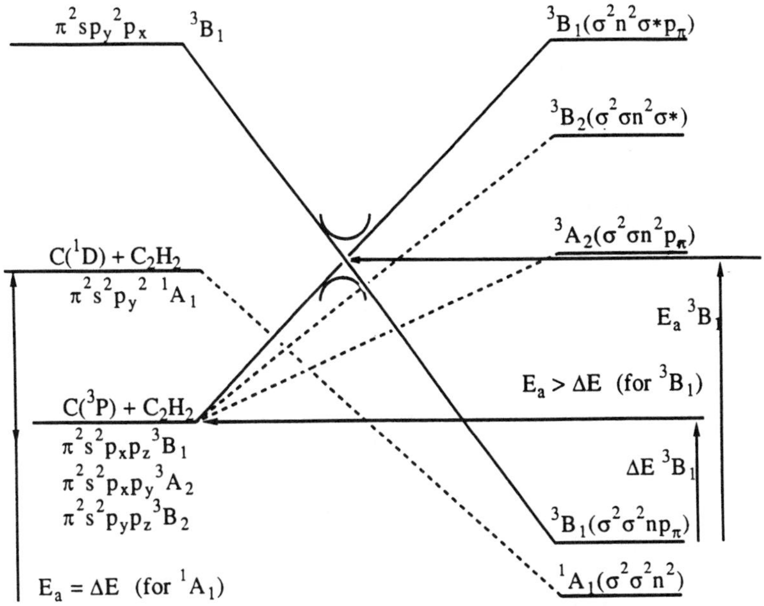

3. a. $$\left\langle \sigma_g | \sigma_g \right\rangle = (2+2S)^{-\frac{1}{2}}\left(1s_A + 1s_B\right)\left|(2+2S)^{-\frac{1}{2}}\left(1s_A + 1s_B\right)\right\rangle$$
$$= (2+2S)^{-1}\left(\left\langle 1s_A|1s_A\right\rangle + \left\langle 1s_A|1s_B\right\rangle + \left\langle 1s_B|1s_A\right\rangle + \left\langle 1s_B|1s_B\right\rangle\right)$$
$$= (0.285)((1.000) + (0.753) + (0.753) + (1.000))$$
$$= 0.999 \approx 1$$

$$\left\langle \sigma_g | \sigma_u \right\rangle = \left\langle (2+2S)^{-\frac{1}{2}}\left(1s_A + 1s_B\right)|(2-2S)^{-\frac{1}{2}}\left(1s_A - 1s_B\right)\right\rangle$$
$$= (2+2S)^{-\frac{1}{2}}(2-2S)^{-\frac{1}{2}}\left(\left\langle 1s_A|1s_A\right\rangle + \left\langle 1s_A|1s_B\right\rangle + \left\langle 1s_B|1s_A\right\rangle + \left\langle 1s_B|1s_B\right\rangle\right)$$
$$= (1.423)(0.534)((1.000) - (0.753) + (0.753) - (1.000))$$
$$= 0$$

$$\left\langle \sigma_u | \sigma_u \right\rangle = \left\langle (2-2S)^{-\frac{1}{2}}\left(1s_A - 1s_B\right)|(2-2S)^{-\frac{1}{2}}\left(1s_A - 1s_B\right)\right\rangle$$
$$= (2-2S)^{-1}\left(\left\langle 1s_A|1s_A\right\rangle - \left\langle 1s_A|1s_B\right\rangle - \left\langle 1s_B|1s_A\right\rangle + \left\langle 1s_B|1s_B\right\rangle\right)$$
$$= (2.024)((1.000) - (0.753) - (0.753) + (1.000))$$
$$= 1.000$$

b. $$\left\langle \sigma_g | h | \sigma_g \right\rangle = \left\langle (2+2S)^{-\frac{1}{2}}\left(1s_A + 1s_B\right)|h|(2+2S)^{-\frac{1}{2}}\left(1s_A + 1s_B\right)\right\rangle$$
$$= (2+2S)^{-1}\left(\left\langle 1s_A|h|1s_A\right\rangle + \left\langle 1s_A|h|1s_B\right\rangle + \left\langle 1s_B|h|1s_A\right\rangle + \left\langle 1s_B|h|1s_B\right\rangle\right)$$
$$= (0.285)((-1.110) + (-0.968) + (-0.968) + (-1.110))$$
$$= -1.184$$

$$\left\langle \sigma_u | h | \sigma_u \right\rangle = \left\langle (2-2S)^{-\frac{1}{2}}\left(1s_A - 1s_B\right)|h|(2-2S)^{-\frac{1}{2}}\left(1s_A - 1s_B\right)\right\rangle$$
$$= (2-2S)^{-1}\left(\left\langle 1s_A|h|1s_A\right\rangle - \left\langle 1s_A|h|1s_B\right\rangle - \left\langle 1s_B|h|1s_A\right\rangle + \left\langle 1s_B|h|1s_B\right\rangle\right)$$
$$= (2.024)((-1.110) + (0.968) + (0.968) + (-1.110))$$
$$= -0.575$$

$$\left\langle \sigma_g\sigma_g | h | \sigma_g\sigma_g \right\rangle \equiv \left\langle gg|gg \right\rangle = (2+2S)^{-1}(2+2S)^{-1}$$

$$\left\langle \left(1s_A + 1s_B\right)\left(1s_A + 1s_B\right)|\left(1s_A + 1s_B\right)\left(1s_A + 1s_B\right)\right\rangle$$

$$= (2+2S)^{-2}\left(\left\langle AA|AA\right\rangle + \left\langle AA|AB\right\rangle + \left\langle AA|BA\right\rangle + \left\langle AA|BB\right\rangle + \right.$$

$$\left\langle AB|AA\right\rangle + \left\langle AB|AB\right\rangle + \left\langle AB|BA\right\rangle + \left\langle AB|BB\right\rangle +$$

$$\left\langle BA|AA\right\rangle + \left\langle BA|AB\right\rangle + \left\langle BA|BA\right\rangle + \left\langle BA|BB\right\rangle +$$

$$\left.\left\langle BB|AA\right\rangle + \left\langle BB|AB\right\rangle + \left\langle BB|BA\right\rangle + \left\langle BB|BB\right\rangle\right)$$

$$= (0.081)\,((0.625) + (0.426) + (0.426) + (0.323) + (0.426) + (0.504) + (0.323) + (0.426) + (0.426) + (0.323) + (0.504) + (0.426) + (0.323) + (0.426) + (0.426) + (0.625)) = 0.564$$

$$\left\langle uu|uu \right\rangle = (2-2S)^{-1}(2-2S)^{-1}\left\langle \left(1s_A - 1s_B\right)\left(1s_A - 1s_B\right)|\left(1s_A - 1s_B\right)\left(1s_A - 1s_B\right)\right\rangle$$

$$= (2-2S)^{-2}\Big(\left\langle AA|AA\right\rangle - \left\langle AA|AB\right\rangle - \left\langle AA|BA\right\rangle + \left\langle AA|BB\right\rangle - \left\langle AB|AA\right\rangle + \left\langle AB|AB\right\rangle + \left\langle AB|BA\right\rangle - \left\langle AB|BB\right\rangle - \left\langle BA|AA\right\rangle + \left\langle BA|AB\right\rangle + \left\langle BA|BA\right\rangle - \left\langle BA|BB\right\rangle + \left\langle BB|AA\right\rangle - \left\langle BB|AB\right\rangle - \left\langle BB|BA\right\rangle + \left\langle BB|BB\right\rangle\Big)$$

$$= (4.100)\,((0.625) - (0.426) - (0.426) + (0.323) - (0.426) + (0.504) + (0.323) - (0.426) - (0.426) + (0.323) + (0.504) - (0.426) + (0.323) - (0.426) - (0.426) + (0.625)) = 0.582$$

$$\left\langle gg|uu \right\rangle = (2+2S)^{-1}(2-2S)^{-1}\left\langle \left(1s_A + 1s_B\right)\left(1s_A + 1s_B\right)|\left(1s_A - 1s_B\right)\left(1s_A - 1s_B\right)\right\rangle$$

$$= (2+2S)^{-1}(2-2S)^{-1}\Big(\left\langle AA|AA\right\rangle - \left\langle AA|AB\right\rangle - \left\langle AA|BA\right\rangle + \left\langle AA|BB\right\rangle + \left\langle AB|AA\right\rangle - \left\langle AB|AB\right\rangle - \left\langle AB|BA\right\rangle + \left\langle AB|BB\right\rangle + \left\langle BA|AA\right\rangle - \left\langle BA|AB\right\rangle - \left\langle BA|BA\right\rangle + \left\langle BA|BB\right\rangle + \left\langle BB|AA\right\rangle - \left\langle BB|AB\right\rangle - \left\langle BB|BA\right\rangle + \left\langle BB|BB\right\rangle\Big)$$

$$= (0.285)\,(2.024)\,((0.625) - (0.426) - (0.426) + (0.323) + (0.426) - (0.504) - (0.323) + (0.426) + (0.426) - (0.323) - (0.504) + (0.426) + (0.323) - (0.426) - (0.426) + (0.625)) = 0.140$$

$$\left\langle gu|gu \right\rangle = (2+2S)^{-1}(2-2S)^{-1}\left\langle \left(1s_A + 1s_B\right)\left(1s_A - 1s_B\right)|\left(1s_A + 1s_B\right)\left(1s_A - 1s_B\right)\right\rangle$$

$$= (2+2S)^{-1}(2-2S)^{-1}\Big(\left\langle AA|AA\right\rangle - \left\langle AA|AB\right\rangle + \left\langle AA|BA\right\rangle - \left\langle AA|BB\right\rangle -$$

$$\left\langle AB|AA \right\rangle + \left\langle AB|AB \right\rangle - \left\langle AB|BA \right\rangle + \left\langle AB|BB \right\rangle +$$

$$\left\langle BA|AA \right\rangle - \left\langle BA|AB \right\rangle + \left\langle BA|BA \right\rangle - \left\langle BA|BB \right\rangle -$$

$$\left\langle BB|AA \right\rangle + \left\langle BB|AB \right\rangle - \left\langle BB|BA \right\rangle + \left\langle BB|BB \right\rangle)$$

$$= (0.285)\ (2.024)\ ((0.625) - (0.426) + (0.426) - (0.323) -$$
$$(0.426) + (0.504) - (0.323) + (0.426) +$$
$$(0.426) - (0.323) + (0.504) - (0.426) -$$
$$(0.323) + (0.426) - (0.426) + (0.625))$$
$$= 0.557$$

Note, that

$$\left\langle gg|gu \right\rangle = \left\langle uu|ug \right\rangle = 0$$

from symmetry considerations, but this can be easily verified. For example,

$$\left\langle gg|gu \right\rangle = (2+2S)^{-\frac{1}{2}}(2-2S)^{-\frac{3}{2}}\left\langle \left(1s_A + 1s_B\right)\left(1s_A + 1s_B\right)|\left(1s_A + 1s_B\right)\left(1s_A - 1s_B\right)\right\rangle$$

$$= (2+2S)^{-\frac{1}{2}}(2-2S)^{-\frac{3}{2}}(\left\langle AA|AA \right\rangle - \left\langle AA|AB \right\rangle + \left\langle AA|BA \right\rangle - \left\langle AA|BB \right\rangle +$$

$$\left\langle AB|AA \right\rangle - \left\langle AB|AB \right\rangle + \left\langle AB|BA \right\rangle - \left\langle AB|BB \right\rangle +$$

$$\left\langle BA|AA \right\rangle - \left\langle BA|AB \right\rangle + \left\langle BA|BA \right\rangle - \left\langle BA|BB \right\rangle +$$

$$\left\langle BB|AA \right\rangle - \left\langle BB|AB \right\rangle + \left\langle BB|BA \right\rangle - \left\langle BB|BB \right\rangle)$$

$$= (0.534)\ (2.880)\ ((0.625) - (0.426) + (0.426) - (0.323) +$$
$$(0.426) - (0.504) + (0.323) - (0.426) +$$
$$(0.426) - (0.323) + (0.504) - (0.426) +$$
$$(0.323) - (0.426) + (0.426) - (0.625))$$
$$= 0.000$$

c. We can now set up the configuration interaction Hamiltonian matrix. The elements are evaluated by using the Slater-Condon rules as shown in the text.

$$H_{11} = \left\langle \sigma_g\alpha\sigma_g\beta|H|\sigma_g\alpha\sigma_g\beta \right\rangle$$

$$= 2f_{\sigma_g\sigma_g} + g_{\sigma_g\sigma_g\sigma_g\sigma_g}$$
$$= 2(-1.184) + 0.564 = -1.804$$

$$H_{21} = H_{12} = \left\langle \sigma_g\alpha\sigma_g\beta|H|\sigma_u\alpha\sigma_u\beta \right\rangle$$

$$= g_{\sigma_g\sigma_g\sigma_u\sigma_u}$$
$$= 0.140$$

$$H_{22} = \left\langle \sigma_u\alpha\sigma_u\beta|H|\sigma_u\alpha\sigma_u\beta \right\rangle$$

$$= 2f_{\sigma_u\sigma_u} + g_{\sigma_u\sigma_u\sigma_u\sigma_u}$$
$$= 2(-0.575) + 0.582 = -0.568$$

d. Solving this eigenvalue problem:

$$\begin{vmatrix} -1.804 - \varepsilon & 0.140 \\ 0.140 & -0.568 - \varepsilon \end{vmatrix} = 0$$

$$(-1.804 - \varepsilon)\,(-0.568 - \varepsilon) - (0.140)^2 = 0$$

$$1.025 + 1.804\varepsilon + 0.568\varepsilon + \varepsilon^2 - 0.0196 = 0$$

$$\varepsilon^2 + 2.372\varepsilon + 1.005 = 0$$

$$\varepsilon = \frac{-2.372 \pm \sqrt{(2.372)^2 - 4(1)(1.005)}}{(2)(1)}$$

$$= -1.186 \pm 0.634$$

$$= 1.820, \text{ and } -0.552.$$

Solving for the coefficients:

$$\begin{bmatrix} -1.804 - \varepsilon & 0.140 \\ 0.140 & -0.568 - \varepsilon \end{bmatrix} \begin{bmatrix} C_1 \\ C_2 \end{bmatrix} = \begin{bmatrix} 0 \\ 0 \end{bmatrix}$$

For the first eigenvalue this becomes:

$$\begin{bmatrix} -1.804 + 1.820 & 0.140 \\ 0.140 & -0.568 + 1.820 \end{bmatrix} \begin{bmatrix} C_1 \\ C_2 \end{bmatrix} = \begin{bmatrix} 0 \\ 0 \end{bmatrix}$$

$$\begin{bmatrix} 0.016 & 0.140 \\ 0.140 & 1.252 \end{bmatrix} \begin{bmatrix} C_1 \\ C_2 \end{bmatrix} = \begin{bmatrix} 0 \\ 0 \end{bmatrix}$$

$$(0.140)\,(C_1) + (1.252)\,(C_2) = 0$$

$$C_1 = -8.943$$

$C_2 C_1^2 + C_2^2 = 1$ (from normalization)

$$(-8.943 C_2)^2 + C_2^2 = 1$$

$$80.975 C_2^2 = 1$$

$$C_2 = 0.111,\ C_1 = -0.994$$

For the second eigenvalue this becomes:

$$\begin{bmatrix} -1.804+0.552 & 0.140 \\ 0.140 & -0.568+0.552 \end{bmatrix} \begin{bmatrix} C_1 \\ C_2 \end{bmatrix} = \begin{bmatrix} 0 \\ 0 \end{bmatrix}$$

$$\begin{bmatrix} -1.252 & 0.140 \\ 0.140 & -0.016 \end{bmatrix} \begin{bmatrix} C_1 \\ C_2 \end{bmatrix} = \begin{bmatrix} 0 \\ 0 \end{bmatrix}$$

$$(-1.252)(C_1) + (0.140)(C_2) = 0$$

$$C_1 = 0.112\, C_2$$

$$C_1^2 + C_2^2 = 1 \text{ (from normalization)}$$

$$(0.112C_2)^2 + C_2^2 = 1$$

$$1.0125C_2^2 = 1$$

$$C_2 = 0.994,\ C_1 = 0.111$$

e. The polarized orbitals, $R_\pm$, are given by:

$$R_\pm = \sigma_g \pm \sqrt{\frac{C_2}{C_1}}\sigma_u$$

$$R_\pm = \sigma_g \pm \sqrt{\frac{0.111}{0.994}}\sigma_u$$

$$R_\pm = \sigma_g \pm 0.334\sigma_u$$

$$R_+ = \sigma_g + 0.334\sigma_u \quad \text{(left polarized)}$$

$$R_- = \sigma_g - 0.334\sigma_u \quad \text{(right polarized)}$$

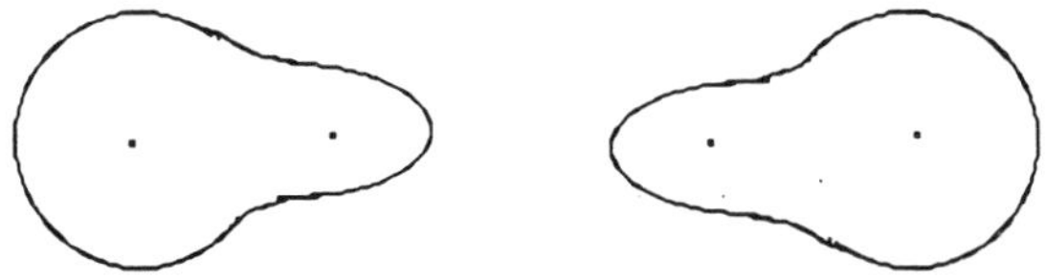

R_+ Left Polarized $\qquad$ R_- Right Polarized

SECTION

4

MOLECULAR ROTATION AND VIBRATION

CHAPTER

13

Treating the full internal nuclear-motion dynamics of a polyatomic molecule is complicated. It is conventional to examine the rotational movement of a hypothetical "rigid" molecule as well as the vibrational motion of a non-rotating molecule, and to then treat the rotation-vibration couplings using perturbation theory.

I. ROTATIONAL MOTIONS OF RIGID MOLECULES

In chapter 3 and appendix G the energy levels and wavefunctions that describe the rotation of rigid molecules are described. Therefore, in this chapter these results will be summarized briefly and emphasis will be placed on detailing how the corresponding rotational Schrödinger equations are obtained and the assumptions and limitations underlying them.

A. Linear Molecules

1. The Rotational Kinetic Energy Operator

As given in chapter 3, the Schrödinger equation for the angular motion of a rigid (i.e., having fixed bond length R) diatomic molecule is

$$-\hbar^2/2\mu\left\{(R^2 \sin\theta)^{-1}\partial/\partial\theta(\sin\theta\partial/\partial\theta) + (R^2 \sin^2\theta)^{-1}\partial^2/\partial\phi^2\right\}\psi = E\psi$$

or

$$L^2\psi/2\mu R^2 = E\psi.$$

The Hamiltonian in this problem contains only the kinetic energy of rotation; no potential energy is present because the molecule is undergoing unhindered "free rotation." The angles θ and ϕ

describe the orientation of the diatomic molecule's axis relative to a laboratory-fixed coordinate system, and μ is the reduced mass of the diatomic molecule $\mu = m_1 m_2/(m_1 + m_2)$.

2. The Eigenfunctions and Eigenvalues

The eigenvalues corresponding to each eigenfunction are straightforward to find because H_{rot} is proportional to the L^2 operator whose eigenvalues have already been determined. The resultant rotational energies are given as:

$$E_J = \hbar^2 J(J + 1)/(2\mu R^2) = BJ(J + 1)$$

and are independent of M. Thus each energy level is labeled by J and is $2J + 1$-fold degenerate (because M ranges from $-J$ to J). The rotational constant B (defined as $\hbar^2/2\mu R^2$) depends on the molecule's bond length and reduced mass. Spacings between successive rotational levels (which are of spectroscopic relevance because angular momentum selection rules often restrict ΔJ to 1, 0, and -1) are given by

$$\Delta E = B(J + 1)(J + 2) - BJ(J + 1) = 2B(J + 1).$$

Within this "rigid rotor" model, the absorption spectrum of a rigid diatomic molecule should display a series of peaks, each of which corresponds to a specific $J \rightarrow J + 1$ transition. The energies at which these peaks occur should grow linearally with J. An example of such a progression of rotational lines is shown in the figure below.

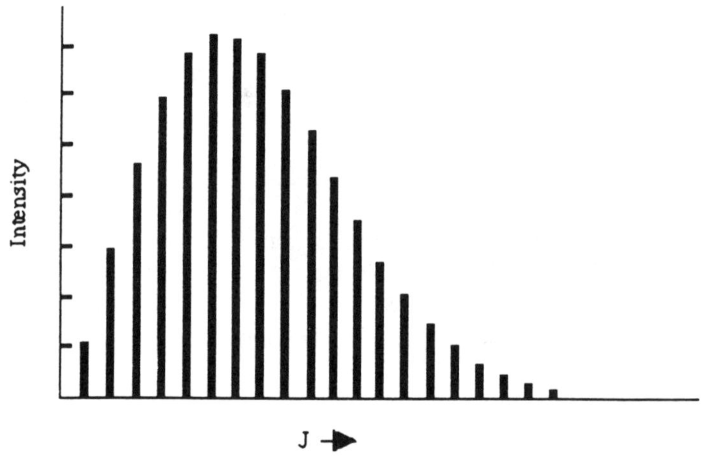

The energies at which the rotational transitions occur appear to fit the $\Delta E = 2B(J + 1)$ formula rather well. The intensities of transitions from level J to level $J + 1$ vary strongly with J primarily because the population of molecules in the absorbing level varies with J. These populations P_J are given, when the system is at equilibrium at temperature T, in terms of the degeneracy $(2J + 1)$ of the Jth level and the energy of this level $BJ(J + 1)$:

$$P_J = Q^{-1}(2J + 1) \exp(-BJ(J + 1)/kT) ,$$

where Q is the rotational partition function:

$$Q = \sum_J (2J + 1) \exp(-BJ(J + 1)/kT) .$$

For low values of J, the degeneracy is low and the $\exp(-BJ(J+1)/kT)$ factor is near unity. As J increases, the degeneracy grows linearly but the $\exp(-BJ(J+1)/kT)$ factor decreases more rapidly. As a result, there is a value of J, given by taking the derivative of $(2J+1)\exp(-BJ(J+1)/kT)$ with respect to J and setting it equal to zero,

$$2J_{max} + 1 = \sqrt{2kT/B}$$

at which the intensity of the rotational transition is expected to reach its maximum. The eigenfunctions belonging to these energy levels are the spherical harmonics $Y_{L,M}(\theta,\phi)$ which are normalized according to

$$\int_0^{\pi}\left(\int_0^{2\pi}(Y^*_{L,M}(\theta,\phi)Y_{L',M'}(\theta,\phi)\sin\theta d\theta d\phi)\right) = \delta_{L,L'}\delta_{M,M'}.$$

These functions are identical to those that appear in the solution of the angular part of Hydrogen-like atoms. The above energy levels and eigenfunctions also apply to the rotation of rigid linear polyatomic molecules; the only difference is that the moment of inertia I entering into the rotational energy expression is given by

$$I = \sum_a m_a R_a^2$$

where m_a is the mass of the ath atom and R_a is its distance from the center of mass of the molecule. This moment of inertia replaces μR^2 in the earlier rotational energy level expressions.

B. Non-Linear Molecules

1. *The Rotational Kinetic Energy Operator*

The rotational kinetic energy operator for a rigid polyatomic molecule is shown in appendix G to be

$$H_{rot} = J_a^2/2I_a + J_b^2/2I_b + J_c^2/2I_c$$

where the $I_k(k = a,b,c)$ are the three principal moments of inertia of the molecule (the eigenvalues of the moment of inertia tensor). This tensor has elements in a cartesian coordinate system $(K,K' = X,Y,Z)$ whose origin is located at the center of mass of the molecule that are computed as:

$$I_{K,K} = \sum_j m_j(R_j^2 - R_{K,j}^2) \text{ (for } K = K')$$

$$I_{K,K'} = -\sum_j m_j R_{K,j} R_{K',j} \text{ (for } K \neq K').$$

The components of the quantum mechanical angular momentum operators along the three principal axes are:

$$J_a = -i\hbar\cos\chi\,[\cot\theta\partial/\partial\chi - (\sin\theta)^{-1}\partial/\partial\phi] -\!\!- i\hbar\sin\chi\partial/\partial\theta$$
$$J_b = i\hbar\sin\chi[\cot\theta\partial/\partial\chi - (\sin\theta)^{-1}\partial/\partial\phi] -\!\!- i\hbar\cos\chi\partial/\partial\theta$$
$$J_c = -\,i\hbar\partial/\partial\chi.$$

The angles θ, φ, and χ are the Euler angles needed to specify the orientation of the rigid molecule relative to a laboratory-fixed coordinate system. The corresponding square of the total angular momentum operator J^2 can be obtained as

$$J^2 = J_a^2 + J_b^2 + J_c^2$$
$$= -\hbar^2\partial^2/\partial\theta^2 - \hbar^2 \cot\theta\partial/\partial\theta - \hbar^2(1/\sin\theta)(\partial^2/\partial\phi^2 + \partial^2/\partial\chi^2 - 2\cos\theta\partial^2/\partial\phi\partial\chi),$$

and the component along the lab-fixed Z axis J_Z is

$$-i\hbar\partial/\partial\phi.$$

2. The Eigenfunctions and Eigenvalues for Special Cases

a. Spherical Tops When the three principal moment of inertia values are identical, the molecule is termed a **spherical top.** In this case, the total rotational energy can be expressed in terms of the total angular momentum operator J^2

$$H_{rot} = J^2/2I\ .$$

As a result, the eigenfunctions of H_{rot} are those of J^2 (and J_a as well as J_Z both of which commute with J^2 and with one another; J_Z is the component of $\boldsymbol{J}$ along the lab-fixed Z-axis and commutes with J_a because $J_Z = -\ i\hbar\partial/\partial\phi$ and $J_\alpha = -\ i\hbar\partial/\partial\chi$ act on different angles). The energies associated with such eigenfunctions are

$$E(J,K,M) = \hbar^2 J(J+1)/2I^2,$$

for all K (i.e., J_a quantum numbers) ranging from $-J$ to J in unit steps and for all M (i.e., J_Z quantum numbers) ranging from $-J$ to J. Each energy level is therefore $(2J+1)^2$ degenerate because there are $2J + 1$ possible K values and $2J + 1$ possible M values for each J. The eigenfunctions of J^2, J_Z and J_a,$|J, M, K>$ are given in terms of the set of rotation matrices $D_{J,M,K}$:

$$|J,M,K> = \sqrt{\frac{2J+1}{8\pi^2}}\, D^*_{J,M,K}(\theta,\phi,\chi)$$

which obey

$$J^2|J,M,K> = \hbar^2 J(J+1)|J,M,K>,$$
$$J_a|J,M,K> = \hbar K|J,M,K>,$$
$$J_Z|J,M,K> = \hbar M|J,M,K>.$$

b. Symmetric tops Molecules for which two of the three principal moments of inertia are equal are called **symmetric tops.** Those for which the unique moment of inertia is smaller than the other two are termed **prolate** symmetric tops; if the unique moment of inertia is larger than the others, the molecule is an **oblate** symmetric top.

Again, the rotational kinetic energy, which is the full rotational Hamiltonian, can be written in terms of the total rotational angular momentum operator J^2 and the component of angular momentum along the axis with the unique principal moment of inertia:

$$H_{rot} = J^2/2I + J_a^2\{1/2I_a - 1/2I\},\ \text{for prolate tops}$$
$$H_{rot} = J^2/2I + J_c^2\{1/2I_c - 1/2I\},\ \text{for oblate tops.}$$

As a result, the eigenfunctions of H_{rot} are those of J^2 and J_a or J_c (and of J_Z), and the corresponding energy levels are:

$$E(J,K,M) = \hbar^2 J(J+1)/2I^2 + \hbar^2 K^2\{1/2I_a - 1/2I\},$$

for prolate tops

$$E(J,K,M) = \hbar^2 J(J+1)/2I^2 + \hbar^2 K^2\{1/2I_c - 1/2I\},$$

for oblate tops, again for K and M (i.e., J_a or J_c and J_Z quantum numbers, respectively) ranging from $-J$ to J in unit steps. Since the energy now depends on K, these levels are only $2J+1$ degenerate due to the $2J+1$ different M values that arise for each J value. The eigenfunctions $|J,M,K\rangle$ are the same rotation matrix functions as arise for the spherical-top case.

c. Asymmetric tops The rotational eigenfunctions and energy levels of a molecule for which all three principal moments of inertia are distinct (a so-called **asymmetric top**) can not easily be expressed in terms of the angular momentum eigenstates and the J, M, and K quantum numbers. However, given the three principal moments of inertia I_a, I_b, and I_c, a matrix representation of each of the three contributions to the rotational Hamiltonian

$$H_{rot} = J_a^2/2I_a + J_b^2/2I_b + J_c^2/2I_c$$

can be formed within a basis set of the $\{|J,M,K\rangle\}$ rotation matrix functions. This matrix will not be diagonal because the $|J,M,K\rangle$ functions are not eigenfunctions of the asymmetric top H_{rot}. However, the matrix can be formed in this basis and subsequently brought to diagonal form by finding its eigenvectors $\{C_{n,J,M,K}\}$ and its eigenvalues $\{E_n\}$. The vector coefficients express the asymmetric top eigenstates as

$$\Psi_n(\theta,\phi,\chi) = \sum_{J,M,K} C_{n,J,M,K}|J,M,K\rangle.$$

Because the total angular momentum J^2 still commutes with H_{rot}, each such eigenstate will contain only one J-value, and hence Ψ_n can also be labeled by a J quantum number:

$$\Psi_{n,J}(\theta,\phi,\chi) = \sum_{M,K} C_{n,J,M,K}|J,M,K\rangle.$$

To form the only non-zero matrix elements of H_{rot} within the $|J,M,K\rangle$ basis, one can use the following properties of the rotation-matrix functions (see, for example, Zare's book on angular momentum):

$$\langle J,M,K|J_a^2|J,M,K\rangle = \langle J,M,K|J_b^2|J,M,K\rangle = 1/2\langle J,M,K|J^2 - J_c^2|J,M,K\rangle = \hbar^2[J(J+1) - K^2],$$
$$\langle J,M,K|J_c^2|J,M,K\rangle = \hbar^2 K^2,$$
$$\langle J,M,K|J_a^2|J,M,K\pm 2\rangle = -\langle J,M,K|J_b^2|J,M,K\pm 2\rangle$$
$$= \hbar^2[J(J+1) - K(K\pm 1)]^{1/2}[J(J+1) - (K\pm 1)(K\pm 2)]^{1/2},$$
$$\langle J,M,K|J_c^2|J,M,K\pm 2\rangle = 0.$$

Each of the elements of J_c^2, J_a^2, and J_b^2 must, of course, be multiplied, respectively, by $1/2I_c$, $1/2I_a$, and $1/2I_b$ and summed together to form the matrix representation of H_{rot}. The diagonalization of this matrix then provides the asymmetric top energies and wavefunctions.

II. VIBRATIONAL MOTION WITHIN THE HARMONIC APPROXIMATION

The simple harmonic motion of a diatomic molecule was treated in chapter 1, and will not be repeated here. Instead, emphasis is placed on polyatomic molecules whose electronic energy's dependence on the $3N$ cartesian coordinates of its N atoms can be written (approximately) in terms of a Taylor series expansion about a stable local minimum. We therefore assume that the molecule of interest exists in an electronic state for which the geometry being considered is stable (i.e., not subject to spontaneous geometrical distortion).

The Taylor series expansion of the electronic energy is written as:

$$V(q_k) = V(0) + \sum_k (\partial V/\partial q_k)q_k + 1/2 \sum_{j,k} q_j H_{j,k}\, q_k + \ldots,$$

where $V(0)$ is the value of the electronic energy at the stable geometry under study, q_k is the *displacement* of the kth cartesian coordinate away from this starting position, $(\partial V/\partial q_k)$ is the gradient of the electronic energy along this direction, and the $H_{j,k}$ are the second derivative or *Hessian* matrix elements along these directions $H_{j,k} = (\partial^2 V/\partial q_j \partial q_k)$. If the starting geometry corresponds to a stable species, the gradient terms will all vanish (meaning this geometry corresponds to a minimum, maximum, or saddle point), and the Hessian matrix will possess $3N - 5$ (for linear species) or $3N - 6$ (for non-linear molecules) positive eigenvalues and five or six zero eigenvalues (corresponding to three translational and two or three rotational motions of the molecule). If the Hessian has one negative eigenvalue, the geometry corresponds to a transition state (these situations are discussed in detail in chapter 20).

From now on, we assume that the geometry under study corresponds to that of a stable minimum about which vibrational motion occurs. The treatment of unstable geometries is of great importance to chemistry, but this chapter deals with vibrations of stable species. For a good treatment of situations under which geometrical instability is expected to occur, see chapter 2 of the text *Energetic Principles of Chemical Reactions* by J. Simons. A discussion of how local minima and transition states are located on electronic energy surfaces is provided in chapter 20 of the present text.

A. The Newton Equations of Motion for Vibration

1. The Kinetic and Potential Energy Matrices

Truncating the Taylor series at the quadratic terms (assuming these terms dominate because only small displacements from the equilibrium geometry are of interest), one has the so-called **harmonic** potential:

$$V(q_k) = V(0) + 1/2 \sum_{j,k} q_j H_{j,k} q_k.$$

The classical mechanical equations of motion for the $3N \{q_k\}$ coordinates can be written in terms of the above potential energy and the following kinetic energy function:

$$T = 1/2 \sum_j m_j \dot{q}_j^2 ,$$

where $\dot{q}_j$ denotes the time rate of change of the coordinate q_j and m_j is the mass of the atom on which the jth cartesian coordinate resides. The Newton equations thus obtained are:

$$m_j \ddot{q}_j = - \sum_k H_{j,k} q_k$$

where the force along the jth coordinate is given by minus the derivative of the potential V along this coordinate $(\partial V/\partial q_j) = \Sigma_k H_{j,k} q_k$ within the harmonic approximation. These classical equations can more compactly be expressed in terms of the time evolution of a set of so-called **mass-weighted** cartesian coordinates defined as:

$$x_j = q_j (m_j)^{1/2},$$

in terms of which the Newton equations become

$$\ddot{x}_j = - \sum_k H'_{j,k} x_k$$

and the *mass-weighted Hessian* matrix elements are

$$H'_{j,k} = H_{j,k} (m_j m_k)^{-1/2}.$$

2. The Harmonic Vibrational Energies and Normal Mode Eigenvectors

Assuming that the x_j undergo some form of sinusoidal time evolution:

$$x_j(t) = x_j(0) \cos(\omega t),$$

and substituting this into the Newton equations produces a matrix eigenvalue equation:

$$\omega^2 x_j = \sum_k H'_{j,k} x_k$$

in which the eigenvalues are the squares of the so-called **normal mode** vibrational frequencies and the eigenvectors give the amplitudes of motion along each of the $3N$ mass weighted cartesian coordinates that belong to each mode. Within this harmonic treatment of vibrational motion, the total vibrational energy of the molecule is given as

$$E(v_1, v_2, \ldots v_{3N-5 \text{ or } 6}) = \sum_{j=1}^{3N-5 \text{ or } 6} \hbar\omega_j (v_j + 1/2),$$

a sum of $3N - 5$ or $3N - 6$ independent contributions one for each normal mode. The corresponding total vibrational wavefunction

$$\Psi(x_1, x_2, \ldots x_{3N-5 \text{ or } 6}) = \prod \psi_{v_j}(x_j)$$

is a product of $3N - 5$ or $3N - 6$ harmonic oscillator functions $\psi_{v_j}(x_j)$, one for each normal mode. Within this picture, the energy gap between one vibrational level and another in which one of the v_j quantum numbers is increased by unity (the origin of this "selection rule" is discussed in chapter 15) is

$$\Delta E_{v_j} \rightarrow {}_{v_j+1} = \hbar\omega_j$$

The harmonic model thus predicts that the "fundamental" $(v = 0 \rightarrow v = 1)$ and "hot band" $(v = 1 \rightarrow v = 2)$ transitions should occur at the same energy, and the overtone $(v = 0 \rightarrow v = 2)$ transition should occur at exactly twice this energy.

B. The Use of Symmetry

1. Symmetry Adapted Modes

It is often possible to simplify the calculation of the normal mode frequencies and eigenvectors by exploiting molecular point group symmetry. For molecules that possess symmetry, the electronic potential $V(q_j)$ displays symmetry with respect to displacements of symmetry equivalent cartesian coordinates. For example, consider the water molecule at its C_{2v} equilibrium geometry as illustrated in the figure below. A very small movement of the H_2O molecule's left H atom in the positive x direction (Δx_L) produces the same change in V as a correspondingly small displacement of the right H atom in the negative x direction ($-\Delta x_R$). Similarly, movement of the left H in the positive y direction (Δy_L) produces an energy change identical to movement of the right H in the positive y direction (Δy_R).

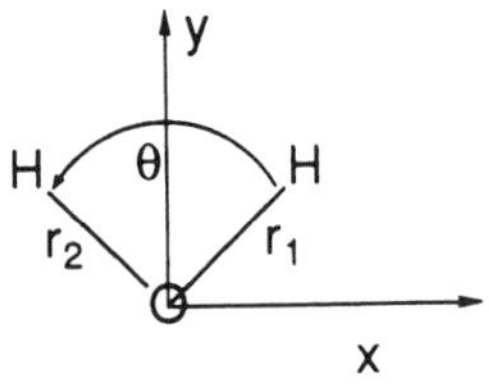

The equivalence of the pairs of cartesian coordinate displacements is a result of the fact that the displacement vectors are connected by the point group operations of the C_{2v} group. In particular, reflection of Δx_L through the yz plane produces $-\Delta x_R$, and reflection of Δy_L through this same plane yields Δy_R.

More generally, it is possible to combine sets of cartesian displacement coordinates $\{q_k\}$ into so-called symmetry adapted coordinates $\{Q_{\Gamma_j}\}$, where the index Γ labels the irreducible representation and j labels the particular combination of that symmetry. These symmetry adapted coordinates can be formed by applying the point group projection operators to the individual cartesian displacement coordinates.

To illustrate, again consider the H_2O molecule in the coordinate system described above. The $3N = 9$ mass weighted cartesian displacement coordinates $(X_L, Y_L, Z_L, X_O, Y_O, Z_O, X_R, Y_R, Z_R)$ can be symmetry adapted by applying the following four projection operators:

$$P_{a_1} = 1 + \sigma_{yz} + \sigma_{xy} + C_2$$
$$P_{b_1} = 1 + \sigma_{yz} - \sigma_{xy} - C_2$$
$$P_{b_2} = 1 - \sigma_{yz} + \sigma_{xy} - C_2$$
$$P_{a_2} = 1 - \sigma_{yz} - \sigma_{xy} + C_2$$

to each of the nine original coordinates. Of course, one will *not* obtain $9 \times 4 = 36$ independent symmetry adapted coordinates in this manner; many identical combinations will arise, and only nine will be independent.

The independent combination of a_1 symmetry (normalized to produce vectors of unit length) are

$$Q_{a_1,1} = 2^{-1/2}[X_L - X_R]$$
$$Q_{a_1,2} = 2^{-1/2}[Y_L + Y_R]$$
$$Q_{a_1,3} = [Y_O]$$

Those of b_2 symmetry are

$$Q_{b_2,1} = 2^{-1/2}[X_L + X_R]$$
$$Q_{b_2,2} = 2^{-1/2}[Y_L - Y_R]$$
$$Q_{b_2,3} = [X_O],$$

and the combinations

$$Q_{b_1,1} = 2^{-1/2}[Z_L + Z_R]$$
$$Q_{b_1,2} = [Z_O]$$

are of b_1 symmetry, whereas

$$Q_{a_2,1} = 2^{-1/2}[Z_L - Z_R]$$

is of a_2 symmetry.

2. Point Group Symmetry of the Harmonic Potential

These nine $Q_{\Gamma,j}$ are expressed as unitary transformations of the original mass weighted cartesian coordinates:

$$Q_{\Gamma,j} = \sum_k C_{\Gamma,j,k} X_k$$

These transformation coefficients $\{C_{\Gamma,j,k}\}$ can be used to carry out a unitary transformation of the 9×9 mass-weighted Hessian matrix. In so doing, we need only form blocks

$$H_{j,l}^{\Gamma} = \sum_{kk'} C_{\Gamma,j,k} H_{k,k'} (m_k m_k')^{-1/2} \; C_{\Gamma,l,k}'$$

within which the symmetries of the two modes are identical. The off-diagonal elements

$$H_{jl}^{\Gamma\Gamma'} = \sum_{kk'} C_{\Gamma,j,k} H_{k,k'} (m_k m_{k'})^{-1/2} C_{\Gamma',l,k'}$$

vanish because the potential $V(q_j)$ (and the full vibrational Hamiltonian $H = T + V$) commutes with the C_{2V} point group symmetry operations. As a result, the 9×9 mass-weighted Hessian eigenvalue problem can be subdivided into two 3×3 matrix problems (of a_1 and b_2 symmetry), one 2×2 matrix of b_1 symmetry and one 1×1 matrix of a_2 symmetry. The eigenvalues of each of these blocks provide the squares of the harmonic vibrational frequencies, the eigenvectors provide the normal mode displacements as linear combinations of the symmetry adapted $\{Q_j^\Gamma\}$.

Regardless of whether symmetry is used to block diagonalize the mass-weighted Hessian, six (for non-linear molecules) or five (for linear species) of the eigenvalues will equal zero. The eigenvectors belonging to these zero eigenvalues describe the three translations and two or three rotations of the molecule. For example,

$$\frac{1}{\sqrt{3}}[X_L + X_R + X_O]$$
$$\frac{1}{\sqrt{3}}[Y_L + Y_R + Y_O]$$
$$\frac{1}{\sqrt{3}}[Z_L + Z_R + Z_O]$$

are three translation eigenvectors of b_2, a_1 and b_1 symmetry, and

$$\frac{1}{\sqrt{2}}(Z_L - Z_R)$$

is a rotation (about the Y-axis in the figure shown above) of a_2 symmetry. This rotation vector can be generated by applying the a_2 projection operator to Z_L or to Z_R. The fact that rotation about the Y-axis is of a_2 symmetry is indicated in the right-hand column of the C_{2v} character table of appendix E via the symbol R_Z (n.b., care must be taken to realize that the axis convention used in the above figure is different than that implied in the character table; the latter has the Z-axis out of the molecular plane, while the figure calls this the X-axis). The other two rotations are of b_1 and b_2 symmetry (see the C_{2v} character table of appendix E) and involve spinning of the molecule about the X- and Z-axes of the figure drawn above, respectively. So, of the nine cartesian displacements, three are of a_1 symmetry, three of b_2, two of b_1, and one of a_2. Of these, there are three translations (a_1, b_2, and b_1) and three rotations (b_2, b_1, and a_2). This leaves two vibrations of a_1 and one of b_2 symmetry. For the H_2O example treated here, the three non-zero eigenvalues of the mass-weighted Hessian are therefore of a_1, b_2, and a_1 symmetry. They describe the symmetric and asymmetric stretch vibrations and the bending mode, respectively as illustrated below.

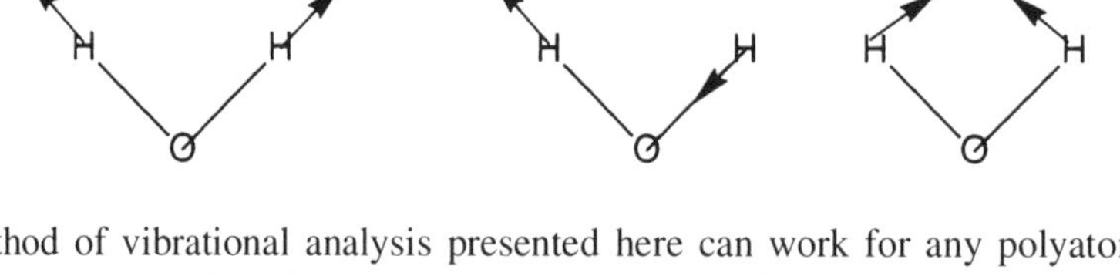

The method of vibrational analysis presented here can work for any polyatomic molecule. One knows the mass-weighted Hessian and then computes the non-zero eigenvalues which then provide the squares of the normal mode vibrational frequencies. Point group symmetry can be used to block diagonalize this Hessian and to label the vibrational modes according to symmetry.

III. ANHARMONICITY

The electronic energy of a molecule, ion, or radical at geometries near a stable structure can be expanded in a Taylor series in powers of displacement coordinates as was done in the preceding section of this chapter. This expansion leads to a picture of uncoupled harmonic vibrational energy levels

$$E(v_1 \cdots v_{3N-5\text{ or }6}) = \sum_{j=1}^{3N-5\text{ or }6} \hbar\omega_j(v_j + 1/2)$$

and wavefunctions

$$\psi(x_1 \cdots x_{3N-5\text{ or }6}) = \prod_{j=1}^{3N-5\text{ or }6} \psi_{v_j}(x_j).$$

The spacing between energy levels in which one of the normal-mode quantum numbers increases by unity

$$\Delta E_{v_j} = E(\cdots v_j + 1 \cdots) - E(\cdots v_j \cdots) = \hbar\omega_j$$

is predicted to be independent of the quantum number v_j. This picture of evenly spaced energy levels

$$\Delta E_0 = \Delta E_1 = \Delta E_2 = \cdots$$

is an incorrect aspect of the harmonic model of vibrational motion, and is a result of the quadratic model for the potential energy surface $V(x_j)$.

A. The Expansion of $E(v)$ in Powers of $(v + 1/2)$

Experimental evidence clearly indicates that significant deviations from the harmonic oscillator energy expression occur as the quantum number v_j grows. In chapter 1 these deviations were explained in terms of the diatomic molecule's true potential $V(R)$ deviating strongly from the harmonic $1/2k(R - R_e)^2$ potential at higher energy (and hence larger $|R - R_e|$) as shown in the following figure.

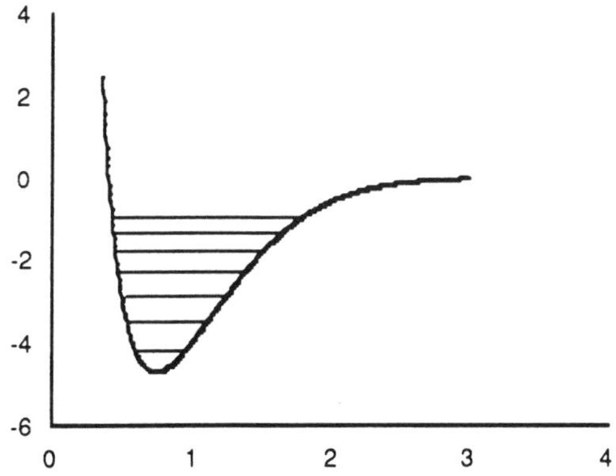

At larger bond lengths, the true potential is "softer" than the harmonic potential, and eventually reaches its asymptote which lies at the dissociation energy D_e above its minimum. This negative deviation of the true $V(R)$ from $1/2k(R - R_e)^2$ causes the true vibrational energy levels to lie below the harmonic predictions. It is convention to express the experimentally observed vibrational energy levels, along each of the $3N - 5$ or 6 independent modes, as follows:

$$E(v_j) = \hbar[\omega_j(v_j + 1/2) - (\omega x)_j(v_j + 1/2)^2 + (\omega y)_j\,(v_j + 1/2)^3 + (\omega z)_j(v_j + 1/2)^4 + \cdots]$$

The first term is the harmonic expression. The next is termed the first anharmonicity; it (usually) produces a negative contribution to $E(v_j)$ that varies as $(v_j + 1/2)^2$. The spacings between successive $v_j \rightarrow v_j + 1$ energy levels are then given by:

$$\Delta E_{v_j} = E(v_j + 1) - E(v_j) = \hbar[\omega_j - 2(\omega x)_j(v_j + 1) + \cdots]$$

A plot of the spacing between neighboring energy levels versus v_j should be linear for values of v_j where the harmonic and first anharmonic terms dominate. The slope of such a plot is expected to be $-2\hbar(\omega x)_j$ and the small $-v_j$ intercept should be $\hbar[\omega_j - 2(\omega x)_j]$. Such a plot of experimental data, which clearly can be used to determine the ω_j and $(\omega x)_j$ parameters of the vibrational mode of study, is shown in the figure below.

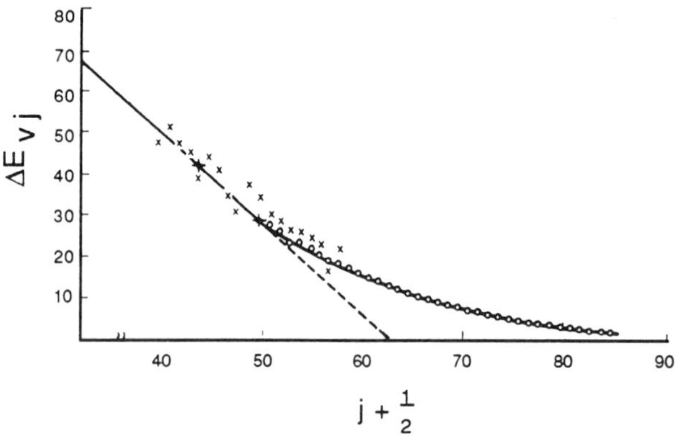

B. The Birge-Sponer Extrapolation

These so-called Birge-Sponer plots can also be used to determine dissociation energies of molecules. By linearly extrapolating the plot of experimental ΔE_{v_j} values to large v_j values, one can find the value of v_j at which the spacing between neighboring vibrational levels goes to zero. This value v_j,max specifies the quantum number of the last bound vibrational level for the particular potential energy function $V(x_j)$ of interest. The dissociation energy D_e can then be computed by adding to $1/2\hbar\omega_j$ (the zero point energy along this mode) the sum of the spacings between neighboring vibrational energy levels from $v_j = 0$ to $v_j = v_j max$:

$$D_e = 1/2\hbar\omega_j + \sum_{v_j=0}^{v_j\max} \Delta E_{v_j}.$$

Since experimental data are not usually available for the entire range of v_j values (from 0 to $v_{j,\max}$), this sum must be computed using the anharmonic expression for ΔE_{vj}:

$$\Delta E_{v_j} = \hbar[\omega_j - 2(\omega x)_j(v_j + 1/2) + \cdots].$$

Alternatively, the sum can be computed from the Birge-Sponer graph by measuring the area under the straight-line fit to the graph of ΔE_{v_j} or v_j from $v_j = 0$ to $v_j = v_{j,\text{max}}$.

SECTION SUMMARY

This completes our introduction to the subject of rotational and vibrational motions of molecules (which applies equally well to ions and radicals). The information contained in this section is used again in section 5 where photon-induced transitions between pairs of molecular electronic, vibrational, and rotational eigenstates are examined. More advanced treatments of the subject matter of this section can be found in the text by Wilson, Decius, and Cross, as well as in Zare's text on angular momentum.

Exercises, Problems, and Solutions

EXERCISES

1. Consider the molecules CCl_4, $CHCl_3$, and CH_2Cl_2.

   a. What kind of rotor are they (symmetric top, etc.; do not bother with oblate, or near-prolate, etc.)

   b. Will they show pure rotational spectra?

   c. Assume that ammonia shows a pure rotational spectrum. If the rotational constants are 9.44 and 6.20 cm^{-1}, use the energy expression:

   $$E = (A - B)K^2 + BJ(J + 1),$$

   to calculate the energies (in cm^{-1}) of the first three lines (i.e., those with lowest K, J quantum number for the absorbing level) in the absorption spectrum (ignoring higher order terms in the energy expression).

2. The molecule $^{11}B^{16}O$ has a vibrational frequency $\omega_e = 1885\ cm^{-1}$, a rotational constant $B_e = 1.78 cm^{-1}$, and a bond energy from the bottom of the potential well of $D_e^0 = 8.28$ eV. Use integral atomic masses in the following:

   a. In the approximation that the molecule can be represented as a Morse oscillator, calculate the bond length, R_e in angstroms, the centrifugal distortion constant, D_e in cm^{-1}, the anharmonicity constant, $\omega_e x_e$ in cm^{-1}, the zero-point corrected bond energy, D_0^0 in eV, the vibration rotation interaction constant, α_e in cm^{-1}, and the vibrational state specific rotation constants, B_0 and B_1 in cm^{-1}. Use the vibration-rotation energy expression for a Morse oscillator:

$E = \hbar\omega_e(v + 1/2) - \hbar\omega_e x_e(v + 1/2)^2 + B_v J(J+1) - D_e J^2(J + 1)^2$, where

$$B_v = B_e - \alpha_e(v + 1/2),$$

$$\alpha_e = \frac{-6B_e^2}{\hbar\omega_e} + \frac{6\sqrt{B_e^3 \hbar\omega_e x_e}}{\hbar\omega_e},$$

and

$$D_e = \frac{4B_e^3}{\hbar\omega_e^2}.$$

b. Will this molecule show a pure rotation spectrum? A vibration-rotation spectrum? Assume that it does, what are the energies (in cm^{-1}) of the first three lines in the P branch ($\Delta v = +1, \Delta J = -1$) of the fundamental absorption?

3. Consider trans-$C_2H_2Cl_2$. The vibrational normal modes of this molecule are shown below. What is the symmetry of the molecule? Label each of the modes with the appropriate irreducible representation.

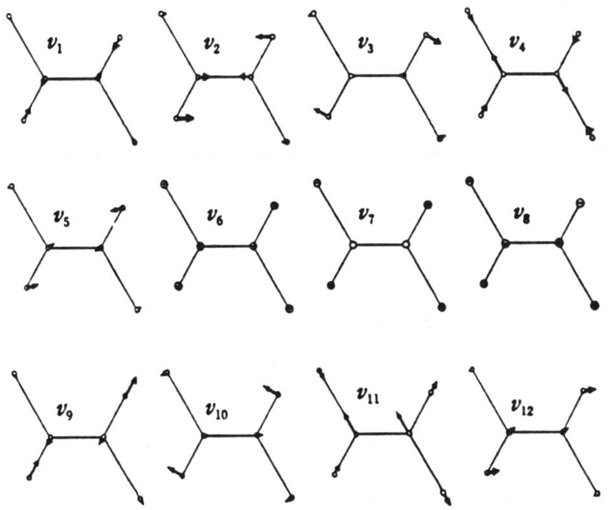

PROBLEMS

1. Suppose you are given two molecules (one is CH_2 and the other is CH_2^- but you don't know which is which). Both molecules have C_{2v} symmetry. The CH bond length of molecule I is 1.121 Å and for molecule II it is 1.076 Å. The bond angle of molecule I is 104° and for molecule II it is 136°.

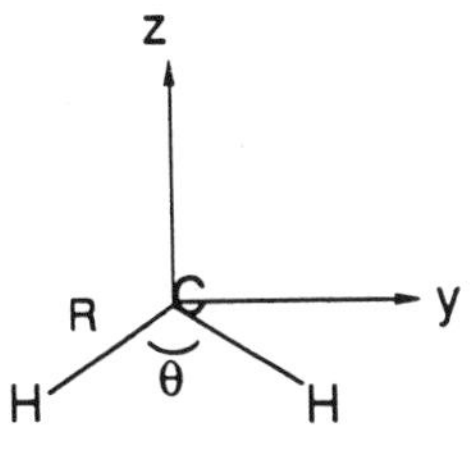

a. Using a coordinate system centered on the C nucleus as shown above (the molecule is in the *YZ* plane), compute the moment of inertia tensors of both species (I and II). The definitions of the components of the tensor are, for example:

$$I_{xx} = \sum_j m_j(y_j^2 + z_j^2) - M(Y^2 + Z^2)$$

$$I_{xy} = -\sum_j m_j x_j y_j - MXY$$

Here, m_j is the mass of the nucleus *j*, *M* is the mass of the entire molecule, and *X*, *Y*, *Z* are the coordinates of the center of mass of the molecule. Use Å for distances and amu's for masses.

b. Find the principal moments of inertia $I_a < I_b < I_c$ for both compounds (in amu Å^2 units) and convert these values into rotational constants *A*, *B*, and *C* in cm^{-1} using, for example,

$$A = h(8\pi^2 c I_a)^{-1}.$$

c. Both compounds are "nearly prolate tops" whose energy levels can be well approximated using the prolate top formula:

$$E = (A - B)K^2 + BJ(J + 1),$$

if one uses for the *B* constant the average of the *B* and *C* values determined earlier. Thus, take *B* and *C* values (for each compound) and average them to produce an effective *B* constant to use in the above energy formula. Write down (in cm^{-1} units) the energy formula for both species. What values are *J* and *K* allowed to assume? What is the degeneracy of the level labeled by a given *J* and *K*?

d. Draw a picture of both compounds and show the directions of the three principle axes (*a*,*b*,*c*). On these pictures show the kind of rotational motion associated with the quantum number *K*.

e. Given that the electronic *transition* moment vector $\vec{\mu}$ connecting species I and II is directed along the *Y* axis, what are the selection rules *J* and *K*?

f. Suppose you are given the photoelectron spectrum of CH_2^-. In this spectrum $J_j = J_i + 1$ transitions are called *R*-branch absorptions and those obeying $J_j = J_i - 1$ are called *P*-branch transitions. The spacing between lines can increase or decrease as functions of J_i depending on the changes in the moment of inertia for the transition. If spacings grow closer and closer, we say that the spectrum exhibits a so-called band head formation. In the photoelectron spectrum that you are given, a rotational analysis of the vibrational lines in this spectrum is carried out and it is found that the *R*-branches show band head formation but the *P*-branches do not. Based on this information, determine which compound I or II is the CH_2^- anion. Explain your reasoning.

g. At what *J* value (of the absorbing species) does the band head occur and at what rotational energy difference?

2. Let us consider the vibrational motions of benzene. To consider all of the vibrational modes of benzene we should attach a set of displacement vectors in the *x*, *y*, and *z* directions to each atom in the molecule (giving 36 vectors in all), and evaluate how these transform under the symmetry operations of D_{6h}. For this problem, however, let's only inquire about the C-H stretching vibrations.

a. Represent the C-H stretching motion on each C-H bond by an outward-directed vector on each H atom, designated r_i:

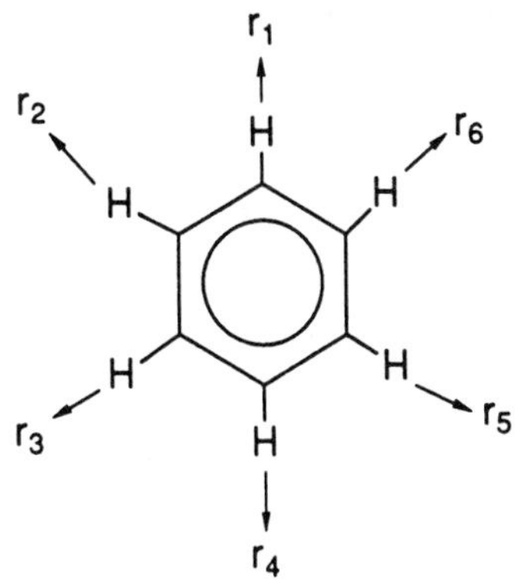

These vectors form the basis for a reducible representation. Evaluate the characters for this reducible representation under the symmetry operations of the D_{6h} group.

b. Decompose the reducible representation you obtained in part (a) into its irreducible components. These are the symmetries of the various C-H stretching vibrational modes in benzene.

c. The vibrational state with zero quanta in each of the vibrational modes (the ground vibrational state) of any molecule always belongs to the totally symmetric representation. For benzene the ground vibrational state is therefore of A_{1g} symmetry. An excited state which has one quantum of vibrational excitation in a mode which is of a given symmetry species has the same symmetry species as the mode which is excited (because the vibrational wave functions are given as Hermite polynomials in the stretching coordinate). Thus, for example, excitation (by one quantum) of a vibrational mode of A_{2u} symmetry gives a wavefunction of A_{2u} symmetry. To resolve the question of what vibrational modes may be excited by the absorption of infrared radiation we must examine the x, y, and z components of the transition dipole integral for initial and final state wave functions ψ_i and ψ_f, respectively:

$|<\psi_f|x|\psi_i>|$, $|<\psi_f|y|\psi_i>|$, and $|<\psi_f|z|\psi_i>|$.

Using the information provided above, which of the C-H vibrational modes of benzene will be infrared-active, and how will the transitions be polarized? How many C-H vibrations will you observe in the infrared spectrum of benzene?

d. A vibrational mode will be active in Raman spectroscopy only if one of the following integrals is nonzero:

$|<\psi_f|xy|\psi_i>|$, $|<\psi_f|xz|\psi_i>|$, $|<\psi_f|yz|\psi_i>|$,
$|<\psi_f|x^2|\psi_i>|$, $|<\psi_f|y^2|\psi_i>|$, and $|<\psi_f|z^2|\psi_i>|$.

Using the fact that the quadratic operators transform according to the irreducible representations:

$(x^2 + y^2, z^2) \Rightarrow A_{1g}$
$(xz, yz) \Rightarrow E_{1g}$
$(x^2 - y^2, xy) \Rightarrow E_{2g}$

Determine which of the C-H vibrational modes will be Raman-active.

e. Are there any of the C-H stretching vibrational motions of benzene which cannot be observed in either infrared or Raman spectroscopy? Give the irreducible representation label for these unobservable modes.

3. In treating the vibrational and rotational motion of a diatomic molecule having reduced mass μ, equilibrium bond length r_e and harmonic force constant k, we are faced with the following radial Schrödinger equation:

$$\frac{-h^2}{2\mu r^2}\frac{d}{dr}\left(r^2\frac{dR}{dr}\right)+\frac{J(J+1)\hbar^2}{2\mu r^2}R+\frac{1}{2}k(r-r_e)^2R=ER$$

a. Show that the substitution $R = r^{-1}F$ leads to:

$$\frac{-h^2}{2\mu}F''+\frac{J(J+1)\hbar^2}{2\mu r^2}F+\frac{1}{2}k(r-r_e)^2F=EF$$

b. Taking $r = r_e + \Delta r$ and expanding $(1+x)^{-2} = 1 - 2x + 3x^2 + \ldots$, show that the so-called vibration-rotation coupling term

$$\frac{J(J+1)\hbar^2}{2\mu r^2}$$

can be approximated (for small Δr) by

$$\frac{J(J+1)\hbar^2}{2\mu r_e^2}\left(1-\frac{2\Delta r}{r_e}+\frac{3\Delta r^2}{r_e^2}\right).$$

Keep terms only through order Δr^2.

c. Show that, through terms of order Δr^2, the above equation for F can be rearranged to yield a new equation of the form:

$$\frac{-\hbar^2}{2\mu}F''+\frac{1}{2}\bar{k}(r-\bar{r}_e)^2F=\left(E-\frac{J(J+1)\hbar^2}{2\mu r_e^2}+\Delta\right)F$$

Give explicit expressions for how the modified force constant $\bar{k}$, bond length $\bar{r}_e$, and energy shift Δ depend on J, k, r_e, and μ.

d. Given the above modified vibrational problem, we can now conclude that the modified energy levels are:

$$E=\hbar\sqrt{\frac{\bar{k}}{\mu}}\left(v+\frac{1}{2}\right)+\frac{J(J+1)\hbar^2}{2\mu r_e^2}-\Delta.$$

Explain how the conclusion is "obvious," how for $J = 0$, $k = \bar{k}$, and $\Delta = 0$, we obtain the usual harmonic oscillator energy levels. Describe how the energy levels would be expected to vary as J increases from zero and explain how these changes arise from changes in k and r_e. Explain in terms of physical forces involved in the rotating-vibrating molecule why r_e and k are changed by rotation.

SOLUTIONS

Exercises

1. a. CCl_4 is tetrahedral and therefore is a spherical top. $CHCl_3$ has C_{3v} symmetry and therefore is a symmetric top. CH_2Cl_2 has C_{2v} symmetry and therefore is an asymmetric top.

b. CCl_4 has such high symmetry that it will not exhibit pure rotational spectra. $CHCl_3$ and CH_2Cl_2 will both exhibit pure rotation spectra.

c. NH_3 is a symmetric top (oblate). Use the given energy expression,

$E = (A - B)K^2 + BJ(J + 1)$,
$A = 6.20\text{cm}^{-1}$, $B = 9.44\text{cm}^{-1}$,

selection rules $\Delta J = \pm 1$, and the fact that $\vec{\mu_0}$ lies along the figure axis such that $\Delta K = 0$, to give:

$\Delta E = 2B(J + 1) = 2B, 4B$, and $6B(J = 0, 1$, and $2)$.

So, lines are at 18.88cm^{-1}, 37.76cm^{-1}, and 56.64cm^{-1}.

2. To convert between cm^{-1} and energy, multiply by

$$hc = (6.62618 \times 10^{-34} J\text{sec})(2.997925 \times 10^{10}\text{cm sec}^{-1}) = 1.9865 \times 10^{23} J\text{cm}.$$

Let all quantities in cm^{-1} be designated with a bar, e.g., $\overline{B_e} = 1.78\text{cm}^{-1}$.

a. $hc\overline{B_e} = \dfrac{\hbar^2}{2\mu R_e^2}$

$$R_e = \frac{\hbar}{\sqrt{2\mu hc\overline{B_e}}},$$

$$\mu = \frac{m_B m_O}{m_B + m_O} = \frac{(11)(16)}{(11 + 16)} \times 1.66056 \times 10^{-27}\text{kg}$$

$$= 1.0824 \times 10^{-26}\text{kg}.$$

$$hc\overline{B_e} = hc(1.78\text{cm}^{-1}) = 3.5359 \times 10^{-23} J$$

$$R_e = \frac{1.05459 \times 10^{-34} J\text{sec}}{\sqrt{(2)1.0824 \times 10^{-26}\text{kg}3.5359 \times 10^{-23} J}}$$

$$R_e = 1.205 \times 10^{-10} m = 1.205\text{Å}$$

$$D_e = \frac{4B_e^3}{\hbar\omega_e^2},\ \overline{D_e} = \frac{4\overline{B_e}^3}{\overline{\omega_e}^2} = \frac{(4)(1.78\text{cm}^{-1})^3}{(1885\text{cm}^{-1})^2} = 6.35 \times 10^{-6}\text{cm}^{-1}$$

$$\omega_e x_e = \frac{\hbar\omega_e^2}{4D_e^0},\ \overline{\omega_e x_e} = \frac{\overline{\omega_e}^2}{4\overline{D_e^0}} = \frac{(1885\text{cm}^{-1})^2}{(4)(66782.2\text{cm}^{-1})} = 13.30\text{cm}^{-1}.$$

$$D_0^0 = D_e^0 - \frac{\hbar\omega_e}{2} + \frac{\hbar\omega_e x_e}{4},\ \overline{D_0^0} = \overline{D_e^0} - \frac{\overline{\omega_e}}{2} + \frac{\overline{\omega_e x_e}}{4}$$

$$= 66782.2 - \frac{1885}{2} + \frac{13.3}{4}$$

$$= 65843.0\text{cm}^{-1} = 8.16eV.$$

$$\alpha_e = \frac{-6B_e^2}{\hbar\omega_e} + 6\frac{\sqrt{B_e^3\hbar\omega_e x_e}}{\hbar\omega_e}$$

$$\overline{\alpha_e} = \frac{-6\overline{B_e}^2}{\overline{\omega_e}} + \frac{6\sqrt{\overline{B_e}^3\overline{\omega_e x_e}}}{\overline{\omega_e}}$$

$$\overline{\alpha_e} = \frac{(-6)(1.78)^2}{(1885)} + \frac{6\sqrt{(1.78)^3(13.3)}}{(1885)} = 0.0175\text{cm}^{-1}.$$

$B_0 = B_e - \alpha_e(1/2)$, $\overline{B_0} = \overline{B_e} - \overline{\alpha_e}(1/2) = 1.78 - 0.0175/2 = 1.77\text{cm}^{-1}$

$B_1 = B_e - \alpha_e(3/2)$, $\overline{B_1} = \overline{B_e} - \overline{\alpha_e}(3/2) = 1.78 - 0.0175(1.5) = 1.75\text{cm}^{-1}$

b. The molecule has a dipole moment and so it should have a pure rotational spectrum. In addition, the dipole moment should change with R and so it should have a vibration rotation spectrum. The first three lines correspond to

$$J = 1 \rightarrow 0, J = 2 \rightarrow 1, J = 3 \rightarrow 2$$
$$E = \hbar\omega_e(v + 1/2) - \hbar\omega_e x_e(v + 1/2)^2 + B_v J(J + 1) - D_e J^2(J + 1)^2$$
$$\Delta E = \hbar\omega_e - 2\hbar\omega_e x_e - B_0 J(J + 1) + B_1 J(J - 1) - 4D_e J^3$$
$$\overline{\Delta E} = \overline{\omega_e} - 2\overline{\omega_e x_e} - \overline{B_0} J(J + 1) + \overline{B_1} J(J - 1) - 4\overline{D_e} J^3$$
$$\overline{\Delta E} = 1885 - 2(13.3) - 1.77J(J + 1) + 1.75J(J - 1) - 4(6.35 \times 10^{-6})J^3$$
$$= 1858.4 - 1.77J(J + 1) + 1.75J(J - 1) - 2.54 \times 10^{-5} J^3$$
$$\overline{\Delta E}(J = 1) = 1854.9\text{cm}^{-1}$$
$$\overline{\Delta E}(J = 2) = 1851.3\text{cm}^{-1}$$
$$\overline{\Delta E}(J = 3) = 1847.7\text{cm}^{-1}$$

3. The $C_2H_2Cl_2$ molecule has a σ_h plane of symmetry (plane of molecule), a C_2 axis ($\perp$ to plane), and inversion symmetry, this results in C_{2h} symmetry. Using C_{2h} symmetry labels the modes can be labeled as follows: ν_1, ν_2, ν_3, ν_4, and ν_5 are a_g, ν_6 and ν_7 are a_u, ν_8 is b_g, and ν_9, ν_{10}, ν_{11}, and ν_{12} are b_u.

Problems

1.

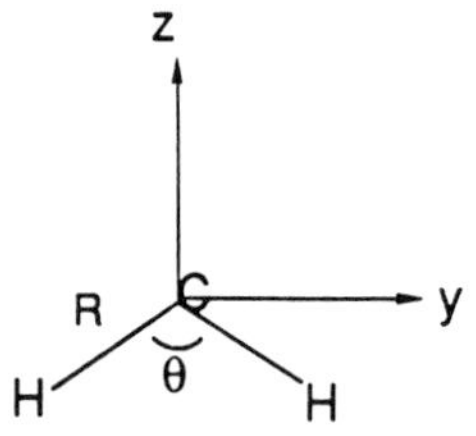

Molecule I	Molecule II
$R_{CH} = 1.121\text{Å}$	$R_{CH} = 1.076\text{Å}$
$\angle\text{HCH} = 104°$	$\angle\text{HCH} = 136°$
$y_H = R\sin(\theta/2) = \pm 0.8834$	$y_H = \pm 0.9976$
$z_H = R\cos(\theta/2) = -0.6902$	$z_H = -0.4031$
Center of Mass(COM): clearly, $X = Y = 0$	
$Z = \dfrac{12(0) - 2R\cos(\theta/2)}{14} = -0.0986$	$Z = -0.0576$

a. $I_{xx} = \sum_j m_j(y_j^2 + z_j^2) - M(Y^2 + Z^2)$

$$I_{xy} = -\sum_j m_j x_j y_j - MXY$$

$I_{xx} = 2(1.121)^2 - 14(-0.0986)^2 = 2.377$

$I_{xx} = 2(1.076)^2 - 14(-0.0576)^2 = 2.269$

$I_{yy} = 2(0.6902)^2 - 14(-0.0986)^2$
$= 0.8167$
$I_{zz} = 2(0.8834)^2$
$= 1.561$
$I_{xz} = I_{yz} = I_{xy} = 0$

$I_{yy} = 2(0.4031)^2 - 14(-0.0576)^2$
$= 0.2786$
$I_{zz} = 2(0.9976)^2$
$= 1.990$

b. Since the moment of inertia tensor is already diagonal, the principal moments of inertia have already been determined to be ($I_\alpha < I_b < I_c$):

$I_{yy} < I_{zz} < I_{xx}$
$0.8167 < 1.561 < 2.377$

$I_{yy} < I_{zz} < I_{xx}$
$0.2786 < 1.990 < 2.269$

Using the formula:

$$A = \frac{h}{8\pi^2 c I_a} = \frac{6.626 \times 10^{-27}}{8\pi^2(3 \times 10^{10})I_a} \times \frac{6.02 \times 10^{23}}{(1 \times 10^{-8})^2}$$

$$A = \frac{16.84}{I_a}\text{cm}^{-1}$$

similarly,

$$B = \frac{16.84}{I_b}\text{cm}^{-1}, \text{ and } C = \frac{16.84}{I_c}\text{cm}^{-1}.$$

So,

Molecule I	**Molecule II**
$y \Rightarrow A = 20.62$	$y \Rightarrow A = 60.45$
$z \Rightarrow B = 10.79$	$z \Rightarrow B = 8.46$
$x \Rightarrow C = 7.08$	$x \Rightarrow C = 7.42$
c. Averaging $B + C$:	
$B = (B + C)/2 = 8.94$	$B = (B + C)/2 = 7.94$
$A - B = 11.68$	$A - B = 52.51$
Using the prolate top formula: $E = (A - B)K^2 + BJ(J + 1)$,	
$E = 11.68K^2 + 8.94J(J + 1)$	$E = 52.51K^2 + 7.94J(J + 1)$

Levels: $J = 0,1,2, \ldots$ and $K = 0,1, \ldots J$

For a given level defined by J and K, there are M_J degeneracies given by:

$$(2J + 1) \times \begin{Bmatrix} 1 \text{ for } K = 0 \\ 2 \text{ for } K \neq 0 \end{Bmatrix}$$

d.

Molecule I

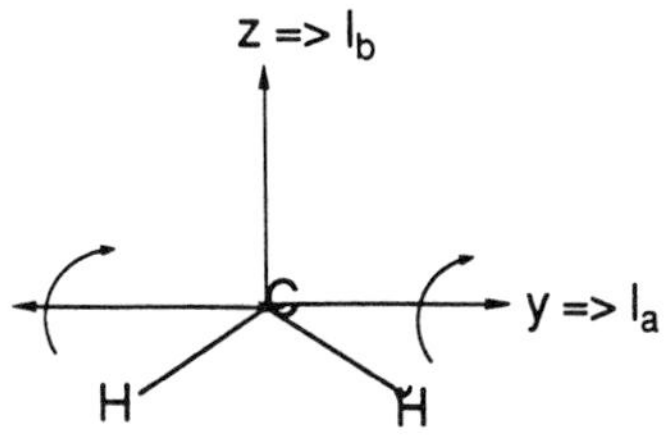

Molecule II

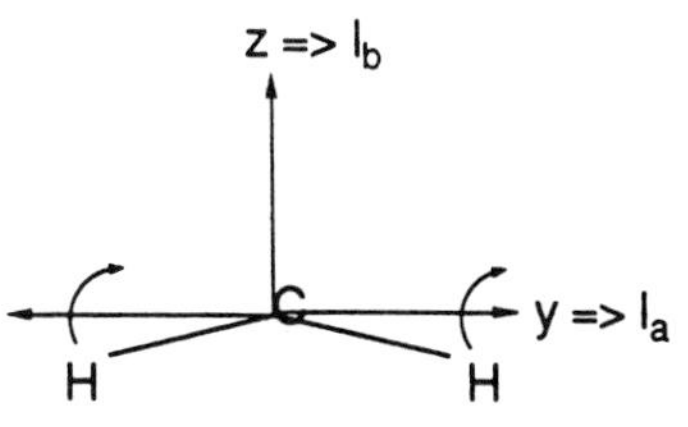

e. Since $\vec{\mu}$ is along Y, $\Delta K = 0$ since K describes rotation about the y axis. Therefore $\Delta J = \pm 1$

f. *Assume* molecule I is CH_2^- and molecule II is CH_2. Then,

$\Delta E = E_{J_j}(CH_2) - E_{J_i}(CH_2^-)$, where:

$E(CH_2) = 52.51K^2 + 7.94J(J+1)$, and $E(CH_2^-) = 11.68K^2 + 8.94J(J+1)$

For R-branches: $J_j = J_i + 1$, $\Delta K = 0$:

$$\Delta E_R = E_{J_j}(CH_2) - E_{J_i}(CH_2^-)$$
$$= 7.94(J_i+1)(J_i+1+1) - 8.94J_i(J_i+1)$$
$$= (J_i+1)\{7.94(J_i+1+1) - 8.94J_i\}$$
$$= (J_i+1)\{(7.94 - 8.94)J_i + 2(7.94)\}$$
$$= (J_i+1)\{-J_i + 15.88\}$$

For P-branches: $J_j = J_i - 1$, $\Delta K = 0$:

$$\Delta E_P = E_{J_j}(CH_2) - E_{J_i}(CH_2^-)$$
$$= 7.94(J_i-1)(J_i-1+1) - 8.94J_i(J_i+1)$$
$$= J_i\{7.94(J_i-1) - 8.94(J_i+1)\}$$
$$= J_i\{(7.94 - 8.94)J_i - 7.94 - 8.94\}$$
$$= J_i\{-J_i - 16.88\}$$

This indicates that the R branch lines occur at energies which grow closer and closer together as J increases (since the $15.88 - J_i$ term will cancel). The P branch lines occur at energies which lie more and more negative (i.e., to the left of the origin). So, you can predict that if molecule I is CH_2^- and molecule II is CH_2 then the R-branch has a band head and the P-branch does not. This is observed therefore our assumption was correct: molecule I is CH_2^- and molecule II is CH_2.

g. The band head occurs when

$$\frac{d(\Delta E_R)}{dJ} = 0.$$
$$\frac{d(\Delta E_R)}{dJ} = \frac{d}{dJ}[(J_i+1)\{-J_i + 15.88\}] = 0$$
$$= \frac{d}{dJ}(-J_i^2 - J_i + 15.88J_i + 15.88) = 0$$
$$= -2J_i + 14.88 = 0$$

$\therefore J_i = 7.44$, so $J = 7$ or 8.

At $J = 7.44$:

$\Delta E_R = (J+1)\{-J + 15.88\}$

$\Delta E_R = (7.44+1)\{-7.44 + 15.88\} = (8.44)(8.44) = 71.2\text{cm}^{-1}$ above the origin.

2. a.

D_{6h}	E	$2C_6$	$2C_3$	C_2	$3C_2'$	$3C_2''$	i	$2S_3$	$2S_6$	σ_h	$3\sigma_d$	$3\sigma_v$		
A_{1g}	1	1	1	1	1	1	1	1	1	1	1	1		x^2+y^2, z^2
A_{2g}	1	1	1	1	−1	−1	1	1	1	1	−1	−1	R_z	
B_{1g}	1	−1	1	−1	1	−1	1	−1	1	−1	1	−1		
B_{2g}	1	−1	1	−1	−1	1	1	−1	1	−1	−1	1		

E_{1g}	2	1	−1	−2	0	0	2	1	−1	−2	0	0	(R_x,R_y)	(xz,yx)
E_{2g}	2	−1	−1	2	0	0	2	−1	−1	2	0	0		(x^2-y^2,xy)
A_{1u}	1	1	1	1	1	1	−1	−1	−1	−1	−1	−1		
A_{2u}	1	1	1	1	−1	−1	−1	−1	−1	−1	1	1	z	
B_{1u}	1	−1	1	−1	1	−1	−1	1	−1	1	−1	1		
B_{2u}	1	−1	1	−1	−1	1	−1	1	−1	1	1	−1		
E_{1u}	2	1	−1	−2	0	0	−2	−1	1	2	0	0	(x,y)	
E_{2u}	2	−1	−1	2	0	0	−2	1	1	−2	0	0		
Γ_{C-H}	6	0	0	0	0	2	0	0	0	6	2	0		

b. The number of irreducible representations may be found by using the following formula:

$$n_{irrep} = \frac{1}{g}\sum_R \chi_{red}(R)\chi_{irrep}(R),$$

where g = the order of the point group (24 for D_{6h}).

$$\begin{aligned} n_{A_{1g}} &= \frac{1}{24}\sum_R \Gamma_{C\text{-}H}(R)\,.\,A_{1g}(R) \\ &= \frac{1}{24}\{(1)(6)(1)+(2)(0)(1)+(2)(0)(1)+(1)(0)(1)+(3)(0)(1)+(3)(2)(1)+(1)(0)(1)+(2)(0)(1) \\ &\quad +(2)(0)(1)+(1)(6)(1)+(3)(2)(1)+(3)(0)(1)\} \\ &= 1 \end{aligned}$$

$$\begin{aligned} n_{A_{2g}} &= \frac{1}{24}\{(1)(6)(1)+(2)(0)(1)+(2)(0)(1)+(1)(0)(1) \\ &\quad +(3)(0)(-1)+(3)(2)(-1)+(1)(0)(1)+(2)(0)(1) \\ &\quad +(2)(0)(1)+(1)(6)(1)+(3)(2)(-1)+(3)(0)(-1)\} \\ &= 0 \end{aligned}$$

$$\begin{aligned} n_{B_{1g}} &= \frac{1}{24}\{(1)(6)(1)+(2)(0)(-1)+(2)(0)(1)+(1)(0)(-1) \\ &\quad +(3)(0)(1)+(3)(2)(-1)+(1)(0)(1)+(2)(0)(-1) \\ &\quad +(2)(0)(1)+(1)(6)(-1)+(3)(2)(1)+(3)(0)(-1)\} \\ &= 0 \end{aligned}$$

$$\begin{aligned} n_{B_{2g}} &= \frac{1}{24}\{(1)(6)(1)+(2)(0)(-1)+(2)(0)(1)+(1)(0)(-1) \\ &\quad +(3)(0)(-1)+(3)(2)(1)+(1)(0)(1)+(2)(0)(-1) \\ &\quad +(2)(0)(1)+(1)(6)(-1)+(3)(2)(-1)+(3)(0)(1)\} \\ &= 0 \end{aligned}$$

$$\begin{aligned} n_{E_{1g}} &= \frac{1}{24}\{(1)(6)(2)+(2)(0)(1)+(2)(0)(-1)+(1)(0)(-2) \\ &\quad +(3)(0)(0)+(3)(2)(0)+(1)(0)(2)+(2)(0)(1) \\ &\quad +(2)(0)(-1)+(1)(6)(-2)+(3)(2)(0)+(3)(0)(0)\} \\ &= 0 \end{aligned}$$

$$n_{E_{2g}} = \frac{1}{24}\{(1)(6)(2) + (2)(0)(-1) + (2)(0)(-1) + (1)(0)(2) + (3)(0)(0) + (3)(2)(0) + (1)(0)(2) + (2)(0)(-1) + (2)(0)(-1) + (1)(6)(2) + (3)(2)(0) + (3)(0)(0)\} = 1$$

$$n_{A_{1u}} = \frac{1}{24}\{(1)(6)(1) + (2)(0)(1) + (2)(0)(1) + (1)(0)(1) + (3)(0)(1) + (3)(2)(1) + (1)(0)(-1) + (2)(0)(-1) + (2)(0)(-1) + (1)(6)(-1) + (3)(2)(-1) + (3)(0)(-1)\} = 0$$

$$n_{A_{2u}} = \frac{1}{24}\{(1)(6)(1) + (2)(0)(1) + (2)(0)(1) + (1)(0)(1) + (3)(0)(-1) + (3)(2)(-1) + (1)(0)(-1) + (2)(0)(-1) + (2)(0)(-1) + (1)(6)(-1) + (3)(2)(1) + (3)(0)(1)\} = 0$$

$$n_{B_{1u}} = \frac{1}{24}\{(1)(6)(1) + (2)(0)(-1) + (2)(0)(1) + (1)(0)(-1) + (3)(0)(1) + (3)(2)(-1) + (1)(0)(-1) + (2)(0)(1) + (2)(0)(-1) + (1)(6)(1) + (3)(2)(-1) + (3)(0)(1)\} = 0$$

$$n_{B_{2u}} = \frac{1}{24}\{(1)(6)(1) + (2)(0)(-1) + (2)(0)(1) + (1)(0)(-1) + (3)(0)(-1) + (3)(2)(1) + (1)(0)(-1) + (2)(0)(1) + (2)(0)(-1) + (1)(6)(1) + (3)(2)(1) + (3)(0)(-1)\} = 1$$

$$n_{E_{1u}} = \frac{1}{24}\{(1)(6)(2) + (2)(0)(1) + (2)(0)(-1) + (1)(0)(-2) + (3)(0)(0) + (3)(2)(0) + (1)(0)(-2) + (2)(0)(-1) + (2)(0)(1) + (1)(6)(2) + (3)(2)(0) + (3)(0)(0)\} = 1$$

$$n_{E_{2u}} = \frac{1}{24}\{(1)(6)(2) + (2)(0)(-1) + (2)(0)(-1) + (1)(0)(2) + (3)(0)(0) + (3)(2)(0) + (1)(0)(-2) + (2)(0)(1) + (2)(0)(1) + (1)(6)(-2) + (3)(2)(0) + (3)(0)(0)\} = 0$$

We see that $\Gamma_{C\text{–}H} = A_{1g} \oplus E_{2g} \oplus B_{2u} \oplus E_{1u}$

c. x and $y \Rightarrow E_{1u}$, $z \Rightarrow A_{2u}$, so, the ground state A_{1g} level can be excited to the degenerate E_{1u} level by coupling through the x or y transition dipoles. Therefore E_{1u} is infrared active and $\perp$ polarized.

d. $(x^2 + y^2, z^2) \Rightarrow A_{1g}$, $(xz,yz) \Rightarrow E_{1g}$, $(x^2 - y^2,xy) \Rightarrow E_{2g}$, so the ground state A_{1g} level can be excited to the degenerate E_{2g} level by coupling through the x^2-y^2 or xy transitions or be excited to the degenerate A_{1g} level by coupling through the xz or yz transitions. Therefore A_{1g} and E_{2g} are Raman active.

e. The B_{2u} mode is not IR or Raman active.

3. a. $\frac{d}{dr}(Fr^{-1}) = F'r^{-1} - r^{-2}F$

$$r^2\frac{d}{dr}(Fr^{-1}) = rF' - F$$

$$\frac{d}{dr}\left(r^2\frac{d}{dr}(Fr^{-1})\right) = F' - F' + rF''$$

So,

$$\frac{-\hbar^2}{2\mu r^2}\frac{d}{dr}\left(r^2\frac{d}{dr}(Fr^{-1})\right) = \frac{-\hbar^2}{2\mu}\frac{F''}{r}.$$

Rewriting the radial Schrödinger equation with the substitution: $R = r^{-1}F$ gives:

$$\frac{-h^2}{2\mu r^2}\frac{d}{dr}\left(r^2\frac{d(Fr^{-1})}{dr}\right) + \frac{J(J+1)\hbar^2}{2\mu r^2}(Fr^{-1}) + \frac{1}{2}k(r - r_e)^2(Fr^{-1}) = E(Fr^{-1})$$

Using the above derived identity gives:

$$\frac{-\hbar^2}{2\mu}\frac{F''}{r} + \frac{J(J+1)\hbar^2}{2\mu r^2}(Fr^{-1}) + \frac{1}{2}k(r - r_e)^2(Fr^{-1}) = E(Fr^{-1})$$

Cancelling out an r^{-1}:

$$\frac{-\hbar^2}{2\mu}F'' + \frac{J(J+1)\hbar^2}{2\mu r^2}F + \frac{1}{2}k(r - r_e)^2F = EF$$

b. $\frac{1}{r^2} = \frac{1}{(r_e+\Delta r)^2} = \frac{1}{r_e^2\left(1 + \frac{\Delta r}{r_e}\right)^2} \approx \frac{1}{r_e^2}\left(1 - \frac{2\Delta r}{r_e} + \frac{3\Delta r^2}{r_e^2}\right)$

So,

$$\frac{J(J+1)\hbar^2}{2\mu r^2} \approx \frac{J(J+1)\hbar^2}{2\mu r_e^2}\left(1 - \frac{2\Delta r}{r_e} + \frac{3\Delta r^2}{r_e^2}\right)$$

c. Using this substitution we now have:

$$\frac{-\hbar^2}{2\mu}F'' + \frac{J(J+1)\hbar^2}{2\mu r_e^2}\left(1 - \frac{2\Delta r}{r_e} + \frac{3\Delta r^2}{r_e^2}\right)F + \frac{1}{2}k(r - r_e)^2F = EF$$

Now, regroup the terms which are linear and quadratic in $\Delta r = r - r_e$:

$$\frac{1}{2}k\Delta r^2 + \frac{J(J+1)\hbar^2}{2\mu r_e^2}\frac{3}{r_e^2}\Delta r^2 - \frac{J(J+1)\hbar^2}{2\mu r_e^2}\frac{2}{r_e}\Delta r$$

$$= \left(\frac{1}{2}k + \frac{J(J+1)\hbar^2}{2\mu r_e^2}\frac{3}{r_e^2}\right)\Delta r^2 - \left(\frac{J(J+1)\hbar^2}{2\mu r_e^2}\frac{2}{r_e}\right)\Delta r$$

Now, we must complete the square:

$$a\Delta r^2 - b\Delta r = a\left(\Delta r - \frac{b}{2a}\right)^2 - \frac{b^2}{4a}.$$

So,

$$\left(\frac{1}{2}k + \frac{J(J+1)\hbar^2}{2\mu r_e^2}\frac{3}{r_e^2}\right)\left(\Delta r - \frac{\frac{J(J+1)\hbar^2}{2\mu r_e^2}\frac{1}{r_e}}{\frac{1}{2}k + \frac{J(J+1)\hbar^2}{2\mu r_e^2}\frac{3}{r_e^2}}\right)^2 - \frac{\left(\frac{J(J+1)\hbar^2}{2\mu r_e^2}\frac{1}{r_e}\right)^2}{\frac{1}{2}k + \frac{J(J+1)\hbar^2}{2\mu r_e^2}\frac{3}{r_e^2}}$$

Now, redefine the first term as $\frac{1}{2}\bar{k}$, second term as $(r - \bar{r}_e)^2$, and the third term as $-\Delta$ giving:

$$\frac{1}{2}\bar{k}\left(r - \bar{r}_e\right)^2 - \Delta$$

From:

$$\frac{-\hbar^2}{2\mu}F'' + \frac{J(J+1)\hbar^2}{2\mu r_e^2}\left(1 - \frac{2\Delta r}{r_e} + \frac{3\Delta r^2}{r_e^2}\right)F + \frac{1}{2}k(r - r_e)^2F = EF,$$

$$\frac{-\hbar^2}{2\mu}F'' + \frac{J(J+1)\hbar^2}{2\mu r_e^2}F + \left(\frac{J(J+1)\hbar^2}{2\mu r_e^2}\left(-\frac{2\Delta r}{r_e} + \frac{3\Delta r^2}{r_e^2}\right) + \frac{1}{2}k\Delta r^2\right)F = EF,$$

and making the above substitution results in:

$$\frac{-\hbar^2}{2\mu}F'' + \frac{J(J+1)\hbar^2}{2\mu r_e^2}F + \left(\frac{1}{2}\bar{k}\left(r - \bar{r}_e\right)^2 - \Delta\right)F = EF,$$

or,

$$\frac{-\hbar^2}{2\mu}F'' + \frac{1}{2}\bar{k}(r - \bar{r}_e)^2F = \left(E - \frac{J(J+1)\hbar^2}{2\mu r_e^2} + \Delta\right)F.$$

d. Since the above is nothing but a harmonic oscillator differential equation in x with force constant $\bar{k}$ and equilibrium bond length $\bar{r}_e$, we know that:

$\frac{-\hbar^2}{2\mu}F'' + \frac{1}{2}\bar{k}(r - \bar{r}_e)^2F = \varepsilon F$, has energy levels:

$$\varepsilon = \hbar\sqrt{\frac{\bar{k}}{\mu}}\left(v + \frac{1}{2}\right),\ v = 0,1,2,\ldots$$

So,

$$E + \Delta - \frac{J(J+1)\hbar^2}{2\mu r_e^2} = \varepsilon$$

tells us that:

$$E = \hbar\sqrt{\frac{k}{\mu}}\left(v + \frac{1}{2}\right) + \frac{J(J+1)\hbar^2}{2\mu r_e^2} - \Delta.$$

As J increases, r_e increases because of the centrifugal force pushing the two atoms apart. On the other hand k increases which indicates that the molecule finds it more difficult to stretch against both the centrifugal and Hooke's law (spring) Harmonic force field. The total energy level (labeled by J and v) will equal a rigid rotor component

$$\frac{J(J+1)\hbar^2}{2\mu r_e^2}$$

plus a Harmonic oscillator part

$$\hbar\sqrt{\frac{k}{\mu}}\left(v + \frac{1}{2}\right)$$

(which has a force constant k which increases with J).

SECTION

5

TIME-DEPENDENT PROCESSES

CHAPTER

14

The interaction of a molecular species with electromagnetic fields can cause transitions to occur among the available molecular energy levels (electronic, vibrational, rotational, and nuclear spin). Collisions among molecular species likewise can cause transitions to occur. Time-dependent perturbation theory and the methods of molecular dynamics can be employed to treat such transitions.

I. THE PERTURBATION DESCRIBING INTERACTIONS WITH ELECTROMAGNETIC RADIATION

The full N-electron non-relativistic Hamiltonian H discussed earlier in this text involves the kinetic energies of the electrons and of the nuclei and the mutual coulombic interactions among these particles

$$H = \sum_{a=1,M} (-\hbar^2/2m_a)\nabla_a^2 + \sum_j [(-\hbar^2/2m_e)\nabla_j^2 - \sum_a Z_a e^2/r_{j,a}]$$

$$+ \sum_{j<k} e^2/r_{j,k} + \sum_{a<b} Z_a Z_b e^2/R_{a,b}.$$

When an electromagnetic field is present, this is not the correct Hamiltonian, but it can be modified straightforwardly to obtain the proper H.

A. The Time-Dependent Vector $A(\boldsymbol{r},t)$ Potential

The only changes required to achieve the Hamiltonian that describes the same system in the presence of an electromagnetic field are to replace the momentum operators $\boldsymbol{P}_a$ and $\boldsymbol{p}_j$ for the nuclei and electrons, respectively, by $(\boldsymbol{P}_a - (Z_a e/c)\boldsymbol{A}(R_a,t))$ and $(\boldsymbol{p}_j - (e/c)\boldsymbol{A}(r_j,t))$. Here $Z_a e$ is the charge on the ath nucleus, $-e$ is the charge of the electron, and c is the speed of light.

The vector potential $\boldsymbol{A}$ depends on time t and on the spatial location $\boldsymbol{r}$ of the particle in the following manner:

$$\boldsymbol{A}(\boldsymbol{r},t) = 2\boldsymbol{A}_o \cos(\omega t - \boldsymbol{k}\bullet\boldsymbol{r}).$$

The circular frequency of the radiation ω (radians per second) and the wave vector $\boldsymbol{k}$ (the magnitude of $\boldsymbol{k}$ is $|\boldsymbol{k}| = 2\pi/\lambda$, where λ is the wavelength of the light) control the temporal and spatial oscillations of the photons. The vector $\boldsymbol{A}_o$ characterizes the strength (through the magnitude of $\boldsymbol{A}_o$) of the field as well as the direction of the $\boldsymbol{A}$ potential; the direction of propagation of the photons is given by the unit vector $\boldsymbol{k}/|\boldsymbol{k}|$. The factor of 2 in the definition of $\boldsymbol{A}$ allows one to think of $\boldsymbol{A}_0$ as measuring the strength of both $\exp(i(\omega t - \boldsymbol{k}\bullet\boldsymbol{r}))$ and $\exp(-i(\omega t - \boldsymbol{k}\bullet\boldsymbol{r}))$ components of the $\cos(\omega t - \boldsymbol{k}\bullet\boldsymbol{r})$ function.

B. The Electric $E(r,t)$ and Magnetic $H(r,t)$ Fields

The electric $\boldsymbol{E}(\boldsymbol{r},t)$ and magnetic $\boldsymbol{H}(\boldsymbol{r},t)$ fields of the photons are expressed in terms of the vector potential $\boldsymbol{A}$ as

$$\boldsymbol{E}(\boldsymbol{r},t) = -1/c\ \partial\boldsymbol{A}/\partial t = \omega/c\ 2\,\boldsymbol{A}_o \sin(\omega t - \boldsymbol{k}\bullet\boldsymbol{r})$$
$$\boldsymbol{H}(\boldsymbol{r},t) = \nabla\times\boldsymbol{A} = \boldsymbol{k}\times\boldsymbol{A}_o\, 2\sin(\omega t - \boldsymbol{k}\bullet\boldsymbol{r}).$$

The $\boldsymbol{E}$ field lies parallel to the $\boldsymbol{A}_o$ vector, and the $\boldsymbol{H}$ field is perpendicular to $\boldsymbol{A}_o$; both are perpendicular to the direction of propagation of the light $\boldsymbol{k}/|\boldsymbol{k}|$. $\boldsymbol{E}$ and $\boldsymbol{H}$ have the same phase because they both vary with time and spatial location as $\sin(\omega t - \boldsymbol{k}\bullet\boldsymbol{r})$. The relative orientations of these vectors are shown below.

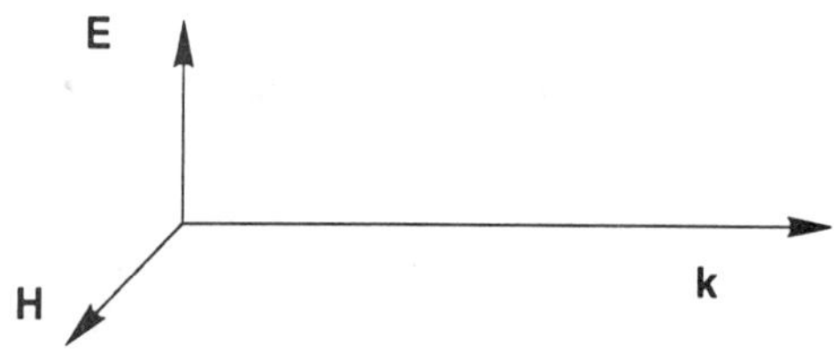

C. The Resulting Hamiltonian

Replacing the nuclear and electronic momenta by the modifications shown above in the kinetic energy terms of the full electronic and nuclear-motion hamiltonian results in the following *additional* factors appearing in *H:*

$$H_{int} = \sum_j \left\{(ie\hbar/m_e c)\boldsymbol{A}(r_j,t)\bullet\nabla_j + (e^2/2m_e c^2)|\boldsymbol{A}(r_j,t)|^2\right\}$$
$$+\sum_a \left\{(iZ_a e\hbar/m_a c)\boldsymbol{A}(R_a,t)\bullet\nabla_a + (Z_a^2 e^2/2m_a c^2)|\boldsymbol{A}(R_a,t)|^2\right\}$$

These so-called interaction perturbations H_{int} are what induces transitions among the various electronic/vibrational/rotational states of a molecule. The one-electron additive nature of H_{int} plays an important role in determining the kind of transitions that H_{int} can induce. For example, it causes the most intense electronic transitions to involve excitation of a single electron from one orbital to another (recall the Slater-Condon rules).

II. TIME-DEPENDENT PERTURBATION THEORY

A. The Time-Dependent Schrödinger Equation

The mathematical machinery needed to compute the rates of transitions among molecular states induced by such a time-dependent perturbation is contained in time-dependent perturbation theory (TDPT). The development of this theory proceeds as follows. One first assumes that one has in-hand *all* of the eigenfunctions $\{\Phi_k\}$ and eigenvalues $\{E_k^0\}$ that characterize the Hamiltonian H^0 of the molecule in the absence of the external perturbation:

$$H^0\Phi_k = E_k^0\Phi_k.$$

One then writes the time-dependent Schrödinger equation

$$i\hbar\partial\Psi/\partial t = (H^0 + H_{int})\Psi$$

in which the full Hamiltonian is explicitly divided into a part that governs the system in the absence of the radiation field and H_{int} which describes the interaction with the field.

B. Perturbative Solution

By treating H^0 as of zeroth order (in the field strength $|\mathbf{A_0}|$) , expanding Ψ order-by-order in the field-strength parameter:

$$\Psi = \Psi^0 + \Psi^1 + \Psi^2 + \Psi^3 + \ldots,$$

realizing that H_{int} contains terms that are both first- and second-order in $|\mathbf{A_0}|$

$$H_{int}^1 = \sum_j \{(ie\hbar/m_e c)\mathbf{A}(r_j,t)\bullet\nabla_j\}$$
$$+ \sum_a \{(iZ_a e\hbar/m_a c)\mathbf{A}(R_a,t)\bullet\nabla_a\},$$

$$H_{int}^2 = \sum_j \{(e^2/2m_e c^2)|\mathbf{A}(r_j,t)|^2\}$$
$$+ \sum_a \{(Z_a^2 e^2/2m_a c^2)|\mathbf{A}(R_a,t)|^2\},$$

and then collecting together all terms of like power of $|\mathbf{A_0}|$, one obtains the set of time-dependent perturbation theory equations. The lowest order such equations read:

$$i\hbar\partial\Psi^0/\partial t = H^0\Psi^0$$
$$i\hbar\partial\Psi^1/\partial t = (H^0\Psi^1 + H_{int}^1\Psi^0)$$
$$i\hbar\partial\Psi^2/\partial t = (H^0\Psi^2 + H_{int}^2\Psi^0 + H_{int}^1\Psi^1).$$

The zeroth order equations can easily be solved because H^0 is independent of time. Assuming that at $t = -\infty, \Psi = \Psi_i$ (we use the index i to denote the initial state), this solution is:

$\Psi^0 = \Phi_i \exp(-iE_i^0 t/\hbar)$.

The first-order correction to Ψ^0,Ψ^1 can be found by

i. expanding Ψ^1 in the complete set of zeroth-order states $\{\Phi_f\}$:

$$\Psi^1 = \sum_f \Phi_f <\Phi_f|\Psi^1> = \sum_f \Phi_f C_f^1,$$

ii. using the fact that

$$H^0\Phi_f = E_f^0 \Phi_f,$$

and

iii. substituting all of this into the equation that Ψ^1 obeys. The resultant equation for the coefficients that appear in the first-order equation can be written as

$$i\hbar \partial C_f^1/\partial t = \sum_k \{E_k^0 C_k^1 \delta_{f,k}\} + <\Phi_f|H_{int}^1|\Phi_i> \exp(-iE_i^0 t/\hbar),$$

or

$$i\hbar \partial C_f^1/\partial t = E_f^0 C_f^1 + <\Phi_f|H_{int}^1|\Phi_i> \exp(-iE_i^0 t/\hbar).$$

Defining

$$C_f^1(t) = D_f^1(t) \exp(-iE_f^0 t/\hbar),$$

this equation can be cast in terms of an easy-to-solve equation for the D_f^1 coefficients:

$$i\hbar \partial D_f^1/\partial t = <\Phi_f|H_{int}^1|\Phi_i> \exp(i[E_f^0 - E_i^0]t/\hbar).$$

Assuming that the electromagnetic field $\boldsymbol{A}(\boldsymbol{r},t)$ is turned on at $t = 0$, and remains on until $t = T$, this equation for D_f^1 can be integrated to yield:

$$D_f^1(t) = (i\hbar)^{-1} \int_0^T <\Phi_f|H_{int}^1|\Phi_i> \exp(i[E_f^0 - E_i^0]t'/\hbar)dt'.$$

C. Application to Electromagnetic Perturbations

1. First-Order Fermi-Wentzel "Golden Rule"

Using the earlier expressions for H_{int}^1 and for $\boldsymbol{A}(\boldsymbol{r},t)$

$$H_{int}^1 = \sum_j \{(ie\hbar/m_e c)\boldsymbol{A}(r_j,t)\bullet\nabla_j\} + \sum_a \{(iZ_a e\hbar/m_a c)\boldsymbol{A}(R_a,t)\bullet\nabla_a\}$$

and

$$2\boldsymbol{A}_o \cos(\omega t - \boldsymbol{k}\bullet\boldsymbol{r}) = \boldsymbol{A}_o \{\exp[i(\omega t - \boldsymbol{k}\bullet\boldsymbol{r})] + \exp[-i(\omega t - \boldsymbol{k}\bullet\boldsymbol{r})]\},$$

it is relatively straightforward to carry out the above time integration to achieve a final expression for $D_f^1(t)$, which can then be substituted into $C_f^1(t) = D_f^1(t) \exp(-iE_f^0 t/\hbar)$ to obtain the final expression for the first-order estimate of the probability amplitude for the molecule appearing in the state

$\Phi_f \exp(-iE_f^0 t/\hbar)$ after being subjected to electromagnetic radiation from $t = 0$ until $t = T$. This final expression reads:

$$C_f^1(T) = (i\hbar)^{-1} \exp(-iE_f^0 T/\hbar)\{<\Phi_f|\sum_j \{(ie\hbar/m_e c) \exp[-i\boldsymbol{k}\bullet\boldsymbol{r}_j\}]\boldsymbol{A}_0\bullet\nabla_j$$

$$+ \sum_a (iZ_a e\hbar/m_a c) \exp[-i\boldsymbol{k}\bullet\boldsymbol{R}_a]\boldsymbol{A}_0\bullet\nabla_a|\Phi_i>\} \frac{\exp(i(\omega + \omega_{f,i})T) - 1}{i(\omega + \omega_{f,i})}$$

$$+ (i\hbar)^{-1}\exp(-iE_f^0 T/\hbar) \{<\Phi_f|\sum_j \{(ie\hbar/m_e c) \exp[i\boldsymbol{k}\bullet\boldsymbol{r}_j]\boldsymbol{A}_0\bullet\nabla_j$$

$$+ \sum_a (iZ_a e\hbar/m_a c) \exp[i\boldsymbol{k}\bullet\boldsymbol{R}_a]\boldsymbol{A}_0\bullet\nabla_a|\Phi_i>\} \frac{\exp(i(-\omega + \omega_{f,i})T) - 1}{i(-\omega + \omega_{f,i})},$$

where

$$\omega_{f,i} = [E_f^0 - E_i^0]/\hbar$$

is the resonance frequency for the transition between "initial" state Φ_i and "final" state Φ_f.

Defining the time-independent parts of the above expression as

$$\alpha_{f,i} = <\Phi_f|\sum_j \{(e/m_e c) \exp[-i\boldsymbol{k}\bullet\boldsymbol{r}_j]\boldsymbol{A}_0\bullet\nabla_j + \sum_a (Z_a e/m_a c) \exp[-i\boldsymbol{k}\bullet\boldsymbol{R}_a]\boldsymbol{A}_0\bullet\nabla_a|\Phi_i>,$$

this result can be written as

$$C_f^1(T) = \exp(-iE_f^0 T/\hbar) \left\{\alpha_{f,i} \frac{\exp(i(\omega + \omega_{f,i})T) - 1}{i(\omega + \omega_{f,i})} + \alpha_{f,i}^* \frac{\exp(-i(\omega - \omega_{f,i})T) - 1}{-i(\omega - \omega_{f,i})}\right\}.$$

The modulus squared $|C_f^1(T)|^2$ gives the probability of finding the molecule in the final state Φ_f at time T, given that it was in Φ_i at time $t = 0$. If the light's frequency ω is tuned close to the transition frequency $\omega_{f,i}$ of a particular transition, the term whose denominator contains $(\omega - \omega_{f,i})$ will dominate the term with $(\omega + \omega_{f,i})$ in its denominator. Within this "near-resonance" condition, the above probability reduces to:

$$|C_f^1(T)|^2 = 2|\alpha_{f,i}|^2 \frac{(1 - \cos((\omega - \omega_{f,i})T))}{(\omega - \omega_{f,i})^2}$$

$$= 4|\alpha_{f,i}|^2 \frac{\sin^2(1/2(\omega - \omega_{f,i})T)}{(\omega - \omega_{f,i})^2}.$$

This is the final result of the first-order time-dependent perturbation theory treatment of light-induced transitions between states Φ_i and Φ_f.

The so-called sine-function

$$\frac{\sin^2(1/2(\omega - \omega_{f,i})T)}{(\omega - \omega_{f,i})^2}$$

as shown in the figure below is strongly peaked near $\omega = \omega_{f,i}$, and displays secondary maxima (of decreasing amplitudes) near $\omega = \omega_{f,i} + 2n\pi/T$, $n = 1,2, \ldots$. In the $T \to \infty$ limit, this function becomes narrower and narrower, and the area under it

$$\int_{-\infty}^{\infty} \frac{\sin^2(1/2(\omega - \omega_{f,i})T)}{(\omega - \omega_{f,i})^2} d\omega = T/2 \int_{-\infty}^{\infty} \frac{\sin^2(1/2(\omega - \omega_{f,i})T)}{1/4T^2(\omega - \omega_{f,i})^2} d\omega T/2$$

$$= T/2 \int_{-\infty}^{\infty} \frac{\sin^2(x)}{x^2} dx = \pi T/2$$

grows with T. Physically, this means that when the molecules are exposed to the light source for long times (large T), the sinc function emphasizes ω values near $\omega_{f,i}$ (i.e., the on-resonance ω values). These properties of the sinc function will play important roles in what follows.

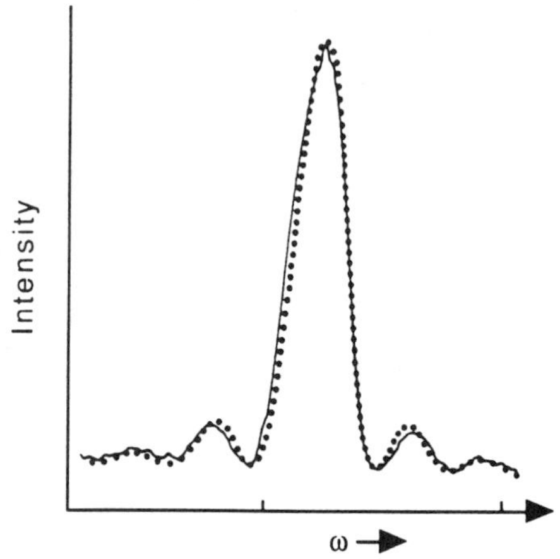

In most experiments, light sources have a "spread" of frequencies associated with them; that is, they provide photons of various frequencies. To characterize such sources, it is common to introduce the spectral source function $g(\omega)d\omega$ which gives the probability that the photons from this source have frequency somewhere between ω and $\omega + d\omega$. For narrow-band lasers, $g(\omega)$ is a sharply peaked function about some "nominal" frequency ω_o; broader band light sources have much broader $g(\omega)$ functions.

When such non-monochromatic light sources are used, it is necessary to average the above formula for $|C_f^1(T)|^2$ over the $g(\omega)d\omega$ probability function in computing the probability of finding the molecule in state Φ_f after time T, given that it was in Φ_i up until $t = 0$, when the light source was turned on. In particular, the proper expression becomes:

$$\left| C_f^1(T) \right|^2_{ave} = 4|\alpha_{f,i}|^2 \int g(\omega) \frac{\sin^2(1/2(\omega - \omega_{f,i})T)}{(\omega - \omega_{f,i})^2} d\omega$$

$$= 2|\alpha_{f,i}|^2 T \int_{-\infty}^{\infty} g(\omega) \frac{\sin^2(1/2(\omega - \omega_{f,i})T)}{1/4T^2(\omega - \omega_{f,i})^2} d\omega T/2.$$

If the light-source function is "tuned" to peak near $\omega = \omega_{f,i}$, and if $g(\omega)$ is much broader (in ω-space) than the

$$\frac{\sin^2(1/2(\omega - \omega_{f,i})T)}{(\omega - \omega_{f,i})^2}$$

function, $g(\omega)$ can be replaced by its value at the peak of the

$$\frac{\sin^2(1/2(\omega - \omega_{f,i})T)}{(\omega - \omega_{f,i})^2}$$

function, yielding:

$$\left|C_f^1(T)\right|^2_{ave} = 2g(\omega_{f,i})|\alpha_{f,i}|^2 T \int_{-\infty}^{\infty} \frac{\sin^2(1/2(\omega - \omega_{f,i})T)}{1/4T^2(\omega - \omega_{f,i})^2} d\omega T/2$$

$$= 2g(\omega_{f,i})|\alpha_{f,i}|^2 T \int_{-\infty}^{\infty} \frac{\sin^2(x)}{x^2} dx = 2\pi g(\omega_{f,i})|\alpha_{f,i}|^2 T .$$

The fact that the *probability* of excitation from Φ_i to Φ_f grows linearly with the time T over which the light source is turned on implies that the *rate* of transitions between these two states is constant and given by:

$$R_{i,f} = 2\pi g(\omega_{f,i})|\alpha_{f,i}|^2;$$

this is the so-called first-order Fermi-Wentzel "golden rule" expression for such transition rates. It gives the rate as the square of a transition matrix element between the two states involved, of the first order perturbation multiplied by the light source function $g(\omega)$ evaluated at the transition frequency $\omega_{f,i}$.

2. Higher Order Results

Solution of the second-order time-dependent perturbation equations,

$$i\hbar\partial\Psi^2/\partial t = (H^0\Psi^2 + H^2_{int}\Psi^0 + H^1_{int}\Psi^1)$$

which will not be treated in detail here, gives rise to two distinct types of contributions to the transition probabilities between Φ_i and Φ_f:

i. There will be matrix elements of the form

$$<\Phi_f|\sum_j \left\{(e^2/2m_e c^2)|\mathbf{A}(r_j,t)|^2\right\} + \sum_a \left\{(Z_a^2 e^2/2m_a c^2)|\mathbf{A}(R_a,t)|^2\right\}|\Phi_i>$$

arising when H^2_{int} couples Φ_i to Φ_f.

ii. There will be matrix elements of the form

$$\sum_k <\Phi_f|\sum_j \left\{(ie\hbar/m_e c)\mathbf{A}(r_j,t)\bullet\nabla_j\right\} + \sum_a \left\{(iZ_a e\hbar/m_a c)\mathbf{A}(R_a,t)\bullet\nabla_a\right\}|\Phi_k>$$

$$<\Phi_k|\sum_j \left\{(ie\hbar/m_e c)\mathbf{A}(r_j,t)\bullet\nabla_j\right\} + \sum_a \left\{(iZ_a e\hbar/m_a c)\mathbf{A}(R_a,t)\bullet\nabla_a\right\}|\Phi_i>$$

arising from expanding

$$H^1_{int}\Psi^1 = \sum_k C^1_k H^1_{int}|\Phi_k>$$

and using the earlier result for the first-order amplitudes C_k^1. Because both types of second-order terms vary quadratically with the $\boldsymbol{A}(\boldsymbol{r},t)$ potential, and because $\boldsymbol{A}$ has time dependence of the form $\cos(\omega t - \boldsymbol{k}\bullet\boldsymbol{r})$, these terms contain portions that vary with time as $\cos(2\omega t)$. As a result, transitions between initial and final states Φ_i and Φ_f whose transition frequency is $\omega_{f,i}$ can be induced when $2\omega = \omega_{f,i}$; in this case, one speaks of coherent two-photon induced transitions in which the electromagnetic field produces a perturbation that has twice the frequency of the "nominal" light source frequency ω.

D. The "Long-Wavelength" Approximation

To make progress in further analyzing the first-order results obtained above, it is useful to consider the wavelength λ of the light used in most visible/ultraviolet, infrared, or microwave spectroscopic experiments. Even the shortest such wavelengths (ultraviolet) are considerably longer than the spatial extent of all but the largest molecules (i.e., polymers and biomolecules for which the approximations we introduce next are not appropriate).

In the definition of the essential coupling matrix element $\alpha_{f,i}$

$$\alpha_{f,i} = <\Phi_f|\sum_j (e/m_e c)\exp[-i\boldsymbol{k}\bullet\boldsymbol{r}_j]\boldsymbol{A}_0\bullet\nabla_j + \sum_a (Z_a e/m_a c)\exp[-i\boldsymbol{k}\bullet\boldsymbol{R}_a]\boldsymbol{A}_0\bullet\nabla_a|\Phi_i>,$$

the factors $\exp[-i\boldsymbol{k}\bullet\boldsymbol{r}_j]$ and $\exp[-i\boldsymbol{k}\bullet\boldsymbol{R}_a]$ can be expanded as:

$$\exp[-i\boldsymbol{k}\bullet\boldsymbol{r}_j] = 1 + (-i\boldsymbol{k}\bullet\boldsymbol{r}_j) + 1/2(-i\boldsymbol{k}\bullet\boldsymbol{r}_j)^2 + \ldots$$
$$\exp[-i\boldsymbol{k}\bullet\boldsymbol{R}_a] = 1 + (-i\boldsymbol{k}\bullet\boldsymbol{R}_a) + 1/2(-i\boldsymbol{k}\bullet\boldsymbol{R}_a)^2 + \ldots.$$

Because $|\boldsymbol{k}| = 2\pi/\lambda$, and the scales of $\boldsymbol{r}_j$ and $\boldsymbol{R}_a$ are of the dimension of the molecule, $\boldsymbol{k}\bullet\boldsymbol{r}_j$ and $\boldsymbol{k}\bullet\boldsymbol{R}_a$ are less than unity in magnitude, within this so-called "long-wavelength" approximation.

1. *Electric Dipole Transitions*

Introducing these expansions into the expression for $\alpha_{f,i}$ gives rise to terms of various powers in $1/\lambda$. The lowest order terms are:

$$\alpha_{f,i}(E1) = <\Phi_f|\sum_j (e/m_e c)\boldsymbol{A}_0\bullet\nabla_j + \sum_a (Z_a e/m_a c)\boldsymbol{A}_0\bullet\nabla_a|\Phi_i>$$

and are called "electric dipole" terms, and are denoted $E1$. To see why these matrix elements are termed $E1$, we use the following identity (see chapter 1) between the momentum operator $-i\hbar\nabla$ and the corresponding position operator $\boldsymbol{r}$:

$$\nabla_j = -(m_e/\hbar^2)[H,\boldsymbol{r}_j]$$
$$\nabla_a = -(m_a/h^2)[H,\boldsymbol{R}_a]\,.$$

This derives from the fact that H contains ∇_j and ∇_a in its kinetic energy operators (as ∇_a^2 and ∇_j^2).

Substituting these expressions into the above $\alpha_{f,i}(E1)$ equation and using

$$H^0_{i\,\mathrm{or}\,f} = E^0_{i\,\mathrm{or}\,f}\,\Phi_{i\,\mathrm{or}\,f},$$

one obtains:

$$\alpha_{f,i}(E1) = (E_f^0 - E_i^0)\boldsymbol{A_0}\bullet\langle\Phi_f|\sum_j (e/\hbar^2c)\boldsymbol{r}_j + \sum_a (Z_ae/\hbar^2c)\boldsymbol{R}_a|\Phi_i\rangle$$

$$= \omega_{f,i}\boldsymbol{A}_0\bullet\langle\Phi_f|\sum_j (e/\hbar c)\boldsymbol{r}_j + \sum_a (Z_ae/\hbar c)\boldsymbol{R_a}|\Phi_i\rangle$$

$$= (\omega_{f,i}/\hbar c)\boldsymbol{A}_0\bullet\langle\Phi_f|\mu|\Phi_i\rangle,$$

where μ is the electric dipole moment operator for the electrons and nuclei:

$$\mu = \sum_j e\boldsymbol{r}_j + \sum_a Z_ae\boldsymbol{R_a}.$$

The fact that the $E1$ approximation to $\alpha_{f,i}$ contains matrix elements of the electric dipole operator between the initial and final states makes it clear why this is called the electric dipole contribution to $\alpha_{f,i}$; within the $E1$ notation, the E stands for electric moment and the 1 stands for the first such moment (i.e., the dipole moment).

Within this approximation, the overall rate of transitions is given by:

$$R_{i,f} = 2\pi g(\omega_{f,i})|\alpha_{f,i}|^2$$
$$= 2\pi g(\omega_{f,i})(\omega_{f,i}/\hbar c)^2|\boldsymbol{A_0}\bullet\langle\Phi_f|\mu|\Phi_i\rangle|^2.$$

Recalling that $\boldsymbol{E}(\boldsymbol{r},t) = -1/c\partial \boldsymbol{A}/\partial t = \omega/c\,\boldsymbol{A}_o \sin(\omega t - \boldsymbol{k}\bullet\boldsymbol{r})$, the magnitude of $\boldsymbol{A}_0$ can be replaced by that of $\boldsymbol{E}$, and this rate expression becomes

$$R_{i,f} = (2\pi/\hbar^2)g(\omega_{f,i})|\boldsymbol{E_0}\bullet\langle\Phi_f|\mu|\Phi_i\rangle|^2.$$

This expresses the widely used $E1$ approximation to the Fermi-Wentzel golden rule.

2. Magnetic Dipole and Electric Quadrupole Transitions

When $E1$ predictions for the rates of transitions between states vanish (e.g., for symmetry reasons as discussed below), it is essential to examine higher order contributions to $\alpha_{f,i}$. The next terms in the above long-wavelength expansion vary as $1/\lambda$ and have the form:

$$\alpha_{f,i}(E2 + M1) = \langle\Phi_f|\sum_j (e/m_ec)[-i\boldsymbol{k}\bullet\boldsymbol{r}_j]\boldsymbol{A}_0\bullet\nabla_j + \sum_a (Z_ae/m_ac)[-i\boldsymbol{k}\bullet\boldsymbol{R_a}]\boldsymbol{A_0}\bullet\nabla_a|\Phi_i\rangle.$$

For reasons soon to be shown, they are called electric quadrupole ($E2$) and magnetic dipole ($M1$) terms. Clearly, higher and higher order terms can be so generated. Within the long-wavelength regime, however, successive terms should decrease in magnitude because of the successively higher powers of $1/\lambda$ that they contain.

To further analyze the above $E2 + M1$ factors, let us label the propagation direction of the light as the z-axis (the axis along which $\boldsymbol{k}$ lies) and the direction of $\boldsymbol{A}_0$ as the x-axis. These axes are so-called "lab-fixed" axes because their orientation is determined by the direction of the light source and the direction of polarization of the light source's $\boldsymbol{E}$ field, both of which are specified by laboratory conditions. The molecule being subjected to this light can be oriented at arbitrary angles relative to these lab axes.

With the x, y, and z axes so defined, the above expression for $\alpha_{f,i}(E2 + M1)$ becomes

$$\alpha_{f,i}(E2 + M1) = -\,i(A_0 2\pi/\lambda)\langle\Phi_f|\sum_j (e/m_ec)z_j\partial/\partial x_j + \sum_a (Z_ae/m_ac)z_a\partial/\partial x_a|\Phi_i\rangle.$$

Now writing (for both z_j and z_a)

$$z\partial/\partial x = 1/2(z\partial/\partial x - x\partial/\partial z + z\partial/\partial x + x\partial/\partial z),$$

and using

$$\nabla_j = -(m_e/\hbar^2)[H,\boldsymbol{r}_j]$$
$$\nabla_a = -(m_a/\hbar^2)[H,\boldsymbol{R}_a],$$

the contributions of $1/2(z\partial/\partial x + x\partial/\partial z)$ to $\alpha_{f,i}(E2 + M1)$ can be rewritten as

$$\alpha_{f,i}(E2) = -i\frac{(A_0 e 2\pi\omega_{f,i})}{c\lambda\hbar} <\Phi_f| \sum_j z_j x_j + \sum_a Z_a z_a x_a |\Phi_i>.$$

The operator

$$\sum_i z_i x_j + \sum_a Z_a z_a x_a$$

that appears above is the z,x element of the electric quadrupole moment operator $Q_{z,x}$; it is for this reason that this particular component is labeled $E2$ and denoted the electric quadrupole contribution.

The remaining $1/2(z\partial/\partial x - x\partial/\partial z)$ contribution to $\alpha_{f,i}(E2 + M1)$ can be rewritten in a form that makes its content more clear by first noting that

$$1/2(z\partial/\partial x - x\partial/\partial z) = (i/2\hbar)(zp_x - xp_z) = (i/2\hbar)L_y$$

contains the y-component of the angular momentum operator. Hence, the following contribution to $\alpha_{f,i}(E2 + M1)$ arises:

$$\alpha_{f,i}(M1) = \frac{A_0 2\pi e}{2\lambda c\hbar} <\Phi_f| \sum_j L_{y_j}/m_e + \sum_a Z_a L_{y_a}/m_a |\Phi_i>.$$

The magnetic dipole moment of the electrons about the y axis is

$$\mu_{y,electrons} = \sum_j (e/2m_e c)L_{y_j};$$

that of the nuclei is

$$\mu_{y,nuclei} = \sum_a (Z_a e/2m_a c)L_{y_a}.$$

The $\alpha_{f,i}(M1)$ term thus describes the interaction of the magnetic dipole moments of the electrons and nuclei with the magnetic field (of strength $|H| = A_0 k$) of the light (which lies along the y axis):

$$\alpha_{f,i}(M1) = \frac{|H|}{\hbar} <\Phi_f|\mu_{y,electrons} + \mu_{y,nuclei}|\Phi_i>.$$

The total rate of transitions from Φ_i to Φ_f is given, through first-order in perturbation theory, by

$$R_{i,f} = 2\pi g(\omega_{f,i})|\alpha_{f,i}|^2,$$

where $\alpha_{f,i}$ is a sum of its $E1$, $E2$, $M1$, etc. pieces. In the next chapter, molecular symmetry will be shown to be of use in analyzing these various pieces. It should be kept in mind that the contributions caused by $E1$ terms will dominate, within the long-wavelength approximation, unless symmetry causes these terms to vanish. It is primarily under such circumstances that consideration of $M1$ and $E2$ transitions is needed.

III. THE KINETICS OF PHOTON ABSORPTION AND EMISSION

A. The Phenomenological Rate Laws

Before closing this chapter, it is important to emphasize the context in which the transition rate expressions obtained here are most commonly used. The perturbative approach used in the above development gives rise to various contributions to the overall rate coefficient for transitions from an initial state Φ_i to a final state Φ_f. These contributions include the electric dipole, magnetic dipole, and electric quadrupole first order terms as well contributions arising from second (and higher) order terms in the perturbation solution.

In principle, once the rate expression

$$R_{i,f} = 2\pi g(\omega_{f,i})|\alpha_{f,i}|^2$$

has been evaluated through some order in perturbation theory and including the dominant electromagnetic interactions, one can make use of these *state-to-state rates*, which are computed on a per-molecule basis, to describe the time evolution of the populations of the various energy levels of the molecule under the influence of the light source's electromagnetic fields.

For example, given two states, denoted i and f, between which transitions can be induced by photons of frequency $\omega_{f,i}$, the following kinetic model is often used to describe the time evolution of the numbers of molecules n_i and n_f in the respective states:

$$\frac{dn_i}{dt} = -R_{i,f}\, n_i + R_{f,i} n_f$$

$$\frac{dn_f}{dt} = -R_{f,i} n_f + R_{i,f}\, n_i.$$

Here, $R_{i,f}$ and $R_{f,i}$ are the rates (per molecule) of transitions for the $i ==> f$ and $f ==> i$ transitions respectively. As noted above, these rates are proportional to the intensity of the light source (i.e., the photon intensity) at the resonant frequency and to the square of a matrix element connecting the respective states. This matrix element square is $|\alpha_{i,f}|^2$ in the former case and $|\alpha_{f,i}|^2$ in the latter. Because the perturbation operator whose matrix elements are $\alpha_{i,f}$ and $\alpha_{f,i}$ is Hermitian (this is true through all orders of perturbation theory and for all terms in the long-wavelength expansion), these two quantities are complex conjugates of one another, and, hence $|\alpha_{i,f}|^2 = |\alpha_{f,i}|^2$, from which it follows that $R_{i,f} = R_{f,i}$. This means that the state-to-state absorption and stimulated emission rate coefficients (i.e., the rate per molecule undergoing the transition) are identical. This result is referred to as the principle of **microscopic reversibility**.

Quite often, the states between which transitions occur are members of *levels* that contain more than a single state. For example, in rotational spectroscopy a transition between a state in the $J = 3$ level of a diatomic molecule and a state in the $J = 4$ level involve such states; the respective levels are $2J + 1 = 7$ and $2J + 1 = 9$ fold degenerate, respectively.

To extend the above kinetic model to this more general case in which degenerate levels occur, one uses the number of molecules in each **level** (N_i and N_f for the two levels in the above example) as the time dependent variables. The kinetic equations then governing their time evolution can be obtained by summing the state-to-state equations over all states in each level

$$\sum_{i \text{ in level } I} \left(\frac{dn_i}{dt} \right) = \frac{dN_I}{dt}$$

$$\sum_{f \text{ in level } F} \left(\frac{dn_f}{dt} \right) = \frac{dN_F}{dt}$$

and realizing that each state within a given level can undergo transitions to all states within the other level (hence the total rates of production and consumption must be summed over all states to or from which transitions can occur). This generalization results in a set of rate laws for the populations of the respective levels:

$$\frac{dN_i}{dt} = -g_f R_{i,f} N_i + g_i R_{f,i} N_f$$

$$\frac{dN_f}{dt} = -g_i R_{f,i} N_f + g_f R_{i,f} N_i.$$

Here, g_i and g_f are the degeneracies of the two levels (i.e., the number of states in each level) and the $R_{i,f}$ and $R_{f,i}$, which are equal as described above, are the state-to-state rate coefficients introduced earlier.

B. Spontaneous and Stimulated Emission

It turns out (the development of this concept is beyond the scope of this text) that the rate at which an excited level can emit photons and decay to a lower energy level is dependent on two factors: (i) the rate of **stimulated** photon emission as covered above, and (ii) the rate of **spontaneous** photon emission. The former rate $g_f R_{i,f}$ (per molecule) is proportional to the light intensity $g(\omega_{f,i})$ at the resonance frequency. It is conventional to separate out this intensity factor by defining an intensity independent rate coefficient $B_{i,f}$ for this process as:

$$g_f R_{i,f} = g(\omega_{f,i}) B_{i,f}.$$

Clearly, $B_{i,f}$ embodies the final-level degeneracy factor g_f, the perturbation matrix elements, and the 2π factor in the earlier expression for $R_{i,f}$. The spontaneous rate of transition from the excited to the lower level is found to be *independent* of photon intensity, because it deals with a process that does not require collision with a photon to occur, and is usually denoted $A_{i,f}$. The rate of photon-stimulated upward transitions from state f to state i ($g_i R_{f,i} = g_i R_{i,f}$ in the present case) is also proportional to $g(\omega_{f,i})$, so it is written by convention as:

$$g_i R_{f,i} = g(\omega_{f,i}) B_{f,i}.$$

An important relation between the $B_{i,f}$ and $B_{f,i}$ parameters exists and is based on the identity $R_{i,f} = R_{f,i}$ that connects the state-to-state rate coefficients:

$$\frac{(B_{i,f})}{(B_{f,i})} = \frac{(g_f R_{i,f})}{(g_i R_{f,i})} = \frac{g_f}{g_i}.$$

This relationship will prove useful in the following sections.

C. Saturated Transitions and Transparency

Returning to the kinetic equations that govern the time evolution of the populations of two levels connected by photon absorption and emission, and adding in the term needed for spontaneous emission, one finds (with the initial level being of the lower energy):

$$\frac{dN_i}{dt} = -gB_{i,f}N_i + (A_{f,i} + gB_{f,i})N_f$$

$$\frac{dN_f}{dt} = -(A_{f,i} + gB_{f,i})N_f + gB_{i,f}N_i$$

where $g = g(\omega)$ denotes the light intensity at the resonance frequency.

At steady state, the populations of these two levels are given by setting

$$\frac{dN_i}{dt} = \frac{dN_f}{dt} = 0:$$

$$\frac{N_f}{N_i} = \frac{(gB_{i,f})}{(A_{f,i} + gB_{f,i})}.$$

When the light source's intensity is so large as to render $gB_{f,i} >> A_{f,i}$ (i.e., when the rate of spontaneous emission is small compared to the stimulated rate), this population ratio reaches $(B_{i,f}/B_{f,i})$, which was shown earlier to equal (g_f/g_i). In this case, one says that the populations have been ***saturated*** by the intense light source. Any further increase in light intensity will result in *zero* increase in the rate at which photons are being absorbed. Transitions that have had their populations saturated by the application of intense light sources are said to display optical ***transparency*** because they are unable to absorb (or emit) any further photons because of their state of saturation.

D. Equilibrium and Relations between *A* and *B* Coefficients

When the molecules in the two levels being discussed reach *equilibrium* (at which time the

$$\frac{dN_i}{dt} = \frac{dN_f}{dt} = 0$$

also holds) with a photon source that itself is in equilibrium characterized by a temperature T, we must have:

$$\frac{N_f}{N_i} = \frac{g_f}{g_i}\exp(-(E_f - E_i)/kT) = \frac{g_f}{g_i}\exp(-\hbar\omega/kT)$$

where g_f and g_i are the degeneracies of the states labeled f and i. The photon source that is characterized by an equilibrium temperature T is known as a **black body** radiator, whose intensity profile $g(\omega)$ (in erg cm^{-3}sec) is know to be of the form:

$$g(\omega) = \frac{2(\hbar\omega)^3}{\pi c^3\hbar^2}(\exp(\hbar\omega/kT) - 1)^{-1}.$$

Equating the kinetic result that must hold at equilibrium:

$$\frac{N_f}{N_i} = \frac{(gB_{i,f})}{(A_{f,i} + gB_{f,i})}$$

to the thermodynamic result:

$$\frac{N_f}{N_i} = \frac{g_f}{g_i}\exp(\hbar\omega/kT) ,$$

and using the above black body $g(\omega)$ expression and the identity

$$\frac{(B_{i,f})}{(B_{f,i})} = \frac{g_f}{g_i},$$

one can solve for the $A_{f,i}$ rate coefficient in terms of the $B_{f,i}$ coefficient. Doing so yields:

$$A_{f,i} = B_{f,i}\frac{2(\hbar\omega)^3}{\pi c^3\hbar^2}.$$

E. Summary

In summary, the so-called **Einstein *A* and *B* rate coefficients** connecting a lower-energy initial state i and a final state f are related by the following conditions:

$$B_{i,f} = \frac{g_f}{g_i}B_{f,i}$$

and

$$A_{f,i} = \frac{2(\hbar\omega)^3}{\pi c^3\hbar^2}B_{f,i}.$$

These phenomenological level-to-level rate coefficients are related to the state-to-state $R_{i,f}$ coefficients derived by applying perturbation theory to the electromagnetic perturbation through

$$g_f R_{i,f} = g(\omega_{f,i})B_{i,f}.$$

The A and B coefficients can be used in a kinetic equation model to follow the time evolution of the populations of the corresponding levels:

$$\frac{dN_i}{dt} = -gB_{i,f}N_i + (A_{f,i} + gB_{f,i})N_f$$

$$\frac{dN_f}{dt} = -(A_{f,i} + gB_{f,i})N_f + gB_{i,f}N_i.$$

These equations possess steady state solutions

$$\frac{N_f}{N_i} = \frac{(gB_{i,f})}{(A_{f,i} + gB_{f,i})}$$

which, for large $g(\omega)$, produce saturation conditions:

$$\frac{N_f}{N_i} = \frac{(B_{i,f})}{(B_{f,i})} = \frac{g_f}{g_i}.$$

CHAPTER

15 The tools of time-dependent perturbation theory can be applied to transitions among electronic, vibrational, and rotational states of molecules.

I. ROTATIONAL TRANSITIONS

Within the approximation that the electronic, vibrational, and rotational states of a molecule can be treated as independent, the total molecular wavefunction of the "initial" state is a product

$$\Phi_i = \psi_{ei}\chi_{vi}\phi_{ri}$$

of an electronic function ψ_{ei}, a vibrational function χ_{vi}, and a rotational function ϕ_{ri}. A similar product expression holds for the "final" wavefunction Φ_f.

In microwave spectroscopy, the energy of the radiation lies in the range of fractions of a cm^{-1} through several cm^{-1}; such energies are adequate to excite rotational motions of molecules but are not high enough to excite any but the weakest vibrations (e.g., those of weakly bound Van der Waals complexes). In rotational transitions, the electronic and vibrational states are thus left unchanged by the excitation process; hence $\psi_{ei} = \psi_{ef}$ and $\chi_{vi} = \chi_{vf}$.

Applying the first-order electric dipole transition rate expressions

$$R_{i,f} = 2\pi g(\omega_{f,i})|\alpha_{f,i}|^2$$

obtained in chapter 14 to this case requires that the $E1$ approximation

$$R_{i,f} = (2\pi/\hbar^2)g(\omega_{f,i})|\mathbf{E}_0\bullet\langle\Phi_f|\mu|\Phi_i\rangle|^2$$

be examined in further detail. Specifically, the electric dipole matrix elements $\langle\Phi_f|\mu|\Phi_i\rangle$ with

$$\mu = \sum_j e\boldsymbol{r}_j + \sum_a Z_a e\boldsymbol{R}_a$$

must be analyzed for Φ_i and Φ_f being of the product form shown above.

The integrations over the electronic coordinates contained in $<\Phi_f|\mu|\Phi_i>$, as well as the integrations over vibrational degrees of freedom yield "expectation values" of the electric dipole moment operator because the electronic and vibrational components of Φ_i and Φ_f are identical:

$$<\psi_{ei}|\mu|\psi_{ei}> = \mu(\boldsymbol{R})$$

is the dipole moment of the initial electronic state (which is a function of the internal geometrical degrees of freedom of the molecule, denoted $\boldsymbol{R}$); and

$$<\chi_{vi}|\mu(\boldsymbol{R})|\chi_{vi}> = \mu_{ave}$$

is the vibrationally averaged dipole moment for the particular vibrational state labeled χ_{vi}. The vector μ_{ave} has components along various directions and can be viewed as a vector "locked" to the molecule's internal coordinate axis (labeled *a, b, c* as below).

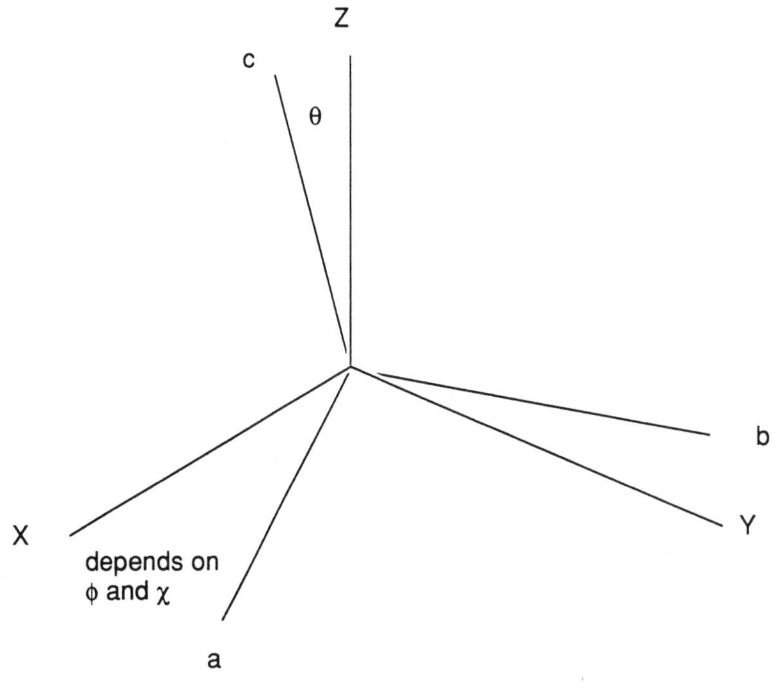

The rotational part of the $<\Phi_f|\mu|\Phi_i>$ integral is not of the expectation value form because the initial rotational function ϕ_{ir} is not the same as the final ϕ_{fr}. This integral has the form:

$$<\phi_{ir}|\mu_{ave}|\phi_{fr}> = \int (Y^*_{L,M}(\theta,\phi)\mu_{ave}Y_{L',M'}(\theta,\phi)\ \sin\theta d\theta d\phi)$$

for linear molecules whose initial and final rotational wavefunctions are $Y_{L,M}$ and $Y_{L',M'}$, respectively, and

$$<\phi_{ir}|\mu_{ave}|\phi_{fr}> = \sqrt{\frac{2L+1}{8\pi^2}}\sqrt{\frac{2L'+1}{8\pi^2}}\int (D_{L,M,K}(\theta,\phi,\chi)\mu_{ave}D^*_{L',M',K'}(\theta,\phi,\chi)\ \sin\theta d\theta d\phi d\chi)$$

for spherical or symmetric top molecules (here,

$$\sqrt{\frac{2L+1}{8\pi^2}}D^*_{L,M,K}(\theta,\phi,\chi)$$

are the normalized rotational wavefunctions described in chapter 13 and in appendix G). The angles θ, ϕ, and χ refer to how the molecule-fixed coordinate system is oriented with respect to the space-fixed X, Y, Z axis system.

A. Linear Molecules

For linear molecules, the vibrationally averaged dipole moment μ_{ave} lies along the molecular axis; hence its orientation in the lab-fixed coordinate system can be specified in terms of the same angles (θ and ϕ) that are used to describe the rotational functions $Y_{L,M}(\theta,\phi)$. Therefore, the three components of the $<\phi_{ir}|\mu_{ave}|\phi_{fr}>$ integral can be written as:

$$<\phi_{ir}|\mu_{ave}|\phi_{fr}>_x = \mu\int (Y^*_{L,M}(\theta,\phi)\ \sin\theta\cos\phi Y_{L',M'}(\theta,\phi)\ \sin\theta d\theta d\phi)$$

$$<\phi_{ir}|\mu_{ave}|\phi_{fr}>_y = \mu\int (Y^*_{L,M}(\theta,\phi)\ \sin\theta\sin\phi Y_{L',M'}(\theta,\phi)\ \sin\theta d\theta d\phi)$$

$$<\phi_{ir}|\mu_{ave}|\phi_{fr}>_z = \mu\int (Y^*_{L,M}(\theta,\phi)\ \cos\theta Y_{L',M'}(\theta,\phi)\ \sin\theta d\theta d\phi),$$

where μ is the magnitude of the averaged dipole moment. **If the molecule has no dipole moment**, all of the above electric dipole integrals vanish and the **intensity** of $E1$ rotational transitions **is zero.**

The three E1 integrals can be further analyzed by noting that $\cos\theta \propto Y_{1,0}$; $\sin\theta\cos\phi \propto Y_{1,1} + Y_{1,-1}$; and $\sin\theta\sin\phi \propto Y_{1,1} - Y_{1,-1}$ and using the angular momentum coupling methods illustrated in appendix G. In particular, the result given in that appendix:

$$D_{j,m,m'}D_{l,n,n'} = \sum_{J,M,M'} <J,M|j,m;l,n>\ <j,m';l,n'|J,M'>D_{J,M,M'}$$

when multiplied by $D^*_{J,M,M'}$ and integrated over $\sin\theta d\theta d\phi d\chi$, yields:

$$\int (D^*_{J,M,M'}D_{j,m,m'}D_{l,n,n'}\ \sin\theta d\theta d\phi d\chi) = \frac{8\pi^2}{2J+1} <J,M|j,m;l,n>\ <j,m';l,n'|J,M'>$$

$$= 8\pi^2\begin{pmatrix} j\ l\ J \\ m\ n\ -M \end{pmatrix}\begin{pmatrix} j\ l\ J \\ m'\ n'\ -M' \end{pmatrix}(-1)^{M+M'}.$$

To use this result in the present linear-molecule case, we note that the $D_{J,M,K}$ functions and the $Y_{J,M}$ functions are related by:

$$Y_{J,M}(\theta,\phi) = \sqrt{(2J+1)/4\pi}\, D^*_{J,M,0}(\theta,\phi,\chi).$$

The normalization factor is now $\sqrt{(2J+1)/4\pi}$ rather than $\sqrt{(2J+1)/8\pi^2}$ because the $Y_{J,M}$ are no longer functions of χ, and thus the need to integrate over $0 \le \chi \le 2\pi$ disappears. Likewise, the χ-dependence of $D^*_{J,M,K}$ disappears for $K = 0$.

We now use these identities in the three $E1$ integrals of the form

$$\mu \int (Y^*_{L,M}(\theta,\phi) Y_{1,m}(\theta,\phi) Y_{L',M'}(\theta,\phi) \sin\theta d\theta d\phi),$$

with $m = 0$ being the Z-axis integral, and the Y- and X-axis integrals being combinations of the $m = 1$ and $m = -1$ results. Doing so yields:

$$\mu \int (Y^*_{L,M}(\theta,\phi) Y_{1,m}(\theta,\phi) Y_{L',M'}(\theta,\phi) \sin\theta d\theta d\phi)$$

$$= \mu \sqrt{\frac{2L+1}{4\pi} \frac{2L'+1}{4\pi} \frac{3}{4\pi}} \int (D_{L,M,0} D^*_{1,m,0} D^*_{L',M',0} \sin\theta d\theta d\phi d\chi / 2\pi).$$

The last factor of $1/2\pi$ is inserted to cancel out the integration over $d\chi$ that, because all K-factors in the rotation matrices equal zero, trivially yields 2π. Now, using the result shown above expressing the integral over three rotation matrices, these $E1$ integrals for the linear-molecule case reduce to:

$$\mu \int (Y^*_{L,M}(\theta,\phi) Y_{1,m}(\theta,\phi) Y_{L',M'}(\theta,\phi) \sin\theta d\theta d\phi)$$

$$= \mu \sqrt{\frac{2L+1}{4\pi} \frac{2L'+1}{4\pi} \frac{3}{4\pi} \frac{8\pi^2}{2\pi}} \begin{pmatrix} L' & 1 & L \\ M' & m & -M \end{pmatrix} \begin{pmatrix} L' & 1 & L \\ 0 & 0 & -0 \end{pmatrix} (-1)^M$$

$$= \mu \sqrt{(2L+1)(2L'+1) \frac{3}{4\pi}} \begin{pmatrix} L' & 1 & L \\ M' & m & -M \end{pmatrix} \begin{pmatrix} L' & 1 & L \\ 0 & 0 & -0 \end{pmatrix} (-1)^M.$$

Applied to the z-axis integral (identifying $m = 0$), this result therefore vanishes unless:

$$M = M'$$

and

$$L = L' + 1 \text{ or } L' - 1.$$

Even though angular momentum coupling considerations would allow $L = L'$ (because coupling two angular momenta with $j = 1$ and $j = L'$ should give $L' + 1$, L', and $L' - 1$), the 3-j symbol

$$\begin{pmatrix} L' & 1 & L \\ 0 & 0 & -0 \end{pmatrix}$$

vanishes for the $L = L'$ case since 3-j symbols have the following symmetry

$$\begin{pmatrix} L' & 1 & L \\ M' & m & -M \end{pmatrix} = (-1)^{L+L'+1} \begin{pmatrix} L' & 1 & L \\ -M' & -m & M \end{pmatrix}$$

with respect to the M, M', and m indices. Applied to the

$$\begin{pmatrix} L' & 1 & L \\ 0 & 0 & -0 \end{pmatrix}$$

3-j symbol, this means that this particular 3-j element vanishes for $L = L'$ since $L + L' + 1$ is odd and hence $(-1)^{L+L'+1}$ is -1.

Applied to the x- and y-axis integrals, which contain $m = \pm 1$ components, this same analysis yields:

$$\mu\sqrt{(2L+1)(2L'+1)\frac{3}{4\pi}}\begin{pmatrix} L' & 1 & L \\ M' & \pm 1 & -M \end{pmatrix}\begin{pmatrix} L' & 1 & L \\ 0 & 0 & -0 \end{pmatrix}(-1)^M$$

which then requires that

$M = M' \pm 1$ and

$L = L' + 1, L' - 1,$

with $L = L'$ again being forbidden because of the second 3-j symbol.

These results provide so-called "**selection rules**" because they limit the L and M values of the final rotational state, given the L',M' values of the initial rotational state. In the figure shown below, the $L = L' + 1$ absorption spectrum of NO at 120°K is given. The intensities of the various peaks are related to the populations of the lower-energy rotational states which are, in turn, proportional to $(2L' + 1)\exp(-L'(L' + 1)\hbar^2/8\pi^2 IkT)$. Also included in the intensities are so-called **line strength factors** that are proportional to the squares of the quantities:

$$\mu\sqrt{(2L+1)(2L'+1)\frac{3}{4\pi}}\begin{pmatrix} L' & 1 & L \\ M' & m & -M \end{pmatrix}\begin{pmatrix} L' & 1 & L \\ 0 & 0 & -0 \end{pmatrix}(-1)^M$$

which appear in the $E1$ integrals analyzed above (recall that the rate of photon absorption $R_{i,f} = (2\pi/\hbar^2) g(\omega_{f,i}) |\mathbf{E}_0 \bullet \langle\Phi_f|\mu|\Phi_i\rangle|^2$ involves the squares of these matrix elements). The book by Zare gives an excellent treatment of line strength factors' contributions to rotation, vibration, and electronic line intensities.

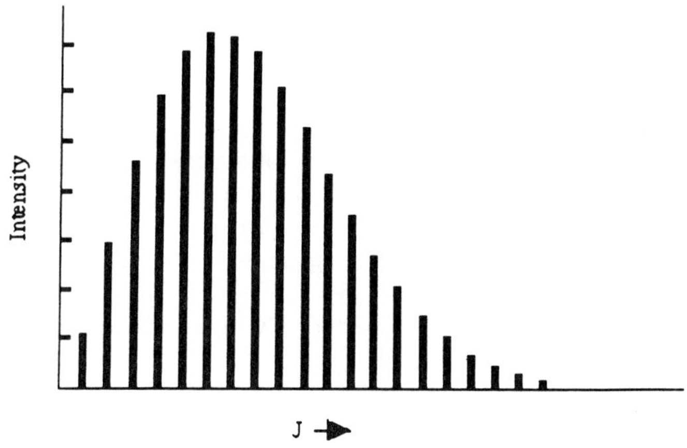

B. Non-Linear Molecules

For molecules that are non-linear and whose rotational wavefunctions are given in terms of the spherical or symmetric top functions $D^*_{L,M,K}$ the dipole moment μ_{ave} can have components along any or all three of the molecule's internal coordinates (e.g., the three molecule-fixed coordinates that describe the orientation of the principal axes of the moment of inertia tensor). For a spherical top molecule, $|\mu_{ave}|$ vanishes, so $E1$ transitions do not occur.

For symmetric top species, μ_{ave} lies along the symmetry axis of the molecule, so the orientation of μ_{ave} can again be described in terms of θ and ϕ, the angles used to locate the orientation of the molecule's symmetry axis relative to the lab-fixed coordinate system. As a result, the $E1$ integral again can be decomposed into three pieces:

$$<\phi_{ir}|\mu_{ave}|\phi_{fr}>_x = \mu\int (D_{L,M,K}(\theta,\phi,\chi) \cos\theta\cos\phi\, D^*_{L',M',K'}(\theta,\phi,\chi) \sin\theta d\theta d\phi d\chi)$$

$$<\phi_{ir}|\mu_{ave}|\phi_{fr}>_y = \mu\int (D_{L,M,K}(\theta,\phi,\chi) \cos\theta\sin\phi\, D^*_{L',M',K'}(\theta,\phi,\chi) \sin\theta d\theta d\phi d\chi)$$

$$<\phi_{ir}|\mu_{ave}|\phi_{fr}>_z = \mu\int (D_{L,M,K}(\theta,\phi,\chi) \cos\theta\, D^*_{L',M',K'}(\theta,\phi,\chi) \sin\theta d\theta d\phi d\chi).$$

Using the fact that $\cos\theta \propto D^*_{1,0,0}$; $\sin\theta\cos\phi \propto D^*_{1,1,0} + D^*_{1,-1,0}$; and $\sin\theta\sin\phi \propto D^*_{1,1,0} - D^*_{1,-1,0}$, and the tools of angular momentum coupling, allows these integrals to be expressed, as above, in terms of products of the following 3-j symbols:

$$\begin{pmatrix} L' & 1 & L \\ M' & m & -M \end{pmatrix}\begin{pmatrix} L' & 1 & L \\ K' & 0 & -K \end{pmatrix},$$

from which the following selection rules are derived:

$L = L' + 1, L', L' - 1$ (but not $L = L' = 0$),
$K = K'$,
$M = M' + m$,

with $m = 0$ for the Z-axis integral and $m = \pm 1$ for the X- and Y-axis integrals. In addition, if $K = K' = 0$, the $L = L'$ transitions are also forbidden by the second 3-j symbol vanishing.

II. VIBRATION-ROTATION TRANSITIONS

When the initial and final electronic states are identical but the respective vibrational and rotational states are not, one is dealing with transitions between vibration-rotation states of the molecule. These transitions are studied in infrared (IR) spectroscopy using light of energy in the 30cm^{-1} (far IR) to 5000cm^{-1} range. The electric dipole matrix element analysis still begins with the electronic dipole moment integral $<\psi_{ei}|\mu|\psi_{ei}> = \mu(\boldsymbol{R})$, but the integration over internal vibrational coordinates no longer produces the vibrationally averaged dipole moment. Instead one forms the vibrational **transition dipole** integral:

$$<\chi_{vf}|\mu(\boldsymbol{R})|\chi_{vi}> = \mu_{f,i}$$

between the initial χ_i and final χ_f vibrational states.

A. The Dipole Moment Derivatives

Expressing $\mu(\boldsymbol{R})$ in a power series expansion about the equilibrium bond length position (denoted $\boldsymbol{R}_e$ collectively and $R_{a,e}$ individually):

$$\mu(\boldsymbol{R}) = \mu(\boldsymbol{R}_e) + \sum_a \partial\mu/\partial R_a(R_a - R_{a,e}) + \ldots,$$

substituting into the $<\chi_{vf}|\mu(\boldsymbol{R})|\chi_{vi}>$ integral, and using the fact that χ_i and χ_f are orthogonal (because they are eigenfunctions of vibrational motion on the same electronic surface and hence of the same vibrational Hamiltonian), one obtains:

$$<\chi_{vf}|\mu(\boldsymbol{R})|\chi_{vi}> = \mu(\boldsymbol{R}_e)<\chi_{vf}|\chi_{vi}> + \sum_a \partial\mu/\partial\boldsymbol{R}_a<\chi_{vf}|(R_a - R_{a,e})|\chi_{vi}> + \ldots$$

$$= \sum_a (\partial\mu/\partial R_a)<\chi_{vf}|(R_a - R_{a,e})|\chi_{vi}> + \ldots.$$

This result can be interpreted as follows:

1. Each independent vibrational mode of the molecule contributes to the $\mu_{f,i}$ vector an amount equal to $(\partial\mu/\partial R_a)<\chi_{vf}|(R_a - R_{a,e})|\chi_{vi}> + \ldots.$
2. Each such contribution contains one part $(\partial\mu/\partial R_a)$ that depends on how the molecule's dipole moment function varies with vibration along that particular mode (labeled a),
3. and a second part $<\chi_{vf}|(R_a - R_{a,e})|\chi_{vi}>$ that depends on the character of the initial and final vibrational wavefunctions.

If the vibration does not produce **modulation of the dipole moment** (e.g., as with the symmetric stretch vibration of the CO_2 molecule), its infrared intensity vanishes because $(\partial\mu/\partial R_a) = 0$. One says that such transitions are infrared "inactive."

B. Selection Rules on the Vibrational Quantum Number in the Harmonic Approximation

If the vibrational functions are described within the harmonic oscillator approximation, it can be shown that the $<\chi_{vf}|(R_a - R_{a,e})|\chi_{vi}>$ integrals vanish unless $vf = vi + 1$, $vi - 1$ (and that these integrals are proportional to $(vi + 1)^{1/2}$ and $(vi)^{1/2}$ in the respective cases). Even when χ_{vf} and χ_{vi} are rather non-harmonic, it turns out that such $\Delta v = \pm 1$ transitions have the largest $<\chi_{vf}|(R_a - R_{a,e})|\chi_{vi}>$ integrals and therefore the highest infrared intensities. For these reasons, transitions that correspond to $\Delta v = \pm 1$ are called "**fundamental**"; those with $\Delta v = \pm 2$ are called "first **overtone**" transitions.

In summary then, vibrations for which the molecule's dipole moment is modulated as the vibration occurs (i.e., for which $(\partial\mu/\partial R_a)$ is non-zero) *and* for which $\Delta v = \pm 1$ tend to have large infrared intensities; overtones of such vibrations tend to have smaller intensities, and those for which $(\partial\mu/\partial R_a) = 0$ have no intensity.

C. Rotational Selection Rules for Vibrational Transitions

The result of all of the vibrational modes' contributions to

$$\sum_a (\partial\mu/\partial R_a)<\chi_{vf}|(R_a - R_{a,e})|\chi_{vi}>$$

is a vector μ_{trans} that is termed the vibrational "transition dipole" moment. This is a vector with components along, in principle, all three of the internal axes of the molecule. For each particular vibrational transition (i.e., each particular χ_i and χ_f) its orientation in space depends only on the orientation of the molecule; it is thus said to be locked to the molecule's coordinate frame. As such, its orientation relative to the lab-fixed coordinates (which is needed to effect a derivation of rotational selection rules as was done earlier in this chapter) can be described much as was done above for the vibrationally averaged dipole moment that arises in purely rotational transitions. There are, however, important differences in detail. In particular,

1. For a linear molecule μ_{trans} can have components either along (e.g., when stretching vibrations are excited; these cases are denoted σ-cases) or perpendicular to (e.g., when bending vibrations are excited; they are denoted π cases) the molecule's axis.
2. For symmetric top species, μ_{trans} need not lie along the molecule's symmetry axis; it can have components either along or perpendicular to this axis.
3. For spherical tops, μ_{trans} will vanish whenever the vibration does not induce a dipole moment in the molecule. Vibrations such as the totally symmetric a_1 C-H stretching motion in CH_4 do not induce a dipole moment, and are thus infrared inactive; non-totally-symmetric vibrations can also be inactive if they induce no dipole moment.

As a result of the above considerations, the angular integrals

$$<\phi_{ir}|\mu_{trans}|\phi_{fr}> = \int (Y^*_{L,M}(\theta,\phi)\mu_{trans}Y_{L',M'}(\theta,\phi)\ \sin\theta d\theta d\phi)$$

and

$$<\phi_{ir}|\mu_{trans}|\phi_{fr}> = \int (D_{L,M,K}(\theta,\phi,\chi)\mu_{trans}D^*_{L',M',K'}(\theta,\phi,\chi)\ \sin\theta d\theta d\phi d\chi)$$

that determine the rotational selection rules appropriate to vibrational transitions produce similar, but not identical, results as in the purely rotational transition case.

The derivation of these selection rules proceeds as before, with the following additional considerations. The transition dipole moment's μ_{trans} components along the lab-fixed axes must be related to its molecule-fixed coordinates (that are determined by the nature of the vibrational transition as discussed above). This transformation, as given in Zare's text, reads as follows:

$$(\mu_{trans})_m = \sum_k D^*_{l,m,k}(\theta,\phi,\chi)(\mu_{trans})_k$$

where $(\mu_{trans})_m$ with $m = 1, 0, -1$ refer to the components along the lab-fixed (X, Y, Z) axes and $(\mu_{trans})_k$ with $k = 1, 0, -1$ refer to the components along the molecule-fixed (a, b, c) axes.

This relationship, when used, for example, in the symmetric or spherical top $E1$ integral:

$$<\phi_{ir}|\mu_{trans}|\phi_{fr}> = \int (D_{L,M,K}(\theta,\phi,\chi)\mu_{trans}D^*_{L',M',K'}(\theta,\phi,\chi)\ \sin\theta d\theta d\phi d\chi)$$

gives rise to products of 3-j symbols of the form:

$$\begin{pmatrix} L' & 1 & L \\ M' & m & -M \end{pmatrix}\begin{pmatrix} L' & 1 & L \\ K' & k & -K \end{pmatrix}.$$

The product of these 3-j symbols is nonvanishing only under certain conditions that provide the rotational selection rules applicable to vibrational lines of symmetric and spherical top molecules.

Both 3-j symbols will vanish unless

$$L = L' + 1,\ L',\ \text{or}\ L' - 1.$$

In the special case in which $L = L' = 0$ (and hence with $M = M' = 0 = K = K'$, which means that $m = 0 = k$), these 3-j symbols again vanish. Therefore, transitions with

$$L = L' = 0$$

are again **forbidden.** As usual, the fact that the lab-fixed quantum number m can range over $m = 1, 0, -1$, requires that

$$M = M' + 1,\ M',\ M'-1.$$

The selection rules for ΔK depend on the nature of the vibrational transition, in particular, on the component of μ_{trans} along the molecule-fixed axes. For the second 3-j symbol to not vanish, one must have

$$K = K' + k,$$

where $k = 0$, 1, and -1 refer to these molecule-fixed components of the transition dipole. Depending on the nature of the transition, various k values contribute.

1. Symmetric Tops

In a symmetric top molecule such as NH_3, if the transition dipole lies along the molecule's symmetry axis, only $k = 0$ contributes. Such vibrations preserve the molecule's symmetry relative to this symmetry axis (e.g., the totally symmetric N-H stretching mode in NH_3). The additional selection rule $\Delta K = 0$ is thus obtained. Moreover, for $K = K' = 0$, all transitions with $\Delta L = 0$ vanish because the second 3-j symbol vanishes. In summary, one has:

$\Delta K = 0$; $\Delta M = \pm 1, 0$; $\Delta L = \pm 1, 0$ (but $L = L' = 0$ is forbidden and all $\Delta L = 0$ are forbidden for $K = K' = 0$)

for symmetric tops with vibrations whose transition dipole lies along the symmetry axis.

If the transition dipole lies perpendicular to the symmetry axis, only $k = \pm 1$ contribute. In this case, one finds

$\Delta K = \pm 1$; $\Delta M = \pm 1, 0$; $\Delta L = \pm 1, 0$ (neither $L = L' = 0$ nor $K = K' = 0$ can occur for such transitions, so there are no additional constraints).

2. Linear Molecules

When the above analysis is applied to a diatomic species such as HCl, only $k = 0$ is present since the only vibration present in such a molecule is the bond stretching vibration, which has σ symmetry. Moreover, the rotational functions are spherical harmonics (which can be viewed as $D^{*}_{L',M',K'}(\theta,\phi,\chi)$ functions with $K' = 0$), so the K and K' quantum numbers are identically zero. As a result, the product of 3-j symbols

$$\begin{pmatrix} L' & 1 & L \\ M' & m & -M \end{pmatrix}\begin{pmatrix} L' & 1 & L \\ K' & k & -K \end{pmatrix}$$

reduces to

$$\begin{pmatrix} L' & 1 & L \\ M' & m & -M \end{pmatrix} \begin{pmatrix} L' & 1 & L \\ 0 & 0 & 0 \end{pmatrix},$$

which will vanish unless

$$L = L' + 1, L' - 1,$$

but *not* $L = L'$(since parity then causes the second 3-j symbol to vanish), and

$$M = M' + 1, M', M' - 1.$$

The $L = L' + 1$ transitions are termed ***R*-branch** absorptions and those obeying $L = L' - 1$ are called ***P*-branch** transitions. Hence, the selection rules

$$\Delta M = \pm 1, 0; \Delta L = \pm 1$$

are identical to those for purely rotational transitions.

When applied to linear polyatomic molecules, these same selection rules result if the vibration is of σ symmetry (i.e., has $k = 0$). If, on the other hand, the transition is of π symmetry (i.e., has $k = \pm 1$), so the transition dipole lies perpendicular to the molecule's axis, one obtains:

$$\Delta M = \pm 1, 0; \Delta L = 1, 0.$$

These selection rules are derived by realizing that in addition to $k = \pm 1$, one has:

1. a linear-molecule rotational wavefunction that in the $v = 0$ vibrational level is described in terms of a rotation matrix $D_{L',M',0}(\theta,\phi,\chi)$ with no angular momentum along the molecular axis, $K' = 0$;
2. a $v = 1$ molecule whose rotational wavefunction must be given by a rotation matrix $D_{L,M,1}(\theta,\phi,\chi)$ with one unit of angular momentum about the molecule's axis, $K = 1$. In the latter case, the angular momentum is produced by the degenerate π vibration itself. As a result, the selection rules above derive from the following product of 3-j symbols:

$$\begin{pmatrix} L' & 1 & L \\ M' & m & -M \end{pmatrix} \begin{pmatrix} L' & 1 & L \\ 0 & 1 & -1 \end{pmatrix},$$

Because $\Delta L = 0$ transitions are allowed for π vibrations, one says that π vibrations possess ***Q*-branches** in addition to their *R*- and *P*-branches (with $\Delta L = 1$ and -1, respectively).

In the figure shown below, the $v = 0 ==> v = 1$ (fundamental) vibrational absorption spectrum of HCl is shown. Here the peaks at lower energy (to the right of the figure) belong to *P*-branch transitions and occur at energies given approximately by:

$$E = \hbar\omega_{stretch} + (h^2/8\pi^2 I)\,((L-1)L - L(L+1))$$
$$= \hbar\omega_{stretch} - 2(h^2/8\pi^2 I)L.$$

The *R*-branch transitions occur at higher energies given approximately by:

$$E = \hbar\omega_{stretch} + (h^2/8\pi^2 I)\,((L+1)(L+2) - L(L+1))$$
$$= \hbar\omega_{stretch} + 2(h^2/8\pi^2 I)\,(L+1).$$

The absorption that is "missing" from the figure below lying slightly below 2900cm^{-1} is the ***Q*-branch** transition for which $L = L'$; it is absent because the selection rules forbid it.

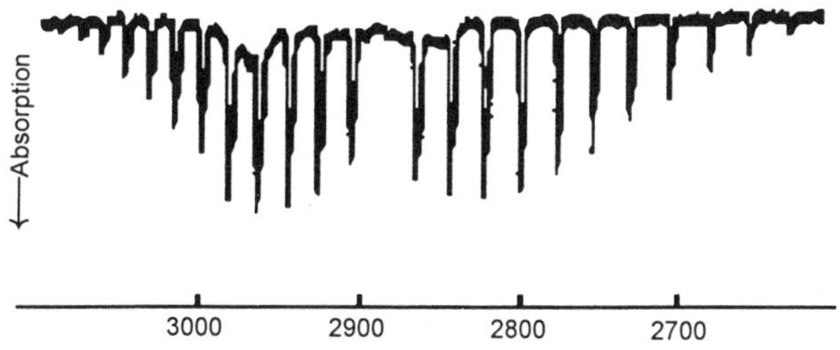

It should be noted that the spacings between the experimentally observed peaks in HCl are not constant as would be expected based on the above P- and R-branch formulas. This is because the moment of inertia appropriate for the $v = 1$ vibrational level is different than that of the $v = 0$ level. These effects of vibration-rotation coupling can be modeled by allowing the $v = 0$ and $v = 1$ levels to have rotational energies written as

$$E = \hbar\omega_{stretch}(v + 1/2) + (h^2/8\pi^2 I_v)\ (L(L + 1))$$

where v and L are the vibrational and rotational quantum numbers. The P- and R-branch transition energies that pertain to these energy levels can then be written as:

$$E_P = \hbar\omega_{stretch} - [(h^2/8\pi^2 I_1) + (h^2/8\pi^2 I_0)]L + [(h^2/8\pi^2 I_1) - (h^2/8\pi^2 I_0)]L^2$$
$$E_R = \hbar\omega_{stretch} + 2(h^2/8\pi^2 I_1) + [3(h^2/8\pi^2 I_1) - (h^2/8\pi^2 I_0)]L + [(h^2/8\pi^2 I_1) - (h^2/8\pi^2 I_0)]L^2.$$

Clearly, these formulas reduce to those shown earlier in the $I_1 = I_0$ limit.

If the vibrationally averaged bond length is longer in the $v = 1$ state than in the $v = 0$ state, which is to be expected, I_1 will be larger than I_0, and therefore $[(h^2/8\pi^2 I_1) - (h^2/8\pi^2 I_0)]$ will be negative. In this case, the *spacing* between neighboring P-branch lines will increase as shown above for HCl. In contrast, the fact that $[(h^2/8\pi^2 I_1) - (h^2/8\pi^2 I_0)]$ is negative causes the *spacing* between neighboring R-branch lines to decrease, again as shown for HCl.

III. ELECTRONIC-VIBRATION-ROTATION TRANSITIONS

When electronic transitions are involved, the initial and final states generally differ in their electronic, vibrational, and rotational energies. Electronic transitions usually require light in the 5000cm^{-1} to $100{,}000\text{cm}^{-1}$ regime, so their study lies within the domain of visible and ultraviolet spectroscopy. Excitations of inner-shell and core orbital electrons may require even higher energy photons, and under these conditions, $E2$ and $M1$ transitions may become more important because of the short wavelength of the light involved.

A. The Electronic Transition Dipole and Use of Point Group Symmetry

Returning to the expression

$$R_{i,f} = (2\pi/\hbar^2)g(\omega_{f,i})|\mathbf{E}_0 \bullet \langle\Phi_f|\mu|\Phi_i\rangle|^2$$

for the rate of photon absorption, we realize that the electronic integral now involves

$$<\psi_{ef}|\mu|\psi_{ei}> = \mu_{f,i}(\boldsymbol{R}),$$

a transition dipole matrix element between the initial ψ_{ei} and final ψ_{ef} electronic wavefunctions. This element is a function of the internal vibrational coordinates of the molecule, and again is a vector locked to the molecule's internal axis frame.

Molecular point-group symmetry can often be used to determine whether a particular transition's dipole matrix element will vanish and, as a result, the electronic transition will be "forbidden" and thus predicted to have zero intensity. If the direct product of the symmetries of the initial and final electronic states ψ_{ei} and ψ_{ef} do not match the symmetry of the electric dipole operator (which has the symmetry of its x, y, and z components; these symmetries can be read off the right most column of the character tables given in appendix E), the matrix element will vanish.

For example, the formaldehyde molecule H_2CO has a ground electronic state (see chapter 11) that has $^1A_1$ symmetry in the C_{2v} point group. Its π ==> π^* singlet excited state also has $^1A_1$ symmetry because both the π and π^* orbitals are of b_1 symmetry. In contrast, the lowest n ==> π^* singlet excited state is of $^1A_2$ symmetry because the highest energy oxygen centered n orbital is of b_2 symmetry and the π^* orbital is of b_1 symmetry, so the Slater determinant in which both the n and π^* orbitals are singly occupied has its symmetry dictated by the $b_2 \, x \, b_1$ direct product, which is A_2.

The π ==> π^* transition thus involves ground ($^1A_1$) and excited ($^1A_1$) states whose direct product ($A_1 \, x \, A_1$) is of A_1 symmetry. This transition thus requires that the electric dipole operator possess a component of A_1 symmetry. A glance at the C_{2v} point group's character table shows that the molecular z-axis is of A_1 symmetry. Thus, if the light's electric field has a non-zero component along the C_2 symmetry axis (the molecule's z-axis), the π ==> π^* transition is predicted to be allowed. Light polarized along either of the molecule's other two axes cannot induce this transition.

In contrast, the n ==> π^* transition has a ground-excited state direct product of $b_2 \times b_1 = A_2$ symmetry. The C_{2v}'s point group character table clearly shows that the electric dipole operator (i.e., its x, y, and z components in the molecule-fixed frame) has no component of A_2 symmetry; thus, light of no electric field orientation can induce this n ==> π^* transition. We thus say that the n ==> π^* transition is $E1$ forbidden (although it is $M1$ allowed).

Beyond such electronic symmetry analysis, it is also possible to derive vibrational and rotational selection rules for electronic transitions that are $E1$ allowed. As was done in the vibrational spectroscopy case, it is conventional to expand $\mu_{f,i}(\boldsymbol{R})$ in a power series about the equilibrium geometry of the initial electronic state (since this geometry is more characteristic of the molecular structure prior to photon absorption):

$$\mu_{f,i}(\boldsymbol{R}) = \mu_{f,i}(\boldsymbol{R}_e) + \sum_a \partial\mu_{f,i}/\partial R_a(R_a - R_{a,e}) + \ldots .$$

B. The Franck-Condon Factors

The first term in this expansion, when substituted into the integral over the vibrational coordinates, gives $\mu_{f,i}(\boldsymbol{R}_e)<\chi_{vf}|\chi_{vi}>$, which has the form of the electronic transition dipole multiplied by the "overlap integral" between the initial and final vibrational wavefunctions. The $\mu_{f,i}(\boldsymbol{R}_e)$ factor was discussed above; it is the electronic E1 transition integral evaluated at the equilibrium geometry of the absorbing state. Symmetry can often be used to determine whether this integral vanishes, as a result of which the $E1$ transition will be "forbidden."

Unlike the vibration-rotation case, the vibrational overlap integrals $<\chi_{vf}|\chi_{vi}>$ do not necessarily vanish because χ_{vf} and χ_{vi} are no longer eigenfunctions of the same vibrational Hamiltonian. χ_{vf} is an eigenfunction whose potential energy is the *final* electronic state's energy surface; χ_{vi} has the *initial* electronic state's energy surface as its potential. The squares of these $<\chi_{vf}|\chi_{vi}>$ integrals, which are what eventually enter into the transition rate expression

$$R_{i,f} = (2\pi/\hbar^2)g(\omega_{f,i})|\mathbf{E}_0\bullet<\Phi_f|\mu|\Phi_i>|^2,$$

are called "**Franck-Condon factors.**" Their relative magnitudes play strong roles in determining the relative intensities of various vibrational "bands" (i.e., peaks) within a particular electronic transition's spectrum.

Whenever an electronic transition causes a large change in the geometry (bond lengths or angles) of the molecule, the Franck-Condon factors tend to display the characteristic "broad progression" shown below when considered for one initial-state vibrational level *vi* and various final-state vibrational levels *vf:*

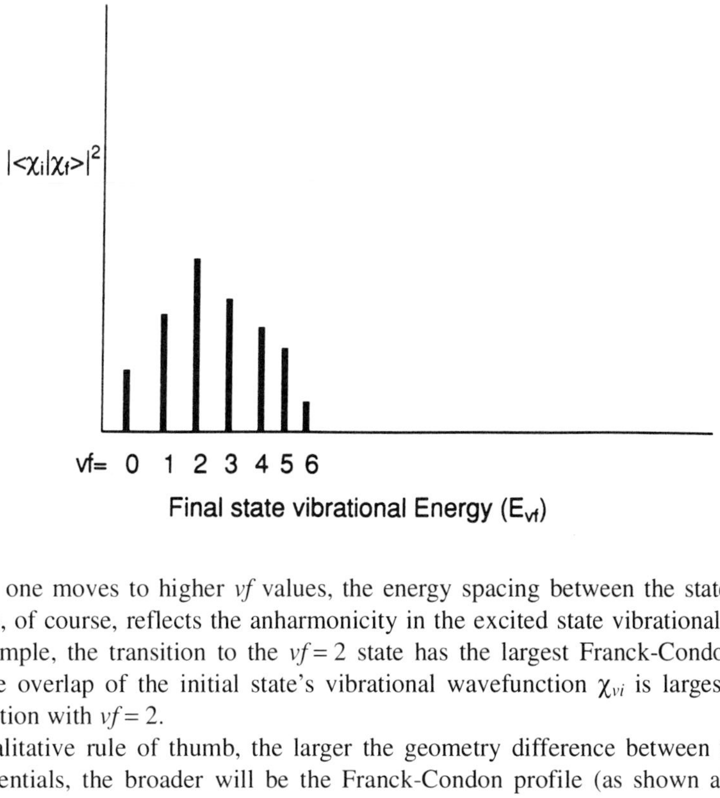

Notice that as one moves to higher *vf* values, the energy spacing between the states ($E_{vf} - E_{vf-1}$) decreases; this, of course, reflects the anharmonicity in the excited state vibrational potential. For the above example, the transition to the $vf = 2$ state has the largest Franck-Condon factor. This means that the overlap of the initial state's vibrational wavefunction χ_{vi} is largest for the final state's χ_{vf} function with $vf = 2$.

As a qualitative rule of thumb, the larger the geometry difference between the initial and final state potentials, the broader will be the Franck-Condon profile (as shown above) and the larger the *vf* value for which this profile peaks. Differences in harmonic frequencies between the two states can also broaden the Franck-Condon profile, although not as significantly as do geometry differences.

For example, if the initial and final states have very similar geometries and frequencies along the mode that is excited when the particular electronic excitation is realized, the following type of Franck-Condon profile may result:

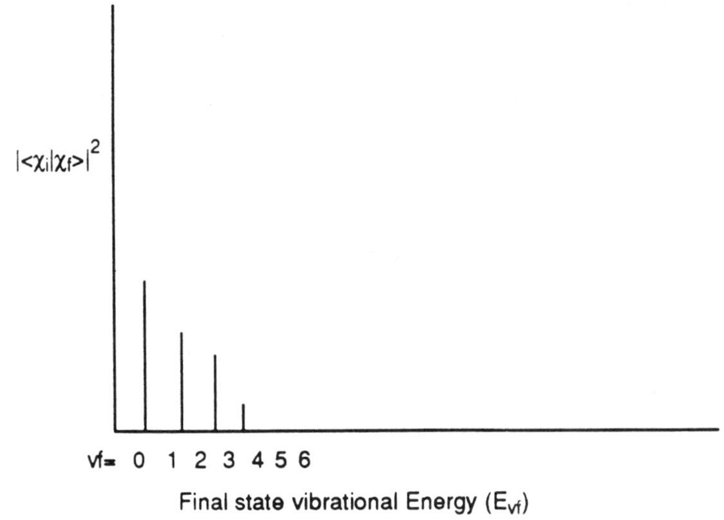

In contrast, if the initial and final electronic states have very different geometries and/or vibrational frequencies along some mode, a very broad Franck-Condon envelope peaked at high-*vf* will result as shown below:

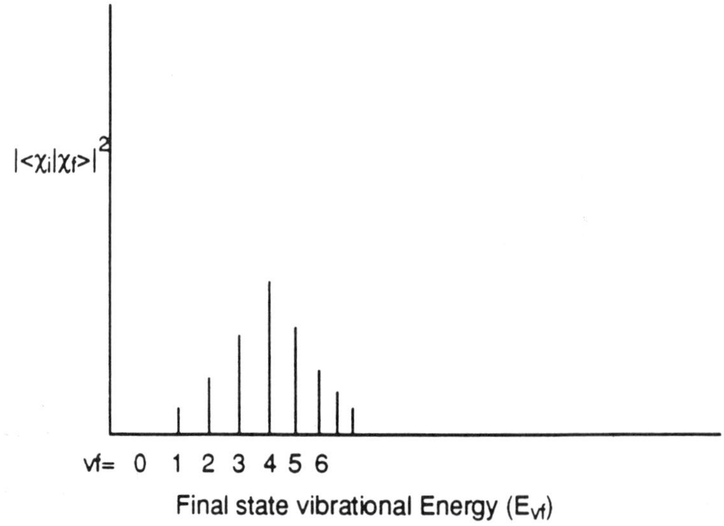

C. Vibronic Effects

The second term in the above expansion of the transition dipole matrix element

$$\sum_a \partial\mu_{f,i}/\partial R_a(R_a - R_{a,e})$$

can become important to analyze when the first term $\mu_{f,i}(\boldsymbol{R_e})$ vanishes (e.g., for reasons of symmetry). This dipole derivative term, when substituted into the integral over vibrational coordinates gives

$$\sum_a \partial\mu_{f,i}/\partial R_a <\chi_{vf}|(R_a - R_{a,e})|\chi_{vi}>.$$

Transitions for which $\mu_{f,i}(\boldsymbol{R_e})$ vanishes but for which $\partial\mu_{f,i}/\partial R_a$ does not for the ath vibrational mode are said to derive intensity through "vibronic coupling" with that mode. The intensities of such modes are dependent on how strongly the electronic dipole integral varies along the mode (i.e, on $\partial\mu_{f,i}/\partial R_a$) as well as on the magnitude of the vibrational integral $<\chi_{vf}|(R_a - R_{a,e})|\chi_{vi}>$.

An example of an E1 forbidden but "vibronically allowed" transition is provided by the singlet $n \Longrightarrow \pi^*$ transition of H_2CO that was discussed earlier in this section. As detailed there, the ground electronic state has ${}^1A_1$ symmetry, and the $n \Longrightarrow \pi^*$ state is of ${}^1A_2$ symmetry, so the $E1$ transition integral $<\psi_{ef}|\mu|\psi_{ei}>$ vanishes for all three (x, y, z) components of the electric dipole operator μ. However, vibrations that are of b_2 symmetry (e.g., the H-C-H asymmetric stretch vibration) can induce intensity in the $n \Longrightarrow \pi^*$ transition as follows:

1. For such vibrations, the b_2 mode's $vi = 0$ to $vf = 1$ vibronic integral $<\chi_{vf}|(R_a - R_{a,e})|\chi_{vi}>$ will be non-zero and probably quite substantial (because, for harmonic oscillator functions these "fundamental" transition integrals are dominant- see earlier);
2. Along these same b_2 modes, the electronic transition dipole integral *derivative* $\partial\mu_{f,i}/\partial R_a$ will be non-zero, even though the integral itself $\mu_{f,i}(\boldsymbol{R_e})$ vanishes when evaluated at the initial state's equilibrium geometry.

To understand why the derivative $\partial\mu_{f,i}/\partial R_a$ can be non-zero for distortions (denoted R_a) of b_2 symmetry, consider this quantity in greater detail:

$$\partial\mu_{f,i}/\partial R_a = \partial<\psi_{ef}|\mu|\psi_{ei}>/\partial R_a$$
$$= <\partial\psi_{ef}/\partial R_a|\mu|\psi_{ei}> + <\psi_{ef}|\mu|\partial\psi_{ei}/\partial R_a> + <\psi_{ef}|\partial\mu/\partial R_a|\psi_{ei}>.$$

The third integral vanishes because the derivative of the dipole operator itself

$$\mu = \sum_i e\boldsymbol{r}_j + \sum_a Z_a e\boldsymbol{R}_a$$

with respect to the coordinates of atomic centers, yields an operator that contains only a sum of scalar quantities (the elementary charge e and the magnitudes of various atomic charges Z_a); as a result and because the integral over the electronic wavefunctions $<\psi_{ef}|\psi_{ei}>$ vanishes, this contribution yields zero. The first and second integrals need not vanish by symmetry because the wavefunction derivatives $\partial\psi_{ef}/\partial R_a$ and $\partial\psi_{ei}/\partial R_a$ do *not* possess the same symmetry as their respective wavefunctions ψ_{ef} and ψ_{ei}. In fact, it can be shown that the symmetry of such a derivative is given by the direct product of the symmetries of its wavefunction and the symmetry of the vibrational mode that gives rise to the $\partial/\partial R_a$. For the H_2CO case at hand, the b_2 mode vibration can induce in the excited ${}^1A_2$ state a derivative component (i.e., $\partial\psi_{ef}/\partial R_a$) that is of ${}^1B_1$ symmetry) and this same vibration can induce in the ${}^1A_1$ ground state a derivative component of ${}^1B_2$ symmetry.

As a result, the contribution

$<\partial\psi_{ef}/\partial R_a|\mu|\psi_{ei}>$ to $\partial\mu_{f,i}/\partial R_a$

arising from vibronic coupling within the *excited* electronic state can be expected to be non-zero for components of the dipole operator μ that are of $(\partial\psi_{ef}/\partial R_a \times \psi_{ei}) = (B_1 \times A_1) = B_1$ symmetry. Light polarized along the molecule's x-axis gives such a b_1 component to μ (see the C_{2v} character table in appendix E). The second contribution $<\psi_{ef}|\mu|\partial\psi_{ei}/\partial R_a>$ can be non-zero for components of μ that are of $(\psi_{ef} \times \partial\psi_{ei}/\partial R_a) = (A_2 \times B_2) = B_1$ symmetry; again, light of x-axis polarization can induce such a transition.

In summary, electronic transitions that are $E1$ forbidden by symmetry can derive significant (e.g., in H_2CO the singlet $n ==> \pi*$ transition is rather intense) intensity through vibronic coupling. In such coupling, one or more vibrations (either in the initial or the final state) cause the respective electronic wavefunction to acquire (through $\partial\psi/\partial R_a$) a symmetry component that is different than that of ψ itself. The symmetry of $\partial\psi/\partial R_a$, which is given as the direct product of the symmetry of ψ and that of the vibration, can then cause the electric dipole integral $<\psi'|\mu|\partial\psi/\partial R_a>$ to be non-zero even when $<\psi'|\mu|\psi>$ is zero. Such vibronically allowed transitions are said to derive their intensity through vibronic borrowing.

D. Rotational Selection Rules for Electronic Transitions

Each vibrational peak within an electronic transition can also display rotational structure (depending on the spacing of the rotational lines, the resolution of the spectrometer, and the presence or absence of substantial line broadening effects such as those discussed later in this chapter). The selection rules for such transitions are derived in a fashion that parallels that given above for the vibration-rotation case. The major difference between this electronic case and the earlier situation is that the vibrational transition dipole moment μ_{trans} appropriate to the former is replaced by $\mu_{f,i}(\boldsymbol{R}_e)$ for conventional (i.e., non-vibronic) transitions or

$$\frac{\partial\mu_{f,i}}{\partial R_a}$$

(for vibronic transitions).

As before, when $\mu_{f,i}(\boldsymbol{R}_e)$ (or $\partial\mu_{f,i}/\partial R_a$) lies along the molecular axis of a linear molecule, the transition is denoted σ and $k = 0$ applies; when this vector lies perpendicular to the axis it is called π and $k = \pm1$ pertains. The resultant ***linear-molecule*** rotational selection rules are the same as in the vibration-rotation case:

$\Delta L = \pm1$, and $\Delta M = \pm1, 0$ (for σ transitions).
$\Delta L = \pm1, 0$ and $\Delta M = \pm1, 0$ (for π transitions).

In the latter case, the $L = L' = 0$ situation does not arise because a π transition has one unit of angular momentum along the molecular axis which would preclude both L and L' vanishing.

For **non-linear molecules** of the spherical or symmetric top variety, $\mu_{f,i}(\boldsymbol{R}_e)$ (or $\partial\mu_{f,i}/\partial R_a$) may be aligned along or perdendicular to a symmetry axis of the molecule. The selection rules that result are

$\Delta L = \pm1, 0$; $\Delta M = \pm1, 0$; and $\Delta K = 0$ ($L = L' = 0$ is not allowed and all $\Delta L = 0$

are forbidden when $K = K' = 0$)

which applies when $\mu_{f,i}(\boldsymbol{R}_e)$ or $\partial\mu_{f,i}/\partial R_a$ lies along the symmetry axis, and

$\Delta L = \pm1, 0$; $\Delta M = \pm1, 0$; and $\Delta K = \pm1$ ($L = L' = 0$ is not allowed)

which applies when $\mu_{f,i}(\boldsymbol{R}_e)$ or $\partial\mu_{f,i}/\partial R_a$ lies perpendicular to the symmetry axis.

IV. TIME CORRELATION FUNCTION EXPRESSIONS FOR TRANSITION RATES

The first-order $E1$ "golden-rule" expression for the rates of photon-induced transitions can be recast into a form in which certain specific physical models are easily introduced and insights are easily gained. Moreover, by using so-called equilibrium averaged time correlation functions, it is possible to obtain rate expressions appropriate to a large number of molecules that exist in a distribution of initial states (e.g., for molecules that occupy many possible rotational and perhaps several vibrational levels at room temperature).

A. State-to-State Rate of Energy Absorption or Emission

To begin, the expression obtained earlier

$$R_{i,f} = (2\pi/\hbar^2) g(\omega_{f,i}) |\boldsymbol{E}_0 \bullet \langle \Phi_f | \mu | \Phi_i \rangle|^2,$$

that is appropriate to transitions between a particular initial state Φ_i and a specific final state Φ_f, is rewritten as

$$R_{i,f} = (2\pi/\hbar^2) \int g(\omega) |\boldsymbol{E}_0 \bullet \langle \Phi_f | \mu | \Phi_i \rangle|^2 \delta(\omega_{f,i} - \omega) d\omega.$$

Here, the $\delta(\omega_{f,i} - \omega)$ function is used to specifically enforce the "resonance condition" that resulted in the time-dependent perturbation treatment given in chapter 14; it states that the photons' frequency ω must be resonant with the transition frequency $\omega_{f,i}$. It should be noted that by allowing ω to run over positive and negative values, the photon absorption (with $\omega_{f,i}$ positive and hence ω positive) and the stimulated emission case (with $\omega_{f,i}$ negative and hence ω negative) are both included in this expression (as long as $g(\omega)$ is defined as $g(|\omega|)$ so that the negative-ω contributions are multiplied by the light source intensity at the corresponding positive ω value).

The following integral identity can be used to replace the δ-function:

$$\delta(\omega_{f,i} - \omega) = \frac{1}{2\pi} \int_{-\infty}^{\infty} \exp[i(\omega_{f,i} - \omega)t] dt$$

by a form that is more amenable to further development. Then, the state-to-state rate of transition becomes:

$$R_{i,f} = (1/\hbar^2) \int g(\omega) |\boldsymbol{E}_0 \bullet \langle \Phi_f | \mu | \Phi_i \rangle|^2 \int_{-\infty}^{\infty} \exp[i(\omega_{f,i} - \omega)t] dt d\omega.$$

B. Averaging over Equilibrium Boltzmann Population of Initial States

If this expression is then multiplied by the **equilibrium probability** ρ_i that the molecule is found in the state Φ_i and summed over all such initial states and summed over all final states Φ_f that can be reached from Φ_i with photons of energy $\hbar\omega$, the *equilibrium averaged rate of photon absorption* by the molecular sample is obtained:

$$R_{eq.ave.} = (1/\hbar^2) \sum_{i,f} \rho_i \int g(\omega) |\boldsymbol{E}_0 \bullet \langle \Phi_f | \mu | \Phi_i \rangle|^2 \int_{-\infty}^{\infty} \exp[i(\omega_{f,i} - \omega)t] dt d\omega.$$

This expression is appropriate for an ensemble of molecules that can be in various initial states Φ_i with probabilities ρ_i. The corresponding result for transitions that originate in a particular state (Φ_i) but end up in any of the "allowed" (by energy and selection rules) final states reads:

$$R_{statei.} = (1/\hbar^2) \sum_f \int g(\omega) |\boldsymbol{E}_0 \bullet \langle \Phi_f | \mu | \Phi_i \rangle|^2 \int_{-\infty}^{\infty} \exp[i(\omega_{f,i} - \omega)t] dt d\omega.$$

For a canonical ensemble, in which the number of molecules, the temperature, and the system volume are specified, ρ_i takes the form:

$$\rho_i = \frac{g_i \exp(-E_i^0/kT)}{Q}$$

where Q is the canonical partition function of the molecules and g_i is the degeneracy of the state Φ_i whose energy is E_i^0.

In the above expression for $R_{eq.ave.}$, a double sum occurs. Writing out the elements that appear in this sum in detail, one finds:

$$\sum_{i,f} \rho_i \boldsymbol{E}_0 \bullet \langle \Phi_i | \mu | \Phi_f \rangle \boldsymbol{E}_0 \bullet \langle \Phi_f | \mu | \Phi_i \rangle \exp i(\omega_{f,i})t.$$

In situations in which one is interested in developing an expression for the intensity arising from transitions to *all* allowed final states, the sum over these final states can be carried out explicitly by first writing

$$\langle \Phi_f | \mu | \Phi_i \rangle \exp i(\omega_{f,i})t = \langle \Phi_f | \exp(iHt/\hbar) \mu \exp(-iHt/\hbar) | \Phi_i \rangle$$

and then using the fact that the set of states $\{\Phi_k\}$ are complete and hence obey

$$\sum_k |\Phi_k\rangle\langle\Phi_k| = 1.$$

The result of using these identities as well as the **Heisenberg definition** of the time-dependence of the dipole operator

$$\mu(t) = \exp(iHt/\hbar)\mu \exp(-iHt/\hbar),$$

is:

$$\sum_i \rho_i \langle \Phi_i | \boldsymbol{E}_0 \bullet \mu \boldsymbol{E}_0 \bullet \mu(t) | \Phi_i \rangle.$$

In this form, one says that the time dependence has been reduce to that of an equilibrium averaged **time correlation function** involving the component of the dipole operator along the external electric field at $t = 0$ ($\boldsymbol{E}_0 \bullet \mu$) and this component at a different time t ($\boldsymbol{E}_0 \bullet \mu(t)$).

C. Photon Emission and Absorption

If $\omega_{f,i}$ is positive (i.e., in the photon absorption case), the above expression will yield a non-zero contribution when multiplied by $\exp(-i\omega t)$ and integrated over positive ω-values. If $\omega_{f,i}$ is negative (as for stimulated photon emission), this expression will contribute, again when multiplied by $\exp(-i\omega t)$, for negative ω-values. In the latter situation, ρ_i is the equilibrium probability of finding the molecule in the (excited) state from which emission will occur; this probability can be related to that of the lower state ρ_f by

$$\rho_{excited} = \rho_{lower} \exp[-(E^0_{excited} - E^0_{lower})/kT]$$
$$= \rho_{lower} \exp[-\hbar\omega/kT].$$

In this form, it is important to realize that the excited and lower states are treated as individual *states*, not as levels that might contain a degenerate set of states.

The absorption and emission cases can be combined into a single *net* expression for the rate of photon absorption by recognizing that the latter process leads to photon production, and thus must be entered with a negative sign. The resultant expression for the *net rate of decrease of photons* is:

$$R_{eq.ave.net} = (1/\hbar^2) \sum_i \rho_i(1 - \exp(-\hbar\omega/kT)) \int\int g(\omega) <\Phi_i|(\boldsymbol{E}_0 \bullet \mu)\boldsymbol{E}_0 \bullet \mu(t)|\Phi_i> \exp(-i\omega t) d\omega dt.$$

D. The Line Shape and Time Correlation Functions

Now, it is convention to introduce the so-called "line shape" function $I(\omega)$:

$$I(\omega) = \sum_i \rho_i \int <\Phi_i|(\boldsymbol{E}_0 \bullet \mu)\boldsymbol{E}_0 \bullet \mu(t)|\Phi_i>) \exp(-i\omega t) dt$$

in terms of which the net photon absorption rate is

$$R_{eq.ave.net} = (1/\hbar^2)(1 - \exp(-\hbar\omega/kT)) \int g(\omega) I(\omega) d\omega.$$

As stated above, the function

$$C(t) = \sum_i \rho_i <\Phi_i|(\boldsymbol{E}_0 \bullet \mu)\boldsymbol{E}_0 \bullet \mu(t)|\Phi_i>$$

is called the equilibrium averaged **time correlation function** of the component of the electric dipole operator along the direction of the external electric field $\boldsymbol{E}_0$. Its Fourier transform is $I(\omega)$, the **spectral line shape** function. The convolution of $I(\omega)$ with the light source's $g(\omega)$ function, multiplied by $(1 - \exp(-\hbar\omega/kT))$, the correction for stimulated photon emission, gives the net rate of photon absorption.

E. Rotational, Translational, and Vibrational Contributions to the Correlation Function

To apply the time correlation function machinery to each particular kind of spectroscopic transition, one proceeds as follows:

1. For purely **rotational transitions**, the initial and final electronic and vibrational states are the same. Moreover, the electronic and vibrational states are not summed over in the analog of the above development because one is interested in obtaining an expression for a particular $\chi_{iv}\psi_{ie} ==> \chi_{fv}\psi_{fe}$ electronic-vibrational transition's lineshape. As a result, the sum over final states contained in the expression (see earlier)

$$\sum_{i,f} \rho_i \boldsymbol{E}_0 \bullet \langle \Phi_i | \mu | \Phi_f \rangle \boldsymbol{E}_0 \bullet \langle \Phi_f | \mu(t) | \Phi_i \rangle \exp i(\omega_{f,i})t$$

applies only to summing over final rotational states. In more detail, this can be shown as follows:

$$\sum_{i,f} \rho_i \boldsymbol{E}_0 \bullet \langle \Phi_i | \mu | \Phi_f \rangle \boldsymbol{E}_0 \bullet \langle \Phi_f | \mu(t) | \Phi_i \rangle$$

$$= \sum_{i,f} \rho_i \boldsymbol{E}_0 \bullet \langle \phi_{ir} \chi_{iv} \psi_{ie} | \mu | \phi_{fr} \chi_{iv} \psi_{ie} \rangle \boldsymbol{E}_0 \bullet \langle \phi_{fr} \chi_{iv} \psi_{ie} | \mu(t) | \phi_{ir} \chi_{iv} \psi_{ie} \rangle$$

$$= \sum_{i,f} \rho_{ir} \rho_{iv} \rho_{ie} \boldsymbol{E}_0 \bullet \langle \phi_{ir} \chi_{iv} | \mu(\boldsymbol{R}) | \phi_{fr} \chi_{iv} \rangle \boldsymbol{E}_0 \bullet \langle \phi_{fr} \chi_{iv} | \mu(\boldsymbol{R},t) | \phi_{ir} \chi_{iv} \rangle$$

$$= \sum_{i,f} \rho_{ir} \rho_{iv} \rho_{ie} \boldsymbol{E}_0 \bullet \langle \phi_{ir} | \boldsymbol{\mu_{ave.iv}} | \phi_{fr} \rangle \boldsymbol{E}_0 \bullet \langle \phi_{fr} | \boldsymbol{\mu_{ave.iv}}(t) | \phi_{ir} \rangle$$

$$= \sum_{i} \rho_{ir} \rho_{iv} \rho_{ie} \boldsymbol{E}_0 \bullet \langle \phi_{ir} | \mu_{ave.iv} \boldsymbol{E}_0 \bullet \mu_{ave.iv}(t) | \phi_{ir} \rangle.$$

In moving from the second to the third lines of this derivation, the following identity was used:

$$\langle \phi_{fr} \chi_{iv} \psi_{ie} | \mu(t) | \phi_{ir} \chi_{iv} \psi_{ie} \rangle = \langle \phi_{fr} \chi_{iv} \psi_{ie} | \exp(iHt/\hbar) \mu \exp(-iHt/\hbar) | \phi_{ir} \chi_{iv} \psi_{ie} \rangle$$
$$= \langle \phi_{fr} \chi_{iv} \psi_{ie} | \exp(iH_{v,r}t/\hbar) \mu(\boldsymbol{R}) \exp(-iH_{v,r}t/\hbar) | \phi_{ir} \chi_{iv} \psi_{ie} \rangle,$$

where H is the full (electronic plus vibrational plus rotational) Hamiltonian and $H_{v,r}$ is the vibrational and rotational Hamiltonian for motion on the electronic surface of the state ψ_{ie} whose dipole moment is $\mu(\boldsymbol{R})$. From the third line to the fourth, the (approximate) separation of rotational and vibrational motions in $H_{v,r}$

$$H_{v,r} = H_v + H_r$$

has been used along with the fact that χ_{iv} is an eigenfunction of H_v :

$$H_v \chi_{iv} = E_{iv} \chi_{iv}$$

to write

$$\langle \chi_{iv} | \mu(\boldsymbol{R},t) | \chi_{iv} \rangle = \exp(iH_r t/\hbar) \langle \chi_{iv} | \exp(iH_v t/\hbar) \mu(\boldsymbol{R}) \exp(-iH_v t/\hbar) | \chi_{iv} \rangle \exp(-iH_r t/\hbar)$$
$$= \exp(iH_r t/\hbar) \langle \chi_{iv} | \exp(iE_{iv} t/\hbar) \mu(\boldsymbol{R}) \exp(-iE_{iv} t/\hbar) | \chi_{iv} \rangle \exp(-iH_r t/\hbar)$$
$$= \exp(iH_r t/\hbar) \langle \chi_{iv} | \mu(\boldsymbol{R}) | \chi_{iv} \rangle \exp(-iH_r t/\hbar)$$
$$= \mu_{ave.iv}(t).$$

In effect, μ is replaced by the vibrationally averaged electronic dipole moment $\mu_{ave,iv}$ for each initial vibrational state that can be involved, and the time correlation function thus becomes:

$$C(t) = \sum_{i} \rho_{ir} \rho_{iv} \rho_{ie} \langle \phi_{ir} | (\boldsymbol{E_0} \bullet \mu_{ave,iv}) \boldsymbol{E}_0 \bullet \mu_{ave,iv}(t) | \phi_{ir} \rangle,$$

where $\mu_{ave,iv}(t)$ is the averaged dipole moment for the vibrational state χ_{iv} at time t, given that it was $\mu_{ave,iv}$ at time $t = 0$. The time dependence of $\mu_{ave,iv}(t)$ is induced by the rotational Hamiltonian H_r, as shown clearly in the steps detailed above:

$$\mu_{ave,iv}(t) = \exp(iH_r t/\hbar) \langle \chi_{iv} | \mu(\boldsymbol{R}) | \chi_{iv} \rangle \exp(-iH_r t/\hbar).$$

In this particular case, the equilibrium average is taken over the initial rotational states whose probabilities are denoted ρ_{ir}, any initial vibrational states that may be populated, with probabilities ρ_{iv}, and any populated electronic states, with probabilities ρ_{ie}.

2. For **vibration-rotation transitions** within a single electronic state, the initial and final electronic states are the same, but the initial and final vibrational and rotational states differ. As a result, the sum over final states contained in the expression

$$\sum_{i,f} \rho_i \boldsymbol{E}_0 \bullet \langle \Phi_i | \mu | \Phi_f \rangle \boldsymbol{E}_0 \bullet \langle \Phi_f | \mu | \Phi_i \rangle \exp i(\omega_{f,i})t$$

applies only to summing over final vibrational and rotational states. Paralleling the development made in the pure rotation case given above, this can be shown as follows:

$$\sum_{i,f} \rho_i \boldsymbol{E}_0 \bullet \langle \Phi_i | \mu | \Phi_f \rangle \boldsymbol{E}_0 \bullet \langle \Phi_f | \mu(t) | \Phi_i \rangle$$

$$= \sum_{i,f} \rho_i \boldsymbol{E}_0 \bullet \langle \phi_{ir}\chi_{iv}\psi_{ie} | \mu | \phi_{fr}\chi_{fv}\psi_{ie} \rangle \boldsymbol{E}_0 \bullet \langle \phi_{fr}\chi_{fv}\psi_{ie} | \mu(t) | \phi_{ir}\chi_{iv}\psi_{ie} \rangle$$

$$= \sum_{i,f} \rho_{ir}\rho_{iv}\rho_{ie} \boldsymbol{E}_0 \bullet \langle \phi_{ir}\chi_{iv} | \mu(\boldsymbol{R}) | \phi_{fr}\chi_{fv} \boldsymbol{E}_0 \bullet \langle \phi_{fr}\chi_{fv} | \mu(\boldsymbol{R},t) | \phi_{ir}\chi_{iv} \rangle$$

$$= \sum_{i,f} \rho_{ir}\rho_{iv}\rho_{ie} \boldsymbol{E}_0 \bullet \langle \phi_{ir}\chi_{iv} | \mu(\boldsymbol{R_e}) + \sum_a (R_a - R_{a,eq}) \partial\mu/\partial R_a | \phi_{fr}\chi_{fv} \rangle \boldsymbol{E}_0 \bullet \langle \phi_{fr}\chi_{fv} | \exp(iH_r t/\hbar)(\mu(\boldsymbol{R_e})$$

$$+ \sum_a (R_a - R_{a,eq}) \partial\mu/\partial R_a) \exp(-iH_r t/\hbar) | \phi_{ir}\chi_{iv} \rangle \exp(i\omega_{fv,iv}t)$$

$$= \sum_{ir,iv,ie} \rho_{ir}\rho_{iv}\rho_{ie} \sum_{fv,fr} \sum_a \langle \chi_{iv} | (R_a - R_{a,eq}) | \chi_{fv} \rangle$$

$$\sum_{a'} \langle \chi_{fv} | (R_{a'} - R_{a',eq}) | \chi_{iv} \rangle \exp(i\omega_{fv,iv}t)$$

$$\boldsymbol{E}_0 \bullet \langle \phi_{ir} | \partial\mu/\partial R_a \boldsymbol{E}_0 \bullet \exp(iH_r t/\hbar) \partial\mu/\partial R_{a'} \exp(-iH_r t/\hbar) | \phi_{ir} \rangle$$

$$= \sum_{ir,iv,ie} \rho_{ir}\rho_{iv}\rho_{ie} \sum_{fv,fr} \exp(i\omega_{fv,iv}t)$$

$$\langle \phi_{ir} | (\boldsymbol{E}_0 \bullet \mu_{trans}) \boldsymbol{E}_0 \bullet \exp(iH_r t/\hbar) \mu_{trans} \exp(-iH_r t/\hbar) | \phi_{ir} \rangle,$$

where the vibrational transition dipole matrix element is defined as before

$$\mu_{trans} = \sum_a \langle \chi_{iv} | (R_a - R_{a,eq}) | \chi_{fv} \rangle \partial\mu/\partial R_a,$$

and derives its time dependence above from the rotational Hamiltonian:

$$\mu_{trans}(t) = \exp(iH_r t/\hbar) \mu_{trans} \exp(-iH_r t/\hbar).$$

The corresponding final expression for the time correlation function $C(t)$ becomes:

$$C(t) = \sum_i \rho_{ir}\rho_{iv}\rho_{ie} \langle \phi_{ir} | (\boldsymbol{E}_0 \bullet \mu_{trans}) \boldsymbol{E}_0 \bullet \mu_{trans}(t) | \phi_{ir} \rangle \exp(i\omega_{fv,iv}t).$$

The net rate of photon absorption remains:

$$R_{eq.ave.net} = (1/\hbar^2)(1 - exp(-\hbar\omega))\int g(\omega)I(\omega)d\omega,$$

where $I(\omega)$ is the Fourier transform of $C(t)$.

The expression for $C(t)$ clearly contains two types of time dependences: (i) the $\exp(i\omega_{fv,iv}t)$, upon Fourier transforming to obtain $I(\omega)$, produces δ-function "spikes" at frequencies $\omega = \omega_{fv,iv}$ equal to the spacings between the initial and final vibrational states, and (ii) rotational motion time dependence that causes $\mu_{trans}(t)$ to change with time. The latter appears in the form of a correlation function for the component of μ_{trans} along $\boldsymbol{E_0}$ at time $t = 0$ and this component at another time t. The convolution of both these time dependences determines the from of $I(\omega)$.

3. For **electronic-vibration-rotation transitions**, the initial and final electronic states are different as are the initial and final vibrational and rotational states. As a result, the sum over final states contained in the expression

$$\sum_{i,f} \rho_i \boldsymbol{E}_0\bullet\langle\Phi_i|\mu|\Phi_f\rangle \boldsymbol{E}_0\bullet\langle\Phi_f|\mu|\Phi_i\rangle \exp i(\omega_{f,i})t$$

applies to summing over final electronic, vibrational, and rotational states. Paralleling the development made in the pure rotation case given above, this can be shown as follows:

$$\sum_{i,f} \rho_i \boldsymbol{E}_0\bullet\langle\Phi_i|\mu|\Phi_f\rangle \boldsymbol{E}_0\bullet\langle\Phi_f|\mu(t)|\Phi_i\rangle$$

$$= \sum_{i,f} \rho_i \boldsymbol{E}_0\bullet\langle\phi_{ir}\chi_{iv}\psi_{ie}|\mu|\phi_{fr}\chi_{fv}\psi_{fe}\rangle \boldsymbol{E}_0\bullet\langle\phi_{fr}\chi_{fv}\psi_{fe}|\mu(t)|\phi_{ir}\chi_{iv}\psi_{ie}\rangle$$

$$= \sum_{i,f} \rho_{ir}\rho_{iv}\rho_{ie} \boldsymbol{E}_0\bullet\langle\phi_{ir}\chi_{iv}|\mu_{i,f}(\boldsymbol{R})|\phi_{fr}\chi_{fv}\boldsymbol{E}_0\bullet\langle\phi_{fr}\chi_{fv}|\mu_{i,f}(\boldsymbol{R},t)|\phi_{ir}\chi_{iv}\rangle$$

$$= \sum_{i,f} \rho_{ir}\rho_{iv}\rho_{ie} \boldsymbol{E}_0\bullet\langle\phi_{ir}|\mu_{i,f}(\boldsymbol{R}_e)|\phi_{fr}\rangle|\langle\chi_{iv}\chi_{fv}\rangle|^2$$

$$\boldsymbol{E}_0\bullet\langle\phi_{fr}| \exp(iH_rt/\hbar)\mu_{i,f}(\boldsymbol{R}_e) \exp(-iH_rt/\hbar)|\phi_{ir}\rangle \exp(i\omega_{fv,iv}t + i\Delta E_{i,f}\, t/\hbar)$$

$$= \sum_{i,f} \rho_{ir}\rho_{iv}\rho_{ie}\langle\phi_{ir}|\boldsymbol{E}_0\bullet\mu_{i,f}(\boldsymbol{R}_e)\boldsymbol{E}_0\bullet\mu_{i,f}(\boldsymbol{R}_e,t)|\phi_{ir}\rangle|\langle\chi_{iv}|\chi_{fv}\rangle|^2 \exp(i\omega_{fv,iv}t + i\Delta E_{i,f}\, t/\hbar),$$

where

$$\mu_{i,f}(\boldsymbol{R}_e,t) = \exp(iH_rt/\hbar)\mu_{i,f}(\boldsymbol{R}_e) \exp(-iH_rt/\hbar)$$

is the electronic transition dipole matrix element, evaluated at the equilibrium geometry of the absorbing state, that derives its time dependence from the rotational Hamiltonian H_r as in the time correlation functions treated earlier.

This development thus leads to the following definition of $C(t)$ for the electronic, vibration, and rotation case:

$$C(t) = \sum_{i,f} \rho_{ir}\rho_{iv}\rho_{ie}\langle\phi_{ir}|\boldsymbol{E}_0\bullet\mu_{i,f}(\boldsymbol{R}_e)\boldsymbol{E}_0\bullet\mu_{i,f}(\boldsymbol{R}_e,t)|\phi_{ir}\rangle|\langle\chi_{iv}|\chi_{fv}\rangle|^2 \exp(i\omega_{fv,iv}t + i\Delta E_{i,f}\, t/\hbar)$$

but the net rate of photon absorption remains:

$$R_{eq.ave.net} = (1/\hbar^2)(1 - \exp(-\hbar\omega/kT))\int g(\omega)I(\omega)d\omega.$$

Here, $I(\omega)$ is the Fourier transform of the above $C(t)$ and $\Delta E_{i,f}$ is the adiabatic electronic energy difference (i.e., the energy difference between the $v = 0$ level in the final electronic state and the $v = 0$ level in the initial electronic state) for the electronic transition of interest. The above $C(t)$ clearly contains Franck-Condon factors as well as time dependence $\exp(i\omega_{fv,iv}t + i\Delta E_{i,f}\, t/\hbar)$ that produces δ-function spikes at each electronic-vibrational transition frequency and rotational time dependence contained in the time correlation function quantity

$$<\phi_{ir}|\boldsymbol{E}_0\bullet\mu_{i,f}(\boldsymbol{R}_e)\boldsymbol{E}_0\bullet\mu_{i,f}(\boldsymbol{R}_e,t)|\phi_{ir}>.$$

To summarize, the line shape function $I(\omega)$ produces the net rate of photon absorption

$$R_{eq.ave.net} = (1/\hbar^2)(1 - \exp(-h\omega/kT))\int g(\omega)I(\omega)d\omega\ .$$

in all of the above cases, and $I(\omega)$ is the Fourier transform of a corresponding time-dependent $C(t)$ function in all cases. However, the pure rotation, vibration-rotation, and electronic-vibration-rotation cases differ in the form of their respective $C(t)$'s. Specifically,

$$C(t) = \sum_i \rho_{ir}\rho_{iv}\rho_{ie}<\phi_{ir}|(\boldsymbol{E}_0\bullet\mu_{ave,iv})\boldsymbol{E}_0\bullet\mu_{ave,iv}(t)|\phi_{ir}>$$

in the pure rotational case,

$$C(t) = \sum_i \rho_{ir}\rho_{iv}\rho_{ie}<\phi_{ir}|(\boldsymbol{E}_0\bullet\mu_{trans})\boldsymbol{E}_0\bullet\mu_{trans}(t)|\phi_{ir}> \exp(i\omega_{fv,iv}t)$$

in the vibration-rotation case, and

$$C(t) = \sum_{i,f} \rho_{ir}\rho_{iv}\rho_{ie}<\phi_{ir}|\boldsymbol{E}_0\bullet\mu_{i,f}(\boldsymbol{R}_e)\boldsymbol{E}_0\bullet\mu_{i,f}(\boldsymbol{R}_e,t)|\phi_{ir}>|<\chi_{iv}|\chi_{fv}>|^2 \exp(i\omega_{fv,iv}t + \Delta E_{i,f}\, t/\hbar)$$

in the electronic-vibration-rotation case.

All of these time correlation functions contain time dependences that arise from rotational motion of a dipole-related vector (i.e., the vibrationally averaged dipole $\mu_{ave,iv}(t)$, the vibrational transition dipole $\mu_{trans}(t)$, or the electronic transition dipole $\mu_{i,f}(\boldsymbol{R}_e,t)$) and the latter two also contain oscillatory time dependences (i.e., $\exp(i\omega_{fv,iv}t)$ or $\exp(i\omega_{fv,iv}t + i\Delta E_{i,f}\, t/\hbar)$) that arise from vibrational or electronic-vibrational energy level differences. In the treatments of the following sections, consideration is given to the rotational contributions under circumstances that characterize, for example, dilute gaseous samples where the collision frequency is low and liquid-phase samples where rotational motion is better described in terms of diffusional motion.

F. Line Broadening Mechanisms

If the rotational motion of the molecules is assumed to be entirely unhindered (e.g., by any environment or by collisions with other molecules), it is appropriate to express the time dependence of each of the dipole time correlation functions listed above in terms of a "free rotation" model. For example, when dealing with diatomic molecules, the electronic-vibrational-rotational $C(t)$ appropriate to a specific electronic-vibrational transition becomes:

$$C(t) = (q_r q_v q_e q_t)^{-1} \sum_J (2J+1)\exp(-h^2 J(J+1)/(8\pi^2 IkT))\exp(-h\nu_{vib} iv/kT)$$

$$g_{ie} \langle\phi_J | \boldsymbol{E}_0 \bullet \mu_{i,f}(\boldsymbol{R}_e) \boldsymbol{E}_0 \bullet \mu_{i,f}(\boldsymbol{R}_e,t) | \phi_J\rangle |\langle\chi_{iv}\chi_{fv}\rangle|^2 \exp(i[h\nu_{vib}]t + i\Delta E_{i,f} t/\hbar).$$

Here,

$$q_r = (8\pi^2 IkT/h^2)$$

is the rotational partition function (I being the molecule's moment of inertia $I = \mu R_e^2$, and $h^2 J(J+1)/(8\pi^2 I)$ the molecule's rotational energy for the state with quantum number J and degeneracy $2J+1$)

$$q_v = \exp(-h\nu_{vib}/2kT)(1 - \exp(-h\nu_{vib}/kT))^{-1}$$

is the vibrational partition function (ν_{vib} being the vibrational frequency), g_{ie} is the degeneracy of the initial electronic state,

$$q_t = (2\pi mkT/h^2)^{3/2} V$$

is the translational partition function for the molecules of mass m moving in volume V, and $\Delta E_{i,f}$ is the adiabatic electronic energy spacing.

The functions $\langle\phi_J | \boldsymbol{E}_0 \bullet \mu_{i,f}(\boldsymbol{R}_e) \boldsymbol{E}_0 \bullet \mu_{i,f}(\boldsymbol{R}_e,t) | \phi_J\rangle$ describe the time evolution of the dipole-related vector (the electronic transition dipole in this case) for the rotational state J. In a "free-rotation" model, this function is taken to be of the form:

$$\langle\phi_J | \boldsymbol{E}_0 \bullet \mu_{i,f}(\boldsymbol{R}_e) \boldsymbol{E}_0 \bullet \mu_{i,f}(\boldsymbol{R}_e,t) | \phi_J\rangle = \langle\phi_J | \boldsymbol{E}_0 \bullet \mu_{i,f}(\boldsymbol{R}_e) \boldsymbol{E}_0 \bullet \mu_{i,f}(\boldsymbol{R}_e,0) | \phi_J\rangle \cos\frac{hJ(J+1)t}{4\pi I},$$

where

$$\frac{hJ(J+1)}{4\pi I} = \omega_J$$

is the rotational frequency (in cycles per second) for rotation of the molecule in the state labeled by J. This oscillatory time dependence, combined with the $\exp(i\omega_{fv,iv} t + i\Delta E_{i,f} t/\hbar)$ time dependence arising from the electronic and vibrational factors, produce, when this $C(t)$ function is Fourier transformed to generate $I(\omega)$ a series of δ-function "peaks" whenever

$$\omega = \omega_{fv,iv} + \Delta E_{i,f}/\hbar \pm \omega_J.$$

The intensities of these peaks are governed by the

$$(q_r q_v q_e q_t)^{-1} \sum_J (2J+1)\exp(-h^2 J(J+1)/(8\pi^2 IkT))\exp(-h\nu_{vib} iv/kT) g_{ie}$$

Boltzmann population factors as well as by the $|\langle\chi_{iv}|\chi_{fv}\rangle|^2$ Franck-Condon factors and the $\langle\phi_J | \boldsymbol{E}_0 \bullet \mu_{i,f}(\boldsymbol{R}_e) \boldsymbol{E}_0 \bullet \mu_{i,f}(\boldsymbol{R}_e,0) | \phi_J\rangle$ terms.

This same analysis can be applied to the pure rotation and vibration-rotation $C(t)$ time dependences with analogous results. In the former, δ-function peaks are predicted to occur at

$$\omega = \pm\omega_J$$

and in the latter at

$$\omega = \omega_{fv,iv} \pm \omega_J;$$

with the intensities governed by the time independent factors in the corresponding expressions for $C(t)$.

In experimental measurements, such sharp δ-function peaks are, of course, not observed. Even when very narrow band width laser light sources are used (i.e., for which $g(\omega)$ is an extremely narrowly peaked function), spectral lines are found to possess finite widths. Let us now discuss several sources of line broadening, some of which will relate to deviations from the "unhindered" rotational motion model introduced above.

1. Doppler Broadening

In the above expressions for $C(t)$, the averaging over initial rotational, vibrational, and electronic states is explicitly shown. There is also an average over the translational motion implicit in all of these expressions. Its role has not (yet) been emphasized because the molecular energy levels, whose spacings yield the characteristic frequencies at which light can be absorbed or emitted, do not depend on translational motion. However, the frequency of the electromagnetic field experienced by moving molecules does depend on the velocities of the molecules, so this issue must now be addressed.

Elementary physics classes express the so-called **Doppler shift** of a wave's frequency induced by movement either of the light source or of the molecule (Einstein tells us these two points of view must give identical results) as follows:

$$\omega_{observed} = \omega_{nominal}(1 + v_z/c)^{-1} \approx \omega_{nominal}(1 - v_z/c + \ldots).$$

Here, $\omega_{nominal}$ is the frequency of the unmoving light source seen by unmoving molecules, v_z is the velocity of relative motion of the light source and molecules, c is the speed of light, and $\omega_{observed}$ is the Doppler shifted frequency (i.e., the frequency seen by the molecules). The second identity is obtained by expanding, in a power series, the $(1 + v_z/c)^{-1}$ factor, and is valid in truncated form when the molecules are moving with speeds significantly below the speed of light.

For all of the cases considered earlier, a $C(t)$ function is subjected to Fourier transformation to obtain a spectral lineshape function $I(\omega)$, which then provides the essential ingredient for computing the net rate of photon absorption. In this Fourier transform process, the variable ω is assumed to be the frequency of the electromagnetic field *experienced by the molecules*. The above considerations of Doppler shifting then leads one to realize that the correct functional form to use in converting $C(t)$ to $I(\omega)$ is:

$$I(\omega) = \int C(t) \exp(-it\omega(1 - v_z/c))dt,$$

where ω is the nominal frequency of the light source.

As stated earlier, within $C(t)$ there is also an equilibrium average over translational motion of the molecules. For a gas-phase sample undergoing random collisions and at thermal equilibrium, this average is characterized by the well known Maxwell-Boltzmann velocity distribution:

$$(m/2\pi kT)^{3/2} \exp(-m(v_x^2 + v_y^2 + v_z^2)/2kT)dv_x dv_y dv_z.$$

Here m is the mass of the molecules and v_x, v_y, and v_z label the velocities along the lab-fixed cartesian coordinates.

Defining the z-axis as the direction of propagation of the light's photons and carrying out the averaging of the Doppler factor over such a velocity distribution, one obtains:

$$\int_{-\infty}^{\infty} \exp(-it\omega(1 - v_z/c))(m/2\pi kT)^{3/2} \exp(-m(v_x^2 + v_y^2 + v_z^2)/2kT)dv_x dv_y dv_z$$

$$= \exp(-i\omega t)\int_{-\infty}^{\infty} (m/2\pi kT)^{1/2} \exp(i\omega t v_z/c) \exp(-mv_z^2/2kT)dv_z$$

$$= \exp(-i\omega t) \exp(-\omega^2 t^2 kT/(2mc^2)).$$

This result, when substituted into the expressions for $C(t)$, yields expressions identical to those given for the three cases treated above *but* with one modification. The translational motion average need no longer be considered in each $C(t)$; instead, the earlier expressions for $C(t)$ must each be multiplied by a factor $\exp(-\omega^2 t^2 kT/(2mc^2))$ that embodies the translationally averaged Doppler shift. The spectral line shape function $I(\omega)$ can then be obtained for each $C(t)$ by simply Fourier transforming:

$$I(\omega) = \int_{-\infty}^{\infty} \exp(-i\omega t)C(t)dt.$$

When applied to the rotation, vibration-rotation, or electronic-vibration-rotation cases within the "unhindered" rotation model treated earlier, the Fourier transform involves integrals of the form:

$$I(\omega) = \int_{-\infty}^{\infty} \exp(-i\omega t) \exp(-\omega^2 t^2 kT/(2mc^2)) \exp(i(\omega_{fv,iv} + \Delta E_{i,f}/\hbar \pm \omega_J)t)dt.$$

This integral would arise in the electronic-vibration-rotation case; the other two cases would involve integrals of the same form but with the $\Delta E_{i,f}/\hbar$ absent in the vibration-rotation situation and with $\omega_{fv,iv} + \Delta E_{i,f}/\hbar$ missing for pure rotation transitions. All such integrals can be carried out analytically and yield:

$$I(\omega) = \sqrt{\frac{2mc^2\pi}{\omega^2 kT}} \exp[-(\omega - \omega_{fv,iv} - \Delta E_{i,f}/\hbar \pm \omega_J)^2 mc^2/(2\omega^2 kT)].$$

The result is a series of **Gaussian** "peaks" in ω-space, centered at:

$$\omega = \omega_{fv,iv} + \Delta E_{i,f}/\hbar \pm \omega_J$$

with widths (σ) determined by

$$\sigma^2 = \omega^2 kT/(mc^2),$$

given the temperature T and the mass of the molecules m. The hotter the sample, the faster the molecules are moving on average, and the broader is the distribution of Doppler shifted frequencies experienced by these molecules. The net result then of the Doppler effect is to produce a line shape function that is similar to the "unhindered" rotation model's series of δ-functions but with each δ-function peak broadened into a Gaussian shape.

2. Pressure Broadening

To include the effects of collisions on the rotational motion part of any of the above $C(t)$ functions, one must introduce a model for how such collisions change the dipole-related vectors that enter into $C(t)$. The most elementary model used to address collisions applies to gaseous samples which are assumed to undergo unhindered rotational motion until struck by another molecule at which time a randomizing "kick" is applied to the dipole vector and after which the molecule returns to its unhindered rotational movement.

The effects of such collisionally induced kicks are treated within the so-called **pressure broadening** (sometimes called collisional broadening) model by modifying the free-rotation correlation function through the introduction of an exponential damping factor $\exp(-|t|/\tau)$:

$$<\phi_J|\mathbf{E}_0\bullet\mu_{i,f}(\mathbf{R}_e)\mathbf{E}_0\bullet\mu_{i,f}(\mathbf{R}_e,0)|\phi_J> \cos\frac{hJ(J+1)t}{4\pi I}$$

$$==> \int <\phi_J|\mathbf{E}_0\bullet\mu_{i,f}(\mathbf{R}_e)\mathbf{E}_0\bullet\mu_{i,f}(\mathbf{R}_e,0)|\phi_J> \cos\frac{hJ(J+1)t}{4\pi I}\exp(-|t|/\tau).$$

This damping function's time scale parameter τ is assumed to characterize the average time between collisions and thus should be inversely proportional to the collision frequency. Its magnitude is also related to the effectiveness with which collisions cause the dipole function to deviate from its unhindered rotational motion (i.e., related to the collision strength). In effect, the exponential damping causes the time correlation function

$$<\phi_J|\mathbf{E}_0\bullet\mu_{i,f}(\mathbf{R}_e)\mathbf{E}_0\bullet\mu_{i,f}(\mathbf{R}_e,t)|\phi_J>$$

to "lose its memory" and to decay to zero; this "memory" point of view is based on viewing

$$<\phi_J|\mathbf{E}_0\bullet\mu_{i,f}(\mathbf{R}_e)\mathbf{E}_0\bullet\mu_{i,f}(\mathbf{R}_e,t)|\phi_J>$$

as the projection of $\mathbf{E}_0\bullet\mu_{i,f}(\mathbf{R}_e,t)$ along its $t = 0$ value $\mathbf{E}_0\bullet\mu_{i,f}(\mathbf{R}_e,0)$ as a function of time t.

Introducing this additional $\exp(-|t|/\tau)$ time dependence into $C(t)$ produces, when $C(t)$ is Fourier transformed to generate $I(\omega)$,

$$I(\omega) = \int_{-\infty}^{\infty} \exp(-i\omega t)\exp(-|t|/\tau)\exp(-\omega^2 t^2 kT/(2mc^2))\exp(i(\omega_{fv,iv} + \Delta E_{i,f}/\hbar \pm \omega_J)t)dt.$$

In the limit of very small Doppler broadening, the $(\omega^2 t^2 kT/(2mc^2))$ factor can be ignored (i.e., $\exp(-\omega^2 t^2 kT/(2mc^2))$ set equal to unity), and

$$I(\omega) = \int_{-\infty}^{\infty} \exp(-i\omega t)\exp(-|t|/\tau)\exp(i(\omega_{fv,iv} + \Delta E_{i,f}/\hbar \pm \omega_J)t)dt$$

results. This integral can be performed analytically and generates:

$$I(\omega) = \frac{1}{4\pi}\left\{\frac{1/\tau}{(1/\tau)^2 + (\omega - \omega_{fv,iv} - \Delta E_{i,f}/\hbar \pm \omega_J)^2} + \frac{1/\tau}{(1/\tau)^2 + (\omega + \omega_{fv,iv} + \Delta E_{i,f}/\hbar \pm \omega_J)^2}\right\},$$

a pair of **Lorentzian** peaks in ω -space centered again at

$$\omega = \pm[\omega_{fv,iv} + \Delta E_{i,f}/\hbar \pm \omega_J].$$

The full width at half height of these Lorentzian peaks is $2/\tau$. One says that the individual peaks have been pressure or collisionally broadened.

When the Doppler broadening can not be neglected relative to the collisional broadening, the above integral

$$I(\omega) = \int_{-\infty}^{\infty} \exp(-i\omega t)\exp(-|t|/\tau)\exp(-\omega^2 t^2 kT/(2mc^2))\exp(i(\omega_{fv,iv} + \Delta E_{i,f}/\hbar \pm \omega_J)t)dt$$

is more difficult to perform. Nevertheless, it can be carried out and again produces a series of peaks centered at

$$\omega = \omega_{fv,iv} + \Delta E_{i,f}/\hbar \pm \omega_J$$

but whose widths are determined both by Doppler and pressure broadening effects. The resultant line shapes are thus no longer purely Lorentzian nor Gaussian (which are compared in the figure below for both functions having the same full width at half height and the same integrated area), but have a shape that is called a **Voight** shape.

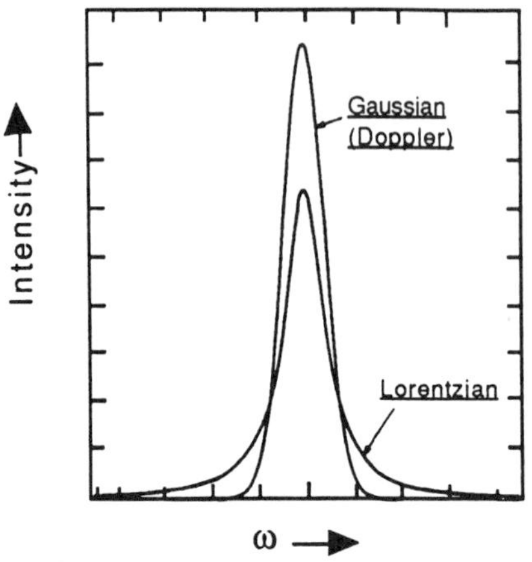

3. Rotational Diffusion Broadening

Molecules in liquids and very dense gases undergo frequent collisions with the other molecules; that is, the mean time between collisions is short compared to the rotational period for their unhindered rotation. As a result, the time dependence of the dipole related correlation function can no longer be modeled in terms of free rotation that is interrupted by (infrequent) collisions and Dopler shifted. Instead, a model that describes the incessant buffeting of the molecule's dipole by surrounding molecules becomes appropriate. For liquid samples in which these frequent collisions cause the molecule's dipole to undergo angular motions that cover all angles (i.e., in contrast to a frozen glass or solid in which the molecule's dipole would undergo strongly perturbed pendular motion about some favored orientation), the so-called **rotational diffusion** model is often used.

In this picture, the rotation-dependent part of $C(t)$ is expressed as:

$$<\phi_J|\boldsymbol{E}_0\bullet\mu_{i,f}(\boldsymbol{R}_e)\boldsymbol{E}_0\bullet\mu_{i,f}(\boldsymbol{R}_e,t)|\phi_J> = <\phi_J|\boldsymbol{E}_0\bullet\mu_{i,f}(\boldsymbol{R}_e)\boldsymbol{E}_0\bullet\mu_{i,f}(\boldsymbol{R}_e,0)|\phi_J> \exp(-2D_{rot}|t|),$$

where D_{rot} is the *rotational diffusion constant* whose magnitude details the time decay in the averaged value of $\boldsymbol{E}_0\bullet\mu_{i,f}(\boldsymbol{R}_e,t)$ at time t with respect to its value at time $t = 0$; the larger D_{rot}, the faster is this decay.

As with pressure broadening, this exponential time dependence, when subjected to Fourier transformation, yields:

$$I(\omega) = \int_{-\infty}^{\infty} \exp(-i\omega t) \exp(-2D_{rot}|t|) \exp(-\omega^2 t^2 kT/(2mc^2)) \exp(i(\omega_{fv,iv} + \Delta E_{i,f}/\hbar \pm \omega_J)t)dt.$$

Again, in the limit of very small Doppler broadening, the $(\omega^2 t^2 kT/(2mc^2))$ factor can be ignored (i.e., $\exp(-\omega^2 t^2 kT/(2mc^2))$ set equal to unity), and

$$I(\omega) = \int_{-\infty}^{\infty} \exp(-i\omega t) \exp(-2D_{rot}|t|) \exp(i(\omega_{fv,iv} + \Delta E_{i,f}/\hbar \pm \omega_J)t)dt$$

results. This integral can be evaluated analytically and generates:

$$I(\omega) = \frac{1}{4\pi}\left\{\frac{2D_{rot}}{(2D_{rot})^2 + (\omega - \omega_{fv,iv} - \Delta E_{i,f}/\hbar \pm \omega_J)^2} + \frac{2D_{rot}}{(2D_{rot})^2 + (\omega + \omega_{fv,iv} + \Delta E_{i,f}/\hbar \pm \omega_J)^2}\right\},$$

a pair of **Lorentzian** peaks in ω-space centered again at

$$\omega = \pm[\omega_{fv,iv} + \Delta E_{i,f}/\hbar \pm \omega_J].$$

The full width at half height of these Lorentzian peaks is $4D_{rot}$. In this case, one says that the individual peaks have been broadened via rotational diffusion. When the Doppler broadening can not be neglected relative to the collisional broadening, the above integral

$$I(\omega) = \int_{-\infty}^{\infty} \exp(-i\omega t) \exp(-2D_{rot}|t|) \exp(-\omega^2 t^2 kT/(2mc^2)) \exp(i(\omega_{fv,iv} + \Delta E_{i,f}/\hbar \pm \omega_J)t)dt.$$

is more difficult to perform. Nevertheless, it can be carried out and again produces a series of peaks centered at

$$\omega = \pm[\omega_{fv,iv} + \Delta E_{i,f}/\hbar \pm \omega_J]$$

but whose widths are determined both by Doppler and rotational diffusion effects.

4. Lifetime or Heisenberg Homogeneous Broadening

Whenever the absorbing species undergoes one or more processes that depletes its numbers, we say that it has a finite lifetime. For example, a species that undergoes unimolecular dissociation has a finite lifetime, as does an excited state of a molecule that decays by spontaneous emission of a photon. Any process that depletes the absorbing species contributes another source of time dependence for the dipole time correlation functions $C(t)$ discussed above. This time dependence is usually modeled by appending, in a multiplicative manner, a factor $\exp(-|t|/\tau)$. This, in turn modifies the line shape function $I(\omega)$ in a manner much like that discussed when treating the rotational diffusion case:

$$I(\omega) = \int_{-\infty}^{\infty} \exp(-i\omega t) \exp(-|t|/\tau) \exp(-\omega^2 t^2 kT/(2mc^2)) \exp(i(\omega_{fv,iv} + \Delta E_{i,f}/\hbar \pm \omega_J)t)dt.$$

Not surprisingly, when the Doppler contribution is small, one obtains:

$$I(\omega) = \frac{1}{4\pi}\left\{\frac{1/\tau}{(1/\tau)^2 + (\omega - \omega_{fv,iv} - \Delta E_{i,f}/\hbar \pm \omega_J)^2} + \frac{1/\tau}{(1/\tau)^2 + (\omega + \omega_{fv,iv} + \Delta E_{i,f}/\hbar \pm \omega_J)^2}\right\}.$$

In these Lorentzian lines, the parameter τ describes the kinetic decay lifetime of the molecule. One says that the spectral lines have been **lifetime or Heisenberg broadened** by an amount proportional to $1/\tau$. The latter terminology arises because the finite lifetime of the molecular states can be viewed as producing, via the Heisenberg uncertainty relation $\Delta E \Delta t > \hbar$, states whose energy is "uncertain" to within an amount ΔE.

5. Site Inhomogeneous Broadening

Among the above line broadening mechanisms, the pressure, rotational diffusion, and lifetime broadenings are all of the **homogeneous** variety. This means that each molecule in the sample is affected in exactly the same manner by the broadening process. For example, one does not find some molecules with short lifetimes and others with long lifetimes, in the Heisenberg case; the entire ensemble of molecules is characterized by a single lifetime.

In contrast, Doppler broadening is **inhomogeneous** in nature because each molecule experiences a broadening that is characteristic of its particular nature (velocity v_z in this case). That is, the fast molecules have their lines broadened more than do the slower molecules. Another important example of inhomogeneous broadening is provided by so-called **site broadening.** Molecules imbedded in a liquid, solid, or glass do not, at the instant of photon absorption, all experience exactly the same interactions with their surroundings. The distribution of instantaneous "solvation" environments may be rather "narrow" (e.g., in a highly ordered solid matrix) or quite "broad" (e.g., in a liquid at high temperature). Different environments produce different energy level splittings $\omega = \omega_{fv,iv} + \Delta E_{i,f}/\hbar \pm \omega_J$ (because the initial and final states are "solvated" differently by the surroundings) and thus different frequencies at which photon absorption can occur. The distribution of energy level splittings causes the sample to absorb at a range of frequencies as illustrated in the figure below where homogeneous and inhomogeneous line shapes are compared.

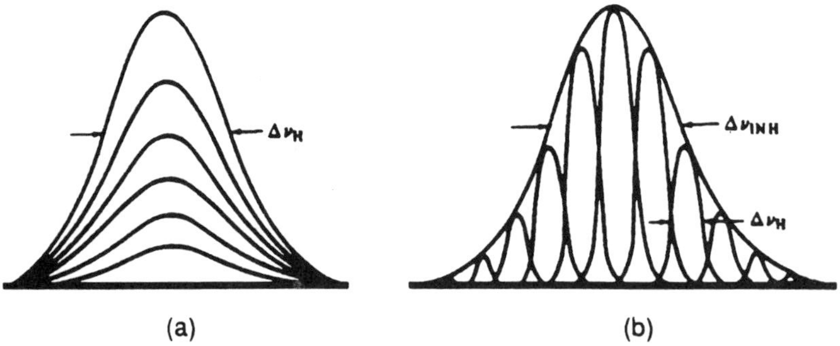

Homogeneous (a) and inhomogeneous (b) band shapes having inhomogeneous width $\Delta\nu_{INH}$, and homogeneous width $\Delta\nu_H$.

The spectral line shape function $I(\omega)$ is further broadened when site inhomogeneity is present and significant. These effects can be modeled by convolving the kind of $I(\omega)$ function that results from Doppler, lifetime, rotational diffusion, and pressure broadening with a Gaussian distribution $P(\Delta E)$ that describes the inhomogeneous distribution of energy level splittings:

$$I(\omega) = \int I^0(\omega;\Delta E) P(\Delta E) d\Delta E.$$

Here $I^0(\omega;\Delta E)$ is a line shape function such as those described earlier each of which contains a set of frequencies (e.g., $\omega = \omega_{fv,iv} + \Delta E_{i,f}/\hbar \pm \omega_J = \omega + \Delta E/\hbar$) at which absorption or emission occurs.

A common experimental test for inhomogeneous broadening involves **hole burning.** In such experiments, an intense light source (often a laser) is tuned to a frequency ω_{burn} that lies within the spectral line being probed for inhomogeneous broadening. Then, a second tunable light source is used to scan through the profile of the spectral line, and, for example, an absorption spectrum is recorded. Given an absorption profile as shown below in the absence of the intense burning light source:

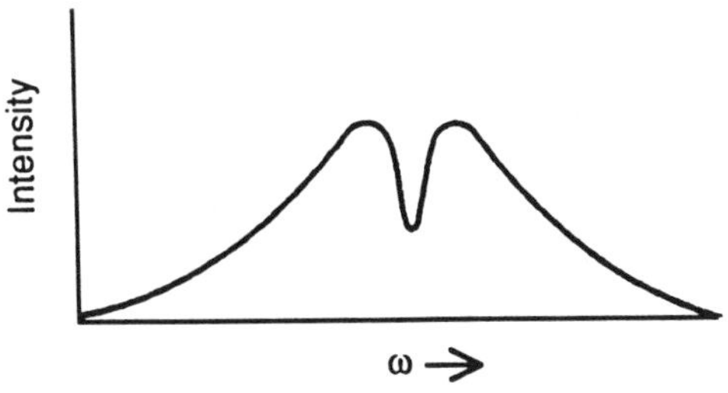

one expects to see a profile such as that shown below:

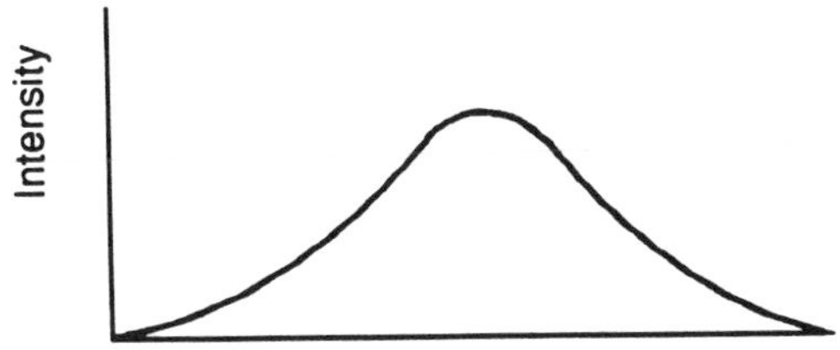

if inhomogeneous broadening is operative.

The interpretation of the change in the absorption profile caused by the bright light source proceeds as follows:

1. In the ensemble of molecules contained in the sample, some molecules will absorb at or near the frequency of the bright light source ω_{burn}; other molecules (those whose environments do not produce energy level splittings that match ω_{burn}) will not absorb at this frequency.
2. Those molecules that do absorb at ω_{burn} will have their transition saturated by the *intense* light source, thereby rendering this frequency region of the line profile transparent to *further* absorption.
3. When the "probe" light source is scanned over the line profile, it will induce absorptions for those molecules whose local environments did not allow them to be saturated by the ω_{burn} light. The absorption profile recorded by this probe light source's detector thus will match that of the original line profile, *until*
4. the probe light source's frequency matches ω_{burn}, upon which no absorption of the probe source's photons will be recorded because molecules that absorb in this frequency regime have had their transition saturated.

5. Hence, a "**hole**" will appear in the spectrum recorded by the probe light source's detector in the region of ω_{burn}.

Unfortunately, the technique of hole burning does not provide a fully reliable method for identifying inhomogeneously broadened lines. If a hole is observed in such a burning experiment, this provides ample evidence, but if one is not seen, the result is not definitive. In the latter case, the transition may not be strong enough (i.e., may not have a large enough "rate of photon absorption") for the intense light source to saturate the transition to the extent needed to form a hole.

This completes our introduction to the subject of molecular spectroscopy. More advanced treatments of many of the subjects treated here as well as many aspects of modern experimental spectroscopy can be found in the text by Zare on angular momentum as well as in Steinfeld's text *Molecules and Radiation*, 2nd edition, by J. I. Steinfeld, MIT Press (1985).

CHAPTER

16

Collisions among molecules can also be viewed as a problem in time-dependent quantum mechanics. The perturbation is the "interaction potential," and the time dependence arises from the movement of the nuclear positions.

The simplest and most widely studied problems in chemical reaction dynamics involve describing the unimolecular motion or bimolecular collision of a system in a well characterized electronic state. Referring back to the discussion of chapter 3, we recall that the motion of the nuclei are governed by a Schrödinger equation

$$[E_j(R)\Xi_j^0(R) + T\Xi_j^0(R)] = E\Xi_j^0(R)$$

in which the electronic energy $E_j(R)$ assumes the role of the potential upon which movement occurs. This treatment of the nuclear motion is based on the Born-Oppenheimer approximation (see chapter 3 for details) which assumes that coupling to nearby electronic states can be ignored. These assumptions are valid only when the energy surface of interest $E_j(R)$ is not crossed or closely approached by another electronic energy surface $E_k(R)$. When the electronic states are so widely spaced, it is proper to speak of the movement of the molecule(s) on the electronic surface $E_j(R)$, and to use either classical or quantum mechanical methods to follow such movements.

To simplify the notation throughout this chapter, the above Schrödinger equation appropriate to movement on a single electronic energy surface will be written as follows:

$$[T + V(R)]\Xi(R) = E\Xi(R),$$

where T denotes the kinetic energy operator for *all 3N* of the geometrical coordinates (collectively denoted R) needed to specify the location of the N nuclei, $V(R)$ is the electronic energy as a function of these coordinates, and $\Xi(R)$ is the nuclear-motion wavefunction.

For example, when diatomic species are considered, V is a function of the radial coordinate describing the distance between the two nuclei, T contains derivatives with respect to radial as well as two angular coordinates (those pertaining to rotation or relative angular motion of the two nuclei), and R refers to these radial and angular coordinates. For a triatomic species such as H_2O, V is a function of two O-H bond lengths and the H-O-H angle, and R refers to these three internal coordinates as well as the three angle coordinates needed to specify the orientation of the H_2O molecule in space relative to a space-fixed coordinate system (e.g., three Euler angles used in chapter 3 to treat rotation of spherical and symmetric top molecules).

In chapters 1 and 3 and in all of section 4, such nuclear-motion Schrödinger equations were used to treat the *bound* vibrational motions of molecules (i.e., the movement of the nuclei when the energy available is not adequate to rupture one or more of the bonds in the molecule). These same Schrödinger equations also apply to the scattering of the constituent nuclei (e.g., the vibration-rotation motion of HCl is treated by the same Schrödinger equation as the scattering of an H atom and a Cl atom). The primary difference between these two situations lies in the total energy (E) available: in the former, E lies below the dissociation asymptote of the ground-state HCl electronic potential energy; in the latter E is higher than this asymptote (e.g., see the potential curve shown below with some of its bound state energies and a state in the continuum).

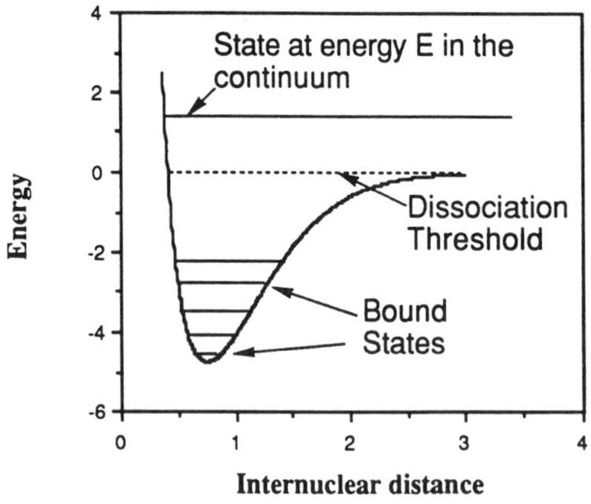

The different energies appropriate to bound-state and scattering situations affect the boundary conditions appropriate to the nuclear-motion wavefunctions in the large internuclear distance region. For the HCl example at hand, the bound-state vibrational wavefunctions $\Xi(R,\theta,\phi)$ decay exponentially (see chapter 1) for large R because such R-values lie in the classically forbidden region of R-space where $E - V(R)$ is negative. In contrast, the scattering wavefunctions for this same $V(R)$ potential and the same HCl molecule need not decay in the large-R region. As illustrated explicitly below for a model problem, this difference in large-R boundary conditions causes major differences in the eigenvalue spectrum of the Hamiltonian in these two cases. In particular, the bound-state energy levels of HCl are discrete (i.e., quantized) but the scattering states are not (i.e., an H atom and a Cl atom may collide with arbitrary relative translational energy).

Let us now examine how the Schrödinger equation is solved for cases in which E lies above the dissociation energy of $V(R)$ by considering a few simple model problems that can be solved exactly.

I. ONE-DIMENSIONAL SCATTERING

Atom-atom scattering on a single Born-Oppenheimer energy surface can be reduced to a one-dimensional Schrödinger equation by separating the radial and angular parts of the three-dimensional Schrödinger equation in the same fashion as used for the Hydrogen atom in chapter 1. The resultant equation for the radial part $\Psi(R)$ of the wavefunction can be written as:

$$-(\hbar^2/2\mu)R^{-2}\partial/\partial R(R^2\partial\psi/\partial R) + L(L+1)\hbar^2/(2\mu R^2)\Psi + V(R)\Psi = E\Psi,$$

where L is the quantum number that labels the angular momentum of the colliding particles whose reduced mass is μ.

Defining $\Psi(R) = R\Psi(R)$ and substituting into the above equation gives the following equation for Ψ:

$$-(\hbar^2/2\mu)\partial^2\Psi/\partial R^2 + L(L+1)\hbar^2/(2\mu R^2)\Psi + V(R)\Psi = E\Psi.$$

The combination of the "centrifugal potential" $L(L+1)\hbar^2/(2\mu R^2)$ and the electronic potential $V(R)$ thus produces a total "effective potential" for describing the radial motion of the system.

The simplest reasonable model for such an effective potential is provided by the "square well" potential illustrated below. This model $V(R)$ could, for example, be applied to the $L = 0$ scattering of two atoms whose bond dissociation energy is D_e and whose equilibrium bond length for this electronic surface lies somewhere between $R = 0$ and $R = R_{max}$.

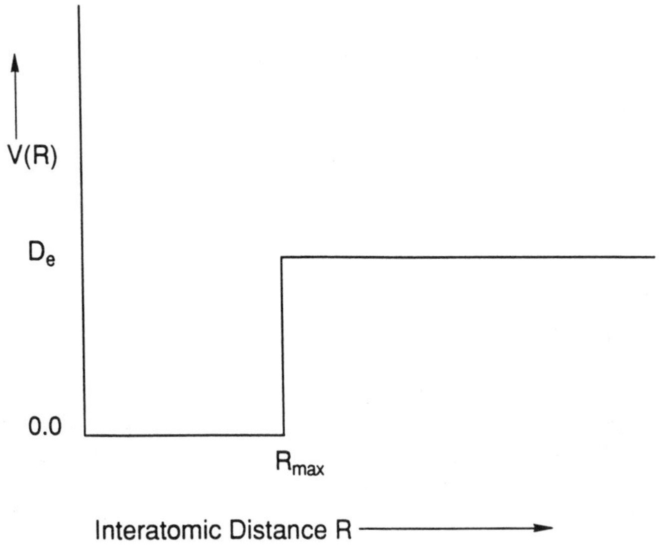

The piecewise constant nature of this particular $V(R)$ allows exact solutions to be written both for bound and scattering states because the Schrödinger equation

$$-(\hbar^2/2\mu)d^2\Psi/dR^2 = E\Psi \text{ (for } 0 \leq R \leq R_{max})$$
$$-(\hbar^2/2\mu)d^2\Psi/dR^2 + D_e\Psi = E\Psi \ (R_{max} \leq R < \infty)$$

admits simple sinusoidal solutions.

A. Bound States

The **bound states** are characterized by having $E < D_e$. For the inner region, the two solutions to the above equation are

$$\Psi_1(R) = A \sin(kR)$$

and

$$\Psi_2(R) = B \cos(kR)$$

where

$$k = \sqrt{2\mu E/\hbar^2}$$

is termed the "local wave number" because it is related to the momentum values for the $\exp(\pm ikR)$ components of such a function:

$$-i\hbar\partial \exp(\pm ikR)/\partial R = \hbar k \exp(\pm ikR).$$

The $\cos(kR)$ solution must be excluded (i.e., its amplitude B in the general solution of the Schrödinger equation must be chosen equal to 0.0) because this function does not vanish at $R = 0$, where the potential moves to infinity and thus the wavefunction must vanish. This means that only the

$$\Psi = A \sin(kR)$$

term remains for this inner region.

Within the asymptotic region ($R > R_{max}$) there are also two solutions to the Schrödinger equation:

$$\Psi_3 = C \exp(-\kappa R)$$

and

$$\Psi_4 = D \exp(\kappa R)$$

where

$$\kappa = \sqrt{2\mu(D_e - E)/\hbar^2}.$$

Clearly, one of these functions is a decaying function of R for large R and the other Ψ_4 grows exponentially for large R. The latter's amplitude D must be set to zero because this function generates a probability density that grows larger and larger as R penetrates deeper and deeper into the classically forbidden region (where $E < V(R)$).

To connect Ψ_1 in the inner region to Ψ_3 in the outer region, we use the fact that Ψ and $d\Psi/dR$ must be continuous except at points R where $V(R)$ undergoes an infinite discontinuity (see chapter 1). Continuity of Ψ at R_{max} gives:

$$A \sin(kR_{max}) = C \exp(-\kappa R_{max}),$$

and continuity of $d\Psi/dR$ at R_{max} yields

$$Ak \cos(kR_{max}) = -\kappa C \exp(-\kappa R_{max}).$$

These two equations allow the ratio C/A as well as the energy E (which appears in κ and in k) to be determined:

$$A/C = -\kappa/k \exp(-\kappa R_{max})/\cos(kR_{max}).$$

The condition that determines E is based on the well known requirement that the determinant of coefficients must vanish for homogeneous linear equations to have no-trivial solutions (i.e., not $A = C = 0$) :

$$\det\begin{pmatrix} \sin(kR_{max}) & -\exp(-\kappa R_{max}) \\ k\cos(kR_{max}) & \kappa \exp(-\kappa R_{max}) \end{pmatrix} = 0$$

The vanishing of this determinant can be rewritten as

$$\kappa \sin(kR_{max}) \exp(-\kappa R_{max}) + k \cos(kR_{max}) \exp(-\kappa R_{max}) = 0$$

or

$$\tan(kR_{max}) = -k/\kappa.$$

When employed in the expression for A/C, this result gives

$$A/C = \exp(-\kappa R_{max})/\sin(kR_{max}).$$

For very large D_e compared to E, the above equation for E reduces to the familiar "particle in a box" energy level result since k/κ vanishes in this limit, and thus $\tan(kR_{max}) = 0$, which is equivalent to $\sin(kR_{max}) = 0$, which yields the familiar $E = n^2h^2/(8\mu R_{max}^2)$ and $C/A = 0$, so $\Psi = A \sin(kR)$.

When D_e is not large compared to E, the full transcendental equation $\tan(kR_{max}) = -k/\kappa$ must be solved numerically or graphically for the eigenvalues E_n, $n = 1,2,3,\ldots$. These energy levels, when substituted into the definitions for k and κ give the wavefunctions:

$$\Psi = A \sin(kR) \qquad (\text{for } 0 \le R \le R_{max})$$
$$\Psi = A \sin(kR_{max}) \exp(\kappa R_{max}) \exp(-\kappa R) \qquad (\text{for } R_{max} \le R < \infty).$$

The one remaining unknown A can be determined by requiring that the modulus squared of the wavefunction describe a probability density that is normalized to unity when integrated over all space:

$$\int_0^\infty |\Psi|^2 dR = 1.$$

Note that this condition is equivalent to

$$\int_0^\infty |\Psi|^2 R^2 dR = 1$$

which would pertain to the original radial wavefunction. In the case of an infinitely deep potential well, this normalization condition reduces to

$$\int_0^{R_{max}} A^2 \sin^2(kR) dR = 1$$

which produces

$$A = \sqrt{\frac{2}{R_{max}}}\ .$$

B. Scattering States

The **scattering states** are treated in much the same manner. The functions Ψ_1 and Ψ_2 arise as above, and the amplitude of Ψ_2 must again be chosen to vanish because Ψ must vanish at $R = 0$ where the potential moves to infinity. However, in the exterior region ($R > R_{max}$), the two solutions are now written as:

$$\Psi_3 = C \exp(ik'R)$$
$$\Psi_4 = D \exp(-ik'R)$$

where the large-R local wavenumber

$$k' = \sqrt{2\mu(E - D_e)/\hbar^2}$$

arises because $E > D_e$ for scattering states.

The conditions that Ψ and $d\Psi/dR$ be continuous at R_{max} still apply:

$$A \sin(kR_{max}) = C \exp(ik'R_{max}) + D \exp(-ik'R_{max})$$

and

$$kA \cos(kR_{max}) = ik'C \exp(ik'R_{max}) - ik'D \exp(-ik'R_{max}).$$

However, these two equations (in *three* unknowns A, C, and D) can no longer be solved to generate eigenvalues E and amplitude ratios. There are now three amplitudes as well as the E value but only these two equations plus a normalization condition to be used. The result is that the energy no longer is specified by a boundary condition; it can take on any value. We thus speak of scattering states as being "in the continuum" because the allowed values of E form a continuum beginning at $E = D_e$ (since the zero of energy is defined in this example as at the bottom of the potential well).

The $R > R_{max}$ components of Ψ are commonly referred to as "incoming"

$$\Psi_{in} = D \exp(-ik'R)$$

and "outgoing"

$$\Psi_{out} = C \exp(ik'R)$$

because their radial momentum eigenvalues are $-\hbar k'$ and $\hbar k'$, respectively. It is a common convention to define the amplitude D so that the **flux** of incoming particles is unity. Choosing

$$D = \sqrt{\frac{\mu}{\hbar k'}}$$

produces an incoming wavefunction whose current density is:

$$S(R) = -i\hbar/2\mu[\Psi_{in}^*(d/dR\Psi_{in}) - (d\Psi_{in}/dR)^*\Psi_{in}]$$
$$= |D|^2(-i\hbar/2\mu)[-2ik']$$
$$= -1.$$

This means that there is one unit of current density moving inward (this produces the minus sign) for all values of R at which Ψ_{in} is an appropriate wavefunction (i.e., $R > R_{max}$). This condition takes the place of the probability normalization condition specified in the bound-state case when the modulus squared of the total wavefunction is required to be normalized to unity over all space. Scattering wavefunctions can not be so normalized because they do not decay at large R; for this reason, the flux normalization condition is usually employed. The magnitudes of the outgoing (C) and short range (A) wavefunctions relative to that of the incoming function (D) then provide information about the scattering and "trapping" of incident flux by the interaction potential.

Once D is so specified, the above two boundary matching equations are written as a set of two inhomogeneous linear equations in two unknowns (A and C):

$$A \sin(kR_{max}) - C \exp(ik'R_{max}) = D \exp(-ik'R_{max})$$

and

$$kA \cos(kR_{max}) - ik'C \exp(ik'R_{max}) = -ik'D \exp(-ik'R_{max})$$

or

$$\begin{pmatrix} \sin(kR_{max}) & -\exp(ik'R_{max}) \\ k\cos(kR_{max}) & -ik'\exp(ik'R_{max}) \end{pmatrix} \begin{pmatrix} A \\ C \end{pmatrix} = \begin{pmatrix} D\exp(-ik'R_{max}) \\ -ik'D\exp(-ik'R_{max}) \end{pmatrix}.$$

Non-trivial solutions for A and C will exist except when the determinant of the matrix on the left side vanishes:

$$-ik' \sin(kR_{max}) + k\cos(kR_{max}) = 0,$$

which can be true only if

$$\tan(kR_{max}) = ik'/k.$$

This equation is not obeyed for any (real) value of the energy E, so solutions for A and C in terms of the specified D can always be found.

In summary, specification of unit incident flux is made by choosing D as indicated above. For any collision energy $E > D_e$, the 2×1 array on the right hand side of the set of linear equations written above can be formed, as can the 2×2 matrix on the left side. These linear equations can then be solved for A and C. The overall wavefunction for this E is then given by:

$$\Psi = A\sin(kR) \qquad (\text{for } 0 \le R \le R_{max})$$
$$\Psi = C\exp(ik'R) + D\exp(-ik'R) \qquad (\text{for } R_{max} \le R < \infty).$$

C. Shape Resonance States

If the angular momentum quantum number L in the effective potential introduced earlier is non-zero, this potential has a repulsive component at large R. This repulsion can combine with short-range attractive interactions due, for example, to chemical bond forces, to produce an effective potential that one can model in terms of simple piecewise functions shown below.

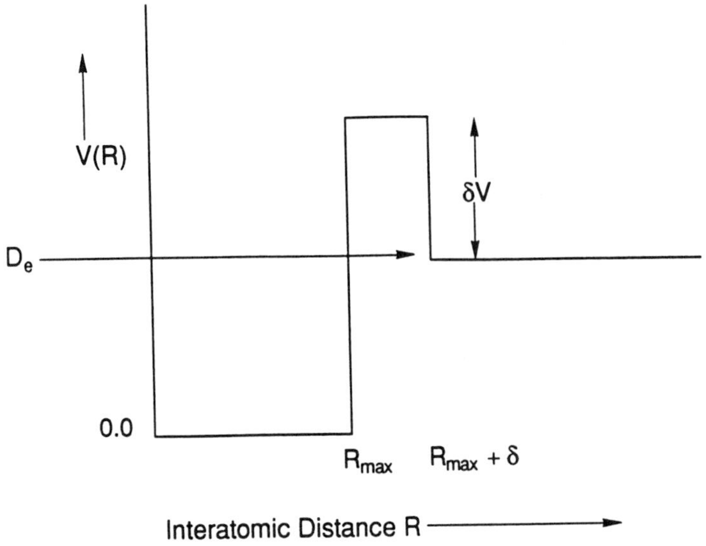

Again, the piecewise nature of the potential allows the one-dimensional Schrödinger equation to be solved analytically. For energies below D_e, one again finds bound states in much the same way as illustrated above (but with the exponentially decaying function $\exp(-\kappa' R)$ used in the region $R_{max} \le R \le R_{max} + \delta$, with

$$\kappa' = \sqrt{2\mu(D_e + \delta V - E)/\hbar^2}\ .$$

For energies lying above $D_e + \delta V$, scattering states occur and the four amplitudes of the functions ($\sin(kR)$, $\exp(\pm ik'''R)$ with $k''' = \sqrt{2\mu(-D_e - \delta V + E)/\hbar^2}$, and $\exp(ik'R)$) appropriate to each R-region are determined in terms of the amplitude of the incoming asymptotic function $D\exp(-ik'R)$ from the four equations obtained by matching Ψ and $d\Psi/dR$ at R_{max} and at $R_{max} + \delta$.

For energies lying in the range $D_e < E < D_e + \delta V$, a qualitatively different class of scattering function exists. These so-called **shape resonance** states occur at energies that are determined by the condition that the amplitude of the wavefunction within the barrier (i.e., for $0 \le R \le R_{max}$) be large so that incident flux successfully tunnels through the barrier and builds up, through constructive interference, large probability amplitude there. Let us now turn our attention to this specific energy regime.

The piecewise solutions to the Schrödinger equation appropriate to the shape-resonance case are easily written down:

$$\Psi = A\sin(kR) \qquad (\text{for } 0 \le R \le R_{max})$$
$$\Psi = B_+\exp(\kappa' R) + B_-\exp(-\kappa' R) \qquad (\text{for } R_{max} \le R \le R_{max} + \delta)$$
$$\Psi = C\exp(ik'R) + D\exp(-ik'R) \qquad (\text{for } R_{max} + \delta \le R < \infty).$$

Note that both exponentially growing and decaying functions are acceptable in the $R_{max} \le R \le R_{max} + \delta$ region because this region does not extend to $R = \infty$.

There are four amplitudes (A, B_+, B_-, and$C$) that must be expressed in terms of the specified amplitude D of the incoming flux. Four equations that can be used to achieve this goal result when Ψ and $d\Psi/dR$ are matched at R_{max} and at $R_{max} + \delta$:

$$A \sin(kR_{max}) = B_+ \exp(\kappa' R_{max}) + B_- \exp(-\kappa' R_{max}),$$
$$Ak \cos(kR_{max}) = \kappa' B_+ \exp(\kappa' R_{max}) - \kappa' B_- \exp(-\kappa' R_{max}),$$
$$B_+ \exp(\kappa'(R_{max} + \delta)) + B_- \exp(-\kappa'(R_{max} + \delta)) = C \exp(ik'(R_{max} + \delta)) + D \exp(-ik'(R_{max} + \delta)),$$
$$\kappa' B_+ \exp(\kappa'(R_{max} + \delta)) - \kappa' B_- \exp(-\kappa'(R_{max} + \delta)) = ik'C \exp(ik'(R_{max} + \delta)) - ik'D \exp(-ik'(R_{max} + \delta)).$$

It is especially instructive to consider the value of A/D that results from solving this set of four equations in four unknowns because the modulus of this ratio provides information about the relative amount of amplitude that exists inside the centrifugal barrier in the attractive region of the potential compared to that existing in the asymptotic region as incoming flux.

The result of solving for A/D is:

$$A/D = 4\kappa' \exp(-ik'(R_{max} + \delta)) \{\exp(\kappa'\delta)(ik' - \kappa')\, (\kappa' \sin(kR_{max}) + k \cos(kR_{max}))/ik' \\ + \exp(-\kappa'\delta)(ik' + \kappa')\, (\kappa' \sin(kR_{max}) - k \cos(kR_{max}))/ik'\}^{-1}.$$

Further, it is instructive to consider this result under conditions of a high (large $D_e + \delta V - E$) and thick (large δ) barrier. In such a case, the "tunnelling factor" $\exp(-\kappa'\delta)$ will be very small compared to its counterpart $\exp(\kappa'\delta)$, and so

$$A/D = 4\frac{ik'\kappa'}{ik' - \kappa'} exp(-ik'(R_{max} + \delta)) exp(-\kappa'\delta)\{\kappa' sin(kR_{max}) + k cos(kR_{max})\}^{-1}.$$

The $\exp(-\kappa'\delta)$ factor in A/D causes the magnitude of the wavefunction inside the barrier to be small in most circumstances; we say that incident flux must tunnel through the barrier to reach the inner region and that $\exp(-\kappa'\delta)$ gives the probability of this tunnelling. The magnitude of the A/D factor could become large if the collision energy E is such that

$$\kappa' \sin(kR_{max}) + k \cos(kR_{max})$$

is small. In fact, if

$$\tan(kR_{max}) = -k/\kappa'$$

this denominator factor in A/D will vanish and A/D will become infinite. Note that the above condition is similar to the energy quantization condition

$$\tan(kR_{max}) = -k/\kappa$$

that arose when bound states of a finite potential well were examined earlier in this chapter. There is, however, an important difference. In the bound-state situation

$$k = \sqrt{2\mu E/\hbar^2}$$

and

$$\kappa = \sqrt{2\mu(D_e - E)/\hbar^2}\ ;$$

in this shape-resonance case, k is the same, but

$$\kappa' = \sqrt{2\mu(D_e + \delta V - E)/\hbar^2})$$

rather than κ occurs, so the two $\tan(kR_{max})$ equations are not identical.

However, in the case of a very high barrier (so that κ' is much larger than k), the denominator

$$\kappa' \sin(kR_{max}) + k \cos(kR_{max}) \cong \kappa' \sin(kR_{max})$$

in *A*/*D* can become small if

$$\sin(kR_{max}) \cong 0.$$

This condition is nothing but the energy quantization condition that would occur for the particle-in-a-box potential shown below.

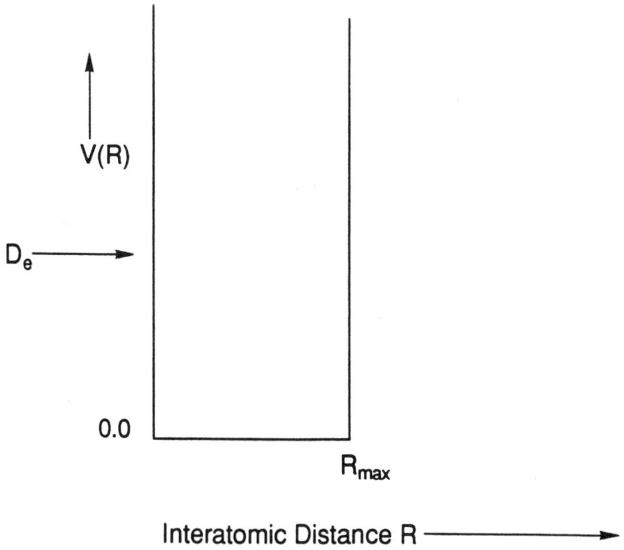

This potential is identical to the true effective potential for $0 \le R \le R_{max}$, but extends to infinity beyond R_{max}; the barrier and the dissociation asymptote displayed by the true potential are absent.

In summary, when a barrier is present on a potential energy surface, at energies above the dissociation asymptote D_e but below the top of the barrier ($D_e + \delta V$ here), one can expect shape-resonance states to occur at "special" scattering energies E. These so-called resonance energies can often be approximated by the bound-state energies of a potential that is identical to the potential of interest in the inner region ($0 \le R \le R_{max}$ here) but that extends to infinity beyond the top of the barrier (i.e., beyond the barrier, it does not fall back to values below E).

The chemical significance of shape resonances is great. Highly rotationally excited molecules may have more than enough total energy to dissociate (D_e), but this energy may be "stored" in the rotational motion, and the vibrational energy may be less than D_e. In terms of the above model, high angular momentum may produce a significant barrier in the effective potential, but the system's vibrational energy may lie significantly below D_e. In such a case, and when viewed in terms of motion on an angular momentum modified effective potential, the lifetime of the molecule with respect to dissociation is determined by the rate of tunnelling through the barrier.

For the case at hand, one speaks of "rotational predissociation" of the molecule. The lifetime τ can be estimated by computing the frequency ν at which flux existing inside R_{max} strikes the barrier at R_{max}

$$\nu = \frac{\hbar k}{2\mu R_{max}} \text{ (sec}^{-1}\text{)}$$

and then multiplying by the probability P that flux tunnels through the barrier from R_{max} to $R_{max} + \delta$:

$$P = \exp(-2\kappa'\delta).$$

The result is that

$$\tau^{-1} = \frac{\hbar k}{2\mu R_{max}} \exp(-2\kappa'\delta)$$

with the energy E entering into k and κ' being determined by the resonance condition: $(\kappa' \sin(kR_{max}) + k \cos(kR_{max}))$ = minimum.

Although the examples treated above involved piecewise constant potentials (so the Schrödinger equation and the boundary matching conditions could be solved exactly), many of the characteristics observed carry over to more chemically realistic situations. As discussed, for example, in *Energetic Principles of Chemical Reactions*, J. Simons, Jones and Bartlett, Portola Valley, Calif. (1983), one can often model chemical reaction processes in terms of:

1. motion along a "reaction coordinate" (s) from a region characteristic of reactant materials where the potential surface is positively curved in all direction and all forces (i.e., gradients of the potential along all internal coordinates) vanish,
2. to a transition state at which the potential surface's curvature along s is negative while all other curvatures are positive and all forces vanish,
3. onward to product materials where again all curvatures are positive and all forces vanish.

Within such a "reaction path" point of view, motion transverse to the reaction coordinate s is often modelled in terms of local harmonic motion although more sophisticated treatments of the dynamics is possible. In any event, this picture leads one to consider motion along a single degree of freedom (s), with respect to which much of the above treatment can be carried over, coupled to transverse motion along all other internal degrees of freedom taking place under an entirely positively curved potential (which therefore produces restoring forces to movement away from the "streambed" traced out by the reaction path s).

II. MULTICHANNEL PROBLEMS

When excited electronic states are involved, couplings between two or more electronic surfaces may arise. Dynamics occuring on an excited-state surface may evolve in a way that produces flux on another surface. For example, collisions between an electronically excited $1s2s(^3S)$ He atom and a ground-state $1s^2(^1S)$ He atom occur on a potential energy surface that is repulsive at large R (due to the repulsive interaction between the closed-shell $1s^2$ He and the large $2s$ orbital) but attractive at smaller R (due to the $\sigma^2\sigma*^1$ orbital occupancy arising from the three $1s$-derived electrons). The ground-state potential energy surface for this system (pertaining to two $1s^2(^1S)$ He atoms is repulsive at small R values (because of the $\sigma^2\sigma*^2$ nature of the electronic state). In this case, there are two Born-Oppenheimer electronic-nuclear motion states that are degenerate and thus need to be combined to achieve a proper description of the dynamics:

$$\Psi_1 = |\sigma^2\sigma*^2| = \Psi_{grnd.}(R,\theta,\phi)$$

pertaining to the ground electronic state and the scattering state $\Psi_{grnd.}$ on this energy surface, and

$$\Psi_2 = |\sigma^2\sigma*^1 2\sigma^1| = \Psi_{ex.}(R,\theta,\phi)$$

Pertaining to the excited electronic state and the nuclear-motion state $\Psi_{ex.}$ on this energy surface. Both of these wavefunctions can have the same energy E; the former has high nuclear-motion energy and low electronic energy, while the latter has higher electronic energy and lower nuclear-motion energy.

A simple model that can be used to illustrate the two-state couplings that arise in such cases is introduced through the two one-dimensional piecewise potential surfaces shown below.

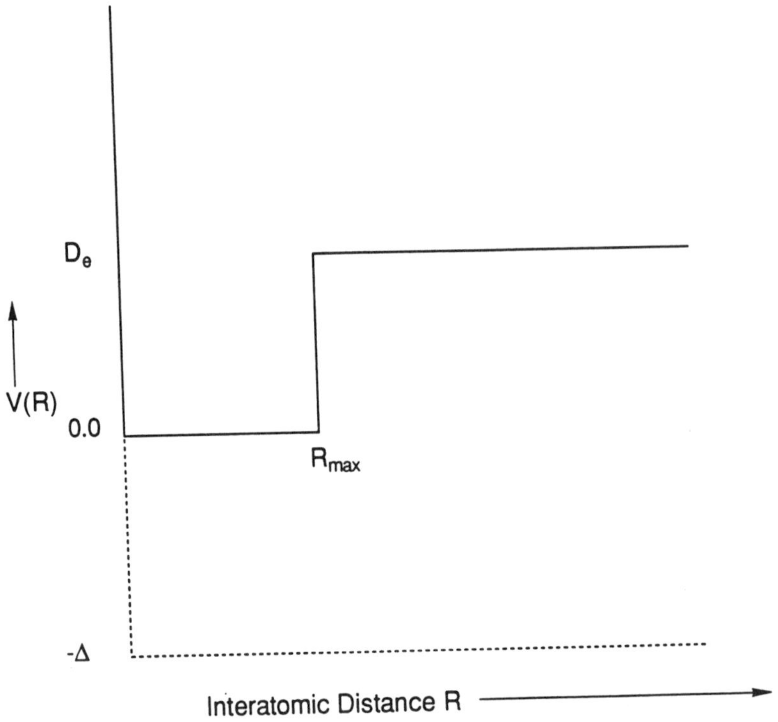

The dashed energy surface

$$V(R) = -\Delta \qquad (\text{for } 0 \le R < \infty)$$

provides a simple representation of a repulsive lower-energy surface, and the solid-line plot represents the excited-state surface that has a well of depth D_e and whose well lies Δ above the ground-state surface.

In this case, and for energies lying above zero (for $E < 0$, only nuclear motion on the lower energy dashed surface is "open" (i.e., accessible)) yet below D_e, the nuclear-motion wavefunction can have amplitudes belonging to both surfaces. That is, the total (electronic and nuclear) wavefunction consists of two portions that can be written as:

$$\Psi = A\phi \sin(kR) + \phi'' A'' \sin(k''R) \qquad (\text{for } 0 \le R \le R_{\max})$$

and

$$\Psi = A\phi \sin(kR_{\max}) \exp(\kappa R_{\max}) \exp(-\kappa R) + \phi'' A'' \sin(k''R) \qquad (\text{for } R_{\max} \le R < \infty),$$

where ϕ and ϕ'' denote the electronic functions belonging to the upper and lower energy surfaces, respectively. The wavenumbers k and k'' are defined as:

$$k = \sqrt{2\mu E/\hbar^2}$$
$$k'' = \sqrt{2\mu(E+\Delta)/\hbar^2}$$

and κ is as before

$$\kappa = \sqrt{2\mu(D_e - E)/\hbar^2}.$$

For the lower-energy surface, only the $\sin(k''R)$ function is allowed because the $\cos(k''R)$ function does not vanish at $R = 0$.

A. The Coupled Channel Equations

In such cases, the relative amplitudes (A and A'') of the nuclear motion wavefunctions on each surface must be determined by substituting the above "two-channel" wavefunction (the word channel is used to denote separate asymptotic states of the system; in this case, the ϕ and ϕ'' electronic states) into the full Schrödinger equation. In chapter 3, the couplings among Born-Oppenheimer states were so treated and resulted in the following equation:

$$[(E_j(R) - E)\Xi_j(R) + T\,\Xi_j(R)] = -\sum_i \{<\Psi_j|T\,|\Psi_i>(R)\Xi_i(R)$$
$$+ \sum_{a=1,M} (-\hbar^2/m_a) < \Psi_j|\nabla_a|\Psi_i > (R) \cdot \nabla_a\Xi_i(R)\}$$

where $E_j(R)$ and $\Xi_j(R)$ denote the electronic energy surfaces and nuclear-motion wavefunctions, Ψ_j denote the corresponding electronic wavefunctions, and the ∇_a represent derivatives with respect to the various coordinates of the nuclei. Changing to the notation used in the one-dimensional model problem introduced above, these so-called **coupled-channel** equations read:

$$[(-\Delta - E) - \hbar^2/2\mu d^2/dR^2]A''\sin(k''R)$$
$$= -\{<\phi''| - \hbar^2/2\mu d^2/dR^2|\phi''>A''\sin(k''R)$$
$$+ (-\hbar^2/\mu)<\phi''|d/dR|\phi>d/dRA\sin(kR)\} \qquad (\text{for } 0 \le R \le R_{max}),$$

$$[(-\Delta - E) - \hbar^2/2\mu d^2/dR^2]A''\sin(k''R)$$
$$= -\{<\phi''| - \hbar^2/2\mu d^2/dR^2|\phi''>A''\sin(k''R)$$
$$+ (-\hbar^2/\mu)<\phi''|d/dR|\phi>d/dRA\phi\sin(kR_{max})\exp(\kappa R_{max})\exp(-\kappa R)\} \qquad (\text{for } R_{max} \le R < \infty);$$

when the index j refers to the ground-state surface ($V(R) = -\Delta$, for $0 < R < \infty$), and

$$[(0 - E) - \hbar^2/2\mu d^2/dR^2]A\sin(kR) = -\{<\phi| - \hbar^2/2\mu d^2/dR^2|\phi>A\sin(kR)$$
$$+ (-\hbar^2/\mu)<\phi|d/dR|\phi''>d/dRA''\sin(k''R)\} \qquad (\text{for } 0 \le R \le R_{max}),$$

$$[(D_e - E) - \hbar^2/2\mu d^2/dR^2]A\sin(kR_{max})\exp(\kappa R_{max})\exp(-\kappa R)$$
$$= -\{<\phi|-\hbar^2/2\mu d^2/dR^2|\phi>A\sin(kR_{max})\exp(\kappa R_{max})\exp(-\kappa R)$$
$$+ (-\hbar^2/\mu)<\phi|d/dR|\phi''>d/dRA''\sin(k''R)\} \qquad (\text{for } R_{max} \le R < \infty)$$

when the index j refers to the excited-state surface (where $V(R) = 0$, for $0 < R \le R_{max}$ and $V(R) = D_e$ for $R_{max} \le R < \infty$).

Clearly, if the right-hand sides of the above equations are ignored, one simply recaptures the Schrödinger equations describing motion on the separate potential energy surfaces:

$$[(-\Delta - E) - \hbar^2/2\mu d^2/dR^2]A'' \sin(k''R) = 0 \qquad \text{(for } 0 \leq R \leq R_{\max}\text{)},$$
$$[(-\Delta - E) - \hbar^2/2\mu d^2/dR^2]A'' \sin(k''R) = 0 \qquad \text{(for } R_{\max} \leq R < \infty\text{)};$$

that describe motion on the lower-energy surface, and

$$[(0 - E) - \hbar^2/2\mu d^2/dR^2]A \sin(kR) = 0 \qquad \text{(for } 0 \leq R \leq R_{\max}\text{)},$$
$$[(D_e - E) - \hbar^2/2\mu d^2/dR^2]A \sin(kR_{\max}) \exp(\kappa R_{\max}) \exp(-\kappa R) = 0 \qquad \text{(for } R_{\max} \leq R < \infty\text{)}$$

describing motion on the upper surface on which the bonding interaction occurs. The terms on the right-hand sides provide the couplings that cause the true solutions to the Schrödinger equation to be combinations of solutions for the two separate surfaces.

In applications of the coupled-channel approach illustrated above, coupled sets of second order differential equations (two in the above example) are solved by starting with a specified flux in one of the channels and a chosen energy E. For example, one might specify the amplitude A to be unity to represent preparation of the system in a bound vibrational level (with $E < D_e$) of the excited electronic-state potential. One would then choose E to be one of the eigenenergies of that potential. Propagation methods could be used to solve the coupled differential equations subject to these choices of E and A. The result would be the determination of the amplitude A'' of the wavefunction on the ground-state surface. The ratio A''/A provides a measure of the strength of coupling between the two Born-Oppenheimer states.

B. Perturbative Treatment

Alternatively, one can treat the coupling between the two states via time dependent perturbation theory. For example, by taking $A = 1.0$ and choosing E to be one of the eigenenergies of the excited-state potential, one is specifying that the system is initially (just prior to $t = 0$) prepared in a state whose wavefunction is:

$$\Psi^0_{ex} = \phi \sin(kR) \qquad \text{(for } 0 \leq R \leq R_{\max}\text{)}$$
$$\Psi^0_{ex} = \phi \sin(kR_{\max}) \exp(\kappa R_{\max}) \exp(-\kappa R) \qquad \text{(for } R_{\max} \leq R < \infty\text{)}.$$

From $t = 0$ on, the coupling to the other state

$$\Psi^0_{grnd} = \phi'' \sin(k''R) \qquad \text{(for } 0 \leq R < \infty\text{)}$$

is induced by the "perturbation" embodied in the terms on the right-hand side of the coupled-channel equations.

Within this time dependent perturbation theory framework, the rate of transition of probability amplitude from the initially prepared state (on the excited state surface) to the ground-state surface is proportional to the square of the perturbation matrix elements between these two states:

$$\text{Rate} \ \alpha \left| \int_0^{R_{\max}} \sin(kR)\langle\phi|d/dR|\phi''\rangle(d/dR \sin(k''R))dR \right.$$
$$\left. + \int_{R_{\max}}^{\infty} \sin(kR_{\max}) \exp(\kappa R_{\max}) \exp(-\kappa R)\langle\phi|d/dR|\phi''\rangle(d/dR \sin(k''R))dR \right|^2$$

The matrix elements occurring here contain two distinct parts:

$$<\phi|d/dR|\phi''>$$

has to do with the electronic state couplings that are induced by radial movement of the nuclei; and both

$$\sin(kR)d/dR\,\sin(k''R)$$

and

$$\sin(kR_{max})\,\exp(\kappa R_{max})\,\exp(-\kappa R)d/dR\,\sin(k''R)$$

relate to couplings between the two nuclear-motion wavefunctions induced by these same radial motions. For a transition to occur, both the electronic and nuclear-motion states must undergo changes. The initially prepared state (the bound state on the upper electronic surface) has high electronic and low nuclear-motion energy, while the state to which transitions may occur (the scattering state on the lower electronic surface) has low electronic energy and higher nuclear-motion energy.

Of course, in the above example, the integrals over R can be carried out if the electronic matrix elements $<\phi|d/dR|\phi''>$ can be handled. In practical chemical applications (for an introductory treatment see *Energetic Principles of Chemical Reactions*, J. Simons, Jones and Bartlett, Portola Valley, Calif. (1983)), the evaluation of these electronic matrix elements is a formidable task that often requires computation intensive techniques such as those discussed in section 6.

Even when the electronic coupling elements are available (or are modelled or parameterized in some reasonable manner), the solution of the coupled-channel equations that govern the nuclear motion is a demanding task. For the purposes of this text, it suffices to note that:

1. couplings between motion on two or more electronic states can and do occur;
2. these couplings are essential to treat whenever the electronic energy difference (i.e., the spacing between pairs of Born-Oppenheimer potential surfaces) is small (i.e., comparable to vibrational or rotational energy level spacings);
3. there exists a rigorous theoretical framework in terms of which one can evaluate the rates of so-called **radiationless transitions** between pairs of such electronic, vibrational, rotational states. Expressions for such transitions involve (a) electronic matrix elements $<\phi|d/dR|\phi''>$ that depend on how strongly the electronic states are modulated by movement (hence the d/dR) of the nuclei, and (b) nuclear-motion integrals connecting the initial and final nuclear-motion wavefunctions, which also contain d/dR because they describe the "recoil" of the nuclei induced by the electronic transition.

C. Chemical Relevance

As presented above, the most obvious situation of multichannel dynamics arises when electronically excited molecules undergo radiationless relaxation (e.g., internal conversion when the spin symmetry of the two states is the same or intersystem crossing when the two states differ in spin symmetry). These subjects are treated in some detail in the text *Energetic Principles of Chemical Reactions*, J. Simons, Jones and Bartlett, Portola Valley, Calif. (1983) where radiationless transitions arising in photochemistry and polyatomic molecule reactivity are discussed.

Let us consider an example involving the chemical reactivity of electronically excited alkaline earth or $d^{10}s^2$ transition metal atoms with H_2 molecules. The particular case for $Cd^* + H_2 \rightarrow CdH + H$ has been studied experimentally and theoretically. In such systems, the potential energy surface connecting to ground-state $Cd(^1S) + H_2$ becomes highly repulsive as the collision partners approach (see the depiction provided in the figure shown below). The three surfaces that correlate with the $Cd(^1P) + H_2$ species prepared by photo-excitation of $Cd(^1S)$ behave quite differently as functions of the Cd-to-H_2 distance because in each the singly occupied $6p$ orbital assumes a different orientation relative to the H_2 molecule's bond axis. For (near) C_{2v} orientations, these states are labeled $^1B_2$, $^1B_1$, and $^1A_1$; they have the $6p$ orbital directed as shown in the second figure, respectively. The corresponding triplet surfaces that derive from $Cd(^3P) + H_2$ behave, as functions of the Cd-to-H_2 distance (R) in similar manner, except they are shifted to lower energy because $Cd(^3P)$ lies below $Cd(^1P)$ by ca.37kcal/mol.

Collisions between $Cd(^1P)$ and H_2 can occur on any of the three surfaces mentioned above. Flus on the $^1A_1$ surface is primarily reflected (at low collision energies characteristic of the thermal experiments) because this surface is quite repulsive at large R. Flux on the $^1B_1$ surface can proceed in to quite small R (ca. 2.4Å) before repulsive forces on this surface reflect it. At geometries near $R = 2.0$Å and $r_{HH} = 0.88$Å, the highly repulsive $^3A_1$ surface intersects this $^1B_1$ surface from below. At and near this intersection, a combination of spin-orbit coupling (which is large for Cd) and non-adiabatic coupling may induce flux to evolve onto the $^3A_1$ surface, after which fragmentation to $Cd(^3P) + H_2$ could occur.

In contrast, flux on the $^1B_2$ surface propogates inward under attractive forces to $R = 2.25$Å and $r_{HH} = 0.79$Å where it may evolve onto the $^3A_1$ surface which intersects from below. At and near this intersection, a combination of spin-orbit coupling (which is large for Cd) and non-adiabatic coupling may induce flux to evolve onto the $^3A_1$ surface, after which fragmentation to $Cd(^3P) + H_2$ could occur. Flux that continues to propogate inward to smaller R values experiences even stronger attractive forces that lead, near $R = 1.69$Å and $r_{HH} = 1.54$Å, to an intersection with the $^1A_1$ surface that connects to $Cd(^1S) + H_2$. Here, non-adiabatic couplings may cause flux to evolve onto the $^1A_1$ surface which may then lead to formation of ground state $Cd(^1S) + H_2$ or $Cd(^1S) + H + H$, both of which are energetically possible. Processes in which electronically excited atoms produce ground-state atoms through such collisions and surface hopping are termed "electronic quenching."

The nature of the non-adiabatic couplings that arise in the two examples given above are quite different. In the former case, when the $^1B_1$ and $^3A_1$ surfaces are in close proximity to one another, the first-order coupling element:

$$<\Psi(^1B_1)|\nabla j|\Psi(^3A_1)>$$

is non-zero only for nuclear motions (i.e.,∇j) of $b_1 \times a_1 = b_1$ symmetry. For the CdH_2 collision complex being considered in (or near) C_{2v} symmetry, such a motion corresponds to rotational motion of the nuclei about an axis lying parallel to the H-H bond axis. In contrast, to couple the $^3A_1$ and $^1B_2$ electronic states through an element of the form

$$<\Psi(^1B_2)|\nabla j|\Psi(^3A_1)>,$$

the motion must be of $b_2 \times a_1 = b_2$ symmetry. This movement corresponds to asymmetric vibrational motion of the two Cd-H interatomic coordinates.

The implications of these observations are clear. For example, in so-called half-collision experiments in which a van der Waals CdH_2 complex is probed, internal rotational motion would be expected to enhance $^1B_1 \rightarrow {}^3A_1$ quenching, whereas asymmetric vibrational motion should enhance the $^1B_2 \rightarrow {}^3A_1$ process.

Moreover, the production of ground-state $Cd(^1S) + H_2$ via $^1B_2 \rightarrow {}^1A_1$ surface hopping (near $R = 1.69$Å and $r_{HH} = 1.54$Å) should also be enhanced by asymmetric vibrational excitation. The $^1B_2$ and $^1A_1$ surfaces also provide, through their non-adiabatic couplings, a "gateway" to formation of the asymmetric bond cleavage products $CdH(^2\Sigma) + H$. It can be shown that the curvature (i.e., second energy derivative) of a potential energy surface consists of two parts: (i) one part that is always positive, and (ii) a second that can be represented in terms of the non-adiabatic coupling elements between the two surfaces and the energy gap ΔE between the two surfaces. Applied to the two states at hand, this second contributor to the curvature of the $^1B_2$ surface is:

$$\frac{|\langle\Psi(^1B_2)|\nabla j|\Psi(^1A_1)\rangle|^2}{E(^1B_2) - E(^1A_1)}.$$

Clearly, when the $^1A_1$ state is higher in energy but strongly non-adiabatically coupled to the $^1B_2$ state, negative curvature along the asymmetric b_2 vibrational mode is expected for the $^1B_2$ state. When the $^1A_1$ state is lower in energy, negative curvature along the b_2 vibrational mode is expected for the $^1A_1$ state (because the above expression also expresses the curvature of the $^1A_1$ state).

Therefore, in the region of close-approach of these two states, state-to-state surface hopping can be facile. Moreover, one of the two states (the lower lying at each geometry) will likely possess negative curvature along the b_2 vibrational mode. It is this negative curvature that causes movement away from C_{2v} symmetry to occur spontaneously, thus leading to the $CdH(^2\Sigma) + H$ reaction products.

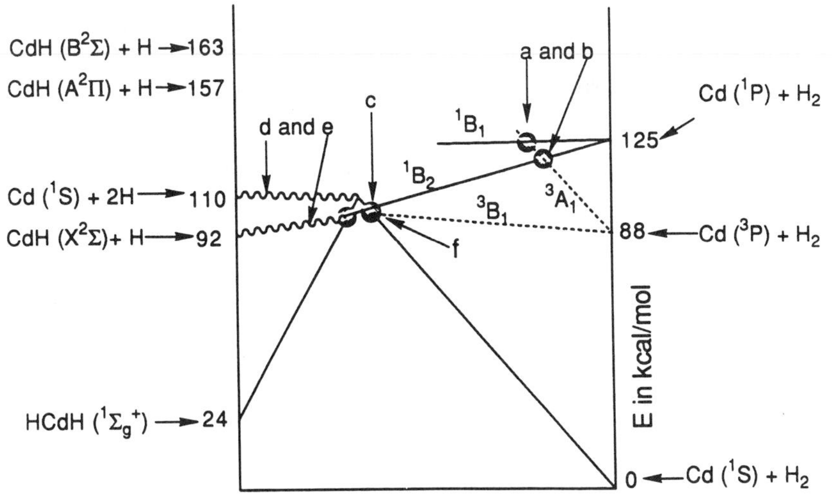

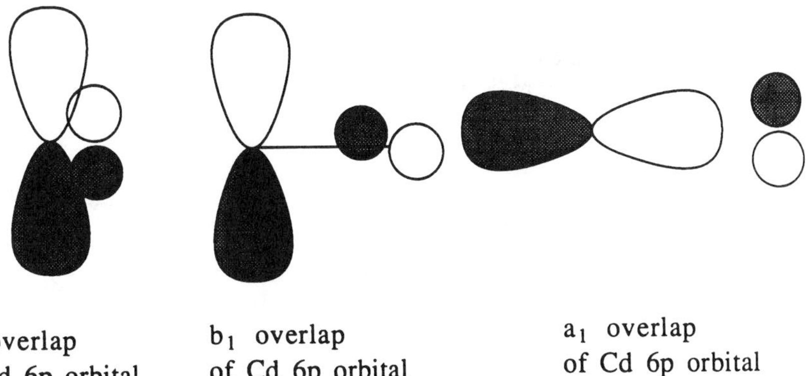

b_2 overlap of Cd 6p orbital and H_2 σ_g orbital

b_1 overlap of Cd 6p orbital and H_2 σ_g orbital

a_1 overlap of Cd 6p orbital and H_2 σ_g orbital

Coupled-state dynamics can also be used to describe situations in which vibrational rather than electronic-state transitions occur. For example, when van der Waals complexes such as HCl · · · Ar undergo so-called vibrational predissociation, one thinks in terms of movement of the Ar atom relative to the center of mass of the HCl molecule playing the role of the R coordinate above, and the vibrational state of HCl as playing the role of the quantized (electronic) state in the above example.

In such cases, a vibrationally excited HCl molecule (e.g., in $v = 1$) to which an Ar atom is attached via weak van der Waals attraction transfers its vibrational energy to the Ar atom, subsequently dropping to a lower (e.g., $v = 0$) vibrational level. Within the two-coupled-state model introduced above, the upper energy surface pertains to Ar in a bound vibrational level (having dissociation energy D_e) with HCl in an excited vibrational state (Δ being the $v = 0$ to $v = 1$ vibrational energy gap), and the lower surface describes an Ar atom that is free from the HCl molecule that is itself in its $v = 0$ vibrational state. In this case, the coordinate R is the Ar-to-HCl distance.

In analogy with the electronic-nuclear coupling example discussed earlier, the rate of transition from HCl($v = 1$) bound to Ar to HCl($v = 0$) plus a free Ar atom depends on the strength of coupling between the Ar · · · HCl relative motion coordinate (R) and the HCl internal vibrational coordinate. The $<\phi|d/dR|\phi''>$ coupling elements in this case are integrals over the HCl vibrational coordinate x involving the $v = 0(\phi)$ and $v = 1(\phi'')$ vibrational functions. The integrals over the R coordinate in the earlier expression for the rate of radiationless transitions now involve integration over the distance R between the Ar atom and the center of mass of the HCl molecule.

This completes our discussion of dynamical processes in which more than one Born-Oppenheimer state is involved. There are many situations in molecular spectroscopy and chemical dynamics where consideration of such coupled-state dynamics is essential. These cases are characterized by

1. total energies E which may be partitioned in two or more ways among the internal degrees of freedom (e.g., electronic and nuclear motion or vibrational and ad-atom in the above examples),
2. Born-Oppenheimer potentials that differ in energy by a small amount (so that energy transfer from the other degree(s) of freedom is facile).

III. CLASSICAL TREATMENT OF NUCLEAR MOTION

For all but very elementary chemical reactions (e.g., D + HH → HD + H or F + HH → FH + H) or scattering processes (e.g., CO(v,J) + He → CO(v',J') + He), the above fully quantal coupled channel equations simply can not be solved even when modern supercomputers are employed. Fortunately, the Schrödinger equation can be replaced by a simple classical mechanics treatment of nuclear motions under certain circumstances.

For motion of a particle of mass μ along a direction R, the primary condition under which a classical treatment of nuclear motion is valid

$$\frac{\lambda}{4\pi}\frac{1}{p}\left|\frac{dp}{dR}\right| << 1$$

relates to the fractional change in the *local momentum* defined as:

$$p = \sqrt{2\mu(E - E_j(R))}$$

along R within the $3N - 5$ or $3N - 6$ dimensional internal coordinate space of the molecule, as well as to the *local de Broglie wavelength*

$$\lambda = \frac{2\pi\hbar}{|p|}\ .$$

The inverse of the quantity

$$\frac{1}{p}\left|\frac{dp}{dR}\right|$$

can be thought of as the length over which the momentum changes by 100%. The above condition then states that the local de Broglie wavelength must be short with respect to the distance over which the potential changes appreciably. Clearly, whenever one is dealing with heavy nuclei that are moving fast (so $|p|$ is large), one should anticipate that the local de Broglie wavelength of those particles may be short enough to meet the above criteria for classical treatment.

It has been determined that for potentials characteristic of typical chemical bonding (whose depths and dynamic range of interatomic distances are well known), and for all but low-energy motions (e.g., zero-point vibrations) of light particles such as Hydrogen and Deuterium nuclei or electrons, the local de Broglie wavelengths are often short enough for the above condition to be met (because of the large masses μ of non-Hydrogenic species) except when their velocities approach zero (e.g., near classical turning points). It is therefore common to treat the nuclear-motion dynamics of molecules that do not contain H or D atoms in a purely classical manner, and to apply so-called semi-classical corrections near classical turning points. The motions of H and D atomic centers usually require quantal treatment except when their kinetic energies are quite high.

A. Classical Trajectories

To apply classical mechanics to the treatment of nuclear-motion dynamics, one solves Newtonian equations

$$m_k \frac{d^2X_k}{dt^2} = -\frac{dE_j}{dX_k}$$

where X_k denotes one of the $3N$ cartesian coordinates of the atomic centers in the molecule, m_k is the mass of the atom associated with this coordinate, and

$$\frac{dE_j}{dX_k}$$

is the derivative of the potential, which is the electronic energy $E_j(R)$, along the kth coordinate's direction. Starting with coordinates $\{X_k(0)\}$ and corresponding momenta $\{P_k(0)\}$ at some initial time $t = 0$, and given the ability to compute the force

$$-\frac{dE_j}{dX_k}$$

at any location of the nuclei, the Newton equations can be solved (usually on a computer) using finite-difference methods:

$$X_k(t+\delta t) = X_k(t) + P_k(t)\delta t/m_k$$

$$P_k(t+\delta t) = P_k(t) - \frac{dE_j}{dX_k}(t)\delta t.$$

In so doing, one generates a sequence of coordinates $\{X_k(t_n)\}$ and momenta $\{P_k(t_n)\}$, one for each "time step" t_n. The histories of these coordinates and momenta as functions of time are called "**classical trajectories.**" Following them from early times, characteristic of the molecule(s) at "reactant" geometries, through to late times, perhaps characteristic of "product" geometries, allows one to monitor and predict the fate of the time evolution of the nuclear dynamics. Even for large molecules with many atomic centers, propagation of such classical trajectories is feasible on modern computers *if* the forces

$$-\frac{dE_j}{dX_k}$$

can be computed in a manner that does not consume inordinate amounts of computer time.

In section 6, methods by which such force calculations are performed using first-principles quantum mechanical methods (i.e., so-called *ab initio* methods) are discussed. Suffice it to say that these calculations are often the rate limiting step in carrying out classical trajectory simulations of molecular dynamics. The large effort involved in the *ab initio* determination of electronic energies and their gradients

$$-\frac{dE_j}{dX_k}$$

motivate one to consider using empirical "force field" functions $V_j(R)$ in place of the *ab initio* electronic energy $E_j(R)$. Such model potentials $V_j(R)$, are usually constructed in terms of easy to compute and to differentiate functions of the interatomic distances and valence angles that appear in the molecule. The parameters that appear in the attractive and repulsive parts of these potentials are usually chosen so the potential is consistent with certain experimental data (e.g., bond dissociation energies, bond lengths, vibrational energies, torsion energy barriers).

For a large polyatomic molecule, the potential function V usually contains several distinct contributions:

$$V = V_{bond} + V_{bend} + V_{vanderWaals} + V_{torsion} + V_{electrostatic}.$$

Here V_{bond} gives the dependence of V on stretching displacements of the bonds (i.e., interatomic distances between pairs of bonded atoms) and is usually modeled as a harmonic or Morse function for each bond in the molecule:

$$V_{bond} = \sum_J 1/2k_J(R_J - R_{eq,J})^2$$

or

$$V_{bond} = \sum_J D_{e,J}(1 - \exp(-a_J(R_J - R_{eq,J})))^2$$

where the index J labels the bonds and the k_J,a_J and $R_{eq,J}$ are the force constant and equilibrium bond length parameters for the Jth bond.

V_{bend} describes the bending potentials for each triplet of atoms (ABC) that are bonded in a A-B-C manner; it is usually modeled in terms of a harmonic potential for each such bend:

$$V_{bend} = \sum_J 1/2k_J^0(\theta_J - \theta_{eq,J})^2.$$

The $\theta_{eq,J}$ and k_J^0 are the equilibrium angles and force constants for the Jth angle.

$V_{vanderWaals}$ represents the van der Waals interactions between all pairs of atoms that are not bonded to one another. It is usually written as a sum over all pairs of such atoms (labeled J and K) of a Lennard-Jones 6,12 potential:

$$V_{vanderWaals} = \sum_{J<K} [a_{J,K}(R_{J,K})^{-12} - b_{J,K}(R_{J,K})^{-6}]$$

where $a_{J,K}$ and $b_{J,K}$ are parameters relating to the repulsive and dispersion attraction forces, respectively for the Jth and K^{th} atoms.

$V_{torsion}$ contributions describe the dependence of V on angles of rotation about single bonds. For example, rotation of a CH_3 group around the single bond connecting the carbon atom to another group may have an angle dependence of the form:

$$V_{torsion} = V_0(1 - \cos(3\theta))$$

where θ is the torsion rotation angle, and V_0 is the magnitude of the interaction between the C-H bonds and the group on the atom bonded to carbon.

$V_{electrostatic}$ contains the interactions among polar bonds or other polar groups (including any charged groups). It is usually written as a sum over pairs of atomic centers (J and K) of Coulombic interactions between fractional charges $\{Q_J\}$ (chosen to represent the bond polarity) on these atoms:

$$V_{electrostatic} = \sum_{J<K} Q_J Q_K / R_{J,K}$$

Although the total potential V as written above contains many components, each is a relatively simple function of the cartesian positions of the atomic centers. Therefore, it is relatively straightforward to evaluate V and its gradient along all $3N$ cartesian directions in a computationally efficient manner. For this reason, the use of such empirical force fields in so-called **molecular mechanics** simulations of classical dynamics is widely used for treating large organic and biological molecules.

B. Initial Conditions

No single trajectory can be used to simulate chemical reaction or collisions that relate to realistic experiments. To generate classical trajectories that are characteristic of particular experiments, one must choose many initial conditions (coordinates and momenta) the *collection* of which is representative of the experiment. For example, to use an ***ensemble*** of trajectories to simulate a molecular beam collision between H and Cl atoms at a collision energy E, one must follow many classical trajectories that have a range of "impact parameters" (b) from zero up to some maximum value b_{max} beyond which the H $\cdots$ Cl interaction potential vanishes. The figure shown below describes the impact parameter as the distance of closest approach that a trajectory would have if no attractive or repulsive forces were operative.

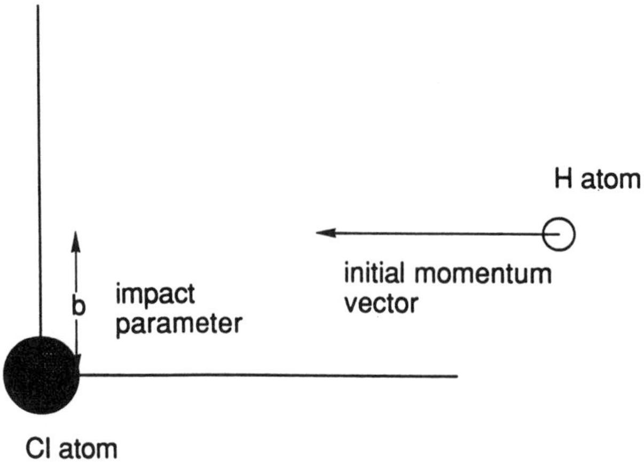

Moreover, if the energy resolution of the experiment makes it impossible to fix the collision energy closer than an amount δE, one must run collections of trajectories for values of E lying within this range.

If, in contrast, one wishes to simulate thermal reaction rates, one needs to follow trajectories with various E values and various impact parameters b from initiation at $t = 0$ to their conclusion (at which time the chemical outcome is interrogated). Each of these trajectories must have their outcome weighted by an amount proportional to a Boltzmann factor $\exp(-E/RT)$, where R is the ideal gas constant and T is the temperature because this factor specifies the probability that a collision occurs with kinetic energy E.

As the complexity of the molecule under study increases, the number of parameters needed to specify the initial conditions also grows. For example, classical trajectories that relate to $F + H_2 \rightarrow HF + H$ need to be specified by providing (i) an impact parameter for the F to the center of mass of H_2, (ii) the relative translational energy of the F and H_2, (iii) the radial momentum and coordinate of the H_2 molecule's bond length, and (iv) the angular momentum of the H_2 molecule as well as the angle of the H-H bond axis relative to the line connecting the F atom to the center of mass of the H_2 molecule. Many such sets of initial conditions must be chosen and the resultant classical trajectories followed to generate an ensemble of trajectories pertinent to an experimental situation.

It should be clear that even the classical mechanical simulation of chemical experiments involves considerable effort because no single trajectory can represent the experimental situation. Many trajectories, each with different initial conditions selected so they represent, as an ensemble, the experimental conditions, must be followed and the outcome of all such trajectories must be averaged over the probability of realizing each specific initial condition.

C. Analyzing Final Conditions

Even after classical trajectories have been followed from $t = 0$ until the outcomes of the collisions are clear, one needs to properly relate the fate of each trajectory to the experimental situation. For the $F + H_2 \rightarrow HF + H$ example used above, one needs to examine each trajectory to determine, for example, (i) whether HF + H products are formed or non-reactive collision to produce $F + H_2$ has occurred, (ii) the amount of rotational energy and angular momentum that is contained in the HF product molecule, (iii) the amount of relative translational energy that remains in the H + FH products, and (iv) the amount of vibrational energy that ends up in the HF product molecule.

Because classical rather than quantum mechanical equations are used to follow the time evolution of the molecular system, there is no guarantee that the amount of energy or angular momentum found in degrees of freedom for which these quantities should be quantized will be so. For example, $F + H_2 \rightarrow HF + H$ trajectories may produce HF molecules with internal vibrational energy that is not a half integral multiple of the fundamental vibrational frequency ω of the HF bond. Also, the rotational angular momentum of the HF molecule may not fit the formula $J(J+1)h^2/(8\pi^2 I)$, where I is HF's moment of inertia.

To connect such purely classical mechanical results more closely to the world of quantized energy levels, a method know as "binning" is often used. In this technique, one assigns the outcome of a classical trajectory to the particular quantum state (e.g., to a vibrational state v or a rotational state J of the HF molecule in the above example) whose quantum energy is closest to the classically determined energy. For the HF example at hand, the classical vibrational energy $E_{cl,vib}$ is simply used to define, as the closest integer, a vibrational quantum number v according to:

$$v = \frac{(E_{cl,vib})}{\hbar\omega} - 1/2.$$

Likewise, a rotational quantum number J can be assigned as the closest integer to that determined by using the classical rotational energy $E_{cl,rot}$ in the formula:

$$J = 1/2\left\{(1 + 32\pi^2 I E_{cl,rot}/h^2)^{1/2} - 1\right\}$$

which is the solution of the quadratic equation $J(J+1)h^2/8\pi^2 I = E_{cl,rot}$.

By following many trajectories and assigning vibrational and rotational quantum numbers to the product molecules formed in each trajectory, one can generate histograms giving the frequency with which each product molecule quantum state is observed for the ensemble of trajectories used to simulate the experiment of interest. In this way, one can approximately extract product-channel quantum state distributions from classical trajectory simulations.

IV. WAVEPACKETS

In an attempt to combine the attributes and strengths of classical trajectories, which allow us to "watch" the motions that molecules undergo, and quantum mechanical wavefunctions, which are needed if interference phenomena are to be treated, a hybrid approach is sometimes used. A popular and rather successful such point of view is provided by so called **coherent state wavepackets**.

A quantum mechanical wavefunction $\psi(\boldsymbol{x}|\boldsymbol{X},\boldsymbol{P})$ that is a function of all pertinent degrees of freedom (denoted collectively by $\boldsymbol{x}$) and that depends on two sets of parameters (denoted $\boldsymbol{X}$ and $\boldsymbol{P}$, respectively) is defined as follows:

$$\psi(\boldsymbol{x}|\boldsymbol{X},\boldsymbol{P}) = \prod_{k=1}^{N} (2\pi\langle\Delta x_k\rangle^2)^{-1/2} \exp\left\{iP_k x_k/\hbar - (x_k - X_k)^2/(4\langle\Delta x_k\rangle^2)\right\}.$$

Here, $\langle\Delta x_k\rangle^2$ is the uncertainty

$$\langle\Delta x_k\rangle^2 = \int |\psi|^2 (x_k - X_k)^2 d\boldsymbol{x}$$

along the kth degree of freedom for this wavefunction, defined as the mean squared displacement away from the average coordinate

$$\int |\psi|^2 x_k d\mathbf{x} = X_k.$$

So, the parameter X_k specifies the *average value* of the coordiate x_k. In like fashion, it can be shown that the parameter P_k is equal to the *average value* of the momentum along the kth coordinate:

$$\int \psi^*(-i\hbar\partial/\partial x_k \psi) d\mathbf{x} = P_k.$$

The uncertainty in the momentum along each coordinate:

$$<\Delta p_k>^2 = \int \psi^*(-i\hbar\partial/\partial x_k - P_k)^2 \psi d\mathbf{x}$$

is given, for functions of the coherent state form, in terms of the coordinate uncertainty as

$$<\Delta p_k>^2 <\Delta x_k>^2 = \hbar^2/4.$$

Of course, the general Heisenberg uncertainty condition

$$<\Delta p_k>^2 <\Delta x_k>^2 \geq \hbar^2/4$$

limits the coordinate and momentum uncertainty products for arbitrary wavefunctions. The coherent state wave packet functions are those for which this *uncertainty* product *is minimum*. In this sense, coherent state wave packets are seen to be as close to classical as possible since in classical mechanics there are no limits placed on the resolution with which one can observe coordinates and momenta.

These wavepacket functions are employed as follows in the most straightforward treatments of combined quantal/classical mechanics:

1. Classical trajectories are used, as discribed in greater detail above, to generate a series of coordinates $X_k(t_n)$ and momenta $P_k(t_n)$ at a sequence of times denoted $\{t_n\}$.
2. These classical coordinates and momenta are used to *define* a wavepacket function as written above, whose X_k and P_k parameters are taken to be the coordinates and momenta of the classical trajectory. In effect, the wavepacket moves around "riding" the classical trajectory's coordinates and momenta as time evolves.
3. At any time t_n, the quantum mechanical properties of the system are computed by forming the expectation values of the corresponding quantum operators for a wavepacket wavefunction of the form given above with X_k and P_k given by the classical coordinates and momenta at that time t_n.

Such wavepackets are, of course, simple approximations to the true quantum mechanical functions of the system because they do not obey the Schrödinger equation appropriate to the system. The should be expected to provide accurate representations to the true wavefunctions for systems that are more classical in nature (i.e., when the local de Broglie wave lengths are short compared to the range over which the potentials vary appreciably). For species containing light particles (e.g., electrons or H atoms) or for low kinetic energies, the local de Broglie wave lengths will not satisfy such criteria, and these approaches can be expected to be less reliable. For further information about the use of coherent state wavepackets in molecular dynamics and molecular spectroscopy, see E. J. Heller, *Acc. Chem. Res. 14*, 368 (1981).

This completes our treatment of the subjects of molecular dynamics and molecular collisions. Neither its depth not its level was at the research level; rather, we intended to provide the reader with an introduction to many of the theoretical concepts and methods that arise when applying either the quantum Schrödinger equation or classical Newtonian mechanics to chemical reaction dynamics. Essentially none of the experimental aspects of this subject (e.g., molecular beam methods for preparing "cold" molecules, laser pump-probe methods for preparing reagents in specified quantum states and observing products in such states) have been discussed. An excellent introduction to both the experimental and theoretical foundations of modern chemical and collision dynamics is provided by the text *Molecular Reaction Dynamics and Chemical Reactivity* by R. D. Levine and R. B. Bernstein, Oxford Univ. Press (1987).

Exercises, Problems, and Solutions

EXERCISES

1. Time-dependent perturbation theory provides an expression for the radiative lifetime of an excited electronic state, given by τ_R:

$$\tau_R = \frac{3\hbar^4 c^3}{4(E_i - E_f)^3 |\mu_{fi}|^2},$$

where i refers to the excited state, f refers to the lower state, and μ_{fi} is the transition dipole.

a. Evaluate the z-component of the transition dipole for the $2p_z \rightarrow 1s$ transition in a hydrogenic atom of nuclear charge Z, given:

$$\psi_{1s} = \frac{1}{\sqrt{\pi}}\left(\frac{Z}{a_0}\right)^{\frac{3}{2}} e^{\frac{-Zr}{a_0}}, \text{ and } \psi_{2p_z} = \frac{1}{4\sqrt{2\pi}}\left(\frac{Z}{a_0}\right)^{\frac{5}{2}} r \cos\theta e^{\frac{-Zr}{2a_0}}.$$

Express your answer in units of ea_0.

b. Use symmetry to demonstrate that the x- and y-components of μ_{fi} are zero, i.e.,

$<2p_z|ex|1s> = <2p_z|ey|1s> = 0.$

c. Calculate the radiative lifetime τ_R of a hydrogenlike atom in its $2p_z$ state. Use the relation

$$e^2 = \frac{\hbar^2}{m_e a_0}$$

to simplify your results.

2. Consider a case in which the complete set of states $|\phi_K|$ for a Hamiltonian is known.

a. If the system is initially in the state m at time $t = 0$ when a constant perturbation V is suddenly turned on, find the probability amplitudes $C_k^{(2)}(t)$ and $C_m^{(2)}(t)$, to second order in V, that describe the system being in a different state k or the same state m at time t.

b. If the perturbation is turned on adiabatically, what are $C_k^{(2)}(t)$ and $C_m^{(2)}(t)$? Here, consider that the initial time is $t_0 \rightarrow -\infty$, and the potential is $Ve^{\eta}t$, where the positive parameter η is allowed to approach zero $\eta \rightarrow 0$ in order to describe the adiabatically (i.e., slowly) turned on perturbation.

c. Compare the results of parts (a) and (b) and explain any differences.

d. Ignore first-order contributions (assume they vanish) and evaluate the transition rates

$$\frac{d}{dt}|C_k^{(2)}(t)|^2$$

for the results of part (b) by taking the limit $\eta \rightarrow 0^+$, to obtain the adiabatic results.

3. If a system is initially in a state m, conservation of probability requires that the total probability of transitions out of state m be obtainable from the decrease in the probability of being in state m. Prove this to the lowest order by using the results of exercise 2, i.e., show that:

$$|C_m|^2 = 1 - \sum_{k \neq m} |C_k|^2.$$

PROBLEMS

1. Consider an interaction or perturbation which is carried out suddenly (instantaneously, e.g., within an interval of time Δt which is small compared to the natural period ω_{nm}^{-1} corresponding to the transition from state m to state n), and after that is turned off adiabatically (i.e., extremely slowly as $Ve^{\eta}t$). The transition probability in this case is given as:

$$T_{nm} \approx \frac{|<n|V|m>|^2}{\hbar^2 \omega_{nm}^2}$$

where V corresponds to the maximum value of the interaction when it is turned on. This formula allows one to calculate the transition probabilities under the action of sudden perturbations which are small in absolute value whenever perturbation theory is applicable. Let's use this "sudden approximation" to calculate the probability of excitation of an electron under a sudden change of the charge of the nucleus. Consider the reaction:

$${}_1^3H \rightarrow {}_2^3He^+ + e^-,$$

and assume the tritium atom has its electron initially in a $1s$ orbital.

a. Calculate the transition probability for the transition $1s \rightarrow 2s$ for this reaction using the above formula for the transition probability.

b. Suppose that at time $t = 0$ the system is in a state which corresponds to the wavefunction φ_m, which is an eigenfunction of the operator H_0. At $t = 0$,the sudden change of the Hamiltonian occurs (now denoted as H and remains unchanged). Calculate the same $1s \rightarrow 2s$ transition probability as in part (a), only this time as the square of the magnitude of the coefficient, $A_{1s,2s}$ using the expansion:

$$\Psi(r,0) = \varphi_m(r) = \sum_n A_{nm}\psi_n(r), \text{ where}$$

$$A_{nm} = \int \varphi_m(r)\psi_n(r)d^3r$$

Note, that the eigenfunctions of H are ψ_n with eigenvalues E_n. Compare this "exact" value with that obtained by perturbation theory in part (a).

2. The methyl iodide molecule is studied using microwave (pure rotational) spectroscopy. The following integral governs the rotational selection rules for transitions labeled $J,M,K \rightarrow J',M',K'$:

$$I = \langle D^{J'}_{M'K'}|\vec{\varepsilon}\cdot\vec{\mu}|D^{J}_{MK}\rangle.$$

The dipole moment $\vec{\mu}$ lies along the molecule's C_3 symmetry axis. Let the electric field of the light $\vec{\varepsilon}$ define the lab-fixed Z-direction.

a. Using the fact that $\cos\beta = D^{1*}_{00}$, show that

$$I = 8\pi^2\mu\varepsilon(-1)^{(M+K)}\begin{pmatrix} J' & 1 & J \\ M & 0 & M \end{pmatrix}\begin{pmatrix} J' & 1 & J \\ K & 0 & K \end{pmatrix}\delta_{M'M}\delta_{K'K}$$

b. What restrictions does this result place on $\Delta J = J' - J$? Explain physically why the K quantum number can not change.

3. Consider the molecule BO.

a. What are the total number of possible electronic states which can be formed by combination of ground state B and O atoms?

b. What electron configurations of the molecule are likely to be low in energy? Consider all reasonable orderings of the molecular orbitals. What are the states corresponding to these configurations?

c. What are the bond orders in each of these states?

d. The true ground state of BO is $^2\Sigma$. Specify the $+/-$ and u/g symmetries for this state.

e. Which of the excited states you derived above will radiate to the $^2\Sigma$ ground state? Consider electric dipole, magnetic dipole, and electric quadrupole radiation.

f. Does ionization of the molecule to form a cation lead to a stronger, weaker, or equivalent bond strength?

g. Assuming that the energies of the molecular orbitals do not change upon ionization, what are the ground state, the first excited state, and the second excited state of the positive ion?

h. Considering only these states, predict the structure of the photoelectron spectrum you would obtain for ionization of BO.

4.

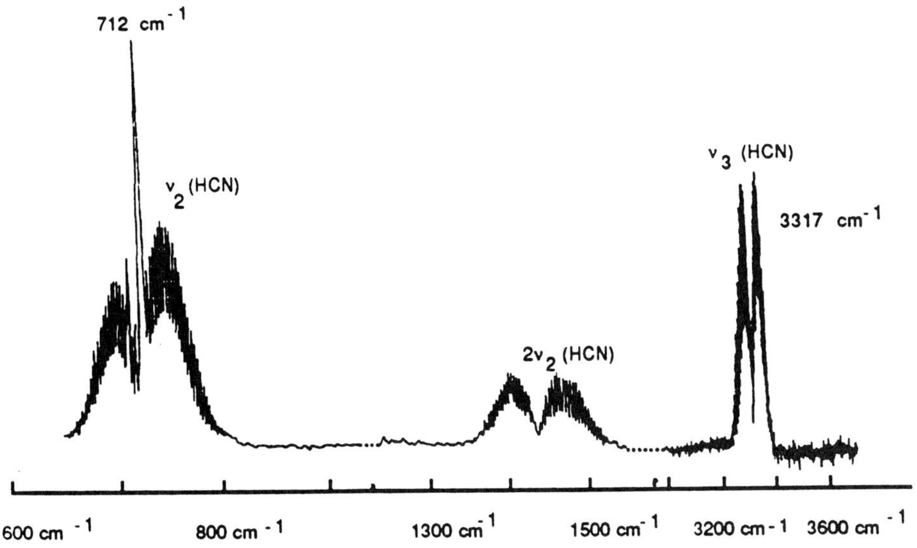

The above figure shows part of the infrared absorption spectrum of HCN gas. The molecule has a CH stretching vibration, a bending vibration, and a CN stretching vibration.

a. Are any of the vibrations of linear HCN degenerate?

b. To which vibration does the group of peaks between 600cm^{-1} and 800cm^{-1} belong?

c. To which vibration does the group of peaks between 3200cm^{-1} and 3400cm^{-1} belong?

d. What are the symmetries (σ,π,δ) of the CH stretch, CN stretch, and bending vibrational motions?

e. Starting with HCN in its 0,0,0 vibrational level, which fundamental transitions would be infrared active under parallel polarized light (i.e., z-axis polarization):

i. 000 $\rightarrow$ 001 ?
ii. 000 $\rightarrow$ 100 ?
iii. 000 $\rightarrow$ 010 ?

f. Which transitions would be active when perpendicular polarized light is used?

g. Why does the 712cm^{-1} transition have a Q-branch, whereas that near 3317cm^{-1} has only P- and R-branches?

SOLUTIONS

Exercises

1. a. Evaluate the z-component of μ_{fi}:

$$\mu_{fi} = \langle 2p_z|er\cos\theta|1s\rangle, \text{ where } \psi_{1s} = \frac{1}{\sqrt{\pi}}\left(\frac{Z}{a_0}\right)^{\frac{3}{2}} e^{\frac{-Zr}{a_0}}, \text{ and } \psi_{2p_z} = \frac{1}{4\sqrt{2\pi}}\left(\frac{Z}{a_0}\right)^{\frac{5}{2}} r\cos\theta\, e^{\frac{-Zr}{2a_0}}.$$

$$\mu_{fi} = \frac{1}{4\sqrt{2\pi}}\left(\frac{Z}{a_0}\right)^{\frac{5}{2}}\frac{1}{\sqrt{\pi}}\left(\frac{Z}{a_0}\right)^{\frac{3}{2}} < r\cos\theta\, e^{\frac{-Zr}{2a_0}} | er\cos\theta | e^{\frac{-Zr}{a_0}} >$$

$$= \frac{1}{4\pi\sqrt{2}}\left(\frac{Z}{a_0}\right)^{4} < r\cos\theta\, e^{\frac{-Zr}{2a_0}} | er\cos\theta | e^{\frac{-Zr}{a_0}} >$$

$$= \frac{e}{4\pi\sqrt{2}}\left(\frac{Z}{a_0}\right)^{4}\int_0^{\infty} r^2 dr \int_0^{\pi} \sin\theta\, d\theta \int_0^{2\pi} d\varphi \left(r^2 e^{\frac{-Zr}{2a_0}} e^{\frac{-Zr}{a_0}}\right)\cos^2\theta$$

$$= \frac{e}{4\pi\sqrt{2}}\, 2\pi\left(\frac{Z}{a_0}\right)^{4}\int_0^{\infty}\left(r^4 e^{\frac{-3Zr}{2a_0}}\right) dr \int_0^{\pi} \sin\theta\cos^2\theta\, d\theta$$

Using integral equation (4) to integrate over r and equation (17) to integrate over θ we obtain:

$$= \frac{e}{4\pi\sqrt{2}}\, 2\pi\left(\frac{Z}{a_0}\right)^{4}\frac{4!}{\left(\frac{3Z}{2a_0}\right)^{5}}\left(\frac{-1}{3}\right)\cos^3\theta\Big|_0^{\pi}$$

$$= \frac{e}{4\pi\sqrt{2}}\, 2\pi\left(\frac{Z}{a_0}\right)^{4}\frac{2^5 a_0^5 4!}{3^5 Z^5}\left(\frac{-1}{3}\right)\left((-1)^3 - (1)^3\right)$$

$$= \frac{e}{\sqrt{2}}\frac{2^8 a_0}{3^5 Z} = \frac{ea_0}{Z}\frac{2^8}{\sqrt{2}\,3^5} = 0.7449\,\frac{ea_0}{Z}$$

b. Examine the symmetry of the integrands for $<2p_z|ex|1s>$ and $<2p_z|ey|1s>$. Consider reflection in the xy plane:

Function	**Symmetry**
$2p_z$	-1
x	$+1$
$1s$	$+1$
y	$+1$

Under this operation the integrand of $<2p_z|ex|1s>$ is $(-1)(1)(1) = -1$ (it is antisymmetric) and hence $<2p_z|ex|1s> = 0$. Similarly, under this operation the integrand of $<2p_z|ey|1s>$ is $(-1)(1)(1) = -1$ (it is also antisymmetric) and hence $<2p_z|ey|1s> = 0$.

c. $$\tau_R = \frac{3\hbar^4 c^3}{4(E_i - E_f)^3 |\mu_{fi}|^2},$$

$$E_i = E_{2pz} = -\frac{1}{4}Z^2\left(\frac{e^2}{2a_0}\right)$$

$$E_f = E_{1s} = -Z^2\left(\frac{e^2}{2a_0}\right)$$

$$E_i - E_f = \frac{3}{8}\left(\frac{e^2}{a_0}\right)Z^2$$

Making the substitutions for $E_i - E_f$ and $|\mu_{fi}|$ in the expression for τ_R we obtain:

$$\tau_R = \frac{3\hbar^4 c^3}{4\left(\frac{3}{8}\left(\frac{e^2}{a_0}\right)Z^2\right)^3\left(\left(\frac{ea_0}{Z}\right)\frac{2^8}{\sqrt{2}3^5}\right)^2},$$

$$= \frac{3\hbar^4 c^3}{4\frac{3^3}{8^3}\left(\frac{e^6}{a_0^3}\right)Z^6\left(\frac{e^2a_0^2}{Z^2}\right)\frac{2^{16}}{(2)3^{10}}},$$

$$= \frac{\hbar^4 c^3 3^8 a_0}{e^8 Z^4 2^8},$$

Inserting $e^2 = \frac{\hbar^2}{m_e a_0}$ we obtain:

$$\tau_R = \frac{\hbar^4 c^3 3^8 a_0 m_e^4 a_0^4}{\hbar^8 Z^4 2^8} = \frac{3^8}{2^8}\frac{c^3 a_0^5 m_e^4}{\hbar^4 Z^4}$$

$$= 25.6289\,\frac{c^3 a_0^5 m_e^4}{\hbar^4 Z^4}$$

$$= 25{,}6289\left(\frac{1}{Z^4}\right)\frac{(2.998\times 10^{10}\text{cm sec}^{-1})^3(0.529177\times 10^{-8}\text{cm})^5(9.109\times 10^{-28}\text{g})^4}{(1.0546\times 10^{-27}\text{g cm}^2\text{sec}^{-1})^4}$$

$$= 1.595\times 10^{-9}\text{sec}\left(\frac{1}{Z^4}\right)$$

So, for example:

Atom	τ_R
H	1.595 ns
He^+	99.7 ps
Li^{+2}	19.7 ps
Be^{+3}	6.23 ps
Ne^{+9}	159 fs

2. a. $H = H_0 + \lambda H'(t),\ H'(t) = V\theta(t),\ H_0\varphi_k = E_k\varphi_k,\ \omega_k = E_k/\hbar$

$$i\hbar\frac{\partial\psi}{\partial t} = H\psi$$

$$let\ \psi(r,t) = i\hbar\sum_j c_j(t)\varphi_j e^{-i\omega_j t}$$

and insert into the above expression:

$$i\hbar\sum_j\left[\dot{c}_j - i\omega_j c_j\right]e^{-i\omega_j t}\varphi_j = i\hbar\sum_j c_j(t)e^{-i\omega_j t}(H_0 + \lambda H'(t))\varphi_j$$

$$\sum_j\left[i\hbar\dot{c}_j + E_j c_j - c_j E_j - c_j\lambda H'\right]e^{-i\omega_j t}\varphi_j = 0$$

$$\sum_j\left[i\hbar\dot{c}_j<m|j> - c_j\lambda<m|H'|j>\right]e^{-i\omega_j t} = 0$$

$$i\hbar\dot{c}_m e^{-i\omega_m t} = \sum_j c_j\lambda H'_{mj}e^{-i\omega_j t}$$

So,

$$\dot{c}_m = \frac{1}{i\hbar}\sum_j c_j\lambda H'_{mj}e^{-i(\omega_{jm})t}$$

Going back a few equations and multiplying from the left by φ_k instead of φ_m we obtain:

$$\sum_j\left[i\hbar\dot{c}_j<k|j> - c_j\lambda<k|H'|j>\right]e^{-i\omega_j t} = 0$$

$$i\hbar\dot{c}_k e^{-i\omega_k t} = \sum_j c_j\lambda H'_{kj}e^{-i\omega_j t}$$

So,

$$\dot{c}_k = \frac{1}{i\hbar}\sum_j c_j\lambda H'_{kj}e^{-i(\omega_{jk})t}$$

Now, let:

$$c_m = c_m^{(0)} + c_m^{(1)}\lambda + c_m^{(2)}\lambda^2 + \ldots$$
$$c_k = c_k^{(0)} + c_k^{(1)}\lambda + c_k^{(2)}\lambda^2 + \ldots$$

and substituting into above we obtain:

$$\dot{c}_m^{(0)} + \dot{c}_m^{(1)}\lambda + \dot{c}_m^{(2)}\lambda^2 + \ldots = \frac{1}{i\hbar}\sum_j\left[c_j^{(0)} + c_j^{(1)}\lambda + c_j^{(2)}\lambda^2 + \ldots\right]\lambda H'_{mj}e^{-i(\omega_{jm})t}$$

first order:

$$\dot{c}_m^{(0)} = 0 \Rightarrow c_m^{(0)} = 1$$

second order:

$$\dot{c}_m^{(1)} = \frac{1}{i\hbar} \sum_j c_j^{(0)} H'_{mj} e^{-i(\omega_{jm})t}$$

$(n+1)^{\text{st}}$ order:

$$\dot{c}_m^{(n)} = \frac{1}{i\hbar} \sum_j c_j^{(n-1)} H'_{mj} e^{-i(\omega_{jm})t}$$

Similarly:
first order:

$$\dot{c}_k^{(0)} = 0 \Rightarrow c_{k\neq m}^{(0)} = 0$$

second order:

$$\dot{c}_k^{(1)} = \frac{1}{i\hbar} \sum_j c_j^{(0)} H'_{kj} e^{-i(\omega_{jk})t}$$

$(n+1)^{\text{st}}$ order:

$$\dot{c}_k^{(n)} = \frac{1}{i\hbar} \sum_j c_j^{(n-1)} H'_{kj} e^{-i(\omega_{jk})t}$$

So,

$$\dot{c}_m^{(1)} = \frac{1}{i\hbar} c_m^{(0)} H'_{mm} e^{-i(\omega_{mm})t} = \frac{1}{i\hbar} H'_{mm}$$

$$c_m^{(1)}(t) = \frac{1}{i\hbar} \int_0^t dt' V_{mm} = \frac{V_{mm} t}{i\hbar}$$

and similarly,

$$\dot{c}_k^{(1)} = \frac{1}{i\hbar} c_m^{(0)} H'_{km} e^{-i(\omega_{mk})t} = \frac{1}{i\hbar} H'_{km} e^{-i(\omega_{mk})t}$$

$$c_k^{(1)}(t) = \frac{1}{i\hbar} V_{km} \int_0^t dt' e^{-i(\omega_{mk})t'} = \frac{V_{km}}{\hbar\omega_{mk}} \left[e^{-i(\omega_{mk})t} - 1 \right]$$

$$\dot{c}_m^{(2)} = \frac{1}{i\hbar} \sum_j c_j^{(1)} H'_{mj} e^{-i(\omega_{jm})t}$$

$$\dot{c}_m^{(2)} = \sum_{j\neq m} \frac{1}{i\hbar} \frac{V_{jm}}{\hbar\omega_{mj}} \left[e^{-i(\omega_{mj})t} - 1 \right] H'_{mj} e^{-i(\omega_{jm})t} + \frac{1}{i\hbar} \frac{V_{mm} t}{i\hbar} H'_{mm}$$

$$c_m^{(2)} = \sum_{j\neq m} \frac{1}{i\hbar}\frac{V_{jm}V_{mj}}{\hbar\omega_{mj}} \int_0^t dt' e^{-i(\omega_{jm})t'}\left[e^{-i(\omega_{mj})t'} - 1\right] - \frac{V_{mm}V_{mm}}{\hbar^2}\int_0^t t'dt'$$

$$= \sum_{j\neq m} \frac{V_{jm}V_{mj}}{i\hbar^2\omega_{mj}} \int_0^t dt'\left[1 - e^{-i(\omega_{jm})t'}\right] - \frac{|V_{mm}|^2}{\hbar^2}\frac{t^2}{2}$$

$$= \sum_{j\neq m} \frac{V_{jm}V_{mj}}{i\hbar^2\omega_{mj}}\left(t - \frac{e^{-i(\omega_{jm})t} - 1}{-i\omega_{jm}}\right) - \frac{|V_{mm}|^2}{\hbar^2}\frac{t^2}{2}$$

$$= \sum_{j\neq m}{}' \frac{V_{jm}V_{mj}}{\hbar^2\omega_{mj}^2}\left(e^{-i(\omega_{jm})t} - 1\right) + \sum_{j\neq m}{}' \frac{V_{jm}V_{mj}}{i\hbar^2\omega_{mj}}t - \frac{|V_{mm}|^2t^2}{2\hbar^2}$$

Similarly,

$$\dot{c}_k^{(2)} = \frac{1}{i\hbar}\sum_j c_j^{(1)} H'_{kj} e^{-i(\omega_{jk})t}$$

$$= \sum_{j\neq m} \frac{1}{i\hbar}\frac{V_{jm}}{\hbar\omega_{mj}}\left[e^{-i(\omega_{mj})t} - 1\right]H'_{kj}e^{-i(\omega_{jk})t} + \frac{1}{i\hbar}\frac{V_{mm}t}{i\hbar}H'_{km}e^{-i(\omega_{mk})t}$$

$$c_k^{(2)}(t) = \sum_{j\neq m}{}' \frac{V_{jm}V_{kj}}{i\hbar^2\omega_{mj}}\int_0^t dt' e^{-i(\omega_{jk})t'}\left[e^{-i(\omega_{mj})t'} - 1\right] - \frac{V_{mm}V_{km}}{\hbar^2}\int_0^t t'dt' e^{-i(\omega_{mk})t'}$$

$$= \sum_{j\neq m}{}' \frac{V_{jm}V_{kj}}{i\hbar^2\omega_{mj}}\left(\frac{e^{-i(\omega_{mj}+\omega_{jm})t} - 1}{-i\omega_{mk}} - \frac{e^{-i(\omega_{jk})t} - 1}{-i\omega_{jk}}\right) - \frac{V_{mm}V_{km}}{\hbar^2}\left[e^{-i(\omega_{mk})t'}\left(\frac{t'}{-i\omega_{mk}} - \frac{1}{-(i\omega_{mk})^2}\right)\right]_{(0)}^{t}$$

$$= \sum_{j\neq m}{}' \frac{V_{jm}V_{kj}}{\hbar^2\omega_{mj}}\left(\frac{e^{-i(\omega_{mk})t} - 1}{\omega_{mk}} - \frac{e^{-i(\omega_{jk})t} - 1}{\omega_{jk}}\right) + \frac{V_{mm}V_{km}}{\hbar^2\omega_{mk}}\left[e^{-i(\omega_{mk})t'}\left(\frac{t'}{i} - \frac{1}{\omega_{mk}}\right)\right]_{(0)}^{t}$$

$$= \sum_{j\neq m}{}' \frac{V_{jm}V_{kj}}{E_m - E_j}\left(\frac{e^{-i(\omega_{mk})t} - 1}{E_m - E_k} - \frac{e^{-i(\omega_{jk})t} - 1}{E_j - E_k}\right) + \frac{V_{mm}V_{km}}{\hbar(E_m - E_k)}\left[e^{-i(\omega_{mk})t}\left(\frac{t}{i} - \frac{1}{\omega_{mk}}\right) + \frac{1}{\omega_{mk}}\right]$$

So, the overall amplitudes c_m, and c_k, to second order are:

$$c_m(t) = 1 + \frac{V_{mm}t}{i\hbar} + \sum_{j\neq m}{}' \frac{V_{jm}V_{mj}}{i\hbar(E_m - E_j)}t + \sum_{j\neq m}{}' \frac{V_{jm}V_{mj}}{\hbar^2(E_m - E_j)^2}\left(e^{-i(\omega_{jm})t} - 1\right) - \frac{|V_{mm}|^2t^2}{2\hbar^2}$$

$$c_k(t) = \frac{V_{km}}{(E_m - E_k)}\left[e^{-i(\omega_{mk})t} - 1\right] + \frac{V_{mm}V_{km}}{(E_m - E_k)^2}\left[1 - e^{-i(\omega_{mk})t}\right] + \frac{V_{mm}V_{km}}{(E_m - E_k)}\frac{t}{\hbar i}e^{-i(\omega_{mk})t} +$$

$$\sum_{j\neq m}{}' \frac{V_{jm}V_{kj}}{E_m - E_j}\left(\frac{e^{-i(\omega_{mk})t} - 1}{E_m - E_k} - \frac{e^{-i(\omega_{jk})t} - 1}{E_j - E_k}\right)$$

b. The perturbation equations still hold:

$$\dot{c}_m^{(n)} = \frac{1}{i\hbar} \sum_j c_j^{(n-1)} H'_{mj} e^{-i(\omega_{jm})t}; \quad \dot{c}_k^{(n)} = \frac{1}{i\hbar} \sum_j c_j^{(n-1)} H'_{kj} e^{-i(\omega_{jk})t}$$

So, $c_m^{(0)} = 1$ and $c_k^{(0)} = 0$

$$\dot{c}_m^{(1)} = \frac{1}{i\hbar} H'_{mm}$$

$$c_m^{(1)} = \frac{1}{i\hbar} V_{mm} \int_{-\infty}^{t} dt' e^{\eta t} = \frac{V_{mm} e^{\eta t}}{i\hbar\eta}$$

$$\dot{c}_k^{(1)} = \frac{1}{i\hbar} H'_{km} e^{-i(\omega_{mk})t}$$

$$c_k^{(1)} = \frac{1}{i\hbar} V_{km} \int_{-\infty}^{t} dt' e^{-i(\omega_{mk}+\eta)t'}) = \frac{V_{km}}{i\hbar(-i\omega_{mk} + \eta)} \left[e^{-i(\omega_{mk}+\eta)t} \right]$$

$$= \frac{V_{km}}{E_m - E_k + i\hbar\eta} \left[e^{-i(\omega_{mk}+\eta)t} \right]$$

$$\dot{c}_m^{(2)} = \sum_{j\neq m}{}' \frac{1}{i\hbar} \frac{V_{jm}}{E_m - E_j + i\hbar\eta} e^{-i(\omega_{mj}+\eta)t} V_{mj} e^{\eta t} e^{-i(\omega_{jm})t} + \frac{1}{i\hbar} \frac{V_{mm} e^{\eta t}}{i\hbar\eta} V_{mm} e^{\eta t}$$

$$c_m^{(2)} = \sum_{j\neq m}{}' \frac{1}{i\hbar} \frac{V_{jm} V_{mj}}{E_m - E_j + i\hbar\eta} \int_{-\infty}^{t} e^{2\eta t'} dt' - \frac{|V_{mm}|^2}{\hbar^2\eta} \int_{-\infty}^{t} e^{2\eta t'} dt'$$

$$= \sum_{j\neq m}{}' \frac{V_{jm} V_{mj}}{i\hbar 2\eta (E_m - E_j + i\hbar\eta)} e^{2\eta t} - \frac{|V_{mm}|^2}{2\hbar^2\eta^2} e^{2\eta t}$$

$$\dot{c}_k^{(2)} = \sum_{j\neq m}{}' \frac{1}{i\hbar} \frac{V_{jm}}{E_m - E_j + i\hbar\eta} e^{-i(\omega_{mj}+\eta)t} H'_{kj} e^{-i(\omega_{jk})t} + \frac{1}{i\hbar} \frac{V_{mm} e^{\eta t}}{i\hbar\eta} H'_{km} e^{-i(\omega_{mk})t}$$

$$c_k^{(2)} = \sum_{j\neq m}{}' \frac{1}{i\hbar} \frac{V_{jm} V_{kj}}{E_m - E_j + i\hbar\eta} \int_{-\infty}^{t} e^{-i(\omega_{mk}+2\eta)t'} dt' - \frac{V_{mm} V_{km}}{\hbar^2\eta} \int_{-\infty}^{t} e^{-i(\omega_{mk}+2\eta)t'} dt'$$

$$= \sum_{j\neq m}{}' \frac{V_{jm} V_{kj} e^{-i(\omega_{mk}+2\eta)t}}{(E_m - E_j + i\hbar\eta)(E_m - E_k + 2i\hbar\eta)} - \frac{V_{mm} V_{km} e^{-i(\omega_{mk}+2\eta)t}}{i\hbar\eta(E_m - E_k + 2i\hbar\eta)}$$

Therefore, to second order:

$$c_m(t) = 1 + \frac{V_{mm} e^{\eta t}}{i\hbar\eta} + \sum_j \frac{V_{jm} V_{mj}}{i\hbar 2\eta (E_m - E_j + i\hbar\eta)} e^{2\eta t}$$

$$c_k(t) = \frac{V_{km}}{i\hbar(-i\omega_{mk} + \eta)} \left[e^{-i(\omega_{mk}+\eta)t} \right] + \sum_j \frac{V_{jm} V_{kj} e^{-i(\omega_{mk}+2\eta)t}}{(E_m - E_j + i\hbar\eta)(E_m - E_k + 2i\hbar\eta)}$$

c. In part (a) the $c^{(2)}(t)$ grow linearly with time (for $V_{mm} = 0$), while in part b they remain finite for $\eta > 0$. The result in part a is due to the sudden turning on of the field.

d. $$|c_k(t)|^2 = \left| \sum_j \frac{V_{jm}V_{kj}e^{-i(\omega_{mk}+2\eta)t}}{(E_m - E_j + i\hbar\eta)(E_m - E_k + 2i\hbar\eta)} \right|^2$$

$$= \sum_{jj'} \frac{V_{kj}V_{kj'}V_{jm}V_{j'm}e^{-i(\omega_{mk}+2\eta)t}e^{i(\omega_{mk}+2\eta)t}}{(E_m - E_j + i\hbar\eta)(E_m - E_{j'} - i\hbar\eta)(E_m - E_k + 2i\hbar\eta)(E_m - E_k - 2i\hbar\eta)}$$

$$= \sum_{jj'} \frac{V_{kj}V_{kj'}V_{jm}V_{j'm}e^{4\eta t}}{[(E_m - E_j)(E_m - E_{j'}) + i\hbar\eta(E_j - E_{j'}) + \hbar^2\eta^2]((E_m - E_k)^2 + 4\hbar^2\eta^2)}$$

$$\frac{d}{dt}|c_k(t)|^2 = \sum_{jj'} \frac{4\eta V_{kj}V_{kj'}V_{jm}V_{j'm}}{[(E_m - E_j)(E_m - E_{j'}) + i\hbar\eta(E_j - E_{j'}) + \hbar^2\eta^2]((E_m - E_k)^2 + 4\hbar^2\eta^2)}$$

Now, look at the limit as $\eta \to 0^+$:

$$\frac{d}{dt}|c_k(t)|^2 \neq 0 \text{ when } E_m = E_k$$

$$\lim_{\eta \to 0^+} \frac{4\eta}{((E_m - E_k)^2 + 4\hbar^2\eta^2)} \quad \alpha\, \delta(E_m - E_k)$$

So, the final result is the second-order golden rule expression:

$$\frac{d}{dt}|c_k(t)|^2 \frac{2\pi}{\hbar} \delta(E_m - E_k) \lim_{\eta \to 0^+} \left| \sum_j \frac{V_{jm}V_{kj}}{(E_j - E_m - i\hbar\eta)} \right|^2$$

3. For the sudden perturbation case:

$$|c_m(t)|^2 = 1 + \sum_j{}' \frac{V_{jm}V_{mj}}{(E_m - E_j)^2}\left[e^{-i(\omega_{jm})t} - 1 + e^{i(\omega_{jm})t} - 1\right] + O(V^3)$$

$$|c_m(t)|^2 = 1 + \sum_j{}' \frac{V_{jm}V_{mj}}{(E_m - E_j)^2}\left[e^{-i(\omega_{jm})t} + e^{i(\omega_{jm})t} - 2\right] + O(V^3)$$

$$|c_k(t)|^2 = \frac{V_{km}V_{mk}}{(E_m - E_k)^2}\left[-e^{-i(\omega_{mk})t} - e^{i(\omega_{mk})t} + 2\right] + O(V^3)$$

$$1 - \sum_{k \neq m}{}' |c_k(t)|^2 = 1 - \sum_k{}' \frac{V_{km}V_{mk}}{(E_m - E_k)^2}\left[-e^{-i(\omega_{mk})t} - e^{i(\omega_{mk})t} + 2\right] + O(V^3)$$

$$= 1 + \sum_k{}' \frac{V_{km}V_{mk}}{(E_m - E_k)^2}\left[e^{-i(\omega_{mk})t} + e^{i(\omega_{mk})t} - 2\right] + O(V^3)$$

$\therefore$ to order V^2,

$$|c_m(t)|^2 = 1 - \sum_k{}' |c_k(t)|^2,$$ with no assumptions made regarding V_{mm}.

For the adiabatic perturbation case:

$$|c_m(t)|^2 = 1 + \sum_{j\neq m}{}' \left[\frac{V_{jm}V_{mj}e^{2\eta t}}{i\hbar 2\eta(E_m - E_j + i\hbar\eta)} + \frac{V_{jm}V_{mj}e^{2\eta t}}{-i\hbar 2\eta(E_m - E_j - i\hbar\eta)}\right] + O(V^3)$$

$$= 1 + \sum_{j\neq m}{}' \frac{1}{i\hbar 2\eta}\left[\frac{1}{(E_m - E_j + i\hbar\eta)} - \frac{1}{(E_m - E_j - i\hbar\eta)}\right] V_{jm}V_{mj}e^{2\eta t} + O(V^3)$$

$$= 1 + \sum_{j\neq m}{}' \frac{1}{i\hbar 2\eta}\left[\frac{-2i\hbar\eta}{(E_m - E_j)^2 + \hbar^2\eta^2}\right] V_{jm}V_{mj}e^{2\eta t} + O(V^3)$$

$$= 1 - \sum_{j\neq m}{}' \left[\frac{V_{jm}V_{mj}e^{2\eta t}}{(E_m - E_j)^2 + \hbar^2\eta^2}\right] + O(V^3)$$

$$|c_k(t)|^2 = \frac{V_{km}V_{mk}}{(E_m - E_k)^2 + \hbar^2\eta^2}e^{2\eta t} + O(V^3)$$

$\therefore$ to order V^2,

$$|c_m(t)|^2 = 1 - \sum_k{}' |c_k(t)|^2,$$ with no assumptions made regarding V_{mm} for this case as well.

Problems

1. a. $T_{nm} \approx \dfrac{|\langle n|V|m\rangle|^2}{\hbar^2\omega_{nm}^2}$

evaluating $\langle 1s|V|2s\rangle$ (using only the radial portions of the $1s$ and $2s$ wavefunctions since the spherical harmonics will integrate to unity) where $V = (e^2,r)$:

$$\langle 1s|V|2s\rangle = \int 2\left(\frac{Z}{a_0}\right)^{\frac{3}{2}} e^{\frac{-Zr}{a_0}} \frac{1}{r}\frac{1}{\sqrt{2}}\left(\frac{Z}{a_0}\right)^{\frac{3}{2}}\left(1 - \frac{Zr}{2a_0}\right)e^{\frac{-Zr}{2a_0}} r^2 dr$$

$$\langle 1s|V|2s\rangle = \frac{2}{\sqrt{2}}\left(\frac{Z}{a_0}\right)^3\left[\int re^{\frac{-3Zr}{2a_0}}dr - \int \frac{Zr^2}{2a_0}e^{\frac{-3Zr}{2a_0}}dr\right]$$

Using integral equation (4) for the two integrations we obtain:

$$\langle 1s|V|2s\rangle = \frac{2}{\sqrt{2}}\left(\frac{Z}{a_0}\right)^3\left[\frac{1}{\left(\frac{3Z}{2a_0}\right)^2} - \left(\frac{Z}{2a_0}\right)\frac{2}{\left(\frac{3Z}{2a_0}\right)^3}\right]$$

$$<1s|V|2s> = \frac{2}{\sqrt{2}}\left(\frac{Z}{a_0}\right)^3\left[\frac{2^2a_0^2}{3^2Z^2} - \frac{2^3a_0^2}{3^3Z^2}\right]$$

$$<1s|V|2s> = \frac{2}{\sqrt{2}}\left(\frac{Z}{a_0}\right)^3\left[\frac{(3)2^2a_0^2 - 2^3a_0^2}{3^3Z^2}\right] = \frac{8Z}{\sqrt{2}27a_0}$$

Now,

$$E_n = -\frac{Z^2e^2}{n^22a_0},\ E_{1s} = -\frac{Z^2e^2}{2a_0},\ E_{2s} = -\frac{Z^2e^2}{8a_0},\ E_{2s} - E_{1s} = \frac{3Z^2e^2}{8a_0}$$

So,

$$T_{nm} = \frac{\left(\frac{8Z}{\sqrt{2}27a_0}\right)^2}{\left(\frac{3Z^2}{8a_0}\right)^2} = \frac{2^6Z^22^6a_0^2}{(2)3^8a_0^2Z^4} = \frac{2^{11}}{3^8Z^2} = 0.312 \quad (\text{for } Z = 1)$$

b. $\varphi_m(r) = \varphi_{1s} = 2\left(\frac{Z}{a_0}\right)^{\frac{3}{2}} e^{\frac{-Zr}{a_0}} Y_{00}$

The orthogonality of the spherical harmonics results in only s-states having non-zero values for A_{nm}. We can then drop the Y_{00} (integrating this term will only result in unity) in determining the value of $A_{1s,2s}$.

$$\psi_n(r) = \psi_{2s} = \frac{1}{\sqrt{2}}\left(\frac{Z}{a_0}\right)^{\frac{3}{2}}\left(1 - \frac{Zr}{2a_0}\right)e^{\frac{-Zr}{2a_0}}$$

Remember for φ_{1s} $Z = 1$ and for ψ_{2s} $Z = 2$

$$A_{nm} = \int 2\left(\frac{Z}{a_0}\right)^{\frac{3}{2}} e^{\frac{-Zr}{a_0}} \frac{1}{\sqrt{2}}\left(\frac{Z+1}{a_0}\right)^{\frac{3}{2}}\left(1 - \frac{(Z+1)r}{2a_0}\right)e^{\frac{-(Z+1)r}{2a_0}} r^2dr$$

$$A_{nm} = \frac{2}{\sqrt{2}}\left(\frac{Z}{a_0}\right)^{\frac{3}{2}}\left(\frac{Z+1}{a_0}\right)^{\frac{3}{2}}\int e^{\frac{-(3Z+1)r}{2a_0}}\left(1 - \frac{(Z+1)r}{2a_0}\right)r^2dr$$

$$A_{nm} = \frac{2}{\sqrt{2}}\left(\frac{Z}{a_0}\right)^{\frac{3}{2}}\left(\frac{Z+1}{a_0}\right)^{\frac{3}{2}}\left[\int r^2e^{\frac{-(3Z+1)r}{2a_0}}dr - \int\frac{(Z+1)r^3}{2a_0}e^{\frac{-(3Z+1)r}{2a_0}}dr\right]$$

Evaluating these integrals using integral equation (4) we obtain:

$$A_{nm} = \frac{2}{\sqrt{2}}\left(\frac{Z}{a_0}\right)^{\frac{3}{2}}\left(\frac{Z+1}{a_0}\right)^{\frac{3}{2}}\left[\frac{2}{\left(\frac{3Z+1}{2a_0}\right)^3} - \left(\frac{Z+1}{2a_0}\right)\frac{(3)(2)}{\left(\frac{3Z+1}{2a_0}\right)^4}\right]$$

$$A_{nm} = \frac{2}{\sqrt{2}}\left(\frac{Z}{a_0}\right)^{\frac{3}{2}}\left(\frac{Z+1}{a_0}\right)^{\frac{3}{2}}\left[\frac{2^4 a_0^3}{(3Z+1)^3} - (Z+1)\frac{(3)2^4\, a_0^3}{(3Z+1)^4}\right]$$

$$A_{nm} = \frac{2}{\sqrt{2}}\left(\frac{Z}{a_0}\right)^{\frac{3}{2}}\left(\frac{Z+1}{a_0}\right)^{\frac{3}{2}}\left[\frac{-2^5 a_0^3}{(3Z+1)^4}\right]$$

$$A_{nm} = -2\frac{\left[2^3 Z(Z+1)\right]^{\frac{3}{2}}}{(3Z+1)^4}$$

The transition probability is the square of this amplitude:

$$T_{nm} = \left(-2\frac{\left[2^3 Z(Z+1)\right]^{\frac{3}{2}}}{(3Z+1)^4}\right)^2 = \frac{2^{11}Z^3(Z+1)^3}{(3Z+1)^8} = 0.25 \quad (\text{for } Z = 1).$$

The difference in these two results (parts a and b) will become negligible at large values of Z when the perturbation becomes less significant as in the case of $Z = 1$.

2. a. $\vec{\varepsilon}$ is along Z (lab fixed), and $\vec{\mu}$ is along z (the C-I molecule fixed bond). The angle between Z and z is β:

$$\vec{\varepsilon} \cdot \vec{\mu} = e\mu \cos\beta = e\mu D^{1*}_{00}(\alpha b\gamma)$$

So,

$$I = \langle D^{J'}_{M'K'}|\vec{\varepsilon} \cdot \vec{\mu}|D^{J}_{MK}\rangle = \int D^{J'}_{M'K'}\vec{\varepsilon} \cdot \vec{\mu} D^{J}_{MK} \sin\beta d\beta d\gamma\delta\alpha$$

$$= \varepsilon\mu\int D^{J'}_{M'K'}D^{1*}_{00}D^{J}_{MK} \sin\beta d\beta d\gamma d\alpha.$$

Now use:

$$D^{J'*}_{M'n'}D^{1*}_{00} = \sum_{jmn} \langle J'M'10|jm\rangle^* D^{j*}_{mn}\langle jn|J'K'10\rangle^*,$$

to obtain:

$$I = \varepsilon\mu\sum_{jmn} \langle J'M'10|jm\rangle^*\langle jn|J'K'10\rangle^* \int D^{j*}_{mn}D^{J}_{MK} \sin\beta d\beta d\gamma d\alpha.$$

Now use:

$$\int D^{j*}_{mn}D^{J}_{MK} \sin\beta d\beta d\gamma d\alpha = \frac{8\pi^2}{2J+1}\delta_{Jj}\delta_{Mm}\delta_{Kn},$$

to obtain:

$$I = \varepsilon\mu\frac{8\pi^2}{2J+1}\sum_{jmn} \langle J'M'10|jm\rangle^* \langle jn|J'K'10\rangle^* \delta_{Jj}\delta_{Mm}\delta_{Kn}$$

$$= \varepsilon\mu\frac{8\pi^2}{2J+1}\langle J'M'10|JM\rangle\langle JK|J'K'10\rangle.$$

We use:

$$\langle JK|J'K'10\rangle = \sqrt{2J+1}(-i)^{(J'-1+K)}\begin{pmatrix} J' & 1 & J \\ K' & 0 & K \end{pmatrix}$$

and,

$$\langle J'M'10|JM\rangle = \sqrt{2J+1}(-i)^{(J'-1+M)}\begin{pmatrix} J' & 1 & J \\ M' & 0 & M \end{pmatrix}$$

to give:

$$I = \varepsilon\mu\frac{8\pi^2}{2J+1}\sqrt{2J+1}(-i)^{(J'-1+M)}\begin{pmatrix} J' & 1 & J \\ M' & 0 & M \end{pmatrix}\sqrt{2J+1}(-i)^{(J'-1+K)}\begin{pmatrix} J' & 1 & J \\ K' & 0 & K \end{pmatrix}$$

$$= \varepsilon\mu 8\pi^2(-i)^{(J'-1+M+J'-1+K)}\begin{pmatrix} J' & 1 & J \\ M' & 0 & M \end{pmatrix}\begin{pmatrix} J' & 1 & J \\ K' & 0 & K \end{pmatrix}$$

$$= \varepsilon\mu 8\pi^2(-i)^{(M+K)}\begin{pmatrix} J' & 1 & J \\ M' & 0 & M \end{pmatrix}\begin{pmatrix} J' & 1 & J \\ K' & 0 & K \end{pmatrix}$$

The 3-J symbols vanish unless: $K' + 0 = K$ and $M' + 0 = M$. So,

$$I = \varepsilon\mu 8\pi^2(-i)^{(M+K)}\begin{pmatrix} J' & 1 & J \\ M & 0 & M \end{pmatrix}\begin{pmatrix} J' & 1 & J \\ K & 0 & K \end{pmatrix}\delta_{M'M}\delta_{K'K}.$$

b. $\begin{pmatrix} J' & 1 & J \\ M & 0 & M \end{pmatrix}$ and $\begin{pmatrix} J' & 1 & J \\ K & 0 & K \end{pmatrix}$ vanish unless $J' = J+1, J, J-1$

$\therefore \Delta J = \pm 1, 0$

The K quantum number can not change because the dipole moment lies along the molecule's C_3 axis and the light's electric field thus can exert no torque that twists the molecule about this axis. As a result, the light can not induce transitions that excite the molecule's spinning motion about this axis.

3. a. B atom: $1s^2 2s^2 2p^1, {}^2P$ ground state $L = 1, S = \frac{1}{2}$, gives a degeneracy $((2L+1)(2S+1))$ of 6.
O atom: $1s^2 2s^2 2p^4, {}^3P$ ground state $L = 1, S = 1$, gives a degeneracy $((2L+1)(2S+1))$ of 9. The total number of states formed is then (6)(9) = 54.

b. We need only consider the p orbitals to find the low-lying molecular states:

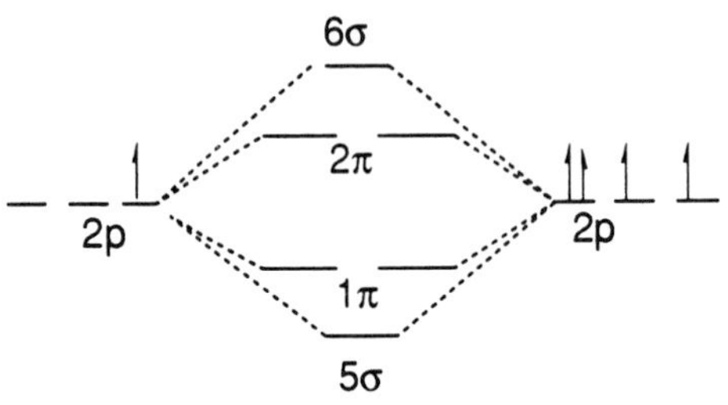

Which, in reality look like this:

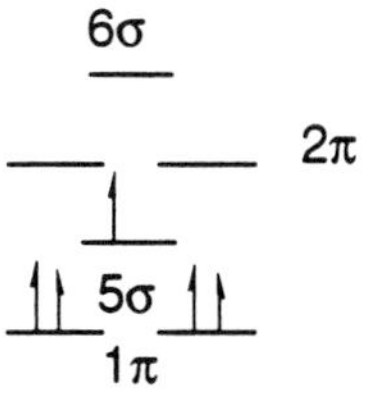

This is the correct ordering to give a $^2\Sigma^+$ ground state. The only low-lying electron configurations are $1\pi^3 5\sigma^2$ or $1\pi^4 5\sigma^1$. These lead to $^2\Pi$ and $^2\Sigma^+$ states, respectively.

c. The bond orders in both states are 2½.

d. The $^2\Sigma$ is + and g/u cannot be specified since this is a heteronuclear molecule.

e. Only one excited state, the $^2\Pi$, is spin-allowed to radiate to the $^2\Sigma^+$. Consider symmetries of transition moment operators that arise in the $E1$, $E2$ and $M1$ contributions to the transition rate electric dipole allowed: $z \rightarrow \Sigma^+$, $x,y \rightarrow \Pi$, ∴ the $^2\Pi \rightarrow {}^2\Sigma^+$ is electric dipole allowed via a perpendicular band. Magnetic dipole allowed: $R_z \rightarrow \Sigma^-$, $R_{x,y} \rightarrow \Pi$, ∴ the $^2\Pi \rightarrow {}^2\Sigma^+$ is magnetic dipole allowed. Electric quadrupole allowed: $x^2 + y^2, z^2 \rightarrow \Sigma^+, xy, yz \rightarrow \Pi$, $x^2 - y^2, xy \rightarrow \Delta$ ∴ the $^2\Pi \rightarrow {}^2\Sigma^+$ is electric quadrupole allowed as well.

f. Since ionization will remove a bonding electron, the BO^+ bond is weaker than the BO bond.

g. The ground state BO^+ is $^1\Sigma^+$ corresponding to a $1\pi^4$ electron configuration. An electron configuration of $1\pi^3 5\sigma^1$ leads to a $^3\Pi$ and a $^1\Pi$ state. The $^3\Pi$ will be lower in energy. A $1\pi^2 5\sigma^2$ configuration will lead to higher lying states of $^3\Sigma^-, {}^1\Delta$, and $^1\Sigma^+$.

h. There should be three bands corresponding to formation of BO^+ in the $^1\Sigma^+$, $^3\Pi$, and $^1\Pi$ states. Since each of these involves removing a bonding electron, the Franck-Conden integrals will be appreciable for several vibrational levels, and thus a vibrational progression should be observed.

4. a. The bending (π) vibration is degenerate.

b.

```
H---C≡N
  ⇑
  bending  fundamental
```

c.

```
H---C≡N
  ⇑
  stretching  fundamental
```

d. CH stretch (ν_3 in figure) is σ, CN stretch is σ, and HCN (ν_2 in figure) bend is π.

e. Under $z(\sigma)$ light the CN stretch and the CH stretch can be excited, since $\psi_0 = \sigma$, $\psi_1 = \sigma$ and $z = \sigma$ provides coupling.

f. Under $x,y(\pi)$ light the HCN bend can be excited, since $\psi_0 = \sigma$, $\psi_1 = \pi$ and $x,y = \pi$ provides coupling.

g. The bending vibration is active under (x,y) perpendicular polarized light. $\Delta J = 0, \pm 1$ are the selection rules for $\perp$ transitions. The CH stretching vibration is active under (z) $||$ polarized light. $\Delta J = \pm 1$ are the selection rules for $||$ transitions.

SECTION

6

MORE QUANTITATIVE ASPECTS OF ELECTRONIC STRUCTURE CALCULATIONS

CHAPTER

17

Electrons interact via pairwise Coulomb forces; within the "orbital picture" these interactions are modelled by less difficult to treat "averaged" potentials. The difference between the true Coulombic interactions and the averaged potential is not small, so to achieve reasonable (ca.1 kcal/mol) chemical accuracy, high-order corrections to the orbital picture are needed.

The discipline of computational *ab initio* quantum chemistry is aimed at determining the electronic energies and wavefunctions of atoms, molecules, radicals, ions, solids, and all other chemical species. The phrase *ab initio* implies that one attempts to solve the Schrödinger equation from first principles, treating the molecule as a collection of positive nuclei and negative electrons moving under the influence of coulombic potentials, and not using any prior knowledge about this species' chemical behavior.

To make practical use of such a point of view requires that approximations be introduced; the full Schrödinger equation is too difficult to solve exactly for any but simple model problems. These approximations take the form of physical concepts (e.g., orbitals, configurations, quantum numbers, term symbols, energy surfaces, selection rules, etc.) that provide useful means of organizing and interpreting experimental data and computational methods that allow quantitative predictions to be made.

Essentially all *ab initio* quantum chemistry methods use, as a starting point from which improvements are made, a picture in which the electrons interact via a one-electron additive potential. These so-called **mean-field** potentials

$$V_{mf}(\mathbf{r}) = \sum_j V_{mf}(\mathbf{r}_j)$$

provide descriptions of atomic and molecular structure that are approximate. Their predictions must be improved to achieve reasonably accurate solutions to the true electronic Schrödinger equation. In so doing, three constructs that characterize essentially all *ab initio* quantum chemical methods are employed: **orbitals, configurations, and electron correlation.** Since the electronic kinetic energy

$$T = \sum_j T_j$$

operator is one-electron additive, the mean-field Hamiltonian $H^0 = T + V_{mf}$ is also of this form. The additivity of H^0 implies that the mean-field wavefunctions $\{\Psi_k^0\}$ can be formed in terms of products of functions $\{\phi_k\}$ of the coordinates of the individual electrons, and that the corresponding energies $\{E_k^0\}$ are additive. Thus, it is the ansatz that V_{mf} is separable that leads to the concept of **orbitals**, which are the one-electron functions $\{\phi_j\}$. These orbitals are found by solving the one-electron Schrödinger equations:

$$(T_1 + V_{mf}(\boldsymbol{r}_1))\, \phi_j(\boldsymbol{r}_1) = \varepsilon_j \phi_j(\boldsymbol{r}_1);$$

the eigenvalues $\{\varepsilon_j\}$ are called **orbital energies.**

Because each of the electrons also possesses intrinsic spin, the one-electron functions $\{\phi_j\}$ used in this construction are taken to be eigenfunctions of $(T_1 + V_{mf}(\boldsymbol{r}_1))$ multiplied by either α or β. This set of functions is called the set of mean-field **spin-orbitals.**

Given the complete set of solutions to this one-electron equation, a complete set of N-electron mean-field wavefunctions can be written down. Each Ψ_k^0 is constructed by forming an antisymmetrized product of N spin-orbitals chosen from the set of $\{\phi_j\}$, allowing each spin-orbital in the list to be a function of the coordinates of one of the N electrons (e.g,

$$\Psi_k^0 = |\phi_{k1}(r_1)\, \phi_{k2}(r_2)\, \phi_{k3}(r_3) \ldots \phi_{kN-1}(\boldsymbol{r}_{N-1})\, \phi_{kN}(r_N)|,$$

as above). The corresponding mean field energy is evaluated as the sum over those spin-orbitals that appear in Ψ_k^0:

$$E_k^0 = \sum_{j=1,N} \varepsilon_{kj}.$$

By choosing to place N electrons into specific spin-orbitals, one has specified a **configuration.** By making other choices of which N ϕ_j to occupy, one describes other configurations. Just as the one-electron mean-field Schrödinger equation has a complete set of spin-orbital solutions $\{\phi_j$ and $\varepsilon_j\}$, the N-electron mean-field Schrödinger equation has a complete set of N-electron *configuration state functions* (CSFs) Ψ_k^0 and energies E_k^0.

I. ELECTRON CORRELATION REQUIRES MOVING BEYOND A MEAN-FIELD MODEL

To improve upon the mean-field picture of electronic structure, one must move beyond the single-configuration approximation. It is essential to do so to achieve higher accuracy, but it is also important to do so to achieve a *conceptually* correct view of chemical electronic structure. However, it is very disconcerting to be told that the familiar $1s^22s^22p^2$ description of the carbon atom is inadequate and that instead one must think of the ^3P ground state of this atom as a "mixture" of $1s^22s^22p^2$, $1s^22s^23p^2$, $1s^22s^23d^2$, $2s^23s^22p^2$ (and any other configurations whose angular momenta can be coupled to produce $L = 1$ and $S = 1$).

Although the picture of configurations in which N electrons occupy N spin-orbitals may be very familiar and useful for systematizing electronic states of atoms and molecules, these constructs are approximations to the true states of the system. They were introduced when the mean-field approximation was made, and neither orbitals nor configurations describe the proper eigenstates $\{\Psi_k, E_k\}$. The inclusion of instantaneous spatial correlations among electrons is necessary to achieve a more accurate description of atomic and molecular electronic structure.

No single spin-orbital product wavefunction is capable of treating electron correlation to *any* extent; its product nature renders it incapable of doing so.

II. MOVING FROM QUALITATIVE TO QUANTITATIVE MODELS

The preceding chapters introduced, in a qualitative manner, many of the concepts which are used in applying quantum mechanics to electronic structures of atoms and molecules. Atomic, bonding, non-bonding, antibonding, Rydberg, hybrid, and delocalized orbitals and the configurations formed by occupying these orbitals were discussed. Spin and spatial symmetry as well as permutational symmetry were treated, and properly symmetry-adapted configuration state functions were formed. The Slater-Condon rules were shown to provide expressions for Hamiltonian matrix elements (and those involving any one- or two-electron operator) over such CSFs in terms of integrals over the orbitals occupied in the CSFs. Orbital, configuration, and state correlation diagrams were introduced to allow one to follow the evolution of electronic structures throughout a "reaction path."

Section 6 addresses the *quantitative and computational implementation* of many of the above ideas. It is not designed to address all of the state-of-the-art methods which have been, and are still being, developed to calculate orbitals and state wavefunctions. The rapid growth in computer hardware and software power and the evolution of new computer architectures makes it difficult, if not impossible, to present an up-to-date overview of the techniques that are presently at the cutting edge in computational chemistry. Nevertheless, this section attempts to describe the essential elements of several of the more powerful and commonly used methods; it is likely that many of these elements will persist in the next generation of computational chemistry techniques although the details of their implementation will evolve considerably. The text by Szabo and Ostlund provides excellent insights into many of the theoretical methods treated in this section.

III. ATOMIC UNITS

The electronic Hamiltonian is expressed, in this section, in so-called atomic units (aus)

$$H_e = \sum_j \{(-1/2)\nabla_j^2 - \Sigma_a Z_a/r_{j,a}\} + \sum_{j<k} 1/r_{j,k}.$$

These units are introduced to remove all $\hbar$, e, and m_e factors from the equations.

To effect this unit transformation, one notes that the kinetic energy operator scales as r_j^{-2} whereas the coulombic potentials scale as r_j^{-1} and as $r_{j,k}^{-1}$. So, if each of the distances appearing in the cartesian coordinates of the electrons and nuclei were expressed as a unit of length a_0 multiplied by a dimensionless length factor, the kinetic energy operator would involve terms of the form $(-\hbar^2/2(a_0)^2 m_e)\nabla_j^2$, and the coulombic potentials would appear as $Z_a e^2/(a_0)r_{j,a}$ and $e^2/(a_0)r_{j,k}$. A factor of e^2/a_0 (which has units of energy since a_0 has units of length) can then be removed from the coulombic and kinetic energies, after which the kinetic energy terms appear as $(-\hbar^2/2(e^2a_0)m_e)\nabla_j^2$ and the potential energies appear as $Z_a/r_{j,a}$ and $1/r_{j,k}$. Then, choosing $a_0 = \hbar^2/e^2m_e$ changes the kinetic energy terms into $-\frac{1}{2}\nabla_j^2$; as a result, the entire electronic Hamiltonian takes the form given above in which no e^2, m_e, or $\hbar^2$ factors appear. The value of the so-called Bohr radius $a_0 = \hbar^2/e^2m_e$ is 0.529Å, and the so-called Hartree energy unit e^2/a_0, which factors out of H_e, is 27.21 eV or 627.51 kcal/mol.

CHAPTER

18

The single Slater determinant wavefunction (properly spin and symmetry adapted) is the starting point of the most common mean-field potential. It is also the origin of the molecular orbital concept.

I. OPTIMIZATION OF THE ENERGY FOR A MULTICONFIGURATION WAVEFUNCTION

A. The Energy Expression

The most straightforward way to introduce the concept of optimal molecular orbitals is to consider a trial wavefunction of the form which was introduced earlier in chapter 9.II. The expectation value of the Hamiltonian for a wavefunction of the multiconfigurational form

$$\Psi = \sum_I C_I \Phi_I,$$

where Φ_I is a space- and spin-adapted CSF which consists of determinental wavefunctions $|\phi_{I1}\phi_{I2}\phi_{I3} \ldots \phi_{IN}|$, can be written as:

$$E = \sum_{I,J=1,M} C_I C_J \langle\Phi_I|H|\Phi_J\rangle.$$

The spin- and space-symmetry of the Φ_I determine the symmetry of the state Ψ whose energy is to be optimized.

In this form, it is clear that E is a quadratic function of the CI amplitudes C_J; it is a quartic functional of the spin-orbitals because the Slater-Condon rules express each $\langle\Phi_I|H|\Phi_J\rangle$ *CI* matrix element in terms of one- and two-electron integrals $\langle\phi_i|f|\phi_j\rangle$ and $\langle\phi_i\phi_j|g|\phi_k\phi_l\rangle$ over these spin-orbitals.

B. Application of the Variational Method

The *variational* method can be used to optimize the above expectation value expression for the electronic energy (i.e., to make the functional stationary) as a function of the CI coefficients C_J and the LCAO-MO coefficients $\{C_{\nu,i}\}$ that characterize the spin-orbitals. However, in doing so the set of $\{C_{\nu,i}\}$ can not be treated as entirely independent variables. The fact that the spin-orbitals $\{\phi_i\}$ are assumed to be orthonormal imposes a set of constraints on the $\{C_{\nu,i}\}$:

$$<\phi_i|\phi_j> = \delta_{i,j} = \sum_{\mu,\nu} C^*_{\mu,i}<\chi_\mu|\chi_\nu>C_{\nu,j}.$$

These constraints can be enforced within the variational optimization of the energy function mentioned above by introducing a set of Lagrange multipliers $\{\varepsilon_{i,j}\}$, one for each constraint condition, and subsequently differentiating

$$E - \sum_{i,j} \varepsilon_{i,j}[\delta_{i,j} - \sum_{\mu,\nu} C^*_{\mu,i}<\chi_\mu|\chi_\nu>C_{\nu,j}]$$

with respect to each of the $C_{\nu,i}$ variables.

C. The Fock and Secular Equations

Upon doing so, the following set of equations is obtained (early references to the derivation of such equations include A. C. Wahl, *J. Chem. Phys. 41*, 2600 (1964) and F. Grein and T. C. Chang, *Chem. Phys. Lett. 12*, 44 (1971); a more recent overview is presented by R. Shepard, p 63, in *Adv. in Chem. Phys. LXIX*, K. P. Lawley, Ed., Wiley-Interscience, New York (1987); the subject is also treated in the textbook *Second Quantization Based Methods in Quantum Chemistry*, P. Jφrgensen and J. Simons, Academic Press, New York (1981)):

$$\sum_{J=1,M} H_{I,J}C_J = EC_I, = 1,2, \ldots M, \text{ and}$$

$$F\phi_i = \sum_j \varepsilon_{i,j}\phi_j,$$

where the $\varepsilon_{i,j}$ are Lagrange multipliers.

The first set of equations govern the $\{C_J\}$ amplitudes and are called the CI-secular equations. The second set determine the LCAO-MO coefficients of the spin-orbitals $\{\phi_j\}$ and are called the Fock equations. The Fock operator F is given in terms of the one- and two-electron operators in H itself as well as the so-called one- and two-electron density matrices $\gamma_{i,j}$ and $\Gamma_{i,j,k,l}$ which are defined below. These density matrices reflect the averaged occupancies of the various spin orbitals in the CSFs of Ψ. The resultant expression for F is:

$$F\phi_i = \sum_j \gamma_{i,j}h\phi_j + \sum_{j,k,l} \Gamma_{i,j,k,l}J_{j,l}\phi_k,$$

where h is the one-electron component of the Hamiltonian (i.e., the kinetic energy operator and the sum of coulombic attractions to the nuclei). The operator $J_{j,l}$ is defined by:

$$J_{j,l}\phi_k(r) = \int \phi^*_j(r')\phi_l(r')1/|r - r'|d\tau'\phi_k(r),$$

where the integration denoted $d\tau'$ is over the spatial and spin coordinates. The so-called spin integration simply means that the α or β spin function associated with ϕ_l must be the same as the α or β spin function associated with ϕ_j or the integral will vanish. This is a consequence of the orthonormality conditions $<\alpha|\alpha> = <\beta|\beta> = 1$, $<\alpha|\beta> = <\beta|\alpha> = 0$.

D. One- and Two-Electron Density Matrices

The density matrices introduced above can most straightforwardly be expressed in terms of the CI amplitudes and the nature of the orbital occupancies in the CSFs of Ψ as follows:

1. $\gamma_{i,i}$ is the sum over all CSFs, in which ϕ_i is occupied, of the square of the C_I coefficient of that CSF:

$$\gamma_{i,i} = \sum_I \text{ (with } \phi_i \text{ occupied) } C_I^2.$$

2. $\gamma_{i,j}$ is the sum over pairs of CSFs which differ by a single spin-orbital occupancy (i.e., one having ϕ_i occupied where the other has ϕ_j occupied after the two are placed into maximal coincidence—the sign factor (sign) arising from bringing the two to maximal coincidence is attached to the final density matrix element):

$$\gamma_{i,j} = \sum_{I,J} \text{ (sign) (with } \phi_i \text{ occupied in } I \text{ where } \phi_j \text{ is in } J)\ C_I C_J.$$

The two-electron density matrix elements are given in similar fashion:

3. $\Gamma_{i,j,i,j} = \sum_I$ (with both ϕ_i and ϕ_j occupied) $C_I C_I$;

4. $\Gamma_{i,j,j,i} = -\sum_I$ (with both ϕ_i and ϕ_j occupied) $C_I C_I = -\Gamma_{i,j,i,j}$

(it can be shown, in general that $\Gamma_{i,j,k,l}$ is odd under exchange of i and j, odd under exchange of k and l and even under $(i,j) <=> (k,l)$ exchange; this implies that $\Gamma_{i,j,k,l}$ vanishes if $i = j$ or $k = l$) ;

5. $\Gamma_{i,j,k,j} = \sum_{I,J}$ (sign) (with ϕ_j in both I and J and ϕ_i in I where ϕ_k is in J)$C_I C_J$

$$= \Gamma_{j,i,j,k} = -\Gamma_{i,j,j,k} = -\Gamma_{j,i,k,j};$$

6. $\Gamma_{i,j,k,l} = \sum_{I,J}$ (sign) (with ϕ_i in I where ϕ_k is in J and ϕ_j in I where ϕ_l is in J)$C_I C_J$

$$= \Gamma_{j,i,l,k} = -\Gamma_{j,i,k,l} = -\Gamma_{i,j,l,k} = \Gamma_{j,i,l,k}.$$

These density matrices are themselves quadratic functions of the CI coefficients and they reflect all of the permutational symmetry of the determinental functions used in constructing Ψ; they are a compact representation of all of the Slater-Condon rules as applied to the particular CSFs which appear in Ψ. They contain all information about the spin-orbital occupancy of the

CSFs in Ψ. The one- and two-electron integrals $<\phi_i|f|\phi_j>$ and $<\phi_i\phi_j|g|\phi_k\phi_l>$ contain all of the information about the magnitudes of the kinetic and Coulombic interaction energies.

II. THE SINGLE-DETERMINANT WAVEFUNCTION

The simplest trial function of the form given above is the single Slater determinant function:

$$\Psi = |\phi_1\phi_2\phi_3 \ldots \phi_N|.$$

For such a function, the CI part of the energy minimization is absent (the classic papers in which the SCF equations for closed- and open-shell systems are treated are C. C. J. Roothaan, *Rev. Mod. Phys. 23*, 69 (1951); *32*, 179 (1960)) and the density matrices simplify greatly because only one spin-orbital occupancy is operative. In this case, the orbital optimization conditions reduce to:

$$F\phi_i = \sum_j \varepsilon_{i,j}\phi_j,$$

where the so-called Fock operator F is given by

$$F\phi_i = h\phi_i + \sum_{j(occupied)} [J_j - K_j]\phi_i.$$

The coulomb (J_j) and exchange (K_j) operators are defined by the relations:

$$J_j\phi_i = \int \phi*_j(r')\phi_j(r')1/|r - r'|d\tau'\phi_i(r), \text{ and}$$

$$K_j\phi_i = \int \phi*_j(r')\phi_i(r')1/|r - r'|d\tau'\phi_j(r).$$

Again, the integration implies integration over the spin variables associated with the ϕ_j (and, for the exchange operator, ϕ_i), as a result of which the exchange integral vanishes unless the spin function of ϕ_j is the same as that of ϕ_i; the coulomb integral is non-vanishing no matter what the spin functions of ϕ_j and ϕ_i.

The sum over coulomb and exchange interactions in the Fock operator runs only over those spin-orbitals that are occupied in the trial Ψ. Because a unitary transformation among the orbitals that appear in Ψ leaves the determinant unchanged (this is a property of determinants $-\det(UA) = \det(U)\det(A) = 1\det(A)$, if U is a unitary matrix), it is possible to choose such a unitary transformation to make the $\varepsilon_{i,j}$ matrix diagonal. Upon so doing, one is left with the so-called *canonical Hartree-Fock equations:*

$$F\phi_i = \varepsilon_i\phi_j,$$

where ε_i is the diagonal value of the $\varepsilon_{i,j}$ matrix after the unitary transformation has been applied; that is, ε_i is an eigenvalue of the $\varepsilon_{i,j}$ matrix. These equations are of the eigenvalue-eigenfunction form with the Fock operator playing the role of an effective one-electron Hamiltonian and the ϕ_i playing the role of the one-electron eigenfunctions.

It should be noted that the Hartree-Fock equations $F\phi_i = \varepsilon_i\phi_j$ possess solutions for the spin-orbitals which appear in Ψ (the so-called *occupied* spin-orbitals) as well as for orbitals which are not occupied in Ψ (the so-called *virtual* spin-orbitals). In fact, the F operator is hermitian, so it possesses a complete set of orthonormal eigenfunctions; only those which appear in Ψ appear in the coulomb and exchange potentials of the Fock operator. The physical meaning of the occupied and virtual orbitals will be clarified later in this chapter (section VII.A)

III. THE UNRESTRICTED HARTREE-FOCK SPIN IMPURITY PROBLEM

As formulated above in terms of spin-orbitals, the Hartree-Fock (HF) equations yield orbitals that do not guarantee that Ψ possesses proper spin symmetry. To illustrate the point, consider the form of the equations for an open-shell system such as the Lithium atom Li. If $1s\alpha$, $1s\beta$, and $2s\alpha$ spin-orbitals are chosen to appear in the trial function Ψ, then the Fock operator will contain the following terms:

$$F = h + J_{1s\alpha} + J_{1s\beta} + J_{2s\alpha} - [K_{1s\alpha} + K_{1s\beta} + K_{2s\alpha}].$$

Acting on an α spin-orbital $\phi_{\kappa\alpha}$ with F and carrying out the spin integrations, one obtains

$$F\phi_{\kappa\alpha} = h\phi_{\kappa\alpha} + (2J_{1s} + J_{2s})\phi_{\kappa\alpha} - (K_{1s} + K_{2s})\phi_{\kappa\alpha}.$$

In contrast, when acting on a β spin-orbital, one obtains

$$F\phi_{\kappa\beta} = h\phi_{\kappa\beta} + (2J_{1s} + J_{2s})\phi_{\kappa\beta} - (K_{1s})\phi_{\kappa\beta}.$$

Spin-orbitals of α and β type do *not* experience the same exchange potential in this model, which is clearly due to the fact that Ψ contains two α spin-orbitals and only one β spin-orbital.

One consequence of the spin-polarized nature of the effective potential in F is that the optimal $1s\alpha$ and $1s\beta$ spin-orbitals, which are themselves solutions of $F\phi_i = \varepsilon_i\phi_i$, do not have identical orbital energies (i.e., $\varepsilon_{1s\alpha} \neq \varepsilon_{1s\beta}$) and are not spatially identical to one another (i.e., $\phi_{1s\alpha}$ and $\phi_{1s}\beta$ do not have identical LCAO-MO expansion coefficients). This resultant spin polarization of the orbitals in Ψ gives rise to spin impurities in Ψ. That is, the determinant $|1s\alpha 1s'\beta 2s\alpha|$ is not a pure doublet spin eigenfunction although it is an S_z eigenfunction with $M_s = 1/2$; it contains both $S = 1/2$ and $S = 3/2$ components. If the $1s\alpha$ and $1s'\beta$ spin-orbitals were spatially identical, then $|1s\alpha 1s'\beta 2s\alpha|$ would be a pure spin eigenfunction with $S = 1/2$.

The above single-determinant wavefunction is commonly referred to as being of the **unrestricted Hartree-Fock** (UHF) type because no restrictions are placed on the spatial nature of the orbitals which appear in Ψ. In general, UHF wavefunctions are not of pure spin symmetry for any open-shell system. Such a UHF treatment forms the starting point of early versions of the widely used and highly successful Gaussian 70 through Gaussian-8X series of electronic structure computer codes which derive from J. A. Pople and co-workers (see, for example, M. J. Frisch, J. S. Binkley, H. B. Schlegel, K Raghavachari, C. F. Melius, R. L. Martin, J. J. P. Stewart, F. W. Bobrowicz, C. M. Rohlfing, L. R. Kahn, D. J. Defrees, R. Seeger, R. A. Whitehead, D. J. Fox, E. M. Fleuder, and J. A. Pople, *Gaussian 86*, Carnegie-Mellon Quantum Chemistry Publishing Unit, Pittsburgh, Penn. (1984)).

The inherent spin-impurity problem is sometimes "fixed" by using the orbitals which are obtained in the UHF calculation to subsequently form a properly spin-adapted wavefunction. For the above Li atom example, this amounts to forming a new wavefunction (after the orbitals are obtained via the UHF process) using the techniques detailed in section 3 and appendix G:

$$\Psi = 1/\sqrt{2}\,[|1s\alpha 1s'\beta 2s\alpha| - |1s\beta 1s'\alpha 2s\alpha|].$$

This wavefunction is a pure $S = 1/2$ state. This prescription for avoiding spin contamination (i.e., carrying out the UHF calculation and then forming a new spin-pure Ψ) is referred to as *spin-projection*.

It is, of course, possible to first form the above spin-pure Ψ as a trial wavefunction and to then determine the orbitals $1s$, $1s'$, and $2s$ which minimize its energy; in so doing, one is dealing with a spin-pure function from the start. The problem with carrying out this process, which is referred to as a *spin-adapted* Hartree-Fock calculation, is that the resultant $1s$ and $1s'$ orbitals still do not have identical spatial attributes. Having a set of orbitals (1*s*, 1*s*', 2*s*, and the virtual orbitals) that form a non-orthogonal set (1*s* and 1*s*' are neither identical nor orthogonal) makes it difficult to progress beyond the single-configuration wavefunction as one often wishes to do. That is, it is difficult to use a spin-adapted wavefunction as a starting point for a correlated-level treatment of electronic motions.

Before addressing head-on the problem of how to best treat orbital optimization for open-shell species, it is useful to examine how the HF equations are solved in practice in terms of the LCAO-MO process.

IV. THE LCAO-MO EXPANSION

The HF equations $F\phi_i = \varepsilon_i\phi_i$ comprise a set of integro-differential equations; their differential nature arises from the kinetic energy operator in h, and the coulomb and exchange operators provide their integral nature. The solutions of these equations must be achieved iteratively because the J_i and K_i operators in F depend on the orbitals ϕ_i which are to be solved for. Typical iterative schemes begin with a "guess" for those ϕ_i which appear in Ψ, which then allows F to be formed. Solutions to $F\phi_i = \varepsilon_i\phi_i$ are then found, and those ϕ_i which possess the space and spin symmetry of the occupied orbitals of Ψ and which have the proper energies and nodal character are used to generate a new F operator (i.e., new J_i and K_i operators). The new F operator then gives new ϕ_i and ε_i via solution of the new $F\phi_i = \varepsilon_i\phi_i$ equations. This iterative process is continued until the ϕ_i and ε_i do not vary significantly from one iteration to the next, at which time one says that the process has converged. This iterative procedure is referred to as the Hartree-Fock *self-consistent field* (SCF) procedure because iteration eventually leads to coulomb and exchange potential fields that are consistent from iteration to iteration.

In practice, solution of $F\phi_i = \varepsilon_i\phi_i$ as an integro-differential equation can be carried out only for atoms (C. Froese-Fischer, *Comp. Phys. Commun. 1*, 152 (1970)) and linear molecules (P. A. Christiansen and E. A. McCullough, *J. Chem. Phys. 67*, 1877 (1977)) for which the angular parts of the ϕ_i can be exactly separated from the radial because of the axial- or full-rotation group symmetry (e.g., $\phi_i = Y_{l,m}R_{n,l}(r)$ for an atom and $\phi_i = \exp(im\phi)\, R_{n,l,m}(r,\theta)$ for a linear molecule). In such special cases, $F\phi_i = \varepsilon_i\phi_i$ gives rise to a set of coupled equations for the $R_{n,l}(r)$ or $R_{n,l,m}(r,\theta)$ which can and have been solved. However, for non-linear molecules, the HF equations have not yet been solved in such a manner because of the three-dimensional nature of the ϕ_i and of the potential terms in F.

In the most commonly employed procedures used to solve the HF equations for non-linear molecules, the ϕ_i are expanded in a basis of functions χ_μ according to the LCAO-MO procedure:

$$\phi_i = \sum_\mu C_{\mu,i}\chi_\mu .$$

Doing so then reduces $F\phi_i = \varepsilon_i\phi_i$ to a matrix eigenvalue-type equation of the form:

$$\sum_\nu F_{\mu,\nu}C_{\nu,i} = \varepsilon_i \sum_\nu S_{\mu,\nu}C_{\nu,i},$$

where $S_{\mu,\nu} = \langle\chi_\mu|\chi_\nu\rangle$ is the overlap matrix among the atomic orbitals (aos) and

$$F_{\mu,\nu} = \langle\chi_\mu|h|\chi_\nu\rangle + \sum_{\delta,\kappa} [\gamma_{\delta,\kappa}\langle\chi_\mu\chi_\delta|g|\chi_\nu\chi_\kappa\rangle - \gamma_{\delta,\kappa}^{ex}\langle\chi_\mu\chi_\delta|g|\chi_\kappa\chi_\nu\rangle]$$

is the matrix representation of the Fock operator in the ao basis. The coulomb and exchange-density matrix elements in the ao basis are:

$$\gamma_{\delta,\kappa} = \sum_{i(occupied)} C_{\delta,i}C_{\kappa,i}, \text{ and}$$

$$\gamma_{\delta,\kappa}{}^{ex} = \sum_{i(occ., \text{ and } same\ spin)} C_{\delta,i}C_{\kappa,i},$$

where the sum in $\gamma_{\delta,\kappa}{}^{ex}$ runs over those occupied spin-orbitals whose m_s value is equal to that for which the Fock matrix is being formed (for a closed-shell species, $\gamma_{\delta,\kappa}{}^{ex} = 1/2\ \gamma_{\delta,\kappa}$).

It should be noted that by moving to a matrix problem, one does not remove the need for an iterative solution; the $F_{\mu,\nu}$ matrix elements depend on the $C_{\nu,i}$ LCAO-MO coefficients which are, in turn, solutions of the so-called Roothaan matrix Hartree-Fock equations

$$\sum_{\nu} F_{\mu,\nu}C_{\nu,i} = \varepsilon_i \sum_{\nu} S_{\mu,\nu}C_{\nu,i}.$$

One should also note that, just as $F\phi_i = \varepsilon_i\phi_j$ possesses a complete set of eigenfunctions, the matrix $F_{\mu,\nu}$, whose dimension M is equal to the number of atomic basis orbitals used in the LCAO-MO expansion, has M eigenvalues ε_i and M eigenvectors whose elements are the $C_{\nu,i}$. Thus, there are occupied and virtual molecular orbitals (mos) each of which is described in the LCAO-MO form with $C_{\nu,i}$ coefficients obtained via solution of

$$\sum_{\nu} F_{\mu,\nu}C_{\nu,i} = \varepsilon_i \sum_{\nu} S_{\mu,\nu}C_{\nu,i}.$$

V. ATOMIC ORBITAL BASIS SETS

A. STOs and GTOs

The basis orbitals commonly used in the LCAO-MO-SCF process fall into two classes:

1. Slater-type orbitals

$$\chi_{n,l,m}(r,\theta,\phi) = N_{n,l,m,\zeta}Y_{l,m}(\theta,\phi)r^{n-1}e^{-\zeta r},$$

which are characterized by quantum numbers n, l, and m and exponents (which characterize the "size" of the basis function) ζ. The symbol $N_{n,l,m,\zeta}$ denotes the normalization constant.

2. Cartesian Gaussian-type orbitals

$$\chi_{a,b,c}(r,\theta,\phi) = N'_{a,b,c,\alpha}x^ay^bz^c \exp(-\alpha r^2),$$

characterized by quantum numbers a, b, and c which detail the angular shape and direction of the orbital and exponents α which govern the radial "size" of the basis function. For example, orbitals with a, b, and c values of 1,0,0 or 0,1,0 or 0,0,1 are p_x, p_y, and p_z orbitals; those with a,b,c values of 2,0,0 or 0,2,0 or 0,0,2 and 1,1,0 or 0,1,1 or 1,0,1 span the space of five d orbitals and one s orbital (the sum of the 2,0,0 and 0,2,0 and 0,0,2 orbitals is an s orbital because $x^2 + y^2 + z^2 = r^2$ is independent of θ and ϕ).

For both types of orbitals, the coordinates r, θ, and φ refer to the position of the electron relative to a set of axes attached to the center on which the basis orbital is located. Although Slater-type orbitals (STOs) are preferred on fundamental grounds (e.g., as demonstrated in appendices A and B, the hydrogen atom orbitals are of this form and the exact solution of the many-electron Schrödinger equation can be shown to be of this form (in each of its coordinates) near the nuclear centers), STOs are used primarily for atomic and linear-molecule calculations because the multi-center integrals $<\chi_a\chi_b|g|\chi_c\chi_d>$ (each basis orbital can be on a separate atomic center) which arise in polyatomic-molecule calculations can not efficiently be performed when STOs are employed. In contrast, such integrals can routinely be done when Gaussian-type orbitals (GTOs) are used. This fundamental advantage of GTOs has lead to the dominance of these functions in molecular quantum chemistry.

To understand why integrals over GTOs can be carried out when analogous STO-based integrals are much more difficult, one must only consider the orbital products ($\chi_a\chi_c(r_1)$ and $\chi_b\chi_d(r_2)$) which arise in such integrals. For orbitals of the GTO form, such products involve $\exp(-\alpha_a(\boldsymbol{r} - R_a)^2)\exp(-\alpha_c(\boldsymbol{r} - \boldsymbol{R}_c)^2)$. By completing the square in the exponent, this product can be rewritten as follows:

$$\exp(-\alpha_a(\boldsymbol{r} - R_a)^2)\exp(-\alpha_c(\boldsymbol{r} - R_c)^2) = \exp(-(\alpha_a + \alpha_c)(\boldsymbol{r} - R')^2)\exp(-\alpha'(\boldsymbol{R}_a - \boldsymbol{R}_c)^2),$$

where

$$\boldsymbol{R}' = [\alpha_a\boldsymbol{R}_a + \alpha_c\boldsymbol{R}_c]/(\alpha_a + \alpha_c) \text{ and}$$
$$\alpha' = \alpha_a\alpha_c/(\alpha_a + \alpha_c).$$

Thus, the product of two GTOs on different centers is equal to a single other GTO at a center $\boldsymbol{R}'$ between the two original centers. As a result, even a four-center two-electron integral over GTOs can be written as, at most, a two-center two-electron integral; it turns out that this reduction in centers is enough to allow all such integrals to be carried out. A similar reduction does not arise for STOs because the product of two STOs can not be rewritten as a new STO at a new center.

To overcome the primary weakness of GTO functions, that they have incorrect behavior near the nuclear centers (i.e., their radial derivatives vanish at the nucleus whereas the derivatives of STOs are non-zero), it is common to combine two, three, or more GTOs, with combination coefficients which are fixed and *not* treated as LCAO-MO parameters, into new functions called contracted GTOs or CGTOs. Typically, a series of tight, medium, and loose GTOs (i.e., GTOs with large, medium, and small α values, respectively) are multiplied by so-called contraction coefficients and summed to produce a CGTO which appears to possess the proper "cusp" (i.e., non-zero slope) at the nuclear center (although even such a combination can not because each GTO has zero slope at the nucleus).

B. Basis Set Libraries

Much effort has been devoted to developing sets of STO or GTO basis orbitals for main-group elements and the lighter transition metals. This ongoing effort is aimed at providing standard basis set libraries which:

1. Yield reasonable chemical accuracy in the resultant wavefunctions and energies.
2. Are cost effective in that their use in practical calculations is feasible.
3. Are relatively transferrable in the sense that the basis for a given atom is flexible enough to be used for that atom in a variety of bonding environments (where the atom's hybridization and local polarity may vary).

C. The Fundamental Core and Valence Basis

In constructing an atomic orbital basis to use in a particular calculation, one must choose from among several classes of functions. First, the size and nature of the primary core and valence basis must be specified. Within this category, the following choices are common:

1. A *minimal basis* in which the number of STO or CGTO orbitals is equal to the number of core and valence atomic orbitals in the atom.
2. A *double-zeta* (DZ) basis in which twice as many STOs or CGTOs are used as there are core and valence atomic orbitals. The use of more basis functions is motivated by a desire to provide additional variational flexibility to the LCAO-MO process. This flexibility allows the LCAO-MO process to generate molecular orbitals of variable diffuseness as the local electronegativity of the atom varies. Typically, double-zeta bases include pairs of functions with one member of each pair having a smaller exponent (ζ or α value) than in the minimal basis and the other member having a larger exponent.
3. A *triple-zeta* (TZ) basis in which three times as many STOs or CGTOs are used as the number of core and valence atomic orbitals.
4. Dunning has developed CGTO bases which range from approximately DZ to substantially beyond TZ quality (T. H. Dunning, *J. Chem. Phys. 53*, 2823 (1970); T. H. Dunning and P. J. Hay in *Methods of Electronic Structure Theory*, H. F. Schaefer, III Ed., Plenum Press, New York (1977))). These bases involve contractions of primitive GTO bases which Huzinaga had earlier optimized (S. Huzinaga, *J. Chem. Phys. 42*, 1293 (1965)) for use as uncontracted functions (i.e., for which Huzinaga varied the α values to minimize the energies of several electronic states of the corresponding atom). These Dunning bases are commonly denoted, for example, as follows for first-row atoms: (10*s*,6*p*/5*s*,4*p*), which means that 10 *s*-type primitive GTOs have been contracted to produce five separate *s*-type CGTOs and that six primitive *p*-type GTOs were contracted to generate four separate *p*-type CGTOs. More recent basis sets from the Dunning group are given in T. Dunning, *J. Chem. Phys. 90*, 1007 (1990).
5. Even-tempered basis sets (M. W. Schmidt and K. Ruedenberg, *J. Chem. Phys. 71*, 3961 (1979)) consist of GTOs in which the orbital exponents α_k belonging to series of orbitals consist of geometrical progressions: $\alpha_k = a\beta^k$, where a and β characterize the particular set of GTOs.
6. STO-3G bases were employed some years ago (W. J. Hehre, R. F. Stewart, and J. A. Pople, *J. Chem. Phys. 51*, 2657 (1969)) but are less popular recently. These bases are constructed by least squares fitting GTOs to STOs which have been optimized for various electronic states of the atom. When three GTOs are employed to fit each STO, a STO-3G basis is formed.
7. 4-31G, 5-31G, and 6-31G bases (R. Ditchfield, W. J. Hehre, and J. A. Pople, *J. Chem. Phys. 54*, 724 (1971); W. J. Hehre, R. Ditchfield, and J. A. Pople, *J. Chem. Phys. 56*, 2257 (1972); P. C. Hariharan and J. A. Pople, *Theoret. Chim. Acta.* (Berl.) *28*, 213 (1973); R. Krishnan, J. S. Binkley, R. Seeger, and J. A. Pople, *J. Chem. Phys. 72*, 650 (1980)) employ a single CGTO of contraction length 4, 5, or 6 to describe the core orbital. The valence space is described at the DZ level with the first CGTO constructed from three primitive GTOs and the second CGTO built from a single primitive GTO.

The values of the orbital exponents (ζs or αs) and the GTO-to-CGTO contraction coefficients needed to implement a particular basis of the kind described above have been tabulated in several journal articles and in computer data bases (in particular, in the data base contained in the book *Handbook of Gaussian Basis Sets: A. Compendium for Ab initio Molecular Orbital Calculations*, R. Poirer, R. Kari, and I. G. Csizmadia, Elsevier Science Publishing Co., Inc., New York, (1985)).

Several other sources of basis sets for particular atoms are listed in the Table shown below (here JCP and JACS are abbreviations for the *Journal of Chemical Physics* and the *Journal of The American Chemical Society,* respectively).

Literature Reference	**Basis Type**	**Atoms**
Hehre, W.J.; Stewart , R.F.; Pople, J.A. *JCP 51*, 2657 (1969). Hehre, W.J.; Ditchfield, R.; Stewart, R.F.; Pople, J.A. *JCP 52*, 2769 (1970).	STO-3G	H-Ar
Binkley, J.S.; Pople, J.A.; Hehre, W.J. *JACS 102*, 939 (1980).	3-21G	H-Ne
Gordon, M.S.; Binkley, J.S.; Pople, J.A.; Pietro, W.J.; Hehre, W.J. *JACS 104*, 2797 (1982).	3-21G	Na-Ar
Dobbs, K.D.; Hehre, W.J. J. *Comput. Chem. 7*, 359 (1986).	3-21G	K , Ca , Ga
Dobbs, K.D.; Hehre, W.J. J. *Comput. Chem. 8*, 880 (1987).	3-21G	Sc-Zn
Ditchfield, R.; Hehre, W.J.; Pople, J.A. *JCP 54*, 724 (1971).	6-31G	H
Dill, J.D.; Pople, J.A. *JCP 62*, 2921 (1975).	6-31G	Li , B
Binkley, J.S.; Pople, J.A. *JCP 66*, 879 (1977).	6-31G	Be
Hehre, W.J.; Ditchfield, R.; Pople, J.A. *JCP 56*, 2257 (1972).	6-31G	C-F
Francl, M.M.; Pietro, W.J.; Hehre, W.J.; Binkley, J.S.; Gordon, M.S.; DeFrees, D.J.; Pople, J.A. *JCP 77*, 3654 (1982).	6-31G	Na-Ar
Dunning, T. *JCP 53*, 2823 (1970).	(4*s*/2*s*) (4*s*/3*s*) 9*s*5*p*/3*s*2*p*) 9*s*5*p*/4*s*2*p*) (9*s*5*p*/5*s*3*p*)	H H B-F B-F B-F
Dunning, T. *JCP 55*, 716 (1971).	(5*s*/3*s*) (10*s*/4*s*) (10*s*/5*s*) (10*s*6*p*/5*s*3*p*) (10*s*6*p*/5*s*4*p*)	H Li Be B-Ne B-Ne
Krishnan, R.; Binkley, J.S.; Seeger , R.; Popl, J.A. *JCP 72*, 650 (1980).	6-311G	H-Ne

Literature Reference	Basis Type	Atoms
Dunning, unpublished VDZ.	$(4s/2s)$ $(9s5p/3s2)$ $(12s8p/4s3p)$	H Li, Be, C-Ne Na-Ar
Dunning, unpublished VTZ.	$(5s/3s)$ $(6s/3s)$ $(12s6p/4s3p)$ $(17s10p/5s4p)$	H H Li, Be, C-Ne Mg-Ar
Dunning, unpublished VQZ.	$(7s/4s)$ $(8s/4s)$ $(16s7p/5s4p)$	H H B-Ne
Dunning, T. *JCP 90*, 1007 (1989). (pVDZ , pVTZ, pVQZ correlation-consistent)	$(4s1p/2s1p)$ $(5s2p1d/3s2p1d)$ $(6s3p1d1f/4s3p2d1f)$ $(9s4p1d/3s2p1d)$ $(10s5p2d1f/4s3p2d1f)$ $(12s6p3d2f1g/5s4p3d2f1g)$	H H H B-Ne B-Ne B-Ne
Huzinaga, S.; Klobukowski, M.; Tatewaki, H. Can. *J. Chem. 63*, 1812 (1985).	$(14s/2s)$ $(14s9p/2s1p)$ $(16s9p/3s1p)$ $(16s11p/3s2p)$	Li, Be B-Ne Na-Mg Al-Ar
Huzinaga, S.; Klobukowski, M. *THEOCHEM. 44*, 1 (1988).	$(14s10p/2s1p)$ $(17s10p/3s1p)$ $(17s13p/3s2p)$ $(20s13p/4s2p)$ $(20s13p10d/4s2p1d)$ $(20s14p9d/4s3d1d)$	B-Ne Na-Mg Al-Ar K-Ca Sc-Zn Ga
McLean, A.D.; Chandler, G.S. *JCP 72*, 5639 (1980).	$(12s8p/4s2p)$ $(12s8p/5s3p)$ $(12s8p/6s4p)$ $(12s9p/6s4p)$ $(12s9p/6s5p)$	Na-Ar, P^-,S^-,Cl^- Na-Ar, P^-,S^-,Cl^- Na-Ar, P^-,S^-,Cl^- Na-Ar, P^-,S^-,Cl^- Na-Ar, P^-,S^-,Cl^-
Dunning, T.H.Jr.; Hay, P.J. chapter 1 in *Methods of Electronic Structure Theory,* Schaefer, H.F., III, Ed., Plenum Press, N.Y., 1977.	$(11s7p/6s4p)$	Al-Cl
Hood, D.M.; Pitzer, R.M.; Schaefer, H.F., III *JCP 71*, 705 (1979).	$(14s11p6d/10s8p3d)$	Sc-Zn
Schmidt, M.W.; Ruedenberg, K. *JCP 71*, 3951 (1979). (regular even-tempered)	$([N]s)$, $N = 3\text{-}10$ $([2N]s)$, $N = 3\text{-}10$ $([2N]s)$, $N = 3\text{-}14$ $([2N]s[N]p)$, $N = 3\text{-}11$ $([2N]s[N]p)$, $N = 3\text{-}13$ $([2N]s[N]p)$, $N = 4\text{-}12$ $([2N\text{-}6]s[N]p)$, $N = 7\text{-}15$	H He Li , Be B, N-Ne C Na, Mg Al-Ar

D. Polarization Functions

In addition to the fundamental core and valence basis described above, one usually adds a set of so-called *polarization functions* to the basis. Polarization functions are functions of one higher angular momentum than appears in the atom's valence orbital space (e.g, *d*-functions for C, N, and O and *p*-functions for H). These polarization functions have exponents (ζ or α) which cause their radial sizes to be similar to the sizes of the primary valence orbitals (i.e., the polarization *p* orbitals of the H atom are similar in size to the 1s orbital). Thus, they are *not* orbitals which provide a description of the atom's valence orbital with one higher l-value; such higher-l valence orbitals would be radially more diffuse and would therefore require the use of STOs or GTOs with smaller exponents.

The primary purpose of polarization functions is to give additional angular flexibility to the LCAO-MO process in forming the valence molecular orbitals. This is illustrated below where polarization d_π orbitals are seen to contribute to formation of the bonding π orbital of a carbonyl group by allowing polarization of the Carbon atom's p_π orbital toward the right and of the Oxygen atom's p_π orbital toward the left.

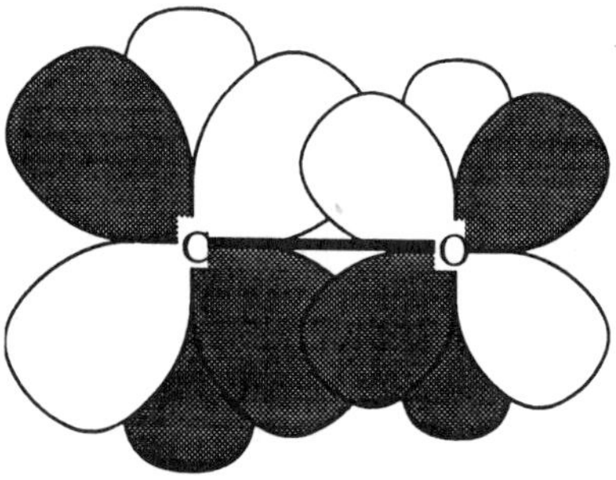

Polarization functions are essential in strained ring compounds because they provide the angular flexibility needed to direct the electron density into regions between bonded atoms.

Functions with higher l-values and with "sizes" more in line with those of the lower l-orbitals are also used to introduce additional angular correlation into the calculation by permitting polarized orbital pairs (see chapter 10) involving higher angular correlations to be formed. Optimal polarization functions for first and second row atoms have been tabulated (B. Roos and P. Siegbahn, *Theoret. Chim. Acta* (Berl.) *17*, 199 (1970); M. J. Frisch, J. A. Pople, and J. S. Binkley, *J. Chem. Phys. 80*, 3265 (1984)).

E. Diffuse Functions

When dealing with anions or Rydberg states, one must augment the above basis sets by adding so-called diffuse basis orbitals. The conventional valence and polarization functions described above do not provide enough radial flexibility to adequately describe either of these cases. Energy-optimized diffuse functions appropriate to anions of most lighter main group elements have been tabulated in the literature (an excellent source of Gaussian basis set information is provided in *Handbook of Gaussian Basis Sets*, R. Poirier, R. Kari, and I. G. Csizmadia, Elsevier, Amsterdam (1985)) and in data bases. Rydberg diffuse basis sets are usually created by adding to conventional valence-plus-polarization bases sequences of primitive GTOs whose exponents are smaller than that (call it α_{diff}) of the most diffuse GTO which contributes strongly to the valence CGTOs. As a "rule of thumb," one can generate a series of such diffuse orbitals which are liniarly independent yet span considerably different regions of radial space by introducing primitive GTOs whose exponents are $\alpha_{diff}/3$, $\alpha_{diff}/9$, $\alpha_{diff}/27$, etc.

Once one has specified an atomic orbital basis for each atom in the molecule, the LCAO-MO procedure can be used to determine the $C_{\nu,i}$ coefficients that describe the occupied and virtual orbitals in terms of the chosen basis set. It is important to keep in mind that the basis orbitals are *not* themselves the true orbitals of the isolated atoms; even the proper atomic orbitals are combinations (with atomic values for the $C_{\nu,i}$ coefficients) of the basis functions. For example, in a minimal-basis-level treatment of the Carbon atom, the 2*s* atomic orbital is formed by combining, with opposite sign to achieve the radial node, the two CGTOs (or STOs); the more diffuse *s*-type basis function will have a larger $C_{i,\nu}$ coefficient in the 2*s* atomic orbital. The 1*s* atomic orbital is formed by combining the same two CGTOs but with the same sign and with the less diffuse basis function having a larger $C_{\nu,i}$ coefficient. The LCAO-MO-SCF process itself determines the magnitudes and signs of the $C_{\nu,i}$.

VI. THE ROOTHAAN MATRIX SCF PROCESS

The matrix SCF equations introduced earlier

$$\sum_{\nu} F_{\mu,\nu}C_{\nu,i} = \varepsilon_i \sum_{\nu} S_{\mu,\nu}C_{\nu,i}$$

must be solved both for the occupied and virtual orbitals' energies ε_i and $C_{\nu,i}$ values. Only the occupied orbitals' $C_{\nu,i}$ coefficients enter into the Fock operator

$$F_{\mu,\nu} = \langle\chi_\mu|h|\chi_\nu\rangle + \sum_{\delta,\kappa} [\gamma_{\delta,\kappa}\langle\chi_\mu\chi_\delta|g|\chi_\nu\chi_\kappa\rangle - \gamma_{\delta,\kappa}^{ex}\langle\chi_\mu\chi_\delta|g|\chi_k\chi_\nu\rangle],$$

but both the occupied and virtual orbitals are solutions of the SCF equations. Once atomic basis sets have been chosen for each atom, the *one- and two-electron integrals* appearing in $F_{\mu,\nu}$ must be evaluated. Doing so is a time consuming process, but there are presently several highly efficient computer codes which allow such integrals to be computed for *s*, *p*, *d*, *f*, and even *g*, *h*, and *i* basis functions. After executing one of these "*integral packages*" for a basis with a total of *N* functions, one has available (usually on the computer's hard disk) of the order of $N^2/2$ one-electron and $N^4/8$ two-electron integrals over these atomic basis orbitals (the factors of 1/2 and 1/8 arise from permutational symmetries of the integrals). When treating extremely large atomic orbital basis sets (e.g., 200 or more basis functions), modern computer programs calculate the requisite integrals but never store them on the disk. Instead, their contributions to $F_{\mu,\nu}$ are accumulated "on the fly" after which the integrals are discarded.

To begin the SCF process, one must input to the computer routine which computes $F_{\mu,\nu}$ *initial "guesses"* for the $C_{\nu,i}$ values corresponding to the occupied orbitals. These initial guesses are typically made in one of the following ways:

1. If one has available $C_{\nu,i}$ values for the system from an SCF calculation performed earlier at a nearby molecular geometry, one can use these $C_{\nu,i}$ values to begin the SCF process.
2. If one has $C_{\nu,i}$ values appropriate to fragments of the system (e.g., for C and O atoms if the CO molecule is under study or for CH_2 and O if H_2CO is being studied), one can use these.

3. If one has no other information available, one can carry out one iteration of the SCF process in which the two-electron contributions to $F_{\mu,\nu}$ are ignored (i.e., take $F_{\mu,\nu} = \langle\chi_\mu|h|\chi_\nu\rangle$) and use the resultant solutions to

$$\sum_\nu F_{\mu,\nu} C_{\nu,i} = \varepsilon_i \sum_\nu S_{\mu,\nu} C_{\nu,i}$$

as initial guesses for the $C_{\nu,i}$. Using only the one-electron part of the Hamiltonian to determine initial values for the LCAO-MO coefficients may seem like a rather severe step; it is, and the resultant $C_{\nu,i}$ values are usually far from the converged values which the SCF process eventually produces. However, the initial $C_{\nu,i}$ obtained in this manner have proper symmetries and nodal patterns because the one-electron part of the Hamiltonian has the same symmetry as the full Hamiltonian.

Once initial guesses are made for the $C_{\nu,i}$ of the occupied orbitals, the full $F_{\mu,\nu}$ matrix is formed and new ε_i and $C_{\nu,i}$ values are obtained by solving

$$\sum_\nu F_{\mu,\nu} C_{\nu,i} = \varepsilon_i \sum_\nu S_{\mu,\nu} C_{\nu,i}.$$

These new orbitals are then used to form a new $F_{\mu,\nu}$ matrix from which new ε_i and $C_{\nu,i}$ are obtained. This iterative process is carried on until the ε_i and $C_{\nu,i}$ do not vary (within specified tolerances) from iteration to iteration, at which time one says that the SCF process has converged and reached self-consistency.

As presented, the Roothaan SCF process is carried out in a fully *ab initio* manner in that all one- and two-electron integrals are computed in terms of the specified basis set; no experimental data or other input is employed. As described in appendix F, it is possible to introduce approximations to the coulomb and exchange integrals entering into the Fock matrix elements that permit many of the requisite $F_{\mu,\nu}$ elements to be evaluated in terms of experimental data or in terms of a small set of "fundamental" orbital-level coulomb interaction integrals that can be computed in an *ab initio* manner. This approach forms the basis of so-called "semi-empirical" methods. Appendix F provides the reader with a brief introduction to such approaches to the electronic structure problem and deals in some detail with the well known Hückel and CNDO-level approximations.

VII. OBSERVATIONS ON ORBITALS AND ORBITAL ENERGIES

A. The Meaning of Orbital Energies

The physical content of the Hartree-Fock orbital energies can be seen by observing that $F\phi_i = \varepsilon_i\phi_i$ implies that ε_i can be written as:

$$\varepsilon_i = \langle\phi_i|F|\phi_i\rangle = \langle\phi_i|h|\phi_i\rangle + \sum_{j(occupied)} \langle\phi_i|J_j - K_j|\phi_i\rangle$$

$$= \langle\phi_i|h|\phi_i\rangle + \sum_{j(occupied)} [J_{i,j} - K_{i,j}].$$

In this form, it is clear that ε_i is equal to the average value of the kinetic energy plus coulombic attraction to the nuclei for an electron in ϕ_i plus the sum over all of the spin-orbitals occupied in Ψ of coulomb minus exchange interactions between ϕ_i and these occupied spin-orbitals. If ϕ_i itself is an occupied spin-orbital, the term $[J_{i,i} - K_{i,i}]$ disappears and the latter sum represents the coulomb minus exchange interaction of ϕ_i with all of the $N - 1$ *other* occupied spin-orbitals. If ϕ_i is a virtual spin-orbital, this cancellation does not occur, and one obtains the coulomb minus exchange interaction of ϕ_i with all N of the occupied spin-orbitals.

In this sense, the orbital energies for occupied orbitals pertain to interactions which are appropriate to a total of N electrons, while the orbital energies of virtual orbitals pertain to a system with $N + 1$ electrons. It is this fact that makes SCF virtual orbitals not optimal (in fact, not usually very good) for use in subsequent correlation calculations where, for instance, they are used, in combination with the occupied orbitals, to form polarized orbital pairs as discussed in chapter 12. To correlate a pair of electrons that occupy a valence orbital requires double excitations into a virtual orbital that is not too dislike in size. Although the virtual SCF orbitals themselves suffer these drawbacks, the space they span can indeed be used for treating electron correlation. To do so, it is useful to recombine (in a unitary manner to preserve orthonormality) the virtual orbitals to "focus" the correlating power into as few orbitals as possible so that the multiconfigurational wavefunction can be formed with as few CSFs as possible. Techniques for effecting such reoptimization or improvement of the virtual orbitals are treated later in this text.

B. Koopmans' Theorem

Further insight into the meaning of the energies of occupied and virtual orbitals can be gained by considering the following model of the vertical (i.e., at fixed molecular geometry) detachment or attachment of an electron to the original N-electron molecule:

1. In this model, *both* the parent molecule and the species generated by adding or removing an electron are treated at the single-determinant level.
2. In this model, the Hartree-Fock orbitals of the parent molecule are used to describe both the parent and the species generated by electron addition or removal. It is said that such a model neglects "*orbital relaxation*" which would accompany the electron addition or removal (i.e., the reoptimization of the spin-orbitals to allow them to become appropriate to the daughter species).

Within this simplified model, the energy difference between the daughter and the parent species can be written as follows (ϕ_k represents the particular spin-orbital that is added or removed):

1. For electron detachment:

$$E^{N-1} - E^{N} = \langle|\phi_1\phi_2 \ldots \phi_{k-1} \ldots \phi_N|H|\phi_1\phi_2 \ldots \phi_{k-1} \ldots \phi_N|\rangle -$$

$$\langle|\phi_1\phi_2 \ldots \phi_{k-1}\phi_k \ldots \phi_N|H||\phi_1\phi_2 \ldots \phi_{k-1}\phi_k \ldots \phi_N|\rangle$$

$$= - \langle\phi_k|h|\phi_k\rangle - \sum_{j=(1,\, k-1,\, k+1,\, N)} [J_{k,j} - K_{k,j}] = -\varepsilon_k;$$

2. For electron attachment:

$$E^{N} - E^{N+1} = \langle|\phi_1\phi_2 \ldots \phi_N|H|\phi_1\phi_2 \ldots \phi_N|\rangle - \langle|\phi_1\phi_2 \ldots \phi_N\phi_k|H||\phi_1\phi_2 \ldots \phi_N\phi_k|\rangle$$

$$= - \langle\phi_k|h|\phi_k\rangle - \sum_{j=(1,N)} [J_{k,j} - K_{k,j}] = -\varepsilon_k.$$

So, within the limitations of the single-determinant, frozen-orbital model set forth, the ionization potentials (IPs) and electron affinities (EAs) are given as the negative of the occupied and virtual spin-orbital energies, respectively. This statement is referred to as Koopmans' theorem (T. Koopmans, *Physica 1*, 104 (1933)); it is used extensively in quantum chemical calculations as a means for estimating IPs and EAs and often yields results that are at least qualitatively correct (i.e., ±0.5eV).

C. Orbital Energies and the Total Energy

For the N-electron species whose Hartree-Fock orbitals and orbital energies have been determined, the total SCF electronic energy can be written, by using the Slater-Condon rules, as:

$$E = \sum_{i(occupied)} \langle\phi_i|h|\phi_i\rangle + \sum_{i>j(occupied)} [J_{i,j} - K_{i,j}].$$

For this same system, the sum of the orbital energies of the occupied spin-orbitals is given by:

$$\sum_{i(occupied)} \varepsilon_i = \sum_{i(occupied)} \langle\phi_i|h|\phi_i\rangle + \sum_{i,j(occupied)} [J_{i,j} - K_{i,j}].$$

These two seemingly very similar expressions differ in a very important way; the sum of occupied orbital energies, when compared to the total energy, double counts the coulomb minus exchange interaction energies. Thus, within the Hartree-Fock approximation, the sum of the occupied orbital energies is *not* equal to the total energy. The total SCF energy can be computed in terms of the sum of occupied orbital energies by taking one-half of

$$\sum_{i(occupied)} \varepsilon_i$$

and then adding to this one-half of

$$\sum_{i(occupied)} \langle\phi_i|h|\phi_i\rangle:$$

$$E = 1/2[\sum_{i(occupied)} \langle\phi_i|h|\phi_i\rangle + \sum_{i(occupied)} \varepsilon_i].$$

The fact that the sum of orbital energies is not the total SCF energy also means that as one attempts to develop a qualitative picture of the energies of CSFs along a reaction path, as when orbital and configuration correlation diagrams are constructed, one must be careful not to equate the sum of orbital energies with the total configurational energy; the former is higher than the latter by an amount equal to the sum of the coulomb minus exchange interactions.

D. The Brillouin Theorem

The condition that the SCF energy $\langle|\phi_1 \ldots \phi_N|H|\phi_1 \ldots \phi_N|\rangle$ be stationary with respect to variations $\delta\phi_i$ in the occupied spin-orbitals (that preserve orthonormality) can be written

$$\langle|\phi_1 \ldots \delta\phi_i \ldots \phi_N|H|\phi_1 \ldots \phi_i \ldots \phi_N|\rangle = 0.$$

The infinitesimal variation of ϕ_i can be expressed in terms of its (small) components along the other occupied ϕ_j and along the virtual ϕ_m as follows:

$$\delta\phi_i = \sum_{j=occ} U_{ij}\phi_j + \sum_m U_{im}\phi_m.$$

When substituted into $|\phi_1 \ldots \delta\phi_i \ldots \phi_N|$, the terms

$$\sum_{j'=occ} |\phi_1 \ldots \phi_j \ldots \phi_N| U_{ij}$$

vanish because ϕ_j already appears in the original Slater determinant $|\phi_1 \ldots \phi_N|$, so $|\phi_1 \ldots \phi_j \ldots \phi_N|$ contains ϕ_j twice. Only the sum over virtual orbitals remains, and the stationary property written above becomes

$$\sum_m U_{im} < |\phi_1 \ldots \phi_m \ldots \phi_N| H |\phi_1 \ldots \phi_i \ldots \phi_N| > = 0.$$

The Slater-Condon rules allow one to express the Hamiltonian matrix elements appearing here as

$$< |\phi_1 \ldots \phi_m \ldots \phi_N| H |\phi_1 \ldots \phi_i \ldots \phi_N| > = <\phi_m|h|\phi_i> + \sum_{j=occ,\, \neq i} <\phi_m|[J_j - K_j]|\phi_i>,$$

which (because the term with $j = i$ can be included since it vanishes) is equal to the following element of the Fock operator: $<\phi_m|F|\phi_i> = \varepsilon_i\delta_{im} = 0$. This result proves that Hamiltonian matrix elements between the SCF determinant and those that are singly excited relative to the SCF determinant vanish because they reduce to Fock-operator integrals connecting the pair of orbitals involved in the "excitation." This stability property of the SCF energy is known as the Brillouin theorem (i.e., that $|\phi_1\phi_i\phi_N|$ and $|\phi_1 \ldots \phi_m \ldots \phi_N|$ have zero Hamiltonian matrix elements *if* the ϕs are SCF orbitals). It is exploited in quantum chemical calculations in two manners:

1. When multiconfiguration wavefunctions are formed from SCF spin-orbitals, it allows one to neglect Hamiltonian matrix elements between the SCF configuration and those that are "singly excited" in constructing the secular matrix.
2. A so-called generalized Brillouin theorem (GBT) arises when one deals with energy optimization for a multiconfigurational variational trial wavefunction for which the orbitals and C_I mixing coefficients are simultaneously optimized. This GBT causes certain Hamiltonian matrix elements to vanish, which, in turn, simplifies the treatment of electron correlation for such wavefunctions. This matter is treated in more detail later in this text.

CHAPTER

19

Corrections to the mean-field model are needed to describe the instantaneous Coulombic interactions among the electrons. This is achieved by including more than one Slater determinant in the wavefunction.

Much of the development of the previous chapter pertains to the use of a single Slater determinant trial wavefunction. As presented, it relates to what has been called the unrestricted Hartree-Fock (UHF) theory in which each spin-orbital ϕ_i has its own orbital energy ε_i and LCAO-MO coefficients $C_{\nu,i}$; there may be different $C_{\nu,i}$ for α spin-orbitals than for β spin-orbitals. Such a wavefunction suffers from the spin contamination difficulty detailed earlier.

To allow for a properly spin- and space-symmetry adapted trial wavefunction and to permit Ψ to contain more than a single CSF, methods which are more flexible than the single-determinant HF procedure are needed. In particular, it may be necessary to use a combination of determinants to describe such a proper symmetry function. Moreover, as emphasized earlier, whenever two or more CSFs have similar energies (i.e., Hamiltonian expectation values) and can couple strongly through the Hamiltonian (e.g., at avoided crossings in configuration correlation diagrams), the wavefunction must be described in a multiconfigurational manner to permit the wavefunction to evolve smoothly from reactants to products. Also, whenever dynamical electron correlation effects are to be treated, a multiconfigurational Ψ must be used; in this case, CSFs that are *doubly excited* relative to one or more of the essential CSFs (i.e., the dominant CSFs that are included in the so-called *reference wavefunction*) are included to permit polarized-orbital-pair formation.

Multiconfigurational functions are needed not only to account for electron correlation but also to permit orbital readjustments to occur. For example, if a set of SCF orbitals is employed in forming a multi-CSF wavefunction, the variational condition that the energy is stationary with respect to variations in the LCAO-MO coefficients is no longer obeyed (i.e., the SCF energy functional is stationary when SCF orbitals are employed, but the MC-energy functional is generally not stationary if SCF orbitals are employed). For such reasons, it is important to include CSFs that are *singly excited* relative to the dominant CSFs in the reference wavefunction.

That singly excited CSFs allow for orbital relaxation can be seen as follows. Consider a wavefunction consisting of one CSF $|\phi_1 \ldots \phi_i \ldots \phi_N|$ to which singly excited CSFs of the form $|\phi_1 \ldots \phi_m \ldots \phi_N|$ have been added with coefficients $C_{i,m}$:

$$\Psi = \sum_m C_{i,m}|\phi_1 \ldots \phi_m \ldots \phi_N| + |\phi_1 \ldots \phi_i \ldots \phi_N| .$$

All of these determinants have all of their columns equal except the ith column; therefore, they can be combined into a single new determinant:

$$\Psi = |\phi_1 \ldots \phi_i' \ldots \phi_N|,$$

where the relaxed orbital ϕ_i' is given by

$$\phi_i' = \phi_i + \sum_m C_{i,m}\phi_m.$$

The sum of CSFs that are singly excited in the ith spin-orbital with respect to $|\phi_1 \ldots \phi_i \ldots \phi_N|$ is therefore seen to allow the spin-orbital ϕ_i to relax into the new spin-orbital ϕ_i'. It is in this sense that singly excited CSFs allow for orbital reoptimization.

In summary, doubly excited CSFs are often employed to permit polarized orbital pair formation and hence to allow for electron correlations. Singly excited CSFs are included to permit orbital relaxation (i.e., orbital reoptimization) to occur.

I. DIFFERENT METHODS

There are numerous procedures currently in use for determining the "best" wavefunction of the form:

$$\Psi = \sum_I C_I\Phi_I,$$

where Φ_I is a spin- and space-symmetry adapted CSF consisting of determinants of the form $|\phi_{I1}\phi_{I2}\phi_{I3} \ldots \phi_{IN}|$. Excellent overviews of many of these methods are included in *Modern Theoretical Chemistry,* Vols. 3 and 4, H. F. Schaefer, III, ed., Plenum Press, New York (1977) and in *Advances in Chemical Physics*, Vols. LXVII and LXIX, K. P. Lawley, ed., Wiley-Interscience, New York (1987). Within the present chapter, these two key references will be denoted MTC, Vols. 3 and 4, and ACP, Vols. 67 and 69, respectively.

In all such trial wavefunctions, there are two fundamentally different kinds of parameters that need to be determined—the CI coefficients C_I and the LCAO-MO coefficients describing the ϕ_{Ik}. The most commonly employed methods used to determine these parameters include:

1. The **multiconfigurational self-consistent field** (MCSCF) method in which the expectation value $\langle\Psi|H|\Psi\rangle/\langle\Psi|\Psi\rangle$ is treated variationally and simultaneously made stationary with respect to variations in the C_I and $C_{\nu,i}$ coefficients subject to the constraints that the spin-orbitals and the full N-electron wavefunction remain normalized:

$$\langle\phi_i|\phi_j\rangle = \delta_{i,j} = \sum_{\nu,\mu} C_{\nu,i}S_{\nu,\mu}C_{\mu,i} \sum_{\nu,\mu} C_{\mu,i}, \text{ and}$$

$$\sum_I C_I^2 = 1.$$

The articles by H.-J. Werner and by R. Shepard in ACP Vol. 69 provide up to date reviews of the status of this approach. The article by A. C. Wahl and G. Das in MTC Vol. 3 covers the "earlier" history on this topic. F. W. Bobrowicz and W. A. Goddard, III provide, in MTC Vol. 3, an overview of the GVB approach, which, as discussed in chapter 12, can be viewed as a specific kind of MCSCF calculation.

2. The **configuration interaction** (CI) method in which the LCAO-MO coefficients are determined first (and independently) via either a single-configuration SCF calculation or an MCSCF calculation using a small number of CSFs. The CI coefficients are subsequently determined by making the expectation value $<\Psi|H|\Psi>/<\Psi|\Psi>$ stationary with respect to variations in the C_I only. In this process, the optimizations of the orbitals and of the CSF amplitudes are done in separate steps. The articles by I. Shavitt and by B. O. Ross and P. E. M. Siegbahn in MTC, Vol. 3 give excellent early overviews of the CI method.

3. The **Møller-Plesset perturbation method** (MPPT) uses the single-configuration SCF process (usually the UHF implementation) to first determine a set of LCAO-MO coefficients and, hence, a set of orbitals that obey $F\phi_i = \varepsilon_i\phi_i$. Then, using an unperturbed Hamiltonian equal to the sum of these Fock operators for each of the N electrons

$$H^0 = \sum_{i=1,N} F(i),$$

perturbation theory (see appendix D for an introduction to time-independent perturbation theory) is used to determine the C_I amplitudes for the CSFs. The MPPT procedure is also referred to as the many-body perturbation theory (MBPT) method. The two names arose because two different schools of physics and chemistry developed them for somewhat different applications. Later, workers realized that they were identical in their working equations when the UHF H^0 is employed as the unperturbed Hamiltonian. In this text, we will therefore refer to this approach as MPPT/MBPT.

The amplitude for the so-called *reference* CSF used in the SCF process is taken as unity and the other CSFs' amplitudes are determined, relative to this one, by Rayleigh-Schrödinger perturbation theory using the full N-electron Hamiltonian minus the sum of Fock operators $H - H^0$ as the perturbation. The Slater-Condon rules are used for evaluating matrix elements of $(H - H^0)$ among these CSFs. The essential features of the MPPT/MBPT approach are described in the following articles: J. A. Pople, R. Krishnan, H. B. Schlegel, and J. S. Binkley, *Int. J. Quantum Chem. 14*, 545 (1978); R. J. Bartlett and D. M. Silver, *J. Chem. Phys. 62*, 3258 (1975); R. Krishnan and J. A. Pople, *Int. J. Quantum* Chem. *14*, 91 (1978).

4. The **Coupled-Cluster method** expresses the CI part of the wavefunction in a somewhat different manner (the early work in chemistry on this method is described in J. Cizek, *J. Chem. Phys. 45*, 4256 (1966); J. Paldus, J. Cizek, and I. Shavitt, *Phys. Rev. A5*, 50 (1972); R. J. Bartlett and G. D. Purvis, *Int. J. Quantum Chem. 14*, 561 (1978); G. D. Purvis and R. J. Bartlett, *J. Chem. Phys. 76*, 1910 (1982)):

$$\Psi = \exp(T)\Phi,$$

where Φ is a single CSF (usually the UHF single determinant) which has been used to independently determine a set of spin-orbitals and LCAO-MO coefficients via the SCF process. The operator T generates, when acting on Φ, single, double, etc. "excitations" (i.e., CSFs in which one, two, etc. of the occupied spin-orbitals in Φ have been replaced by virtual spin-orbitals). T is commonly expressed in terms of operators that effect such spin-orbital removals and additions as follows:

$$T = \sum_{i,m} t_i^m m^+ i + \sum_{i,j,m,n} t_{i,j}^{m,n} m^+ n^+ j i + \ldots,$$

where the operator m^+ is used to denote *creation* of an electron in virtual spin-orbital ϕ_m and the operator j is used to denote *removal* of an electron from occupied spin-orbital ϕ_j.

The t_i^m, $t_{i,j}^{m,n}$, etc. amplitudes, which play the role of the CI coefficients in CC theory, are determined through the set of equations generated by projecting the Schrödinger equation in the form

$$\exp(-T) H \exp(T)\Phi = E\Phi$$

against CSFs which are single, double, etc. excitations relative to Φ. For example, for double excitations $\Phi_{i,j}^{m,n}$ the equations read:

$$<\Phi_{i,j}^{m,n}| \exp(-T) H \exp(T)|\Phi> = E<\Phi_{i,j}^{m,n}|\Phi> = 0;$$

zero is obtained on the right hand side because the excited CSFs $|\Phi_{i,j}^{m,n}>$ are orthogonal to the reference function $|\Phi>$. The elements on the left hand side of the CC equations can be expressed, as described below, in terms of one- and two-electron integrals over the spin-orbitals used in forming the reference and excited CSFs.

A. Integral Transformations

All of the above methods require the evaluation of one- and two-electron integrals over the N atomic *orbital basis:* $<\chi_a|f|\chi_b>$ and $<\chi_a\chi_b|g|\chi_c\chi_d>$. Eventually, all of these methods provide their working equations and energy expressions in terms of one- and two-electron integrals over the N final *molecular orbitals*: $<\phi_i|f|\phi_j>$ and $<\phi_i\phi_j|g|\phi_k\phi_l>$. The mo-based integrals can only be evaluated by *transforming* the AO-based integrals as follows:

$$<\phi_i\phi_j|g|\phi_k\phi_l> = \sum_{a,b,c,d} C_{a,i}C_{b,j}C_{c,k}C_{d,l} <\chi_a\chi_b|g|\chi_c\chi_d>$$

and

$$<\phi_i|f|\phi_j> = \sum_{a,b} C_{a,i}C_{b,j}<\chi_a|f|\chi_b>.$$

It would seem that the process of evaluating all N^4 of the $<\phi_i\phi_j|g|\phi_k\phi_l>$, each of which requires N^4 additions and multiplications, would require computer time proportional to N^8. However, it is possible to perform the full transformation of the two-electron integral list in a time that scales as N^5. This is done by first performing a transformation of the $<\chi_a\chi_b|g|\chi_c\chi_d>$ to an intermediate array labeled $<\chi_a\chi_b|g|\chi_c\phi_l>$ as follows:

$$<\chi_a\chi_b|g|\chi_c\phi_l> = \sum_{d} C_{d,l}<\chi_a\chi_b|g|\chi_c\chi_d>.$$

This partial transformation requires N^5 multiplications and additions. The list $<\chi_a\chi_b|g|\chi_c\phi_l>$ is then transformed to a second-level transformed array $<\chi_a\chi_b|g|\phi_k\phi_l>$:

$$<\chi_a\chi_b|g|\phi_k\phi_l> = \sum_{c} C_{c,k}<\chi_a\chi_b|g|\chi_c\phi_l>,$$

which requires another N^5 operations. This sequential, one-index-at-a-time transformation is repeated four times until the final $<\phi_i\phi_j|g|\phi_k\phi_l>$ array is in hand. The entire transformation done this way requires $\mathbf{4}N^5$ multiplications and additions.

Once the requisite one- and two-electron integrals are available in the molecular orbital basis, the multiconfigurational wavefunction and energy calculation can begin. These transformations consume a large fraction of the computer time used in most such calculations, and represent a severe bottleneck to progress in applying *ab initio* electronic structure methods to larger systems.

B. Configuration List Choices

Once the requisite one- and two-electron integrals are available in the molecular orbital basis, the multiconfigurational wavefunction and energy calculation can begin. Each of these methods has its own approach to describing the configurations $\{\Phi_J\}$ included in the calculation and how the $\{C_J\}$ amplitudes and the total energy E is to be determined.

The *number of configurations* (N_C) varies greatly among the methods and is an important factor to keep in mind when planning to carry out an *ab initio* calculation. Under certain circumstances (e.g., when studying Woodward-Hoffmann forbidden reactions where an avoided crossing of two configurations produces an activation barrier), it may be essential to use more than one electronic configuration. Sometimes, one configuration (e.g., the SCF model) is adequate to capture the qualitative essence of the electronic structure. In all cases, many configurations will be needed if highly accurate treatment of electron-electron correlations are desired.

The value of N_C determines how much computer time and memory is needed to solve the N_C-dimensional

$$\sum_J H_{I,J} C_J = EC_I$$

secular problem in the CI and MCSCF methods. Solution of these matrix eigenvalue equations requires computer time that scales as N_C^2 (if few eigenvalues are computed) to N_C^3 (if most eigenvalues are obtained).

So-called *complete-active-space* (CAS) methods form *all* CSFs that can be created by distributing N valence electrons among P valence orbitals. For example, the eight non-core electrons of H_2O might be distributed, in a manner that gives $M_S = 0$, among six valence orbitals (e.g., two lone-pair orbitals, two OH σ bonding orbitals, and two OH σ^* antibonding orbitals). The number of configurations thereby created is 225. If the same eight electrons were distributed among 10 valence orbitals 44,100 configurations results; for 20 and 30 valence orbitals, 23,474,025 and 751,034,025 configurations arise, respectively. Clearly, practical considerations dictate that CAS-based approaches be limited to situations in which a few electrons are to be correlated using a few valence orbitals. The primary advantage of CAS configurations is discussed below in section II.C.

II. STRENGTHS AND WEAKNESSES OF VARIOUS METHODS

A. Variational Methods such as MCSCF, SCF, and CI Produce Energies That are Upper Bounds, but These Energies Are not Size-Extensive

Methods that are based on making the energy functional $<\Psi|H|\Psi>/<\Psi|\Psi>$ stationary (i.e., variational methods) yield *upper bounds* to the lowest energy of the symmetry which characterizes the

CSFs which comprise Ψ. These methods also can provide approximate excited-state energies and wavefunctions (e.g., in the form of other solutions of the secular equation

$$\sum_J H_{I,J} C_J = E C_I$$

that arises in the CI and MCSCF methods). Excited-state energies obtained in this manner can be shown to "bracket" the true energies of the given symmetry in that between any two approximate energies obtained in the variational calculation, there exists at least one true eigenvalue. This characteristic is commonly referred to as the "bracketing theorem" (E. A. Hylleraas and B. Undheim, *Z. Phys. 65*, 759 (1930); J. K. L. MacDonald, *Phys. Rev. 43*, 830 (1933)). These are strong attributes of the variational methods, as is the long and rich history of developments of analytical and computational tools for efficiently implementing such methods (see the discussions of the CI and MCSCF methods in MTC and ACP).

However, all variational techniques suffer from at least one serious drawback; they are not **size-extensive** (J. A. Pople, page 51 in *Energy, Structure, and Reactivity*, D. W. Smith and W. B. McRae, eds., Wiley, New York (1973)). This means that the energy computed using these tools can not be trusted to scale with the size of the system. For example, a calculation performed on two CH_3 species at large separation may not yield an energy equal to twice the energy obtained by performing the same kind of calculation on a single CH_3 species. Lack of size-extensivity precludes these methods from use in extended systems (e.g., solids) where errors due to improper scaling of the energy with the number of molecules produce nonsensical results.

By carefully adjusting the kind of variational wavefunction used, it is possible to circumvent size-extensivity problems for selected species. For example, a CI calculation on Be_2 using *all* ${}^1\Sigma_g$ CSFs that can be formed by placing the four valence electrons into the orbitals $2\sigma_g$, $2\sigma_u$, $3\sigma_g$, $3\sigma_u$, $1\pi_u$, and $1\pi_g$ can yield an energy equal to twice that of the Be atom described by CSFs in which the two valence electrons of the Be atom are placed into the $2s$ and $2p$ orbitals in all ways consistent with a ${}^1S$ symmetry. Such special choices of configurations give rise to what are called *complete-active-space* (CAS) MCSCF or CI calculations (see the article by B. O. Roos in ACP for an overview of this approach).

Let us consider an example to understand why the CAS choice of configurations works. The ${}^1S$ ground state of the Be atom is known to form a wavefunction that is a strong mixture of CSFs that arise from the $2s^2$ and $2p^2$ configurations:

$$\Psi_{Be} = C_1|1s^2 2s^2| + C_2|1s^2 2p^2|,$$

where the latter CSF is a short-hand representation for the proper spin- and space-symmetry adapted CSF

$$|1s^2 2p^2| = 1/\sqrt{3}\,[|1s\alpha 1s\beta 2p_0\alpha 2p_0\beta| - |1s\alpha 1s\beta 2p_1\alpha 2p_{-1}\beta| - |1s\alpha 1s\beta 2p_{-1}\alpha 2p_1\beta|].$$

The reason the CAS process works is that the Be_2 CAS wavefunction has the flexibility to dissociate into the product of two CAS Be wavefunctions:

$$\Psi = \Psi_{Bea}\Psi_{Beb}$$
$$= \{C_1|1s^2 2s^2| + C_2|1s^2 2p^2|\}_a \{C_1|1s^2 2s^2| + C_2|1s^2 2p^2|\}_b,$$

where the subscripts a and b label the two Be atoms, because the four electron CAS function distributes the four electrons in all ways among the $2s_a$, $2s_b$, $2p_a$, and $2p_b$ orbitals. In contrast, if the Be_2 calculation had been carried out using only the following CSFs: $|1\sigma_g^2 1\sigma_u^2 2\sigma_g^2 2\sigma_u^2|$ and all single and double excitations relative to this (dominant) CSF, which is a very common type of CI procedure to follow, the Be_2 wavefunction would not have contained the particular CSFs $|1s^2 2p^2|_a|1s^2 2p^2|_b$ because these CSFs are four-fold excited relative to the $|1\sigma_g^2 1\sigma_u^2 2\sigma_g^2 2\sigma_u^2|$ "reference" CSF.

In general, one finds that if the "monomer" uses CSFs that are K-fold excited relative to its dominant CSF to achieve an accurate description of its electron correlation, a size-extensive variational calculation on the "dimer" will require the inclusion of CSFs that are 2K-fold excited relative to the dimer's dominant CSF. To perform a size-extensive variational calculation on a species containing M monomers therefore requires the inclusion of CSFs that are $M \times K$-fold excited relative to the M-mer's dominant CSF.

B. Non-Variational Methods Such as MPPT/MBPT and CC Do not Produce Upper Bounds, but Yield Size-Extensive Energies

In contrast to variational methods, perturbation theory and coupled-cluster methods achieve their energies from a "*transition formula*" $<\Phi|H|\Psi>$ rather than from an expectation value $<\Psi|H|\Psi>$. It can be shown (H. P. Kelly, *Phys. Rev. 131*, 684 (1963)) that this difference allows non-variational techniques to yield size-extensive energies. This can be seen in the MPPT/MBPT case by considering the energy of two non-interacting Be atoms. The reference CSF is $\Phi = |1s_a^2 2s_a^2 1s_b^2 2s_b^2|$; the Slater-Condon rules limit the CSFs in Ψ which can contribute to

$$E = <\Phi|H|\Psi> = <\Phi|H| \sum_J C_J \Phi_J>,$$

to be Φ itself and those CSFs that are singly or doubly excited relative to Φ. These "excitations" can involve atom a, atom b, or both atoms. However, any CSFs that involve excitations on both atoms (e.g., $|1s_a^2 2s_a 2p_a 1s_b^2 2s_b 2p_b|$) give rise, via the SC rules, to one- and two-electron integrals over orbitals on both atoms; these integrals (e.g., $<2s_a 2p_a|g|2s_b 2p_b>$) vanish if the atoms are far apart, as a result of which the contributions due to such CSFs vanish in our consideration of size-extensivity. Thus, only CSFs that are excited on one or the other atom contribute to the energy:

$$E = <\Phi_a \Phi_b|H| \sum_{Ja} C_{Ja} \Phi^*_{Ja} \Phi_b + \sum_{Jb} C_{Jb} \Phi_a \Phi^*_{Jb}>,$$

where Φ_a and Φ_b as well as Φ^*_{Ja} and Φ^*_{Jb} are used to denote the a and b parts of the reference and excited CSFs, respectively.

This expression, once the SC rules are used to reduce it to one- and two-electron integrals, is of the additive form required of any size-extensive method:

$$E = <\Phi_a|H| \sum_{Ja} C_{Ja} \Phi_{Ja}> + <\Phi_b|H| \sum_{Jb} C_{Jb} \Phi_{Jb}>,$$

and will yield a size-extensive energy *if* the equations used to determine the C_{Ja} and C_{Jb} amplitudes are themselves separable. In MPPT/MBPT, these amplitudes are expressed, in first order, as:

$$C_{Ja} = <\Phi_a \Phi_b|H|\Phi^*_{Ja} \Phi_b>/[E^0_a + E^0_b - E^*_{Ja} - E^0_b]$$

(and analogously for C_{Jb}). Again using the SC rules, this expression reduces to one that involves only atom a:

$$C_{Ja} = <\Phi_a|H|\Phi^*_{Ja}>/[E^0_a - E^*_{Ja}] \ .$$

The additivity of E and the separability of the equations determining the C_J coefficients make the MPPT/MBPT energy size-extensive. This property can also be demonstrated for the Coupled-Cluster energy (see the references given above in chapter 19.I.4). However, size-extensive methods have at least one serious weakness; their energies do *not* provide upper bounds to the true energies of the system (because their energy functional is not of the expectation-value form for which the upper-bound property has been proven).

C. Which Method Is Best?

At this time, it may not possible to say which method is preferred for applications where all are practical. Nor is it possible to assess, in a way that is applicable to most chemical species, the accuracies with which various methods predict bond lengths and energies or other properties. However, there are reasons to recommend some methods over others in specific cases.

For example, certain applications require a size-extensive energy (e.g., extended systems that consist of a large or macroscopic number of units or studies of weak intermolecular interactions), so MBPT/MPPT or CC or CAS-based MCSCF are preferred. Moreover, certain chemical reactions (e.g., Woodward-Hoffmann forbidden reactions) and certain bond-breaking events require two or more "essential" electronic configurations. For them, single-configuration-based methods such as conventional CC and MBTP/MPPT should not be used; MCSCF or CI calculations would be better. Very large molecules, in which thousands of atomic orbital basis functions are required, may be impossible to treat by methods whose effort scales as N^4 or higher; density functional methods would be better to use then.

For all calculations, the choice of atomic orbital basis set must be made carefully, keeping in mind the N^4 scaling of the one- and two-electron integral evaluation step and the N^5 scaling of the two-electron integral transformation step. Of course, basis functions that describe the essence of the states to be studied are essential (e.g., Rydberg or anion states require diffuse functions, and strained rings require polarization functions).

As larger atomic basis sets are employed, the size of the CSF list used to treat dynamic correlation increases rapidly. For example, most of the above methods use singly and doubly excited CSFs for this purpose. For large basis sets, the number of such CSFs, N_C, scales as the number of electrons squared, n_e^2, times the number of basis functions squared, N^2. Since the effort needed to solve the CI secular problem varies as N_C^2 or N_C^3, a dependence as strong as N^4 to N^6 can result. To handle such large CSF spaces, all of the multiconfigurational techniques mentioned in this paper have been developed to the extent that calculations involving of the order of 100 to 5,000 CSFs are routinely performed and calculations using 10,000, 100,000, and even several million CSFs are practical.

Other methods, most of which can be viewed as derivatives of the techniques introduced above, have been and are still being developed. This ongoing process has been, in large part, stimulated by the explosive growth in computer power and change in computer architecture that has been realized in recent years. All indications are that this growth pattern will continue, so *ab initio* quantum chemistry will likely have an even larger impact on future chemistry research and education (through new insights and concepts).

III. FURTHER DETAILS ON IMPLEMENTING MULTICONFIGURATIONAL METHODS

A. The MCSCF Method

The simultaneous optimization of the LCAO-MO and CI coefficients performed within an MCSCF calculation is a quite formidable task. The variational energy functional is a quadratic function of the CI coefficients, and so one can express the stationary conditions for these variables in the secular form:

$$\sum_J H_{I,J} C_J = E C_I.$$

However, E is a quartic function of the $C_{\nu,i}$ coefficients because each matrix element $<\Phi_I|H|\Phi_J>$ involves one- and two-electron integrals over the mos ϕ_i, and the two-electron integrals depend quartically on the $C_{\nu,i}$ coefficients. The stationary conditions with respect to these $C_{\nu,i}$ parameters must be solved iteratively because of this quartic dependence.

It is well known that minimization of a function (E) of several non-linear parameters (the $C_{\nu,i}$) is a difficult task that can suffer from poor convergence and may locate local rather than global minima. In an MCSCF wavefunction containing many CSFs, the energy is only weakly dependent on the orbitals that are weakly occupied (i.e., those that appear in CSFs with small C_I values); in contrast, E is strongly dependent on the $C_{\nu,i}$ coefficients of those orbitals that appear in the CSFs with larger C_I values. One is therefore faced with minimizing a function of many variables (there may be as many $C_{\nu,i}$ as the square of the number of orbital basis functions) that depends strongly on several of the variables and weakly on many others. This is a very difficult job.

For these reasons, in the MCSCF method, the number of CSFs is usually kept to a small to moderate number (e.g., a few to several hundred) chosen to describe essential correlations (i.e., configuration crossings, proper dissociation) and important dynamical correlations (those electron-pair correlations of angular, radial, left-right, etc. nature that arise when low-lying "virtual" orbitals are present). In such a compact wavefunction, only spin-orbitals with reasonably large occupations (e.g., as characterized by the diagonal elements of the one-particle density matrix $\gamma_{i,j}$) appear. As a result, the energy functional is expressed in terms of variables on which it is strongly dependent, in which case the non-linear optimization process is less likely to be pathological.

Such a compact MCSCF wavefunction is designed to provide a good description of the set of strongly occupied spin-orbitals and of the CI amplitudes for CSFs in which only these spin-orbitals appear. It, of course, provides no information about the spin-orbitals that are not used to form the CSFs on which the MCSCF calculation is based. As a result, the MCSCF energy is invariant to a unitary transformation among these "virtual" orbitals.

In addition to the references mentioned earlier in ACP and MTC, the following papers describe several of the advances that have been made in the MCSCF method, especially with respect to enhancing its rate and range of convergence: E. Dalgaard and P. Jørgensen, *J. Chem. Phys. 69*, 3833 (1978); H. J. Aa. Jensen, P. Jørgensen, and H. Ågren, *J. Chem. Phys. 87*, 457 (1987); B. H. Lengsfield, III and B. Liu, *J. Chem. Phys. 75*, 478 (1981).

B. The Configuration-Interaction Method

In the CI method, one usually attempts to realize a high-level treatment of electron correlation. A set of orthonormal molecular orbitals are first obtained from an SCF or MCSCF calculation (usually involving a small to moderate list of CSFs). The LCAO-MO coefficients of these orbitals are *no longer considered* as variational parameters in the subsequent CI calculation; only the C_I coefficients are to be further optimized.

The CI wavefunction

$$\Psi = \sum_J C_J \Phi_J$$

is most commonly constructed from CSFs Φ_J that include:

1. All of the CSFs in the SCF (in which case only a single CSF is included) or MCSCF wavefunction that was used to generate the molecular orbitals ϕ_i. This set of CSFs are referred to as spanning the '*reference space*' of the subsequent CI calculation, and the particular combination of these CSFs used in this orbital optimization (i.e., the SCF or MCSCF wavefunction) is called the *reference function*.
2. CSFs that are generated by carrying out single-, double-, triple-, etc. level "excitations" (i.e., orbital replacements) relative to reference CSFs. CI wavefunctions limited to include contributions through various levels of excitation (e.g., single, double, etc.) are denoted S (singly excited), D (doubly), SD (singly and doubly), SDT (singly, doubly, and triply), and so on.

The orbitals from which electrons are removed and those into which electrons are excited can be restricted to focus attention on correlations among certain orbitals. For example, if excitations out of core electrons are excluded, one computes a total energy that contains no correlation corrections for these core orbitals. Often it is possible to so limit the nature of the orbital excitations to focus on the energetic quantities of interest (e.g., the CC bond breaking in ethane requires correlation of the σ_{CC} orbital but the 1*s* Carbon core orbitals and the CH bond orbitals may be treated in a non-correlated manner).

Clearly, the number of CSFs included in the CI calculation can be far in excess of the number considered in typical MCSCF calculations; CI wavefunctions including 5,000 to 50,000 CSFs are routinely used, and functions with one to *several million* CSFs are within the realm of practicality (see, for example, J. Olsen, B. Roos, Poul Jørgensen, and H. J. Aa. Jensen, *J. Chem. Phys. 89*, 2185 (1988) and J. Olsen, P. Jørgensen, and J. Simons, *Chem. Phys. Letters 169*, 463 (1990)).

The need for such large CSF expansions should not come as a surprise once one considers that (i) each electron pair requires *at least* two CSFs (let us say it requires P of them, on average, a dominant one and $P-1$ others which are doubly excited) to form polarized orbital pairs, (ii) there are of the order of $N(N-1)/2 = X$ electron pairs in an atom or molecule containing N electrons, and (iii) that the number of terms in the CI wavefunction scales as P^X. So, for an H_2O molecule containing ten electrons, there would be P^{55} terms in the CI expansion. This is 3.6×10^{16} terms if $P = 2$ and 1.7×10^{26} terms if $P = 3$. Undoubtedly, this is an over estimate of the number of CSFs needed to describe electron correlation in H_2O, but it demonstrates how rapidly the number of CSFs can grow with the number of electrons in the system.

The $H_{I,J}$ matrices that arise in CI calculations are evaluated in terms of one- and two-electron integrals over the molecular orbitals using the equivalent of the Slater-Condon rules. For large CI calculations, the full $H_{I,J}$ matrix is not actually evaluated and stored in the computer's memory (or on its disk); rather, so-called "direct CI" methods (see the article by Roos and Siegbahn in MTC) are used to compute and immediately sum contributions to the sum

$$\sum_J H_{I,J} C_J$$

in terms of integrals, density matrix elements, and approximate values of the C_J amplitudes. Iterative methods (see, for example, E. R. Davidson, *J. Comput. Phys. 17*, 87 (1975)), in which approximate values for the C_J coefficients and energy E are refined through sequential application of

$$\sum_J H_{I,J}$$

to the preceding estimate of the C_J vector, are employed to solve these large CI matrix eigenvalue problems.

C. The MPPT/MBPT Method

In the MPPT/MBPT method, once the reference CSF is chosen and the SCF orbitals belonging to this CSF are determined, the wavefunction Ψ and energy E are determined in an order-by-order manner. This is one of the primary strengths of the MPPT/MBPT technique; it does not require one to make further (potentially arbitrary) choices once the basis set and dominant (SCF) configuration are specified. In contrast to the MCSCF and CI treatments, one need not make choices of CSFs to include in or exclude from Ψ. The MPPT/MBPT perturbation equations determine what CSFs must be included through any particular order.

For example, the first-order wavefunction correction Ψ^1 (i.e., $\Psi = \Phi + \Psi^1$ through first order) is given by:

$$\begin{aligned}\Psi^1 &= -\sum_{i<j,m<n} <\Phi_{i,j}^{m,n}|H - H^0|\Phi>[\varepsilon_m - \varepsilon_i + \varepsilon_n - \varepsilon_j]^{-1}|\Phi_{i,j}^{m,n}> \\ &= -\sum_{i<j,m<n} [<i,j|g|m,n> - <i,j|g|n,m>][\varepsilon_m - \varepsilon_i + \varepsilon_n - \varepsilon_j]^{-1}|\Phi_{i,j}^{m,n}>\end{aligned}$$

where the SCF orbital energies are denoted ε_k and $\Phi_{i,j}^{m,n}$ represents a CSF that is *doubly excited* relative to Φ. Thus, only doubly excited CSFs contribute to the *first-order wavefunction*; as a result, the energy E is given through second order as:

$$\begin{aligned}E &= <\Phi|H^0|\Phi> + <\Phi|H - H^0|\Phi> + <\Phi|H - H^0|\Psi^1> \\ &= <\Phi|H|\Phi> - \sum_{i<j,m<n} |<\Phi_{i,j}^{m,n}|H - H^0|\Phi>|^2/[\varepsilon_m - \varepsilon_i + \varepsilon_n - \varepsilon_j] \\ &= E_{SCF} - \sum_{i<j,m<n} |<i,j|g|m,n> - <i,j|g|n,m>|^2/[\varepsilon_m - \varepsilon_i + \varepsilon_n - \varepsilon_j] \\ &= E^0 + E^1 + E^2.\end{aligned}$$

These contributions have been expressed, using the SC rules, in terms of the two-electron integrals $<i,j|g|m,n>$ coupling the excited spin-orbitals to the spin-orbitals from which electrons were excited as well as the orbital energy differences $[\varepsilon_m - \varepsilon_i + \varepsilon_n - \varepsilon_j]$ accompanying such excitations. In this form, it becomes clear that major contributions to the correlation energy of the pair of occupied orbitals $\phi_i\phi_j$ are made by double excitations into virtual orbitals $\phi_m\phi_n$ that have large coupling (i..e., large $<i,j|g|m,n>$ integrals) and small orbital energy gaps, $[\varepsilon_m - \varepsilon_i + \varepsilon_n - \varepsilon_j]$.

In higher order corrections to the wavefunction and to the energy, contributions from CSFs that are singly, triply, etc. excited relative to Φ appear, and additional contributions from the doubly excited CSFs also enter. It is relatively common to carry MPPT/MBPT calculations (see the references given above in chapter 19.I.3 where the contributions of the Pople and Bartlett groups to the development of MPPT/MBPT are documented) through to third order in the energy (whose evaluation can be shown to require only Ψ^0 and Ψ^1). The entire GAUSSIAN-8X series of programs, which have been used in thousands of important chemical studies, calculate E through third order in this manner.

In addition to being size-extensive and not requiring one to specify input beyond the basis set and the dominant CSF, the MPPT/MBPT approach is able to include the effect of *all* CSFs (that contribute to any given order) without having to find any eigenvalues of a matrix. This is an important advantage because matrix eigenvalue determination, which is necessary in MCSCF and CI calculations, requires computer time in proportion to the third power of the dimension of the $H_{I,J}$ matrix. Despite all of these advantages, it is important to remember the primary disadvantages of the MPPT/MBPT approach; its energy is not an upper bound to the true energy and it may not be able to treat cases for which two or more CSFs have equal or nearly equal amplitudes because it obtains the amplitudes of all but the dominant CSF from perturbation theory formulas that assume the perturbation is "small."

D. The Coupled-Cluster Method

The implementation of the CC method begins much as in the MPPT/MBPT case; one selects a reference CSF that is used in the SCF process to generate a set of spin-orbitals to be used in the subsequent correlated calculation. The set of working equations of the CC technique given above in chapter 19.I.4 can be written explicitly by introducing the form of the so-called cluster operator T,

$$T = \sum_{i,m} t_i^m m^+ i + \sum_{i,j,m,n} t_{i,j}^{m,n} m^+ n^+ j i + \ldots,$$

where the combination of operators $m^+ i$ denotes *creation* of an electron in virtual spin-orbital ϕ_m and *removal* of an electron from occupied spin-orbital ϕ_i to generate a single excitation. The operation $m^+ n^+ j i$ therefore represents a double excitation from $\phi_i\phi_j$ to $\phi_m\phi_n$. Expressing the cluster operator T in terms of the amplitudes t_i^m, $t_{i,j}^{m,n}$, etc. for singly, doubly, etc. excited CSFs, and expanding the exponential operators in $\exp(-T)H\exp(T)$ one obtains:

$$<\Phi_i^m|H + [H,T] + 1/2[[H,T],T] + 1/6[[[H,T],T],T] + 1/24[[[[H,T],T],T],T]|\Phi> = 0;$$

$$<\Phi_{i,j}^{m,n}|H + [H,T] + 1/2[[H,T],T] + 1/6[[[H,T],T],T] + 1/24[[[[H,T],T],T],T]|\Phi> = 0;$$

$$<\Phi_{i,j,k}^{m,n,p}|H + [H,T] + 1/2[[H,T],T] + 1/6[[[H,T],T],T] + 1/24[[[[H,T],T],T],T]|\Phi> = 0,$$

and so on for higher-order excited CSFs. It can be shown, because of the one- and two- electron operator nature of H, that the expansion of the exponential operators truncates exactly at the fourth power; that is terms such as [[[[[H,T],T],T],T],T] and higher commutators vanish identically (this is demonstrated in chapter 4 of *Second Quantization Based Methods in Quantum Chemistry*, P. Jørgensen and J. Simons, Academic Press, New York (1981)).

As a result, the exact CC equations are *quartic equations* for the t_i^m, $t_{i,j}^{m,n}$, etc. amplitudes. Although it is a rather formidable task to evaluate all of the commutator matrix elements appearing in the above CC equations, it can be and has been done (the references given above to Purvis and Bartlett are especially relevant in this context). The result is to express each such matrix element, via the Slater-Condon rules, in terms of one- and two-electron integrals over the spin-orbitals used in determining Φ, including those in Φ itself and the "virtual" orbitals not in Φ.

In general, these quartic equations must then be solved in an iterative manner and are susceptible to convergence difficulties that are similar to those that arise in MCSCF-type calculations. In any such iterative process, it is important to start with an approximation (to the t amplitudes, in this case) which is reasonably close to the final converged result. Such an approximation is often achieved, for example, by neglecting all of the terms that are non-linear in the t amplitudes (because these amplitudes are assumed to be less than unity in magnitude). This leads, for the CC working equations obtained by projecting onto the doubly excited CSFs, to:

$$<i,j|g|m,n>' + [\varepsilon_m - \varepsilon_i + \varepsilon_n - \varepsilon_j]t_{i,j}^{m,n} + \sum_{i',j',m',n'} <\Phi_{i,j}^{m,n}|H - H^0|\Phi_{i',j'}^{m',n'}>t_{i',j'}^{m',n'} = 0,$$

where the notation $<i,j|g|m,n>'$ is used to denote the two-electron integral difference $<i,j|g|m,n> - <i,j|g|n,m>$. If, in addition, the factors that couple different doubly excited CSFs are ignored (i.e., the sum over i',j',m',n'), the equations for the t amplitudes reduce to the equations for the CSF amplitudes of the first-order MPPT/MBPT wavefunction:

$$t_{i,j}^{m,n} = - <i,j|g|m,n>'/[\varepsilon_m - \varepsilon_i + \varepsilon_n - \varepsilon_j] \ .$$

As Bartlett and Pople have both demonstrated, there is, in fact, close relationship between the MPPT/MBPT and CC methods when the CC equations are solved iteratively starting with such an MPPT/MBPT-like initial "guess" for these double-excitation amplitudes.

The CC method, as presented here, suffers from the same drawbacks as the MPPT/MBPT approach; its energy is not an upper bound and it may not be able to accurately describe wavefunctions which have two or more CSFs with approximately equal amplitude. Moreover, solution of the non-linear CC equations may be difficult and slowly (if at all) convergent. It has the same advantages as the MPPT/MBPT method; its energy is size-extensive, it requires no large matrix eigenvalue solution, and its energy and wavefunction are determined once one specifies the basis and the dominant CSF.

E. Density Functional or X-alpha (X_α) Methods

These approaches provide alternatives to the conventional tools of quantum chemistry. The family of approaches that often are referred to as X_α-type methods is realizing a renewed importance and popularity in quantum chemistry, and a density functional module already appears in Gaussian 9X.

The CI, MCSCF, MPPT/MBPT, and CC methods move beyond the single-configuration picture in a rather straightforward manner; they add to the expansion of the wavefunction more configurations whose amplitudes they each determine in their own way. This can lead to a very large number of CSFs in the correlated wavefunction, and, as a result, a large need for computer resources.

The density functional approaches are different. Here one solves a set of orbital-level equations

$$\left[-\hbar^2/2m_e\nabla^2 - \sum_A Z_A e^2/|\mathbf{r} - \mathbf{R}_A| + \int \rho(\mathbf{r}')e^2/|\mathbf{r} - \mathbf{r}'|d\mathbf{r}' + U_{X\alpha}(\mathbf{r})\right]\phi_i = \varepsilon_i\phi_i$$

in which the orbitals $\{\phi_i\}$ "feel" potentials due to the nuclear centers (labeled A and having charges Z_A), due to coulombic interaction with the *total* electron density $\rho(\mathbf{r}')$, and due to a so-called correlation-exchange potential which will be denoted $U_{X\alpha}(\mathbf{r}')$. The particular electronic state for which the calculation is being performed is specified by forming a corresponding density $\rho(\mathbf{r}')$.

This potential $U_{X\alpha}(\mathbf{r}')$ must remove the "self interaction" of the electron in $\phi_i(\mathbf{r})$ that is incorrectly included in

$$\int \rho(\mathbf{r}')e^2/|\mathbf{r}-\mathbf{r}'|d\mathbf{r}'\phi_i(\mathbf{r}).$$

The self-interaction is included in this integral because $\rho(\mathbf{r}')$ represents the *total* electron density at $\mathbf{r}'$, including that due to the electron in ϕ_i. To compensate for this incorrect evaluation of the electron-electron interactions, $U_{X\alpha}$ must include terms which remove the self-interactions.

The fundamental theory that underlies the density functional methods dates back to the work of Hohenberg and Kohn (P. Hohenberg and W. Kohn, *Phys. Rev. 136*, B864 (1964)) who showed that the *ground-state* energy E_0 of an N-electron system can be expressed as a *functional* (not necessarily a function) of the electron density $\rho(\mathbf{r})$ of that system. A great deal of work has subsequently been devoted to (i) finding a functional relationship for $E_0[\rho]$ in terms of $\rho(\mathbf{r})$, and (ii) finding computational schemes to compute $\rho(\mathbf{r})$ and then E_0 in an efficient manner.

Researchers studying the idealized "uniform electron gas" found that the exchange and correlation energy (per electron) for this system could be written exactly as a *function* of the electron density ρ of the system: $E_{exchange} = W(\rho)$. This motivated workers to suggest that this exact result could be used to define, for each point in $\mathbf{r}$-space, a correlation-exchange potential, $W(\mathbf{r}) = W(\rho(\mathbf{r}))$, and that $W(\mathbf{r})$ could be used to generate the exchange energy by integrating over all points $\mathbf{r}$:

$$E_{exchange} = \int \rho(\mathbf{r})W(\rho(\mathbf{r}))\, d\mathbf{r}.$$

These studies of the uniform electron gas influenced early work by Slater and coworkers (J. C. Slater, *Quantum Theory of Molecules and Solids*, Vol. 4, McGraw-Hill, New York (1974); K. H. Johnson, *Adv. Quantum Chem. 7*, 143 (1973)) who focused on single-configuration descriptions where the electron density is straightforwardly evaluated in terms of the orbitals $\{\phi_j\}$:

$$\rho(\mathbf{r}) = \sum_j n_j|\phi_j(\mathbf{r})|^2.$$

Here $n_j = 0$, 1, or 2 is the occupation number of the orbital ϕ_j in the state being studied. Combining the single-configuration ansatz with the approximation that the exchange contributions to $E_0[\rho]$ be represented by the "uniform electron gas" (ueg) value

$$W_{ueg}(\mathbf{r}) = -9/4\rho(\mathbf{r})(3\rho(\mathbf{r})/8\pi)^{1/3},$$

Slater arrived at an orbital-level equation for the $\{\phi_j\}$ that define ρ:

$$\left[-\hbar^2/2m_e\nabla^2 - \sum_A Z_Ae^2/|\mathbf{r}-\mathbf{R}A| + \int \rho(\mathbf{r}')e^2/|\mathbf{r}-\mathbf{r}'|\, d\mathbf{r}' + U_{ueg}(\mathbf{r})\right]\phi_i = \varepsilon_i\phi_i\,,$$

where

$$U_{ueg}(\mathbf{r}) = -9/2(3\rho(\mathbf{r})/8\pi)^{1/3}.$$

However, Slater found the uniform-electron-gas estimate of the exchange function overestimated the magnitude of the exchange energy when compared to results of *ab initio* calculations. He therefore introduced an empirical scaling of the expression for $U_{ueg}(\mathbf{r})$ and proposed that

$$U_{X\alpha}(\mathbf{r}) = -9\alpha/2(3\rho(\mathbf{r})/8\pi)^{1/3}$$

be used in its place. Slater named the scaling parameter α, which is the origin of the name of the $X\alpha$ procedure (the X denoting "exchange").

Although Slater's work emphasized a single-configuration picture of electronic structure where the evaluation of $\rho(\boldsymbol{r})$ as a sum

$$\sum_j n_j|\phi_j(\boldsymbol{r})|^2$$

is straightforward, the fundamental theory behind density functional methods applies to exact energies if an exact $E_0[\rho]$ functional and an exact density $\rho(\boldsymbol{r})$ are used.

Within the X_α modification of the local uniform gas density functional (sometimes called the local density approximation (LDA))

$$U_{X\alpha}(\boldsymbol{r}) = -9\alpha/2(3\rho(\boldsymbol{r})/8\pi)^{1/3}$$

it has empirically been found that using a value of α near 2/3 produces orbital and total energies that most closely reproduce high quality *ab initio* and experimental values. Moreover, Kohn and Sham (W. Kohn and L. J. Sham, *Phys. Rev. B13*, 4274 (1965)) have shown that the value $\alpha = 2/3$ can be derived from a variational theory as the optimal value.

Most of the early applications of density functional methods solved equations of the form:

$$\left[-\hbar^2/2m_e\nabla^2 - \sum_A Z_Ae^2/|\boldsymbol{r} - \boldsymbol{R}\boldsymbol{A}| + \int \rho(\boldsymbol{r}')e^2/|\boldsymbol{r} - \boldsymbol{r}'|\, d\boldsymbol{r}' + U_{X\alpha}(\boldsymbol{r})\right]\phi_i = \varepsilon_i\phi_i$$

in terms of functions expressed as "scattered waves" defined locally in spherical regions of space surrounding each atom in the molecule or solid. In this so-called "*muffin tin*" view, the volume of space available to electrons in the molecule is divided into atom-centered spheres, regions lying outside such atom-centered spheres yet inside a so-called "outer sphere" surrounding the molecule, and a region lying beyond the outer sphere. In the atom-centered regions and within the outer sphere, the orbital-level equations are easily solved (the angular and radial parts separate because the potential is assumed to be spherically averaged and thus angle-independent). These scattered-wave solutions are then matched at the boundaries among the atom-centered spheres and the outer sphere.

The muffin-tin potential approach was shown to suffer when applied to molecules of low symmetry. Therefore, it became more common to not introduce the local muffin-tin potentials but simply to solve the density functional orbital equations using atomic basis sets such as those utilized in conventional *ab initio* quantum chemistry. The individual orbitals $\phi_j(\boldsymbol{r})$ can easily be expanded in terms of the atomic orbital basis as in the LCAO-MO expansion; it is also straightforward to express

$$\rho(\boldsymbol{r}) = \sum_j n_j|\phi_j(\boldsymbol{r})|^2,$$

or even the multiconfigurational analog

$$\rho(\boldsymbol{r}) = \sum_{i,j} \gamma_{ij}\phi_i{*}(\boldsymbol{r})\phi_j(\boldsymbol{r}),$$

in like fashion. However, to express the

$$U_{ueg}(\boldsymbol{r}) = -9/2(3\rho(\boldsymbol{r})/8\pi)^{1/3}$$

in terms of an atomic orbital basis requires considerably more effort; several approaches to this problem exist and are in common use.

In the simplest version of density functional theory, the total energy E is computed, as

$$E = \sum_j n_j \varepsilon_j,$$

where ε_j are the orbital energies obtained from the X_α-like equations. Although the sum-of-orbital-energies does not correctly represent the total energy in the CI, MCSCF, MPPT/MBPT, or CC approaches, it can be shown that it is reasonably correct for these methods. A more accurate alternative is to compute E as

$$E = \sum_j n_j \langle \phi_j | -\hbar^2/2m_e \nabla^2 | \phi_j \rangle - \int \rho(\boldsymbol{r}) Z_A e^2 / |\boldsymbol{r} - \boldsymbol{R}_A| \, d\boldsymbol{r}$$

$$+ 1/2 \int \rho(\boldsymbol{r}) \rho(\boldsymbol{r}') e^2 / |\boldsymbol{r} - \boldsymbol{r}'| d\boldsymbol{r} d\boldsymbol{r}' - \int 9/4 \rho(\boldsymbol{r}) (3\rho(\boldsymbol{r})/8\pi)^{1/3} d\boldsymbol{r}.$$

Once the orbitals are found by solving the above equations, the kinetic energy integrals, the electron-nuclear attraction integral, and the coulomb integral can be evaluated using conventional *ab initio* basis set tools. As stated above, the exchange correlation integral is more difficult to treat because $[\rho(\boldsymbol{r})]^{1/3}$ appears; $\rho(\boldsymbol{r})$ is easily expanded in a basis, but $[\rho(\boldsymbol{r})]^{1/3}$ requires a separate expansion.

Various density functional approaches are under active development (see, for example, R. O. Jones in *Adv. Chem. Phys.* LXVII, 413 (1987); B. I. Dunlap in *Adv. Chem. Phys.* LXIX, 287 (1987); J. P. Dahl and J. Avery (Eds.) *Local Density Approximations in Quantum Chemistry and Solid State Physics*, Plenum, New York (1984); R. G. Parr, *Ann. Rev. Phys. Chem. 34*, 631 (1983); D. R. Salahub, S. H. Lampson, and R. P. Messmer, *Chem. Phys. Lett. 85*, 430 (1982); T. Ziegler, A. Rauk, and E. J. Baerends, *Theor. Chim. Acta* (Berl.) *43*, 261 (1977); A. Becke, *J. Chem. Phys. 76*, 6037 (1983); D. A. Case, *Ann. Rev. Phys. Chem. 33*, 151 (1982)) and thus appear in the literature in many forms; refs. 39 describe many of the historically important developments as well as several of the more recent advances. Because the computational effort involved in these approaches scales much less strongly with basis set size than for conventional (SCF, MCSCF, CI, etc.) methods, density functional methods offer great promise and are likely to contribute much to quantum chemistry in the next decade.

CHAPTER

20

Many physical properties of a molecule can be calculated as expectation values of a corresponding quantum mechanical operator. The evaluation of other properties can be formulated in terms of the "response" (i.e., derivative) of the electronic energy with respect to the application of an external field perturbation.

I. CALCULATIONS OF PROPERTIES OTHER THAN THE ENERGY

There are, of course, properties other than the energy that are of interest to the practicing chemist. Dipole moments, polarizabilities, transition probabilities among states, and vibrational frequencies all come to mind. Other properties that are of importance involve operators whose quantum numbers or symmetry indices label the state of interest. Angular momentum and point group symmetries are examples of the latter properties; for these quantities the properties are precisely specified once the quantum number or symmetry label is given (e.g., for a ^3P state, the average value of L^2 is $<^3P|L^2|^3P> = \hbar^2 1(1+1) = 2\hbar^2$).

Although it may be straightforward to specify what property is to be evaluated, often computational difficulties arise in carrying out the calculation. For some *ab initio* methods, these difficulties are less severe than for others. For example, to compute the electric dipole transition matrix element $<\Psi_2|\,\boldsymbol{r}|\Psi_1>$ between two states Ψ_1 and Ψ_2, one must evaluate the integral involving the one-electron dipole operator

$$\boldsymbol{r} = \sum_j e\boldsymbol{r}_j - \sum_a eZ_a\boldsymbol{R}_a;$$

here the first sum runs over the N electrons and the second sum runs over the nuclei whose charges are denoted Z_a. To evaluate such transition matrix elements in terms of the Slater-Condon rules is relatively straightforward as long as Ψ_1 and Ψ_2 are expressed in terms of Slater determinants involving a single set of orthonormal spin-orbitals. If Ψ_1 and Ψ_2, have been obtained, for example, by carrying out separate MCSCF calculations on the two states in question, the energy optimized spin-orbitals for one state will not be the same as the optimal spin-orbitals for the second state. As a result, the determinants in Ψ_1 and those in Ψ_2 will involve spin-orbitals that are not orthonormal to one another. Thus, the SC rules can not immediately be applied. Instead, a transformation of the spin-orbitals of Ψ_1 and Ψ_2 to a single set of orthonormal functions must be carried out. This then expresses Ψ_1 and Ψ_2 in terms of new Slater determinants over this new set of orthonormal spin-orbitals, after which the SC rules can be exploited.

In contrast, if Ψ_1 and Ψ_2 are obtained by carrying out a CI calculation using a single set of orthonormal spin-orbitals (e.g., with Ψ_1 and Ψ_2 formed from two different eigenvectors of the resulting secular matrix), the SC rules can immediately be used to evaluate the transition dipole integral.

A. Formulation of Property Calculations as Responses

Essentially all experimentally measured properties can be thought of as arising through the *response* of the system to some externally applied perturbation or disturbance. In turn, the calculation of such properties can be formulated in terms of the response of the energy E or wavefunction Ψ to a perturbation. For example, molecular dipole moments μ are measured, via electric-field deflection, in terms of the change in energy

$$\Delta E = \mu \bullet \boldsymbol{E} + 1/2\boldsymbol{E} \bullet \alpha \bullet \boldsymbol{E} + 1/6\boldsymbol{E} \bullet \boldsymbol{E} \bullet \boldsymbol{E} \bullet \beta + \ldots$$

caused by the application of an external electric field $\boldsymbol{E}$ which is spatially inhomogeneous, and thus exerts a force

$$\boldsymbol{F} = -\nabla \Delta E$$

on the molecule proportional to the dipole moment (good treatments of response properties for a wide variety of wavefunction types (i.e., SCF, MCSCF, MPPT/MBPT, etc.) are given in *Second Quantization Based Methods in Quantum Chemistry*, P. Jørgensen and J. Simons, Academic Press, New York (1981) and in *Geometrical Derivatives of Energy Surfaces and Molecular Properties*, P. Jørgensen and J. Simons, eds., NATO ASI Series, Vol. 166, D. Reidel, Dordrecht (1985)).

To obtain expressions that permit properties other than the energy to be evaluated in terms of the state wavefunction Ψ, the following strategy is used:

1. The perturbation $V = H - H^0$ appropriate to the particular property is identified. For dipole moments (μ), polarizabilities (α), and hyperpolarizabilities (β), V is the interaction of the nuclei and electrons with the external electric field

$$V = \sum_a Z_a e \boldsymbol{R}_a \bullet E - \sum_j e \boldsymbol{r}_j \bullet E.$$

For vibrational frequencies, one needs the derivatives of the energy E with respect to deformation of the bond lengths and angles of the molecule, so V is the sum of all changes in the electronic Hamiltonian that arise from displacements $\delta \boldsymbol{R}_a$ of the atomic centers

$$V = \sum_a (\nabla_{\boldsymbol{R}_a} H) \prod \delta \boldsymbol{R}_a.$$

2. A power series expansion of the state energy E, computed in a manner consistent with how Ψ is determined (i.e., as an expectation value for SCF, MCSCF, and CI wavefunctions or as

$<\Phi|H|\Psi>$ for MPPT/MBPT or as $<\Phi| \exp(-T)H \exp(T)|\Phi>$ for CC wavefunctions), is carried out in powers of the perturbation V:

$$E = E^0 + E^{(1)} + E^{(2)} + E^{(3)} + \ldots$$

In evaluating the terms in this expansion, the dependence of $H = H^0 + V$ *and* of Ψ (which is expressed as a solution of the SCF, MCSCF, . . . , or CC equations for H *not* for H^0) must be included.

3. The desired physical property must be extracted from the power series expansion of ΔE in powers of V.

B. The MCSCF Response Case

1. The Dipole Moment

To illustrate how the above developments are carried out and to demonstrate how the results express the desired quantities in terms of the original wavefunction, let us consider, for an MCSCF wavefunction, the response to an external electric field. In this case, the Hamiltonian is given as the conventional one- and two-electron operators H^0 to which the above one-electron electric dipole perturbation V is added. The MCSCF wavefunction Ψ and energy E are assumed to have been obtained via the MCSCF procedure with $H = H^0 + \lambda V$, where λ can be thought of as a measure of the strength of the applied electric field.

The terms in the expansion of $E(\lambda)$ in powers of λ:

$$E = E(\lambda = 0) + \lambda(dE/d\lambda)_0 + 1/2\lambda^2(d^2E/d\lambda^2)_0 + \ldots$$

are obtained by writing the total derivatives of the MCSCF energy functional with respect to λ and evaluating these derivatives at $\lambda = 0$ (which is indicated by the subscript $(..)_0$ on the above derivatives):

$$E(\lambda = 0) = <\Psi(\lambda = 0)|H^0|\Psi(\lambda = 0)> = E^0,$$

$$(dE/d\lambda)_0 = <\Psi(\lambda = 0)|V|\Psi(\lambda = 0)> + 2\sum_J (\partial C_J/\partial\lambda)_0<\partial\Psi/\partial C_J|H^0|\Psi(\lambda = 0)>$$

$$+ 2\sum_{i,a} (\partial C_{a,i}/\partial\lambda)_0<\partial\Psi/\partial C_{a,i}|H^0|\Psi(\lambda = 0)>$$

$$+ 2\sum_{\nu} (\partial\chi_\nu/\partial\lambda)_0<\partial\Psi/\partial\chi_\nu|H^0|\Psi(\lambda = 0)>,$$

and so on for higher-order terms. The factors of 2 in the last three terms come through using the hermiticity of H^0 to combine terms in which derivatives of Ψ occur.

The first-order correction can be thought of as arising from the response of the wavefunction (as contained in its LCAO-MO and CI amplitudes and basis functions χ_ν) plus the response of the Hamiltonian to the external field. Because the MCSCF energy functional has been made stationary with respect to variations in the C_J and $C_{i,a}$ amplitudes, the second and third terms above vanish:

$$\partial E/\partial C_J = 2<\partial\Psi/\partial C_J|H^0|\Psi(\lambda = 0)> = 0,$$

$$\partial E/\partial C_{a,i} = 2<\partial\Psi/\partial C_{a,i}|H^0|\Psi(\lambda = 0)> = 0.$$

If, as is common, the atomic orbital bases used to carry out the MCSCF energy optimization are not explicitly dependent on the external field, the third term also vanishes because $(\partial\chi_\nu/\partial\lambda)_0 = 0$. Thus for the MCSCF case, the first-order response is given as the average value of the perturbation over the wavefunction with $\lambda = 0$:

$$(dE/d\lambda)_0 = <\Psi(\lambda = 0)|V|\Psi(\lambda = 0)>.$$

For the external electric field case at hand, this result says that the field-dependence of the state energy will have a linear term equal to

$$<\Psi(\lambda = 0)|V|\Psi(\lambda = 0)> = <\Psi|\sum_a Z_a e \boldsymbol{R_a} \bullet \boldsymbol{e} - \sum_j e\boldsymbol{r_j} \bullet \boldsymbol{e}|\Psi>,$$

where $\boldsymbol{e}$ is a unit vector in the direction of the applied electric field (the magnitude of the field λ having already been removed in the power series expansion). Since the dipole moment is determined experimentally as the energy's slope with respect to field strength, this means that the dipole moment is given as:

$$\mu = <\Psi|\sum_a Z_a e \boldsymbol{R_a} - \sum_j e\boldsymbol{r_j}|\Psi>.$$

2. The Geometrical Force

These same techniques can be used to determine the response of the energy to displacements $\delta\boldsymbol{R_a}$ of the atomic centers. In such a case, the perturbation is

$$V = \sum_a \delta\boldsymbol{R_a} \bullet \nabla_{R_a}\left(-\sum_i Z_a e^2/|r_i - R_a|\right)$$
$$= -\sum_a Z_a e^2 \delta\boldsymbol{R_a} \bullet \sum_i (\boldsymbol{r}_i - \boldsymbol{R}_a)/|r_i - R_a|^3.$$

Here, the one-electron operator

$$\sum_i (\boldsymbol{r}_i - \boldsymbol{R_a})/|r_i - R_a|^3$$

is referred to as the "Hellmann-Feynman" force operator; it is the derivative of the Hamiltonian with respect to displacement of center-a in the x, y, or z direction.

The expressions given above for $E(\lambda = 0)$ and $(dE/d\lambda)_0$ can once again be used, but with the Hellmann-Feynman form for V. Once again, for the MCSCF wavefunction, the variational optimization of the energy gives

$$<\partial\Psi/\partial C_J|H^0|\Psi(\lambda = 0)> = <\partial\Psi/\partial C_{a,i}|H^0|\Psi(\lambda = 0)> = 0.$$

However, because the atomic basis orbitals are attached to the centers, and because these centers are displaced in forming V, it is no longer true that $(\partial\chi_\nu/\partial\lambda)_0 = 0$; the variation in the wavefunction caused by movement of the basis functions now contributes to the first-order energy response. As a result, one obtains

$$(dE/d\lambda)_0 = -\sum_a Z_a e^2 \delta\boldsymbol{R_a} \bullet <\Psi|\sum_i (\boldsymbol{r}_i - \boldsymbol{R_a})/|r_i - R_a|^3|\Psi>$$
$$+ 2\sum_a \delta\boldsymbol{R_a} \bullet \sum_\nu (\nabla_{R_a}\chi_\nu)_0 <\partial\Psi/\partial\chi_\nu|H^0|\Psi(\lambda = 0)>.$$

The first contribution to the *force*

$$F_a = -Z_a e^2 \langle \Psi | \sum_i (r_i - R_a)/|r_i - R_a|^3 |\Psi\rangle$$
$$+ 2 \sum_\nu (\nabla_{R_a} \chi_\nu)_0 \langle \partial\Psi/\partial\chi_\nu | H^0 | \Psi(\lambda = 0)\rangle$$

along the x, y, and z directions for center-a involves the expectation value, with respect to the MCSCF wavefunction with $\lambda = 0$, of the Hellmann-Feynman force operator. The second contribution gives the forces due to infinitesimal displacements of the basis functions on center-a.

The evaluation of the latter contributions can be carried out by first realizing that

$$\Psi = \sum_J C_J |\phi_{J1}\phi_{J2}\phi_{J3} \ldots \phi_{Jn} \ldots \phi_{JN}|$$

with

$$\phi_j = \sum_\mu C_{\mu,j}\, \chi_\mu$$

involves the basis orbitals through the LCAO-MO expansion of the ϕ_js. So the derivatives of the basis orbitals contribute as follows:

$$\sum_\nu (\nabla_{R_a} \chi_\nu) \langle \partial\Psi/\partial\chi_\nu | = \sum_J \sum_{j,\nu} C_J C_{n,j} \langle |\phi_{J1}\phi_{J2}\phi_{J3} \ldots \nabla_{R_a} \chi_\nu \ldots \phi_{JN}|.$$

Each of these factors can be viewed as combinations of CSFs with the same C_J and $C_{\nu,j}$ coefficients as in Ψ but with the jth spin-orbital involving basis functions that have been differentiated with respect to displacement of center-a. It turns out that such derivatives of Gaussian basis orbitals can be carried out analytically (giving rise to new Gaussians with one higher and one lower l-quantum number).

When substituted into

$$\sum_\nu (\nabla_{R_a} \chi_\nu)_0 \langle \partial\Psi/\partial\chi_\nu | H^0 | \Psi(\lambda = 0)\rangle,$$

these basis derivative terms yield

$$\sum_\nu (\nabla_{R_a} \chi_\nu)_0 \langle \partial\Psi/\partial\chi_\nu | H^0 | \Psi(\lambda = 0)\rangle = \sum_J \sum_{j,\nu} C_J C_{\nu,j} \langle |\phi_{J1}\phi_{J2}\phi_{J3} \ldots \nabla_{R_a} \chi_\nu \ldots \phi_{JN}| H^0 | \Psi\rangle,$$

whose evaluation via the Slater-Condon rules is straightforward. It is simply the expectation value of H^0 with respect to Ψ(with the same density matrix elements that arise in the evaluation of Ψ's energy) *but* with the one- and two-electron integrals over the atomic basis orbitals involving one of these differentiated functions:

$$\langle \chi_\mu\chi_\nu | g | \chi_\gamma\chi_\delta\rangle \Rightarrow \nabla_{R_a} \langle \chi_\mu\chi_\nu | g | \chi_\gamma\chi_\delta\rangle = \langle \nabla_{R_a} \chi_\mu\chi_\nu | g | \chi_\gamma\chi_\delta\rangle$$
$$+ \langle \chi_\mu \nabla_{R_a} \chi_\nu | g | \chi_\gamma\chi_\delta\rangle + \langle \chi_\mu\chi_\nu | g | \nabla_{R_a} \chi_\gamma\chi_\delta\rangle + \langle \chi_\mu\chi_\nu | g | \chi_\gamma \nabla_{R_a} \chi_\delta\rangle .$$

In summary, the force F_a felt by the nuclear framework due to a displacement of center-a along the x, y, or z axis is given as

$$F_a = -Z_a e^2 \langle \Psi | \sum_i (r_i - R_a)/|r_i - R_a|^3 |\Psi\rangle + (\nabla_{R_a} \langle \Psi | H^0 | \Psi\rangle),$$

where the second term is the energy of Ψ but with all atomic integrals replaced by integral derivatives:

$$<\chi_\mu\chi_\nu|g|\chi_\gamma\chi_\delta> \Rightarrow \nabla_{R_a} <\chi_\mu\chi_\nu|g|\chi_\gamma\chi_\delta>.$$

C. Responses for Other Types of Wavefunctions

It should be stressed that the MCSCF wavefunction yields especially compact expressions for responses of E with respect to an external perturbation because of the variational conditions

$$<\partial\Psi/\partial C_J|H^0|\Psi(\lambda=0)> = <\partial\Psi/\partial C_{a,i}|H^0|\Psi(\lambda=0)> = 0$$

that apply. The SCF case, which can be viewed as a special case of the MCSCF situation, also admits these simplifications. However, the CI, CC, and MPPT/MBPT cases involve additional factors that arise because the above variational conditions do not apply (in the CI case, $<\partial\Psi/\partial C_J|H^0|\Psi(\lambda=0)> = 0$ still applies, but the orbital condition $<\partial\Psi/\partial C_{a,i}|H^0|\Psi(\lambda=0)> = 0$ does not because the orbitals are not varied to make the CI energy functional stationary).

Within the CC, CI, and MPPT/MBPT methods, one must evaluate the so-called responses of the C_I and $C_{a,i}$ coefficients $(\partial C_J/\partial\lambda)_0$ and $(\partial C_{a,i}/\partial\lambda)_0$ that appear in the full energy response as (see above)

$$2\sum_J (\partial C_J/\partial\lambda)_0<\partial\Psi/\partial C_J|H^0|\Psi(\lambda=0)> + 2\sum_{i,a} (\partial C_{a,i}/\partial\lambda)_0<\partial\Psi/\partial C_{a,i}|H^0|\Psi(\lambda=0)>.$$

To do so requires solving a set of response equations that are obtained by differentiating whatever equations govern the C_I and $C_{a,i}$ coefficients in the particular method (e.g., CI, CC, or MPPT/MBPT) with respect to the external perturbation. In the geometrical derivative case, this amounts to differentiating with respect to x, y, and z displacements of the atomic centers. These response equations are discussed in *Geometrical Derivatives of Energy Surfaces and Molecular Properties*, P. Jørgensen and J. Simons, Eds., NATO ASI Series, Vol. 166, D. Reidel, Dordrecht (1985). Their treatment is somewhat beyond the scope of this text, so they will not be dealt with further here.

D. The Use of Geometrical Energy Derivatives

1. Gradients as Newtonian Forces

The first energy derivative is called the gradient $\boldsymbol{g}$ and is the negative of the force $\boldsymbol{F}$ (with components along the ath center denoted $\boldsymbol{F}_a$) experienced by the atomic centers $\boldsymbol{F} = -\boldsymbol{g}$. These forces, as discussed in chapter 16, can be used to carry out classical trajectory simulations of molecular collisions or other motions of large organic and biological molecules for which a quantum treatment of the nuclear motion is prohibitive.

The second energy derivatives with respect to the x, y, and z directions of centers a and b (for example, the x, y component for centers a and b is $H_{ax,by} = (\partial^2E/\partial x_a\partial y_b)_0$) form the Hessian matrix $\boldsymbol{H}$. The elements of $\boldsymbol{H}$ give the local curvatures of the energy surface along the 3**N** cartesian directions.

The gradient and Hessian can be used to systematically locate local minima (i.e., stable geometries) and transition states that connect one local minimum to another. At each of these stationary points, all forces and thus all elements of the gradient $\boldsymbol{g}$ vanish. At a local minimum, the $\boldsymbol{H}$ matrix has five or six zero eigenvalues corresponding to translational and rotational displacements of the molecule (five for linear molecules; six for non-linear species) and $3N$–5 or $3N$–6 positive eigenvalues. At a transition state, $\boldsymbol{H}$ has one negative eigenvalue, five or six zero eigenvalues, and $3N$–6 or $3N$–7 positive eigenvalues.

2. Transition State Rate Coefficients

The transition state theory of Eyring or its extensions due to Truhlar and co-workers (see, for example, D. G. Truhlar and B. C. Garrett, *Ann. Rev. Phys. Chem. 35*, 159 (1984)) allow knowledge of the Hessian matrix at a transition state to be used to compute a rate coefficient k_{rate} appropriate to the chemical reaction for which the transition state applies.

More specifically, the geometry of the molecule at the transition state is used to compute a rotational partition function $Q^{\dagger}_{rot}$ in which the principal moments of inertia I_a, I_b, and I_c (see chapter 13) are those of the transition state (the † symbol is, by convention, used to label the transition state):

$$Q^{\dagger}_{rot} = \prod_{n=a,b,c} \sqrt{\frac{8\pi^2 I_n kT}{h^2}},$$

where k is the Boltzmann constant and T is the temperature in °K.

The eigenvalues $\{\omega_\alpha\}$ of the mass weighted Hessian matrix (see below) are used to compute, for each of the $3N$–7 vibrations with real and positive ω_α values, a vibrational partition function that is combined to produce a transition-state vibrational partition function:

$$Q^{\dagger}_{vib} = \prod_{\alpha=1,3N-7} \frac{\exp(-\hbar\omega_\alpha/2kT)}{1-\exp(-\hbar\omega_\alpha/kT)}.$$

The electronic partition function of the transition state is expressed in terms of the activation energy (the energy of the transition state relative to the electronic energy of the reactants) $E^{\dagger}$ as:

$$Q^{\dagger}_{electronic} = \omega^{\dagger} \exp(-E^{\dagger}/kT)$$

where $\omega^{\dagger}$ is the degeneracy of the electronic state at the transition state geometry.

In the original Eyring version of transition state theory (TST), the rate coefficient k_{rate} is then given by:

$$k_{rate} = \frac{kT}{\hbar}\omega^{\dagger} \exp(-E^{\dagger}/kT)\frac{Q^{\dagger}_{rot}Q^{\dagger}_{vib}}{Q_{reactants}},$$

where $Q_{reactants}$ is the conventional partition function for the reactant materials.

For example, in a bimolecular reaction such as:

$$F + H_2 \rightarrow FH + H,$$

the reactant partition function

$$Q_{reactants} = Q_F Q_{H_2}$$

is written in terms of the translational and electronic (the degeneracy of the ^2P state produces the 2(3) overall degeneracy factor) partition functions of the F atom

$$Q_F = \left(\frac{2\pi m_F kT}{h^2}\right)^{3/2} 2(3)$$

and the translational, electronic, rotational, and vibrational partition functions of the H_2 molecule

$$Q_{H_2} = \left(\frac{2\pi m_{H_2} kT}{h^2}\right)^{3/2} \frac{8\pi^2 I_{H_2} kT}{2h^2} \frac{\exp(-\hbar\omega_{H_2}/2kT)}{1-\exp(-\hbar\omega_{H_2}/kT)}.$$

The factor of 2 in the denominator of the H_2 molecule's rotational partition function is the "symmetry number" that must be inserted because of the identity of the two H nuclei.

The overall rate coefficient k_{rate} (with units $\sec^{-1}$ because this is a rate per collision pair) can thus be expressed entirely in terms of energetic, geometrical, and vibrational information about the reactants and the transition state. Even within the extensions to Eyring's original model, such is the case. The primary difference in the more modern theories is that the transition state is identified not as the point on the potential energy surface at which the gradient vanishes and there is one negative Hessian eigenvalue. Instead, a so-called variational transition state (see the above reference by Truhlar and Garrett) is identified. The geometry, energy, and local vibrational frequencies of this transition state are then used to compute, much like outlined above, k_{rate}.

3. Harmonic Vibrational Frequencies

It is possible (see, for example, J. Nichols, H. L. Taylor, P. Schmidt, and J. Simons, *J. Chem. Phys.* *92*, 340 (1990) and references therein) to remove from $\boldsymbol{H}$ the zero eigenvalues that correspond to rotation and translation and to thereby produce a Hessian matrix whose eigenvalues correspond only to internal motions of the system. After doing so, the number of negative eigenvalues of $\boldsymbol{H}$ can be used to characterize the nature of the stationary point (local minimum or transition state), and $\boldsymbol{H}$ can be used to evaluate the local harmonic vibrational frequencies of the system.

The relationship between $\boldsymbol{H}$ and vibrational frequencies can be made clear by recalling the classical equations of motion in the Lagrangian formulation:

$$d/dt(\partial L/\partial \dot{q}_j) - (\partial L/\partial q_j) = 0,$$

where q_j denotes, in our case, the $3N$ cartesian coordinates of the N atoms, and $\dot{q}_j$ is the velocity of the corresponding coordinate. Expressing the Lagrangian L as kinetic energy minus potential energy and writing the potential energy as a local quadratic expansion about a point where $\boldsymbol{g}$ vanishes, gives

$$L = 1/2 \sum_j m_j \dot{q}_j^2 - E(0) - 1/2 \sum_{j,k} q_j H_{j,k} q_k.$$

Here, $E(0)$ is the energy at the stationary point, m_j is the mass of the atom to which q_j applies, and the $H_{j,k}$ are the elements of $\boldsymbol{H}$ along the x, y, and z directions of the various atomic centers.

Applying the Lagrangian equations to this form for L gives the equations of motion of the q_j coordinates:

$$m_j \ddot{q}_j = -\sum_k H_{j,k} q_k.$$

To find solutions that correspond to local harmonic motion, one assumes that the coordinates q_j oscillate in time according to

$$q_j(t) = q_j \cos(\omega t).$$

Substituting this form for $q_j(t)$ into the equations of motion gives

$$m_j\omega^2 q_j = \sum_k H_{j,k} q_k.$$

Defining

$$q_j' = q_j(m_j)^{1/2}$$

and introducing this into the above equation of motion yields

$$\omega^2 q_j' = \sum_k H'_{j,k} q_k',$$

where

$$H'_{j,k} = H_{j,k}(m_j m_k)^{-1/2}$$

is the so-called *mass-weighted Hessian* matrix.

The squares of the desired harmonic vibrational frequencies ω^2 are thus given as eigenvalues of the mass-weighted Hessian $\boldsymbol{H}'$:

$$\boldsymbol{H}'\boldsymbol{q}'_\alpha = \omega_\alpha^2 \boldsymbol{q}'_\alpha$$

The corresponding eigenvector, $\{q'_{\alpha,j}\}$ gives, when multiplied by $m_j^{-1/2}$, the atomic displacements that accompany that particular harmonic vibration. At a transition state, one of the ω_α^2 will be negative and $3N$–6 or $3N$–7 will be positive.

4. Reaction Path Following

The Hessian and gradient can also be used to trace out "streambeds" connecting local minima to transition states. In doing so, one utilizes a local harmonic description of the potential energy surface

$$E(\boldsymbol{x}) = E(\boldsymbol{0}) + \boldsymbol{x}\bullet\boldsymbol{g} + 1/2\boldsymbol{x}\bullet\boldsymbol{H}\bullet\boldsymbol{x} + \ldots,$$

where $\boldsymbol{x}$ represents the (small) step away from the point $\boldsymbol{x} = \boldsymbol{0}$ at which the gradient $\boldsymbol{g}$ and Hessian $\boldsymbol{H}$ have been evaluated. By expressing $\boldsymbol{x}$ and $\boldsymbol{g}$ in terms of the eigenvectors $\boldsymbol{v}_\alpha$ of $\boldsymbol{H}$

$$\boldsymbol{H}\boldsymbol{v}_\alpha = \lambda_\alpha \boldsymbol{v}_\alpha,$$

$$\boldsymbol{x} = \sum_\alpha \langle \boldsymbol{v}_\alpha | \boldsymbol{x} \rangle \boldsymbol{v}_\alpha = \sum_\alpha x_\alpha \boldsymbol{v}_\alpha,$$

$$\boldsymbol{g} = \sum_\alpha \langle \boldsymbol{v}_\alpha | \boldsymbol{g} \rangle \boldsymbol{v}_\alpha = \sum_\alpha g_\alpha \boldsymbol{v}_\alpha,$$

the energy change $E(\boldsymbol{x}) - E(\boldsymbol{0})$ can be expressed in terms of a sum of independent changes along the eigendirections:

$$E(\boldsymbol{x}) - E(\boldsymbol{0}) = \sum_\alpha [x_\alpha g_\alpha + 1/2 x_\alpha^2 \lambda_\alpha] + \ldots$$

Depending on the signs of g_α and of λ_α, various choices for the displacements x_α will produce increases or decreases in energy:

1. If λ_α is positive, then a step x_α "along" g_α (i.e., one with $x_\alpha g_\alpha$ positive) will generate an energy increase. A step "opposed to" g_α will generate an energy decrease if it is short enough that $x_\alpha g_\alpha$ is larger in magnitude than $1/2 x_\alpha^2 \lambda_\alpha$, otherwise the energy will increase.

2. If λ_α is negative, a step opposed to g_α will generate an energy decrease. A step along g_α will give an energy increase if it is short enough for $x_\alpha g_\alpha$ to be larger in magnitude than $1/2x_\alpha^2\lambda_\alpha$, otherwise the energy will decrease.

Thus, to proceed downhill in all directions (such as one wants to do when searching for local minima), one chooses each x_α in opposition to g_α and of small enough length to guarantee that the magnitude of $x_\alpha g_\alpha$ exceeds that of $1/2x_\alpha^2\lambda_\alpha$ for those modes with $\lambda_\alpha > 0$. To proceed uphill along a mode with $\lambda_\alpha' < 0$ and downhill along all other modes with $\lambda_\alpha > 0$, one chooses x_α' along g_α' with x_α' short enough to guarantee that $x_\alpha' g_\alpha'$ is larger in magnitude than $1/2x_\alpha^{2\prime}\lambda_\alpha'$, and one chooses the other x_α opposed to g_α and short enough that $x_\alpha g_\alpha$ is larger in magnitude than $1/2x_\alpha^2\lambda_\alpha$.

Such considerations have allowed the development of highly efficient potential energy surface "walking" algorithms (see, for example, J. Nichols, H. L. Taylor, P. Schmidt, and J. Simons, *J. Chem. Phys.* 92, 340 (1990) and references therein) designed to trace out streambeds and to locate and characterize, via the local harmonic frequencies, minima and transition states. These algorithms form essential components of most modern *ab initio*, semi-empirical, and empirical computational chemistry software packages.

II. *AB INITIO,* SEMI-EMPIRICAL AND EMPIRICAL FORCE FIELD METHODS

A. *Ab Initio* Methods

Most of the techniques described in this chapter are of the *ab initio* type. This means that they attempt to compute electronic state energies and other physical properties, as functions of the positions of the nuclei, from first principles without the use or knowledge of experimental input. Although perturbation theory or the variational method may be used to generate the working equations of a particular method, and although finite atomic orbital basis sets are nearly always utilized, these approximations do not involve "fitting" to known experimental data. They represent approximations that can be systematically improved as the level of treatment is enhanced.

B. Semi-Empirical and Fully Empirical Methods

Semi-empirical methods, such as those outlined in appendix F, use experimental data or the results of *ab initio* calculations to determine some of the matrix elements or integrals needed to carry out their procedures. Totally empirical methods attempt to describe the internal electronic energy of a system as a function of geometrical degrees of freedom (e.g., bond lengths and angles) in terms of analytical "force fields" whose parameters have been determined to "fit" known experimental data on some class of compounds. Examples of such parameterized force fields were presented in section III.A of chapter 16.

C. Strengths and Weaknesses

Each of these tools has advantages and limitations. *Ab initio* methods involve intensive computation and therefore tend to be limited, for practical reasons of computer time, to smaller atoms, molecules, radicals, and ions. Their CPU time needs usually vary with basis set size (M) as at least M^4; correlated methods require time proportional to at least M^5 because they involve transformation of the atomic-orbital-based two-electron integrals to the molecular orbital basis. As computers continue to advance in power and memory size, and as theoretical methods and algorithms continue to improve, *ab initio* techniques will be applied to larger and more complex species. When dealing with systems in which qualitatively new electronic environments and/or new bonding types arise, or excited electronic states that are unusual, *ab initio* methods are essential. Semi-empirical or empirical methods would be of little use on systems whose electronic properties have not been included in the data base used to construct the parameters of such models.

On the other hand, to determine the stable geometries of large molecules that are made of conventional chemical units (e.g., CC, CH, CO, etc. bonds and steric and torsional interactions among same), fully empirical force-field methods are usually quite reliable and computationally very fast. Stable geometries and the relative energetic stabilities of various conformers of large macromolecules and biopolymers can routinely be predicted using such tools if the system contains only conventional bonding and common chemical building blocks. These empirical potentials usually do not contain sufficient flexibility (i.e., their parameters and input data do not include enough knowledge) to address processes that involve rearrangement of the electronic configurations. For example, they can not treat:

1. Electronic transitions, because knowledge of the optical oscillator strengths and of the energies of excited states is absent in most such methods;
2. Concerted chemical reactions involving simultaneous bond breaking and forming, because to do so would require the force-field parameters to evolve from those of the reactant bonding to those for the product bonding as the reaction proceeds;
3. Molecular properties such as dipole moment and polarizability, although in certain fully empirical models, bond dipoles and lone-pair contributions have been incorporated (although again only for conventional chemical bonding situations).

Semi-empirical techniques share some of the strengths and weaknesses of *ab initio* and of fully empirical methods. They treat at least the valence electrons explicitly, so they are able to address questions that are inherently electronic such as electronic transitions, dipole moments, polarizability, and bond breaking and forming. Some of the integrals involving the Hamiltonian operator and the atomic basis orbitals are performed *ab initio;* others are obtained by fitting to experimental data. The computational needs of semi-empirical methods lie between those of the *ab initio* methods and the force-field techniques. As with the empirical methods, they should never be employed when qualitatively new electronic bonding situations are encountered because the data base upon which their parameters were determined contain, by assumption, no similar bonding cases.

Exercises, Problems, and Solutions

REVIEW EXERCISES

1. Contrast Slater-type orbitals (STOs) with Gaussian-type orbitals (GTOs).

EXERCISES

1. By expanding the molecular orbitals $\{\phi_k\}$ as linear combinations of atomic orbitals $\{\chi_\mu\}$,

$$\phi_k = \sum_\mu c_{\mu k}\chi_\mu$$

show how the canonical Hartree-Fock (HF) equations:

$$F\phi_i = \varepsilon_i\phi_j$$

reduce to the matrix eigenvalue-type equation of the form given in the text:

$$\sum_\nu F_{\mu\nu}C_{\nu i} = \varepsilon_i \sum_\nu S_{\mu\nu}C_{\nu i}$$

where:

$$F_{\mu\nu} = \left\langle \chi_\mu | h | \chi_\nu \right\rangle + \sum_{\delta\kappa} \left[\gamma_{\delta\kappa} \left\langle \chi_\mu \chi_\delta | g | \chi_\nu \chi_\kappa \right\rangle - \gamma^{ex}_{\delta\kappa} \left\langle \chi_\mu \chi_\delta | g | \chi_\kappa \chi_\nu \right\rangle \right],$$

$$S_{\mu\nu} = \left\langle \chi_\mu | \chi_\nu \right\rangle, \; \gamma_{\delta\kappa} = \sum_{i=occ} C_{\delta i} C_{\kappa i},$$

and

$$\gamma^{ex}_{\delta\kappa} = \sum_{i=occ \text{ and } same\ spin} C_{\delta i} C_{\kappa i}.$$

Note that the sum over i in $\gamma_{\delta\kappa}$ and $\gamma^{ex}_{\delta\kappa}$ is a sum over spin orbitals. In addition, show that this Fock matrix can be further reduced for the closed shell case to:

$$F_{\mu\nu} = \left\langle \chi_\mu | h | \chi_\nu \right\rangle + \sum_{\delta\kappa} P_{\delta\kappa} \left[\left\langle \chi_\mu \chi_\delta | g | \chi_\nu \chi_\kappa \right\rangle - \frac{1}{2} \left\langle \chi_\mu \chi_\delta | g | \chi_\kappa \chi_\nu \right\rangle \right],$$

where the charge bond order matrix, P, is defined to be:

$$P_{\delta\kappa} = \sum_{i=occ} 2 C_{\delta i} C_{\kappa i},$$

where the sum over i here is a sum over orbitals not spin orbitals.

2. Show that the HF total energy for a closed-shell system may be written in terms of integrals over the orthonormal HF orbitals as:

$$E(SCF) = 2 \sum_{k}^{occ} \left\langle \phi_k | h | \phi_k \right\rangle + \sum_{kl}^{occ} \left[2 \left\langle kl | g | kl \right\rangle - \left\langle kl | g | lk \right\rangle \right] + \sum_{\mu>\nu} \frac{Z_\mu Z_\nu}{R_{\mu\nu}}.$$

3. Show that the HF total energy may alternatively be expressed as:

$$E(SCF) = \sum_{k}^{occ} \left(\varepsilon_k + \left\langle \phi_k | h | \phi_k \right\rangle \right) + \sum_{\mu>\nu} \frac{Z_\mu Z_\nu}{R_{\mu\nu}},$$

where the ε_k refer to the HF orbital energies.

PROBLEMS

1. This problem will be concerned with carrying out an SCF calculation for the HeH^+ molecule in the

$$^1\Sigma_g^+ (1\sigma^2)$$

ground state. The one- and two-electron integrals (in atomic units) needed to carry out this SCF calculation at $R = 1.4$ a.u. using Slater type orbitals with orbital exponents of 1.6875 and 1.0 for the He and H, respectively are:

$S_{11} = 1.0$	$S_{22} = 1.0$	$S_{12} = 0.5784$
$h_{11} = -2.6442$	$h_{22} = -1.7201$	$h_{12} = -1.5113$
$g_{1111} = 1.0547$	$g_{1121} = 0.4744$	$g_{1212} = 0.5664$
$g_{2211} = 0.2469$	$g_{2221} = 0.3504$	$g_{2222} = 0.6250$

where 1 refers to $1s_{He}$ and 2 to $1s_H$. Note that the two-electron integrals are given in Dirac notation. Parts a–d should be done by hand. Any subsequent parts can make use of the QMIC software provided.

a. Using $\phi_1 \approx 1s_{He}$ for the initial guess of the occupied molecular orbital, form a 2×2 Fock matrix. Use the equation derived above in question 1 for $F_{\mu\nu}$.

b. Solve the Fock matrix eigenvalue equations given above to obtain the orbital energies and an improved occupied molecular orbital. In so doing, note that

$$\langle \phi_1 | \phi_1 \rangle = 1 = C_1^T S C_1$$

gives the needed normalization condition for the expansion coefficients of the ϕ_1 in the atomic orbital basis.

c. Determine the total SCF energy using the result of exercise 3 above at this step of the iterative procedure. When will this energy agree with that obtained by using the alternative expression for E(SCF) given in exercise 2?

d. Obtain the new molecular orbital, ϕ_1, from the solution of the matrix eigenvalue problem (part b).

e. A new Fock matrix and related total energy can be obtained with this improved choice of molecular orbital, ϕ_1. This process can be continued until a convergence criterion has been satisfied. Typical convergence criteria include: no significant change in the molecular orbitals or the total energy (or both) from one iteration to the next. Perform this iterative procedure for the HeH^+ system until the difference in total energy between two successive iterations is less than 10^{-5} a.u.

f. Show, by comparing the difference between the SCF total energy at one iteration and the converged SCF total energy, that the convergence of the above SCF approach is primarily linear (or first order).

g. Is the SCF total energy calculated at each iteration of the above SCF procedure (via exercise 3) an upper bound to the exact ground-state total energy?

h. Using the converged self-consistent set of molecular orbitals, ϕ_1 and ϕ_2, calculate the one- and two-electron integrals in the molecular orbital basis. Using the equations for E(SCF) in exercises 2 and 3 calculate the converged values of the orbital energies making use of these integrals in the mo basis.

i. Does this SCF wavefunction give rise (at $R \to \infty$) to proper dissociation products?

2. This problem will continue to address the same HeH^+ molecular system as above, extending the analysis to include "correlation effects." We will use the one- and two-electron integrals (same geometry) in the *converged* (to 10^{-5} au) SCF molecular orbital basis which we would have obtained after seven iterations above. The *converged* mos you would have obtained in problem 1 are:

$$\phi_1 = \begin{bmatrix} -0.89997792 \\ -0.15843012 \end{bmatrix} \qquad \phi_2 = \begin{bmatrix} -0.83233180 \\ 1.21558030 \end{bmatrix}$$

a. Carry out a two configuration CI calculation using the $1\sigma^2$ and $2\sigma^2$ configurations first by obtaining an expression for the CI matrix elements $H_{ij}(i,j = 1\sigma^2, 2\sigma^2)$ in terms of one- and two-electron integrals, and secondly by showing that the resultant CI matrix is (ignoring the nuclear repulsion term):

$$\begin{bmatrix} -4.2720 & 0.1261 \\ 0.1261 & -2.0149 \end{bmatrix}$$

b. Obtain the two CI energies and eigenvectors for the matrix found in part a.

c. Show that the lowest energy CI wavefunction is equivalent to the following two-determinant (single configuration) wavefunction:

$$\frac{1}{2}\left[\left|\left(a^{\frac{1}{2}}\phi_1 + b^{\frac{1}{2}}\phi_2\right)\alpha\left(a^{\frac{1}{2}}\phi_1 - b^{\frac{1}{2}}\phi_2\right)\beta\right| + \left|\left(a^{\frac{1}{2}}\phi_1 - b^{\frac{1}{2}}\phi_2\right)\alpha\left(a^{\frac{1}{2}}\phi_1 + b^{\frac{1}{2}}\phi_2\right)\beta\right|\right]$$

involving the polarized orbitals: $a^{\frac{1}{2}}\phi_1 \pm b^{\frac{1}{2}}\phi_2$, where $a = 0.9984$ and $b = 0.0556$.

d. Expand the CI list to 3 configurations by adding the $1\sigma 2\sigma$ to the original $1\sigma^2$ and $2\sigma^2$ configurations of part a above. First, express the proper singlet spin-coupled $1\sigma 2\sigma$ configuration as a combination of Slater determinants and then compute all elements of this 3×3 matrix.

e. Obtain all eigenenergies and corresponding normalized eigenvectors for this CI problem.

f. Determine the excitation energies and transition moments for HeH^+ using the full CI result of part e above. The nonvanishing matrix elements of the dipole operator $\boldsymbol{r}(x,y,z)$ in the atomic basis are:

$$\left\langle 1s_H \middle| z \middle| 1s_{He} \right\rangle = 0.2854 \text{ and } \left\langle 1s_H \middle| z \middle| 1s_H \right\rangle = 1.4 .$$

First determine the matrix elements of $\boldsymbol{r}$ in the SCF orbital basis then determine the excitation energies and transition moments from the ground state to the two excited singlet states of HeH^+.

g. Now turning to perturbation theory, carry out a RSPT calculation of the first-order wavefunction $|1\sigma^2>^{(1)}$ for the case in which the zeroth-order wavefunction is taken to be the $1\sigma^2$ Slater determinant. Show that the first-order wavefunction is given by:

$$|1\sigma^2>^{(1)} = -0.0442|2\sigma^2> .$$

h. Why does the $|1\sigma 2\sigma>$ configuration not enter into the first-order wavefunction?

i. Normalize the resultant wavefunction that contains zeroth- plus first-order parts and compare it to the wavefunction obtained in the two-configuration CI study of part b.

j. Show that the second-order RSPT correlation energy, $E^{(2)}$, of HeH^+ is -0.0056 a.u. How does this compare with the correlation energy obtained from the two-configuration CI study of part b?

3. Using the QMIC programs, calculate the SCF energy of HeH^+ using the same geometry as in problem 1 and the STO3G basis set provided in the QMIC basis set library. How does this energy compare to that found in problem 1? Run the calculation again with the 3-21G basis basis provided. How does this energy compare to the STO3G and the energy found using STOs in problem 1?

4. Generate SCF potential energy surfaces for HeH^+ and H_2 using the QMIC software provided. Use the 3-21G basis set and generate points for geometries of R = 1.0, 1.2, 1.4, 1.6, 1.8, 2.0, 2.5, and 10.0. Plot the energies vs. geometry for each system. Which system dissociates properly?

5. Generate CI potential energy surfaces for the 4 states of H_2 resulting from a CAS calculation with two electrons in the lowest two SCF orbitals ($1\sigma_g$ and $1\sigma_u$). Use the same geometries and basis set as in problem

4. Plot the energies vs. geometry for each system. Properly label and characterize each of the states (e.g., repulsive, dissociate properly, etc.).

SOLUTIONS

Review Exercises

1. Slater-type orbitals (STOs) are "hydrogen-like" in that they have a normalized form of:

$$\left(\frac{2\zeta}{a_o}\right)^{n+\frac{1}{2}}\left(\frac{1}{(2n)!}\right)^{\frac{1}{2}} r^{n-1} e^{\left(\frac{-\zeta r}{a_o}\right)} Y_{l,m}(\theta,\phi),$$

whereas Gaussian-type orbitals GTOs have the form:

$$N r^l e^{\left(-\alpha r^2\right)} Y_{l,m}(\theta,\phi),$$

although in most quantum chemistry computer programs they are specified in so-called "cartesian" form as:

$$N' x^a y^b z^c e^{\left(-\alpha r^2\right)},$$

where a, b, and c are quantum numbers each ranging from zero upward in unit steps. So, STOs give "better" overall energies and properties that depend on the shape of the wavefunction near the nuclei (e.g., Fermi contact ESR hyperfine constants) but they are more difficult to use (two-electron integrals are more difficult to evaluate; especially the four-center variety which have to be integrated numerically). GTOs on the other hand are easier to use (more easily integrable) but improperly describe the wavefunction near the nuclear centers because of the so-called cusp condition (they have zero slope at $R = 0$, whereas $1s$ STOs have non-zero slopes there).

Exercises

1. $F\phi_i = \varepsilon_i\phi_j = h\phi_i + \sum_j \left[J_j - K_j\right]\phi_i$

Let the closed shell Fock potential be written as:

$$V_{ij} = \sum_k \left(2\left\langle ik|jk\right\rangle - \left\langle ik|kj\right\rangle\right),$$

and the $1e^-$ component as:

$$h_{ij} = \left\langle \phi_i| - \frac{1}{2}\nabla^2 - \sum_A \frac{Z_A}{|r - R_A|}|\phi_j\right\rangle,$$

and the delta as: $\delta_{ij} = \left\langle i|j\right\rangle$, so that: $h_{ij} + V_{ij} = \delta_{ij}\varepsilon_i$

using:

$$\phi_i = \sum_\mu C_{\mu i}\chi_\mu,\ \phi_j = \sum_\nu C_{\nu j}\chi_\nu,\ \text{and}\ \phi_k = \sum_\gamma C_{\gamma k}\chi_\gamma,$$

and transforming from the mo to ao basis we obtain:

$$V_{ij} = \sum_{k\mu\gamma\nu\kappa} C_{\mu i}C_{\gamma k}C_{\nu j}C_{\kappa k}\left(2\left\langle \mu\gamma|\nu\kappa \right\rangle - \left\langle \mu\gamma|\kappa\nu \right\rangle\right)$$

$$= \sum_{k\mu\gamma\nu\kappa} (C_{\gamma k}C_{\kappa k})\,(C_{\mu i}C_{\nu j})\left(2\left\langle \mu\gamma|\nu\kappa \right\rangle - \left\langle \mu\gamma|\kappa\nu \right\rangle\right)$$

$$= \sum_{\mu\nu} (C_{\mu i}C_{\nu j})V_{\mu\nu} \text{ where,}$$

$$V_{\mu\nu} = \sum_{\gamma\kappa} P_{\gamma\kappa}\left(2\left\langle \mu\gamma|\nu\kappa \right\rangle - \left\langle \mu\gamma|\kappa\nu \right\rangle\right), \text{ and } P_{\gamma\kappa} = \sum_{k} (C_{\gamma k}C_{\kappa k}),$$

$$h_{ij} = \sum_{\mu\nu} (C_{\mu i}C_{\nu j})h_{\mu\nu}, \text{ where}$$

$$h_{\mu\nu} = \left\langle \chi_\mu| - \frac{1}{2}\nabla^2 - \sum_A \frac{Z_A}{|r - R_A|}|\chi_\nu \right\rangle, \text{ and}$$

$$\delta_{ij} = \left\langle i|j \right\rangle = \sum_{\mu\nu} (C_{\mu i}S_{\mu\nu}C_{\nu j}).$$

So, $h_{ij} + V_{ij} = \delta_{ij}\varepsilon_j$ becomes:

$$\sum_{\mu\nu} (C_{\mu i}C_{\nu j})h_{\mu\nu} + \sum_{\mu\nu} (C_{\mu i}C_{\nu j})V_{\mu\nu} = \sum_{\mu\nu} (C_{\mu i}S_{\mu\nu}C_{\nu j})\varepsilon_j,$$

$$\sum_{\mu\nu} (C_{\mu i}S_{\mu\nu}C_{\nu j})\varepsilon_j - \sum_{\mu\nu} (C_{\mu i}C_{\nu j})h_{\mu\nu} - \sum_{\mu\nu} (C_{\mu i}C_{\nu j})V_{\mu\nu} = 0 \text{ for all } i,j$$

$$\sum_{\mu\nu} C_{\mu i}\left[\varepsilon_j S_{\mu\nu} - h_{\mu\nu} - V_{\mu\nu}\right]C_{\nu j} = 0 \text{ for all } i,j$$

Therefore,

$$\sum_{\nu}\left[h_{\mu\nu} + V_{\mu\nu} - \varepsilon_j S_{\mu\nu}\right]C_{\nu j} = 0$$

This is FC = SCE.

2. The Slater-Condon rule for zero (spin-orbital) difference with N electrons in N spin orbitals is:

$$E = \left\langle |H + G| \right\rangle = \sum_{i}^{N} \left\langle \phi_i|h|\phi_i \right\rangle + \sum_{i>j}^{N}\left(\left\langle \phi_i\phi_j|g|\phi_i\phi_j \right\rangle - \left\langle \phi_i\phi_j|g|\phi_j\phi_i \right\rangle\right)$$

$$= \sum_i h_{ii} + \sum_{i>j}\left(g_{ijij} - g_{ijji}\right)$$

$$= \sum_i h_{ii} + \frac{1}{2}\sum_{ij}\left(g_{ijij} - g_{ijji}\right)$$

If all orbitals are doubly occupied and we carry out the spin integration we obtain:

$$E = 2\sum_{i}^{occ} h_{ii} + \sum_{ij}^{occ}\left(2g_{ijij} - g_{ijji}\right)$$

where i and j now refer to orbitals (not spin-orbitals).

3. If the occupied orbitals obey $F\phi_k = \varepsilon_k\phi_k$, then the expression for E in problem 2 above can be rewritten as

$$E = \sum_{i}^{occ}\left(h_{ii} + \sum_{j}^{occ}\left(2g_{ijij} - g_{ijji}\right)\right) + \sum_{i}^{occ} h_{ii}$$

We recognize the closed-shell Fock operator expression and rewrite this as:

$$E = \sum_{i}^{occ} F_{ii} + \sum_{i}^{occ} h_{ii} = \sum_{i}^{occ}\left(\varepsilon_i + h_{ii}\right)$$

Problems

1. We will use the QMIC software to do this problem. Lets just start from the beginning. Get the starting "guess" mo coefficients on disk. Using the program MOCOEFS it asks us for the first and second mo vectors. We input 1, 0 for the first mo (this means that the first mo is 1.0 times the He $1s$ orbital plus 0.0 times the H $1s$ orbital; this bonding mo is more likely to be heavily weighted on the atom having the higher nuclear charge) and 0, 1 for the second. Our beginning mo-ao array looks like:

$$\begin{bmatrix} 1.0 & 0.0 \\ 0.0 & 1.0 \end{bmatrix}$$

and is placed on disk in a file we choose to call "mocoefs.dat." We also put the ao integrals on disk using the program RW_INTS. It asks for the unique one- and two-electron integrals and places a canonical list of these on disk in a file we choose to call "ao_integrals.dat." At this point it is useful for us to step back and look at the set of equations which we wish to solve: $FC = SCE$. The QMIC software does not provide us with a so-called generalized eigenvalue solver (one that contains an overlap matrix; or metric), so in order to use the diagonalization program that is provided we must transform this equation ($FC = SCE$) to one that looks like ($F'C' = C'E$) . We do that in the following manner:

Since S is symmetric and positive definite we can find an $S^{-\frac{1}{2}}$ such that $S^{-\frac{1}{2}}S^{+\frac{1}{2}} = 1$, $S^{-\frac{1}{2}}S = S^{+\frac{1}{2}}$, etc. rewrite $FC = SCE$ by inserting unity between FC and multiplying the whole equation on the left by $S^{-\frac{1}{2}}$. This gives:

$$S^{-\frac{1}{2}}FS^{-\frac{1}{2}}S^{+\frac{1}{2}}C = S^{-\frac{1}{2}}SCE = S^{+\frac{1}{2}}CE.$$

Letting: $F' = S^{-\frac{1}{2}}FS^{-\frac{1}{2}}$, $C' = S^{+\frac{1}{2}}C$,
and inserting these expressions above give:

$$F'C' = C'E$$

Note, that to get the next iterations mo coefficients we must calculate C fromC':

$C' = S^{+\frac{1}{2}}C$, so, multiplying through on the left by $S^{-\frac{1}{2}}$ gives:

$$S^{-\frac{1}{2}}C' = S^{-\frac{1}{2}}S^{+\frac{1}{2}}C = C$$

This will be the method we will use to solve our fock equations. Find $S^{-1/2}$ by using the program FUNCT_MAT (this program generates a function of a matrix). This program will ask for the elements of the S array and write to disk a file (name of your choice . . . a good name might be "shalf") containing the $S^{-1/2}$ array. Now we are ready to begin the iterative Fock procedure.

a. Calculate the Fock matrix, F, using program FOCK which reads in the mo coefficients from "mocoefs.dat" and the integrals from "ao_integrals.dat" and writes the resulting Fock matrix to a user specified file (a good filename to use might be something like "fock1").

b. Calculate $F' = S^{-1/2}FS^{-1/2}$ using the program UTMATU which reads in F and $S^{-1/2}$ from files on the disk and writes F' to a user specified file (a good filename to use might be something like "fock1p"). Diagonalize F' using the program DIAG. This program reads in the matrix to be diagonalized from a user specified filename and writes the resulting eigenvectors to disk using a user specified filename (a good filename to use might be something like "coef1p"). You may wish to choose the option to write the eigenvalues (Fock orbital energies) to disk in order to use them at a later time in program FENERGY. Calculate C by back transforming e.g., $C = S^{-1/2}C'$. This is accomplished by using the program MATXMAT which reads in two matrices to be multiplied from user specified files and writes the product to disk using a user specified filename (a good filename to use might be something like "mocoefs.dat").

c. The QMIC program FENERGY calculates the total energy, using the result of exercises 2 and 3;

$$\sum_{kl} 2\langle k|h|k\rangle + 2\langle kl|kl\rangle - \langle kl|lk\rangle + \sum_{\mu>\nu} \frac{Z_\mu Z_\nu}{R_{\mu\nu}}, \text{ and}$$

$$\sum_{k} \varepsilon_k + \langle k|h|k\rangle + \sum_{\mu>\nu} \frac{Z_\mu Z_\nu}{R_{\mu\nu}}.$$

This is the conclusion of one iteration of the Fock procedure . . . you may continue by going back to part a and proceeding onward.

d.–e. Results for the successful convergence of this system using the supplied QMIC software is as follows (this is a lot of bloody detail but will give the user assurance that they are on the right track; alternatively one could switch to the QMIC program SCF and allow that program to iteratively converge the Fock equations):

The one-electron AO integrals:

$$\begin{bmatrix} -2.644200 & -1.511300 \\ -1.511300 & -1.720100 \end{bmatrix}$$

The two-electron AO integrals:

1 1 1 1	1.054700
2 1 1 1	0.4744000
2 1 2 1	0.5664000
2 2 1 1	0.2469000
2 2 2 1	0.3504000
2 2 2 2	0.6250000

The "initial" MO-AO coefficients:

$$\begin{bmatrix} 1.000000 & 0.000000 \\ 0.000000 & 1.000000 \end{bmatrix}$$

AO overlap matrix (S):

$$\begin{bmatrix} 1.000000 & 0.578400 \\ 0.578400 & 1.000000 \end{bmatrix}$$

$$S^{-\frac{1}{2}} \begin{bmatrix} 1.168032 & -0.3720709 \\ -0.3720709 & 1.168031 \end{bmatrix}$$

************** ITERATION 1 **************

The charge bond order matrix:

$$\begin{bmatrix} 1.000000 & 0.0000000 \\ 0.0000000 & 0.0000000 \end{bmatrix}$$

The Fock matrix (F):

$$\begin{bmatrix} -1.589500 & -1.036900 \\ -1.036900 & -0.8342001 \end{bmatrix}$$

$$S^{-\frac{1}{2}}FS^{-\frac{1}{2}} \begin{bmatrix} -1.382781 & -0.5048679 \\ -0.5048678 & -0.4568883 \end{bmatrix}$$

The eigenvalues of this matrix (Fock orbital energies) are:

[−1.604825 −0.2348450]

Their corresponding eigenvectors ($C' = S^{+\frac{1}{2}} \times C$) are:

$$\begin{bmatrix} -0.9153809 & -0.4025888 \\ -0.4025888 & 0.9153810 \end{bmatrix}$$

The "new" MO-AO coefficients ($C = S^{-\frac{1}{2}} \times C'$) :

$$\begin{bmatrix} -0.9194022 & -0.8108231 \\ -0.1296498 & 1.218985 \end{bmatrix}$$

The one-electron MO integrals:

$$\begin{bmatrix} -2.624352 & -0.1644336 \\ -0.1644336 & -1.306845 \end{bmatrix}$$

The two-electron MO integrals:

1 1 1 1	0.9779331
2 1 1 1	0.1924623
2 1 2 1	0.5972075
2 2 1 1	0.1170838
2 2 2 1	-0.0007945194
2 2 2 2	0.6157323

The closed-shell Fock energy from formula:

$$\sum_{kl} 2<k|h|k> + 2<kl|kl> - <kl|lk> + \sum_{\mu>\nu} \frac{Z_\mu Z_n}{R_{\mu\nu}} = -2.84219933$$

from formula:

$$\sum_{k} \varepsilon_k + <k|h|k> + \sum_{\mu>\nu} \frac{Z_\mu Z_n}{R_{\mu\nu}} = -2.80060530$$

the difference is: –0.04159403

************** ITERATION 2 **************

The charge bond order matrix:

$$\begin{bmatrix} 0.8453005 & 0.1192003 \\ 0.1192003 & 0.01680906 \end{bmatrix}$$

The Fock matrix:

$$\begin{bmatrix} -1.624673 & -1.083623 \\ -1.083623 & -0.8772071 \end{bmatrix}$$

$$S^{-\frac{1}{2}}FS^{-\frac{1}{2}} \quad \begin{bmatrix} -1.396111 & -0.5411037 \\ -0.5411037 & -0.4798213 \end{bmatrix}$$

The eigenvalues of this matrix (Fock orbital energies) are:

[–1.646972 –0.2289599]

Their corresponding eigenvectors ($C' = S^{+\frac{1}{2}} \times C$) are:

$$\begin{bmatrix} -0.9072427 & -0.4206074 \\ -0.4206074 & 0.9072427 \end{bmatrix}$$

The "new" MO-AO coefficients ($C = S^{-\frac{1}{2}} \times C'$):

$$\begin{bmatrix} -0.9031923 & -0.8288413 \\ -0.1537240 & 1.216184 \end{bmatrix}$$

The one-electron MO integrals:

$$\begin{bmatrix} -2.617336 & -0.1903475 \\ -0.1903475 & -1.313861 \end{bmatrix}$$

The two-electron MO integrals:

1 1 1 1	0.9626070
2 1 1 1	0.1949828
2 1 2 1	0.6048143
2 2 1 1	0.1246907
2 2 2 1	0.003694540
2 2 2 2	0.6158437

The closed-shell Fock energy from formula:

$$\sum_{kl} 2<k|h|k> + 2<kl|kl> - <kl|lk> + \sum_{\mu>\nu} \frac{Z_\mu Z_\nu}{R_{\mu\nu}} = -2.84349298$$

from formula:

$$\sum_{k} \varepsilon_k + <k|h|k> + \sum_{\mu>\nu} \frac{Z_\mu Z_n}{R_{\mu\nu}} = 2.83573675$$

the difference is: –0.00775623

************** ITERATION 3 **************

The charge bond order matrix:

$$\begin{bmatrix} 0.8157563 & 0.1388423 \\ 0.1388423 & 0.02363107 \end{bmatrix}$$

The Fock matrix:

$$\begin{bmatrix} -1.631153 & -1.091825 \\ -1.091825 & -0.8853514 \end{bmatrix}$$

$$S^{-\frac{1}{2}} F S^{-\frac{1}{2}} \quad \begin{bmatrix} -1.398951 & -0.5470731 \\ -0.5470730 & -0.4847007 \end{bmatrix}$$

The eigenvalues of this matrix (Fock orbital energies) are:

[–1.654745 –0.2289078]

Their corresponding eigenvectors ($C' = S^{+\frac{1}{2}} \times C$) are:

$$\begin{bmatrix} -0.9058709 & -0.4235546 \\ -0.4235545 & 0.9058706 \end{bmatrix}$$

The "new" MO-AO coefficients ($C = S^{-\frac{1}{2}} \times C'$):

$$\begin{bmatrix} -0.9004935 & -0.8317733 \\ -0.1576767 & 1.215678 \end{bmatrix}$$

The one-electron MO integrals:

$$\begin{bmatrix} -2.616086 & -0.1945811 \\ -0.1945811 & -1.315112 \end{bmatrix}$$

The two-electron MO integrals:

1 1 1 1	0.9600707
2 1 1 1	0.1953255
2 1 2 1	0.6060572
2 2 1 1	0.1259332
2 2 2 1	0.004475587
2 2 2 2	0.6158972

The closed-shell Fock energy from formula:

$$\sum_{kl} 2<k|h|k> + 2<kl|kl> - <kl|lk> + \sum_{\mu>\nu} \frac{Z_\mu Z_\nu}{R_{\mu\nu}} = -2.84353018$$

from formula:

$$\sum_{k} \varepsilon_k + <k|h|k> + \sum_{\mu>\nu} \frac{Z_\mu Z_\nu}{R_{\mu\nu}} = -2.84225941$$

the difference is: –0.00127077

************** ITERATION 4 **************

The charge bond order matrix:

$$\begin{bmatrix} 0.8108885 & 0.1419869 \\ 0.1419869 & 0.02486194 \end{bmatrix}$$

The Fock matrix:

$$\begin{bmatrix} -1.632213 & -1.093155 \\ -1.093155 & -0.8866909 \end{bmatrix}$$

$$S^{-\frac{1}{2}} F S^{-\frac{1}{2}} \quad \begin{bmatrix} -1.399426 & -0.5480287 \\ -0.5480287 & -0.4855191 \end{bmatrix}$$

The eigenvalues of this matrix (Fock orbital energies) are:

[–1.656015 –0.2289308]

Their corresponding eigenvectors ($C' = S^{+\frac{1}{2}} \times C$) are:

$$\begin{bmatrix} -0.9056494 & -0.4240271 \\ -0.4240271 & 0.9056495 \end{bmatrix}$$

The "new" MO-AO coefficients ($C = S^{-\frac{1}{2}} \times C'$):

$$\begin{bmatrix} -0.9000589 & -0.8322428 \\ -0.1583111 & 1.215595 \end{bmatrix}$$

The one-electron MO integrals:

$$\begin{bmatrix} -2.615881 & -0.1952594 \\ -0.1952594 & -1.315315 \end{bmatrix}$$

The two-electron MO integrals:

1 1 1 1	0.9596615
2 1 1 1	0.1953781
2 1 2 1	0.6062557
2 2 1 1	0.1261321
2 2 2 1	0.004601604
2 2 2 2	0.6159065

The closed-shell Fock energy from formula:

$$\sum_{kl} 2\langle k|h|k\rangle + 2\langle kl|kl\rangle - \langle kl|lk\rangle + \sum_{\mu>\nu} \frac{Z_\mu Z_\nu}{R_{\mu\nu}} = -2.84352922$$

from formula:

$$\sum_{k} \varepsilon_k + \langle k|h|k\rangle + \sum_{\mu>\nu} \frac{Z_\mu Z_\nu}{R_{\mu\nu}} = -2.84332418$$

the difference is: –0.00020504

************** ITERATION 5 **************

The charge bond order matrix:

$$\begin{bmatrix} 0.8101060 & 0.1424893 \\ 0.1424893 & 0.02506241 \end{bmatrix}$$

The Fock matrix:

$$\begin{bmatrix} -1.632385 & -1.093368 \\ -1.093368 & -0.8869066 \end{bmatrix}$$

$$S^{-\frac{1}{2}}FS^{-\frac{1}{2}} \quad \begin{bmatrix} -1.399504 & -0.5481812 \\ -0.5481813 & -0.4856516 \end{bmatrix}$$

The eigenvalues of this matrix (Fock orbital energies) are:

[–1.656219 –0.2289360]

Their corresponding eigenvectors ($C' = S^{+\frac{1}{2}} \times C$) are:

$$\begin{bmatrix} -0.9056138 & -0.4241026 \\ -0.4241028 & 0.9056141 \end{bmatrix}$$

The "new" MO-AO coefficients ($C = S^{-\frac{1}{2}} \times C'$):

$$\begin{bmatrix} -0.8999892 & -0.8323179 \\ -0.1584127 & 1.215582 \end{bmatrix}$$

The one-electron MO integrals:

$$\begin{bmatrix} -2.615847 & -0.1953674 \\ -0.1953674 & -1.315348 \end{bmatrix}$$

The two-electron MO integrals:

1 1 1 1	0.9595956
2 1 1 1	0.1953862
2 1 2 1	0.6062872
2 2 1 1	0.1261639
2 2 2 1	0.004621811
2 2 2 2	0.6159078

The closed-shell Fock energy from formula:

$$\sum_{kl} 2\langle k|h|k\rangle + 2\langle kl|kl\rangle - \langle kl|lk\rangle + \sum_{\mu>\nu} \frac{Z_\mu Z_\nu}{R_{\mu\nu}} = -2.84352779$$

from formula:

$$\sum_{k} \varepsilon_k + \langle k|h|k\rangle + \sum_{\mu>\nu} \frac{Z_\mu Z_\nu}{R_{\mu\nu}} = -2.84349489$$

the difference is: –0.00003290

************** ITERATION 6 **************

The charge bond order matrix:

$$\begin{bmatrix} 0.8099805 & 0.1425698 \\ 0.1425698 & 0.02509460 \end{bmatrix}$$

The Fock matrix:

$$\begin{bmatrix} -1.632412 & -1.093402 \\ -1.093402 & -0.8869413 \end{bmatrix}$$

$$S^{-\frac{1}{2}}FS^{-\frac{1}{2}} \quad \begin{bmatrix} -1.399517 & -0.5482056 \\ -0.5482056 & -0.4856730 \end{bmatrix}$$

The eigenvalues of this matrix (Fock orbital energies) are:

[–1.656253 –0.2289375]

Their corresponding eigenvectors ($C' = S^{+\frac{1}{2}} \times C$) are:

$$\begin{bmatrix} -0.9056085 & -0.4241144 \\ -0.4241144 & 0.9056086 \end{bmatrix}$$

The "new" MO-AO coefficients ($C = S^{-\frac{1}{2}} * C'$):

$$\begin{bmatrix} -0.8999786 & -0.8323296 \\ -0.1584283 & 1.215580 \end{bmatrix}$$

The one-electron MO integrals:

$$\begin{bmatrix} -2.615843 & -0.1953846 \\ -0.1953846 & -1.315353 \end{bmatrix}$$

The two-electron MO integrals:

1 1 1 1	0.9595859
2 1 1 1	0.1953878
2 1 2 1	0.6062925
2 2 1 1	0.1261690
2 2 2 1	0.004625196
2 2 2 2	0.6159083

The closed-shell Fock energy from formula:

$$\sum_{kl} 2<k|h|k> + 2<kl|kl> - <kl|lk> + \sum_{\mu>\nu} \frac{Z_\mu Z_\nu}{R_{\mu\nu}} = -2.84352827$$

from formula:

$$\sum_{k} \varepsilon_k + <k|h|k> + \sum_{\mu>\nu} \frac{Z_\mu Z_\nu}{R_{\mu\nu}} = -2.84352398 = -2.84352398$$

the difference is: –0.00000429

************** ITERATION 7 **************

The charge bond order matrix:

$$\begin{bmatrix} 0.8099616 & 0.1425821 \\ 0.1425821 & 0.02509952 \end{bmatrix}$$

The Fock matrix:

$$\begin{bmatrix} -1.632416 & -1.093407 \\ -1.093407 & -0.8869464 \end{bmatrix}$$

$$S^{-\frac{1}{2}}FS^{-\frac{1}{2}} \quad \begin{bmatrix} -1.399519 & -0.5482093 \\ -0.5482092 & -0.4856761 \end{bmatrix}$$

The eigenvalues of this matrix (Fock orbital energies) are:

[–1.656257 –0.2289374]

Their corresponding eigenvectors ($C' = S^{+\frac{1}{2}} \times C$) are:

$$\begin{bmatrix} -0.9056076 & -0.4241164 \\ -0.4241164 & 0.9056077 \end{bmatrix}$$

The "new" MO-AO coefficients ($C = S^{-\frac{1}{2}} \times C'$):

$$\begin{bmatrix} -0.8999770 & -0.8323317 \\ -0.1584310 & 1.215580 \end{bmatrix}$$

The one-electron MO integrals:

$$\begin{bmatrix} -2.615843 & -0.1953876 \\ -0.1953876 & -1.315354 \end{bmatrix}$$

The two-electron MO integrals:

1 1 1 1	0.9595849
2 1 1 1	0.1953881
2 1 2 1	0.6062936
2 2 1 1	0.1261697
2 2 2 1	0.004625696
2 2 2 2	0.6159083

The closed-shell Fock energy from formula:

$$\sum_{kl} 2<k|h|k> + 2<kl|kl> - <kl|lk> + \sum_{\mu>\nu} \frac{Z_\mu Z_\nu}{R_{\mu\nu}} = -2.84352922$$

from formula:

$$\sum_{k} \varepsilon_k + <k|h|k> + \sum_{\mu>\nu} \frac{Z_\mu Z_\nu}{R_{\mu\nu}} = -2.84352827$$

the difference is: –0.00000095

************** ITERATION 8 **************

The charge bond order matrix:

$$\begin{bmatrix} 0.8099585 & 0.1425842 \\ 0.1425842 & 0.02510037 \end{bmatrix}$$

The Fock matrix:

$$\begin{bmatrix} -1.632416 & -1.093408 \\ -1.093408 & -0.8869470 \end{bmatrix}$$

$$S^{-\frac{1}{2}}FS^{-\frac{1}{2}} \quad \begin{bmatrix} -1.399518 & -0.5482103 \\ -0.5482102 & -0.4856761 \end{bmatrix}$$

The eigenvalues of this matrix (Fock orbital energies) are:

[–1.656258 –0.2289368]

Their corresponding eigenvectors ($C' = S^{+\frac{1}{2}} \times C$) are:

$$\begin{bmatrix} -0.9056074 & -0.4241168 \\ -0.4241168 & 0.9056075 \end{bmatrix}$$

The "new" MO-AO coefficients ($C = S^{-\frac{1}{2}} \times C'$):

$$\begin{bmatrix} -0.8999765 & -0.8323320 \\ -0.1584315 & 1.215579 \end{bmatrix}$$

The one-electron MO integrals:

$$\begin{bmatrix} -2.615842 & -0.1953882 \\ -0.1953882 & -1.315354 \end{bmatrix}$$

The two-electron MO integrals:

1 1 1 1	0.9595841
2 1 1 1	0.1953881
2 1 2 1	0.6062934
2 2 1 1	0.1261700
2 2 2 1	0.004625901
2 2 2 2	0.6159081

The closed-shell Fock energy from formula:

$$\sum_{kl} 2<k|h|k> + 2<kl|kl> - <kl|lk> + \sum_{\mu>\nu} \frac{Z_\mu Z_\nu}{R_{\mu\nu}} = -2.84352827$$

from formula:

$$\sum_{k} \varepsilon_k + <k|h|k> + \sum_{\mu>\nu} \frac{Z_\mu Z_n}{R_{\mu\nu}} = -2.84352827 = -2.84352827$$

the difference is: 0.00000000

f. In looking at the energy convergence we see the following:

Iter	Formula 1	Formula 2
1	−2.84219933	−2.80060530
2	−2.84349298	−2.83573675
3	−2.84353018	−2.84225941
4	−2.84352922	−2.84332418
5	−2.84352779	−2.84349489
6	−2.84352827	−2.84352398
7	−2.84352922	−2.84352827
8	-2.84352827	-2.84352827

If you look at the energy differences (SCF at iteration n – SCF converged) and plot this data versus iteration number, and do a fifth-order polynomial fit, we see the following:

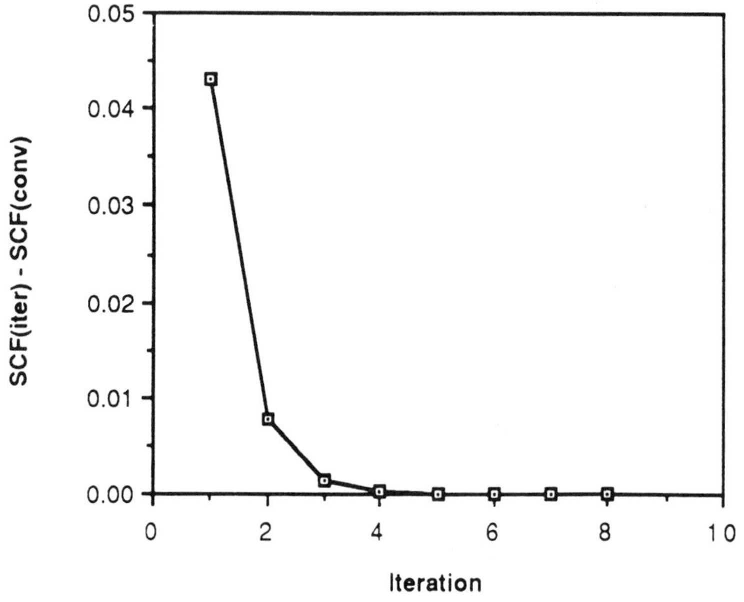

In looking at the polynomial fit we see that the convergence is primarily linear since the coefficient of the linear term is much larger than those of the cubic and higher terms.

g. The converged SCF total energy calculated using the result of exercise 3 is an upper bound to the ground state energy, but, during the iterative procedure it is not. At convergence, the expectation value of the Hamiltonian for the Hartree Fock determinant is given by the equation in exercise 3.

h. The one- and two-electron integrals in the MO basis are given above (see part e, iteration 8). The orbital energies are found using the result of exercise 2 and 3 to be:

$$E(SCF) = \sum_{k} \varepsilon_k + \langle k|h|k\rangle + \sum_{\mu>\nu} \frac{Z_\mu Z_\nu}{R_{\mu\nu}}$$

$$E(SCF) = \sum_{kl} 2\langle k|h|k\rangle + 2\langle kl|kl\rangle - \langle kl|lk\rangle + \sum_{\mu>\nu} \frac{Z_\mu Z_\nu}{R_{\mu\nu}}$$

$$\text{so, } \varepsilon_k = \langle k|h|k\rangle + \sum_{l}^{occ} (2\langle kl|kl\rangle - \langle kl|lk\rangle)$$

$$\begin{aligned}\varepsilon_1 &= h_{11} + 2\langle 11|11\rangle - \langle 11|11\rangle \\ &= -2.615842 + 0.9595841 \\ &= -1.656258 \\ \varepsilon_2 &= h_{22} + 2\langle 21|21\rangle - \langle 21|12\rangle \\ &= -1.315354 + 2*0.6062934 - 0.1261700 \\ &= -0.2289372\end{aligned}$$

i. Yes, the $1\sigma^2$ configuration does dissociate properly because at at $R \to \infty$ the lowest energy state is He + H^+, which also has a $1\sigma^2$ orbital occupancy (i.e., $1s^2$ on He and $1s^0$ on H^+).

2. At convergence the mo coefficients are:

$$\phi_1 = \begin{bmatrix} -0.8999765 \\ -0.1584315 \end{bmatrix} \phi_2 = \begin{bmatrix} -0.8323320 \\ 1.215579 \end{bmatrix}$$

and the integrals in this MO basis are:

$h_{11} = -2.615842$	$h_{21} = -0.1953882$	$h_{22} = -1.315354$
$g_{1111} = 0.9595841$	$g_{2111} = 0.1953881$	$g_{2121} = 0.6062934$
$g_{2211} = 0.1261700$	$g_{2221} = 0.004625901$	$g_{2222} = 0.6159081$

a. $$H = \begin{bmatrix} \langle 1\sigma^2|H|1\sigma^2\rangle & \langle 1\sigma^2|H|2\sigma^2\rangle \\ \langle 2\sigma^2|H|1\sigma^2\rangle & \langle 2\sigma^2|H|2\sigma^2\rangle \end{bmatrix} = \begin{bmatrix} 2h_{11} + g_{1111} & g_{1122} \\ g_{1122} & 2h_{22} + g_{2222} \end{bmatrix}$$

$$= \begin{bmatrix} 2* - 2.615842 + 0.9595841 & 0.1261700 \\ 0.1261700 & 2* -1.315354 + 0.6159081 \end{bmatrix}$$

$$= \begin{bmatrix} -4.272100 & 0.1261700 \\ 0.1261700 & -2.014800 \end{bmatrix}$$

b. The eigenvalues are $E_1 = -4.279131$ and $E_2 = -2.007770$. The corresponding eigenvectors are:

$$C_1 = \begin{bmatrix} -.99845123 \\ 0.05563439 \end{bmatrix}, \quad C_2 = \begin{bmatrix} 0.05563438 \\ 0.99845140 \end{bmatrix}$$

c. $\frac{1}{2}\left[\left|\left(a\frac{1}{2}\phi_1 + b\frac{1}{2}\phi_2\right)\alpha\left(a\frac{1}{2}\phi_1 - b\frac{1}{2}\phi_2\right)\beta\right| + \left|\left(a\frac{1}{2}\phi_1 - b\frac{1}{2}\phi_2\right)\alpha\left(a\frac{1}{2}\phi_1 + b\frac{1}{2}\phi_2\right)\beta\right|\right]$

$$= \frac{1}{2\sqrt{2}}\left[\left(a\frac{1}{2}\phi_1 + b\frac{1}{2}\phi_2\right)\left(a\frac{1}{2}\phi_1 - b\frac{1}{2}\phi_2\right) + \left(a\frac{1}{2}\phi_1 - b\frac{1}{2}\phi_2\right)\left(a\frac{1}{2}\phi_1 + b\frac{1}{2}\phi_2\right)\right](\alpha\beta - \beta\alpha)$$

$$= \frac{1}{\sqrt{2}}\left(a\phi_1\phi_1 - b\phi_2\phi_2\right)(\alpha\beta - \beta\alpha)$$

$$= a\left|\phi_1\alpha\phi_1\beta\right| - b\left|\phi_2\alpha\phi_2\beta\right|.$$

(note from part b, $a = 0.9984$ and $b = 0.0556$)

d. The third configuration $|1\sigma 2\sigma| = 1/\sqrt{2}\,[|1\alpha 2\beta| - |1\beta 2\alpha|]$,
Adding this configuration to the previous 2×2CI results in the following 3×3 "full" CI:

$$H = \begin{bmatrix} \langle 1\sigma^2|H|1\sigma^2\rangle & \langle 1\sigma^2|H|2\sigma^2\rangle & \langle 1\sigma^2|H|1\sigma 2\sigma\rangle \\ \langle 2\sigma^2|H|1\sigma^2\rangle & \langle 2\sigma^2|H|2\sigma^2\rangle & \langle 2\sigma^2|H|1\sigma 2\sigma\rangle \\ \langle 1\sigma 2\sigma|H|1\sigma^2\rangle & \langle 1\sigma 2\sigma|H|2\sigma^2\rangle & \langle 1\sigma 2\sigma|H|1\sigma 2\sigma\rangle \end{bmatrix}$$

$$= \begin{bmatrix} 2h_{11} + g_{1111} & g_{1122} & \frac{1}{\sqrt{2}}\left[2h_{12} + 2g_{2111}\right] \\ g_{1122} & 2h_{22} + g_{2222} & \frac{1}{\sqrt{2}}\left[2h_{12} + 2g_{2221}\right] \\ \frac{1}{\sqrt{2}}\left[2h_{12} + 2g_{2111}\right] & \frac{1}{\sqrt{2}}\left[2h_{12} + 2g_{2221}\right] & h_{11} + h_{22} + g_{2121} + g_{2211} \end{bmatrix}$$

Evaluating the new matrix elements:

$H_{13} = H_{31} = \sqrt{2} * (-0.1953882 + 0.1953881) = 0.0$
$H_{23} = H_{32} = \sqrt{2} * (-0.1953882 + 0.004626) = -0.269778$
$H_{33} = -2.615842 - 1.315354 + 0.606293 + 0.126170$
$= -3.198733$

$$= \begin{bmatrix} -4.272100 & 0.126170 & 0.0 \\ 0.126170 & -2.014800 & -0.269778 \\ 0.0 & -0.269778 & -3.198733 \end{bmatrix}$$

e. The eigenvalues are

$E_1 = -4.279345$, $E_2 = -3.256612$ and $E_3 = -1.949678$.

The corresponding eigenvectors are:

$$C_1 = \begin{bmatrix} -0.99825280 \\ 0.05732290 \\ 0.01431085 \end{bmatrix}, C_2 = \begin{bmatrix} -0.02605343 \\ -0.20969283 \\ -0.97742000 \end{bmatrix}, C_3 = \begin{bmatrix} -0.05302767 \\ -0.97608540 \\ 0.21082004 \end{bmatrix}$$

f. We need the non-vanishing matrix elements of the dipole operator in the mo basis. These can be obtained by calculating them by hand. They are more easily obtained by using the TRANS program. Put the $1e^-$ ao integrals on disk by running the program RW_INTS. In this case you are inserting $z_{11} = 0.0$, $z_{21} = 0.2854$, and $z_{22} = 1.4$ (insert 0.0 for all the $2e^-$ integrals) . . . call the output file "ao_dipole.ints" for example. The converged MO-AO coefficients should be in a file ("mocoefs.dat" is fine). The transformed integrals can be written to a file (name of your choice) for example "mo_dipole.ints." These matrix elements are:

$z_{11} = 0.11652690$, $z_{21} = -0.54420990$, $z_{22} = 1.49117320$

The excitation energies are

$E_2 - E_1 = -3.256612 - -4.279345 = 1.022733$, and $E_3 - E_1 = -1.949678 - -4.279345 = 2.329667$.

Using the Slater-Condon rules to obtain the matrix elements between configurations we get:

$$H_z = \begin{bmatrix} \langle 1\sigma^2|z|1\sigma^2 & \langle 1\sigma^2|z|2\sigma^2\rangle & \langle 1\sigma^2|z|1\sigma 2\sigma\rangle \\ \langle 2\sigma^2|z|1\sigma^2\rangle & \langle 2\sigma^2|z|2\sigma^2\rangle & \langle 2\sigma^2|z|1\sigma 2\sigma\rangle \\ \langle 1\sigma 2\sigma|z|1\sigma^2\rangle & \langle 1\sigma 2\sigma|z|2\sigma^2\rangle & \langle 1\sigma 2\sigma|z|1\sigma 2\sigma\rangle \end{bmatrix}$$

$$= \begin{bmatrix} 2z_{11} & 0 & \frac{1}{\sqrt{2}}\left[2z_{12}\right] \\ 0 & 2z_{22} & \frac{1}{\sqrt{2}}\left[2z_{12}\right] \\ \frac{1}{\sqrt{2}}\left[2z_{12}\right] & \frac{1}{\sqrt{2}}\left[2z_{12}\right] & z_{11} + z_{22} \end{bmatrix}$$

$$= \begin{bmatrix} 0.233054 & 0 & -0.769629 \\ 0 & 2.982346 & -0.769629 \\ -0.769629 & -0.769629 & 1.607700 \end{bmatrix}$$

Now, $\langle \Psi_1|z|\Psi_2\rangle = C_1^T H_z C_2$ (this can be accomplished with the program UTMATU):

$$= \begin{bmatrix} -0.99825280 \\ 0.05732290 \\ 0.01431085 \end{bmatrix}^T \begin{bmatrix} 0.233054 & 0 & -0.769629 \\ 0 & 2.982346 & -0.769629 \\ -0.769629 & -0.769629 & 1.607700 \end{bmatrix} \begin{bmatrix} -0.02605343 \\ -0.20969283 \\ -0.97742000 \end{bmatrix}$$

$= -.757494$ and,

$\langle \Psi_1|z|\Psi_3\rangle = C_1^T H_z C_3$

$$= \begin{bmatrix} -0.99825280 \\ 0.05732290 \\ 0.01431085 \end{bmatrix}^T \begin{bmatrix} 0.233054 & 0 & -0.769629 \\ 0 & 2.982346 & -0.769629 \\ -0.769629 & -0.769629 & 1.607700 \end{bmatrix} \begin{bmatrix} -0.05302767 \\ -0.97608540 \\ 0.21082004 \end{bmatrix}$$

$= 0.014322$

g. Using the converged coefficients the orbital energies obtained from solving the Fock equations are $\varepsilon_1 = -1.656258$ and $\varepsilon_2 = -0.228938$. The resulting expression for the RSPT first-order wavefunction becomes:

$$|1\sigma^2\rangle^{(1)} = -\frac{g_{2211}}{2(\varepsilon_2 - \varepsilon_1)}|2\sigma^2\rangle$$

$$|1\sigma^2\rangle^{(1)} = -\frac{0.126170}{2(-0.228938 + 1.656258)}|2\sigma^2\rangle$$

$$|1\sigma^2\rangle^{(1)} = -0.0441982|2\sigma^2\rangle$$

h. As you can see from part c, the matrix element $\langle 1\sigma^2|H|1\sigma 2\sigma\rangle = 0$ (this is also a result of the Brillouin theorem) and hence this configuration does not enter into the first-order wavefunction.

i. $|0\rangle = |1\sigma^2\rangle - 0.0441982|2\sigma^2\rangle$. To normalize we divide by:

$$\sqrt{\left[1 + (0.0441982)^2\right]} = 1.0009762$$

$|0\rangle = 0.999025|1\sigma^2\rangle - 0.044155|2\sigma^2\rangle$

In the 2×2 CI we obtained:

$|0\rangle = 0.99845123|1\sigma^2\rangle - 0.05563439|2\sigma^2\rangle$.

The expression for the 2nd order RSPT is:

$$E^{(2)} = -\frac{|g_{2211}|^2}{2(\varepsilon_2 - \varepsilon_1)} = -\frac{0.126170^2}{2(-0.228938 + 1.656258)} = -0.005576\text{au}$$

Comparing the 2×2 CI energy obtained to the SCF result we have:

$-4.279131 - (-4.272102) = -0.007029$au

3. STO total energy: –2.8435283
STO3G total energy: –2.8340561
3-21G total energy: –2.8864405

The STO3G orbitals were generated as a best fit of three primitive gaussians (giving 1 CGTO) to the STO. So, STO3G can at best reproduce the STO result. The 3-21G orbitals are more flexible since there are two CGTOs per atom. This gives four orbitals (more parameters to optimize) and a lower total energy.

4.

R	**HeH^+ Energy**	**H_2 Energy**
1.0	–2.812787056	–1.071953297
1.2	–2.870357513	–1.113775015
1.4	–2.886440516	–1.122933507
1.6	–2.886063576	–1.115567684
1.8	–2.880080938	–1.099872589
2.0	–2.872805595	–1.080269098
2.5	–2.856760263	–1.026927710
10.0	–2.835679293	–0.7361705303

Plotting total energy vs. geometry for HeH^+:

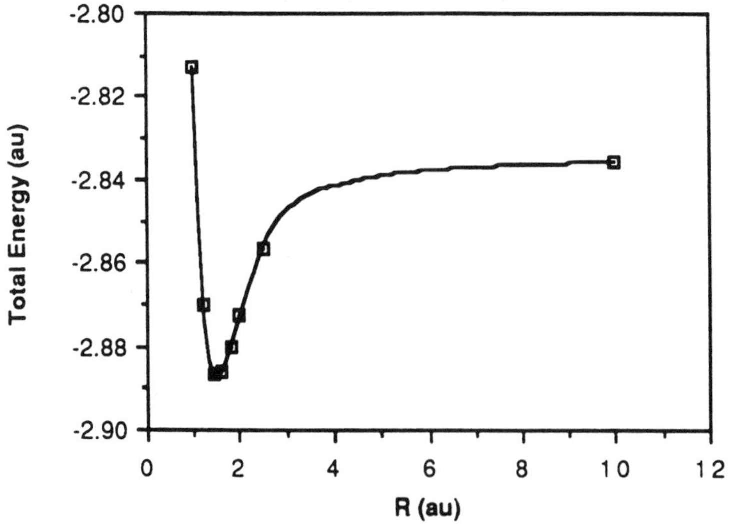

Plotting total energy vs. geometry for H_2:

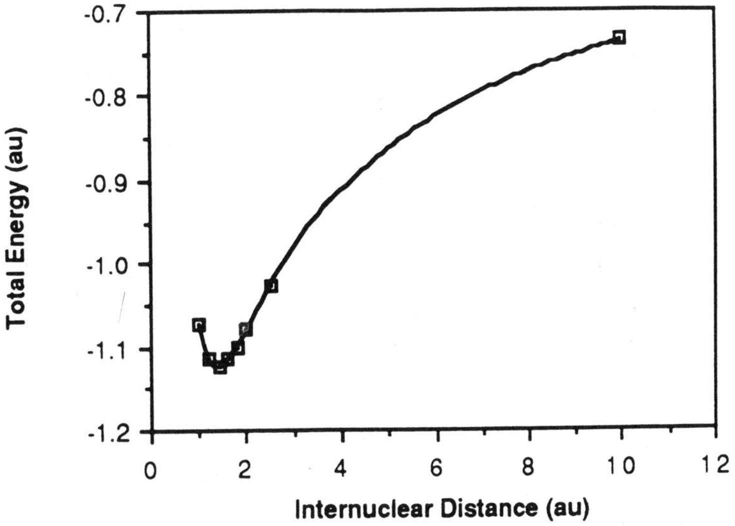

For HeH^+ at $R = 10.0$au, the eigenvalues of the converged Fock matrix and the corresponding converged MO-AO coefficients are:

−.1003571E + 01	−.4961988E + 00	.5864846E + 00	.1981702E + 01
.4579189E + 00	−.8245406E − 05	.1532163E − 04	.1157140E + 01
.6572777E + 00	−.4580946E − 05	−.6822942E − 05	−.1056716E + 01
−.1415438E − 05	.3734069E + 00	.1255539E + 01	−.1669342E − 04
.1112778E − 04	.7173244E + 00	−.1096019E + 01	.2031348E − 04

Notice that this indicates that orbital 1 is a combination of the s functions on He only (dissociating properly to He + H^+).

For H_2 at $R = 10.0$au, the eigenvalues of the converged Fock matrix and the corresponding converged MO-AO coefficients are:

−.2458041E + 00	−.1456223E + 00	.1137235E + 01	.1137825E + 01
.1977649E + 00	−.1978204E + 00	.1006458E + 01	−.7903225E + 00
.5632566E + 00	−.5628273E + 00	−.8179120E + 00	.6424941E + 00
.1976312E + 00	.1979216E + 00	.7902887E + 00	.1006491E + 01
.5629326E + 00	.5631776E + 00	−.6421731E + 00	−.8181460E + 00

Notice that this indicates that orbital 1 is a combination of the *s* functions on both H atoms (dissociating improperly; equal probabilities of H_2 dissociating to two neutral atoms or to a proton plus hydride ion).

5. The H_2 CI result:

R	$^1\Sigma_g^+$	$^3\Sigma_u^+$	$^1\Sigma_u^+$	$^1\Sigma_g^+$
1.0	−1.074970	−0.5323429	−0.3997412	0.3841676
1.2	−1.118442	−0.6450778	−0.4898805	0.1763018
1.4	−1.129904	−0.7221781	−0.5440346	0.0151913
1.6	−1.125582	−0.7787328	−0.5784428	−0.1140074
1.8	−1.113702	−0.8221166	−0.6013855	−0.2190144
2.0	−1.098676	−0.8562555	−0.6172761	−0.3044956
2.5	−1.060052	−0.9141968	−0.6384557	−0.4530645
5.0	−0.9835886	−0.9790545	−0.5879662	−0.5802447
7.5	−0.9806238	−0.9805795	−0.5247415	−0.5246646
10.0	−0.980598	−0.9805982	−0.4914058	−0.4913532

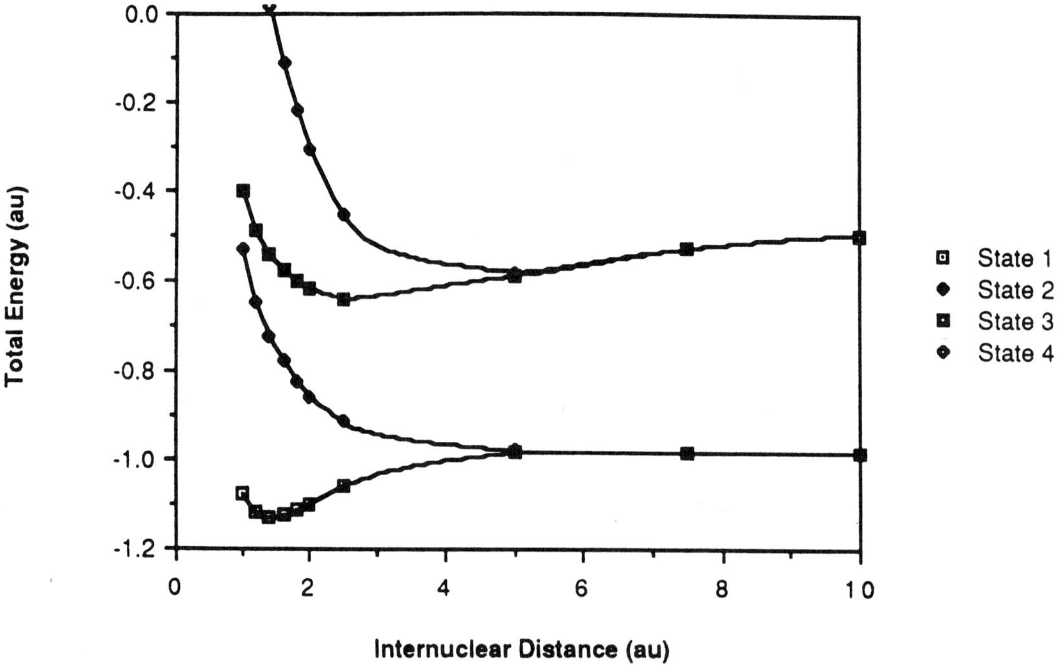

For H_2 at $R = 1.4$au, the eigenvalues of the Hamiltonian matrix and the corresponding determinant amplitudes are:

Determinant	−1.129904	−0.722178	−0.544035	0.015191
$\|1\sigma_g\alpha 1\sigma_g\beta\|$	0.99695	0.00000	0.00000	0.07802
$\|1\sigma_g\beta 1\sigma_u\alpha\|$	0.00000	0.70711	0.70711	0.00000
$\|1\sigma_g\alpha 1\sigma_u\beta\|$	0.00000	0.70711	−0.70711	0.00000
$\|1\sigma_u\alpha 1\sigma_u\beta\|$	−0.07802	0.00000	0.00000	0.99695

This shows, as expected, the mixing of the first $^1\Sigma_g^+(1\sigma_g^2)$ and the second $^1\Sigma_g^+(1\sigma_u^2)$ determinants, the

$$^3\Sigma_u^+ = \left(\frac{1}{\sqrt{2}}\left(|1\sigma_g\beta 1\sigma_u\alpha| + |1\sigma_g\alpha 1\sigma_u\beta|\right)\right), \text{ and the}$$

$$^1\Sigma_u^+ = \left(\frac{1}{\sqrt{2}}\left(|1\sigma_g\beta 1\sigma_u\alpha| - |1\sigma_g\alpha 1\sigma_u\beta|\right)\right).$$

Also notice that the first

$$^1\Sigma_g^+$$

state is the bonding (0.99695 – 0.07802) combination (note specifically the + – combination) and the second

$$^1\Sigma_g^+$$

state is the antibonding combination (note specifically the + + combination). The + + combination always gives a higher energy than the + – combination. Also notice that the first and second states

$$^1\Sigma_g^+ \text{ and } {}^3\Sigma_u^+$$

are dissociating to two neutral atoms and the third and fourth states

$$^1\Sigma_g^+ \text{ and } {}^3\Sigma_u^+$$

are dissociating to proton/anion combinations. The difference in these energies is the ionization potential of H minus the electron affinity of H.

SECTION

7

APPENDICES

APPENDIX

A Mathematics Review

I. MATRICES

A. Vectors

A vector is an object having n-components

$$\boldsymbol{x} = (x_1, x_2, \ldots x_n).$$

These components may represent, for example, the cartesian coordinates of a particle (in which case, $n = 3$) or the cartesian coordinates of N particles (in which case, $n = 3N$) . Alternatively, the vector components may have nothing what so ever to do with cartesian or other coordinate-system positions.

The numbers x_i are called the components of the vector $\boldsymbol{x}$ in the directions of some n elementary unit vectors:

$$\boldsymbol{x} = x_1 \bullet \boldsymbol{e_1} + x_2 \bullet \boldsymbol{e_2} + x_3 \bullet \boldsymbol{e_3} + \ldots + x_n \bullet \boldsymbol{e_n}$$

$$= x_1 \begin{bmatrix} 1 \\ 0 \\ 0 \\ 0 \\ 0 \\ \cdot \\ \cdot \\ \cdot \end{bmatrix} + x_2 \begin{bmatrix} 0 \\ 1 \\ 0 \\ 0 \\ 0 \\ \cdot \\ \cdot \\ \cdot \end{bmatrix} + x_3 \begin{bmatrix} 0 \\ 0 \\ 1 \\ 0 \\ 0 \\ \cdot \\ \cdot \\ \cdot \end{bmatrix} + \ldots + x_n \begin{bmatrix} 0 \\ 0 \\ \cdot \\ \cdot \\ \cdot \\ 0 \\ 0 \\ 1 \end{bmatrix}$$

The unit vectors $\boldsymbol{e}_i$, whose exact definition, meaning and interpretation depend on the particular application at hand, are called basis vectors and form the elements of a *basis*. They are particularly simple to work with because they are *orthogonal*. This means that their dot products vanish $\boldsymbol{e}_i \cdot \boldsymbol{e}_j = 0$, unless $i = j$. If $i = j$, then the scalar or dot product is unity (it is usually convenient, but not necessary, to use bases that are normalized so $\boldsymbol{e}_i \cdot \boldsymbol{e}_i = 1$). The shorthand way of representing this information is to write

$$\boldsymbol{e}_i \cdot \boldsymbol{e}_j = <\boldsymbol{e}_i|\boldsymbol{e}_j> = \delta_{ij},$$

where δ_{ij} is called the Kronecker delta function defined by:

$$\delta_{ij} = 0, \text{ if } i \neq j, \text{ and}$$
$$\delta_{ij} = 1 \text{ if } i = j.$$

The above equation for $\boldsymbol{x}$ provides an example of expressing a vector as a linear combination of other vectors (in this case, the basis vectors). The vector $\boldsymbol{x}$ is expressed as a linear combination of the unit vectors $\boldsymbol{e}_i$, and the numbers x_i are the coefficients in the linear combination. Another way of writing this within the summation notation is:

$$\boldsymbol{x} = \sum_i^n x_i \boldsymbol{e}_i.$$

The idea of a linear combination is an important idea that will be encountered when we discuss how a matrix operator affects a linear combination of vectors.

B. Products of Matrices and Vectors

If $\underline{\mathrm{M}}$ is an $n \times n$ matrix with elements M_{ij}, (the first subscript specifies the row number and the second subscript specifies the column number), then the product $\underline{\mathrm{M}}\ \boldsymbol{x} = \boldsymbol{y}$ is a vector whose components (when subscripts i,j,k, etc. appear, they can take any value $1,2, \ldots n$ unless otherwise specified) are defined as follows:

$$y_k = \sum_j^n M_{kj} x_j$$

The vector components y_k can be understood as either the components of a new vector $\boldsymbol{y}$ in the directions of the original basis $\boldsymbol{e}_i (i = 1,2 \ldots n)$ or as the components of the old vector $\boldsymbol{x}$ in the directions of new bases.

There are always these two ways to view a matrix acting on a vector:

1. The operation can be thought of as transforming the vector into a different vector. This view is called the active view (vector in a different place), and is the interpretation we will use most often.
2. The operation can be thought of as expressing the same vector in terms of a different coordinate system or basis. This view is called the passive view.

Some examples may help to clarify these perspectives:

For the matrix-vector product

$$\begin{bmatrix} a & 0 \\ 0 & 1 \end{bmatrix}\begin{bmatrix} x \\ y \end{bmatrix} = \begin{bmatrix} ax \\ y \end{bmatrix}$$

the active interpretation states that the vector is scaled in the x direction by an amount a. In the passive interpretation, the original vector is written in terms of new bases $(a^{-1}, 0)$ and $(0\ ,\ 1)$:

$$\begin{bmatrix} x \\ y \end{bmatrix} = ax\begin{bmatrix} a^{-1} \\ 0 \end{bmatrix} + y\begin{bmatrix} 0 \\ 1 \end{bmatrix}.$$

As another example, consider the following matrix multiplication:

$$\underline{\mathrm{M}}\boldsymbol{x} = \begin{bmatrix} \cos\theta & -\sin\theta \\ \sin\theta & \cos\theta \end{bmatrix}\begin{bmatrix} x \\ y \end{bmatrix} = \begin{bmatrix} (x\cos\theta - y\sin\theta) \\ (x\sin\theta + y\cos\theta) \end{bmatrix}$$

In the active interpretation, the vector whose cartesian and polar representations are:

$$\boldsymbol{x} = \begin{bmatrix} x \\ y \end{bmatrix} = \begin{bmatrix} r\cos\phi \\ r\sin\phi \end{bmatrix},$$

is rotated by an angle θ to obtain:

$$\begin{aligned} \underline{\mathrm{M}}\boldsymbol{x} &= \begin{bmatrix} \cos\theta & -\sin\theta \\ \sin\theta & \cos\theta \end{bmatrix}\begin{bmatrix} r\cos\phi \\ r\sin\phi \end{bmatrix} \\ &= \begin{bmatrix} (r\cos\phi\cos\theta - r\sin\phi\sin\theta) \\ (r\cos\phi\sin\theta + r\sin\phi\cos\theta) \end{bmatrix} \\ &= \begin{bmatrix} r\cos(\phi + \theta) \\ r\sin(\phi + \theta) \end{bmatrix}. \end{aligned}$$

In the passive interpretation, the original vector $\boldsymbol{x}$ is expressed in terms of a new coordinate system with axes rotated by $-\theta$ with new bases

$$\begin{bmatrix} \cos\theta \\ -\sin\theta \end{bmatrix} \text{and} \begin{bmatrix} \sin\theta \\ \cos\theta \end{bmatrix}.$$

$$\begin{aligned} \begin{bmatrix} x \\ y \end{bmatrix} &= (x\cos\theta - y\sin\theta)\begin{bmatrix} \cos\theta \\ -\sin\theta \end{bmatrix} + (x\sin\theta + y\cos\theta)\begin{bmatrix} \sin\theta \\ \cos\theta \end{bmatrix} \\ &= \begin{bmatrix} x(\cos^2\theta + \sin^2\theta) + y(\sin\theta\cos\theta - \sin\theta\cos\theta) \\ y(\cos^2\theta + \sin^2\theta) + x(\sin\theta\cos\theta - \sin\theta\cos\theta) \end{bmatrix} \\ &= \begin{bmatrix} x \\ y \end{bmatrix} \end{aligned}$$

As a general rule, active transformations and passive transformations are inverses of each other; you can do something *to* a vector or else do the reverse to the coordinate system.

The two pictures can be summarized by the following two equations:

1. $\underline{\mathrm{M}}\boldsymbol{x} = \boldsymbol{y}$ states the active picture, and
2. $\boldsymbol{x} = \underline{\mathrm{M}}^{-1}\boldsymbol{y}$ states the passive picture.

C. Matrices as Linear Operators

Matrices are examples of linear operators for which

$$\underline{M}(a\boldsymbol{x} + b\boldsymbol{y}) = a\underline{M}\boldsymbol{x} + b\underline{M}\boldsymbol{y},$$

which can easily be demonstrated by examining the components:

$$\begin{aligned}[\underline{M}(a\boldsymbol{x} + b\boldsymbol{y})]_i &= \sum_k M_{ik}(ax_k + by_k) \\ &= a\sum_k M_{ik}x_k + b\sum_k M_{ik}y_k \\ &= a(\underline{M}\boldsymbol{x})_i + b(\underline{M}\boldsymbol{y})_i.\end{aligned}$$

One can also see that this property holds for a linear combination of many vectors rather than for the two considered above.

We can visualize how the action of a matrix on arbitrary vectors can be expressed if one knows its action on the elementary basis vectors. Given the expansion of $\boldsymbol{x}$ in the $\boldsymbol{e}_i$,

$$\boldsymbol{x} = \sum_i x_i \boldsymbol{e}_i,$$

one can write

$$\underline{M}\boldsymbol{x} = \sum_i x_i \underline{M}\boldsymbol{e}_i.$$

Using the fact that all of the components of $\boldsymbol{e}_i$ are zero except one, $(\boldsymbol{e}_i)_i = 1$, we see that

$$(M\boldsymbol{e}_i)_k = \sum_j M_{kj}(\boldsymbol{e}_i)_j = M_{ki}$$

This equation tells us that the ith column of a matrix, $\underline{M}$, contains the result of operating on the ith unit vector $\boldsymbol{e}_i$ with the matrix. More specifically, the element M_{ki} in the kth row and i^{th} column is the component of $\underline{M}\boldsymbol{e}_i$ in the direction of the $\boldsymbol{e}_k$ unit vector. As a generalization, we can construct any matrix by first deciding how the matrix affects the elementary unit vectors and then placing the resulting vectors as the columns of the matrix.

II. PROPERTIES OF GENERAL $n \times n$ (SQUARE) MATRICES

The following operations involving square matrices each of the same dimension are useful to express in terms of the individual matrix elements:

1. Sum of matrices; $\underline{A} + \underline{B} = \underline{C}$ if $A_{ij} + B_{ij} = C_{ij}$
2. Scalar multiplication; $c\underline{M} = \underline{N}$ if $cM_{ij} = N_{ij}$
3. Matrix multiplication; $\underline{A}\,\underline{B} = \underline{C}$ if $\sum_k^n A_{ik}B_{kj} = C_{ij}$

4. Determinant of a matrix;

   The determinant is defined inductively

   for $n = 1$ $\quad |A| = det(A) = A_{11}$

   for $n > 1$ $\quad |A| = det(A) = \sum_{i}^{n} A_{ij}\det(a_{ij})(-1)^{(i+j)}$; where

   $j = 1,2, \ldots n$ and a_{ij} is the minor matrix obtained by deleting the ith row and j^{th} column.

5. There are certain special matrices that are important to know:

   a. The zero matrix; $\quad 0_{ij} = 0 \quad$ for $i = 1,2, \ldots n$ and $j = 1,2, \ldots n$

   b. The identity matrix; $\quad I_{ij} = \delta_{ij}$

   (Note $(IM)_{ij} = \sum_{k} \delta_{ik}M_{kj} = M_{ij}$, so $\underline{I}\,\underline{M} = \underline{M}\,\underline{I} = \underline{M}$)

6. Transpose of a matrix; $(M^T)_{ij} = M_{ji}$
7. Complex Conjugate of a Matrix; $(M^*)_{ij} = M_{ij}^*$
8. Adjoint of a Matrix; $(M{=})_{ij} = M_{ji}^* = (M^T)_{ij}^*$
9. Inverse of a Matrix; if $\underline{N}\,\underline{M} = \underline{M}\,\underline{N} = \underline{I}$ then $\underline{N} = \underline{M}^{-1}$
10. Trace (or Character) of Matrix; $Tr(M) = \sum_{i} M_{ii}$ (sum of diagonal elements)

III. SPECIAL KINDS OF SQUARE MATRICES

If a matrix obeys certain conditions, we give it one of several special names. These names include the following:

1. Diagonal Matrix: $D_{ij} = d_i\delta_{ij} = d_j\delta_{ij}$
2. Real Matrix: $\underline{M} = \underline{M}^*$ or $M_{ij} = M_{ij}*$ (real elements)
3. Symmetric Matrix: $\underline{M} = \underline{M}^T$ or $M_{ij} = M_{ji}$ (symmetric about main diagonal)
4. Hermitian Matrix: $\underline{M} = \underline{M}{=}$ or $M_{ij} = M_{ji}^*$
5. Unitary Matrix: $\underline{M}{=} = \underline{M}^{-1}$
6. Real Orthogonal $\underline{M}{=} = \underline{M}^T = \underline{M}^{-1}$

IV. EIGENVALUES AND EIGENVECTORS OF A SQUARE MATRIX

An eigenvector of a matrix, $\underline{M}$, is a vector such that

$$\underline{M}\boldsymbol{v} = \lambda\boldsymbol{v}$$

where λ is called the eigenvalue. An eigenvector thus has the property that when it is multiplied by the matrix, the *direction* of the resultant vector is unchanged. The length, however, is altered by a factor λ. Note that any multiple of an eigenvector is also an eigenvector, which we demonstrate as follows:

$$\underline{M}(a\boldsymbol{v}) = a\underline{M}\boldsymbol{v} = a\lambda\boldsymbol{v} = \lambda(a\boldsymbol{v}).$$

Hence, an eigenvector can be thought of as defining a *direction* in n-dimensional space. The length (normalization) of the eigenvector is arbitrary; we are free to choose the length to be anything we please. Usually we choose the length to be unity (because, in quantum mechanics, our vectors usually represent some wavefunction that we wish to obey a normalization condition).

The basic eigenvalue equation can be rewritten as

$$(\underline{\mathrm{M}} - \lambda\underline{\mathrm{I}})\boldsymbol{v} = 0$$

or, in an element-by-element manner, as

$$\begin{array}{lllll} (M_{11}-\lambda)v_1 & + M_{12}v_2 & + M_{13}v_3 + \cdots & M_{1n}v_n & = 0 \\ M_{21}v_1 & + (M_{22}-\lambda)v_2 & + M_{23}v_3 + \cdots & M_{2n}v_n & = 0 \\ \cdot & \cdot & \cdot & \cdot & \\ \cdot & \cdot & \cdot & \cdot & \\ \cdot & \cdot & \cdot & \cdot & \\ M_{n1}v_1 & + M_{n2}v_2 & + M_{n3}v_3 + \cdots & (M_{nn}-\lambda)v_n & = 0. \end{array}$$

If you try to solve these n equations for all of the elements of the $\boldsymbol{v}$ vector ($v_1 \ldots v_n$), you can eliminate one variable using one equation, a second variable using a second equation, etc., because the equations are linear. For example you could solve for v_1 using the first equation and then substitute for v_1 in the second equation as you solve for v_2, etc. Then when you come to the nth equation, you would have $n - 1$ of the variables expressed in terms of the one remaining variable, v_n.

However, you find that you cannot use the remaining equation to solve for the value of v_n; the last equation is found to appear in the form

$$(C - \lambda)v_n = 0$$

once the v_1,v_2,v_3, and v_{n-1} are expressed in terms of v_n. We should not really have expected to solve for v_n since, as we saw above, the length of the vector $\boldsymbol{v}$ is not determined from the set of eigenvalue equations. You find that the only solution is $v_n = 0$, which then implies that all of the other $v_k = 0$ because you expressed them in terms of v_n, unless the eigenvalue λ is chosen to obey $\lambda = C$.

Upon analyzing what has gone into the C element, one finds that the v_k ($k = 1,2,3, \ldots n - 1$) were eliminated in favor of v_n by successively combining rows of the $(\underline{\mathrm{M}} - \lambda\underline{\mathrm{I}})$ matrix. Thus, $(C - \lambda)$ can vanish if and only if the last row of $(\underline{\mathrm{M}} - \lambda\underline{\mathrm{I}})$ is a linear combination of the other $n - 1$ rows. A theorem dealing with determinants states that the rows of a matrix are linearly dependent (i.e., one row is a linear combination of the rest) if and only if the determinant of the matrix is zero. We can therefore make the eigenvalue equation have a solution $\boldsymbol{v}$ by adjusting λ so the determinant of $(\underline{\mathrm{M}} - \lambda\underline{\mathrm{I}})$ vanishes.

A. Finding the Eigenvalues

In summary, to solve an eigenvalue equation, we first solve the determinantal equation:

$$|M - \lambda I| = 0.$$

Using the definition of a determinant, one can see that expanding the determinant results in an nth-order polynomial in λ

$$a_n(M)\lambda^n + a_{n-1}(M)\lambda^{n-1} + \ldots a_1(M)\lambda + a_o(M) = 0,$$

where the coefficients $a_i(M)$ depend on the matrix elements of M. A theorem in algebra shows that such an equation always has n roots some or all of which may be complex (e.g., a quadratic equation has two roots). Thus there are n different ways of adjusting the parameter λ so the determinant vanishes. Each of these solutions is a candidate for use as λ in subsequently solving for the $\boldsymbol{v}$ vector's coefficients. That is, each λ value has its own vector $\boldsymbol{v}$.

B. Finding the Eigenvectors

One can substitute each of these n λ values ($\lambda_k, k = 1,2,n$) back into the eigenvalue equation, one at a time, and solve for the n eigenvectors $\boldsymbol{v}(k)$. By using one of the λ_k values, the nth equation is guaranteed to be equal to zero, so you use $n - 1$ of the equations to solve for $n - 1$ of the components in terms of the nth one. The eigenvectors will then be determined up to a multiplicative factor which we then fix by requiring normalization:

$$\sum_i v_i^*(k)v_i(k) = \langle \boldsymbol{v}(k) \mid \boldsymbol{v}(k)\rangle = 1.$$

This expression now defines the dot or inner product (Hermitian inner product) for vectors which can have complex valued components. We use this definition so the dot product of a complex valued vector with itself is real.

In summary, we have shown how to solve the n equations:

$$\sum_j M_{ij}v_j(k) = \lambda_k v_i(k);\ k = 1,2, \ldots .\ n.$$

for the eigenvalues of the matrix λ_k and the corresponding normalized eigenvectors $\boldsymbol{v}(k)$, $k = 1, 2, \ldots n$. Now let us work an example that is chosen to illustrate the concepts we have learned as well as an additional complication, that of eigenvalue degeneracy.

C. Examples

Consider the following real symmetric matrix:

$$\underline{M} = \begin{bmatrix} 3 & 0 & 0 \\ 0 & \frac{5}{2} & \frac{1}{2} \\ 0 & \frac{1}{2} & \frac{5}{2} \end{bmatrix}$$

The set of eigenvalue-eigenvector equations has non-trivial ($\boldsymbol{v}(k) = 0$ is "trivial") solutions if

$$|\underline{M} - \lambda_k \underline{I}| = 0.$$

In our example, this amounts to

$$\begin{vmatrix} 3-\lambda_k & 0 & 0 \\ 0 & \frac{5}{2}-\lambda_k & \frac{1}{2} \\ 0 & \frac{1}{2} & \frac{5}{2}-\lambda_k \end{vmatrix} = 0,$$

or,

$$|3-\lambda_k| \begin{vmatrix} \frac{5}{2}-\lambda_k & \frac{1}{2} \\ \frac{1}{2} & \frac{5}{2}-\lambda_k \end{vmatrix} = 0,$$

or,

$$(3-\lambda_k)\bullet\left[\left(\frac{5}{2}-\lambda_k\right)^2 - \frac{1}{4}\right] = 0,$$

or,

$$(3-\lambda_k)\,[\lambda_k^2 - 5\lambda_k + 6] = 0 = (3-\lambda_k)\,(\lambda_k - 3)\,(\lambda_k - 2).$$

There are three real solutions to this cubic equation (why all the solutions are real in this case for which the $\underline{\mathrm{M}}$ matrix is real and symmetric will be made clear later):

1. $\lambda_1 = 3$, 2. $\lambda_2 = 3$, and 3. $\lambda_3 = 2$.

Notice that the eigenvalue $\lambda_k = 3$ appears twice; we say that the eigenvalue $\lambda_k = 3$ is *doubly degenerate*. $\lambda_k = 2$ is a *non-degenerate* eigenvalue.

The eigenvectors $\boldsymbol{v}(k)$ are found by plugging the above values for λ_k into the basic eigenvalue equation

$$\underline{\mathrm{M}}\boldsymbol{v}(k) = \lambda_k \boldsymbol{v}(k)$$

For the non-degenerate eigenvalue ($\lambda_3 = 2$), this gives:

$$\begin{bmatrix} 3 & 0 & 0 \\ 0 & \frac{5}{2} & \frac{1}{2} \\ 0 & \frac{1}{2} & \frac{5}{2} \end{bmatrix} \begin{bmatrix} v_1(3) \\ v_2(3) \\ v_3(3) \end{bmatrix} = 2 \begin{bmatrix} v_1(3) \\ v_2(3) \\ v_3(3) \end{bmatrix}.$$

The following algebraic steps can then be followed:

a. $3v_1(3) = 2v_1(3)$ implies that $v_1(3) = 0$,

b. $\frac{5}{2}v_2(3) + \frac{1}{2}v_3(3) = 2v_2(3)$, and

c. $\frac{1}{2}v_2(3) + \frac{5}{2}v_3(3) = 2v_3(3)$.

The last two equations can not be solved for both $v_2(3)$ and $v_3(3)$. To see the trouble, multiply equation (c) by 5 and subtract it from equation (b) to obtain

$$-12v_3(3) = 2v_2(3) - 10v_3(3),$$

which implies that $v_3(3) = -v_2(3)$. Now, substitute this result into the equation (b) to obtain

$$\frac{5}{2}v_2(3) + \frac{1}{2}(-v_2(3)) = 2v_2(3),$$

or,

$$2v_2(3) = 2v_2(3).$$

This is a trivial identity; it does not allow you to solve for $v_2(3)$. Hence, for this non-degenerate root, one is able to solve for all of the v_j elements in terms of one element that is yet undetermined.

As in all matrix eigenvalue problems, we are able to express $(n - 1)$ elements of the eigenvector $\boldsymbol{v}(k)$ in terms of one remaining element. However, we can never solve for this one last element. So, for convenience, we impose one more constraint (equation to be obeyed) which allows us to solve for the remaining element of $\boldsymbol{v}(k)$. We require that the eigenvectors be *normalized*:

$$<\boldsymbol{v}(k) \mid \boldsymbol{v}(k)> = \sum_a v_a^*(k)v_a(k) = 1.$$

In our example, this means that

$$v_1^2(3) + v_2^2(3) + v_3^2(3) = 1,$$

or,

$$0^2 + v_2^2(3) + (-v_2(3))^2 = 1,$$

which implies that $v_2(3) = \pm\frac{1}{\sqrt{2}}$. So,

$$v_3(3) = \mp\frac{1}{\sqrt{2}},$$

and finally the vector is given by:

$$\boldsymbol{v}(3) = \pm\begin{bmatrix} 0 \\ \frac{1}{\sqrt{2}} \\ -\frac{1}{\sqrt{2}} \end{bmatrix}.$$

Note that even after requiring normalization, there is still an indeterminancy in the sign of $\boldsymbol{v}(3)$. The eigenvalue equation, as we recall, only specifies a direction in space. The sense or sign is not determined. We can choose either sign we prefer.

Finding the first eigenvector was not too difficult. The degenerate eigenvectors are more difficult to find. For $\lambda_1 = \lambda_2 = 3$,

$$\begin{bmatrix} 3 & 0 & 0 \\ 0 & \frac{5}{2} & \frac{1}{2} \\ 0 & \frac{1}{2} & \frac{5}{2} \end{bmatrix} \begin{bmatrix} v_1(1) \\ v_2(1) \\ v_3(1) \end{bmatrix} = 3 \begin{bmatrix} v_1(1) \\ v_2(1) \\ v_3(1) \end{bmatrix}.$$

Again, the algebraic equations can be written as follows:

a. $3v_1(1) = 3v_1(1)$; this tells us nothing!

b. $\frac{5}{2}v_2(1) + \frac{1}{2}v_3(1) = 3v_2(1)$, and

c. $\frac{1}{2}v_2(1) + \frac{5}{2}v_3(1) = 3v_3(1)$.

If we multiply equation (c) by 5 and subtract if from equation (b), we obtain:

$$-12v_3(1) = -15v_3(1) + 3v_2(1),$$

or

$$3v_3(1) = 3v_2(1)\ ,$$

which implies that $v_3(1) = v_2(1)$. So far, all we have is $v_3(1) = v_2(1)$; we don't know $v_1(1)$ nor do we know either $v_3(1)$ or $v_2(1)$, and we have used all three equations. Normalization provides one more equation $v_1^2(1) + v_2^2(1) + (v_2(1))^2 = 1$, but we are still in a situation with more unknowns (2) than equations (1).

One might think that by restricting our eigenvectors to be orthogonal as well as normalized, we might overcome this difficulty (however, such is not the case, as we now show).

For our vectors, the constraint that the nondegenerate vector $\boldsymbol{v}(3)$ be orthogonal to the one we are trying to find $\boldsymbol{v}(1)$, amounts to

$$<\boldsymbol{v}(3)|\boldsymbol{v}(1)> = 0$$

$$v_1(3)^* v_1(1) + v_2(3)^* v_2(1) + v_3(3)^* v_3(1) = 0$$

$$0v_1(1) \pm \left[\frac{1}{\sqrt{2}}v_2(1) - \frac{1}{\sqrt{2}}(v_2(1))\right] = 0.$$

We see that $\boldsymbol{v}(3)$ and $\boldsymbol{v}(1)$ are *already* orthogonal regardless of how $v_2(1)$ and $v_3(1)$ turn out. This is shown below to be guaranteed because $\boldsymbol{v}(1)$ and $\boldsymbol{v}(3)$ have different eigenvalues (i.e., two eigenvectors belonging to different eigenvalues of any symmetric or hermitian matrix must be orthonogonal). Hence, this first attempt at finding additional equations to use has failed.

What about the two degenerate eigenvectors $\boldsymbol{v}(1)$ and $\boldsymbol{v}(2)$? Are they also orthonormal? So far, we know that these two eigenvectors have the structure

$$\boldsymbol{v}(1) = \begin{bmatrix} v_1(1) \\ v_2(1) \\ v_3(1) \end{bmatrix}, \text{ with } 1 = v_1^2(1) + 2v_2^2(1).$$

If we go through all of the above steps for $\boldsymbol{v}(2)$ with $\lambda_2 = 3$, we will find that this vector obeys exactly the same set of equations

$$\boldsymbol{v}(2) = \begin{bmatrix} v_1(2) \\ v_2(2) \\ v_3(2) \end{bmatrix}, \text{ with } 1 = v_1^2(2) + 2v_2^2(2).$$

We showed above that $<\boldsymbol{v}(1)|\boldsymbol{v}(3)> = 0$, and it is easy to show that $<\boldsymbol{v}(2)|\boldsymbol{v}(3)> = 0$ because the elements of $\boldsymbol{v}(2)$, thus far, obey the same equations as $\boldsymbol{v}(1)$.

If we also wish to make the two degenerate eigenvectors orthogonal to one another

$$<\boldsymbol{v}(1)|\boldsymbol{v}(2) = 0,$$

then we obtain additional relationships among our yet-undetermined vector amplitudes. In particular, we obtain

$$v_1(1)v_1(2) + v_2(1)v_2(2) + v_3(1)v_3(2) = 0,$$

or,

$$v_1(1)v_1(2) + 2v_2(1)v_2(2) = 0.$$

Although progress has been made, we still have four *unknowns* $v_1(1)$, $v_2(1)$; $v_1(2)$, $v_2(2)$ and only *three equations:*

$$v_1(1)v_1(2) + 2v_2(1)v_2(2) = 0,$$
$$v_1(1)v_1(1) + 2v_2(1)v_2(1) = 1, \text{ and}$$
$$v_1(2)v_1(2) + 2v_2(2)v_2(2) = 1.$$

It appears as though we are stuck again. We are; but for good reasons. We are trying to find two vectors $\boldsymbol{v}(1)$ and $\boldsymbol{v}(2)$ that are orthonormal and are eigenvectors of $\underline{\text{M}}$ having eigenvalue equal to 3. Suppose that we do find two such vectors. Because $\underline{\text{M}}$ is a linear operator, *any* two vectors generated by taking linear combinations of these two vectors would also be eigenvectors of $\underline{\text{M}}$. There is a degree of freedom, that of recombining $\boldsymbol{v}(1)$ and $\boldsymbol{v}(2)$, which can not be determined by insisting that the two vectors be eigenfunctions. Thus, in this degenerate-eigenvalue case our requirements do not give a *unique* pair of eigenvectors. They just tell us the two-dimensional space in which the acceptable eigenvectors lie. (This difficulty does not arise in nondegenerate cases because one-dimensional spaces have no flexibility other than a sign.)

So to find *an* acceptable pair of vectors, we are free to make an additional choice. For example, we can choose one of the four unknown components of $\boldsymbol{v}(1)$ and $\boldsymbol{v}(2)$ equal to zero. Let us make the choice $v_1(1) = 0$. Then the above equations can be solved for the other elements of $\boldsymbol{v}(1)$ to give

$$v_2(1) = \pm\frac{1}{\sqrt{2}} = v_3(1).$$

The orthogonality between $\boldsymbol{v}(1)$ and $\boldsymbol{v}(2)$ then gives

$$0 = 2\left(\pm\frac{1}{\sqrt{2}}\right)v_2(2),$$

which implies that $v_2(2) = v_3(2) = 0$; the remaining equation involving $\boldsymbol{v}(2)$ then gives $v_1(2) = \pm 1$.

In summary, we have now found a specific solution once the choice $v_1(1) = 0$ is made:

$$\boldsymbol{v}(1) = \pm\begin{bmatrix} 0 \\ \frac{1}{\sqrt{2}} \\ \frac{1}{\sqrt{2}} \end{bmatrix}, \ \boldsymbol{v}(2) = \pm\begin{bmatrix} 1 \\ 0 \\ 0 \end{bmatrix}, \text{ and } \boldsymbol{v}(3) = \pm\begin{bmatrix} 0 \\ \frac{1}{\sqrt{2}} \\ -\frac{1}{\sqrt{2}} \end{bmatrix}$$

Other choices for $v_1(1)$ will yield different specific solutions.

V. PROPERTIES OF EIGENVALUES AND EIGENVECTORS OF HERMITIAN MATRICES

The above example illustrates many of the properties of the matrices that we will most commonly encounter in properties in more detail and to learn about other properties in more detail and to learn about other characteristics that Hermitian matrices have.

A. Outer Product

Given any vector $\boldsymbol{v}$, we can form a square matrix denoted $|v(i)><v(i)|$, whose elements are defined as follows:

$$|v(i)> <v(i)| = \begin{bmatrix} v_1^*(i)v_1(i) & v_1^*(i)v_2(i) & \dots & v_1^*(i)v_n(i) \\ v_2^*(i)v_1(i) & v_2^*(i)v_2(i) & \dots & v_2^*(i)v_n(i) \\ \dots & \dots & \dots & \dots \\ v_n^*(i)v_1(i) & v_n^*(i)v_2(i) & \dots & v_n^*(i)v_n(i) \end{bmatrix}$$

We can use this matrix to project onto the component of a vector in the $\boldsymbol{v}(i)$ direction. For the example we have been considering, if we form the projector onto the $\boldsymbol{v}(1)$ vector, we obtain

$$\begin{bmatrix} 0 \\ \frac{1}{\sqrt{2}} \\ \frac{1}{\sqrt{2}} \end{bmatrix} \begin{bmatrix} 0 & \frac{1}{\sqrt{2}} & \frac{1}{\sqrt{2}} \end{bmatrix} = \begin{bmatrix} 0 & 0 & 0 \\ 0 & \frac{1}{2} & \frac{1}{2} \\ 0 & \frac{1}{2} & \frac{1}{2} \end{bmatrix},$$

for $\boldsymbol{v}(2)$, we get

$$\begin{bmatrix} 1 \\ 0 \\ 0 \end{bmatrix} [\,1\,0\,0\,] = \begin{bmatrix} 1 & 0 & 0 \\ 0 & 0 & 0 \\ 0 & 0 & 0 \end{bmatrix},$$

and for $\boldsymbol{v}(3)$ we find

$$\begin{bmatrix} 0 \\ \frac{1}{\sqrt{2}} \\ -\frac{1}{\sqrt{2}} \end{bmatrix} \begin{bmatrix} 0 & \frac{1}{\sqrt{2}} & -\frac{1}{\sqrt{2}} \end{bmatrix} = \begin{bmatrix} 0 & 0 & 0 \\ 0 & \frac{1}{2} & -\frac{1}{2} \\ 0 & -\frac{1}{2} & \frac{1}{2} \end{bmatrix}.$$

These three projection matrices play important roles in what follows.

B. Completeness Relation or Resolution of the Identity

The set of eigenvectors of any Hermitian matrix form a complete set over the space they span in the sense that the sum of the projection matrices constructed from these eigenvectors gives an exact representation of the identity matrix.

$$\sum_i |v(i)\rangle\langle v(i)| = \underline{I}.$$

For the specific matrix we have been using as an example, this relation reads as follows:

$$\begin{bmatrix} 0 \\ \frac{1}{\sqrt{2}} \\ \frac{1}{\sqrt{2}} \end{bmatrix} \begin{bmatrix} 0 & \frac{1}{\sqrt{2}} & \frac{1}{\sqrt{2}} \end{bmatrix} + \begin{bmatrix} 1 \\ 0 \\ 0 \end{bmatrix} [\, 1\, 0\, 0\,] + \begin{bmatrix} 0 \\ \frac{1}{\sqrt{2}} \\ -\frac{1}{\sqrt{2}} \end{bmatrix} \begin{bmatrix} 0 & \frac{1}{\sqrt{2}} & -\frac{1}{\sqrt{2}} \end{bmatrix} = \begin{bmatrix} 1 & 0 & 0 \\ 0 & 1 & 0 \\ 0 & 0 & 1 \end{bmatrix}.$$

Physically, this means that when you project onto the components of a vector in these three directions, you don't lose any of the vector. This happens because our vectors are orthogonal and complete. The completeness relation means that *any* vector in this three-dimensional space can be written in terms of $\boldsymbol{v}(1)$,$\boldsymbol{v}(2)$, and $\boldsymbol{v}(3)$ (i.e., we can use $\boldsymbol{v}(1)$, $\boldsymbol{v}(2)$, $\boldsymbol{v}(3)$ as a new set of bases instead of $\boldsymbol{e}_1$,$\boldsymbol{e}_2$,$\boldsymbol{e}_3$).

Let us consider an example in which the following vector is expanded or written in terms of our three eigenvectors:

$$\boldsymbol{f} = \begin{bmatrix} 7 \\ -9 \\ 12 \end{bmatrix} = a_1\boldsymbol{v}(1) + a_2\boldsymbol{v}(2) + a_3\boldsymbol{v}(3)$$

The task at hand is to determine the expansion coefficients, the a_i. These coefficients are the projections of the given $\boldsymbol{f}$ vector onto each of the $\boldsymbol{v}(i)$ directions:

$$\langle \boldsymbol{v}(1)|\boldsymbol{f}\rangle = a_1,$$

since,

$$\langle \boldsymbol{v}(1)|\boldsymbol{v}(2)\rangle = \langle \boldsymbol{v}(1)|\boldsymbol{v}(3)\rangle = 0.$$

Using our three $\boldsymbol{v}_i$ vectors and the above $\boldsymbol{f}$ vector, the following three expansion coefficients are obtained:

$$a_1 = \begin{bmatrix} 0 & \frac{1}{\sqrt{2}} & \frac{1}{\sqrt{2}} \end{bmatrix} \begin{bmatrix} 7 \\ -9 \\ 12 \end{bmatrix} = \frac{3}{\sqrt{2}}$$

$$\langle \boldsymbol{v}(2)|\boldsymbol{f}\rangle = 7$$

$$\langle \boldsymbol{v}(3)|\boldsymbol{f}\rangle = -\frac{21}{\sqrt{2}}$$

Therefore, f can be written as:

$$f = \begin{bmatrix} 7 \\ -9 \\ 12 \end{bmatrix} = \frac{3}{\sqrt{2}} \begin{bmatrix} 0 \\ \frac{1}{\sqrt{2}} \\ \frac{1}{\sqrt{2}} \end{bmatrix} + 7 \begin{bmatrix} 1 \\ 0 \\ 0 \end{bmatrix} - \frac{21}{\sqrt{2}} \begin{bmatrix} 0 \\ \frac{1}{\sqrt{2}} \\ -\frac{1}{\sqrt{2}} \end{bmatrix}$$

This works for any vector f, and we could write the process in general in terms of the resolution of the identity as

$$f = If = \sum_k |v(k)\rangle\langle v(k)|f\rangle = \sum_k |v(k)\rangle\, a_k,$$

This is how we will most commonly make use of the completeness relation as it pertains to the eigenvectors of Hermitian matrices.

C. Spectral Resolution of $\underline{M}$

It turns out that not only the identity matrix $\underline{I}$ but also the matrix $\underline{M}$ itself can be expressed in terms of the eigenvalues and eigenvectors. In the so-called spectral representation of $\underline{M}$, we have

$$\underline{M} = \sum_k \lambda_k |\boldsymbol{v}(k)\rangle\langle \boldsymbol{v}(k)| \ .$$

In the example we have been using, the three terms in this sum read

$$3\begin{bmatrix} 0 & 0 & 0 \\ 0 & \frac{1}{2} & \frac{1}{2} \\ 0 & \frac{1}{2} & \frac{1}{2} \end{bmatrix} + 3\begin{bmatrix} 1 & 0 & 0 \\ 0 & 0 & 0 \\ 0 & 0 & 0 \end{bmatrix} + 2\begin{bmatrix} 0 & 0 & 0 \\ 0 & \frac{1}{2} & -\frac{1}{2} \\ 0 & -\frac{1}{2} & \frac{1}{2} \end{bmatrix} = \begin{bmatrix} 3 & 0 & 0 \\ 0 & \frac{5}{2} & \frac{1}{2} \\ 0 & \frac{1}{2} & \frac{5}{2} \end{bmatrix} = \underline{M}$$

This means that a matrix is totally determined if we know its eigenvalues and eigenvectors.

D. Eigenvalues of Hermitian Matrices are Real Numbers

A matrix can be expressed in terms of any complete set of vectors. For $n \times n$ matrices, a complete set is any n linearly independent vectors. For a set of vectors $|\boldsymbol{k}\rangle$, $k = 1,2, \ldots n$, the matrix elements of $\underline{M}$ are denoted M_{jk} or $\langle \boldsymbol{j}|M|\boldsymbol{k}\rangle$. If the matrix is Hermitian then

$$\langle \boldsymbol{j}|M|\boldsymbol{k}\rangle = \langle \boldsymbol{k}|M|\boldsymbol{j}\rangle^*.$$

If the vectors $|\boldsymbol{k}\rangle$ are eigenvectors of $\underline{M}$, that is, if $\underline{M}|\boldsymbol{v}(k)\rangle = M|\boldsymbol{k}\rangle = \lambda_k|\boldsymbol{k}\rangle$, then the eigenvalues are real. This can be shown as follows:

$$\langle \boldsymbol{k}|M|\boldsymbol{k}\rangle = \lambda_k \langle \boldsymbol{k}|\boldsymbol{k}\rangle = \langle \boldsymbol{k}|M|\boldsymbol{k}\rangle^* = \lambda_k^* \langle \boldsymbol{k}|\boldsymbol{k}\rangle,$$

so $\lambda_k = \lambda_k^*$. This is a very important result because it forms the basis of the use of Hermitian operators in quantum mechanics; such operators are used because experiments yield real results, so to connect with experimental reality, only Hermitian operators must occur.

E. Non-Degenerate Eigenvectors of Hermitian Matrices Are Orthogonal

If two eigenvalues are different, $\lambda_k \neq \lambda_j$, then

$$<\boldsymbol{k}|M|\boldsymbol{j}> = \lambda_j <\boldsymbol{k}|\boldsymbol{j}> = <\boldsymbol{j}|M|\boldsymbol{k}>^* = \lambda_k^* <\boldsymbol{j}|\boldsymbol{k}>^* = \lambda_k <\boldsymbol{k}|\boldsymbol{j}>,$$

which implies that $(\lambda_k - \lambda_j) <\boldsymbol{k}|\boldsymbol{j}> = 0$. Since, by assumption, $\lambda_k \neq \lambda_j$, it must be that $<\boldsymbol{k}|\boldsymbol{j}> = 0$. In other words, the eigenvectors are orthogonal. We saw this earlier in our example when we "discovered" that $\boldsymbol{v}(3)$ was automatically orthogonal to $\boldsymbol{v}(1)$ and to $\boldsymbol{v}(2)$.

If one has degenerate eigenvalues, $\lambda_k = \lambda_j$ for some k and j, then the corresponding eigenvectors are not automatically orthogonal to one another (they are orthogonal to other eigenvectors), but the degenerate eigenvectors can always be *chosen* to be orthogonal. We also encountered this in our earlier example.

In all cases then, one can find n orthonormal eigenvectors (remember we required $<\boldsymbol{k}|\boldsymbol{k}> = 1$ as an additional condition so that our amplitudes could be interpreted in terms of probabilities). Since any vector in an n-dimensional space can be expressed as a linear combination of n orthonormal vectors, the eigenvectors form a complete basis set. This is why the so-called resolution of the identity, in which the unit matrix can be expressed in terms of the n eigenvectors of $\underline{M}$, holds for all Hermitian matrices.

F. Diagonalizing a Matrix Using its Eigenvectors

The eigenvectors of $\underline{M}$ can be used to form a matrix that diagonalizes $\underline{M}$. This matrix $\underline{S}$ is defined such that the kth column of $\underline{S}$ contains the elements of $\boldsymbol{v}(k)$

$$S = \begin{bmatrix} v_1(1) & v_1(2) & \cdots & v_1(n) \\ v_2(1) & v_2(2) & \cdots & v_2(n) \\ v_3(1) & v_3(2) & \cdots & v_3(n) \\ \cdot & \cdot & \cdots & \cdot \\ v_n(1) & v_n(2) & \cdots & v_n(n) \end{bmatrix}.$$

Then using the above equation one can write the action of $\underline{M}$ on $\underline{S}$ as

$$\sum_j M_{ij} S_{jk} = \sum_j M_{ij} v_j(k) = \lambda_k v_i(k) = S_{ik}\lambda_k.$$

Now consider another matrix $\underline{\Lambda}$ which is a diagonal matrix with diagonal elements λ_k, i.e.,

$$\Lambda_{ik} = \delta_{ik}\lambda_k.$$

One can easily show using the δ_{jk} matrix that

$$\sum_j S_{ij}\delta_{jk}\lambda_k = S_{ik}\lambda_k,$$

since the only non-zero term in the sum is the one in which $j = k$. Thus, comparing with the previous result for the action of $\underline{M}$ on $\underline{S}$,

$$\sum_j M_{ij}S_{jk} = \sum_j S_{ij}\delta_{jk}\lambda_k.$$

These are just the i,k^{th} matrix elements of the matrix equation

$$\underline{M}\,\underline{S} = \underline{S}\,\underline{\Lambda}.$$

Let us assume that an inverse $\underline{S}^{-1}$ of $\underline{S}$ exists. Then multiply the above equation on the left by $\underline{S}^{-1}$ to obtain

$$\underline{S}^{-1}\underline{M}\,\underline{S} = \underline{S}^{-1}\underline{S}\,\underline{\Lambda} = \underline{I}\,\underline{\Lambda} = \underline{\Lambda}.$$

This identity illustrates a so-called similarity transform of $\underline{M}$ using $\underline{S}$. Since $\underline{\Lambda}$ is diagonal, we say that the similarity transform $\underline{S}$ diagonalizes $\underline{M}$. Notice this would still work if we had used the eigenvectors $\boldsymbol{v}(k)$ in a different order. We would just get $\underline{\Lambda}$ with the diagonal elements in a different order. Note also that the eigenvalues of $\underline{M}$ are the same as those of $\underline{\Lambda}$ since the eigenvalues of $\underline{\Lambda}$ are just the diagonal elements λ_k with eigenvectors $\boldsymbol{e}_k$ (the elementary unit vectors)

$$\underline{\Lambda}\boldsymbol{e}_k = \lambda_k\boldsymbol{e}_k, \text{ or, } (\Lambda e_k)_i = \sum_j \lambda_k\delta_{kj}\delta_{ji} = \lambda_k\delta_{ki} = \lambda_k(\boldsymbol{e}_k)_i.$$

G. The Trace of a Matrix Is the Sum of Its Eigenvalues

Based on the above similarity transform, we can now show that the trace of a matrix (i.e., the sum of its diagonal elements) is independent of the representation in which the matrix is formed, and, in particular, the trace is equal to the sum of the eigenvalues of the matrix. The proof of this theorem proceeds as follows:

$$\sum_k \lambda_k = Tr(\Lambda) = \sum_k (S^{-1}MS)_{kk} = \sum_{kij} (S^{-1}_{ki}M_{ij}S_{jk})$$

$$= \sum_{ij} M_{ij}\sum_k S_{jk}S^{-1}_{ki} = \sum_{ij} M_{ij}\delta_{ji} = \sum_i M_{ii} = Tr(M)$$

H. The Determinant of a Matrix is the Product of Its Eigenvalues

This theorem, $\det(M) = \det(\Lambda)$, can be proven, by using the theorem that $\det(AB) = \det(A)\det(B)$, inductively within the expansion by minors of the determinant

$$\begin{aligned}\lambda_1\lambda_2\lambda_3 \ldots \lambda_n = \det(\Lambda) &= \det(S^{-1}MS) = \det(S^{-1})\det(M)\det(S)\\ &= \det(M)\det(S^{-1})\det(S)\\ &= \det(M)\det(S^{-1}S)\\ &= \det(M)\det(I) = \det(MI) = \det(M).\end{aligned}$$

I. Invariance of Matrix Identities to Similarity Transforms

We will see later that performing a similarity transform expresses a matrix in a different basis (coordinate system). If, for any matrix $\underline{A}$, we call $\underline{S}^{-1}\underline{A}\,\underline{S} = \underline{A}'$, we can show that performing a similarity transform on a matrix equation leaves the form of the equation unchanged. For example, if

$$\underline{A} + \underline{B}\,\underline{C} = \underline{P}\,\underline{Q}\,\underline{R}$$

performing a similarity transform on both sides, and remembering that $\underline{S}^{-1}\underline{S} = \underline{I}$, one finds

$$\underline{S}^{-1}(\underline{A} + \underline{B}\,\underline{C})\underline{S} = \underline{S}^{-1}\underline{P}\,\underline{Q}\,\underline{R}\,\underline{S} = \underline{S}^{-1}\underline{A}\,\underline{S} + \underline{S}^{-1}\underline{B}(\underline{S}\,\underline{S}^{-1})\underline{C}\,\underline{S} = \underline{S}^{-1}\underline{P}(\underline{S}\,\underline{S}^{-1})\underline{Q}\,\underline{S}\,\underline{S}^{-1}\underline{R}\,\underline{S},$$

or,

$$\underline{A}' + \underline{B}'\underline{C}' = \underline{P}'\underline{Q}'\underline{R}'.$$

J. How to Construct $\underline{S}^{-1}$

To form $\underline{S}^{-1}$, recall the eigenvalue equation we began with

$$\underline{M}\boldsymbol{v}(k) = \lambda_k\,\boldsymbol{v}(k).$$

If we multiply on the left by $\underline{S}^{-1}$ and insert $\underline{S}^{-1}\underline{S} = \underline{I}$, we obtain

$$\underline{S}^{-1}\underline{M}(\underline{S}\,\underline{S}^{-1})\boldsymbol{v}(k) = \lambda_k\,\underline{S}^{-1}\boldsymbol{v}(k),$$

or,

$$\underline{\Lambda}\,\boldsymbol{v}(k)' = \lambda_k\,\boldsymbol{v}(k)'.$$

According to the passive picture of matrix operations, $\boldsymbol{v}(k)' = \underline{S}^{-1}\boldsymbol{v}(k)$ is the old vector $\boldsymbol{v}$ expressed in terms of new basis functions which are given by the columns of $\underline{S}$. But the rows of $\underline{S}^{-1}$, or equivalently, the columns of $\underline{S}$ have to be orthonormal vectors (using the definition of an inverse):

$$\sum_j S_{ij}^{-1} S_{jk} = \delta_{ik}.$$

However, the columns of $\underline{S}$ are just the eigenvectors $\boldsymbol{v}(k)$, so the rows of $\underline{S}^{-1}$ must also have something to do with $\boldsymbol{v}(k)$.

Now consider the matrix $S{=}$. Its elements are $(S{=})_{kj} = S_{jk}^{*}$, so $(S{=})_{kj} = v_j^{*}(k)$, as a result of which we can write

$$(S{=}S)_{ik} = \sum_j S{=}_{ij}S_{jk} \qquad = \sum_j \boldsymbol{v}_j^{*}(i)\boldsymbol{v}_j(k) = \langle\boldsymbol{v}(i)|\boldsymbol{v}(k)\rangle = \delta_{ik}.$$

We have therefore found $\underline{S}^{-1}$, and it is $\underline{S}{=}$. We have also proven the important theorem stating that any Hermitian matrix $\underline{M}$, can be diagonalized by a matrix $\underline{S}$, which is unitary ($\underline{S}{=} = \underline{S}^{-1}$) and whose columns are the eigenvectors of $\underline{M}$.

VI. FINDING INVERSES, SQUARE ROOTS, AND OTHER FUNCTIONS OF A MATRIX USING ITS EIGENVECTORS AND EIGENVALUES

Since a matrix is defined by how it affects basis vectors upon which it acts, and since a matrix only changes the lengths, not the directions of its eigenvectors, we can form other matrices that have the same eigenvectors as the original matrix but have different eigenvalues in a straightforward manner. For example, if we want to reverse the effect of $\underline{M}$, we create another matrix that reduces the lengths of the eigenvectors by the same amount that $\underline{M}$ lengthened them. This produces the inverse matrix. We illustrate this by using the same example matrix that we used earlier.

$$\underline{M}^{-1} = \sum_k \lambda_k^{-1} |v(i)\rangle \langle v(i)| \qquad \text{(this is defined only if all } \lambda_k \neq 0)$$

$$\frac{1}{3}\begin{bmatrix} 0 & 0 & 0 \\ 0 & \frac{1}{2} & \frac{1}{2} \\ 0 & \frac{1}{2} & \frac{1}{2} \end{bmatrix} + \frac{1}{3}\begin{bmatrix} 1 & 0 & 0 \\ 0 & 0 & 0 \\ 0 & 0 & 0 \end{bmatrix} + \frac{1}{2}\begin{bmatrix} 0 & 0 & 0 \\ 0 & \frac{1}{2} & -\frac{1}{2} \\ 0 & -\frac{1}{2} & \frac{1}{2} \end{bmatrix} = \begin{bmatrix} \frac{1}{3} & 0 & 0 \\ 0 & \frac{10}{24} & \frac{-2}{24} \\ 0 & \frac{-2}{24} & \frac{10}{24} \end{bmatrix} = \underline{M}^{-1}$$

To show that this matrix obeys $\underline{M}^{-1}\underline{M} = \underline{1}$, we simply multiply the two matrices together:

$$\begin{bmatrix} \frac{1}{3} & 0 & 0 \\ 0 & \frac{10}{24} & \frac{-2}{24} \\ 0 & \frac{-2}{24} & \frac{10}{24} \end{bmatrix} \begin{bmatrix} 3 & 0 & 0 \\ 0 & \frac{5}{2} & \frac{1}{2} \\ 0 & \frac{1}{2} & \frac{5}{2} \end{bmatrix} = \begin{bmatrix} 1 & 0 & 0 \\ 0 & 1 & 0 \\ 0 & 0 & 1 \end{bmatrix}$$

An extension of this result is that a whole family of matrices, each member related to the original $\underline{M}$ matrix, can be formed by combining the eigenvalues and eigenvectors as follows:

$$f(\underline{M}) = \sum_i |v(i)\rangle \langle v(i)| f(\lambda_i),$$

where f is any function for which $f(\lambda_i)$ is defined. Examples include $\exp(\underline{M})$, $\sin(\underline{M})$, $\underline{M}^{1/2}$, and $(\underline{I} - \underline{M})^{-1}$. The matrices so constructed, e.g.,

$$\exp(\underline{M}) = \sum_i \exp(\lambda_i) |v(i)\rangle \langle v(i)|,$$

are proper representations of the functions of $\underline{M}$ in the sense that they give results equal to those of the function of $\underline{M}$ when acting on any eigenvector of $\underline{M}$; because the eigenvectors form complete sets, this implies that they give identical results when acting on any vector. This equivalence can be shown most easily by carrying out a power series expansion of the function of $\underline{M}$ (e.g., of $\exp(\underline{M})$) and allowing each term in the series to act on an eigenvector.

VII. PROJECTORS REVISITED

In hindsight, the relationships developed above for expressing functions of the original matrix in terms of its eigenvectors and eigenvalues are not unexpected because each of the matrices is given in terms of the so-called projector matrices $|\boldsymbol{v}(i)\rangle\langle\boldsymbol{v}(i)|$. As we saw, the matrix

$$\sum_i \lambda_i |\boldsymbol{v}(i)\rangle\langle\boldsymbol{v}(i)|$$

behaves just like $\underline{\mathrm{M}}$, as demonstrated below:

$$\sum_i \lambda_i |\boldsymbol{v}(i)\rangle\langle\boldsymbol{v}(i)|(\boldsymbol{v}(j)\rangle = \sum_i \lambda_i|\boldsymbol{v}(i)\rangle\delta_{ij} = \lambda_j|\boldsymbol{v}(j)\rangle$$

The matrix $|\boldsymbol{v}(i)\rangle\langle\boldsymbol{v}(i)|$ is called a projector onto the space of eigenvector $|\boldsymbol{v}(i)\rangle$ because its action on any vector $|f\rangle$ within the class of admissible vectors (2- or 4-dimensional vectors are in a different class than 3- or 1- or 7- or 96-dimensional vectors)

$$(|\boldsymbol{v}(i)\rangle\langle\boldsymbol{v}(i)|)|f\rangle = |\boldsymbol{v}(i)\rangle(\langle\boldsymbol{v}(i)|f\rangle)$$

gives $|\boldsymbol{v}(i)\rangle$ multiplied by the coefficient of $|f\rangle$ along $|\boldsymbol{v}(i)\rangle$.

This construction in which a vector is used to form a matrix $|\boldsymbol{v}(i)\rangle\langle\boldsymbol{v}(i)|$ is called an "outer product." The projection matrix thus formed can be shown to be *idempotent*, which means that the result of applying it twice (or more times) is identical to the result of applying it once $PP = P$. This property is straightforward to demonstrate. Let us consider the projector $\underline{\mathrm{P}}_{\mathrm{i}} \equiv |\boldsymbol{v}(i)\rangle\langle\boldsymbol{v}(i)|$, which, when applied twice yields

$$|\boldsymbol{v}(i)\rangle\langle\boldsymbol{v}(i)|\;|\boldsymbol{v}(i)\rangle\langle\boldsymbol{v}(i)| = \left|\boldsymbol{v}(i)\rangle\delta_{ii}\langle\boldsymbol{v}(i)\right| = |\boldsymbol{v}(i)\rangle\langle\boldsymbol{v}(i)|.$$

Sets of projector matrices each formed from a member of an orthonormal vector set are mutually *orthogonal*, (i.e., $\underline{\mathrm{P}}_{\mathrm{i}}\underline{\mathrm{P}}_{\mathrm{j}} = 0$ if $i \neq j$, which can be shown as follows:

$$\begin{aligned}|\boldsymbol{v}(i)\rangle\langle\boldsymbol{v}(i)|\;|\boldsymbol{v}(j)\rangle\langle\boldsymbol{v}(j)| &= \left|\boldsymbol{v}(i)\rangle\delta_{ij}\langle\boldsymbol{v}(j)\right| \\ &= |\boldsymbol{v}(i)\rangle 0\langle\boldsymbol{v}(j)| = 0 \qquad (i \neq j)\end{aligned}$$

VIII. HERMITIAN MATRICES AND THE TURNOVER RULE

The eigenvalue equation:

$$\underline{\mathrm{M}}|\boldsymbol{v}(i)\rangle = \lambda_i|\boldsymbol{v}(i)\rangle,$$

which can be expressed in terms of its indices as:

$$\sum_j M_{kj}v_j(i) = \lambda_i v_k(i),$$

is equivalent to (just take the complex conjugate):

$$\sum_j v_j^*(i)M_{kj}^* = \lambda_i v_k^*(i),$$

which, for a Hermitian matrix $\underline{M}$, can be rewritten as:

$$\sum_j v_j^*(i)\underline{M}_{jk} = \lambda_i v_k^*(i),$$

or, equivalently, as:

$$<v(i)|\underline{M} = \lambda_i <v(i)| \,.$$

This means that the $v(i)$, when viewed as column vectors, obey the eigenvalue identity $\underline{M}|v(i)> = m_i|v(i)>$. These same vectors, when viewed as row vectors (and thus complex conjugated), also obey the eigenvalue relation, but in the "turn over" form $<v(i)|\underline{M} = \lambda_i <v(i)|$. For example, in the case we have been studying, the first vector obeys

$$\begin{bmatrix} 0 & \frac{1}{\sqrt{2}} & \frac{1}{\sqrt{2}} \end{bmatrix} \begin{bmatrix} 3 & 0 & 0 \\ 0 & \frac{5}{2} & \frac{1}{2} \\ 0 & \frac{1}{2} & \frac{5}{2} \end{bmatrix} = \begin{bmatrix} 0 & \frac{3}{\sqrt{2}} & \frac{3}{\sqrt{2}} \end{bmatrix}$$
$$= 3 \begin{bmatrix} 0 & \frac{1}{\sqrt{2}} & \frac{1}{\sqrt{2}} \end{bmatrix}$$

As a general rule, a Hermitian matrix operating on a column vector to the right is equivalent to the matrix operating on the complex conjugate of the same vector written as a row vector to the left. For a non-Hermitian matrix, it is the adjoint matrix operating on the complex conjugate row vector that is equivalent to the original matrix acting on the column vector on the right. These two statements are consistent since, for a Hermitian matrix the adjoint is identical to the original matrix (often, Hermitian matrices are called "self-adjoint" matrices for this reason) $\underline{M} = \underline{M}$=.

IX. CONNECTION BETWEEN ORTHONORMAL VECTORS AND ORTHONORMAL FUNCTIONS

For vectors as we have been dealing with, the scalar or dot product is defined as we have seen as follows:

$$<v(i)|v(j)> \equiv \sum_k v_k^*(i) v_k(j).$$

For functions of one or more variable (we denote the variables collectively as x), the generalization of the vector scalar product is

$$<f(i)|f(j)> \equiv \int f_i^*(x) f_j(x) dx.$$

The range of integration and the meaning of the variables x will be defined by the specific problem of interest; for example, in polar coordinates for one particle, $x \rightarrow r,\theta,\phi$, and for N particles in a general coordinate system $x \rightarrow (\mathbf{r}_i, \mathbf{r}_2, \ldots, \mathbf{r}_N)$.

If the functions $f_i(x)$ are orthonormal;

$$<f(i)|f(j)> = \delta_{ij},$$

and complete, then a resolution of the identity holds that is similar to that for vectors

$$\sum_i |f(i)\rangle\langle f(i)| = \sum_i f_i(x) f_i^*(x') = \delta(x - x')$$

where $\delta(x - x')$ is called the Dirac delta function. The δ function is zero everywhere except at x', but has unit area under it. In other words,

$$\int \delta(x - x') g(x') dx' = g(x) = \sum_i |f(i)\rangle\langle f(i)|g\rangle = \sum_i f_i(x) \int f_i^*(x') g(x') dx'$$

$$= \sum_i f_i(x) a_i,$$

where a_i is the projection of $g(x)$ along $f_i(x)$. The first equality can be taken as a definition of $\delta(x - x')$.

X. MATRIX REPRESENTATIONS OF FUNCTIONS AND OPERATORS

As we saw above, if the set of functions $\{f_i(x)\}$ is complete, any function of x can be *expanded* in terms of the $\{f_i(x)\}$

$$g(x) = \sum_i |f(i)\rangle\langle f(i)|\,|g\rangle$$

$$= \sum_i f_i(x) \int f_i^*(x') g(x') dx'$$

The column vector of numbers $a_i \equiv \int f_i^*(x') g(x') dx'$ is called the *representation* of $g(x)$ in the $f_i(x)$ *basis*. Note that this vector may have an infinite number of components because there may be an infinite number of $f_i(x)$.

Suppose that the function $g(x)$ obeys some *operator* equation (e.g., an eigenvalue equation) such as

$$\frac{dg(x)}{dx} = \alpha g(x).$$

Then, if we express $g(x)$ in terms of the $\{f_i(x)\}$, the equation for $g(x)$ can be replaced by a corresponding matrix problem for the a_i vector coefficients:

$$\sum_i \frac{df_i(x)}{dx} \int f_i^*(x') g(x') dx' = \alpha \sum_i f_i(x) \int f_i^*(x') g(x') dx'$$

If we now multiply by $f_j^*(x)$ and integrate over x, we obtain

$$\sum_i \int f_j^*(x) \frac{d}{dx} f_i(x) dx\, a_i = \alpha \sum_i \int f_j^*(x) f_i(x) dx\, a_i.$$

If the $f_i(x)$ functions are orthonormal, this result simplifies to

$$\sum_i \left(\frac{d}{dx}\right)_{ji} a_i = \alpha\, a_j$$

where we have *defined* the *matrix representation* of $\frac{d}{dx}$ in the $\{f_i\}$ basis by

$$\left(\frac{d}{dx}\right)_{ji} \equiv \int f_j^*(x) \frac{d}{dx} f_i(x) dx\ .$$

Remember that a_i is the representation of $g(x)$ in the $\{f_i\}$ basis. So the operator eigenvalue equation is equivalent to the *matrix* eigenvalue problem *if* the functions $\{f_i\}$ form a complete set.

Let us consider an example, that of the derivative operator in the orthonormal basis of Harmonic Oscillator functions. The fact that the solutions of the quantum Harmonic Oscillator, $\psi_n(x)$, are orthonormal and complete means that:

$$\int_{-\infty}^{+\infty} \psi_n^*(x)\psi_m(x)dx = \delta_{mn}.$$

The lowest members of this set are given as

$$\psi_0(x) = \pi^{-\frac{1}{4}} e^{-\frac{x^2}{2}}\ ,$$

$$\psi_1(x) = \pi^{-\frac{1}{4}} 2^{\frac{1}{2}} \times e^{-\frac{x^2}{2}}\ ,$$

$$\psi_2(x) = \pi^{-\frac{1}{4}} 8^{-\frac{1}{2}} (4x^2 - 2) e^{-\frac{x^2}{2}}, \ldots,$$

$$\psi_n(x) = A_n H_n(x) e^{-\frac{x^2}{2}}\ .$$

The derivatives of these functions are

$$\psi_0(x)' = \pi^{-\frac{1}{4}}(-x)e^{-\frac{x^2}{2}}\ ,$$

$$\psi_1(x)' = \pi^{-\frac{1}{4}} 2^{\frac{1}{2}} \left(e^{-\frac{x^2}{2}} - x^2 e^{-\frac{x^2}{2}}\right),$$

$$= 2^{-\frac{1}{2}}\left(\psi_0 - \psi_2\right), \text{ etc.}$$

In general, one finds that

$$\frac{d\psi_n(x)}{dx} = 2^{-\frac{1}{2}}\left(n^{\frac{1}{2}}\psi_{n-1} - (n+1)^{\frac{1}{2}}\psi_{n+1}\right).$$

From this general result, it is clear that the matrix representation of the $\frac{d}{dx}$ operator is given by

$$\underline{\mathrm{D}} = 2^{-\frac{1}{2}} \begin{bmatrix} 0 & 1 & 0 & 0 & \ldots \\ -1 & 0 & \sqrt{2} & 0 & \ldots \\ 0 & -\sqrt{2} & 0 & \sqrt{3} & \ldots \\ 0 & 0 & -\sqrt{3} & 0 & \ldots \\ \ldots & \ldots & \ldots & \ldots & \ldots \end{bmatrix}.$$

The matrix $\underline{\mathrm{D}}$ operates on the unit vectors $\boldsymbol{e}_0 = (1,0,0\ldots)$, $\boldsymbol{e}_1 = (0,1,\ldots.)$ etc. just like

$$\frac{d}{dx}$$

operates on $\psi_n(x)$, because these unit vectors are the representations of the $\psi_n(x)$ in the basis of the $\psi_n(x)$, and $\underline{D}$ is the representation of

$$\frac{d}{dx}$$

in the basis of the $\psi_n(x)$. Since any vector can be represented by a linear combination of the $\boldsymbol{e}_i$ vectors, the matrix $\underline{D}$ operates on any vector $(a_0,a_1,a_2 \ldots)$ just like

$$\frac{d}{dx}$$

operates on any $f(x)$. Note that the matrix is not Hermitian; it is actually antisymmetric. However, if we multiply $\underline{D}$ by $-i\hbar$ we obtain a Hermitian matrix that represents the operator

$$-i\hbar\frac{d}{dx},$$

the momentum operator.

It is easy to see that we can form the matrix representation of any linear operator for any complete basis in any space. To do so, we act on each basis function with the operator and express the resulting function as a linear combination of the original basis functions. The coefficients that arise when we express the operator acting on the functions in terms of the original functions form the the matrix representation of the operator.

It is natural to ask what the eigenvalues and eigenfunctions of the matrix you form through this process mean. If your operator is the Hamiltonian operator, then the matrix eigenvectors and eigenvalues are the representations of the solutions to the Schrodinger equation in this basis. Forming the representation of an operator reduces the solution of the operator eigenvalue equation to a matrix eigenvalue equation.

XI. COMPLEX NUMBERS, FOURIER SERIES, FOURIER TRANSFORMS, BASIS SETS

One of the techniques to which chemists are frequently exposed is Fourier transforms. They are used in NMR and IR spectroscopy, quantum mechanics, and classical mechanics. Just as we expand a function in terms of a complete set of basis function or a vector in terms of a complete set of vectors, the Fourier transform expresses a function $f(\omega)$ of a continuous variable ω in terms of a set of orthonormal functions that are not discretely labeled but which are labeled by a continuous "index" t. These functions are

$$(2\pi)^{-\frac{1}{2}}e^{-i\omega t},$$

and the "coefficients" in the expansion

$$f(\omega) = (2\pi)^{-\frac{1}{2}} \int_{-\infty}^{+\infty} e^{-i\omega t} f(t)dt,$$

are called the Fourier transform $f(t)$ of $f(\omega)$.

The orthonormality of the

$$(2\pi)^{-\frac{1}{2}}e^{-i\omega t}$$

functions will be demonstrated explicitly later. Before doing so however, it is useful to review both complex numbers and basis sets.

A. Complex Numbers

A complex number has a real part and an imaginary part and is usually denoted:

$$z = x + iy.$$

For the complex number z, x is the real part, y is the imaginary part, and $i = \sqrt{-1}$. This is expressed as $x = Re(z)$, $y = Im(z)$. For every complex number z, there is a related one called its complex conjugate $z* = x - iy$.

Complex numbers can be thought of as points in a plane where x and y are the abscissa and ordinate, respectively. This point of view prompts us to introduce polar coordinates r and θ to describe complex numbers, with

$$x = r \cos \theta \qquad r = (x^2 + y^2)^{\frac{1}{2}}$$

or,

$$y = r \sin \theta \qquad \theta = \tan^{-1}\left(\frac{y}{x}\right) + [\pi \text{ (if } x < 0)].$$

Another name for r is the norm of z which is denoted $|z|$. The angle θ is sometimes called the argument of z, $\arg(z)$, or the phase of z.

Complex numbers can be added, subtracted, multiplied and divided like real numbers. For example, the multiplication of z by $z*$ gives:

$$zz* = (x + iy)\,(x - iy) = x^2 + ixy - ixy + y^2 = x^2 + y^2 = r^2$$

Thus $zz* = |z|^2$ is a real number.

An identity due to Euler is important to know and is quite useful. It states that

$$e^{i\theta} = \cos\theta + i\sin\theta.$$

It can be proven by showing that both sides of the identity obey the same differential equation. Here we will only demonstrate its plausibility by Taylor series expanding both sides:

$$e^x = 1 + x + \frac{x^2}{2} + \frac{x^3}{3!} + \frac{x^4}{4!} + \ldots,$$

$$\sin x = x - \frac{x^3}{3!} + \frac{x^5}{5!} + \ldots,$$

and

$$\cos x = 1 - \frac{x^2}{2} + \frac{x^4}{4!} + \ldots.$$

Therefore, the exponential $\exp(i\theta)$ becomes

$$e^{i\theta} = 1 + i\theta + i^2\frac{\theta^2}{2} + i^3\frac{\theta^3}{3!} + i^4\frac{\theta^4}{4!} + i^5\frac{\theta^5}{5!} + \ldots$$

$$= 1 - \frac{\theta^2}{2} + \frac{\theta^4}{4!} + \ldots + i\left(\theta - \frac{\theta^3}{3!} + \frac{\theta^5}{5!} + \ldots\right).$$

The odd powers of θ clearly combine to give the sine function; the even powers give the cosine function, so

$$e^{i\theta} = \cos\theta + i\sin\theta$$

is established. From this identity, it follows that

$$\cos\theta = \frac{1}{2}(e^{i\theta} + e^{-i\theta})$$

and

$$\sin\theta = \frac{1}{2i}(e^{i\theta} - e^{-i\theta}) .$$

It is now possible to express the complex number z as

$$\begin{aligned} z &= x + iy \\ &= r\cos\theta + ir\sin\theta \\ &= r(\cos\theta + i\sin\theta) \\ &= re^{i\theta}. \end{aligned}$$

This form for complex numbers is extremely useful. For instance, it can be used to easily show that

$$zz^* = re^{i\theta}(re^{-i\theta}) = r^2e^{i(\theta-\theta)} = r^2.$$

B. Fourier Series

Now let us consider a function that is periodic in time with period T. Fourier's theorem states that any periodic function can be expressed in a Fourier series as a linear combination (infinite series) of sines and cosines whose frequencies are multiples of a fundamental frequency Ω corresponding to the period:

$$f(t) = \sum_{n=0}^{\infty} a_n\cos(n\Omega t) + \sum_{n=1}^{\infty} b_n\sin(n\Omega t),$$

where

$$\Omega = \frac{2\pi}{T} .$$

The Fourier expansion coefficients are given by projecting $f(t)$ along each of the orthogonal sine and cosine functions:

$$a_o = \frac{1}{T}\int_0^T f(t)dt,$$

$$a_n = \frac{2}{T}\int_0^T f(t)\cos(n\Omega t)dt,$$

$$b_n = \frac{2}{T}\int_0^T f(t)\sin(n\Omega t)dt.$$

The term in the Fourier expansion associated with a_o is a constant giving the average value (i.e., the zero-frequency or DC component) of the function. The terms with $n = 1$ contain the fundamental frequency and higher terms contain the nth harmonics or overtones of the fundamental frequency. The coefficients, a_n and b_n, give the amplitudes of each of these frequencies. Note that if $f(t)$ is an even function (i.e., if $f(t) = f(-t)$), $b_n = 0$ for $n = 1,2,3 \ldots$ so the series only has cosine terms. If $f(t)$ is an odd function (i.e., $f(t) = -f(-t)$), $a_n = 0$ for $n = 0,1,2,3 \ldots$ and the series only has sine terms.

The Fourier series expresses a continuous function as an infinite series of numbers . . . $a_o, a_1, b_1, a_2, b_2, \ldots$. We say that the set of coefficients is a *representation* of the function in the Fourier basis. The expansion in the cos Ωnt and sin Ωnt basis is useful because the basis functions are orthogonal when integrated over the interval 0 to T. An alternative set of functions is sometimes used to carry out the Fourier expansion; namely the

$$\left(\frac{1}{T}\right)^{\frac{1}{2}} \exp(i\Omega nt)$$

functions for n from $-\infty$ to $+\infty$. Their orthogonality can be proven as follows:

$$\frac{1}{T}\int_0^T \psi_n^* \psi_m dt = \frac{1}{T}\int_0^T \exp(i(m-n)\Omega t)dt = 1$$

if $m = n$, and

$$= \frac{1}{T}(i(m-n)\Omega)^{-1}(\exp(i(m-n)\Omega T) - 1) = 0$$

if $m \neq n$.

Let us consider the Fourier representation of $f(t)$ in terms of the complex exponentials introduced above. For an arbitrary periodic function $f(t)$, we can write

$$f(t) = \sum_{-\infty}^{+\infty} c_n e^{in\Omega t}, \text{ where } c_n = \frac{1}{T}\int_0^T f(t)e^{-in\Omega t}dt.$$

This form of the Fourier series is entirely equivalent to the first form and the a_n and b_n can be obtained from the c_n and vice versa. For example, the c_n amplitudes are obtained by projecting $f(t)$ onto exp(inΩt) as:

$$c_n = \frac{1}{T}\int_0^T f(t)(\cos(n\Omega t) - i\sin(n\Omega t))dt$$

$$= \frac{1}{T}\int_0^T f(t)\cos(n\Omega t)dt - \frac{1}{T}\int_0^T f(t)\sin(n\Omega t)dt,$$

but these two integrals are easily recognized as the expansion in the other basis, so

$$c_n = \frac{1}{2}a_n - \frac{1}{2}i\, b_n = \frac{1}{2}(a_n - ib_n).$$

By using complex exponential functions instead of trigonometric functions, we only have one family of basis functions, $e^{in\Omega t}$, but we have twice as many of them. In this form, if $f(t)$ is even, then the c_n are real, but if $f(t)$ is odd, then c_n are imaginary.

It is useful to consider some examples to learn this material. First, let $f(t)$ be the odd function $f(t) = \sin 3t$. Then, one period has elapsed when $3T = 2\pi$, so

$$T = \frac{2\pi}{3}$$

and $\Omega = 3$. For this function, the complex series Fourier expansion coefficients are given by

$$c_0 = \frac{3}{2\pi}\int_0^{\frac{2\pi}{3}} \sin 3t dt = 0.$$

$$c_1 = \frac{3}{2\pi}\int_0^{\frac{2\pi}{3}} \sin 3t e^{-i3t} dt = \frac{3}{2\pi}\int \frac{(e^{i3t} - e^{-i3t})}{2i} e^{-i3t} dt$$

$$= \frac{1}{2i}\frac{3}{2\pi}\int_0^{\frac{2\pi}{3}} (1 - e^{6it}) dt = \frac{3}{4\pi i}\left(\frac{2\pi}{3} - 0\right) = \frac{1}{2i} = -\frac{i}{2}$$

Because $\sin 3t$ is real, it is straightforward to see from the definition of c_n that $c_{-n} = c_n^*$, as a result of which

$$c_{-1} = c_1^* = +\frac{i}{2}.$$

The orthogonality of the $\exp(in\Omega t)$ functions can be used to show that all of the higher c_n coefficients vanish:

$$c_n = 0 \text{ for } n \neq \pm 1.$$

Hence we see that this simple periodic function has just two terms in its Fourier series. In terms of the sine and cosine expansion, one finds for this same $f(t) = \sin 3t$ that $a_n = 0$, $b_n = 0$ for $n \neq 1$, and $b_1 = 1$.

As another example, let $f(t)$ be

$$f(t) = t, \qquad -\pi < t < \pi,$$

and make $f(t)$ periodic with period 2π (so $\Omega = 1$). For this function,

$$c_0 = \frac{1}{2\pi} \int_{-\pi}^{\pi} t dt = 0$$

$$c_n = \frac{1}{2\pi} \int_{-\pi}^{\pi} te^{-int} dt = \frac{1}{2\pi n^2} [e^{-int}(1 + int)] \Big|_{-\pi}^{\pi}$$

$$= \frac{1}{2\pi n^2} (e^{-in\pi}(1 + in\pi(1 - in\pi)))$$

$$= \frac{1}{2\pi n^2} ((-1)^n (1 + in\pi) - (-1)^n (1 - in\pi))$$

$$= \frac{(-1)^n}{2\pi n^2} (2in\pi) = \frac{i}{n} (-1)^n \quad n \neq 0.$$

Note that since t is an odd function, c_n is imaginary, and as $n \rightarrow \infty$, c_n approaches zero.

C. Fourier Transforms

Let us now imagine that the period T becomes very long. This means that the fundamental frequency

$$\Omega = \frac{2\pi}{T}$$

becomes very small and the harmonic frequencies $n\Omega$ are very close together. In the limit of an infinite period one has a non-periodic function (i.e., an arbitrary function). The frequencies needed to represent such a function become continuous, and hence the Fourier sum becomes an integral.

To make this generalization more concrete, we first take the earlier Fourier expression for $f(t)$ and multiply it by unity in the form

$$\frac{\Omega T}{2\pi}$$

to obtain:

$$f(t) = \sum_{-\infty}^{\infty} c_n e^{in\Omega t} \frac{\Omega T}{2\pi}.$$

Now we replace $n\Omega$ (which are frequencies infinitesimally close together for sufficiently long T) by the continuous index ω:

$$f(t) = \frac{1}{2\pi} \sum_{n=-\infty}^{\infty} (c_n T) e^{i\omega t} \Omega.$$

In the limit of long T, this sum becomes an integral because Ω becomes infinitesimally small ($\Omega \rightarrow d\omega$). As T grows, the frequency spacing between n and $n + 1$, which is

$$\frac{2\pi}{T},$$

as well as the frequency associated with a given n-value become smaller and smaller. As a result, the product c_nT evolves into a continuous function of ω which we denote $c(\omega)$. Before, c_n depended on the continuous index n, and represented the contribution to the function $f(t)$ of waves with frequency $n\Omega$. The function $c(\omega)$ is the contribution per unit frequency to $f(t)$ from waves with frequency in the range ω to $\omega + d\omega$. In summary, the Fourier series expansion evolves into an integral that defines the Fourier transformation of $f(t)$:

$$f(t) = \frac{1}{2\pi} \int_{-\infty}^{\infty} c(\omega)e^{-i\omega t}d\omega.$$

It is convenient to define a new function $f(\omega) = (2\pi)^{-\frac{1}{2}}$ where $f(\omega)$ is called the Fourier transform of $f(t)$. The Fourier transform $f(\omega)$ is a representation of $f(t)$ in another basis, that of the orthonormal set of oscillating functions

$$e^{iwt}(2\pi)^{-\frac{1}{2}}:$$

$$f(t) = \left(\frac{1}{2\pi}\right)^{\frac{1}{2}} \int f(\omega)e^{-i\omega t}d\omega.$$

Earlier, for Fourier series, we had the orthogonality relation among the Fourier functions:

$$\frac{1}{T}\int_0^T \psi_n{}^*\psi_m dt = \delta_{nm},$$

but for the continuous variable ω, we have a different kind of orthogonality

$$\int_{-\infty}^{+\infty} \psi^*(\omega_1)\psi(\omega_2)dt = \delta(\omega_1 - \omega_2),$$

where $\psi(\omega_j) = (2\pi)^{-\frac{1}{2}}e^{i\omega_j t}$.

The function $\delta(\omega)$, called the Dirac delta function, is the continuous analog to δ_{nm}. It is zero unless $\omega = o$. If $\omega = o$, $\delta(\omega)$ is infinite, but it is infinite in such a way that the area under the curve is precisely unity. Its most useful definition is that $\delta(\omega)$ is the function which, for arbitrary $f(\omega)$, the following identity holds:

$$\int_{-\infty}^{+\infty} f(\omega)\delta(\omega - \omega')dw = f(\omega').$$

That is, integrating $\delta(\omega - \omega')$ times any function evaluated at ω just gives back the value of the function at ω'.

The Dirac delta function can be expressed in a number of convenient ways, including the following two:

$$\delta(\omega - \omega') = \frac{1}{2\pi} \int_{-\infty}^{+\infty} e^{i(\omega-\omega')t}\,dt$$

$$= \lim_{a\to 0} \sqrt{\frac{\pi}{a}}\, e^{-\frac{(\omega-\omega')^2}{a^2}}$$

As an example of applying the Fourier transform method to a non-periodic function, consider the localized pulse

$$f(t) = \begin{cases} \dfrac{1}{T} & -\dfrac{T}{2} \leq t \leq \dfrac{T}{2} \\ 0 & |t| > \dfrac{T}{2} \end{cases}$$

For this function, the Fourier transform is

$$f(\omega) = (2\pi)^{-\frac{1}{2}} \int_{-\infty}^{\infty} f(t) e^{-i\omega t} dt$$

$$= (2\pi)^{-\frac{1}{2}} \frac{1}{T} \int_{-\frac{T}{2}}^{\frac{T}{2}} e^{-i\omega t} dt$$

$$= (2\pi)^{-\frac{1}{2}} \frac{1}{T} \int_{-\frac{T}{2}}^{\frac{T}{2}} (\cos\omega t - i\sin\omega t) dt$$

$$= (2\pi)^{-\frac{1}{2}} \frac{1}{T} \int_{-\frac{T}{2}}^{\frac{T}{2}} \cos\omega t dt + 0$$

$$= (2\pi)^{-\frac{1}{2}} \frac{1}{\omega T} (\sin\omega t) \Big|_{\frac{-T}{2}}^{\frac{T}{2}}$$

$$= \left(\frac{2}{\pi}\right)^{-\frac{1}{2}} \sin(\omega T/2)/\omega T.$$

Note that $f(\omega)$ has its maximum value of $(2\pi)^{-\frac{1}{2}}$ for $\omega = 0$ and that $f(\omega)$ falls slowly in magnitude to zero as ω increases in magnitude, while oscillating to positive and negative values. However, the primary maximum in $f(\omega)$ near zero-frequency has a width that is inversely proportional to T. This inverse relationship between the width (T) in t-space and the width

$$\left(\frac{2\pi}{T}\right)$$

in ω-space is an example of the uncertainty principle. In the present case it means that if you try to localize a wave in time, you must use a wide range of frequency components.

D. Fourier Series and Transforms in Space

The formalism we have developed for functions of time can also be used for functions of space variables or any other variables for that matter. If $f(x)$ is a periodic function of the coordinate x, with period (repeat distance)

$$\frac{2\pi}{K},$$

then it can be represented in terms of Fourier functions as follows:

$$f(x) = \sum_{n=-\infty}^{\infty} f_n e^{inKx}$$

where the coefficients are

$$f_n = \frac{K}{2\pi} \int_0^{\frac{2\pi}{K}} f(x) e^{-inKx} dx.$$

If $f(x)$ is a non-periodic function of the coordinate x, we write it as

$$f(x) = (2\pi)^{-\frac{1}{2}} \int_{-\infty}^{\infty} f(k) e^{ikx} dk$$

and the Fourier transform is given as

$$f(k) = (2\pi)^{-\frac{1}{2}} \int_{-\infty}^{\infty} f(x) e^{-ikx} dx\ .$$

If f is a function of several spatial coordinates and/or time, one can Fourier transform (or express as Fourier series) simultaneously in as many variables as one wishes. You can even Fourier transform in some variables, expand in Fourier series in others, and not transform in another set of variables. It all depends on whether the functions are periodic or not, and whether you can solve the problem more easily after you have transformed it.

E. Comments

So far we have seen that a periodic function can be expanded in a discrete basis set of frequencies and a non-periodic function can be expanded in a continuous basis set of frequencies. The expansion process can be viewed as expressing a function in a different basis. These basis sets are the collections of solutions to a differential equation called the wave equation. These sets of solutions are useful because they are *complete sets*. Completeness means that *any* arbitrary function can be expressed *exactly* as a linear combination of these functions. Mathematically, completeness can be expressed as

$$1 = \int |\psi(\omega)\rangle \langle\psi(\omega)| d\omega$$

in the Fourier transform case, and

$$1 = \sum_n |\psi_n\rangle \langle\psi_n|$$

in the Fourier series case.

The only limitation on the function expressed is that it has to be a function that has the same boundary properties and depends on the same variables as the basis. You would not want to use Fourier series to express a function that is not periodic, nor would you want to express a three-dimensional vector using a two-dimensional or four-dimensional basis.

Besides the intrinsic usefulness of Fourier series and Fourier transforms for chemists (e.g., in FTIR spectroscopy), we have developed these ideas to illustrate a point that is important in quantum chemistry. Much of quantum chemistry is involved with basis sets and expansions. This has nothing in particular to do with quantum mechanics. Any time one is dealing with linear differential equations like those that govern light (e.g., spectroscopy) or matter (e.g., molecules), the solution can be written as linear combinations of complete sets of solutions.

XII. SPHERICAL COORDINATES

A. Definitions

The relationships among cartesian and spherical polar coordinates are given as follows:

$$z = r\cos\theta \qquad r = \sqrt{x^2 + y^2 + z^2}$$

$$x = r\sin\theta\cos\phi \qquad \theta = \cos^{-1}\left(\frac{z}{\sqrt{x^2 + y^2 + z^2}}\right)$$

$$y = r\sin\theta\sin\phi \qquad \phi = \cos^{-1}\left(\frac{x}{\sqrt{x^2 + y^2}}\right)$$

The ranges of the polar variables are $0 < r < \infty$, $0 \le \theta \le \pi$, $0 \le \phi < 2\pi$.

B. The Jacobian in Integrals

In performing integration over all space, it is necessary to convert the multiple integral from cartesian to spherical coordinates:

$$\int dx \int dy \int dz\, f(x,y,z) \rightarrow \int dr \int d\theta \int d\phi\, f(r,\theta,\phi) \cdot J.$$

where J is the so-called Jacobian of the transformation. J is computed by forming the determinant of the three-by-three matrix consisting of the partial derivatives relating x,y,z to r,θ,ϕ:

$$J = \left|\frac{\partial(x,y,z)}{\partial(r,\theta,\phi)}\right|$$

$$= \begin{vmatrix} \sin\theta\cos\phi & \sin\theta\sin\phi & \cos\theta \\ r\cos\theta\cos\phi & r\cos\theta\sin\phi & -r\sin\theta \\ -r\sin\theta\sin\phi & r\sin\theta\cos\phi & 0 \end{vmatrix}$$

The determinant, J, can be expanded, for example, by the method of diagonals, giving four non-zero terms:

$$
\begin{aligned}
J &= r^2\sin^3\theta\sin^2\phi + r^2\cos^2\theta\sin\theta\cos^2\phi + r^2\sin^3\theta\cos^2\phi + r^2\cos^2\theta\sin\theta\sin^2\phi \\
&= r^2\sin\theta(\sin^2\theta(\sin^2\phi + \cos^2\phi) + \cos^2\theta(\cos^2\phi + \sin^2\phi)) \\
&= r^2\sin\theta.
\end{aligned}
$$

Hence in converting integrals from x,y,z to r,θ,ϕ one writes as a short hand $dxdydz = r^2\sin\theta drd\theta d\phi$.

C. Transforming Operators

In many applications, derivative operators need to be expressed in spherical coordinates. In converting from cartesian to spherical coordinate derivatives, the chain rule is employed as follows:

$$
\begin{aligned}
\frac{\partial}{\partial x} &= \frac{\partial r}{\partial x}\frac{\partial}{\partial r} + \frac{\partial\theta}{\partial x}\frac{\partial}{\partial\theta} + \frac{\partial\phi}{\partial x}\frac{\partial}{\partial\phi} \\
&= \sin\theta\cos\phi\frac{\partial}{\partial r} + \frac{\cos\theta\cos\phi}{r}\frac{\partial}{\partial\theta} - \frac{\sin\phi}{r\sin\theta}\frac{\partial}{\partial\phi}.
\end{aligned}
$$

Likewise

$$\frac{\partial}{\partial y} = \sin\theta\sin\phi\frac{\partial}{\partial r} + \frac{\cos\theta\sin\phi}{r}\frac{\partial}{\partial\theta} + \frac{\cos\phi}{r\sin\theta}\frac{\partial}{\partial\phi},$$

and

$$\frac{\partial}{\partial z} = \cos\theta\frac{\partial}{\partial r} - \frac{\sin\theta}{r}\frac{\partial}{\partial\theta} + 0\frac{\partial}{\partial\phi}$$

Now to obtain an expression for

$$\frac{\partial^2}{\partial x^2}$$

and other second derivatives, one needs to take the following derivatives:

$$
\begin{aligned}
\frac{\partial^2}{\partial x^2} &= \sin\theta\cos\phi\frac{\partial}{\partial r}\left(\sin\theta\cos\phi\frac{\partial}{\partial r} + \frac{\cos\theta\cos\phi}{r}\frac{\partial}{\partial\theta} - \frac{\sin\phi}{r\sin\theta}\frac{\partial}{\partial\phi}\right) \\
&\quad + \frac{\cos\theta\cos\phi}{r}\frac{\partial}{\partial\theta}\left(\sin\theta\cos\phi\frac{\partial}{\partial r} + \frac{\cos\theta\cos\phi}{r}\frac{\partial}{\partial\theta} - \frac{\sin\phi}{r\sin\theta}\frac{\partial}{\partial\phi}\right) \\
&\quad - \frac{\sin\phi}{r\sin\theta}\frac{\partial}{\partial\phi}\left(\sin\theta\cos\phi\frac{\partial}{\partial r} + \frac{\cos\theta\cos\phi}{r}\frac{\partial}{\partial\theta} - \frac{\sin\phi}{r\sin\theta}\frac{\partial}{\partial\phi}\right) \\
&= \sin^2\theta\cos^2\phi\frac{\partial^2}{\partial r^2} + \frac{\cos^2\theta\cos^2\phi}{r^2}\frac{\partial^2}{\partial\theta^2} + \frac{\sin^2\phi}{r^2\sin^2\theta}\frac{\partial^2}{\partial\phi^2}
\end{aligned}
$$

$$+\frac{2\sin\theta\cos\theta\cos^2\phi}{r}\frac{\partial^2}{\partial r\partial\theta}-\frac{2\cos\phi\sin\phi}{r}\frac{\partial^2}{\partial r\partial\phi}-\frac{2\cos\theta\cos\phi\sin\phi}{r^2\sin\theta}\frac{\partial^2}{\partial\theta\partial\phi}$$

$$+\left(\frac{\cos^2\theta\cos^2\phi}{r}+\frac{\sin^2\phi}{r}\right)\frac{\partial}{\partial r}+\left(-\frac{2\sin\theta\cos\theta\cos^2\phi}{r^2}+\frac{\sin^2\phi\cos\theta}{r^2\sin\theta}\right)\frac{\partial}{\partial\theta}$$

$$+\left(\frac{\cos\phi\sin\phi}{r^2}+\frac{\cos^2\theta\cos\phi\sin\phi}{r^2\sin^2\theta}\right)\frac{\partial}{\partial\phi}$$

Analogous steps can be performed for

$$\frac{\partial^2}{\partial y^2} \text{ and } \frac{\partial^2}{\partial z^2}.$$

Adding up the three contributions, one obtains:

$$\nabla^2=\frac{1}{r^2}\frac{\partial}{\partial r}\left(r^2\frac{\partial}{\partial r}\right)+\frac{1}{r^2\sin\theta}\frac{\partial}{\partial\theta}\left(\sin\theta\frac{\partial}{\partial\theta}\right)+\frac{1}{r^2\sin^2\theta}\frac{\partial^2}{\partial\phi^2}.$$

As can be seen by reading Appendix G, the terms involving angular derivatives in ∇^2 are identical to

$$-\frac{L^2}{r^2},$$

where L^2 is the square of the rotational angular momentum operator. Although in this appendix, we choose to treat it as a collection of differential operators that gives rise to differential equations in θ and ϕ to be solved, there are more general tools for treating all such angular momentum operators. These tools are developed in detail in appendix G and are used substantially in the next two appendices.

XIII. SEPARATION OF VARIABLES

In solving differential equations such as the Schrödinger equation involving two or more variables (e.g., equations that depend on three spatial coordinates x, y, and z or r, θ, and ϕ or that depend on time t and several spatial variables denoted $\boldsymbol{r}$), it is sometimes possible to reduce the solution of this one multivariable equation to the solution of several equations each depending on fewer variables. A commonly used device for achieving this goal is the *separation of variables technique*.

This technique is not always successful, but can be used for the type of cases illustrated now. Consider a two-dimensional differential equation that is of second order and of the eigenvalue type:

$$A\,\partial^2/\partial x^2\psi+B\,\partial^2/\partial y^2\,\psi+C\,\partial\psi/\partial x\,\partial\psi/\partial y+D\,\psi=E\,\psi.$$

The solution ψ must be a function of x and y because the differential equation refers to ψ's derivatives with respect to these variables.

The separations of variables device assumes that $\psi(x,y)$ can be written as a *product* of a function of x and a function of y:

$$\psi(x,y)=\alpha(x)\,\beta(y).$$

Inserting this ansatz into the above differential equation and then dividing by $\alpha(x)\ \beta(y)$ produces:

$$A\,\alpha^{-1}\partial^2/\partial x^2\,\alpha + B\,\beta^{-1}\,\partial^2/\partial y^2\,\beta + C\,\alpha^{-1}\,\beta^{-1}\,\partial\alpha/\partial x\,\partial\beta/\partial y + D = E.$$

The key observations to be made are:

1. If A is independent of y, $A\,\alpha^{-1}\,\partial^2/\partial x^2\,\alpha$ must be independent of y.
2. If B is independent of x, $B\,\beta^{-1}\,\partial^2/\partial y^2\,\beta$ must be independent of x.
3. If C vanishes and D does not depend on both x and y, then there are no "cross terms" in the above equations (i.e., terms that contain both x and y). For the sake of argument, let us take D to be dependent on x only for the remainder of this discussion; the case for which D depends on y only is handled in like manner.

Under circumstances for which all three of the above conditions are true, the left-hand side of the above second-order equation in two variables can be written as the sum of

$$A\,\alpha^{-1}\,\partial^2/\partial x^2\,\alpha + D,$$

which is independent of y, and

$$B\,\beta^{-1}\,\partial^2/\partial y^2\,\beta$$

which is independent of x.

The full equation states that the sum of these two pieces must equal the eigenvalue E, which is independent of both x and y.

Because E and $B\,\beta^{-1}\,\partial^2/\partial y^2\,\beta$ are both independent of x, the quantity $A\,\alpha^{-1}\,\partial^2/\partial x^2\,\alpha + D$ (as a whole) can *not* depend on x, since $E - B\,\beta^{-1}\,\partial^2/\partial y^2\,\beta$ must equal $A\,\alpha^{-1}\,\partial^2/\partial x^2\,\alpha + D$ for all values of x and y. That $A\,\alpha^{-1}\,\partial^2/\partial x^2\,\alpha + D$ is independent of x and (by assumption) independent of y allows us to write:

$$A\,\alpha^{-1}\,\partial^2/\partial x^2\,\alpha + D = \varepsilon,$$

a constant.

Likewise, because E and $A\,\alpha^{-1}\,\partial^2/\partial x^2\,\alpha + D$ are independent of y, the quantity $B\,\beta^{-1}$ $\partial^2/\partial y^2\,\beta$ (as a whole) can *not* depend on y, since $E - A\,\alpha^{-1}\,\partial^2/\partial x^2\,\alpha - D$ must equal $B\,\beta^{-1}\,\partial^2/\partial y^2\,\beta$ for all values of x and y. That $B\,\beta^{-1}\,\partial^2/\partial y^2\,\beta$ is independent of y and (by assumption) independent of x allows us to write:

$$B\,\beta^{-1}\,\partial^2/\partial y^2\,\beta = \varepsilon',$$

another constant.

The net result is that we now have *two first-order differential equations* of the eigenvalue form:

$$A\,\partial^2/\partial x^2\,\alpha + D\,\alpha = \varepsilon\,\alpha,$$

and

$$B\,\partial^2/\partial y^2\,\beta = \varepsilon'\,\beta,$$

and the solution of the original equation has been successfully subjected to separation of variables. The two eigenvalues ε and ε' of the separated x- and y-equations must obey:

$$\varepsilon + \varepsilon' = E,$$

which follows by using the two separate (x and y) eigenvalue equations in the full two-dimensional equation that contains E as its eigenvalue.

In summary, when separations of variables can be used, it:

1. reduces one multidimensional differential equation to two or more lower-dimensional differential equations.
2. expresses the eigenvalue of the original equation as a *sum of eigenvalues* (whose values are determined via boundary conditions as usual) of the lower-dimensional problems.
3. expresses the solutions to the original equation as a *product* of solutions to the lower-dimensional equations (i.e., $\psi(x,y) = \alpha(x)\ \beta(y)$, for the example considered above).

APPENDIX

B The Hydrogen Atom Orbitals

In chapter 1 and appendix A, the angular and radial parts of the Schrödinger equation for an electron moving in the potential of a nucleus of charge Z were obtained. These "hydrogen-like" atomic orbitals are proper eigenstates for H, He^{+}, Li^{++}, Be^{+++}, etc. They also serve as useful starting points for treating many-electron atoms and molecules in that they have been found to be good basis functions in terms of which to expand the orbitals of most atoms and ions. In this appendix, we will examine the sizes, energies, and shapes of these orbitals and consider, in particular, how these properties vary with principal quanutm number n, charge Z, and angular quantum numbers l and m.

In chapter 1 and appendix A, it was shown that the total r, θ,ϕ dependence of these so-called hydrogenic orbitals is given as:

$$\psi_{n,l,m} = Y_{l,m}(\theta,\phi)R_{n,l}(r),$$

where the spherical harmonics $Y_{l,m}$ are expressed in terms of the associated Legendre polynomials $P_l^m(\theta)$ as

$$Y_{l,m}(\theta,\phi) = P_l^{|m|}(\cos\theta)(2\pi)^{-1/2}\exp(im\phi),$$

and the radial functions $R_{n,l}(r)$ are given in terms of Laguerre polynomials of order $n-l-1$, $L_{n-l-1}(\rho)$ as follows:

$$R_{n,l}(r) = N_{n,l}\rho^l e^{-\rho/2}L_{n-l-1}(\rho),$$

where $N_{n,l}$ is a normalization constant. Here, the radial coordinate r of the electron is contained in the variable ρ defined as

$$\rho = Zr/a_o n,$$

where a_o is the bohr radius

$$a_o = \hbar^2/\mu e^2 = 0.529 \AA.$$

The energies of these hydrogenic orbitals, relative to an electron infinitely far from the nucleus and with zero kinetic energy, are

$$E = -\mu Z^2 e^4/2\hbar^2 n^2 = -27.21\, Z^2/2n^2 \text{ eV}.$$

The $n = 1$, 2, and 3 wavefunctions are given explicitly as follows:

$n = 1, l = 0, m = 0, \psi = (Z/a_o)^{3/2}(\pi)^{-1/2} \exp(-Zr/a_o)$
$n = 2, l = 0, m = 0, \psi = (Z/a_o)^{3/2}(2\pi)^{-1/2} (1 - Zr/2a_o)/2 \exp(-Zr/2a_o)$
$n = 2, l = 1, m = 0, \psi = (Z/a_o)^{3/2}(2\pi)^{-1/2}\, Zr/4a_o \cos\theta \exp(-Zr/2a_o)$
$n = 2, l = 1, m = \pm 1, \psi = (Z/a_o)^{3/2}(4\pi)^{-1/2}\, Zr/4a_o \sin\theta \exp(\pm i\phi) \exp(-Zr/2a_o)$
$n = 3, l = 0, m = 0, \psi = (Z/a_o)^{3/2}(3\pi)^{-1/2} [27 - 18Zr/a_o + 2(Zr/a_o)^2]/81 \exp(-Zr/3a_o)$
$n = 3, l = 1, m = 0, \psi = (Z/a_o)^{3/2}(2\pi)^{-1/2}\, 2/81 [6Zr/a_o - (Zr/a_o)^2] \cos\theta \exp(-Zr/3a_o)$
$n = 3, l = 1, m = \pm 1, \psi = (Z/a_o)^{3/2}(\pi)^{-1/2}\, 1/81 [6Zr/a_o - (Zr/a_o)^2] \sin\theta$
$\exp(\pm i\phi) \exp(-Zr/3a_o)$
$n = 3, l = 2, m = 0, \psi = (Z/a_o)^{3/2}(6\pi)^{-1/2}\, 1/81 (Zr/a_o)^2 [3\cos^2\theta - 1] \exp(-Zr/3a_o)$
$n = 3, l = 2, m = \pm 1, \psi = (Z/a_o)^{3/2}(\pi)^{-1/2}\, 1/81 (Zr/a_o)^2 \sin\theta \cos\theta \exp(\pm i\phi) \exp(-Zr/3a_o)$
$n = 3, l = 2, m = \pm 2, \psi = (Z/a_o)^{3/2}(4\pi)^{-1/2}\, 1/81 (Zr/a_o)^2 \sin^2\theta \exp(\pm 2i\phi) \exp(-Zr/3a_o)$.

On pages 133–136 of Pauling and Wilson are tabulated the bound-state solutions of this problem for n ranging up to $n = 6$ and l-values up to $l = 5$.

The above $\pm m$ functions are appropriate whenever one wishes to describe orbitals that are eigenfunctions of L_z, the component of orbital angular momentum along the z-axis. The functions with $\pm m$ quantum numbers can be combined in pairs to give the spatially oriented functions with which most chemists are familiar:

$$\psi_+ = 2^{-1/2}[\psi_m + \psi_{-m}]$$

and

$$\psi_- = 2^{-1/2}[\psi_m - \psi_{-m}].$$

For example, when applied to the $2p_{\pm 1}$ functions, one forms the $2p_x$ and $2p_y$ functions; the $3d_{\pm 1}$ functions combine to give the $3d_{xz}$ and $3d_{yz}$ functions; and the $3d_{\pm 2}$ functions combine to give the $3d_{xy}$ and $3d_{x^2-y^2}$ functions. These spatially directed functions are more appropriate for use whenever one is dealing with the interaction of the atom on which these orbitals are located with surrounding "ligands" or bonding partners which are spatially arranged in a manner that destroys the spherical and axial symmetry of the system. One is permitted to combine the $\pm m$ functions in this manner because these pairs of functions are energetically degenerate; any combination of them is therefore of the same energy.

There are several important points to stress about these hydrogenic atomic functions:

1. Their energies vary as $-Z^2/n^2$, so atoms with higher Z values will have more tightly bound electrons (for the same n). In a many-electron atom, one often introduces the concept of an effective nuclear charge Z_{eff}, and takes this to be the full nuclear charge Z minus the number of electrons that occupy orbitals that reside radially "inside" the orbital in question. For example, $Z_{eff} = 6 - 2 = 4$ for the $n = 2$ orbitals of carbon in the $1s^2 2s^2 2p^4$ configuration. This Z_{eff} is then used to qualitatively estimate the relative energetic stability by making use of the $-Z_{eff}^2/n^2$ scaling of the energy.
2. The radial sizes of orbitals are governed by the product of $\exp(-Z_{eff}\, r/na_o)$, an r^l factor which arises from the small-r limit of $\psi_{n,l,m}$, r^1, which enters because the probability density is $|\psi|^2 r^2$, and the highest power of Zr/na_o that appears in the Laguerre polynomial, which is $(Zr/na_o)^{n-l-1}$. This product's r-dependence reduces to $\exp(-Z_{eff}\, r/na_o)\, r^n$, which has a maximum at $r = n^2 a_o/Z_{eff}$. So, orbitals with large Z_{eff} values are smaller and orbitals with larger n-values are larger.
3. The hydrogenic atom energy expression has no l-dependence; the $2s$ and $2p$ orbitals have exactly the same energy, as do the $3s$, $3p$, and $3d$ orbitals. This degree of degeneracy is only present in one-electron atoms and is the result of an additional symmetry (i.e., an additional operator that commutes with the Hamiltonian) that is not present once the atom contains two or more electrons. This additional symmetry is discussed on page 77 of Atkins.
4. The radial part of $\psi_{n,l,m}$ involves a polynomial in r of order $n - l - 1$

$$L_{n-l-1}(r) = \sum_k a_k\, r^k,$$

and the coefficients in these polynomials alternate in sign ($a_k = c_k(-1)^k$). As a result, the radial functions possess nodes (i.e., values of r at which $\psi_{n,l,m}$ vanishes) at points where the polynomial is equal to zero. There are $n - l - 1$ such radial nodes (excluding the nodes at $r = \infty$ and at $r = 0$, the latter of which all but s-orbitals have). A $4s$ orbital has three such radial nodes, and a $3d$ orbital has none. Radial nodes provide a means by which an orbital acquires density closer to the nucleus than its $<r^2>$ value would indicate; such enhanced density near the nucleus differentially stabilizes such orbitals relative to those with the same n but fewer nodes. In the language commonly used in inorganic chemistry, one says that nodes allow an orbital to "penetrate" the underlying inner-shell orbitals.

The presence of radial nodes also indicates that the electron has radial kinetic energy. The $3s$ orbital with 2 radial nodes has more radial kinetic energy than does the $3p$, which, in turn, has more than the $3d$. On the other hand, the $3d$ orbital has the most angular energy $l(l + 1)\hbar^2/2mr^2$ (this is the analog of the rotational energy of a diatomic molecule), the $3p$ has an intermediate amount, and the $3s$ has the least.

APPENDIX

C Quantum Mechanical Operators and Commutation

I. BRA-KET NOTATION

It is conventional to represent integrals that occur in quantum mechanics in a notation that is independent of the number of coordinates involved. This is done because the fundamental structure of quantum chemistry applies to all atoms and molecules, regardless of how many electronic and atom-center coordinates arise. The most commonly used notation, which is referred to as "Dirac" or "bra-ket" notation, can be summarized as follows:

1. The wavefunction itself Ψ is represented as a so-called "ket" $|\Psi\rangle$.
2. The complex conjugate Ψ^* of Ψ is represented as a "bra" $\langle\Psi|$; the complex conjugation is implied by writing $\langle\ |$.
3. The integral, over all of the N coordinates $(q_1 \ldots q_N)$ on which Ψ depends, of the product of Ψ^* and Ψ is represented as a so-called "bra-ket" or bracket:

$$\int \Psi^* \Psi \, dq_1 \ldots dq_N = \langle\Psi|\Psi\rangle.$$

   By convention, the two vertical lines that touch when $\langle\Psi|$ is placed against $|\Psi\rangle$ are merged into a single line in this notation.
4. Integrals involving one function (Ψ^*) and either another function (Φ) or the result of an operator A acting on a function (e.g., $A\Psi$ or $A\Phi$) are denoted as follows:

$$\int \Psi^* \, \Phi \, dq_1 \ldots dq_N = \langle \Psi | \Phi \rangle$$

$$\int \Psi^* \, A\Psi \, dq_1 \ldots dq_N = \langle \Psi | A\Psi \rangle = \langle \Psi | A | \Psi \rangle$$

$$\int \Psi^* \, A\Phi \, dq_1 \ldots dq_N = \langle \Psi | A\Phi \rangle = \langle \Psi | A | \Phi \rangle$$

$$\int (A\Psi)^* \, \Phi \, dq_1 \ldots dq_N = \langle A\Psi | \Phi \rangle.$$

It is merely convention that an "extra" vertical line (e.g., that appearing in $\langle \Psi | A | \Phi \rangle$) is inserted when an operator acting on the ket function appears in the integral.

II. HERMITIAN OPERATORS

In quantum mechanics, physically measurable quantities are represented by Hermitian operators. Such operators $\{R\}$ have matrix representations, in any basis spanning the space of functions on which the $\{R\}$ act, that are Hermitian:

$$\langle \phi_k | R | \phi_1 \rangle = \langle \phi_1 | R | \phi_k \rangle^* = \langle R\phi_k | \phi_l \rangle.$$

The equality of the first and third terms expresses the so-called "turn-over rule"; Hermitian operators can act on the function to their right or, equivalently, on the function to their left.

Operators that do not obey the above identity are not hermitian. For such operators, it is useful to introduce the so-called adjoint operator as follows. If for the operator R, another operator R^+ can be found that obeys

$$\langle \phi_k | R | \phi_1 \rangle = \langle R^+ \phi_k | \phi_l \rangle = \langle \phi_1 | R^+ | \phi_k \rangle^*,$$

for all $\{\phi_k\}$ within the class of functions on which R and R^+ operate, then R^+ is defined to be the *adjoint* of R. With this definition, it should be clear that Hermitian operators are self-adjoint (i.e., they obey $R^+ = R$).

The Hermiticity property guarantees that the eigenvalues $\{\lambda_m\}$ of such operators are real numbers (i.e., not complex) and that the corresponding eigenfunctions $\{f_m\}$, or their representations $\{V_{mk}\}$ within the $\{\phi_k\}$ basis

$$f_m = \sum_k V_{mk} \, \phi_k,$$

corresponding to different eigenvalues, are orthonormal and that the eigenfunctions belonging to degenerate eigenvalues can be made orthonormal.

To prove these claims, start with $R\phi_k = \lambda_k \phi_k$. Multiplying on the left by the complex conjugate of ϕ_k and integrating gives $\langle \phi_k | R | \phi_k \rangle = \lambda_k \langle \phi_k | \phi_k \rangle$. Taking the complex conjugate of this equation and using the Hermiticity property $\langle \phi_k | R | \phi_1 \rangle = \langle \phi_1 | R | \phi_k \rangle^*$ (applied with $k = l$) gives $\lambda_k^* = \lambda_k$.

The orthogonality proof begins with $R\phi_k = \lambda_k \phi_k$, and $R\phi_l = \lambda_l \phi_l$. Multiplying the first of these on the left by $\langle \phi_l |$ and the second by $\langle \phi_k |$ gives $\langle \phi_l | R | \phi_k \rangle = \lambda_k \langle \phi_l | \phi_k \rangle$ and $\langle \phi_k | R | \phi_l \rangle = \lambda_l \langle \phi_k | \phi_l \rangle$. The complex conjugate of the second reads $\langle \phi_k | R | \phi_l \rangle^* = \lambda_l \langle \phi_k | \phi_l \rangle^*$; using the Hermiticity property this reduces to $<\phi_l|R|\phi_k> = \lambda_l <\phi_l|\phi_k>$. If $\lambda_k \neq \lambda_l$, this result can be consistent with $<\phi_l|R|\phi_k> = \lambda_k <\phi_l|\phi_k>$ only if $<\phi_l|\phi_k>$ vanishes.

III. MEANING OF THE EIGENVALUES AND EIGENFUNCTIONS

In quantum mechanics, the eigenvalues of an operator represent the *only* numerical values that can be observed if the physical property corresponding to that operator is measured. Operators for which the eigenvalue spectrum (i.e., the list of eigenvalues) is discrete thus possess discrete spectra when probed experimentally.

For a system in a state ψ *that is an eigenfunction* of R

$$R\psi = \lambda\psi,$$

measurement of the property corresponding to R will yield the value λ. For example, if an electron is in a $2p_{-1}$ orbital and $\mathbf{L}^2$ is measured, the value $L(L+1)\hbar^2 = 2\hbar^2$ (and only this value) will be observed; if $\mathbf{L}_z$ is measured, the value $-1\hbar$ (and only this value) will be observed. If the electron were in a $2p_x$ orbital and $\mathbf{L}^2$ were measured, the value $2\hbar^2$ will be found; however, if $\mathbf{L}_z$ is measured, we can not say that only one value will be observed because the $2p_x$ orbital is not an eigenfunction of $\mathbf{L}_z$ (measurements in such non-eigenfunction situations are discussed below).

In general, if the property R is measured, any one of the eigenvalues $\{\lambda_m\}$ of the operator R may be observed. In a large number of such measurements (i.e., for an ensemble of systems all in states described by ψ that may or may not itself be an eigenfunction of R), the probability or frequency of observing the particular eigenvalue λ_m is given by $|C_m|^2$, where C_m is the expansion coefficient of ψ in the eigenfunctions of R:

$$\psi = \sum_m C_m f_m.$$

In the special case treated earlier in which ψ is an eigenfunction of R, all but one of the C_m vanish; hence the probability of observing various λ_m values is zero except for the *one* state for which C_m is non-zero.

For a measurement that results in the observation of the particular value λ_m, a *subsequent* measurement of R on systems just found to have eigenvalue λ_m will result, with 100% certainty, in observation of this *same* value λ_m. The quantum mechanical interpretation of this experimental observation it to say that the act of measuring the property belonging to the operator R causes the wavefunction to be altered. Once this measurement of R is made, the wavefunction is no longer ψ; it is now f_m for those species for which the value λ_m is observed.

For example (this example and others included in this appendix are also treated more briefly in chapter 1), if the initial ψ discussed above were a so-called superposition state of the form

$$\psi = a(2p_0 + 2p_{-1} - 2p_1) + b(3p_0 - 3p_{-1}), \text{ then:}$$

1. If $\mathbf{L}^2$ were measured, the value $2\hbar^2$ would be observed with probability $3\,|a|^2 + 2\,|b|^2 = 1$, since all of the non-zero C_m coefficients correspond to p-type orbitals for this ψ. After said measurement, the wavefunction would still be this same ψ because this entire ψ is an eigenfunction of $\mathbf{L}^2$.
2. If $\mathbf{L}_z$ were measured for this

$$\psi = a(2p_0 + 2p_{-1} - 2p_1) + b(3p_0 - 3p_{-1}),$$

the values $0\hbar$, $1\hbar$, and $-1\hbar$ would be observed (because these are the only functions with non-zero C_m coefficients for the L_z operator) with respective probabilities $|a|^2 + |b|^2$, $|-a|^2$, and $|a|^2 + |-b|^2$.

3. *After* L_z were measured, if the sub-population for which $-1\hbar$ had been detected were subjected to measurement of L^2 the value $2\hbar^2$ would certainly be found because the *new* wavefunction

$$\psi' = \{a\, 2p_{-1} - b\, 3p_{-1}\}\, (|a|^2 + |b|^2)^{-1/2}$$

is still an eigenfunction of L^2 with this eigenvalue.

4. Again after L_z were measured, if the sub-population for which $-1\hbar$ had been observed and for which the wavefunction is now

$$\psi' = \{a\, 2p_{-1} - b\, 3p_{-1}\}\, (|a|^2 + |b|^2)^{-1/2}$$

were subjected to measurement of the energy (through the Hamiltonian operator), two values would be found. With probability $|a|^2\, (|a|^2 + |b|^2)^{-1}$ the energy of the $2p_{-1}$ orbital would be observed; with probability $|-b|^2\, (|a|^2 + |b|^2)^{-1}$, the energy of the $3p_{-1}$ orbital would be observed.

The *general observation* to make is that, given an initial normalized ψ function, and a physical measurement (with operator R) to be made, one first must express ψ as a linear combination of the complete set of eigenfunctions of that R:

$$\psi = \sum_m C_m f_m.$$

The coefficients C_m tell, through $|C_m|^2$, the probabilities (since ψ is normalized to unity) of observing each of the R eigenvalues λ_m when the measurement is made. Once the measurement is made, that sub-population of the sample on which the experiment was run that gave the particular eigenvalue, say λ_p, now have a wavefunction that no longer is the above ψ; their wavefunction now is f_p.

IV. EXPERIMENTS DO NOT PREPARE ONLY EIGENSTATES

The above remarks should *not* be interpreted to mean that experiments are limited to preparing only eigenstates. Just as one can "pluck" a violin string or "pound" a drum head in any manner, experiments can prepare a system in states that are not pure eigenfunctions (i.e., states that do not contain just one eigenfunction in their expansion). However, *no matter how the state is prepared, it can be interpreted*, via expansion *in the complete set of eigenfunctions* of the operator(s) whose properties are to be measured, as a superposition of eigenfunctions. The superposition amplitudes provide the probabilities of observing each eigenfunction when the measurement is made.

For example, after the drum head has been hit, its shape will evolve spatially and in time in a manner that depends on how it was "prepared" by the initial blow. However, if one carries out an experiment to detect and frequency-analyze the sound that emanates from this drum, thereby measuring differences in the eigen-energies of the system, one finds a set of discrete (quantized) frequencies $\{\omega_k\}$ and corresponding amplitudes $\{A_k\}$. Repeating this experiment after a different "blow" is used to prepare a different initial state of the drum head, one finds the *same* frequencies $\{\omega_k\}$ but *different* amplitudes $\{B_k\}$.

The quantum mechanics interpretation of these observations is that the initial state of the drum head is a superposition of eigenstates:

$$\psi = \sum_n C_n f_n.$$

The $\{C_n\}$ amplitudes are determined by how the drum head is "prepared" in the blow that strikes it. As time evolves, ψ progresses according to the time-dependent Schrödinger equation:

$$\psi(t) = \sum_n C_n f_n \exp(-i\, E_n\, t/\hbar).$$

The frequencies emitted by this system will depend on the probability $|C_n|^2$ of the system being in a particular eigenstate f_n and the energy E_n of each eigenstate.

The frequency spectrum measured for the drum head motion is therefore seen to have variable amplitudes for each observed "peak" because the $|C_n|^2$ vary depending on how the drum head was prepared. In contrast, the frequencies themselves are not characteristic of the preparation process but are properties of the drum head itself; they depend on the eigen-energies of the system, not on how the system is prepared.

This distinction between the characteristic eigenstates of the system with their intrinsic properties and the act of preparing the system in some state that may be a superposition of these eigenstates is essential to keep in mind when applying quantum mechanics to experimental observations.

V. OPERATORS THAT COMMUTE AND THE EXPERIMENTAL IMPLICATIONS

Two Hermitian operators that commute

$$[R,S] = RS - SR = 0$$

can be shown to possess *complete sets* of simultaneous eigenfunctions. That is, one can find complete sets of functions that are eigenfunctions of both R and of S.

The symbol $[R,S]$ is used to denote what is called the *commutator* of the operators R and S. There are several useful identities that can be proven for commutators among operators A, B, C, and D, scalar numbers k, and integers n. Several of these are given as follows:

$$[A,A^n] = 0$$
$$[kA,B] = [A,kB] = k[A,B]$$
$$[A,B+C] = [A,B] + [A,C]$$
$$[A+B,C] = [A,C] + [B,C]$$
$$[A+B,C+D] = [A,C] + [A,D] + [B,C] + [B,D]$$
$$[A,BC] = [A,B]C + B[A,C]$$
$$[AB,C] = [A,C]B + A[B,C]$$
$$[AB,CD] = [A,C]DB + C[A,D]B + A[B,C]D + AC[B,D].$$

The physical implications of the commutation of two operators are very important because they have to do with interfering with one another. For example, the fact that the x coordinate operator $\boldsymbol{x} = x$ and its momentum operator $\boldsymbol{p_x} = -i\hbar\partial/\partial x$ do *not* commute results in the well known Heisenberg uncertainty relationship $\Delta x\, \Delta p_x \geq \hbar/2$ involving measurements of $\boldsymbol{x}$ and $\boldsymbol{p_x}$.

There are two distinct cases that need to be addressed:

A. Operators Involving Different Variables

If the two operators act on *different coordinates* (or, more generally, on different sets of coordinates), then they obviously commute. Moreover, in this case, it is straightforward to find the complete set of eigenfunctions of both operators; one simply forms a product of any eigenfunction (say f_k) of R and any eigenfunction (say g_n) of S. The function $f_k\, g_n$ is an eigenfunction of both R and S:

$$R\, f_k\, g_n = g_n\, (R\, f_k) = g_n\, (\lambda\, f_k) = \lambda\, g_n f_k = \lambda f_k\, g_n;$$
$$S\, f_k\, g_n = f_k\, (S\, g_n) = f_k\, (\mu\, g_n) = \mu\, f_k\, g_n.$$

In these equations use has been made of the fact that g_n and f_k are functions of different sets of coordinates that S and R, respectively, act upon.

Product functions such as $f_k\, g_n$ yield predictable results when measurements are performed. If the property corresponding to R is measured, the value λ is observed, *and* the wavefunction remains $f_k\, g_n$. If S is measured, the value μ is observed, and the wavefunction remains $f_k\, g_n$. For example, the two Hermitian operators $-i\,\partial/\partial\phi$ and $-i\,\partial/\partial r$ clearly commute. An eigenfunction of $-i\,\partial/\partial\phi$ is of the form $\exp(ia\phi)$ and an eigenfunction of $-i\,\partial/\partial r$ is of the form $\exp(ibr)$. The *product* $\exp(ia\phi)\exp(ibr)$ is an eigenfunction of both $-i\,\partial/\partial\phi$ and $-i\,\partial/\partial r$. The corresponding eigenvalues are a and b, respectively; only these values will be observed if measurements of the properties corresponding to $-i\,\partial/\partial\phi$ and $-i\,\partial/\partial r$ are made for a system whose wavefunction is $\exp(ia\phi)\exp(ibr)$.

B. Operators Involving the Same Variables

If the operators R and S act on the *same coordinates* yet still commute, the implications of their commutation are somewhat more intricate to detail.

As a first step, consider the functions $\{g_n\}$ that are eigenfunctions of S with eigenvalues $\{\mu_n\}$. Now, act on g_n with the SR operator and use the fact that $SR = RS$ to obtain

$$S\, R\, g_n = RS\, g_n.$$

Because the $\{g_n\}$ are eigenfunctions of S having eigenvalues $\{\mu_n\}$, this equation further reduces to:

$$S\, R\, g_n = R\, \mu_n g_n = \mu_n\, R\, g_n.$$

This is a *key result*. It shows that the function $(R\, g_n)$ is itself *either* an eigenfunction of S having the same eigenvalue that g_n has *or* it vanishes.

If $R\, g_n$ vanishes, g_n clearly is an eigenfunction of R (since $R\, g_n = 0\, g_n$) and of S. On the other hand, if $R\, g_n$ is non-vanishing, it must be an eigenfunction of S having the same eigenvalue (μ_n) as g_n. If this eigenvalue is non-degenerate (i.e., if g_n is the only function with eigenvalue μ_n), then $R\, g_n$ must be proportional to g_n itself:

$$R\, g_n = c_n\, g_n.$$

This also implies that g_n is an eigenfunction of both R and of S.

Thus far, we can say that functions which are eigenfunctions of S belonging to *non-degenerate* eigenvalues must also be eigenfunctions of R. On the other hand, if the μ_n eigenvalue is degenerate (i.e., there are ω such functions g_n, $g_{n'}$, $g_{n''}$, etc. that are S-eigenfunctions with the same μ_n as their eigenvalue), all that can be said is that $R\, g_n$ is some combination of this ω-fold degenerate manifold of states:

$$R\, g_n = \sum_{n'} c_{n,n'}\, g_{n'}$$

where the sum over n' runs only over the states with S-eigenvalues equal to μ_n. This same conclusion can be reached no matter which particular state g_n among the degenerate manifold we begin with. Therefore, the above equation holds for *all* $\{g_n\}$ that belong to this degenerate group of states.

The constants $c_{n,n'}$ form a square (since we act on all ω states and produce combinations of ω states) Hermitian (since R is Hermitian) matrix; in fact, $c_{n,n'}$ forms the matrix representation of the operator R within the ω-dimensional space of orthonormal functions $\{g_n\}$. As with all Hermitian matrices, a unitary transformation can be employed to bring it to diagonal form. That is, the ω orthonormal $\{g_n\}$ functions can be unitarily combined:

$$G_p = \sum_n U_{p,n}\, g_n$$

to produce ω new orthonormal functions $\{G_p\}$ for which the corresponding matrix elements $c_{p,p'}$, defined by

$$R\, G_p = \sum_n U_{p,n}\, R\, g_n = \sum_{n,n'} U_{p,n}\, c_{n,n'}\, g_n$$

$$= \sum_{p'} \sum_{n,n'} U_{p,n}\, c_{n,n'}\, U^*_{n',p'}\, G_{p'} = \sum_{p'} c_{p,p'}\, G_{p'}$$

are diagonal

$$R\, G_p = c_{p,p}\, G_p.$$

This shows that the set of functions (the G_p in this degenerate case) that are eigenfunctions of S can also be eigenfunctions of R.

C. Summary

In summary, we have shown that if R and S are operators that act on the same set of coordinates (e.g., $-i\partial/\partial x$ and x^2 or $\partial^2/\partial x^2$ and $-i\partial/\partial x$), then an eigenfunction of R (denoted f_k and having eigenvalue λ_k) must either be (i) eigenfunctions of S (if its R-eigenvalue is non-degenerate) or (ii) a member of a degenerate set of R eigenfunctions that can be combined among one another to produce eigenfunctions of S.

An example will help illustrate these points. The p_x, p_y and p_z orbitals are eigenfunctions of the $\boldsymbol{L}^2$ angular momentum operator with eigenvalues equal to $L(L+1)\,\hbar^2 = 2\hbar^2$. Since $\boldsymbol{L}^2$ and $\boldsymbol{L}_z$ commute and act on the same (angle) coordinates, they possess a complete set of simultaneous eigenfunctions. Although the p_x, p_y and p_z orbitals are not eigenfunctions of $\boldsymbol{L}_z$, they can be combined (as above to form the G_p functions) to form three new orbitals: $p_0 = p_z$, $p_1 = 2^{-1/2}[p_x + ip_y]$, and $p_{-1} = 2^{-1/2}[p_x - ip_y]$ that are still eigenfunctions of $\boldsymbol{L}^2$ but are now eigenfunctions of $\boldsymbol{L}_z$ also (with eigenvalues $0\hbar$, $1\hbar$, and $-1\hbar$, respectively).

It should be mentioned that if two operators do not commute, they may still have *some* eigenfunctions in common, but they will not have a complete set of simultaneous eigenfunctions. For example, the L_z and L_x components of the angular momentum operator do not commute; however, a wavefunction with $L = 0$ (i.e., an S-state) is an eigenfunction of both operators.

D. Experimental Significance

We use an example to illustrate the importance of two operators commuting to quantum mechanics' interpretation of experiments. Assume that an experiment has been carried out on an atom to measure its total angular momentum L^2. According to quantum mechanics, only values equal to $L(L+1)\hbar^2$ will be observed. Further assume, for the particular experimental sample subjected to observation, that values of L^2 equal to $2\hbar^2$ and $0\hbar^2$ were detected in relative amounts of 64% and 36%, respectively. This means that the atom's original wavefunction ψ could be represented as:

$$\psi = 0.8\,P + 0.6\,S,$$

where P and S represent the P-state and S-state components of ψ. The squares of the amplitudes 0.8 and 0.6 give the 64% and 36% probabilities mentioned above.

Now assume that a subsequent measurement of the component of angular momentum along the lab-fixed z-axis is to be measured for that subpopulation of the original sample found to be in the P-state. For that population, the wavefunction is now a pure P-function:

$$\psi' = P\,.$$

However, at this stage we have no information about how much of this ψ' is of $m = 1, 0,$ or -1, nor do we know how much $2p, 3p, 4p, \ldots np$ component this state contains.

Because the property corresponding to the operator $\boldsymbol{L_z}$ is about to be measured, we express the above ψ' in terms of the eigenfunctions of $\boldsymbol{L_z}$:

$$\psi' = P = \sum_{m=1,0,-1} C'_m\,P_m.$$

When the measurement of L_z is made, the values $1\hbar$, $0\hbar$, and $-1\hbar$ will be observed with probabilities given by $|C'_1|^2$, $|C'_0|^2$, and $|C'_{-1}|^2$, respectively. For that subpopulation found to have, for example, L_z equal to $-1\hbar$, the wavefunction then becomes

$$\psi'' = P_{-1}.$$

At this stage, we do not know how much of $2p_{-1}, 3p_{-1}, 4p_{-1}, \ldots np_{-1}$ this wavefunction contains. To probe this question another subsequent measurement of the energy (corresponding to the $\boldsymbol{H}$ operator) could be made. Doing so would allow the amplitudes in the expansion of the above

$$\psi'' = P_{-1} = \sum_n C''_n\,nP_{-1}$$

to be found.

The kind of experiment outlined above allows one to find the content of each particular component of an initial sample's wavefunction. For example, the original wavefunction has $0.64\,|C''_n|^2|\,C'_m|^2$ fractional content of the various nP_m functions.

Let us consider another experiment in which an initial sample (with wavefunction ψ) is first subjected to measurement of L_z and then subjected to measurement of L^2 and then of the energy. In this order, one would first find specific values (integer multiples of $\hbar$) of L_z and one would express ψ as

$$\psi = \sum_m D_m\,\psi_m.$$

At this stage, the nature of each ψ_m is unknown (e.g., the ψ_1 function can contain np_1, $n'd_1$, $n''f_1$, etc. components); all that is known is that ψ_m has $m\hbar$ as its L_z value.

Taking that sub-population ($|D_m|^2$ fraction) with a particular $m\hbar$ value for L_z and subjecting it to subsequent measurement of L^2 requires the current wavefunction ψ_m to be expressed as

$$\psi_m = \sum_L D_{L,m}\, \psi_{L,m}.$$

When L^2 is measured the value $L(L+1)\,\hbar^2$ will be observed with probability $|D_{m,L}|^2$, and the wavefunction for that particular sub-population will become

$$\psi'' = \psi_{L,m}.$$

At this stage, we know the value of L and of m, but we do not know the energy of the state. For example, we may know that the present sub-population has $L = 1$, $m = -1$, but we have no knowledge (yet) of how much $2p_{-1}$, $3p_{-1}$, ... np_{-1} the system contains.

To further probe the sample, the above sub-population with $L = 1$ and $m = -1$ can be subjected to measurement of the energy. In this case, the function $\psi_{1,-1}$ must be expressed as

$$\psi_{1,-1} = \sum_n D_n''\, nP_{-1}.$$

When the energy measurement is made, the state nP_{-1} will be found $|D_n''|^2$ fraction of the time.

We now need to explain how the fact that $\boldsymbol{L_z}$, $\boldsymbol{L^2}$, and $\boldsymbol{H}$ all commute with one another (i.e., are *mutually commutative*) makes the series of measurements described above more straightforward than if these operators did not commute. In the first experiment, the fact that they are mutually commutative allowed us to expand the 64% probable $\boldsymbol{L^2}$ eigenstate with $L = 1$ in terms of functions that were eigenfunctions of the operator for which measurement was *about* to be made without destroying our knowledge of the value of L^2. That is, because $\boldsymbol{L^2}$ and $\boldsymbol{L_z}$ can have simultaneous eigenfunctions, the $L = 1$ function can be expanded in terms of functions that are eigenfunctions of *both* $\boldsymbol{L^2}$ and $\boldsymbol{L_z}$. This in turn, allowed us to find experimentally the subpopulation that had, for example a $-1\hbar$ value of L_z while retaining knowledge that the state *remains* an eigenstate of $\boldsymbol{L^2}$ (the state at this time had $L = 1$ *and* $m = -1$ and was denoted P_{-1}) Then, when this P_{-1} state was subjected to energy measurement, knowledge of the energy of the subpopulation could be gained *without* giving up knowledge of the L^2 and L_z information; upon carrying out said measurement, the state became nP_{-1}.

In contrast, if (hypothetically) $\boldsymbol{L^2}$ and $\boldsymbol{L_z}$ did not commute, the $L = 1$ function originally detected with 64% probability would be altered by the subsequent L_z measurement in a manner that *destroys* our knowledge of L^2. The P function could still have been expanded in terms of the eigenfunctions of the property about to be probed (L_z)

$$P = \sum_{m=1,0,-1} C'_m\, \psi_m.$$

However, because $\boldsymbol{L^2}$ and $\boldsymbol{L_z}$ do not commute in this hypothetical example, the states ψ_m that are eigenfunctions of $\boldsymbol{L_z}$ will not, in general, also be eigenfunctions of $\boldsymbol{L^2}$. Hence, when L_z is measured and a particular value (say $-1\hbar$) is detected, the wavefunction becomes

$$\psi''' = \psi_{-1},$$

which is *no longer* an eigenfunction of $\boldsymbol{L^2}$.

The essential observations to be made are:

1. After the first measurement is made (say for operator R), the wavefunction becomes an eigenfunction of R with a well defined R-eigenvalue (say λ): $\psi = \psi(\lambda)$.

2. The eigenfunctions of the second operator S (i.e., the operator corresponding to the measurement about to be made) can be taken to also be eigenfunctions of R if R and S commute. This then allows $\psi(\lambda)$ to be expanded in terms of functions that are both R-eigenfunctions (with eigenvalue λ) and S-eigenfunctions (with various eigenvalues μ_n):

$$\psi(\lambda) = \sum_n C_n \, \psi(\lambda,\mu_n).$$

Upon measurement of S, the wavefunction becomes one of these $\psi(\lambda,\mu_n)$ functions. When the system is in this state, both R- and S-eigenvalues are known precisely; they are λ and μ_n.

3. The eigenfunctions of the second operator S (i.e., the operator corresponding to the measurement about to be made) can *not* be taken to also be eigenfunctions of R if R and S do not commute. The function $\psi(\lambda)$ can still be expanded in terms of functions that are both S-eigenfunctions (with various eigenvalues μ_n):

$$\psi(\lambda) = \sum_n C_n \, \psi(\mu_n).$$

However, because R and S do not commute, these $\psi(\mu_n)$ functions are not, in general, also R-eigenfunctions; they are only S-eigenfunctions. Then, upon measurement of S, the wavefunction becomes one of these $\psi(\mu_n)$ functions. When the system is in this state, the S-eigenvalue is known precisely; it is μ_n. The R-eigenvalue is no longer specified. In fact, the new state $\psi(\mu_n)$ may contain components of all different R-eigenvalues, which can be represented by expanding $\psi(\mu_n)$ in terms of the R-eigenfunctions:

$$\psi(\mu_n) = \sum_k D_k \, \psi(\lambda_k).$$

If R were measured again, after the state has been prepared in $\psi(\mu_n)$, the R-eigenvalues $\{\lambda_k\}$ would be observed with probabilities $\{|D_k|^2\}$, and the wavefunction would, for these respective sub-populations, become $\psi(\lambda_k)$.

It should now be clear that the act of carrying out an experimental measurement disturbs the system in that it causes the system's wavefunction to become an eigenfunction of the operator whose property is measured. If two properties whose corresponding operators commute are measured, the measurement of the second property does *not* destroy knowledge of the first property's value gained in the first measurement. If the two properties do not commute, the second measurement does destroy knowledge of the first property's value. It is thus often said that "measurements for operators that do not commute interfere with one another."

APPENDIX

D Time-Independent Perturbation Theory

Perturbation theory is used in two qualitatively different contexts in quantum chemistry. It allows one to estimate (because perturbation theory is usually employed through some finite order and may not even converge if extended to infinite order) the splittings and shifts in energy levels and changes in wavefunctions that occur when an external field (e.g., electric or magnetic or that due to a surrounding set of "ligands"—a crystal field) is applied to an atom, molecule, ion, or solid whose "unperturbed" states are known. These "perturbations" in energies and wavefunctions are expressed in terms of the (complete) set of unperturbed states. For example, the distortion of the $2s$ orbital of a Li atom due to the application of an external electric field along the y-axis is described by adding to the (unperturbed) $2s$ orbital components of p_y-type orbitals ($2p$, $3p$, etc.):

$$\phi = a\ 2s + \sum_n C_n\ np_y.$$

The effect of adding in the p_y orbitals is to polarize the $2s$ orbital along the y-axis. The amplitudes C_n are determined via the equations of perturbation theory developed below; the change in the energy of the $2s$ orbital caused by the application of the field is expressed in terms of the C_n coefficients and the (unperturbed) energies of the $2s$ and np_y orbitals.

There is another manner in which perturbation theory is used in quantum chemistry that does *not* involve an externally applied perturbation. Quite often one is faced with solving a Schrödinger equation to which no exact solution has been (yet) or can be found. In such cases, one often develops a "model" Schrödinger equation, which in some sense is designed to represent the system whose full Schrödinger equation cannot be solved. The difference between the Hamiltonia of the full and model problems, H and H^0, respectively is used to *define* a perturbation

$V = H - H^0$. Perturbation theory is then employed to approximate the energy levels and wavefunctions of the full H in terms of the energy levels and wavefunctions of the model system (which, by assumption, can be found). The "imperfection" in the model problem is therefore used as the perturbation. The success of such an approach depends strongly on how well the model H^0 represents the true problem (i.e., on how "small" V is). For this reason, much effort is often needed to develop approximate Hamiltonia for which V is small and for which the eigenfunctions and energy levels can be found.

I. STRUCTURE OF TIME-INDEPENDENT PERTURBATION THEORY

A. The Power Series Expansions of the Wavefunction and Energy

Assuming that all wavefunctions Φ_k and energies E_k^0 belonging to the unperturbed Hamiltonian H^0 are known

$$H^0\Phi_k = E_k^0\Phi_k,$$

and given that one wishes to find eigenstates (ψ_k and E_k) of the perturbed Hamiltonian

$$H = H^0 + \lambda V,$$

perturbation theory begins by expressing ψ_k and E_k as power series in the perturbation strength λ:

$$\psi_k = \sum_{n=0,\infty} \lambda^n \psi_k^{(n)}$$

$$E_k = \sum_{n=0,\infty} \lambda^n E_k^{(n)}.$$

Moreover, it is assumed that, as the strength of the perturbation is reduced to zero, ψ_k reduces to one of the unperturbed states Φ_k and that the full content of Φ_k in ψ_k is contained in the first term $\psi_k^{(0)}$. This means that $\psi_k^{(0)} = \Phi_k$ and $E_k^{(0)} = E_k^0$, and so

$$\psi_k = \Phi_k + \sum_{n=1,\infty} \lambda^n \psi_k^{(n)} = \Phi_k + \psi_k'$$

$$E_k = E_k^0 + \sum_{n=1,\infty} \lambda^n E_k^{(n)} = E_k^0 + E_k'.$$

In the above expressions, λ would be proportional to the strength of the electric or magnetic field if one is dealing with an external-field case. When dealing with the situation for which V is the imperfection in the model H^0, λ is equal to unity; in this case, one thinks of formulating and solving for the perturbation expansion for variable λ after which λ is set equal to unity.

B. The Order-by-Order Energy Equations

Equations for the order-by-order corrections to the wavefunctions and energies are obtained by using these power series expressions in the full Schrödinger equation:

$$(H - E_k)\, \psi_k = 0.$$

Multiplying through by $\Phi_k{}^*$ and integrating gives the expression in terms of which the total energy is obtained:

$$<\Phi_k|H|\psi_k> = E_k<\Phi_k|\psi_k> = E_k.$$

Using the fact that Φ_k is an eigenfunction of H^0 and employing the power series expansion of ψ_k allows one to generate the fundamental relationships among the energies $E_k^{(n)}$ and the wavefunctions $\psi_k^{(n)}$:

$$E_k = <\Phi_k|H^0|\psi_k> + <\Phi_k|V|\psi_k> = E_k^0 + <\Phi_k|\lambda V| \sum_{n=0,\infty} \lambda^n \psi_k^{(n)}>.$$

The lowest few orders in this expansion read as follows:

$$E_k = E_k^0 + \lambda<\Phi_k|V|\Phi_k> + \lambda^2<\Phi_k|V|\psi_k^{(1)}> + \lambda^3<\Phi_k|V|\psi_k^{(2)}> + \ldots$$

If the various $\psi_k^{(n)}$ can be found, then this equation can be used to compute the order-by-order energy expansion.

Notice that the first-order energy correction is given in terms of the zeroth-order (i.e., unperturbed) wavefunction as:

$$E_k^{(1)} = <\Phi_k|V|\Phi_k>,$$

the average value of the perturbation taken over Φ_k.

C. The Order-by-Order Wavefunction Equations

To obtain workable expressions for the perturbative corrections to the wavefunction ψ_k, the full Schrödinger equation is first projected against all of the unperturbed eigenstates $\{|\Phi_j>\}$ *other than* the state Φ_k whose perturbative corrections are sought:

$$<\Phi_j|H|\psi_k> = E_k<\Phi_j|\psi_k>, \text{ or}$$
$$<\Phi_j|H^0|\psi_k> + \lambda<\Phi_j|V|\psi_k> = E_k<\Phi_j|\psi_k>, \text{ or}$$
$$<\Phi_j|\psi_k>E_j^0 + \lambda<\Phi_j|V|\psi_k> = E_k<\Phi_j|\psi_k>, \text{ or finally}$$
$$\lambda<\Phi_j|V|\psi_k> = [E_k - E_j^0]<\Phi_j|\psi_k>.$$

Next, each component $\psi_k^{(n)}$ of the eigenstate ψ_k is expanded in terms of the unperturbed eigenstates (as they can be because the $\{\Phi_k\}$ form a complete set of functions):

$$\psi_k = \Phi_k + \sum_{j\neq k} \sum_{n=1,\infty} \lambda^n<\Phi_j|\psi_k^{(n)}>|\Phi_j>.$$

Substituting this expansion for ψ_k into the preceeding equation gives

$$\lambda<\Phi_j|V|\Phi_k> + \sum_{l\neq k} \sum_{n=1,\infty} \lambda^{n+1}<\Phi_l|\psi_k^{(n)}> <\Phi_j|V|\Phi_l> = [E_k - E_j^0] \sum_{n=1,\infty} \lambda^n<\Phi_j|\psi_k^{(n)}>.$$

To extract from this set of *coupled* equations relations that can be solved for the coefficients $<\Phi_j|\psi_k^{(n)}>$, which embodies the desired wavefunction perturbations $\psi_k^{(n)}$, one collects together all terms with like power of λ in the above general equation (in doing so, it is important to keep in mind that E_k itself is given as a power series in λ).

The λ^0 terms vanish, and the first-order terms reduce to:

$$<\Phi_j|V|\Phi_k> = [E_k^0 - E_j^0]<\Phi_j|\psi_k^{(1)}>,$$

which can be solved for the expansion coefficients of the so-called *first-order wavefunction* $\psi_k^{(1)}$:

$$\psi_k^{(1)} = \sum_j <\Phi_j|V|\Phi_k>/[E_k^0 - E_j^0]|\Phi_j>.$$

When this result is used in the earlier expression for the *second-order energy* correction, one obtains:

$$E_k^{(2)} = \sum_j |<\Phi_j|V|\Phi_k>|^2/[E_k^0 - E_j^0] .$$

The terms proportional to λ^2 are as follows:

$$\sum_{l\neq k} <\Phi_l|\psi_k^{(1)}> <\Phi_j|V|\Phi_l> = [E_k^0 - E_j^0]<\Phi_j|\psi_k^{(2)}> + E_k^{(1)}<\Phi_j|\psi_k^{(1)}>.$$

The solution to this equation can be written as:

$$<\Phi_j|\psi_k^{(2)}> = [E_k^0 - E_j^0]^{-1} \sum_{l\neq k} <\Phi_l|\psi_k^{(1)}> \left\{<\Phi_j|V|\Phi_l> - \delta_{j,l}E_k^{(1)}\right\} .$$

Because the expansion coefficients $<\Phi_l|\psi_k^{(1)}>$ of $\psi_k^{(1)}$ are already known, they can be used to finally express the expansion coefficients of $\psi_k^{(2)}$ totally in terms of zeroth-order quantities:

$$<\Phi_j|\psi_k^{(2)}> = [E_k^0 - E_j^0]^{-1} \sum_{l\neq k} \left\{<\Phi_j|V|\Phi_l> - \delta_{j,l}E_k^{(1)}\right\}<\Phi_l|V|\Phi_k>[E_k^0 - E_l^0]^{-1},$$

which then gives

$$\psi_k^{(2)} = \sum_{j\neq k} [E_k^0 - E_j^0]^{-1} \sum_{l\neq k} \left\{<\Phi_j|V|\Phi_l> - \delta_{j,l}E_k^{(1)}\right\}<\Phi_l|V|\Phi_k>[E_k^0 - E_l^0]^{-1}|\Phi_j>.$$

D. Summary

An essential thing to stress concerning the above development of so-called Rayleigh-Schrödinger perturbation theory (RSPT) is that each of the energy corrections $E_k^{(n)}$ and wavefunction corrections $\psi_k^{(n)}$ are expressed in terms of integrals over the unperturbed wavefunctions Φ_k involving the perturbation (i.e., $<\Phi_j|V|\Phi_l>$) and the unperturbed energies E_j^0. As such, these corrections can be symmetry-analyzed to determine, for example, whether perturbations of a given symmetry will or will not affect particular states. For example, if the state under study belongs to a non-degenerate representation in the absence of the perturbation V, then its first-order energy correction $<\Phi_k|V|\Phi_k>$ will be non-zero only if V contains a totally symmetric component (because the direct product of the symmetry of Φ_k with itself is the totally symmetric representation). Such an analysis predicts, for example, that the energy of an s orbital of an atom will be unchanged, in first-order, by the application of an external electric field because the perturbation

$$V = e\mathbf{E} \cdot \mathbf{r}$$

is odd under the inversion operation (and hence can not be totally symmetric). This same analysis, when applied to $E_k^{(2)}$ shows that contributions to the second-order energy of an *s* orbital arise only from unperturbed orbitals ϕ_j that are odd under inversion because only in such cases will the integrals $<s|e\mathbf{E}\cdot\mathbf{r}|\phi_j>$ be non-zero.

II. THE MØLLER-PLESSET PERTURBATION SERIES

A. The Choice of H^0

Let us assume that an SCF calculation has been carried out using the set of *N* spin-orbitals $\{\phi_a\}$ that are occupied in the reference configuration Φ_k to define the corresponding Fock operator:

$$F = h + \sum_{a(occupied)} [J_a - K_a] .$$

Further, we assume that all of the occupied $\{\phi_a\}$ and virtual $\{\phi_m\}$ spin-orbitals and orbital energies have been determined and are available.

This Fock operator is used to define the unperturbed Hamiltonian of Møller-Plesset perturbation theory (MPPT):

$$H^0 = \sum_i F(\mathbf{r}_i).$$

This particular Hamiltonian, when acting on *any* Slater determinant formed by placing *N* electrons into the SCF spin-orbitals, yields a zeroth-order eigenvalue equal to the sum of the orbital energies of the spin-orbitals appearing in that determinant:

$$H^0|\phi_{j1}\phi_{j2}\phi_{j3}\phi_{j4}\ldots\phi_{jN}| = (\varepsilon_{j1}+\varepsilon_{j2}+\varepsilon_{j3}+\varepsilon_{j4}+\ldots+\varepsilon_{jN}|\phi_{j1}\phi_{j2}\phi_{j3}\phi_{j4}\ldots\phi_{jN}|$$

because the spin-orbitals obey

$$F\phi_j = \varepsilon_j\phi_j,$$

where *j* runs over all (*occupied* (*a*, *b*, . . .) and virtual (*m*, *n*, . . .)) spin-orbitals. This result is the MPPT embodiment of $H^0\Phi_k = E_k^0\Phi_k$.

B. The Perturbation V

The perturbation *V* appropriate to this MPPT case is the difference between the full *N*-electronic Hamiltonian and this H^0:

$$V = H - H^0.$$

Matrix elements of *V* among determinental wavefunctions constructed from the SCF spin-orbitals $<\Phi_l|V|\Phi_k>$ can be expressed, using the Slater-Condon rules, in terms of matrix elements over the full Hamiltonian *H*

$$<\Phi_l|V|\Phi_k> = <\Phi_l|H|\Phi_k> - \delta_{k,l}E_k^0,$$

because each such determinant is an eigenfunction of H^0.

C. The MPPT Energy Corrections

Given this particular choice of H^0, it is possible to apply the general RSPT energy and wavefunction correction formulas developed above to generate explicit results in terms of spin-orbital energies and one- and two-electron integrals, $<\phi_i|h|\phi_j>$ and $<\phi_i\phi_j|g|\phi_k\phi_l> = <ij|kl>$, over these spin-orbitals. In particular, the first-order energy correction is given as follows:

$$E_k^{(1)} = <\Phi_k|V|\Phi_k> = <\Phi_k|H|\Phi_k> - \sum_a \varepsilon_a$$

$$= \sum_a \varepsilon_a - \sum_{a<b} [J_{a,b} - K_{a,b}] - \sum_a \varepsilon_a$$

$$= -\sum_{a<b} [J_{a,b} - K_{a,b}] = -\sum_{a<b} [<ab|ab> - <ab|ba>].$$

Thus E_k^0 (the sum of orbital energies) and $E_k^{(1)}$ (the correction for double counting) add up to produce the proper expectation value energy.

The second-order energy correction can be evaluated in like fashion by noting that $<\Phi_k|H|\Phi_l> = 0$ according to the Brillouin theorem for all singly excited Φ_l, and that $<\Phi_k|H|\Phi_l> = <ab|mn> - <ab|nm>$ for doubly excited Φ_l in which excitations from ϕ_a and ϕ_b into ϕ_m and ϕ_n are involved:

$$E_k^{(2)} = \sum_j |<\Phi_j|V|\Phi_k>|^2/[E_k^0 - E_j^0]$$

$$= \sum_{a<b;m<n} |<ab|mn> - <ab|nm>|^2/(\varepsilon_a + \varepsilon_b - \varepsilon_m - \varepsilon_n).$$

D. The Wavefunction Corrections

The first-order MPPT wavefunction can be evaluated in terms of Slater determinants that are excited relative to the SCF reference function Φ_k. Realizing again that the perturbation coupling matrix elements $<\Phi_k|H|\Phi_l>$ are non-zero only for doubly excited CSF's, and denoting such doubly excited Φ_l by $\Phi_{a,b;m,n}$, the first-order wavefunction can be written as:

$$\psi_k^{(1)} = \sum_j <\Phi_j|V|\Phi_k>/[E_k^0 - E_j^0]|\Phi_j>$$

$$= \sum_{a<b;m<n} \Phi_{a,b;m,n}[<ab|mn> - <ab|nm>]/(\varepsilon_a + \varepsilon_b - \varepsilon_m - \varepsilon_n).$$

III. CONCEPTUAL USE OF PERTURBATION THEORY

The first- and second-order RSPT energy and first-order RSPT wavefunction correction expressions form not only a useful computational tool but are also of great use in understanding how strongly a perturbation will affect a particular state of the system. By examining the symmetries of the state of interest Φ_k (this can be an orbital of an atom or molecule, an electronic state of same, or a vibrational/rotational wavefunction of a molecule) and of the perturbation V, one can say whether V will have a significant effect on the energy E_k of Φ_k; if $<\Phi_k|V|\Phi_k>$ is non-zero, the effect can be expected to be significant.

Sometimes the perturbation is of the wrong symmetry to directly (i.e., in a first-order manner) affect E_k. In such cases, one considers whether nearby states $\{\Phi_j, E_j\}$ exist which could couple through V with Φ_k; the second-order energy expression, which contains

$$\sum_j |<\Phi_j|V|\Phi_k>|^2/[E_k^0 - E_j^0]$$

directs one to seek states whose symmetries are contained in the direct product of the symmetries of V and of Φ_k *and* which are close to E_k in energy.

It is through such symmetry and "coupling matrix element" considerations that one can often "guess" whether a given perturbation will have an appreciable effect on the state of interest. The nature of the perturbation is not important to such considerations. It could be the physical interaction that arises as two previously non-interacting atoms are brought together (in which case V would have axial point group symmetry) or it could describe the presence of surrounding ligands on a central transition metal ion (in which case V would carry the symmetry of the "ligand field"). Alternatively, the perturbation might describe the electric dipole interaction of the electrons and nuclei of the atom or molecule with an externally applied electric field $\boldsymbol{E}$, in which case

$$V = -\sum_j e\boldsymbol{r}_j \cdot \boldsymbol{E} + \sum_a Z_a e\boldsymbol{R}_a \cdot \boldsymbol{E}$$

contains components that transform as x, y, and z in the point group appropriate to the system (because the electronic $\boldsymbol{r}_j$ and nuclear $\boldsymbol{R}_a$ coordinate vectors so transform).

APPENDIX

E Point Group Symmetry

It is assumed that the reader has previously learned, in undergraduate inorganic or physical chemistry classes, how symmetry arises in molecular shapes and structures and what symmetry elements are (e.g., planes, axes of rotation, centers of inversion, etc.). For the reader who feels, after reading this appendix, that additional background is needed, the texts by Cotton and EWK, as well as most physical chemistry texts can be consulted. We review and teach here only that material that is of direct application to symmetry analysis of molecular orbitals and vibrations and rotations of molecules. We use a specific example, the ammonia molecule, to introduce and illustrate the important aspects of point group symmetry.

I. THE C_{3v} SYMMETRY GROUP OF AMMONIA—AN EXAMPLE

The ammonia molecule NH_3 belongs, in its ground-state equilibrium geometry, to the C_{3v} point group. Its symmetry *operations* consist of two C_3 rotations, C_3, C_3^2 (rotations by 120° and 240°, respectively about an axis passing through the nitrogen atom and lying perpendicular to the plane formed by the three hydrogen atoms), three vertical reflections, σ_v, σ_v', σ_v'', and the identity operation, E. Corresponding to these six *operations* are symmetry *elements*: the three-fold rotation axis, C_3 and the three symmetry planes σ_v, σ_v' and σ_v'' that contain the three NH bonds and the z-axis (see figure below).

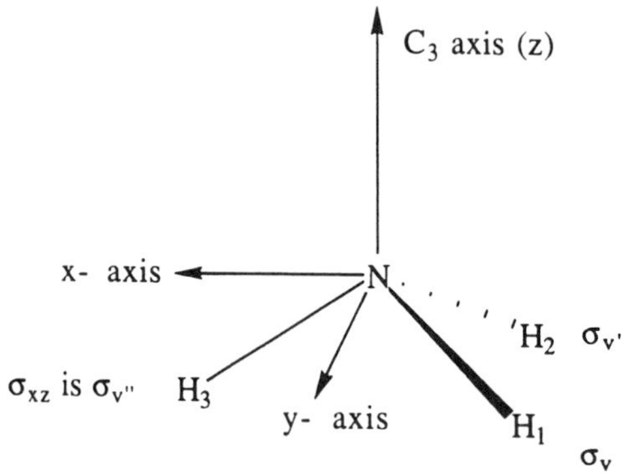

These six symmetry operations form a mathematical *group*. A group is defined as a set of objects satisfying four properties.

1. A combination rule is defined through which two group elements are combined to give a result, which we call the product. The product of two elements in the group must also be a member of the group (i.e., the group is closed under the combination rule).
2. One special member of the group, when combined with any other member of the group, must leave the group member unchanged (i.e., the group contains an identity element).
3. Every group member must have a reciprocal in the group. When any group member is combined with its reciprocal, the product is the identity element.
4. The associative law must hold when combining three group members (i.e., $(AB)C$ must equal $A(BC)$).

The members of symmetry groups are symmetry operations; the combination rule is successive operation. The identity element is the operation of doing nothing at all. The group properties can be demonstrated by forming a multiplication table. Let us label the rows of the table by the first operation and the columns by the second operation. Note that this order is important because most groups are *not commutative*. The C_{3v} group multiplication table is as follows:

	E	C_3	C_3^2	σ_v	σ_v'	σ_v''	**Second operation**
E	E	C_3	C_3^2	σ_v	σ_v'	σ_v''	
C_3	C_3	C_3^2	E	σ_v'	σ_v''	σ_v	
C_3^2	C_3^2	E	C_3	σ_v''	σ_v	σ_v'	
σ_v	σ_v	σ_v''	σ_v'	E	C_3^2	C_3	
σ_v'	σ_v'	σ_v	σ_v''	C_3	E	C_3^2	
σ_v''	σ_v''	σ_v'	σ_v	C_3^2	C_3	E	
First operation							

Note the reflection plane labels do not move. That is, although we start with H_1 in the σ_v plane, H_2 in σ_v'', and H_3 in σ_v'', if H_1 moves due to the first symmetry operation, σ_v remains fixed and a different H atom lies in the σ_v plane.

II. MATRICES AS GROUP REPRESENTATIONS

In using symmetry to help simplify molecular orbital or vibration/rotation energy level calculations, the following strategy is followed:

1. A set of M objects belonging to the constituent atoms (or molecular fragments, in a more general case) is introduced. These objects are the orbitals of the individual atoms (or of the fragments) in the m.o. case; they are unit vectors along the x, y, and z directions located on each of the atoms, and representing displacements along each of these directions, in the vibration/rotation case.
2. Symmetry tools are used to combine these M objects into M new objects each of which belongs to a specific symmetry of the point group. Because the Hamiltonian (electronic in the m.o. case and vibration/rotation in the latter case) commutes with the symmetry operations of the point group, the matrix representation of $\boldsymbol{H}$ within the symmetry adapted basis will be "block diagonal." That is, objects of different symmetry will not interact; only interactions among those of the same symmetry need be considered.

To illustrate such symmetry adaptation, consider symmetry adapting the $2s$ orbital of N and the three $1s$ orbitals of H. We begin by determining how these orbitals transform under the symmetry operations of the C_{3v} point group. The act of each of the six symmetry operations on the four atomic orbitals can be denoted as follows:

$$(S_N,S_1,S_2,S_3) \xrightarrow{E} (S_N,S_1,S_2,S_3)$$
$$\xrightarrow{C_3} (S_N,S_3,S_1,S_2)$$
$$\xrightarrow{C_3^2} (S_N,S_2,S_3,S_1)$$
$$\xrightarrow{\sigma_v} (S_N,S_1,S_3,S_2)$$
$$\xrightarrow{\sigma_v''} (S_N,S_3,S_2,S_1)$$
$$\xrightarrow{\sigma_v'} (S_N,S_2,S_1,S_3)$$

Here we are using the active view that a C_3 rotation rotates the molecule by 120°. The equivalent passive view is that the $1s$ basis functions are rotated −120°. In the C_3 rotation, S_3 ends up where S_1 began, S_1 ends up where S_2 began, and S_2 ends up where S_3 began.

These transformations can be thought of in terms of a matrix multiplying a vector with elements (S_N,S_1,S_2,S_3). For example, if $D^{(4)}(C_3)$ is the representation matrix giving the C_3 transformation, then the above action of C_3 on the four basis orbitals can be expressed as:

$$D^{(4)}(C_3)\begin{bmatrix} S_N \\ S_1 \\ S_2 \\ S_3 \end{bmatrix} = \begin{bmatrix} 1&0&0&0 \\ 0&0&0&1 \\ 0&1&0&0 \\ 0&0&1&0 \end{bmatrix}\begin{bmatrix} S_N \\ S_1 \\ S_2 \\ S_3 \end{bmatrix} = \begin{bmatrix} S_N \\ S_3 \\ S_1 \\ S_2 \end{bmatrix}$$

We can likewise write matrix representations for each of the symmetry operations of the C_{3v} point group:

$$D^{(4)}(C_3^2)=\begin{bmatrix}1&0&0&0\\0&0&1&0\\0&0&0&1\\0&1&0&0\end{bmatrix},\ D^{(4)}(E)=\begin{bmatrix}1&0&0&0\\0&1&0&0\\0&0&1&0\\0&0&0&1\end{bmatrix}$$

$$D^{(4)}(\sigma_v)=\begin{bmatrix}1&0&0&0\\0&1&0&0\\0&0&0&1\\0&0&1&0\end{bmatrix},\ D^{(4)}(\sigma_v')=\begin{bmatrix}1&0&0&0\\0&0&0&1\\0&0&1&0\\0&1&0&0\end{bmatrix}$$

$$D^{(4)}(\sigma_v'')=\begin{bmatrix}1&0&0&0\\0&0&1&0\\0&1&0&0\\0&0&0&1\end{bmatrix}$$

It is easy to verify that a C_3 rotation followed by a σ_v reflection is equivalent to a σ_v' reflection alone. In other words

$$\sigma_v C_3=\sigma_v',\quad \text{or,}\quad \begin{array}{ccc} & S_1 & \\ S_2 & & S_3\end{array}\ \xrightarrow{C_3}\ \begin{array}{ccc} & S_3 & \\ S_1 & & S_2\end{array}\ \xrightarrow{\sigma_v}\ \begin{array}{ccc} & S_3 & \\ S_2 & & S_1\end{array}$$

Note that this same relationship is carried by the matrices:

$$D^{(4)}(\sigma_v)D^{(4)}(C_3)=\begin{bmatrix}1&0&0&0\\0&1&0&0\\0&0&0&1\\0&0&1&0\end{bmatrix}\begin{bmatrix}1&0&0&0\\0&0&0&1\\0&1&0&0\\0&0&1&0\end{bmatrix}=\begin{bmatrix}1&0&0&0\\0&0&0&1\\0&0&1&0\\0&1&0&0\end{bmatrix}=D^{(4)}(\sigma_v')$$

Likewise we can verify that $C_3\sigma_v=\sigma_{v''}$ directly and we can notice that the matrices also show the same identity:

$$D^{(4)}(C_3)D^{(4)}(\sigma_v)=\begin{bmatrix}1&0&0&0\\0&0&0&1\\0&1&0&0\\0&0&1&0\end{bmatrix}\begin{bmatrix}1&0&0&0\\0&1&0&0\\0&0&0&1\\0&0&1&0\end{bmatrix}=\begin{bmatrix}1&0&0&0\\0&0&1&0\\0&1&0&0\\0&0&0&1\end{bmatrix}=D^{(4)}(\sigma_v'').$$

In fact, one finds that the six matrices, $D^{(4)}(R)$, when multiplied together in all 36 possible ways obey the same multiplication table as did the six symmetry operations. We say the matrices form a representation of the group because the matrices have all the properties of the group.

A. Characters of Representations

One important property of a matrix is the sum of its diagonal elements

$$Tr(D)=\sum_i D_{ii}=\chi.$$

χ is called the trace or *character* of the matrix. In the above example

$$\chi(E) = 4$$
$$\chi(C_3) = \chi(C_3^2) = 1$$
$$\chi(\sigma_v) = \chi(\sigma_v') = \chi(\sigma_v'') = 2.$$

The importance of the characters of the symmetry operations lies in the fact that they do not depend on the specific basis used to form them. That is, they are invariant to a unitary or orthorgonal transformation of the objects used to define the matrices. As a result, they contain information about the symmetry operation itself and about the *space* spanned by the set of objects. The significance of this observation for our symmetry adaptation process will become clear later.

Note that the characters of both rotations are the same as are those of all three reflections. Collections of operations having identical characters are called *classes*. Each operation in a *class* of operations has the same *character* as other members of the class. The character of a class depends on the space spanned by the basis of functions on which the symmetry operations act. Above we used (S_N,S_1,S_2,S_3) as a basis.

B. Another Basis and Another Representation

If, alternatively, we use the one-dimensional basis consisting of the 1*s* orbital on the N-atom, we obtain different characters, as we now demonstrate.

The act of the six symmetry operations on this S_N can be represented as follows:

$$S_N \xrightarrow{E} S_N; \quad S_N \xrightarrow{C_3} S_N; \quad S_N \xrightarrow{C_3^2} S_N;$$
$$S_N \xrightarrow{\sigma_v} S_N; \quad S_N \xrightarrow{\sigma_v'} S_N; \quad S_N \xrightarrow{\sigma_v''} S_N.$$

We can represent this group of operations in this basis by the one-dimensional set of matrices:

$$D^{(1)}(E) = 1; \quad D^{(1)}(C_3) = 1; \quad D^{(1)}(C_3^2) = 1,$$
$$D^{(1)}(\sigma_v) = 1; \quad D^{(1)}(\sigma_v'') = 1; \quad D^{(1)}(\sigma_v') = 1.$$

Again we have

$$D^{(1)}(\sigma_v)D^{(1)}(C_3) = 1 \cdot 1 = D^{(1)}(\sigma_v''), \text{ and}$$
$$D^{(1)}(C_3)D^{(1)}(\sigma_v) = 1 \cdot 1 = D^{(1)}(\sigma_v').$$

These six matrices form another representation of the group. In this basis, each character is equal to unity. The representation formed by allowing the six symmetry operations to act on the 1*s* N-atom orbital is clearly not the same as that formed when the same six operations acted on the (S_N,S_1,S_2,S_3) basis. We now need to learn how to further analyze the information content of a specific representation of the group formed when the symmetry operations act on any specific set of objects.

III. REDUCIBLE AND IRREDUCIBLE REPRESENTATIONS

A. A Reducible Representation

Note that every matrix in the four dimensional group representation labeled $D^{(4)}$ has the so-called *block diagonal* form

$$\begin{array}{c|ccc} \boxed{1} & 0 & 0 & 0 \\ \hline 0 & & & \\ 0 & & 3 \times 3 \text{ matrix} & \\ 0 & & & \end{array}$$

This means that these $D^{(4)}$ matrices are really a combination of two separate group representations (mathematically, it is called a *direct sum* representation). We say that $D^{(4)}$ is reducible into a one-dimensional representation $D^{(1)}$ and a three-dimensional representation formed by the 3×3 submatrices which we will call $D^{(3)}$.

$$D^{(3)}(E) = \begin{bmatrix} 1\,0\,0 \\ 0\,1\,0 \\ 0\,0\,1 \end{bmatrix};\ D^{(3)}(C_3) = \begin{bmatrix} 0\,0\,1 \\ 1\,0\,0 \\ 0\,1\,0 \end{bmatrix};\ D^{(3)}(C_3^2) = \begin{bmatrix} 0\,1\,0 \\ 0\,0\,1 \\ 1\,0\,0 \end{bmatrix}$$

$$D^{(3)}(\sigma_v) = \begin{bmatrix} 1\,0\,0 \\ 0\,0\,1 \\ 0\,1\,0 \end{bmatrix};\ D^{(3)}(\sigma_v') = \begin{bmatrix} 0\,0\,1 \\ 0\,1\,0 \\ 1\,0\,0 \end{bmatrix};\ D^{(3)}(\sigma_v'') = \begin{bmatrix} 0\,1\,0 \\ 1\,0\,0 \\ 0\,0\,1 \end{bmatrix}$$

The characters of $D^{(3)}$ are $\chi(E) = 3$, $\chi(2C_3) = 0$, $\chi(3\sigma_v) = 1$. Note that we would have obtained this $D^{(3)}$ representation directly if we had originally chosen to examine the basis (S_1,S_2,S_3); also note that these characters are equal to those of $D^{(4)}$ minus those of $D^{(1)}$.

B. A Change in Basis

Now let us convert to a new basis that is a linear combination of the original S_1,S_2,S_3 basis:

$$T_1 = S_1 + S_2 + S_3$$
$$T_2 = 2S_1 - S_2 - S_3$$
$$T_3 = S_2 - S_3$$

(Don't worry about how we construct T_1, T_2, and T_3 yet. As will be demonstrated later, we form them by using symmetry projection operators defined below) We determine how the "T" basis functions behave under the group operations by allowing the operations to act on the S_j and interpreting the results in terms of the T_i. In particular,

$$\left(T_1,T_2,T_3\right) \xrightarrow{\sigma_v} \left(T_1,T_2,-T_3\right); \qquad \left(T_1,T_2,T_3\right) \xrightarrow{E} \left(T_1,T_2,T_3\right);$$

$$\left(T_1,T_2,T_3\right) \xrightarrow{\sigma_v'} \left(S_3 + S_2 + S_1,\ 2S_3 - S_2 - S_1,\ S_2 - S_1\right) = \left(T_1,\ -\frac{1}{2}T_2 - \frac{3}{2}T_3,\ -\frac{1}{2}T_2 + \frac{1}{2}T_3\right);$$

$$\left(T_1,T_2,T_3\right) \xrightarrow{\sigma_v''} \left(S_2 + S_1 + S_3,\ 2S_2 - S_1 - S_3,\ S_1 - S_3\right) = \left(T_1,\ -\frac{1}{2}T_2 + \frac{3}{2}T_3,\ \frac{1}{2}T_2 + \frac{1}{2}T_3\right);$$

$$\left(T_1,T_2,T_3\right) \xrightarrow{C_3} \left(S_3 + S_1 + S_2,\ 2S_3 - S_1 - S_2,\ S_1 - S_2\right) = \left(T_1,\ -\frac{1}{2}T_2 - \frac{3}{2}T_3,\ \frac{1}{2}T_2 - \frac{1}{2}T_3\right);$$

$$\left(T_1,T_2,T_3\right) \xrightarrow{C_3^2} \left(S_2 + S_3 + S_1,\ 2S_2 - S_3 - S_1,\ S_3 - S_1\right) = \left(T_1,\ -\frac{1}{2}T_2 + \frac{3}{2}T_3,\ -\frac{1}{2}T_2 - \frac{1}{2}T_3\right).$$

So the matrix representations in the new T_i basis are:

$$D^{(3)}(E) = \begin{bmatrix} 1 & 0 & 0 \\ 0 & 1 & 0 \\ 0 & 0 & 1 \end{bmatrix}; \qquad D^{(3)}(C_3) = \begin{bmatrix} 1 & 0 & 0 \\ 0 & -\frac{1}{2} & -\frac{3}{2} \\ 0 & +\frac{1}{2} & -\frac{1}{2} \end{bmatrix};$$

$$D^{(3)}(C_3^2) = \begin{bmatrix} 1 & 0 & 0 \\ 0 & -\frac{1}{2} & +\frac{3}{2} \\ 0 & -\frac{1}{2} & -\frac{1}{2} \end{bmatrix}; \quad D^{(3)}(\sigma_v) = \begin{bmatrix} 1 & 0 & 0 \\ 0 & 1 & 0 \\ 0 & 0 & -1 \end{bmatrix};$$

$$D^{(3)}(\sigma_v') = \begin{bmatrix} 1 & 0 & 0 \\ 0 & -\frac{1}{2} & -\frac{3}{2} \\ 0 & -\frac{1}{2} & +\frac{1}{2} \end{bmatrix}; \quad D^{(3)}(\sigma_v'') = \begin{bmatrix} 1 & 0 & 0 \\ 0 & -\frac{1}{2} & +\frac{3}{2} \\ 0 & +\frac{1}{2} & +\frac{1}{2} \end{bmatrix}.$$

C. Reduction of the Reducible Representation

These six matrices can be verified to multiply just as the symmetry operations do; thus they form another three-dimensional representation of the group. We see that in the T_i basis the matrices are block diagonal. This means that the space spanned by the T_i functions, which is the same space as the S_j span, forms a *reducible representation* that can be decomposed into a one dimensional space and a two dimensional space (via formation of the T_i functions). Note that the characters (traces) of the matrices are not changed by the change in bases. The one-dimensional part of the above reducible three-dimensional representation is seen to be the same as the totally symmetric representation we arrived at before, $D^{(1)}$. The two-dimensional representation that is left can be shown to be *irreducible;* it has the following matrix representations:

$$D^{(2)}(E) = \begin{bmatrix} 1 & 0 \\ 0 & 1 \end{bmatrix}; D^{(2)}(C_3) = \begin{bmatrix} -\frac{1}{2} & -\frac{3}{2} \\ +\frac{1}{2} & -\frac{1}{2} \end{bmatrix}; D^{(2)}(C_3^2) = \begin{bmatrix} -\frac{1}{2} & +\frac{3}{2} \\ -\frac{1}{2} & -\frac{1}{2} \end{bmatrix};$$

$$D^{(2)}(\sigma_v) = \begin{bmatrix} 1 & 0 \\ 0 & -1 \end{bmatrix}; D^{(2)}(\sigma_v') = \begin{bmatrix} -\frac{1}{2} & -\frac{3}{2} \\ -\frac{1}{2} & +\frac{1}{2} \end{bmatrix}; D^{(2)}(\sigma_v'') = \begin{bmatrix} -\frac{1}{2} & +\frac{3}{2} \\ +\frac{1}{2} & +\frac{1}{2} \end{bmatrix}.$$

The characters can be obtained by summing diagonal elements:

$$\chi(E) = 2, \chi(2C_3) = -1, \chi(3\sigma_v) = 0.$$

D. Rotations as a Basis

Another one-dimensional representation of the group can be obtained by taking rotation about the Z-axis (the C_3 axis) as the object on which the symmetry operations act:

$$R_z \xrightarrow{E} R_z; \qquad R_z \xrightarrow{C_3} R_z; \qquad R_z \xrightarrow{C_3^2} R_z';$$

$$R_z \xrightarrow{\sigma_v} -R_z; \qquad R_z \xrightarrow{\sigma_v''} -R_z; \qquad R_z \xrightarrow{\sigma_v'} -R_z.$$

In writing these relations, we use the fact that reflection reverses the sense of a rotation. The matrix representations corresponding to this one-dimensional basis are:

$$D^{(1)}(E) = 1; \qquad D^{(1)}(C_3) = 1; \qquad D^{(1)}(C_3^2) = 1;$$
$$D^{(1)}(\sigma_v) = -1; \qquad D^{(1)}(\sigma_v'') = -1; \qquad D^{(1)}(\sigma_v') = -1.$$

These one-dimensional matrices can be shown to multiply together just like the symmetry operations of the C_{3v} group. They form an *irreducible* representation of the group (because it is one-dimensional, it can not be further reduced). Note that this one-dimensional representation is not identical to that found above for the $1s$ N-atom orbital, or the T_1 function.

E. Overview

We have found three distinct irreducible representations for the C_{3v} symmetry group; two different one-dimensional and one two-dimensional representations. Are there any more? An important theorem of group theory shows that the number of irreducible representations of a group is equal to the number of classes. Since there are three classes of operation, we have found *all* the irreducible representations of the C_{3v} point group. There are no more. The irreducible representations have standard names: the first $D^{(1)}$ (that arising from the T_1 and $1s_N$ orbitals) is called A_1, the $D^{(1)}$ arising from R_z is called A_2 and $D^{(2)}$ is called E (not to be confused with the identity operation E). Thus, our original $D^{(4)}$ representation was a combination of two A_1 representations and one E representation. We say that $D^{(4)}$ is a direct sum representation: $D^{(4)} = 2A_1 \oplus E$. A consequence is that the characters of the combination representation $D^{(4)}$ can be obtained by adding the characters of its constituent irreducible representations.

	E	$2C_3$	$3\sigma_v$
A_1	1	1	1
A_1	1	1	1
E	2	−1	0
$2A_1 \oplus E$	4	1	2

F. How to Decompose Reducible Representations in General

Suppose you were given only the characters (4,1,2). How can you find out how many times A_1, E, and A_2 appear when you reduce $D^{(4)}$ to its irreducible parts? You want to find a linear combination of the characters of A_1, A_2, and E that add up (4,1,2). You can treat the characters of matrices as vectors and take the dot product of A_1 with $D^{(4)}$

$$\begin{bmatrix} 1 & 1 & 1 & 1 & 1 & 1 \\ E & C_3 & & \sigma_v & & \end{bmatrix} \cdot \begin{bmatrix} 4 & E \\ 1 & C_3 \\ 1 & \\ 2 & \sigma_v \\ 2 & \\ 2 & \end{bmatrix} = 4+1+1+2+2+2 = 12.$$

The vector (1,1,1,1,1,1) is not normalized; hence to obtain the component of (4,1,1,2,2,2) along a unit vector in the (1,1,1,1,1,1) direction, one must divide by the norm of (1,1,1,1,1,1); this norm is 6. The result is that the reducible representation contains

$$\frac{12}{6} = 2A_1$$

components. Analogous projections in the E and A_2 directions give components of 1 and 0, respectively. In general, to determine the number n_Γ of times irreducible representation Γ appears in the reducible representation with characters χ_{red}, one calculates

$$n_\Gamma = \frac{1}{g}\sum_R \chi_\Gamma(R)\chi_{red}(R),$$

where g is the order of the group and $\chi_\Gamma(R)$ are the characters of the Γth irreducible representation.

G. Commonly Used Bases

We could take *any* set of functions as a basis for a group representation. Commonly used sets include: coordinates (x,y,z) located on the atoms of a polyatomic molecule (their symmetry treatment is equivalent to that involved in treating a set of p orbitals on the same atoms), quadratic functions such as d orbitals $-xy$, yz, xz, $x^2 - y^2$, z^2, as well as rotations about the x, y, and z axes. The transformation properties of these very commonly used bases are listed in the character tables shown at the end of this appendix.

H. Summary

The basic idea of symmetry analysis is that any basis of orbitals, displacements, rotations, etc. transforms either as one of the irreducible representations or as a direct sum (reducible) representation. Symmetry tools are used to first determine how the basis transforms under action of the symmetry operations. They are then used to decompose the resultant representations into their irreducible components.

IV. ANOTHER EXAMPLE

A. The 2*p* Orbitals of Nitrogen

For a function to transform according to a specific irreducible representation means that the function, when operated upon by a point-group symmetry operator, yields a linear combination of the functions that transform according to that irreducible representation. For example, a $2p_z$ orbital (z is the C_3 axis of NH_3) on the nitrogen atom belongs to the A_1 representation because it yields unity times itself when C_3, C_3^2, σ_v, σ_v', σ_v'' or the identity operation act on it. The factor of 1 means that $2p_z$ has A_1 symmetry since the characters (the numbers listed opposite A_1 and below E, $2C_3$, and $3\sigma_v$ in the C_{3v} character table) of all six symmetry operations are 1 for the A_1 irreducible representation. The $2p_x$ and $2p_y$ orbitals on the nitrogen atom transform as the E representation since C_3, C_3^2, σ_v, σ_v', σ_v'' and the identity operation map $2p_x$ and $2p_y$ among one another. Specifically,

$$C_3\begin{bmatrix} 2p_x \\ 2p_y \end{bmatrix} = \begin{bmatrix} \cos 120° & -\sin 120° \\ \sin 120° & \cos 120° \end{bmatrix} \begin{bmatrix} 2p_x \\ 2p_y \end{bmatrix};$$

$$C_3^2\begin{bmatrix} 2p_x \\ 2p_y \end{bmatrix} = \begin{bmatrix} \cos 240° & -\sin 240° \\ \sin 240° & \cos 240° \end{bmatrix} \begin{bmatrix} 2p_x \\ 2p_y \end{bmatrix};$$

$$E\begin{bmatrix} 2p_x \\ 2p_y \end{bmatrix} = \begin{bmatrix} 1 & 0 \\ 0 & 1 \end{bmatrix} \begin{bmatrix} 2p_x \\ 2p_y \end{bmatrix};$$

$$\sigma_v\begin{bmatrix} 2p_x \\ 2p_y \end{bmatrix} = \begin{bmatrix} -1 & 0 \\ 0 & 1 \end{bmatrix} \begin{bmatrix} 2p_x \\ 2p_y \end{bmatrix};$$

$$\sigma_v'\begin{bmatrix} 2p_x \\ 2p_y \end{bmatrix} = \begin{bmatrix} +\frac{1}{2} & +\frac{\sqrt{3}}{2} \\ +\frac{\sqrt{3}}{2} & -\frac{1}{2} \end{bmatrix} \begin{bmatrix} 2p_x \\ 2p_y \end{bmatrix};$$

$$\sigma_v''\begin{bmatrix} 2p_x \\ 2p_y \end{bmatrix} = \begin{bmatrix} +\frac{1}{2} & -\frac{\sqrt{3}}{2} \\ -\frac{\sqrt{3}}{2} & -\frac{1}{2} \end{bmatrix} \begin{bmatrix} 2p_x \\ 2p_y \end{bmatrix}.$$

The 2×2 matrices, which indicate how each symmetry operation maps $2p_x$ and $2p_y$ into some combinations of $2p_x$ and $2p_y$, are the representation matrices ($D^{(IR)}$) for that particular operation and for this particular irreducible representation (IR). For example,

$$\begin{bmatrix} +\frac{1}{2} & +\frac{\sqrt{3}}{2} \\ +\frac{\sqrt{3}}{2} & -\frac{1}{2} \end{bmatrix} = D^{(E)}(\sigma_v')$$

This set of matrices have the same characters as $D^{(2)}$ above, but the individual matrix elements are different because we used a different basis set (here $2p_x$ and $2p_y$; above it was T_2 and T_3). This illustrates the invariance of the trace to the specific representation; the trace only depends on the space spanned, not on the specific manner in which it is spanned.

B. A Short-Cut

A short-cut device exists for evaluating the trace of such representation matrices (that is, for computing the characters). The diagonal elements of the representation matrices are the projections along each orbital of the effect of the symmetry operation acting on that orbital. For example, a diagonal element of the C_3 matrix is the component of $C_3 2p_y$ along the $2p_y$ direction. More rigorously, it is $\int 2p_y^* C_3 2p_y d\tau$. Thus, the character of the C_3 matrix is the sum of $\int 2p_x^* C_3 2p_y d\tau$ and $\int 2p_x^* C_3 2p_x d\tau$. In general, the character χ of any symmetry operation S can be computed by allowing S to operate on each orbital ϕ_i, then projecting $S\phi_i$ along ϕ_i (i.e., forming $\int \phi_i^* S\phi_i d\tau$), and summing these terms,

$$\sum_i \int \phi_i^* S\phi_i d\tau = \chi(S) \ .$$

If these rules are applied to the $2p_x$ and $2p_y$ orbitals of nitrogen within the C_{3v} point group, one obtains

$$\chi(E) = 2, \chi(C_3) = \chi(C_3^2) = -1, \chi(\sigma_v) = \chi(\sigma_v'') = \chi(\sigma_v') = 0.$$

This set of characters is the same as $D^{(2)}$ above and agrees with those of the E representation for the C_{3v} point group. Hence, $2p_x$ and $2p_y$ belong to or transform as the E representation. This is why (x,y) is to the right of the row of characters for the E representation in the C_{3v} character table. In similar fashion, the C_{3v} character table states that $d_{x^2-y^2}$ and d_{xy} orbitals on nitrogen transform as E, as do d_{xy} and d_{yz}, but d_z^2 transforms as A_1.

Earlier, we considered in some detail how the three $1s_H$ orbitals on the hydrogen atoms transform. Repeating this analysis using the short-cut rule just described, the traces (characters) of the 3×3 representation matrices are computed by allowing E, $2C_3$, and $3\sigma_v$ to operate on $1s_{H1}$, $1s_{H_2}$, and $1s_{Hsub3}$ and then computing the component of the resulting function along the original function. The resulting characters are $\chi(E) = 3, \chi(C_3) = \chi(C_3^2) = 0$, and $\chi(\sigma_v) = \chi(\sigma_v') = \chi(\sigma_v'') = 1$, in agreement with what we calculated before.

Using the orthogonality of characters taken as vectors we can reduce the above set of characters to $A_1 + E$. Hence, we say that our orbital set of three $1s_H$ orbitals forms a *reducible* representation consisting of the sum of A_1 and E *IR*'s. This means that the three $1s_H$ orbitals can be combined to yield one orbital of A_1 symmetry and a *pair* that transform according to the E representation.

V. PROJECTOR OPERATORS: SYMMETRY ADAPTED LINEAR COMBINATIONS OF ATOMIC ORBITALS

To generate the above A_1 and E symmetry-adapted orbitals, we make use of so-called symmetry projection operators P_E and P_{A_1}. These operators are given in terms of linear combinations of products of characters times elementary symmetry operations as follows:

$$P_{A_1} = \sum_S \chi_A(S)S$$

$$P_E = \sum_S \chi_E(S)S$$

where S ranges over C_3, C_3^2, σ_v, σ_v', and σ_v'' and the identity operation. The result of applying P_{A_1} to say $1s_{H_1}$ is

$$\begin{aligned} P_{A_1} 1s_{H_1} &= 1s_{H_1} + 1s_{H_2} + 1s_{H_3} + 1s_{H_2} + 1s_{H_3} + 1s_{H_1} \\ &= 2(1s_{H_1} + 1s_{H_2} + 1s_{H_3}) = \phi_{A_1}, \end{aligned}$$

which is an (unnormalized) orbital having A_1 symmetry. Clearly, this same ϕ_{A_1} would be generated by P_{A_1} acting on $1s_{H_2}$ or $1s_{H_3}$. Hence, only one A_1 orbital exists. Likewise,

$$P_E 1s_{H_1} = 2 \cdot 1s_{H_1} - 1s_{H_2} - 1s_{H_3} \equiv \phi_{E,1},$$

which is *one* of the symmetry adapted orbitals having E symmetry. The other E orbital can be obtained by allowing P_E to act on $1s_{H_2}$ or $1s_{H_3}$:

$$P_E 1s_{H_2} = 2 \cdot 1s_{H_2} - 1s_{H_1} - 1s_{H_3} \equiv \phi_{E,2}$$
$$P_E 1s_{H_3} = 2 \cdot 1s_{H_3} - 1s_{H_1} - 1s_{H_2} = \phi_{E,3}.$$

It might seem as though *three* orbitals having E symmetry were generated, but only two of these are really independent functions. For example, $\phi_{E,3}$ is related to $\phi_{E,1}$ and $\phi_{E,2}$ as follows:

$$\phi_{E,3} = -(\phi_{E,1} + \phi_{E,2}).$$

Thus, only $\phi_{E,1}$ and $\phi_{E,2}$ are needed to span the two-dimensional space of the E representation. If we include $\phi_{E,1}$ in our set of orbitals and require our orbitals to be orthogonal, then we must find numbers a and b such that $\phi'_E = a\phi_{E,2} + b\phi_{E,3}$ is orthogonal to $\phi_{E,1}$: $\int \phi'_E \phi_{E,1} d\tau = 0$. A straightforward calculation gives $a = -b$ or $\phi'_E = a(1s_{H_2} - 1s_{H_3})$, which agrees with what we used earlier to construct the T_i functions in terms of the S_j functions.

VI. SUMMARY

Let us now summarize what we have learned. Any given set of atomic orbitals $\{\phi_i\}$, atom-centered displacements or rotations can be used as a basis for the symmetry operations of the point group of the molecule. The characters $\chi(S)$ belonging to the operations S of this point group within any such space can be found by summing the integrals $\int \phi_i{}^* S \phi_i d\tau$ over all the atomic orbitals (or corresponding unit vector atomic displacements). The resultant characters will, in general, be reducible to a combination of the characters of the irreducible representations $\chi_i(S)$. To decompose the characters $\chi(S)$ of the reducible representation to a sum of characters $\chi_i(S)$ of the irreducible representation

$$\chi(S) = \sum_i n_i \chi_i(S),$$

it is necessary to determine how many times, n_i, the ith irreducible representation occurs in the reducible representation. The expression for n_i is (see the text by Cotton)

$$n_i = \frac{1}{g} \sum_S \chi(S)\, \chi_i(S)$$

in which g is the order of the point group; the total number of symmetry operations in the group (e.g., $g = 6$ for C_{3v}).

For example, the reducible representation $\chi(E) = 3$, $\chi(C_3) = 0$, and $\chi(\sigma_v) = 1$ formed by the three $1s_H$ orbitals discussed above can be decomposed as follows:

$$n_{A_1} = \frac{1}{6}(3 \cdot 1 + 2 \cdot 0 \cdot 1 + 3 \cdot 1 \cdot 1) = 1,$$

$$n_{A_2} = \frac{1}{6}(3 \cdot 1 + 2 \cdot 0 \cdot 1 + 3 \cdot 1 \cdot (-1)) = 0,$$

$$n_E = \frac{1}{6}(3 \cdot 2 + 2 \cdot 0 \cdot (-1) + 3 \cdot 1 \cdot 0) = 1.$$

These equations state that the three $1s_H$ orbitals can be combined to give one A_1 orbital and, since E is degenerate, one *pair* of E orbitals, as established above. With knowledge of the n_i, the symmetry-adapted orbitals can be formed by allowing the projectors

$$P_i = \sum_i \chi_i(S)S$$

to operate on each of the primitive atomic orbitals. How this is carried out was illustrated for the $1s_H$ orbitals in our earlier discussion. These tools allow a symmetry decomposition of any set of atomic orbitals into appropriate symmetry-adapted orbitals.

Before considering other concepts and group-theoretical machinery, it should once again be stressed that these same tools can be used in symmetry analysis of the translational, vibrational and rotational motions of a molecule. The 12 motions of NH_3 (three translations, three rotations, six vibrations) can be described in terms of combinations of displacements of each of the four atoms in each of three (x,y,z) directions. Hence, unit vectors placed on each atom directed in the x, y, and z directions form a basis for action by the operations $\{S\}$ of the point group. In the case of NH_3, the characters of the resultant 12×12 representation matrices form a reducible representation in the C_{2v} point group: $\chi(E) = 12$, $\chi(C_3) = \chi(C_3^2) = 0$, $\chi(\sigma_v) = \chi(\sigma_v') = \chi(\sigma_v'') = 2$. (You should try to prove this. For example, under σ_v, the H_2 and H_3 atoms are are interchanged, so unit vectors on either one will not contribute to the trace. Unit z-vectors on N and H_1 remain unchanged as well as the corresponding y-vectors. However, the x-vectors on N and H_1 are reversed in sign. The total character for σ_v' the H_2 and H_3 atoms are interchanged, so unit vectors on either one will not contribute to the trace. Unit z-vectors on N and H_1 remain unchanged as well as the corresponding y-vectors. However, the x-vectors on N and H_1 are reversed in sign. The total character for σ_v is thus $4 - 2 = 2$. This representation can be decomposed as follows:

$$n_{A_1} = \frac{1}{6}[1 \cdot 1 \cdot 12 + 2 \cdot 1 \cdot 0 + 3 \cdot 1 \cdot 2] = 3,$$

$$n_{A_2} = \frac{1}{6}[1 \cdot 1 \cdot 12 + 2 \cdot 1 \cdot 0 + 3 \cdot (-1) \cdot 2] = 1,$$

$$n_E = \frac{1}{6}[1 \cdot 2 \cdot 12 + 2 \cdot (-1) \cdot 0 + 3 \cdot 0 \cdot 2] = 4.$$

From the information on the right side of the C_{3v} character table, translations of all four atoms in the z, x, and y directions transform as $A_1(z)$ and $E(x,y)$, respectively, whereas rotations about the $z(R_z)$, $x(R_x)$, and $y(R_y)$ axes transform as A_2 and E. Hence, of the 12 motions, three translations have A_1 and E symmetry and three rotations have A_2 and E symmetry. This leaves six vibrations, of which two have A_1 symmetry, none have A_2 symmetry, and two (pairs) have E symmetry. We could obtain symmetry-adapted vibrational and rotational bases by allowing symmetry projection operators of the irreducible representation symmetries to operate on various elementary cartesian (x,y,z) atomic displacement vectors. Both Cotton and Wilson, Decius and Cross show in detail how this is accomplished.

VII. DIRECT PRODUCT REPRESENTATIONS

A. Direct Products in N-Electron Wavefunctions

We now return to the symmetry analysis of orbital products. Such knowledge is important because one is routinely faced with constructing symmetry-adapted N-electron configurations that consist of products of N individual orbitals. A point-group symmetry operator S, when acting on such a product of orbitals, gives the product of S acting on each of the individual orbitals

$$S(\phi_1\phi_2\phi_3 \ldots \phi_N) = (S\phi_1)(S\phi_2)(S\phi_3) \ldots (S\phi_N).$$

For example, reflection of an N-orbital product through the σ_v plane in NH_3 applies the reflection operation to all N electrons.

Just as the individual orbitals formed a basis for action of the point-group operators, the configurations (N-orbital products) form a basis for the action of these same point-group operators. Hence, the various electronic configurations can be treated as functions on which S operates, and the machinery illustrated earlier for decomposing orbital symmetry can then be used to carry out a symmetry analysis of configurations.

Another shortcut makes this task easier. Since the symmetry adapted individual orbitals $\{\phi_i, i = 1, \ldots, M\}$ transform according to irreducible representations, the representation matrices for the N-term products shown above consist of products of the matrices belonging to each ϕ_i. This matrix product is not a simple product but a *direct product*. To compute the characters of the direct product matrices, one multiplies the characters of the individual matrices of the irreducible representations of the N orbitals that appear in the electron configuration. The direct-product representation formed by the orbital products can therefore be symmetry-analyzed (reduced) using the same tools as we used earlier.

For example, if one is interested in knowing the symmetry of an orbital product of the form $a_1^2a_2^2e^2$ (note: lowercase letters are used to denote the symmetry of electronic orbitals) in C_{3v} symmetry, the following procedure is used. For each of the six symmetry operations in the C_{2v} point group, the *product* of the characters associated with each of the *six* spin orbitals (orbital multiplied by α or β spin) is formed

$$\chi(S) = \prod_i \chi_i(S)) = (\chi_{A_1}(S))^2(\chi_{A_2}(S))^2(\chi_E(S))^2.$$

In the specific case considered here, $\chi(E) = 4$, $\chi(2C_3) = 1$, and $\chi(3\sigma_v) = 0$ (you should try this). Notice that the contributions of any doubly occupied nondegenerate orbitals (e.g., a_1^2, and a_2^2) to these direct product characters $\chi(S)$ are unity because for *all* operators $(\chi_k(S))^2 = 1$ for any one-dimensional irreducible representation. As a result, only the singly occupied or degenerate orbitals need to be considered when forming the characters of the reducible direct-product representation $\chi(S)$. For this example this means that the direct-product characters can be determined from the characters $\chi_E(S)$ of the two active (i.e., nonclosed-shell) orbitals—the e^2 orbitals. That is, $\chi(S) = \chi_E(S) \cdot \chi_E(S)$.

From the direct-product characters $\chi(S)$ belonging to a particular electronic configuration (e.g., $a_1^2a_2^2e^2$), one must still decompose this list of characters into a sum of irreducible characters. For the example at hand, the direct-product characters $\chi(S)$ decompose into one A_1, one A_2, and one E representation. This means that the e^2 configuration contains A_1, A_2, and E symmetry elements. Projection operators analogous to those introduced earlier for orbitals can be used to form symmetry-adapted orbital products from the individual basis orbital products of the form $a_1^2a_2^2e_x^m e_y^{m'}$, where m and m' denote the occupation (2, 1, or 0) of the two degenerate orbitals e_x and

e_y. When dealing with indistinguishable particles such as electrons, it is also necessary to further project the resulting orbital products to make them antisymmetric (for Fermions) or symmetric (for Bosons) with respect to interchange of any pair of particles. This step reduces the set of N-electron states that can arise. For example, in the above e^2 configuration case, only ${}^3A_2$, ${}^1A_1$, and ${}^1E$ states arise; the ${}^3E$, ${}^3A_1$, and ${}^1A_2$ possibilities disappear when the antisymmetry projector is applied. In contrast, for an $e^1e'^1$ configuration, all states arise even after the wavefunction has been made antisymmetric. The steps involved in combining the point group symmetry with permutational antisymmetry are illustrated in chapter 10. In appendix III of *Electronic Spectra and Electronic Structure of Polyatomic Molecules*, G. Herzberg, Van Nostrand Reinhold Co., New York, N.Y. (1966), the resolution of direct products among various representations within many point groups are tabulated.

B. Direct Products in Selection Rules

Two states ψ_a and ψ_b that are eigenfunctions of a Hamiltonian $\boldsymbol{H}^o$ in the absence of some external perturbation (e.g., electromagnetic field or static electric field or potential due to surrounding ligands) can be "coupled" by the perturbation $\boldsymbol{V}$ only if the symmetries of $\boldsymbol{V}$ and of the two wavefunctions obey a so-called selection rule. In particular, only if the coupling integral (see appendix D, which deals with time independent perturbation theory)

$$\int \psi_a^* V \psi_b d\tau = V_{a,b}$$

is non-vanishing will the two states be coupled by $\boldsymbol{V}$.

The role of symmetry in determining whether such integrals are non-zero can be demonstrated by noting that the integrand, considered as a whole, must contain a component that is invariant under all of the group operations (i.e., belongs to the totally symmetric representation of the group). In terms of the projectors introduced above in section IV of this appendix, we must have

$$\sum_S \chi_A(S) S \psi_a^* V \psi_b$$

not vanish. Here the subscript A denotes the totally symmetric representation of the group. The symmetry of the product $\psi_a^* V \psi_b$ is, according to what was covered earlier in this section, given by the direct product of the symmetries of ψ_a^* of $\boldsymbol{V}$ and of ψ_b. So, the conclusion is that the integral will vanish unless this triple direct product contains, when it is reduced to its irreducible components, a component of the totally symmetric representation.

To see how this result is used, consider the integral that arises in formulating the interaction of electromagnetic radiation with a molecule within the electric-dipole approximation:

$$\int \psi_a^* \boldsymbol{r} \psi_b d\tau \, .$$

Here $\boldsymbol{r}$ is the vector giving, together with e, the unit charge, the quantum mechanical dipole moment operator

$$\boldsymbol{r} = e \sum_n Z_n \boldsymbol{R}_n - e \sum_j \boldsymbol{r}_j ,$$

where Z_n and $\boldsymbol{R}_n$ are the charge and position of the nth nucleus and $\boldsymbol{r}_j$ is the position of the jth electron. Now, consider evaluating this integral for the singlet $n \rightarrow \pi^*$ transition in formaldehyde. Here, the closed-shell ground state is of $^1A_1$ symmetry and the excited state, which involves promoting an electron from the non-bonding b_2 lone pair orbital on the oxygen into the $\pi^* b_1$ orbital on the CO moiety, is of $^1A_2$ symmetry ($b_1 \times b_2 = a_2$). The direct product of the two wavefunction symmetries thus contains only a_2 symmetry. The three components (x, y, and z) of the dipole operator have, respectively, b_1, b_2, and a_1 symmetry. Thus, the triple direct products give rise to the following possibilities:

$$a_2 \times b_1 = b_2,$$
$$a_2 \times b_2 = b_1,$$
$$a_2 \times a_1 = a_2.$$

There is no component of a_1 symmetry in the triple direct product, so the integral vanishes. This allows us to conclude that the $n \rightarrow \pi^*$ excitation in formaldehyde is electric dipole forbidden.

VIII. SUMMARY

This appendix has reviewed how to make a symmetry decomposition of a basis of atomic orbitals (or cartesian displacements or orbital products) into irreducible representation components. This tool is most helpful when constructing the orbital correlation diagrams that form the basis of the Woodward-Hoffmann rules. We also learned how to form the direct-product symmetries that arise when considering configurations consisting of products of symmetry-adapted spin orbitals. This step is essential for the construction of configuration and state correlation diagrams upon which one ultimately bases a prediction about whether a reaction is allowed or forbidden. Finally, we learned how the direct-product analysis allows one to determine whether or not integrals of products of wavefunctions with operators between them vanish. This tool is of utmost importance in determining selection rules in spectroscopy and for determining the effects of external perturbations on the states of the species under investigation.

CHARACTER TABLES

C_1	E
A	1

C_s	E	σ_h		
A'	1	1	$x,y,R_z,$	x^2,y^2,z^2,xy
A''	1	−1	z,R_x,R_y	yz,xz

C_i	E	i		
A_g	1	1	R_x,R_y,R_z	x^2,y^2,z^2,xy,xz,yz
A_u	1	−1	x,y,z	

C_2	E	C_2		
A	1	1	z,R_z	x^2,y^2,z^2,xy
B	1	−1	x,y,R_x,R_y	yz,xz

D_2	E	$C_2(z)$	$C_2(y)$	$C_2(x)$		
A	1	1	1	1		x^2,y^2,z^2
B_1	1	1	−1	−1	z,R_z	xy
B_2	1	−1	1	−1	y,R_y	xz
B_3	1	−1	−1	1	x,R_x	yz

D_3	E	$2C_3$	$3C_2$		
A_1	1	1	1		x^2+y^2,z^2
A_2	1	1	−1	z,R_z	
E	2	−1	0	$(x,y)\ (R_x,R_y)$	$(x^2-y^2,xy)\ (xz,yz)$

D_4	E	$2C_4$	C_2 $(=C_4^2)$	$2C_2'$	$2C_2''$		
A_1	1	1	1	1	1		x^2+y^2,z^2
A_2	1	1	1	−1	−1	z,R_z	
B_1	1	−1	1	1	−1		x^2-y^2
B_2	1	−1	1	−1	1		xy
E	2	0	−2	0	0	$(x,y)\ (R_x,R_y)$	(xz,yz)

C_{2v}	E	C_2	$\sigma_v(xz)$	$\sigma_v'(yz)$		
A_1	1	1	1	1	z	x^2,y^2,z^2
A_2	1	1	−1	−1	R_z	xy
B_1	1	−1	1	−1	x,R_y	xz
B_2	1	−1	−1	1	y,R_x	yz

C_{3v}	E	$2C_3$	$3\sigma_v$		
A_1	1	1	1	z	x^2+y^2,z^2
A_2	1	1	−1	R_z	
E	2	−1	0	$(x,y)\ (R_x,R_y)$	$(x^2-y^2,xy)\ (xz,yz)$

C_{4v}	E	$2C_4$	C_2	$2\sigma_v$	$2\sigma_d$		
A_1	1	1	1	1	1	z	x^2+y^2,z^2
A_2	1	1	1	−1	−1	R_z	
B_1	1	−1	1	1	−1		x^2-y^2
B_2	1	−1	1	−1	1		xy
E	2	0	−2	0	0	(x,y) (R_x,R_y)	(xz,yz)

C_{2h}	E	C_2	i	σ_h		
A_g	1	1	1	1	R_z	x^2,y^2,z^2,xy
B_g	1	−1	1	−1	R_x,R_y	xz,yz
A_u	1	1	−1	−1	z	
B_u	1	−1	−1	1	x,y	

D_{2h}	E	$C_2(z)$	$C_2(y)$	$C_2(x)$	i	$\sigma(xy)$	$\sigma(xz)$	$\sigma(yz)$		
A_g	1	1	1	1	1	1	1	1		x^2,y^2,z^2
B_{1g}	1	1	−1	−1	1	1	−1	−1	R_z	xy
B_{2g}	1	−1	1	−1	1	−1	1	−1	R_y	xz
B_{3g}	1	−1	−1	1	1	−1	−1	1	R_x	yz
A_u	1	1	1	1	−1	−1	−1	−1		
B_{1u}	1	1	−1	−1	−1	−1	1	1	z	
B_{2u}	1	−1	1	−1	−1	1	−1	1	y	
B_{3u}	1	−1	−1	1	−1	1	1	−1	x	

D_{3h}	E	$2C_3$	$3C_2$	σ_h	$2S_3$	$3\sigma_v$		
A_1'	1	1	1	1	1	1		x^2+y^2,z^2
A_2'	1	1	−1	1	1	−1	R_z	
E'	2	−1	0	2	−1	0	(x,y)	(x^2-y^2,xy)
A_1''	1	1	1	−1	−1	−1		
A_2''	1	1	−1	−1	−1	1	z	
E''	2	−1	0	−2	1	0	(R_x,R_y)	(xz,yz)

D_{4h}	E	$2C_4$	C_2	$2C_2'$	$2C_2''$	i	$2S_4$	σ_h	$2\sigma_v$	$2\sigma_d$		
A_{1g}	1	1	1	1	1	1	1	1	1	1		x^2+y^2, z^2
A_{2g}	1	1	1	−1	−1	1	1	1	−1	−1	R_z	
B_{1g}	1	−1	1	1	−1	1	−1	1	1	−1		x^2-y^2
B_{2g}	1	−1	1	−1	1	1	−1	1	−1	1		xy
E_g	2	0	−2	0	0	2	0	−2	0	0	(R_x, R_y)	(xz, yz)
A_{1u}	1	1	1	1	1	−1	−1	−1	−1	−1		
A_{2u}	1	1	1	−1	−1	−1	−1	−1	1	1	z	
B_{1u}	1	−1	1	1	−1	−1	1	−1	−1	1		
B_{2u}	1	−1	1	−1	1	−1	1	−1	1	−1		
E_u	2	0	−2	0	0	−2	0	2	0	0	(x, y)	

D_{6h}	E	$2C_6$	$2C_3$	C_2	$3C_2'$	$3C_2''$	i	$2S_3$	$2S_6$	σ_h	$3\sigma_d$	$3\sigma_v$		
A_{1g}	1	1	1	1	1	1	1	1	1	1	1	1		x^2+y^2, z^2
A_{2g}	1	1	1	1	−1	−1	1	1	1	1	−1	−1	R_z	
B_{1g}	1	−1	1	−1	1	−1	1	−1	1	−1	1	−1		
B_{2g}	1	−1	1	−1	−1	1	1	−1	1	−1	−1	1		
E_{1g}	2	1	−1	−2	0	0	2	1	−1	−2	0	0	(R_x, R_y)	(xz, yz)
E_{2g}	2	−1	−1	2	0	0	2	−1	−1	2	0	0		(x^2-y^2, xy)
A_{1u}	1	1	1	1	1	1	−1	−1	−1	−1	−1	−1		
A_{2u}	1	1	1	1	−1	−1	−1	−1	−1	−1	1	1	z	
B_{1u}	1	−1	1	−1	1	−1	−1	1	−1	1	−1	1		
B_{2u}	1	−1	1	−1	−1	1	−1	1	−1	1	1	−1		
E_{1u}	2	1	−1	−2	0	0	−2	−1	1	2	0	0	(x, y)	
E_{2u}	2	−1	−1	2	0	0	−2	1	1	−2	0	0		

D_{2d}	E	$2S_4$	C_2	$2C_2'$	$2\sigma_d$		
A_1	1	1	1	1	1		x^2+y^2, z^2
A_2	1	1	1	−1	−1	R_z	
B_1	1	−1	1	1	−1		x^2-y^2
B_2	1	−1	1	−1	1	z	xy
E	2	0	−2	0	0	(x, y) (R_x, R_y)	(xz, yz)

D_{3d}	E	$2C_3$	$3C_2$	i	$2S_6$	$3\sigma_d$		
A_{1g}	1	1	1	1	1	1		x^2+y^2, z^2
A_{2g}	1	1	−1	1	1	−1	R_z	
E_g	2	−1	0	2	−1	0	(R_x,R_y)	(x^2-y^2,xy) (xz,yz)
A_{1u}	1	1	1	−1	−1	−1		
A_{2u}	1	1	−1	−1	−1	1	z	
E_u	2	−1	0	−2	1	0	(x,y)	

S_4	E	S_4	C_2	S_4^3		
A	1	1	1	1	R_z	x^2+y^2, z^2
B	1	−1	1	−1	z	x^2-y^2, xy
E	$\left\{\begin{matrix}1\\1\end{matrix}\right.$	$\begin{matrix}i\\-i\end{matrix}$	$\begin{matrix}-1\\-1\end{matrix}$	$\left.\begin{matrix}-i\\i\end{matrix}\right\}$	(x,y) (R_x,R_y)	(xz,yz)

T	E	$4C_3$	$4C_3^2$	$3C_2$		$\varepsilon = \exp(2\pi i/3)$
A	1	1	1	1		$x^2+y^2+z^2$
E	$\left\{\begin{matrix}1\\1\end{matrix}\right.$	$\begin{matrix}\varepsilon\\\varepsilon^*\end{matrix}$	$\begin{matrix}\varepsilon^*\\\varepsilon\end{matrix}$	$\left.\begin{matrix}1\\-1\end{matrix}\right\}$		$(2z^2-x^2-y^2, x^2-y^2)$
T	3	0	0	−1	(R_x,R_y,R_z) (x,y,z)	(xy,xz,yz)

T_h	E	$4C_3$	$4C_3^2$	$3C_2$	i	$4S_6$	$4S_6^5$	$3\sigma_h$		$\varepsilon = \exp(2\pi i/3)$
A_g	1	1	1	1	1	1	1	1		$x^2+y^2+z^2$
A_u	1	1	1	1	−1	−1	−1	−1		
E_g	$\left\{\begin{matrix}1\\1\end{matrix}\right.$	$\begin{matrix}\varepsilon\\\varepsilon^*\end{matrix}$	$\begin{matrix}\varepsilon^*\\\varepsilon\end{matrix}$	$\begin{matrix}1\\1\end{matrix}$	$\begin{matrix}1\\1\end{matrix}$	$\begin{matrix}\varepsilon\\\varepsilon^*\end{matrix}$	$\begin{matrix}\varepsilon^*\\\varepsilon\end{matrix}$	$\left.\begin{matrix}1\\1\end{matrix}\right\}$		$(2z^2-x^2-y^2, x^2-y^2)$
E_u	$\left\{\begin{matrix}1\\1\end{matrix}\right.$	$\begin{matrix}\varepsilon\\\varepsilon^*\end{matrix}$	$\begin{matrix}\varepsilon^*\\\varepsilon\end{matrix}$	$\begin{matrix}1\\1\end{matrix}$	$\begin{matrix}-1\\-1\end{matrix}$	$\begin{matrix}-\varepsilon\\-\varepsilon^*\end{matrix}$	$\begin{matrix}-\varepsilon^*\\-\varepsilon\end{matrix}$	$\left.\begin{matrix}-1\\-1\end{matrix}\right\}$		
T_g	3	0	0	−1	1	0	0	−1	(R_x,R_y,R_z)	
T_u	3	0	0	−1	−1	0	0	1	(x,y,z)	(xy,xz,yz)

T_d	E	$8C_3$	$3C_2$	$6S_4$	$6\sigma_d$		
A_1	1	1	1	1	1		$x^2+y^2+z^2$
A_2	1	1	1	−1	−1		
E	2	−1	2	0	0		$(2z^2-x^2-y^2, x^2-y^2)$
T_1	3	0	−1	1	−1	(R_x,R_y,R_z)	
T_2	3	0	−1	−1	1	(x,y,z)	(xy,xz,yz)

O	E	$6C_4$	$3C_2$ $(=C_4^2)$	$8C_3$	$6C_2$		
A_1	1	1	1	1	1		$x^2+y^2+z^2$
A_2	1	−1	1	1	−1		
E	2	0	2	−1	0		$(2z^2-x^2-y^2, x^2-y^2)$
T_1	3	1	−1	0	−1	(R_x,R_y,R_z) (x,y,z)	
T_2	3	−1	−1	0	1		(xy,xz,yz)

O_h	E	$8C_3$	$6C_2$	$6C_4$	$3C_2$ $(=C_4^2)$	i	$6S_4$	$8S_6$	$3\sigma_h$	$6\sigma_d$		
A_{1g}	1	1	1	1	1	1	1	1	1	1		$x^2+y^2+z^2$
A_{2g}	1	1	−1	−1	1	1	−1	1	1	−1		
E_g	2	−1	0	0	2	2	0	−1	2	0		$(2z^2-x^2-y^2,$ $x^2-y^2)$
T_{1g}	3	0	−1	1	−1	3	1	0	−1	−1	(R_x,R_y,R_z)	
T_{2g}	3	0	1	−1	−1	3	−1	0	−1	1		(xy,xz,yz)
A_{1u}	1	1	1	1	1	−1	−1	−1	−1	−1		
A_{2u}	1	1	−1	−1	1	−1	1	−1	−1	1		
E_u	2	−1	0	0	2	−2	0	1	−2	0		
T_{1u}	3	0	−1	1	−1	−3	−1	0	1	1	(x,y,z)	
T_{2u}	3	0	1	−1	−1	−3	1	0	1	−1		

$C_{\infty v}$	E	$2C_\infty^\Phi$	. . .	$\infty\sigma_v$		
$A_1 \equiv \Sigma^+$	1	1	. . .	1	z	x^2+y^2, z^2
$A_2 \equiv \Sigma^-$	1	1	. . .	−1	R_z	
$E_1 \equiv \Pi$	2	$2\cos\Phi$	. . .	0	(x,y) (R_x,R_y)	(xz,yz)
$E_2 \equiv \Delta$	2	$2\cos 2\Phi$	. . .	0		(x^2-y^2,xy)
$E_3 \equiv \Phi$	2	$2\cos 3\Phi$	. . .	0		
. . .	. . .	. . .	. . .	. . .		

$D_{\infty h}$	E	$2C_{\infty}^{\Phi}$	...	$\infty\sigma_v$	i	$2S_{\infty}^{\Phi}$	...	∞C_2		
Σ_g^+	1	1	...	1	1	1	...	1		x^2+y^2, z^2
Σ_g^-	1	1	...	−1	1	1	...	−1	R_z	
Π_g	2	$2\cos\Phi$	...	0	2	$-2\cos\Phi$	...	0	(R_x, R_y)	(xz, yz)
Δ_g	2	$2\cos 2\Phi$	...	0	2	$2\cos 2\Phi$	...	0		(x^2-y^2, xy)
...	...	...	...	...	...	...	...	...		
Σ_u^+	1	1	...	1	−1	−1	...	−1	z	
Σ_u^-	1	1	...	−1	−1	−1	...	1		
Π_u	2	$2\cos\Phi$	...	0	−2	$2\cos\Phi$	...	0	(x, y)	
Δ_u	2	$2\cos 2\Phi$	...	0	−2	$-2\cos 2\Phi$	...	0		
...	...	...	...	...	...	...	...	...		

APPENDIX

F Qualitative Orbital Picture and Semi-Empirical Methods

Some of the material contained in the early parts of this appendix appears, in condensed form, near the end of chapter 7. For the sake of completeness and clarity of presentation, this material is repeated and enhanced here.

In the simplest picture of chemical bonding, the valence molecular orbitals ϕ_i are constructed as linear combinations of valence atomic orbitals χ_μ according to the LCAO-MO formula:

$$\phi_i = \sum_\mu C_{i\mu}\chi_\mu.$$

The core electrons are not explicitly included in such a treatment, although their effects are felt through an electrostatic potential V as detailed below. The electrons that reside in the occupied MO's are postulated to experience an effective potential V that has the following properties:

1. V contains contributions from all of the nuclei in the molecule exerting coulombic attractions on the electron, as well as coulombic repulsions and exchange interactions exerted by the other electrons on this electron;
2. As a result of the (assumed) cancellation of electron clouds (i.e., the core, lone-pair, and valence electron clouds (i.e., the core, lone-pair, and valence orbitals) that surround these distant nuclei, the effect of V on any particular MO ϕ_i depends primarily on the atomic charges and local bond polarities of the atoms over which ϕ_i is delocalized.

As a result of these assumptions, qualitative molecular orbital models can be developed in which one assumes that each MO ϕ_i obeys a one-electron Schrödinger equation

$$h\phi_i = \varepsilon_i\phi_i.$$

Here the orbital-level Hamiltonian h contains the kinetic energy of motion of the electron and the potential V mentioned above:

$$[-\hbar^2/2m_e\nabla^2 + V]\phi_i = \varepsilon_i\phi_i.$$

Expanding the MO ϕ_i in the LCAO-MO manner, substituting this expansion into the above Schrödinger equation, multiplying on the left by χ_ν^*, and integrating over the coordinates of the electron generates the following orbital-level eigenvalue problem:

$$\sum_\mu <\chi_\nu| - \hbar^2/2m_e\nabla^2 + V|\chi_\mu> C_{i\mu} = \varepsilon_i \sum_\mu <\chi_\nu|\chi_\mu> C_{i\mu}.$$

If the constituent atomic orbitals $\{\chi_\mu\}$ have been orthonormalized as discussed in chapter 7, the overlap integrals $<\chi_\nu|\chi_\mu>$ reduce to $\delta_{\mu,\nu}$.

In all semi-empirical models, the quantities that explicitly define the potential V are not computed from first principles as they are in so-called *ab initio* methods. Rather, either experimental data or results of *ab initio* calculations are used to determine the parameters in terms of which V is expressed. The various semi-empirical methods discussed below differ in the sophistication used to include electron-electron interactions as well as in the manner experimental data or *ab initio* computational results are used to specify V.

If experimental data is used to parameterize a semi-empirical model, then the model should not be extended beyond the level at which it has been parameterized. For example, orbitals obtained from a semi-empirical theory parameterized such that bond energies, excitation energies, and ionization energies predicted by theory agree with experimental data should not subsequently be used in a configuration interaction (CI) calculation. To do so would be inconsistent because the CI treatment, which is designed to treat dynamical correlations among the electrons, would duplicate what using the experimental data (which already contains mother nature's electronic correlations) to determine the parameters had accomplished.

Alternatively, if results of *ab initio* theory at the single-configuration orbital level are used to define the parameters of a semi-empirical model, it would be proper to use the semi-empirical orbitals in a subsequent higher-level treatment of electron correlations.

A. The Hückel Model

In the most simplified embodiment of the above orbital-level model, the following additional approximations are introduced:

1. The diagonal values $<\chi_\mu| - \hbar^2/2m_e\nabla^2 + V|\chi_\mu>$, which are usually denoted α_μ, are taken to be equal to the energy of an electron in the atomic orbital χ_μ and, as such, are evaluated in terms of atomic ionization energies (IPs) and electron affinities (EAs):

$$<\chi_\mu| - \hbar^2/2m_e\nabla^2 + V|\chi_\mu> = -\mathrm{IP}_\mu,$$

for atomic orbitals that are occupied in the atom, and

$$<\chi_\mu| - \hbar^2/2m_e\nabla^2 + V|\chi_\mu> = -\mathrm{EA}_\mu,$$

for atomic orbitals that are not occupied in the atom.

These approximations assume that contributions in V arising from coulombic attraction to nuclei other than the one on which χ_μ is located, and repulsions from the core, lone-pair, and valence electron clouds surrounding these other nuclei cancel to an extent that $<\chi_\mu|V|\chi_\mu>$ contains only potentials from the atom on which χ_μ sits.

It should be noted that the IPs and EAs of valence-state orbitals are not identical to the experimentally measured IPs and EAs of the corresponding atom, but can be obtained from such information. For example, the $2p$ valence-state IP (VSIP) for a Carbon atom is the energy difference associated with the hypothetical process

$$C(1s^2 2s 2p_x 2p_y 2p_z) ==> C^+(1s^2 2s 2p_x 2p_y).$$

If the energy differences for the "promotion" of C

$$C(1s^2 2s^2 2p_x 2p_y) ==> C(1s^2 2s 2p_x 2p_y 2p_z);\ \Delta E_C$$

and for the promotion of C^+

$$C^+(1s^2 2s^2 2p_x) ==> C^+(1s^2 2s 2p_x 2p_y);\ \Delta E_{C+}$$

are known, the desired VSIP is given by:

$$\mathrm{IP}_{2p_z} = \mathrm{IP}_C + \Delta E_C + -\Delta E_C.$$

The EA of the $2p$ orbital is obtained from the

$$C(1s^2 2s^2 2p_x 2p_y) ==> C^-(1s^2 2s^2 2p_x 2p_y 2p_z)$$

energy gap, which means that $\mathrm{EA}_{2p_z} = \mathrm{EA}_C$. Some common IPs of valence $2p$ orbitals in eV are as follows: C (11.16), N (14.12), N^+ (28.71), O (17.70), O^+ (31.42), F^+ (37.28).

2. The off-diagonal elements $<\chi_\nu| - \hbar^2/2m_e\nabla^2 + V|\chi_\mu>$ are taken as zero if χ_μ and χ_ν belong to the same atom because the atomic orbitals are assumed to have been constructed to diagonalize the one-electron hamiltonian appropriate to an electron moving in that atom. They are set equal to a parameter denoted $\beta_{\mu,\nu}$ if χ_μ and χ_ν reside on neighboring atoms that are chemically bonded. If χ_μ and χ_ν reside on atoms that are not bonded neighbors, then the off-diagonal matrix element is set equal to zero.

3. The geometry dependence of the $\beta_{\mu,\nu}$ parameters is often approximated by assuming that $\beta_{\mu,\nu}$ is proportional to the overlap $S_{\mu,\nu}$ between the corresponding atomic orbitals:

$$\beta_{\mu,\nu} = \beta^o_{\mu,\nu} S_{\mu,\nu}\ .$$

Here $\beta^o_{\mu,\nu}$ is a constant (having energy units) characteristic of the bonding interaction between χ_μ and χ_ν; its value is usually determined by forcing the molecular orbital energies obtained from such a qualitative orbital treatment to yield experimentally correct ionization potentials, bond dissociation energies, or electronic transition energies.

The particular approach described thus far forms the basis of the so-called *Hückel model*. Its implementation requires knowledge of the atomic α_μ and $\beta^0_{\mu,\nu}$ values, which are eventually expressed in terms of experimental data, as well as a means of calculating the geometry dependence of the $\beta_{\mu,\nu}$'s (e.g., some method for computing overlap matrices $S_{\mu,\nu}$).

B. The Extended Hückel Method

It is well known that bonding and antibonding orbitals are formed when a pair of atomic orbitals from neighboring atoms interact. The energy splitting between the bonding and antibonding orbitals depends on the overlap between the pair of atomic orbitals. Also, the energy of the antibonding orbital lies higher above the arithmetic mean $E_{ave} = E_A + E_B$ of the energies of the constituent atomic orbitals (E_A and E_B) than the bonding orbital lies below E_{ave}. If overlap is ignored, as in conventional Hückel theory (except in parameterizing the geometry dependence of $\beta_{\mu,\nu}$), the differential destabilization of antibonding orbitals compared to stabilization of bonding orbitals can not be accounted for.

By parameterizing the off-diagonal Hamiltonian matrix elements in the following overlap-dependent manner:

$$h_{\nu,\mu} = <\chi_\nu| -\hbar^2/2m_e\nabla^2 + V|\chi_\mu> = 0.5K(h_{\mu,\mu} + h_{\nu,\nu})S_{\mu,\nu},$$

and explicitly treating the overlaps among the constituent atomic orbitals $\{\chi_\mu\}$ in solving the orbital-level Schrödinger equation

$$\sum_\mu <\chi_\nu| -\hbar^2/2m_e\nabla^2 + V|\chi_\mu> C_{i\mu} = \varepsilon_i \sum_\mu <\chi_\nu|\chi_\mu> C_{i\mu},$$

Hoffmann introduced the so-called extended Hückel method. He found that a value for $K = 1.75$ gave optimal results when using Slater-type orbitals as a basis (and for calculating the $S_{\mu,\nu}$). The diagonal $h_{\mu,\mu}$ elements are given, as in the conventional Hückel method, in terms of valence-state IPs and EAs. Cusachs later proposed a variant of this parameterization of the off-diagonal elements:

$$h_{\nu,\mu} = 0.5\ K\ (h_{\mu,\mu} + h_{\nu,\nu})\ S_{\mu,\nu}\ (2-|S_{\mu,\nu}|)\ .$$

For first- and second-row atoms, the 1s or (2s, 2p) or (3s, 3p, 3d) valence-state ionization energies (α_μ's), the number of valence electrons (#Elec.) as well as the orbital exponents (e_s, e_p, and e_d) of Slater-type orbitals used to calculate the overlap matrix elements $S_{\mu,\nu}$ corresponding are given below.

Atom	# Elec.	$e_s = e_p$	e_d	α_s(eV)	α_p(eV)	α_d(eV)
H	1	1.3		−13.6		
Li	1	0.650		−5.4	−3.5	
Be	2	0.975		−10.0	−6.0	
B	3	1.300		−15.2	−8.5	
C	4	1.625		−21.4	−11.4	
N	5	1.950		−26.0	−13.4	
O	6	2.275		−32.3	−14.8	
F	7	2.425		−40.0	−18.1	
Na	1	0.733		−5.1	−3.0	
Mg	2	0.950		−9.0	−4.5	
Al	3	1.167		−12.3	−6.5	
Si	4	1.383	1.383	−17.3	−9.2	−6.0
P	5	1.600	1.400	−18.6	−14.0	−7.0
S	6	1.817	1.500	−20.0	−13.3	−8.0
Cl	7	2.033	2.033	−30.0	−15.0	−9.0

In the Hückel or extended Hückel methods no *explicit* reference is made to electron-electron interactions although such contributions are absorbed into the V potential, and hence into the α_μ and $\beta_{\mu,\nu}$ parameters of Hückel theory or the $h_{\mu,\mu}$ and $h_{\mu,\nu}$ parameters of extended Hückel theory.

As electron density flows from one atom to another (due to electronegativity differences), the electron-electron repulsions in various atomic orbitals changes. To account for such charge-density-dependent coulombic energies, one must use an approach that includes explicit reference to interorbital coulomb and exchange interactions. There exists a large family of semi-empirical methods that permit explicit treatment of electronic interactions; some of the more commonly used approaches are discussed below.

C. Semi-Empirical Models that Treat Electron-Electron Interactions

1. The ZDO Approximation

Most methods of this type are based on the so-called zero-differential overlap (ZDO) approximation. Their development begins by using an approximation to the atomic-orbital-based two-electron integrals introduced by Mulliken:

$$<\chi_a\chi_b|g|\chi_c\chi_d> = S_{a,c}S_{b,d}\{\lambda_{a,b} + \lambda_{a,d} + \lambda_{c,b} + \lambda_{c,d}\}/4,$$

where $S_{a,c}$ is the overlap integral between χ_a and χ_c, and

$$\lambda_{a,b} = <\chi_a\chi_b|g|\chi_a\chi_b>$$

is the *coulomb integral* between the charge densities $|\chi_a|^2$ and $|\chi_b|^2$. Then, when the so-called zero-overlap approximation

$$S_{a,c} = \delta_{a,c}$$

is made, the general four-orbital two-electron integral given above reduces to its *zero-differential overlap* value:

$$<\chi_a\chi_b|g|\chi_c\chi_d> = \delta_{a,c}\delta_{b,d}\lambda_{a,b}.$$

This fundamental approximation allows the two-electron integrals that enter into the expression for the Fock matrix elements to be expressed in terms of the set of two-orbital coulomb interaction integrals $\lambda_{a,b}$ as well as experimental or *ab initio* values for valence-state IPs and EAs, as is now illustrated.

2. Resulting Fock Matrices

Using the ZDO approximation, the Fock matrix elements over the valence atomic orbitals (the cores are still treated through an effective electrostatic potential as above)

$$F_{\mu,\nu} = <\chi_\mu|h|\chi_\nu> + \sum_{\delta,\kappa} [\gamma_{\delta,\kappa}<\chi_\mu\chi_\delta|g|\chi_\nu\chi_\kappa> - \gamma_{\delta,\kappa}^{ex}<\chi_\mu\chi_\delta|g|\chi_\kappa\chi_\nu>],$$

reduce, to:

$$F_{\mu,\mu} = <\chi_\mu|h|\chi_\mu> + \sum_{\varepsilon} \gamma_{\varepsilon,\varepsilon}\lambda_{\mu,\varepsilon} - \gamma_{\mu,\mu}^{ex}\lambda_{\mu,\mu},$$

for the diagonal elements and

$$F_{\mu,\nu} = <\chi_\mu|h|\chi_\nu> - \gamma_{\mu,\nu}^{ex}\lambda_{\mu,\nu}$$

for the off-diagonal elements. Here, h represents the kinetic energy $-\hbar^2\nabla^2/2m$ operator plus the sum of the attractive coulombic potential energies to each of the nuclei

$$-\sum_a Z_a e^2/|r - R_a|$$

and the electrostatic repulsions of the core electrons (i.e., all those not explicitly treated as valence in this calculation) around each of the nuclei.

Further reduction of the *diagonal* $F_{\mu,\mu}$ expression is achieved by:

a. Combining terms in the sum

$$\sum_e$$

involving orbitals χ_ε on atomic centers other than where χ_μ sits (atom a) together with the sum of coulomb attractions (which appear in h) over these same centers:

$$\sum_{\varepsilon \text{ (not on atom } a)} \gamma_{\varepsilon,\varepsilon}\lambda_{\mu,\varepsilon} \quad -\sum_{b \text{ (not on atom } a)} \langle\chi_\mu|Z_b e^2|r - R_b|^{-1}|\chi_\mu\rangle \quad = \sum_{b,\,\varepsilon \text{ (not center } a)} (\gamma_{\varepsilon,\varepsilon}\lambda_{\mu,\varepsilon} - V_{\mu,b}).$$

This combination represents the net coulombic interaction of $|\chi_\mu|^2$ with the total electron density (first sum) and the total attractive positive density (second sum) on atoms other than the atom on which χ_μ sits.

b. Recognizing

$$\langle\chi_\mu| - \hbar^2\nabla^2/2m|\chi_\mu\rangle + \sum_{\varepsilon\neq\mu \text{ (on center } a)} \gamma_{\varepsilon,\varepsilon}\lambda_{\mu,\varepsilon} - \langle\chi_\mu|Z_a e^2/|r - R_a|\,|\chi_\mu\rangle = U_{\mu,\mu}$$

as the average value of the atomic Fock operator (i.e., kinetic energy plus attractive colomb potential to that atom's nucleus plus coulomb and exchange interactions with other electrons on that atom) for an electron in χ_μ on the nucleus a. As in Hückel theory, the values of these parameters $U_{\mu,\mu}$, which play the role of the Hückel α_μ, can be determined from atomic valence-state ionization potentials and electron affinities. These quantities, in turn, may be obtained either from experimental data or from the results of *ab initio* calculations.

As a result, the diagonal F matrix elements are given by

$$F_{\mu,\mu} = U_{\mu,\mu} + (\gamma_{\mu,\mu} - \gamma^{ex}_{\mu,\mu})\lambda_{\mu,\mu} + \sum_{b,\varepsilon \text{ (not center } a)} (\gamma_{\varepsilon,\varepsilon}\lambda_{\mu,e} - V_{\mu,b}).$$

The evaluation of the quantities entering into this expression and that for the off-diagonal $F_{\mu,\nu}$ elements differs from one semi-empirical method to another; this topic is covered late in this appendix.

Reduction of the off-diagonal elements involving orbitals χ_μ and χ_ν on the *same* atom (a) is achieved by assuming that the atomic orbitals have been formed in a manner that makes the contributions to $F_{\mu,\nu}$ from atom a vanish

$$<\chi_\mu| - \hbar^2\nabla^2/2m|\chi_\nu> - <\chi_\mu|Z_a e^2/|r - R_a|\,|\chi_\nu> - \gamma_{\mu,\nu}{}^{ex}\lambda_{\mu,\nu} + \sum_{e\ (\text{on atom } a)} \gamma_{\varepsilon,\varepsilon}<\chi_\mu\chi_\varepsilon|g|\chi_\nu\chi_\varepsilon> = 0,$$

and then neglecting, to be consistent with the ZDO assumption, the contributions from atoms other than atom a

$$-\sum_{b\ (\text{not center } a)} <\chi_\mu|Z_b e^2/|r - R_b||\chi_\nu> + \sum_{\varepsilon\ (\text{not on atom } a)} \gamma_{\varepsilon,\varepsilon} <\chi_\mu\chi_\varepsilon|g|\chi_\nu\chi_\varepsilon> = 0.$$

Hence, the off-diagonal F matrix elements vanish, $F_{\mu,\nu} = 0$ for χ_μ and χ_ν both on the same atom (a).

The off-diagonal F matrix elements coupling orbitals from different atoms (a and b) are expressed as

$$F_{\mu,\nu} = <\chi_\mu| - \hbar^2\nabla^2/2m|\chi_\nu> - <\chi_\mu|Z_a e^2/|r - R_a|\,|\chi_\nu>$$
$$-<\chi_\mu|Z_b e^2/|r - R_b|\,|\chi_\nu> - \gamma^{ex}_{\mu,\nu}\lambda_{\mu,\nu} = \beta_{\mu,\nu} - \gamma_{\mu,\nu}{}^{ex}\lambda_{\mu,\nu}$$

Contributions to these elements from atoms other than a and b are neglected, again to be consistent with the ZDO approximation.

Unlike the Hückel and extended Hückel methods, the semi-empirical approaches that explicitly treat electron-electron interactions give rise to Fock matrix element expressions that depend on the atomic-orbital-based density matrix $\gamma_{\mu,\nu}$. This quantity is computed using the LCAO-MO coefficients $\{C_{i,\mu}\}$ of the occupied molecular orbitals from the previous iteration of the

$$\sum_\mu F_{\nu,\mu}C_{i\mu} = \varepsilon_i \sum_\mu <\chi_\nu|\chi_\mu>\, C_{i\mu}$$

equations. In particular,

$$\gamma_{\mu,\nu} = \sum_{i(\text{occupied})} n_i C_{i,\mu}C_{i,\nu},$$

$$\gamma_{\mu,\nu}{}^{ex} = \sum_{i(\text{occupied and of spin } \sigma)} C_{i,\mu}C_{i,\nu}.$$

Here, n_i is the number (0, 1, or 2) of electrons that occupy the ith molecular orbital, and spin σ denotes the spin (α or β) of the orbital whose Fock matrix is being formed. For example, when studying doublet radicals having K doubly occupied orbitals and one half-filled orbital ($K + 1$) in which an α electron resides, σ is α. In this case, $n_i = 2$ for the first K orbitals and $n_i = 1$ for the last occupied orbital. Moreover, the Fock matrix elements $F^\beta_{\mu,\nu}$ for β orbitals contains contributions from $\gamma_{\mu,\nu}{}^{ex}$ that are of the form

$$\gamma_{\mu,\nu}{}^{ex} = \sum_{i=1,K} C_{i,\mu}C_{i,\nu},$$

while the Fock matrix elements $F^\alpha_{\mu,\nu}$ for α orbitals contains

$$\gamma_{\mu,\nu}{}^{ex} = \sum_{i=1,K} C_{i,\mu}C_{i,\nu} + C_{K+1,\mu}C_{K+1,\nu}.$$

For both $F^{\alpha}_{\mu,\nu}$ and $F^{\beta}_{\mu,\nu}$, coulomb contributions arise as

$$\gamma_{\mu,\nu} = 2 \sum_{i=1,K} C_{i,\mu}C_{i,\nu} + C_{K+1,\mu}C_{K+1,\nu}.$$

3. Various Semi-Empirical Methods

a. The Pariser-Parr-Pople (PPP) method for π-orbitals In the PPP method, only the π-orbitals and the corresponding π-electrons are considered. The parameters included in the F matrix

$$F_{\mu,\mu} = U_{\mu,\mu} + (\gamma_{\mu,\mu} - \gamma_{\mu,\mu}{}^{ex})\lambda_{\mu,\mu} + \sum_{b,\varepsilon(\text{not center } a)} (\gamma_{\varepsilon,\varepsilon}\lambda_{\mu,\varepsilon} - V_{\mu,b}).$$

$$F_{\mu,\nu} = \beta_{\mu,\nu} - \gamma_{\mu,\nu}{}^{ex}\lambda_{\mu,\nu}$$

are obtained as follows:

i. The diagonal integrals $\lambda_{a,a}$, which represent the mutual coulomb repulsions between a pair of electrons in the valence-state orbital labeled a, can be estimated, as suggested by Pariser, in terms of the valence-state IP and EA of that orbital:

$$\lambda_{a,a} = \text{IP}_a - \text{EA}_a.$$

Alternatively, these one-center coulomb integrals can be computed from first principles using Slater or Gaussian type orbitals.

ii. The off-diagonal coulomb integrals $\lambda_{a,b}$ are commonly approximated either by the Mataga-Nishimoto formula:

$$\lambda_{a,b} = e^2/(R_{a,b} + x_{a,b}),$$

where

$$x_{a,b} = 2e^2/(\lambda_{a,a} + \lambda_{b,b}),$$

or by the Dewar-Ohno-Klopman expression:

$$\lambda_{a,b} = \frac{e^2}{\sqrt{(R^2_{a,b} + 0.25e^4(1/\lambda_{a,a} + 1/\lambda_{b,b})^2)}}.$$

iii. The valence-state IP's and EA's, $U_{\mu,\mu}$ are evaluated from experimental data or from the results of *ab initio* calculations of the atomic IPs and EAs.

iv. The $\beta_{\mu,\nu}$ integrals are usually chosen to make bond lengths, bond energies, or electronic excitation energies in the molecule agree with experimental data. The geometry dependence of $\beta_{\mu,\nu}$ is often parameterized as in Hückel theory $\beta_{\mu,\nu} = \beta^{o}_{\mu,\nu}S_{\mu,\nu}$, and the overlap is then computed from first principles.

v. The

$$\sum_{b\ (\text{not center } a)} V_{\mu,b}$$

term, which represents the coulombic attraction of an electron in χ_μ to the nucleus at center b, is often approximated as $Z_b\, \lambda_{\mu,\varepsilon}$, where ε labels the one π orbital on center b, and Z_b is the number of π electrons contributed by center b. This parameterization then permits the attractive interaction for center b to be combined with the repulsive interaction to give

$$\sum_{b,\varepsilon\text{ (not center } a)} (\gamma_{\varepsilon,\varepsilon} - Z_b)\lambda_{\mu,\varepsilon}\ .$$

b. All valence electron methods The CNDO, INDO, NDDO, MNDO, and MINDO methods all are defined in terms of an orbital-level Fock matrix with elements

$$F_{\mu,\mu} = U_{\mu,\mu} + (\gamma_{\mu,\mu} - \gamma_{\mu,\mu}{}^{ex})\lambda_{\mu,\mu} + \sum_{b,\varepsilon\text{ (not center } a)} (\gamma_{\varepsilon,\varepsilon}\lambda_{\mu,\varepsilon} - V_{\mu,b}).$$

$$F_{\mu,\nu} = \beta_{\mu,\nu} - \gamma_{\mu,\nu}{}^{ex}\lambda_{\mu,\nu}.$$

They differ among one another in two ways: (i) in the degree to which they employ the ZDO approximation to eliminate two-electron integrals, and (ii) in whether they employ experimental data (MINDO, MNDO, CNDO/S) or results of *ab initio* one-electron calculations (CNDO, INDO, NDDO) to define their parameters.

The CNDO and CNDO/S methods apply the ZDO approximation to all integrals, regardless of whether the orbitals are located on the same atom or not. In the INDO method, which was designed to improve the treatment of spin densities at nuclear centers and to handle singlet-triplet energy differences for open-shell species, exchange integrals $<\chi_a\chi_b|g|\chi_b\chi_a>$ involving orbitals χ_a and χ_b on the same atom are retained. In the NDDO approach, the ZDO approximation is applied only to integrals involving orbitals on two or more different atoms; that is, all one-center integrals are retained. The text *Approximate Molecular Orbital Theory* by J. A. Pople and D. L. Beveridge, McGraw-Hill, New York (1970) gives a treatment of several of these semi-empirical methods beyond the introduction provided here.

To illustrate the differences among the various approaches and to clarify how their parameters are obtained, let us consider two specific and popular choices—CNDO/2 and MINDO.

i. The CNDO/2 and CNDO/S Models. In the CNDO/2 approach as originally implemented, *ab initio* (orbital-level) calculated values of the energies mentioned below are used in determining the requisite parameters. In the later CNDO/S method, experimental values of these energies are employed. Briefly, in any CNDO method:

1. The diagonal integrals $\lambda_{a,a}$, which represent the mutual coulomb repulsions between a pair of electrons in the valence-state orbital labeled a, are calculated in terms of the valence-state IP and EA of that orbital:

   $$\lambda_{a,a} = \mathrm{IP_a} - \mathrm{EA_a}.$$

2. The valence-state IPs and EAs, and hence the $U_{\mu,\mu}$ are evaluated from the results of *ab initio* calculations (CNDO/2) or experimental measurement (CNDO/S) of the atomic IPs and EAs. The expressions used are:

   $$-\mathrm{IP_A} = U_{a,a} + (Z_A - 1)\lambda_{a,a},$$

   for orbitals χ_a, and

   $$-\mathrm{EA_A} = U_{b,b} + Z_A\lambda_{b,b},$$

for orbitals χ_b. Here Z_A is the effective core charge of atom A (the nuclear charge minus the number of "core" electrons not explicitly treated). For first row atoms, several $U_{a,a}$ and $\lambda_{a,a}$ values are tabulated below (all quantities are in eV).

	H	Li	Be	B	C	N	O	F
$U_{s,s}$	−13.6	−5.00	−15.4	−30.37	−50.69	−70.09	−101.3	−129.5
$U_{p,p}$		−3.67	−12.28	−24.7	−41.53	−57.85	−84.28	−108.9
$\lambda_{a,a}$	12.85	3.46	5.95	8.05	10.33	11.31	13.91	15.23

3. The off-diagonal coulomb integrals $\lambda_{a,b}$ are commonly approximated either by the Mataga-Nishimoto formula:

$$\lambda_{a,b} = e^2/(R_{a,b} + x_{a,b}),$$

where

$$x_{a,b} = 2e^2/(\lambda_{a,a} + \lambda_{b,b}),$$

or by the Dewar-Ohno-Klopman expression:

$$\lambda_{a,b} = \frac{e^2}{\sqrt{(R_{a,b}^2 + 0.25e^4(1/\lambda_{a,a} + 1/\lambda_{b,b})^2)}}.$$

4. As in PPP theory, the term

$$\sum_{b,\varepsilon \text{ (not center } a)} (\gamma_{\varepsilon,\varepsilon}\lambda_{\mu,\varepsilon} - V_{\mu,b})$$

is approximated by

$$\sum_{b,\varepsilon \text{ (not center } a)} (\gamma_{\varepsilon,\varepsilon} - Z_b)\lambda_{\mu,\varepsilon},$$

where Z_b is the number of valence electrons contributed by atom b, and χ_ε is one of the valence electrons on atom b.

5. The $\beta_{\mu,\nu}$ parameters are approximated as $\beta_{\mu,\nu} = S_{\mu,\nu}(\beta_a + \beta_b)$, where $S_{\mu,\nu}$ is the overlap between the orbitals χ_μ and χ_ν, and β_a and β_b are atom-dependent parameters given below for first row atoms:

	H	Li	Be	B	C	N	O	F
β_a(eV)	−9	−9	−13	−17	−21	−25	−31	−39

ii. The INDO (and MINDO-Type) Methods. In these methods, the specification of the parameters entering into $F_{\mu,\nu}$ is carried out in the same fashion as in the CNDO/2 approach, except that:

1. The ZDO approximation is made only for two-center integrals; one-center coulomb $\lambda_{a,b} = \langle\chi_a\chi_b|g|\chi_a\chi_b\rangle$ and exchange $\lambda_{a,b}^{ex} = \langle\chi_a\chi_b|g|\chi_b\chi_a\rangle$ integrals are retained. In the INDO approach, the values of these single-atom integrals are determined by requiring the results of the calculation, performed at the Fock-like orbital level, to agree with results of *ab initio* Fock-level calculations. In the MINDO approach, experimental electronic spectra of the particular atom are used to determine these parameters. The "diagonal" values $\lambda_{a,a}$ are determined, as indicated earlier, from valence-state energies (*ab initio* for INDO and experimental for MINDO) of the atom A on which χ_a resides.
2. The values of the $U_{a,a}$ parameters are determined according to the following equations:

$$U_{s,s} = -0.5(\mathrm{IP_H} + \mathrm{EA_H}) - 0.5\lambda_{s,s}$$

for Hydrogen's $1s$ orbital;

$$U_{s,s} = -0.5(\mathrm{IP_s} + \mathrm{EA_s}) - (Z_A - 0.5)\lambda_{s,s} + 1/6(Z_A - 1.5)G^1(s,s)$$

for Boron through Fluorine's $2s$ orbitals; and

$$U_{p,p} = -0.5(\mathrm{IP_p} + \mathrm{EA_p}) - (Z_A - 0.5)\lambda_{p,p} + 2/25(Z_A - 1.5)F^2(p,p) + 1/3G^1(p,p).$$

Here, F^2 and G^1 represent the well known Slater-Condon integrals in terms of which the coulomb and exchange integrals can be expressed:

$$F^k(nl,n'l') = \int_0^\infty\int_0^\infty |R_{nl}(\boldsymbol{r})|^2\,|R_{n'l'}(\boldsymbol{r}')|^2\,2r_<^k/r_>^{k+1}\boldsymbol{r}^2\boldsymbol{r}'^2 d\boldsymbol{r}d\boldsymbol{r}'$$

$$G^k(nl,n'l') = \int_0^\infty\int_0^\infty |R_{nl}(\boldsymbol{r})R_{n'l'}(\boldsymbol{r})|^2\,2r_<^k/r_>^{k+1}\boldsymbol{r}^2\boldsymbol{r}'^2 d\boldsymbol{r}d\boldsymbol{r}'$$

and Z_A is the effective core charge (the nuclear charge minus the number of "core" electrons not explicitly treated in the calculation) of the atom A on which the orbitals in question reside. In the definitions of the integrals, $\boldsymbol{r}_<$ and $\boldsymbol{r}_>$ represent, respectively, the smaller and larger of $\boldsymbol{r}$ and $\boldsymbol{r}'$. Again, *ab initio* calculational data is used in the INDO method, and experimental data in the MINDO method to fix the parameters entering these expressions.

D. Summary

As presented, semi-empirical methods are based on a single-configuration picture of electronic structure. Extensions of such approaches to permit consideration of more than a single important configuration have been made (for excellent overviews, see *Approximate Molecular Orbital Theory* by J. A. Pople and D. L. Beveridge, McGraw-Hill, New York (1970) and *Valence Theory*, second edition, by J. N. Murrell, S. F. A. Kettle, and J. M. Tedder, John Wiley, London (1965)). Pople and co-workers preferred to use data from *ab initio* calculations in developing sets of parameters to use in such methods because they viewed semi-empirical methods as approximations to *ab initio* methods. Others use experimental data to determine parameters because they view semi-empirical methods as models of mother nature.

APPENDIX

G Angular Momentum Operator Identities

I. ORBITAL ANGULAR MOMENTUM

A particle moving with momentum $\boldsymbol{p}$ at a position $\boldsymbol{r}$ relative to some coordinate origin has so-called *orbital* angular momentum equal to $\boldsymbol{L} = \boldsymbol{r} \times \boldsymbol{p}$. The three components of this angular momentum vector in a cartesian coordinate system located at the origin mentioned above are given in terms of the cartesian coordinates of $\boldsymbol{r}$ and $\boldsymbol{p}$ as follows:

$$L_z = xp_y - yp_x,$$
$$L_x = yp_z - zp_y,$$
$$L_y = zp_x - xp_z.$$

Using the fact that the quantum mechanical coordinate operators $\{q_k\} = x, y, z$ as well as the conjugate momentum operators $\{p_j\} = p_x, p_y, p_z$ are Hermitian, it is possible to show that L_x, L_y, and L_z are also Hermitian, as they must be if they are to correspond to experimentally measurable quantities.

Using the fundamental commutation relations among the cartesian coordinates and the cartesian momenta:

$$[q_k, p_j] = q_k p_j - p_j q_k = i\hbar\, \delta_{j,k} \; (j,k = x,y,z),$$

it can be shown that the above angular momentum operators obey the following set of commutation relations:

$$[L_x, L_y] = i\hbar\, L_z,$$
$$[L_y, L_z] = i\hbar\, L_x,$$
$$[L_z, L_x] = i\hbar\, L_y.$$

Although the components of $\boldsymbol{L}$ do not commute with one another, they can be shown to commute with the operator L^2 defined by

$$L^2 = L_x^2 + L_y^2 + L_z^2.$$

This new operator is referred to as the square of the total angular momentum operator.

The commutation properties of the components of $\boldsymbol{L}$ allow us to conclude that complete sets of functions can be found that are eigenfunctions of L^2 and of one, but not more than one, component of $\boldsymbol{L}$. It is convention to select this one component as L_z, and to label the resulting simultaneous eigenstates of L^2 and L_z as $|l,m\rangle$ according to the corresponding eigenvalues:

$$L^2|l,m\rangle = \hbar^2 l(l+1)\, |l,m\rangle,\ l = 0,1,2,3, \ldots .$$
$$L_z|l,m\rangle = \hbar m\, |l,m\rangle,\ m = \pm l, \pm(l-1), \pm(l-2), \ldots \pm(l-(l-1)), 0.$$

That these eigenvalues assume the values specified in these identities is proven in considerable detail below. These eigenfunctions of L^2 and of L_z will not, in general, be eigenfunctions of either L_x or of L_y. This means that any measurement of L_x or L_y will necessarily change the wavefunction if it begins as an eigenfunction of L_z.

The above expressions for L_x, L_y, and L_z can be mapped into quantum mechanical operators by substituting x, y, and z as the corresponding coordinate operators and $-i\hbar\partial/\partial x$, $-i\hbar\partial/\partial y$, and $-i\hbar\partial/\partial z$ for p_x, p_y, and p_z, respectively. The resulting operators can then be transformed into spherical coordinates by using the techniques provided in appendix A, the results of which are:

$$L_z = -i\hbar\, \partial/\partial\varphi,$$
$$L_x = i\hbar \left\{\sin\phi\, \partial/\partial\theta + \cot\theta\, \cos\phi\partial/\partial\phi\right\},$$
$$L_y = -i\hbar \left\{\cos\phi\partial/\partial\theta - \cot\theta\, \sin\phi\, \partial/\partial\phi\right\},$$
$$L^2 = -\hbar^2 \left\{(1/\sin\theta)\partial/\partial\theta\, (sin\theta\, \partial/\partial\theta) + (1/\sin^2\theta)\, \partial^2/\partial\phi^2\right\} .$$

At this point, it should be again stressed that the above form for L^2 appears explicitly when the kinetic energy operator $-\hbar^2/2m\, \nabla^2$ is expressed in spherical coordinates; in particular, the term $L^2/2mr^2$ is what enters. This means that our study of the properties of angular momenta will also help us to understand the angular-motion components of the Hamiltonian for spherically symmetric systems (i.e., those for which the potential V contains no angle dependence, and hence for which the total angle dependence is contained in the kinetic energy term $L^2/2mr^2$).

II. PROPERTIES OF GENERAL ANGULAR MOMENTA

There are many types of angular momenta that one encounters in chemistry. Orbital angular momenta, such as that introduced above, arise in electronic motion in atoms, in atom-atom and electron-atom collisions, and in rotational motion in molecules. Intrinsic spin angular momentum is present in electrons, H^1, H^2, C^{13}, and many other nuclei. In this section, we will deal with the behavior of any and all angular momenta and their corresponding eigenfunctions.

At times, an atom or molecule contains more than one type of angular momentum. The Hamiltonian's interaction potentials present in a particular species may or may not cause these individual angular momenta to be coupled to an appreciable extent (i.e., the Hamiltonian may or may not contain terms that refer simultaneously to two or more of these angular momenta). For example, the NH^- ion, which has a $^2\Pi$ ground electronic state (its electronic configuration is $1s_N^2 2\sigma^2 3\sigma^2 2p_{\pi x}^2 2p_{py}^1$) has electronic spin, electronic orbital, and molecular rotational angular momenta. The full Hamiltonian H contains spin-orbit coupling terms that couple the electronic spin and orbital angular momenta, thereby causing them individually to not commute with H. H also contains terms that couple the ion's rotational and electronic angular momenta, thereby making these quantities no longer "good" quantum numbers (i.e., making the corresponding operators no longer commute with H).

In such cases, the eigenstates of the system can be labeled rigorously only by angular momentum quantum numbers j and m belonging to the total angular momentum $\boldsymbol{J}$. The total angular momentum of a collection of individual angular momenta is defined, component-by-component, as follows:

$$J_k = \sum_i J_k(i),$$

where k labels x, y, and z, and i labels the constituents whose angular momenta couple to produce $\boldsymbol{J}$.

For the remainder of this appendix, we will study eigenfunction-eigenvalue relationships that are characteristic of all angular momenta and which are consequences of the commutation relations among the angular momentum vector's three components. We will also study how one combines eigenfunctions of two or more angular momenta $\{\boldsymbol{J}(i)\}$ to produce eigenfunctions of the the total $\boldsymbol{J}$.

A. Consequences of the Commutation Relations

Any set of three Hermitian operators that obey

$$[J_x, J_y] = i\hbar\, J_z,$$
$$[J_y, J_z] = i\hbar\, J_x,$$
$$[J_z, J_x] = i\hbar\, J_y,$$

will be taken to define an angular momentum $\boldsymbol{J}$, whose square $J^2 = J_x^2 + J_y^2 + J_z^2$ commutes with all three of its components. It is useful to also introduce two combinations of the three fundamental operators:

$$J_\pm = J_x \pm i\, J_y,$$

and to refer to them as *raising* and *lowering operators* for reasons that will be made clear below. These new operators can be shown to obey the following commutation relations:

$$[J^2, J_\pm] = 0,$$
$$[J_z, J_\pm] = \pm\hbar\, J_\pm\ .$$

These two operators are *not* Hermitian operators (although J_x and J_y are), but they are adjoints of one another:

$$J_+^+ = J_-,$$
$$J_-^+ = J_+,$$

as can be shown using the self-adjoint nature of J_x and J_y.

Using only the above commutation properties, it is possible to prove important properties of the eigenfunctions and eigenvalues of J^2 and J_z. Let us assume that we have found a set of simultaneous eigenfunctions of J^2 and J_z; the fact that these two operators commute tells us that this is possible. Let us label the eigenvalues belonging to these functions:

$$J^2 \,|j,m\rangle = \hbar^2 f(j,m)\,|j,m\rangle,$$
$$J_z \,|j,m\rangle = \hbar\, m\,|j,m\rangle,$$

in terms of the quantities m and $f(j,m)$. Although we certainly "hint" that these quantities must be related to certain j and m quantum numbers, we have not yet proven this, although we will soon do so. For now, we view $f(j,m)$ and m simply as symbols that represent the respective eigenvalues. Because both J^2 and J_z are Hermitian, eigenfunctions belonging to different $f(j,m)$ or m quantum numbers must be orthogonal:

$$\langle j,m|j',m'\rangle = \delta_{m,m'}\,\delta_{j,j'}.$$

We now prove several identities that are needed to discover the information about the eigenvalues and eigenfunctions of general angular momenta that we are after. Later in this appendix, the essential results are summarized.

1. There is a Maximum and a Minimum Eigenvalue for J_z

Because all of the components of $\boldsymbol{J}$ are Hermitian, and because the scalar product of any function with itself is positive semi-definite, the following identity holds:

$$\langle j,m|J_x^2 + J_y^2|j,m\rangle = \langle J_x\langle j,m|J_x|j,m\rangle + \langle J_y\langle j,m|J_y|j,m\rangle \geq 0.$$

However, $J_x^2 + J_y^2$ is equal to $J^2 - J_z^2$, so this inequality implies that

$$\langle j,m|J^2 - J_z^2|j,m\rangle = \hbar^2 \{f(j,m) - m^2\} \geq 0,$$

which, in turn, implies that m^2 must be less than or equal to $f(j,m)$. Hence, for any value of the total angular momentum eigenvalue f, the z-projection eigenvalue (m) must have a maximum and a minimum value and both of these must be less than or equal to the total angular momentum squared eigenvalue f.

2. The Raising and Lowering Operators Change the J_z Eigenvalue but not the J^2 Eigenvalue When Acting on $|j,m\rangle$

Applying the commutation relations obeyed by $J_\pm$ to $|j,m\rangle$ yields another useful result:

$$J_z\, J_\pm\,|j,m\rangle - J_\pm\, J_z\,|j,m\rangle = \pm\hbar\, J_\pm\,|j,m\rangle,$$
$$J^2\, J_\pm\,|j,m\rangle - J_\pm\, J^2\,|j,m\rangle = 0.$$

Now, using the fact that $|j,m\rangle$ is an eigenstate of J^2 and of J_z, these identities give

$$J_z\, J_\pm\,|j,m\rangle = (m\hbar \pm\hbar)\, J_\pm\,|j,m\rangle = \hbar(m \pm 1)\,|j,m\rangle,$$
$$J^2\, J_\pm\,|j,m\rangle = \hbar^2 f(j,m)\, J_\pm\,|j,m\rangle.$$

These equations prove that the functions $J_\pm\,|j,m\rangle$ must either themselves be eigenfunctions of J^2 and J_z, with eigenvalues $\hbar^2 f(j,m)$ and $\hbar\,(m+1)$ or $J_\pm\,|j,m\rangle$ must equal zero. In the former case, we see that $J_\pm$ acting on $|j,m\rangle$ generates a new eigenstate with the same J^2 eigenvalue as $|j,m\rangle$ but with one unit of $\hbar$ higher in J_z eigenvalue. It is for this reason that we call $J_\pm$ raising and lowering operators. Notice that, although $J_\pm |j,m>$ is indeed an eigenfunction of J_z with eigenvalue $(m \pm 1)\,\hbar$, $J_\pm |j,m>$ is not identical to the normalized $|j,m \pm 1>$; it is only proportional to $|j,m \pm 1>$:

$$J_\pm |j,m> = C^\pm_{j,m} |j,m \pm 1>.$$

Explicit expressions for these $C^\pm_{j,m}$ coefficients will be obtained below. Notice also that because the $J_\pm |j,m>$, and hence $|j,m \pm 1>$, have the same J^2 eigenvalue as $|j,m>$ (in fact, sequential application of $J_\pm$ can be used to show that all $|j,m'>$, for all m', have this same J^2 eigenvalue), the J^2 eigenvalue $f(j,m)$ must be independent of m. For this reason, f can be labeled by one quantum number j.

3. The J^2 Eigenvalues are Related to the Maximum and Minimum J_z Eigenvalues Which Are Related to One Another

Earlier, we showed that there exists a maximum and a minimum value for m, for any given total angular momentum. It is when one reaches these limiting cases that $J_\pm |j,m> = 0$ applies. In particular,

$$J_+ |j,m_{\text{max}}> = 0,$$
$$J_- |j,m_{\text{min}}> = 0.$$

Applying the following identities:

$$J_- J_+ = J^2 - J_z^2 - \hbar J_z,$$
$$J_+ J_- = J^2 - J_z^2 + \hbar J_z,$$

respectively, to $|j,m_{\text{max}}>$ and $|j,m_{\text{min}}>$ gives

$$\hbar^2 \left\{ f(j,m_{\text{max}}) - m_{\text{max}}^2 - m_{\text{max}} \right\} = 0,$$
$$\hbar^2 \left\{ f(j,m_{\text{min}}) - m_{\text{min}}^2 + m_{\text{min}} \right\} = 0,$$

which immediately gives the J^2 eigenvalue $f(j,m_{\text{max}})$ and $f(j,m_{\text{min}})$ in terms of m_{max} or m_{min}:

$$f(j,m_{\text{max}}) = m_{\text{max}}(m_{\text{max}} + 1)\ ,$$
$$f(j,m_{\text{min}}) = m_{\text{min}}(m_{\text{min}} - 1).$$

So, we now know the J^2 eigenvalues for $|j,m_{\text{max}}>$ and $|j,m_{\text{min}}>$. However, we earlier showed that $|j,m>$ and $|j,m-1>$ have the same J^2 eigenvalue (when we treated the effect of $J_\pm$ on $|j,m>$) and that the J^2 eigenvalue is independent of m. If we therefore define the quantum number j to be m_{max}, we see that the J^2 eigenvalues are given by

$$J^2 |j,m> = \hbar^2 j(j+1) |j,m>.$$

We also see that

$$f(j,m) = j(j+1) = m_{\text{max}}(m_{\text{max}} + 1) = m_{\text{min}}(m_{\text{min}} - 1),$$

from which it follows that

$$m_{\text{min}} = -m_{\text{max}}.$$

4. The j Quantum Number Can Be Integer or Half-Integer

The fact that the m-values run from j to $-j$ in unit steps (because of the property of the $J_\pm$ operators), there clearly can be only integer or half-integer values for j. In the former case, the m

quantum number runs over $-j, -j+1, -j+2, \ldots, -j+(j-1), 0, 1, 2, \ldots j$; in the latter, m runs over $-j, -j+1, -j+2, \ldots -j+(j-1/2), 1/2, 3/2, \ldots j$. Only integer and half-integer values can range from j to $-j$ in steps of unity. Species with integer spin are known as Bosons and those with half-integer spin are called Fermions.

5. More on $J_\pm$ $|j,m>$

Using the above results for the effect of $J_\pm$ acting on $|j,m>$ and the fact that J_+ and J_- are adjoints of one another, allows us to write:

$$\begin{aligned} <j,m|\, J_- J_+\, |j,m> &= <j,m|\, (J^2 - J_z^2 - \hbar J_z)\, |j,m> \\ &= \hbar^2 \{j(j+1) - m(m+1)\} = <J_+ <j,m|\, J_+\, |j,m> = (C^+_{j,m})^2, \end{aligned}$$

where $C^+_{j,m}$ is the proportionality constant between $J_+ |j,m>$ and the normalized function $|j,m+1>$. Likewise, the effect of J_- can be

$$\begin{aligned} &<j,m|\, J_+ J_-\, |j,m> = <j,m|\, J_+\, (J^2 - J_z^2 + \hbar J_z)\, |j,m> \\ &\hbar^2 \{j(j+1) - m(m-1)\} = <J_- <j,m|\, J_-\, |j,m> = (C^-_{j,m})^2, \end{aligned}$$

where $C^-_{j,m}$ is the proportionality constant between $J_- |j,m>$ and the normalized $|j,m-1>$. Thus, we can solve for $C^\pm_{j,m}$ after which the effect of $J_\pm$ on $|j,m>$ is given by:

$$J_\pm |j,m> = \hbar \{j(j+1) - m(m\pm 1)\}^{1/2} |j,m\pm 1>.$$

B. Summary

The above results apply to *any* angular momentum operators. The essential findings can be summarized as follows:

1. J^2 and J_z have complete sets of simultaneous eigenfunctions. We label these eigenfunctions $|j,m>$; they are orthonormal in both their m- and j-type indices: $<j,m|\, j',m'> = \delta_{m,m'}\, \delta_{j,j'}$.
2. These $|j,m>$ eigenfunctions obey:

$$\begin{aligned} &J^2 |j,m> = \hbar^2 j(j+1) |j,m>, \{j = \text{integer or half integer}\}, \\ &J_z |j,m> = \hbar\, m\, |j,m>, \{m = -j, \text{ in steps of 1 to } +j\}. \end{aligned}$$

3. The raising and lowering operators $J_\pm$ act on $|j,m>$ to yield functions that are eigenfunctions of J^2 with the same eigenvalue as $|j,m>$ and eigenfunctions of J_z with eigenvalue of $(m\pm 1)\hbar$:

$$J_\pm |j,m> = \hbar \{j(j+1) - m(m\pm 1)\}^{1/2} |j,m\pm 1>.$$

4. When $J_\pm$ acts on the "extremal" states $|j,j>$ or $|j,-j>$, respectively, the result is zero.

The results given above are, as stated, general. Any and all angular momenta have quantum mechanical operators that obey these equations. It is convention to designate specific kinds of angular momenta by specific letters; however, it should be kept in mind that no matter what letters are used, there are operators corresponding to J^2, J_z, and $J_\pm$ that obey relations as specified above, and there are eigenfunctions and eigenvalues that have all of the properties obtained above. For electronic or collisional orbital angular momenta, it is common to use L^2 and L_z; for electron spin, S^2 and S_z are used; for nuclear spin I^2 and I_z are most common; and for molecular rotational angular momentum, N^2 and N_z are most common (although sometimes J^2 and J_z may be used).

Whenever two or more angular momenta are combined or coupled to produce a "total" angular momentum, the latter is designated by J^2 and J_z.

III. COUPLING OF ANGULAR MOMENTA

If the Hamiltonian under study contains terms that couple two or more angular momenta $\boldsymbol{J}(i)$, then only the components of the total angular momentum

$$\boldsymbol{J} = \sum_i \boldsymbol{J}(i)$$

and $\boldsymbol{J}^2$ will commute with H. It is therefore essential to label the quantum states of the system by the eigenvalues of $\boldsymbol{J}_z$ and $\boldsymbol{J}^2$ and to construct variational trial or model wavefunctions that are eigenfunctions of these total angular momentum operators. The problem of angular momentum coupling has to do with how to combine eigenfunctions of the uncoupled angular momentum operators, which are given as simple products of the eigenfunctions of the individual angular momenta $\prod_i |j_i,m_i\rangle$, to form eigenfunctions of J^2 and J_z.

A. Eigenfunctions of J_z

Because the individual elements of $\boldsymbol{J}$ are formed additively, but J^2 is *not*, it is straightforward to form eigenstates of

$$J_z = \sum_i J_z(i)\ ;$$

simple products of the form $\prod_i |j_i,m_i\rangle$ are eigenfunctions of J_z:

$$J_z \prod_i |j_i,m_i\rangle = \sum_k J_z(k) \prod_i |j_i,m_i\rangle = \sum_k \hbar\, m_k \prod_i |j_i,m_i\rangle,$$

and have J_z eigenvalues equal to the sum of the individual $m_k\hbar$ eigenvalues. Hence, to form an eigenfunction with specified J and M eigenvalues, one must combine only those product states $\prod_i |j_i,m_i\rangle$ whose $m_i h$ sum is equal to the specified M value.

B. Eigenfunctions of J^2; the Clebsch-Gordon Series

The task is then reduced to forming eigenfunctions $|J,M\rangle$, given particular values for the $\{j_i\}$ quantum numbers (e.g., to couple the ^3P states of the Si atom, which are eigenfunctions of L^2 and of S^2, to produce a $^3P_1$ state which is an eigenfunction of J^2, where $J = L + S$). When coupling pairs of angular momenta $\{|j,m\rangle \text{ and } |j',m'\rangle\}$, the total angular momentum states can be written, according to what we determined above, as

$$|J,M\rangle = \sum_{m,m'} C^{J,M}_{j,m;j',m'}\, |j,m\rangle\, |j',m'\rangle,$$

where the coefficients $C^{J,M}_{j,m;j',m'}$ are called vector coupling coefficients (because angular momentum coupling is viewed much like adding two vectors $\boldsymbol{j}$ and $\boldsymbol{j}'$ to produce another vector $\boldsymbol{J}$), and where the sum over m and m' is restricted to those terms for which $m + m' = M$. It is more common to express the vector coupling or so-called **Clebsch-Gordon (CG) coefficients** as $\langle j,m;j'm'|J,M\rangle$ and to view them as elements of a "matrix" whose columns are labeled by the coupled-state J,M quantum numbers and whose rows are labeled by the quantum numbers characterizing the uncoupled "product basis" $j,m;j',m'$. It turns out (see chapter 2 of *Angular Momentum*, by R. N. Zare, John Wiley and Sons, New York, N.Y., (1988)) that this matrix can be shown to be unitary so that the CG coefficients obey:

$$\sum_{m,m'} <j,m;j'm'|J,M>^* <j,m;j'm'|J',M'> = \delta_{J,J'}\, \delta_{M,M'}$$

and

$$\sum_{J,M} <j,n;j'n'|J,M> <j,m;j'm'|J,M>^* = \delta_{n,m}\, \delta_{n',m'}.$$

This unitarity of the CG coefficient matrix allows the inverse of the relation giving coupled functions in terms of the product functions:

$$|J,M> = \sum_{m,m'} <j,m;j'm'|J,M> |j,m>|j',m'>$$

to be written as:

$$|j,m> |j',m'> = \sum_{J,M} <j,m;j'm'|J,M>^* |J,M>$$

$$= \sum_{J,M} <J,M|j,m;j'm'> |J,M>.$$

This result expresses the product functions in terms of the coupled angular momentum functions.

C. Generation of the CG Coefficients

The CG coefficients can be generated in a systematic manner; however, they can also be looked up in books where they have been tabulated (e.g., see Table 2.4 of Zare's book on angular momentum; the reference is given above). Here, we will demonstrate the technique by which the CG coefficients can be obtained, but we will do so for rather limited cases and refer the reader to more extensive tabulations.

The strategy we take is to generate the $|J,J>$ state (i.e., the state with maximum M-value) and to then use J_- to generate $|J,J-1>$, after which the state $|J-1,J-1>$ (i.e., the state with one lower J-value) is constructed by finding a combination of the product states in terms of which $|J,J-1>$ is expressed (because both $|J,J-1>$ and $|J-1,J-1>$ have the same M-value $M = J-1$), which is orthogonal to $|J,J-1>$ (because $|J-1,J-1>$ and $|J,J-1>$ are eigenfunctions of the Hermitian operator J^2 corresponding to different eigenvalues, they must be orthogonal). This same process is then used to generate $|J,J-2>$ $|J-1,J-2>$ and (by orthogonality construction) $|J-2,J-2>$, and so on.

1. The States with Maximum and Minimum M-Values

We begin with the state $|J,J>$ having the highest M-value. This state must be formed by taking the highest m and the highest m' values (i.e., $m = j$ and $m' = j'$), and is given by:

$$|J,J> = |j,j> |j'j'>.$$

Only this one product is needed because only the one term with $m = j$ and $m' = j'$ contributes to the sum in the above CG series. The state

$$|J,-J\rangle = |j,-j\rangle\, |j',-j'\rangle$$

with the minimum M-value is also given as a single product state. Notice that these states have M-values given as $\pm(j + j')$; since this is the maximum M-value, it must be that the J-value corresponding to this state is $J = j + j'$.

2. States with One Lower M-Value but the Same J-Value

Applying J_- to $|J,J\rangle$, and expressing J_- as the sum of lowering operators for the two individual angular momenta:

$$J_- = J_-(1) + J_-(2)$$

gives

$$\begin{aligned} J_-|J,J\rangle &= \hbar\left\{J(J+1) - J(J-1)\right\}^{1/2} |J,J-1\rangle \\ &= (J_-(1) + J_-(2))\, |j,j\rangle\, |j'j'\rangle \\ &= \hbar\left\{j(j+1) - j(j-1)\right\}^{1/2} |j,j-1\rangle\, |j',j'\rangle + \hbar\left\{j'(j'+1) - j'(j'-1)\right\}^{1/2} |j,j\rangle\, |j',j'-1\rangle. \end{aligned}$$

This result expresses $|J,J-1\rangle$ as follows:

$$\begin{aligned} |J,J-1\rangle = [\left\{j(j+1) - j(j-1)\right\}^{1/2} &|j,j-1\rangle\, |j',j'\rangle \\ + \left\{j'(j'+1) - j'(j'-1)\right\}^{1/2} &|j,j\rangle\, |j',j'-1\rangle]\, \left\{J(J+1) - J(J-1)\right\}^{-1/2}; \end{aligned}$$

that is, the $|J,J-1\rangle$ state, which has $M = J - 1$, is formed from the two product states $|j,j-1\rangle\, |j',j'\rangle$ and $|j,j\rangle\, |j',j'-1\rangle$ that have this same M-value.

3. States with One Lower J-Value

To find the state $|J-1,J-1\rangle$ that has the same M-value as the one found above but one lower J-value, we must construct another combination of the two product states with $M = J - 1$ (i.e., $|j,j-1\rangle\, |j',j'\rangle$ and $|j,j\rangle\, |j',j'-1\rangle$) that is orthogonal to the combination representing $|J,J-1\rangle$; after doing so, we must scale the resulting function so it is properly normalized. In this case, the desired function is:

$$\begin{aligned} |J-1,J-1\rangle = [\left\{j(j+1) - j(j-1)\right\}^{1/2} &|j,j\rangle\, |j',j'-1\rangle \\ -\left\{j'(j'+1) - j'(j'-1)\right\}^{1/2} &|j,j-1\rangle\, |j',j'\rangle]\, \left\{J(J+1) - J(J-1)\right\}^{-1/2}. \end{aligned}$$

It is straightforward to show that this function is indeed orthogonal to $|J,J-1\rangle$.

4. States with Even One Lower J-Value

Having expressed $|J,J-1\rangle$ and $|J-1,J-1\rangle$ in terms of $|j,j-1\rangle\, |j',j'\rangle$ and $|j,j\rangle\, |j',j'-1\rangle$, we are now prepared to carry on with this stepwise process to generate the states $|J,J-2\rangle$, $|J-1,J-2\rangle$ and $|J-2,J-2\rangle$ as combinations of the product states with $M = J - 2$. These product states are $|j,j-2\rangle\, |j',j'\rangle$, $|j,j\rangle\, |j',j'-2\rangle$, and $|j,j-1\rangle\, |j',j'-1\rangle$. Notice that there are precisely as many product states whose $m + m'$ values add up to the desired M-value as there are total angular momentum states that must be constructed (there are three of each in this case).

The steps needed to find the state $|J-2, J-2\rangle$ are analogous to those taken above:

a. One first applies J_- to $|J-1, J-1\rangle$ and to $|J, J-1\rangle$ to obtain $|J-1, J-2\rangle$ and $|J, J-2\rangle$, respectively as combinations of $|j, j-2\rangle |j', j'\rangle$, $|j, j\rangle |j', j'-2\rangle$, and $|j, j-1\rangle |j', j'-1\rangle$.
b. One then constructs $|J-2, J-2\rangle$ as a linear combination of the $|j, j-2\rangle |j', j'\rangle$, $|j, j\rangle |j', j'-2\rangle$, and $|j, j-1\rangle |j', j'-1\rangle$ that is orthogonal to the combinations found for $|J-1, J-2\rangle$ and $|J, J-2\rangle$.

Once $|J-2, J-2\rangle$ is obtained, it is then possible to move on to form $|J, J-3\rangle$, $|J-1, J-3\rangle$, and $|J-2, J-3\rangle$ by applying J_- to the three states obtained in the preceding application of the process, and to then form $|J-3, J-3\rangle$ as the combination of $|j, j-3\rangle |j', j'\rangle$, $|j, j\rangle |j', j'-3\rangle$, $|j, j-2\rangle |j', j'-1\rangle$, $|j, j-1\rangle |j', j'-2\rangle$ that is orthogonal to the combinations obtained for $|J, J-3\rangle$, $|J-1, J-3\rangle$, and $|J-2, J-3\rangle$.

Again notice that there are precisely the correct number of product states (four here) as there are total angular momentum states to be formed. In fact, the product states and the total angular momentum states are equal in number and are both members of orthonormal function sets (because $J^2(1)$, $J_z(1)$, $J^2(2)$, and $J_z(2)$ as well as J^2 and J_z are Hermitian operators). This is why the CG coefficient matrix is unitary; because it maps one set of orthonormal functions to another, with both sets containing the same number of functions.

D. An Example

Let us consider an example in which the spin and orbital angular momenta of the Si atom in its ^3P ground state can be coupled to produce various ^3P_J states. In this case, the specific values for j and j' are $j = S = 1$ and $j' = L = 1$. We could, of course take $j = L = 1$ and $j' = S = 1$, but the final wavefunctions obtained would span the same space as those we are about to determine.

The state with highest M-value is the $^3P(M_s = 1, M_L = 1)$ state. As shown in chapter 10 which deals with electronic configurations and states, this particular product wavefunction can be represented by the product of an $\alpha\alpha$ spin function (representing $S = 1, M_s = 1$) and a $3p_1 3p_0$ spatial function (representing $L = 1, M_L = 1$), where the first function corresponds to the first open-shell orbital and the second function to the second open-shell orbital. Thus, the maximum M-value is $M = 2$ and corresponds to a state with $J = 2$:

$$|J = 2, M = 2\rangle = |2,2\rangle = \alpha\alpha\ 3p_1 3p_0.$$

Clearly, the state $|2,-2\rangle$would be given as $\beta\beta\ 3p_{-1} 3p_0$.

The states $|2,1\rangle$ and $|1,1\rangle$ with one lower M-value are obtained by applying $J_- = S_- + L_-$ to $|2,2\rangle$ as follows:

$$\begin{aligned} J_- |2,2\rangle &= \hbar \{2(3) - 2(1)\}^{1/2} |2,1\rangle \\ &= (S_- + L_-)\ \alpha\alpha\ 3p_1 3p_0. \end{aligned}$$

To apply S_- or L_- to $\alpha\alpha\ 3p_1 3p_0$, one must realize that each of these operators is, in turn, a sum of lowering operators for each of the two open-shell electrons:

$$\begin{aligned} S_- &= S_-(1) + S_-(2)\ , \\ L_- &= L_-(1) + L_-(2)\ . \end{aligned}$$

The result above can therefore be continued as

$$(S_- + L_-)\,\alpha\alpha\,3p_13p_0 = \hbar\{1/2(3/2) - 1/2(-1/2)\}^{1/2}\,\beta\alpha\,3p_13p_0$$
$$+ \hbar\{1/2(3/2) - 1/2(-1/2)\}^{1/2}\,\alpha\beta\,3p_13p_0$$
$$+ \hbar\{1(2) - 1(0)\}^{1/2}\,\alpha\alpha\,3p_03p_0$$
$$+ \hbar\{1(2) - 0(-1)\}^{1/2}\,\alpha\alpha\,3p_13p_{-1}.$$

So, the function | 2,1> is given by

$$|2,1\rangle = [\beta\alpha\,3p_13p_0 + \alpha\beta\,3p_13p_0 + \{2\}^{1/2}\,\alpha\alpha\,3p_03p_0 + \{2\}^{1/2}\,\alpha\alpha\,3p_13p_{-1}]/2,$$

which can be rewritten as:

$$|2,1\rangle = [(\beta\alpha + \alpha\beta)3p_13p_0 + \{2\}^{1/2}\,\alpha\alpha\,(3p_03p_0 + 3p_13p_{-1})]/2.$$

Writing the result in this way makes it clear that | 2,1> is a combination of the product states $|S = 1, M_S = 0\rangle\,|L = 1, M_L = 1\rangle$ (the terms containing $|S = 1, M_S = 0\rangle = 2^{-1/2}(\alpha\beta + \beta\alpha)$) and $|S = 1, M_S = 1\rangle\,|L = 1, M_L = 0\rangle$ (the terms containing $|S = 1, M_S = 1\rangle = \alpha\alpha$).

To form the other function with $M = 1$, the | 1,1> state, we must find another combination of $|S = 1, M_S = 0\rangle\,|L = 1, M_L = 1\rangle$ and $|S = 1, M_S = 1\rangle\,|L = 1, M_L = 0\rangle$ that is orthogonal to | 2,1> and is normalized. Since

$$|2,1\rangle = 2^{-1/2}\,[|S = 1, M_S = 0\rangle\,|L = 1, M_L = 1\rangle + |S = 1, M_S = 1\rangle\,|L = 1, M_L = 0\rangle],$$

we immediately see that the requisite function is

$$|1,1\rangle = 2^{-1/2}\,[|S = 1, M_S = 0\rangle\,|L = 1, M_L = 1\rangle - |S = 1, M_S = 1\rangle\,|L = 1, M_L = 0\rangle].$$

In the spin-orbital notation used above, this state is:

$$|1,1\rangle = [(\beta\alpha + \alpha\beta)3p_13p_0 - \{2\}^{1/2}\,\alpha\alpha\,(3p_03p_0 + 3p_13p_{-1})]/2.$$

Thus far, we have found the ${}^3P_J$ states with $J = 2$, $M = 2$; $J = 2$, $M = 1$; and $J = 1$, $M = 1$.

To find the ${}^3P_J$ states with $J = 2$, $M = 0$; $J = 1$, $M = 0$; and $J = 0$, $M = 0$, we must once again apply the J_- tool. In particular, we apply J_- to | 2,1> to obtain | 2,0>, and we apply J_- to | 1,1> to obtain | 1,0>, each of which will be expressed in terms of $|S = 1, M_S = 0\rangle\,|L = 1, M_L = 0\rangle$, $|S = 1, M_S = 1\rangle\,|L = 1, M_L = -1\rangle$, and $|S = 1, M_S = -1\rangle\,|L = 1, M_L = 1\rangle$. The | 0,0> state is then constructed to be a combination of these same product states which is orthogonal to | 2,0> and to | 1,0>. The results are as follows:

$$|J = 2, M = 0\rangle = 6^{-1/2}[2|1,0\rangle\,|1,0\rangle + |1,1\rangle\,|1,-1\rangle + |1,-1\rangle\,|1,1\rangle],$$
$$|J = 1, M = 0\rangle = 2^{-1/2}[|1,1\rangle\,|1,-1\rangle - |1,-1\rangle\,|1,1\rangle],$$
$$|J = 0, M = 0\rangle = 3^{-1/2}[|1,0\rangle\,|1,0\rangle - |1,1\rangle\,|1,-1\rangle - |1,-1\rangle\,|1,1\rangle],$$

where, in all cases, a shorthand notation has been used in which the $|S,M_S\rangle\,|L,M_L\rangle$ product stated have been represented by their quantum numbers with the spin function always appearing first in the product. To finally express all three of these new functions in terms of spin-orbital products it is necessary to give the $|S,M_S\rangle\,|L,M_L\rangle$ products with $M = 0$ in terms of these products. For the spin functions, we have:

$$|S = 1, M_S = 1\rangle = \alpha\alpha,$$
$$|S = 1, M_S = 0\rangle = 2^{-1/2}(\alpha\beta + \beta a).$$
$$|S = 1, M_S = -1\rangle = \beta\beta.$$

For the orbital product function, we have:

$$|L = 1, M_L = 1> = 3p_1 3p_0,$$
$$|L = 1, M_L = 0> = 2^{-1/2}(3p_0 3p_0 + 3p_1 3p_{-1}),$$
$$|L = 1, M_L = -1> = 3p_0 3p_{-1}.$$

E. CG Coefficients and 3-*j* Symbols

As stated above, the CG coefficients can be worked out for any particular case using the raising and lowering operator techniques demonstrated above. Alternatively, as also stated above, the CG coefficients are tabulated (see, for example, Zare's book on angular momentum the reference to which is given earlier in this appendix) for several values of j, j', and J.

An alternative to the CG coefficients is provided by the so-called 3-j coefficients (see section 2.2 of Zare's book), which are defined in terms of the CG coefficients as follows:

$$\begin{pmatrix} j & j' & J \\ m & m' & -M \end{pmatrix} = (-1)^{j-j'-M} <j,m; j',m'|J,M> (2J+1)^{-1/2}.$$

Clearly, these coefficients contain no more or less information than do the CG coefficients. However, both sets of symbols have symmetries under interchange of the j and m quantum number that are more easily expressed in terms of the 3-j symbols. In particular, odd permutations of the columns of the 3-j symbol leave the magnitude unchanged and change the sign by $(-1)^{j+j'+J}$, whereas even permutations leave the value unchanged. Moreover, replacement of all of the m-values (m, m', and M) by their negatives leave the magnitude the same and changes the sign by $(-1)^{j+j'+J}$. Table 2.5 in Zare's book (see above for reference) contains 3-j symbols for $J = 0, 1/2, 1, 3/2$, and 2.

IV. HOW ANGULAR MOMENTUM ARISES IN MOLECULAR QUANTUM CHEMISTRY

A. The Hamiltonian May Commute with Angular Momentum Operators

As is illustrated throughout this text, angular momentum operators often commute with the Hamiltonian of the system. In such cases, the eigenfunctions of the Hamiltonian can be made to also be eigenfunctions of the angular momentum operators. This allows one to label the energy eigenstates by quantum numbers associated with the angular momentum eigenvalues.

1. Electronic Atomic Hamiltonia without Spin-Orbit Coupling

For example, the electronic Hamiltonian of atoms, as treated in chapters 1 and 3 in which only kinetic and coulombic interaction energies are treated, commutes with L^2, and L_z, where

$$L_z = \sum_j L_z(j)$$

and

$$L^2 = L_z^2 + L_x^2 + L_y^2.$$

The fact that H commutes with L_z, L_x, and L_y and hence L^2 is a result of the fact that the total coulombic potential energies among all the electrons and the nucleus is invariant to rotations of *all* electrons about the z, x, or y axes (H does not commute with $L_z(j)$ since if only the jth electron's coordinates are so rotated, the total coulombic potential is altered because inter-electronic distances change). The invariance of the potential to rotations of all electrons is, in turn, related to the spherical nature of the atom. As a result, atomic energy leveis for such a Hamiltonian can be labeled by their *total* L and M quantum numbers.

2. Electronic Linear-Molecule Hamiltonia without Spin-Orbit Coupling

For linear molecules, the coulombic potential is unchanged (because the set of all inter-particle distances are unchanged) by rotations about the molecular axis (the z axis); hence H commutes with L_z. H does not commute with L_x or L_y, and thus not L^2, because the potential is altered by rotations about the x or y axes. As a result, linear-molecule energy levels for such a Hamiltonian can be labeled by their *total* M quantum number, which in this context is usually replaced by the quantum number $\Lambda = |M|$.

3. Spin-Orbit Effects

When spin-orbit couplings are added to the electrostatic Hamiltonian considered in the text, additional terms arise in H. These terms have the form of a one-electron additive operator:

$$H_{SO} = \sum_j \left\{g_e/2m_e^2c^2\right\} r_j^{-1}\ \partial V/\partial R_j\ \boldsymbol{S}(j)\bullet\boldsymbol{L}(j)$$

where V is the total coulombic potential the that electron j feels due to the presence of the other electrons and the nuclei. $\boldsymbol{S}(j)$ and $\boldsymbol{L}(j)$ are the spin- and orbital-angular momentum operators of electron j, and g_e is the electron magnetic moment in Bohr magneton units ($g_e = 2.002319$) . For atoms in which these spin-orbit terms are considered (they are important for "heavy atoms" because $r_j^{-1}\ \partial V/\partial \boldsymbol{R}_j$ varies as Zr_j^{-3} for atoms, whose expectation value varies as Z^4), it turns out that neither L^2 nor S^2 commute with H_{SO}. However, the "combined" angular momentum

$$\boldsymbol{J} = \boldsymbol{L} + \boldsymbol{S}$$
$$J_z = L_z + S_z$$
$$J^2 = J_z^2 + J_x^2 + J_y^2$$

does commute with H_{SO}, and hence with the full $H + H_{SO}$ Hamiltonian including spin-orbit coupling. For this reason, the eigenstates of atoms in which spin-orbit coupling is important can not be labeled by L, M, S, and M_S, but only by J and M_J.

B. The Hamiltonian May Contain Angular Momentum Operators

1. Electronic Hamiltonia for Atoms without Spin-Orbit Effects

There are cases in which the angular momentum operators themselves appear in the Hamiltonian. For electrons moving around a single nucleus, the total kinetic energy operator T has the form:

$$T = \sum_j \left\{-\hbar^2/2m_e\ \nabla_j^2\right\}$$

$$= \sum_j \left\{-\hbar^2/2m_e\ [r_j^{-2}\partial/\partial r_j(r_j^2\partial/\partial r_j) - (r_j^2\sin\theta_j)^{-1}\partial/\partial\theta_j(\sin\theta_j\partial/\partial\theta_j) - (r_j\sin\theta_j)^{-2}\partial^2/\partial\phi_j^2\right\}.$$

The factor $\hbar^2[(\sin\theta_j)^{-1}\partial/\partial\theta_j(\sin\theta_j\partial/\partial\theta_j) + (\sin\theta_j)^{-2}\ \partial^2/\partial\phi_j^2]$ is $L^2(j)$, the square of the angular momentum for the jth electron. In this case, the Hamiltonian contains $L^2(j)$ for the individual electrons, not the total L^2, although it still commutes with the total L^2 (which thus renders L and M good quantum numbers).

2. Linear Rigid-Molecule Rotation

The rotational Hamiltonian for a diatomic molecule as given in chapter 3 is

$$H_{rot} = \hbar^2/2\mu \left\{(R^2\sin\theta)^{-1}\partial/\partial\theta\ (\sin\theta\ \partial/\partial\theta) + (R^2\sin^2\theta)^{-1}\ \partial^2/\partial\phi^2\right\},$$

where μ is the reduced mass of the molecule, and R is its bond length. Again, the square of the total rotational angular momentum operator appears in H_{rot}

$$H_{rot} = L^2/2\mu R^2.$$

In this case, the Hamiltonian both contains and commutes with the total L^2; it also commutes with L_z, as a result of which L and M are both good quantum numbers and the spherical harmonics $Y_{L,M}(\theta,\phi)$ are eigenfunctions of H. These eigenfunctions obey orthogonality relations:

$$\int_0^{\pi}\left(\int_0^{2\pi}(Y^*_{L,M}(\theta,\phi)\ Y_{L',M'}(\theta,\phi)\sin\theta\ d\theta\ d\phi)\right) = \delta_{L,L'}\ \delta_{M,M'}$$

because they are eigenfunctions of two Hermitian operators (L^2 and L_z) with (generally) different eigenvalues.

3. Non-Linear Molecule Rotation

For non-linear molecules, when treated as rigid (i.e., having fixed bond lengths, usually taken to be the equilibrium values or some vibrationally averaged values), the rotational Hamiltonian can be written in terms of rotation about three axes. If these axes (X,Y,Z) are located at the center of mass of the molecule but fixed in space such that they do not move with the molecule, then the rotational Hamiltonian can be expressed as:

$$H_{rot} = 1/2 \sum_{K,K'} \omega_K\, I_{K,K'}\, \omega_{K'}$$

where ω_K is the angular velocity about the Kth axis and

$$I_{K,K} = \sum_j m_j\,(R_j^2 - R_{K,j}^2) \qquad (\text{for } K = K')$$

$$I_{K,K'} = -\sum_j m_j\, R_{K,j}\, R_{K',j} \qquad (\text{for } K \neq K')$$

are the elements of the so-called moment of inertia tensor. This tensor has components along the axes labeled K and K' (each of which runs over X, Y, and Z). The m_j denote the masses of the atoms (labeled j) in the molecule, $R_{K,j}$ is the coordinate of atom j along the K-axis relative to the center of mass of the molecule, and R_j is the distance of atom j from the center of mass

$$\left(R_j^2 = \sum_K (R_{K,j})^2\right).$$

Introducing a new set of axes x, y, z that also have their origin at the center of mass, but that rotate with the molecule, it is possible to reexpress H_{rot} in terms of motions of these axes. It is especially useful to choose a particular set of such molecule-fixed axes, those that cause the moment of inertial tensor to be diagonal. This symmetric matrix can, of course, be made diagonal by first computing $I_{k,k'}$(where k and k' run over x, y, and z) for an arbitrary x, y, z axis choice and then finding the orthogonal transformation (i.e., the eigenvectors of the $\boldsymbol{I}$ matrix) that brings $\boldsymbol{I}$ to diagonal form. Such molecule-fixed axes (which we denote as a, b, and c) in which $\boldsymbol{I}$ is diagonal are called **principal axes**; in terms of them, H_{rot} becomes:

$$H_{rot} = 1/2[I_a\, \omega_a^2 + I_b\, \omega_b^2 + I_c\, \omega_c^2]\ .$$

The angular momentum conjugate to each of these three angular coordinates (each ω is the time rate of change of an angle of rotation about an axis: $\omega = d\,(\text{angle})/dt)$) is obtained, as usual, from the Lagrangian function $L = T - V$ of classical mechanics:

$$p = \partial L/\partial \dot{q} = \partial(\text{Kinetic Energy} - \text{Potential Energy})/\partial(dq/dt)$$

or (using J_a to denote the angular momentum conjugate to ω_a and realizing that since this free rotational motion has no potential energy, $L = T = H_{rot}$)

$$J_a = \frac{\partial H_{rot}}{\partial \omega_a} = I_a \omega_a$$

$$J_b = I_b \omega_b$$

$$J_c = I_c \omega_c.$$

The rotational Hamiltonian can then be written in terms of angular momenta and principal-axis moments of inertia as:

$$H_{rot} = J_a^2/2I_a + J_b^2/2I_b + J_c^2/2I_c.$$

With respect to this principal axis point of view, the rotation of the molecule is described in terms of three angles (it takes three angles to specify the orientation of such a rigid body) that detail the movement of the a, b, and c axes relative to the lab-fixed X, Y, and Z axes. It is convention to call these angles θ'(which can be viewed as the angle between the lab-fixed Z axis and one of the principal axes—say c—in the molecule), ϕ', and χ'. The volume element for integration over these three angles is $\sin\theta\, d\theta\, d\phi\, d\chi$, with ϕ and χ running between 0 and 2π, and θ going from 0 to π. These coordinates are described visually below.

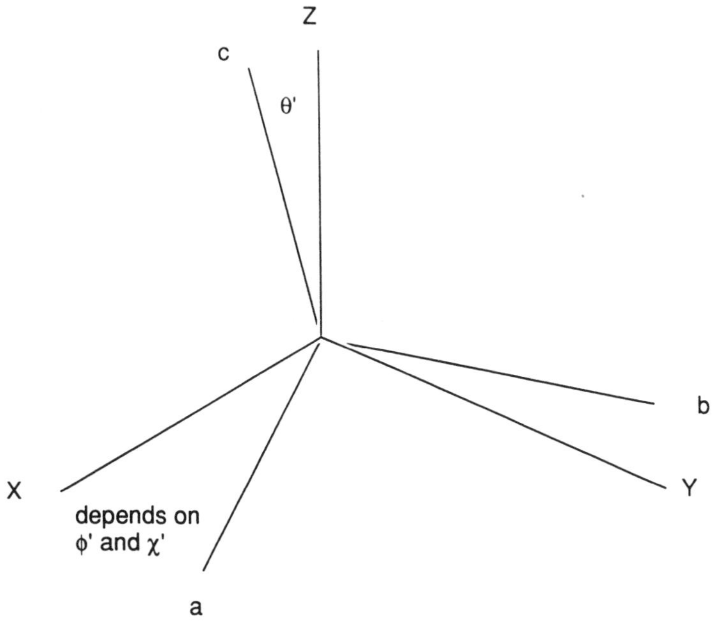

The a, b, c coordinate system can be formed by beginning with the original X, Y, Z system and sequentially:

i. rotating about the Z axis by an amount ϕ', to generate intermediate $X'Y'$, and $Z = Z'$ axes (X' and Y' being rotated by ϕ' relative to X and Y);

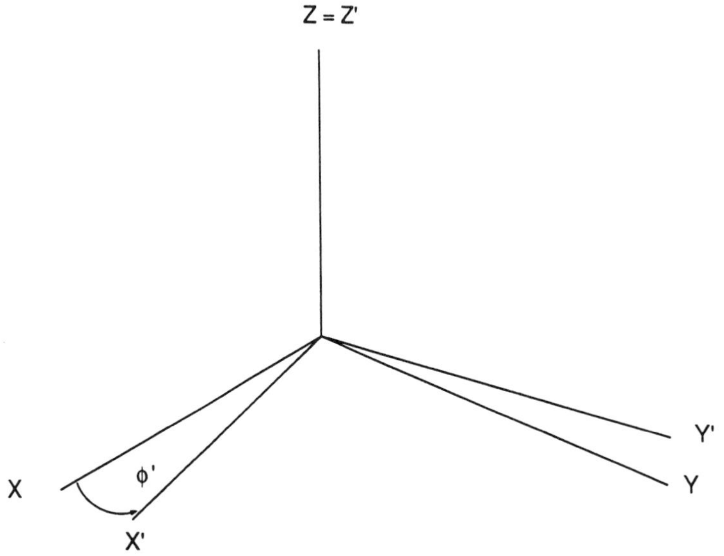

ii. next rotating about the Y' axis by an amount θ', to generate X'',Y', and $Z'' = c$ axes;

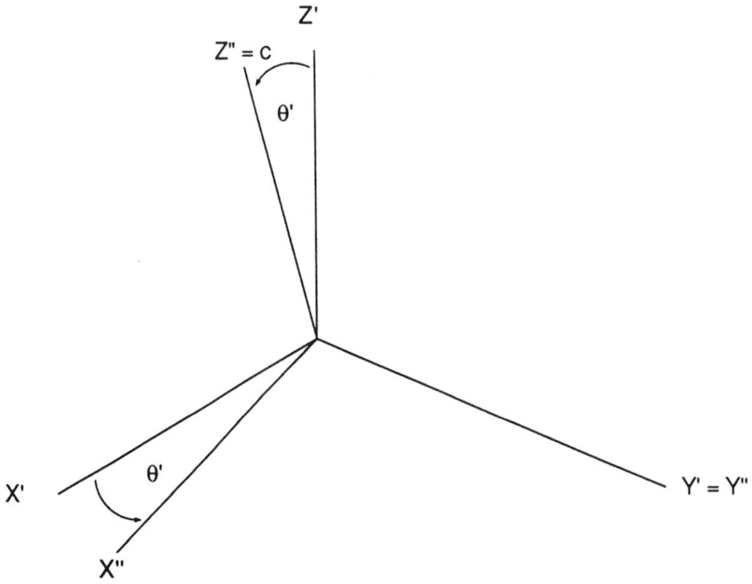

iii. and finally rotating about the new $Z'' = c$ axis by an amount χ' to generate the final $X''' = a$ and $Y''' = b$ axes.

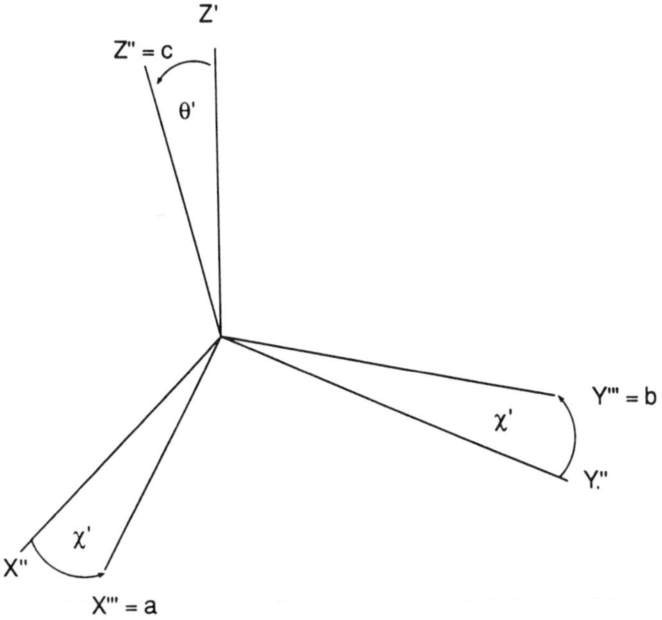

Thus, the original and final coordinates can be depicted as follows:

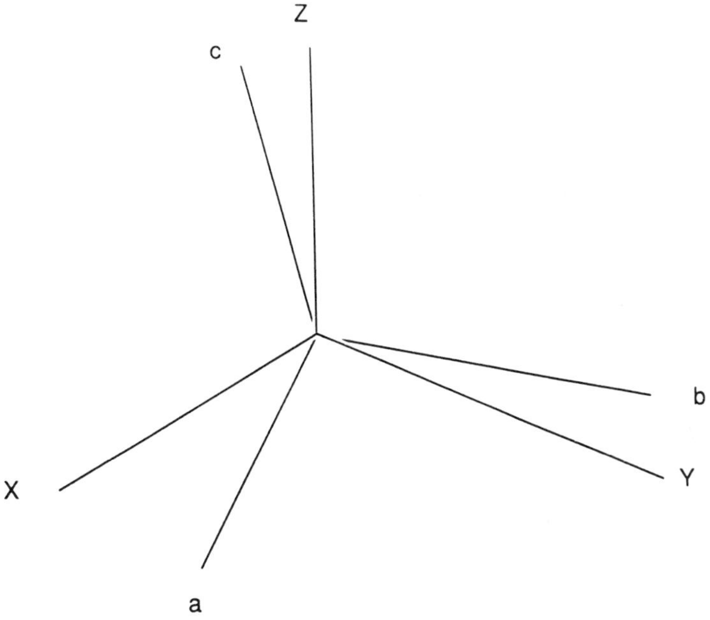

The explicit expressions for the components of the quantum mechanical angular momentum operators along the three new axes are:

$$J_a = -i\hbar \cos\chi\, [\cot\theta\, \partial/\partial\chi - (\sin\theta)^{-1}\partial/\partial\phi] - -i\hbar \sin\chi\, \partial/\partial\theta$$
$$J_b = i\hbar \sin\chi\, [\cot\theta\, \partial/\partial\chi - (\sin\theta)^{-1}\partial/\partial\phi] - -i\hbar \cos\chi\, \partial/\partial\theta$$
$$J_c = -\, i\hbar\, \partial/\partial\chi.$$

The corresponding total angular momentum operator J^2 can be obtained as

$$J^2 = J_a^2 + J_b^2 + J_c^2$$
$$= -\partial^2/\partial\theta^2 - \cot\theta\, \partial/\partial\theta - (1/\sin\theta)\, (\partial^2/\partial\phi^2 + \partial^2/\partial\chi^2 - 2\cos\theta\partial^2/\partial\phi\partial\chi),$$

and the component along the original Z axis J_Z is still $-i\hbar\, \partial/\partial\phi$.

Returning now to the rigid-body rotational Hamiltonian shown above, there are two special cases for which exact eigenfunctions and energy levels can be found using the general properties of angular momentum operators.

a. Spherical and Symmetric Top Energies The special cases for which $I_a = I_b = I_c$ (the **spherical top**) and for which $I_a = I_b > I_c$ (the oblate symmetric top) or $I_a > I_b = I_c$ (the prolate symmetric top) are covered in chapter 3. In the former case, the rotational Hamiltonian can be expressed in terms of $J^2 = J_a^2 + J_b^2 + J_c^2$ because all three moments of inertia are identical:

$$H_{rot} = J^2/2I,$$

as a result of which the eigenfunctions of H_{rot} are those of J^2 (and J_a as well as J_Z both of which commute with J^2 and with one another; J_Z is the component of $\boldsymbol{J}$ along the lab-fixed Z-axis and commutes with J_a because $J_Z = -\, i\hbar\, \partial/\partial\phi$ and $J_a = -\, i\hbar\partial/\partial\chi$ act on different angles). The energies associated with such eigenfunctions are

$$E(J,K,M) = \hbar^2 J(J+1)/2I^2,$$

for all K (i.e., J_a quantum numbers) ranging from $-J$ to J in unit steps and for all M (i.e., J_Z quantum numbers) ranging from $-J$ to J. Each energy level is therefore $\mathbf{(2J+1)^2}$ **degenerate** because there are $2J+1$ possible K values and $2J+1$ M values for each J.

In the **symmetric top** cases, H_{rot} can be expressed in terms of J^2 and the angular momentum along the axis with the unique moment of inertia (denoted the a-axis for prolate tops and the c-axis of oblate tops):

$$H_{rot} = J^2/2I + J_a^2\{1/2I_a - 1/2I\}, \text{ for prolate tops}$$
$$H_{rot} = J^2/2I + J_c^2\{1/2I_c - 1/2I\}, \text{ for oblate tops.}$$

H_{rot}, along with J^2 and J_a (or J_c for oblate tops) and J_Z (the component of $\boldsymbol{J}$ along the lab-fixed Z-axis) form a mutually commutative set of operators. J_Z, which is $-i\hbar\, \partial/\partial\phi$, and J_a (or c) which is $-i\hbar\, \partial/\partial\chi$, commute because they act on different angles. As a result, the eigenfunctions of H_{rot} are those of J^2 and J_a or J_c (and of J_Z), and the corresponding energy levels are:

$$E(J,K,M) = \hbar^2 J(J+1)/2I^2 + \hbar^2 K^2\{1/2I_a - 1/2I\},$$

for prolate tops

$$E(J,K,M) = \hbar^2 J(J+1)/2I^2 + \hbar^2 K^2\{1/2I_c - 1/2I\},$$

for oblate tops, again for K and M (i.e., J_a or J_c and J_Z quantum numbers, respectively) ranging from $-J$ to J in unit steps. Since the energy now depends on K, these levels are only **$2J+1$ degenerate** due to the $2J+1$ different M values that arise for each J value.

b. Spherical- and symmetric top wavefunctions The eigenfunctions of J^2, J_a (or J_c), and J_Z clearly play important roles in polyatomic molecule rotational motion; they are the eigenstates for spherical-top and symmetric-top species, and they can be used as a basis in terms of which to expand the eigenstates of asymmetric-top molecules whose energy levels do not admit an analytical solution. These eigenfunctions $|J,M,K>$ are given in terms of the set of so-called **"rotation matrices,"** which are denoted $D_{J,M,K}$:

$$|J,M,K> = \sqrt{\frac{2J+1}{8\pi^2}}\, D^*_{J,M,K}(\theta,\phi,\chi).$$

They obey

$$J^2\,|J,M,K> = \hbar^2 J(J+1)\,|J,M,K>,$$
$$J_a \text{ (or } J_c \text{ for oblate tops) } |J,M,K> = \hbar\, K\,|J,M,K>,$$
$$J_Z\,|J,M,K> = \hbar\, M\,|J,M,K>.$$

It is demonstrated below why the symmetric and spherical top wavefunctions are given in terms of these $D_{J,M',M}$ functions.

c. Rotation matrices These same rotation matrices arise when the transformation properties of spherical harmonics are examined for transformations that rotate coordinate systems. For example, given a spherical harmonic $Y_{L,M}(\theta, \phi)$ describing the location of a particle in terms of polar angles θ,ϕ within the X, Y, Z axes, one might want to rotate this function by Euler angles θ',ϕ',χ' and *evaluate this rotated function at the same physical point*. As shown in Zare's text on angular momentum, the rotated function $\Omega Y_{L,M}$ evaluated at the angles θ,ϕ can be expressed as follows:

$$\Omega\, Y_{L,M}(\theta,\phi) = \sum_{M'} D_{L,M',M}(\theta',\phi',\chi')\, Y_{L,M'}(\theta,\phi).$$

In this form, one sees why the array $D_{J,M',M}$ is viewed as a unitary matrix, with M' and M as indices, that describes the effect of rotation on the set of functions $\{Y_{L,M}\}$. This mapping from the unrotated set $\{Y_{L,M}\}$ into the rotated set of functions $\{\Omega\, Y_{L,M}\}$ must be **unitary** if the sets $\{\Omega\, Y_{L,M}\}$ and $\{Y_{L,M}\}$ are both orthonormal. The unitary matrix carries an additional index (L in this example) that details the dimension $(2L+1)$ of the space of functions whose transformations are so parameterized. An example, for $L = 1$, of a set of unrotated and rotated functions is shown below.

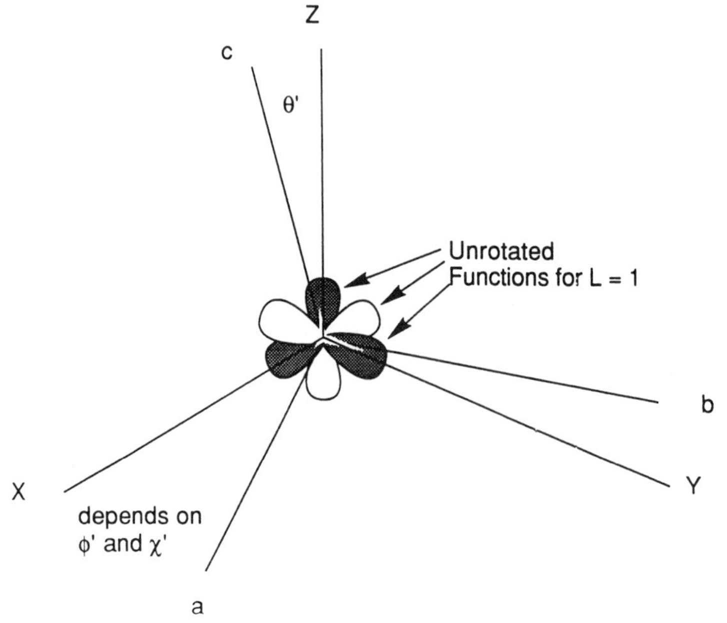

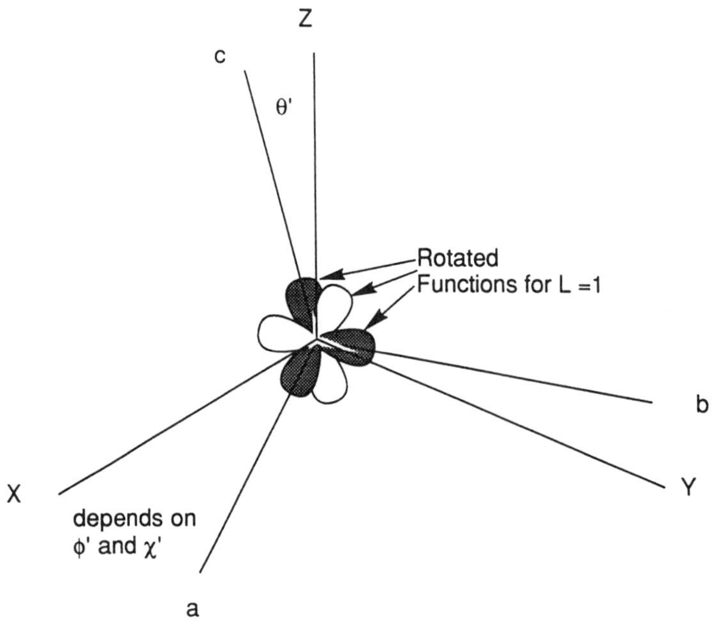

d. Products of rotation matrices An identity that proves very useful when treating coupled angular momenta that are subjected to rotations of the axes with respect to which their eigenfunctions are quantized can be derived by combining the above result:

$$\Omega\, Y_{L,M}(\theta,\phi) = \sum_{M'} D_{L,M',M}(\theta',\phi',\chi')\; Y_{L,M'}(\theta,\phi)$$

and the expression for couping two angular momenta:

$$|J,M\rangle = \sum_{m,n} \langle j,m;l,n|J,M\rangle\, |j,m\rangle\, |l,n\rangle.$$

Applying the rotation Ω to the left and right sides of the equation defining $|J,M\rangle$, gives:

$$\sum_{M'} D_{J,M,M'}(\Omega)\, |J,M'\rangle$$
$$= \sum_{m',n'} \langle j,m;l,n|J,M\rangle\, D_{j,m,m'}(\Omega) D_{l,n,n'}(\Omega)\, |j,m'\rangle\, |l,n'\rangle.$$

Multiplying both sides of this equation by $\langle J,M'|$ and using the orthonormality of the angular momentum eigenfunctions gives:

$$D_{J,M,M'} = \sum_{m',n'} \langle j,m;l,n|J,M\rangle\, D_{j,m,m'} D_{l,n,n'}\, \langle J,M'|j,m';\, l,n'\rangle.$$

This result expresses one $D_{J,M,M'}$ in terms of sums of products of D matrix elements for angular momenta j,m,m' and l,n,n' that can be coupled to form J,M,M'.

If the above series of operations is applied to the angular momentum coupling expression in the form:

$$|j,m\rangle\, |l,n\rangle = \sum_{J,M} \langle J,M|j,m;l,n\rangle\, |J,M\rangle,$$

one can obtain:

$$\sum_{m',n'} D_{j,m,m'}(\Omega) D_{l,n,n'}(\Omega)\, |j,m'\rangle\, |l,n'\rangle = \sum_{J,M} \langle J,M|j,m;l,n\rangle \sum_{M'} D_{J,M,M'}(\Omega)\, |J,M'\rangle.$$

Multiplying by $\langle j,m'|\, \langle l,n'|$ then yields:

$$D_{j,m,m'} D_{l,n,n'} = \sum_{J,M,M'} \langle J,M|j,m;l,n\rangle\, \langle j,m';\, l,n'|J,M'\rangle\, D_{J,M,M'}$$

which expresses the product of two D matrices as a sum of D matrices whose angular momentum indices are related to those of the product.

e. Rigid body rotational wavefunctions as rotation matrices This same analysis can be used to describe how a set of functions $\psi_{J,M}(\theta, \phi, \chi)$ (labeled by a total angular momentum quantum number that determines the number of functions in the set and an M quantum number that labels the Z-axis projection of this angular momentum) that are functions of three coordinates θ, ϕ, χ, transform under rotation. In particular, one obtains a result analogous to the spherical harmonic expression:

$$\Omega\psi_{J,M}(\theta, \phi, \chi) = \sum_{M'} D_{J,M',M}(\theta',\phi',\chi')\, \psi_{J,M'}(\theta,\phi,\chi).$$

Here $\psi_{J,M}(\theta, \phi, \chi)$ is the original unrotated function evaluated at a point whose angular coordinates are θ, ϕ, χ; θ', ϕ', χ' are the Euler angles through which this function is rotated to obtain the rotated function $\Omega\psi_{J,M}$ whose value at the above point is denoted $\Omega\psi_{J,M}(\theta, \phi, \chi)$.

Now, if the angles θ', ϕ', χ' through which the original function is rotated were *chosen* to equal the angular coordinates θ, ϕ, χ of the point discussed here, then the rotated function $\Omega\psi_{J,M}$ evaluated at this point could easily be identified. Its value would be nothing more than the unrotated function $\psi_{J,M}$ evaluated at $\theta = 0, \phi = 0, \chi = 0$. In this case, we can write:

$$\Omega\psi_{J,M}(\theta, \phi, \chi) = \psi_{J,M}(0, 0, 0) = \sum_{M'} D_{J,M',M}(\theta, \phi, \chi)\psi_{J,M'}(\theta, \phi, \chi).$$

Using the unitary nature of the $D_{L,M',M}$ array, this equation can be solved for the $\psi_{J,M'}(\theta, \phi, \chi)$ functions:

$$\psi_{J,M'}(\theta, \phi, \chi) = \sum_{M} D^*_{J,M',M}(\theta, \phi, \chi)\psi_{J,M}(0, 0, 0).$$

This result shows that the functions that describe the rotation of a rigid body through angles θ, ϕ, χ must be a combination of rotation matrices (actually $D^*_{L,M',M}(\theta, \phi, \chi)$ functions). Because of the normalization of the $D_{L,M,M'}(\theta, \phi, \chi)$ functions:

$$\int (D^*_{L',M',K'}(\theta, \phi, \chi)\, D_{L,M,K}(\theta, \phi, \chi) \sin\theta\, d\theta\, d\phi\, d\chi) = \frac{8\pi^2}{2L+1}\,\delta_{L,L'}\delta_{M,M'}\delta_{K,K'}$$

the properly normalized rotational functions that describe spherical or symmetric tops are:

$$|J,M,K> = \sqrt{\frac{2J+1}{8\pi^2}}\, D^*_{J,M,K}(\theta,\phi,\chi)$$

as given above. For asymmetric top cases, the correct eigenstates are combinations of these $\{|J,M,K>\}$ functions:

$$\psi_{J,M}(\theta, \phi, \chi) = \sum_{K} \sqrt{\frac{2J+1}{8\pi^2}}\, D^*_{J,M,K}(\theta, \phi, \chi)\, C_K$$

with amplitudes $\{C_K\}$ determined by diagonalizing the full H_{rot} Hamiltonian within the basis consisting of the set of

$$\sqrt{\frac{2J+1}{8\pi^2}}\, D^*_{J,M,K}(\theta, \phi, \chi)$$

functions.

4. Electronic and Nuclear Zeeman Interactions

When magnetic fields are present, the intrinsic spin angular momenta of the electrons $\boldsymbol{S}(j)$ and of the nuclei $\boldsymbol{I}(k)$ are affected by the field in a manner that produces additional energy contributions to the total Hamiltonian H. The Zeeman interaction of an external magnetic field (e.g., the earth's magnetic field of 4. Gauss or that of a NMR machine's magnet) with such intrinsic spins is expressed in terms of the following contributions to H:

$$H_{zeeman} = (g_e\, e/2m_e c) \sum_j S_z(j)\, H - (e/2m_p c) \sum_k g_k\, I_z(k)\, H.$$

Here g_k is the so-called nuclear g-value of the kth nucleus, H is the strength of the applied field, m_p is the mass of the proton, g_e is the electron magnetic moment, and c is the speed of light. When chemical shieldings (denoted σ_k), nuclear spin-spin couplings (denoted $J_{k,l}$), and electron-nuclear spin couplings (denoted $a_{j,k}$) are considered, the following spin-dependent Hamiltonian is obtained:

$$H = (g_e\, e/2m_e c) \sum_j S_z(j)\, H - (e/2m_p c) \sum_k g_k\, (1 - \sigma_k) I_z(k)\, H$$

$$+ h \sum_{j,k} (a_{j,k}/\hbar^2)\, \boldsymbol{I}(k) \bullet \boldsymbol{S}(j) + h \sum_{k,l} (J_{k,l}/\hbar^2)\, \boldsymbol{I}(k) \bullet \boldsymbol{I}(l).$$

Clearly, the treatment of electron and nuclear spin angular momenta is essential to analyzing the energy levels of such Hamiltonia, which play a central role in NMR and ESR spectroscopy.

APPENDIX

H QMIC Programs

The Quantum Mechanics in Chemistry (QMIC) programs, whose source and executable versions are provided along with the text, are designed to be pedagogical in nature; therefore they are not designed with optimization in mind, and could certainly be improved by interested students or instructors. The software is actually a suite of programs allowing the student to carry out many different types of *ab initio* calculations. The student can perform Hartree-Fock, MP2, or CI calculations, in a single step or by putting together a series of steps, by running the programs provided. The software can be found on the World Wide Web at several locations:

at the University of Utah, located at: http://www.chem.utah.edu

at the Pacific Northwest National Laboratory, located at:
http://www.emsl.pnl.gov:2080/people/bionames/nichols_ja.html

at Oxford University Press, located at: http://www.oup-usa.org

These programs are designed to run in very limited environments (e.g., memory, disk, and CPU power). With the exception of "integral.f", all are written in single precision and use minimal memory (less than 640K) in most instances. The programs are designed for simple systems, i.e., only a few atoms (usually less than eight), and small basis sets. They do not use group symmetry, and they use simple Slater determinants rather than spin-adapted configuration state functions to perform the CI. The programs were all originally developed and run on an IBM RISC System 6000 using AIX v3.2 and Fortran compilers xlf v2 and v3. All routines compile untouched with gnu compilers and utilities for workstations and PCs. The gnu utilities were obtained from the ftp server: ftp.coast.net in directory: simtel/vendors/gnu.

Except for very minor modifications all run untouched when compiled using Language Systems Fortran for the Macintosh. The instrinsics "and", "xor", and "rshift" have to be replaced by their counterparts "iand", "ixor", and "ishft". These intrinsic functions are only used in program hamilton.f, and their replacement functions are detailed and commented in the hamilton program source. No floating point unit has been turned on in the compilation. Because of this, computations on chemical systems with lots of basis functions performed on an old Mac SE can be tiring (the N^5 processes such as the transformation can take as long as a half hour on these systems). Needless to say, all of these run in less than a minute on the fancier workstations. Special thanks goes to Martin Feyereisen (Cray Research) for supplying us with very compact subroutines, which evaluate one- and two-electron integrals in a very simple and straightforward manner. Brief descriptions of each of the programs in QMIC follow.

Current QMIC program limits:

Maximum number of atoms: 8

Maximum number of orbitals: 26

Maximum number of shells: 20

Maximum number of primitives per shell: 7

Maximum orbital angular momentum: 1

Maximum number of active orbitals in the CI: 15

Maximum number of determinants: 350

Maximum matrix size (row or column): 350

QMIC PROGRAM DESCRIPTIONS

Program INTEGRAL This program is designed to calculate one- and two-electron AO integrals and to write them out to disk in canonical order (in Dirac <12|12> convention). It is designed to handle only *S* and *P* orbitals. With the program limitations described above, INTEGRAL memory usage is 542776 bytes.

Program MOCOEFS This program is designed to read in (from the keyboard) the LCAO-MO coefficient matrix and write it out to disk. Alternatively, you can choose to have a unit matrix (as your initial guess) put out to disk. With the program limitations described above, MOCOEFS memory usage is 2744 bytes.

Program FNCT_MAT This program is designed to read in a real square matrix, perform a function on it, and return this new array. Possible functions, using *X* as the input matrix, are:

1. $X^{(-1/2)}$, NOTE: *X* must be real symmetric, and positive definite.
2. $X^{(+1/2)}$, NOTE: *X* must be real symmetrix, and positive definite.
3. $X^{(-1)}$, NOTE: *X* must be real symmetrix, and have non-zero eigenvalues.
4. a power series expansion of a matrix to find the transformation matrix:
   $U = \exp(X) = 1 + X + X{**}2/2! + X{**}3/3! + \ldots + X{**}N/N!$

With the program limitations described above, FNCT_MAT memory usage is 1960040 bytes.

Program FOCK This program is designed to read in the LCAO-MO coefficient matrix, the one- and two-electron AO integrals, and to form a **closed-shell** Fock matrix (i.e., a Fock matrix for species with all doubly occupied orbitals). With the program limitations described above, FOCK memory usage is 255256 bytes.

Program UTMATU This program is designed to read in a real matrix, A, a real transformation matrix, B, perform the transformation: $X = B(\text{transpose}) * A * B$, and output the result. With the program limitations desccribed above, UTMATU memory usage is 1960040 bytes.

Program DIAG This program is designed to read in a real symmetric matrix (but as a square matrix on disk), diagonalize it, and return all the eigenvalues and corresponding eigenvectors. With the program limitations described above, DIAG memory usage is 738540 bytes.

Program MATXMAT This program is designed to read in two real matrices, A and B, and to multiply them together, $AB = A * B$, and output the result. With the program limitations described above, MATXMAT memory usage is 1470040 bytes.

Program FENERGY This program is designed to read in the LCAO-MO coefficient matrix, the one- and two-electron AO integrals (in Dirac <12|12> convention), and the Fock orbital energies. Upon transformation of the one- and two-electron integrals from the AO to the MO basis, the closed-shell Hartree-Fock energy is calculated in two ways: First, the energy is calculated with the MO integrals,

$$\sum_{(k)} 2*\langle k|h|k\rangle + \sum_{(k,l)} (2*\langle k,l|k,l\rangle - \langle k,l|l,k\rangle) + ZuZv/Ruv.$$

Second, the energy is calculated with the Fock orbital energies and one-electron energies in the MO basis,

$$\sum_{(k)} (\varepsilon(k) + \langle k|h|k\rangle) + ZuZv/Ruv.$$

With the program limitations described above, FENERGY memory usage is 1905060 bytes.

Program TRANS This program is designed to read in the LCAO-MO coefficient matrix, the one- and two-electron AO integrals (in Dirac <12|12> convention), and to transform the integrals from the AO to the MO basis, and write these MO integrals to a file. With the program limitations described above, TRANS memory usage is 1905060 bytes.

Program SCF This program is designed to read in the LCAO-MO coefficient matrix (or generate one), the one- and two-electron AO integrals, and form a **closed-shell** FOCK matrix (i.e., a Fock matrix for species with all doubly occupied orbitals). It then solves the Fock equations; iterating until convergence to six significant figures in the energy expression. A modified damping algorithm is used to ensure convergence. With the program limitations described above, SCF memory usage is 259780 bytes.

Program MP2 This program is designed to read in the transformed one- and two-electron integrals and the Fock orbital energies, after which it will compute the second-order Moller Plesset perturbation theory energy (MP2). With the program limitations described above, MP2 memory usage is 250056 bytes.

Program HAMILTON This program is designed to generate or read in a list of determinants. You can generate determinants for a CAS (Complete Active Space) of orbitals or you can input your own list of determinants. Next, if you wish, you may read in the one- and two-electron MO integrals and form a Hamiltonian matrix over the determinants. Finally, if you so choose, you may diagonalize the Hamiltonian matrix constructed over the determinants generated. With the program limitations described above, HAMILTON memory usage is 988784 bytes.

Program RW_INTS This program is designed to read the one- and two-electron AO integrals (in Dirac <12/12> convention) from user input and output them to disk in canonical order. There are no memory limitations associated with program RW_INTS.

QMLIB This is a library of subroutines and functions used by the QMIC programs.

"limits.h" This is an include file containing *all* the parameters that determine memory requirements for the QMIC programs.

Makefile There are a few versions of Makefiles available: a generic Makefile (Makefile.gnu), which works with Gnu make on a unix box; a Makefile (Makefile.486), which was used to make the programs on a 486 PC using other Gnu utilities such as "f2c", "gcc", etc.; and a Makefile (Makefile.mac), which was used on the Macintosh.

BasisLib This is a library that contains Gaussian atomic orbital basis sets for Hydrogen–Neon. The basis sets available to choose from are

1. STO3G by Hehre, Stewart, and Pople, *JCP, 51,* 2657 (1969).
2. 3-21G by Brinkley, Pople, and Hehre, *JACS, 102,* 939 (1980).
3. [3s2p] by Dunning and Hay in *Modern Theoretical Chemistry,* Vol. 3, Henry F. Schaefer, III, (ed.), Plenum Press, N.Y. (1977).

The QMIC software is broken up into the following folders (directories); see the diagram opposite.

Source This folder (directory) contains all Fortran source code, include files, Makefiles, and the master copy of the basis set theory.

Execs This folder (directory) contains all the executables as well as the basis set library file accessed by the "integral" executable (BasisLib). The executables are stored as a self-extracting archive file. The executables require about 1.3Mbytes and cannot, once extracted, be held on a floppy disk (therefore, copy the files to a hard disk before extracting).

Examples This folder (directory) contains input and associated output examples.

```
|------- Doc (potential contributed teaching material)
           |
           |          Source
           |          /
           |          /
           |/
           |/
|/
QMIC --- Examples
           |\
           |\         RS6000
           |          \        /
           |          \!/
|          \          /
           |          Execs ------ Mac
           |          |          \
           |          |          \
           |          |          \
par        |          |          PC486
           |          |
           |          |
           |          |
           |          |
           |          |--- Other platforms as
           |          requested and available
           |
           |
           |------Readme.1st, Readme.2nd
```

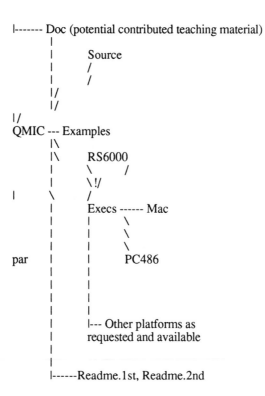

APPENDIX

I Useful Information and Data

I. INTEGRALS AND IDENTITIES

Frequently Encountered Integrals

$$\int_0^\infty e^{-\alpha x^2} dx = \frac{1}{2}\sqrt{\pi/\alpha} \tag{1}$$

$$\int_0^\infty x^{2n} e^{-\alpha x^2} dx = \frac{1 \cdot 3 \cdot 5 \cdot 7 \cdot \ldots (2n-1)}{2^{n+1}\alpha^n} \sqrt{\pi/\alpha} \quad n = 1, 2, \ldots \tag{2}$$

$$\int_0^\infty x^{2n+1} e^{-\alpha x^2} dx = \frac{n!}{2\alpha^{n+1}} \quad n = 0, 1, 2, \ldots \quad 0! \equiv 1 \tag{3}$$

$$\int_0^\infty x^n e^{-\alpha x} dx = \frac{n!}{\alpha^{n+1}} \quad n = 0, 1, 2, \ldots \tag{4}$$

$$\int u\,dv = uv - \int v\,du, \quad \text{Integration by Parts} \tag{5}$$

$$\int u^2 e^{-u^2}\,du = -\frac{1}{2}\,ue^{-u^2} + \frac{1}{2}\int e^{-u^2}\,du \tag{6}$$

$$\frac{2}{\sqrt{\pi}}\int_0^x e^{-t^2}\,dt = erf(x). \quad \text{The } erf(x) \text{ function is called the Error Function} \tag{7}$$

Brief Table of the Error Function

$erf(0) = 0$
$erf(1/2) = 0.5205$
$erf(1) = 0.8427$
$erf(2) = 0.9953$
$erf(\infty) = 1.0$

Integrals Involving Trigonometric Functions

$$\int \sin x\,dx = -\cos x \tag{8}$$

$$\int \cos x\,dx = \sin x \tag{9}$$

$$\int \sin^2 x\,dx = -\frac{1}{4}\sin 2x + \frac{x}{2} \tag{10}$$

$$\int \cos^2 x\,dx = \frac{1}{4}\sin 2x + \frac{x}{2} \tag{11}$$

$$\int \sin^3 x\,dx = \frac{1}{12}\cos 3x - \frac{3}{4}\cos x \tag{12}$$

$$\int \cos^3 x\,dx = \frac{1}{12}\sin 3x + \frac{3}{4}\sin x \tag{13}$$

$$\int x \sin ax\,dx = \frac{1}{a^2}\sin ax - \frac{x}{a}\cos ax \tag{14}$$

$$\int \sin\theta\cos\theta\,d\theta = -\frac{1}{2}\cos^2\theta \tag{15}$$

$$\int x^2 \sin ax\,dx = \frac{2x}{a^2}\sin ax - \frac{a^2x^2 - 2}{a^3}\cos ax \tag{16}$$

$$\int \sin\theta\cos^2\theta\,d\theta = -\frac{1}{3}\cos^3\theta \tag{17}$$

$$\int x\sin^2 ax\,dx = \frac{x^2}{4} - \frac{x\sin 2ax}{4a} - \frac{\cos 2ax}{8a^2} \tag{18}$$

$$\int x^2\sin^2 ax\,dx = \frac{x^3}{6} - \left(\frac{x^2}{4a} - \frac{1}{8a^3}\right)\sin 2ax - \frac{x\cos 2ax}{4a^2} \tag{19}$$

Trigonometric Identities

$$\sin\alpha\sin\beta = \frac{1}{2}\cos(\alpha-\beta) - \frac{1}{2}\cos(\alpha+\beta) \tag{20}$$

$$\cos\alpha\cos\beta = \frac{1}{2}\cos(\alpha-\beta) + \frac{1}{2}\cos(\alpha+\beta) \tag{21}$$

$$\sin\alpha\cos\beta = \frac{1}{2}\sin(\alpha-\beta) + \frac{1}{2}\sin(\alpha+\beta) \quad (22)$$

$$\sin(\alpha\pm\beta) = \sin\alpha\cos\beta \pm \cos\alpha\sin\beta \quad (23)$$

$$\cos(\alpha\pm\beta) = \cos\alpha\cos\beta - (\pm)\sin\alpha\sin\beta \quad (24)$$

$$e^{\pm i\theta} = \cos\theta \pm i\sin\theta \quad (25)$$

$$\cos\theta = \frac{e^{i\theta}+e^{-i\theta}}{2} \quad (26)$$

$$\sin\theta = \frac{e^{i\theta}-e^{-i\theta}}{2i} \quad (27)$$

Series Expansions

$$f(x) = f(a) + f'(a)(x-a) + \frac{1}{2!}f''(a)(x-a)^2 + \frac{1}{3!}f'''(a)(x-a)^3 + \ldots \quad (28)$$

$$e^x = 1 + x + \frac{x^2}{2!} + \frac{x^3}{3!} + \frac{x^4}{4!} + \ldots \quad (29)$$

$$\cos x = 1 - \frac{x^2}{2!} + \frac{x^4}{4!} - \frac{x^6}{6!} + \ldots \quad (30)$$

$$\sin x = x - \frac{x^3}{3!} + \frac{x^5}{5!} - \frac{x^7}{7!} + \ldots \quad (31)$$

$$\frac{1}{1-x} = 1 + x + x^2 + x^3 + x^4 + \ldots, \quad x^2 < 1 \quad (32)$$

$$(1 \pm x)^n = 1 \pm nx + \frac{n(n-1)}{2!}x^2 \pm \frac{n(n-1)(n-2)}{3!}x^3 + \ldots, \quad x^2<1 \quad (33)$$

II. COMMONLY USED UNITS

Unit	Symbol	SI Value
Angstrom	Å	$10^{-10}m$
Micron	μ	$10^{-6}m$
Calorie	cal	4.184 J

Atomic Units: 1 bohr unit of length = 0.529 Å; 1 Hartree unit of energy = 27.21 eV = 627.51 kcal/mol (1 eV = 8067 cm^{-1} = 23.06 kcal/mol)

III. CONVERSION FACTORS FOR ENERGY UNITS

	joule	**$kJ \cdot mol^{-1}$**	**eV**	***au***	**cm^{-1}**	**Hz**
joule	1	6.022×10^{20}	6.242×10^{18}	2.2939×10^{17}	5.035×10^{22}	1.509×10^{33}
$kJ \cdot \mathrm{mol}^{-1}$	1.6661×10^{-21}	1	1.036×10^{-2}	3.089×10^{-4}	83.60	2.506×10^{12}
eV	1.602×10^{-19}	96.48	1	3.675×10^{-2}	8065	2.418×10^{14}
au	4.359×10^{-18}	2625	27.21	1	2.195×10^{5}	6.580×10^{15}
cm^{-1}	1.986×10^{-23}	1.196×10^{-2}	1.240×10^{-4}	4.556×10^{-6}	1	2.998×10^{10}
Hz	6.626×10^{-34}	3.990×10^{-13}	4.136×10^{-15}	1.520×10^{-16}	3.336×10^{-11}	1

IV. ATOMIC UNITS AND THEIR SI EQUIVALENTS

Quantity	Atomic Unit	SI Equivalent
Mass	$m = 1$ (electron mass)	9.1091×10^{-31}kg
Charge	$\|e\| = 1$ (electron charge)	$1.6021 \times 10^{-19}\ C$ $= 1.52 \times 10^{-14} \sqrt{J_m}$
Angular momentum	$\hbar = 1$	$1.0545 \times 10^{-34} J \cdot s$
Length	$a_o = 1(bohr)$ (Bohr radius)	$5.29177 \times 10^{-11} m$
Energy	$= e^2/a_o = 1$ (hartree) (twice the ionization energy of atomic hydrogen)	$4.35944 \times 10^{-18} J$
Time	$\hbar^3/me^4 = 1$ (period of an electron in the first Bohr orbit)	$2.41889 \times 10^{-17} s$
Speed	$e^2/\hbar = 1$ (speed of an electron in the first Bohr orbit)	$2.18764 \times 10^{6} m \cdot s^{-1}$

V. VALUES OF SOME PHYSICAL CONSTANTS

Constant	Symbol	Value
Avogadro's number	N_o	6.02205×10^{23} mol^{-1}
Proton charge	e	$1.60219 \times 10^{-19}\ C$
Planck's constant	h $\hbar$	$6.62618 \times 10^{-34} J \cdot s$ $1.05459 \times 10^{-34} J \cdot s$
Speed of light in vacuum	c	$2.997925 \times 10^{8} m \cdot s^{-1}$
Atomic mass unit	amu	1.66056×10^{-27}kg
Electron rest mass	m_e	9.10953×10^{-31}kg
Proton rest mass	m_p	1.67265×10^{-27}kg
Boltzmann constant	k_b	$1.38066 \times 10^{-23} J \cdot K^{-1}$ $0.69509\ cm^{-1}$
Molar gas constant	R	$8.31441\ J \cdot K^{-1} \cdot$ mol^{-1}
Rydberg constant (for infinite nuclear mass)	R_∞	$2.179914 \times 10^{-23} J$ $1.097373\ cm^{-1}$
First Bohr radius	a_o	$5.29177 \times 10^{-11}\ m$

Index

A and *B* rate coefficients, 323–324
Ab initio
  defined, 401
  methods, 156, 448
Action, 10–12
  on a closed path, 11
  equation, 10–11
Angular momenta. *See also* Spin angular momentum
  coupling, 575–580
  equivalent orbital term symbols, 208–209
  when Hamiltonian commutes with angular momentum operators, 580–581
  when Hamiltonian contains angular momentum operators, 581–591
  non-equivalent orbital term symbols, 207–208
  non-vector coupling, 207
  orbital, 569–570
  properties of, 570–575
  vector coupling, 206–207
Anharmonicity, 291
  Birge-Sponer extrapolation, 292–293
  expansion of *E(v)*, 291–292
Antisymmetry. *See also* Symmetry
  antisymmetrizer operator, 198–199
  general concepts, 198–199
  physical consequences, 200–201
  Slater determinant, 199
Approximation methods
  perturbation theory, 39–41
  variational method, 37–39
Approximations. *See also* Born-Oppenheimer approximation
  ab initio approach, 401, 448–449
  configurations, 402
  electron correlation, 402–403
  empirical and semi-empirical, 448–449
  mean-field potentials, 401, 402
  orbital energies, 402
  orbitals, 402
  spin-orbitals, 402
  zero-differential overlap, 561
Associated Legendre polynomials, 17
Asymmetric tops, 285
Atomic orbitals, 125, 212–213. *See also* Molecular orbitals, One-electron model
  basis set libraries, 412
  Brillouin theorem, 420–421
  and configuration of electrons, 189
  core and valence basis, 413–415
  diffuse functions, 416–417
  directions, 126
  and electronic configurations, 197–198
  equivalent and non-equivalent term symbols, 207–209
  Gaussian-type, 411–415
  and Hückel model, 156–159, 558–561
  interaction topology, 135–138
  Koopmans' theorem, 419–420
  orbital angular momentum, 195
  orbital energies, 418–419
  orbital energies and total energy, 420

Atomic orbitals (*cont'd*)
  point-group symmetry operators, 195
  polarization functions, 416
  and Roothaan process, 417–418
  shapes, 125–126
  sizes and energies, 126–127
  Slater-type, 411–415
  spin angular momentum, 195
  symmetry, 138–140, 146–147
Atomic units, 403–404, 602
Atomic valence, 3
Aufbau principle, 3

Basis
  diffuse, 416–417
  double-zeta, 413
  minimal, 413
  set libraries, 412
  triple-zeta, 413
  valence, 413–415
Birge-Sponer extrapolation, 292–293
Black body radiators, 323–324
BO approximation. *See* Born-Oppenheimer approximation
Bohr radius, 19–20
Bohr-Sommerfeld quantization condition, 12
Boltzmann population, 341–342. *See also* Maxwell-Boltzmann velocity distribution
Bonding, 238–239, 249
  and configuration correlation diagrams, 244–246
  and configuration state functions, 240–244
  heterolytic bond cleavage, 239–240
  homolytic bond cleavage, 239
  and orbital diagrams, 241–242
  and state correlation diagrams, 246–249
Born-Oppenheimer approximation, 43–45, 357, 359
  and vibration/rotation states, 45
  time scale separation, 45
Bound states, 358
  and energy levels, 10
  and one dimensional scattering, 360–362
Boundary conditions
  and free-particle motion, 9–10
Box method, 208–209
Bra-ket notation, 517–518
Brillouin theorem, 420–421

Cartesian coordinates, 4, 5
  mass-weighted, 287
  and spherical, 508
Cartesian Gaussian-type orbitals. *See* Gaussian-type orbitals
CAS methods. *See* Complete-active-space methods
CC method. *See* Coupled-cluster method
CCDs. *See* Configuration correlation diagrams
CGTOs. *See* Gaussian-type orbitals
CI. *See* Configuration interaction
CI method. *See* Configuration interaction method
Classical trajectories, 375–377
Closed path, 11
Coherent state wavepackets, 379–381
Commutation of operators, 521–526
Commuting, 30–33
  mutually, 31, 33
Complete-active-space methods, 427
Complex numbers, 500–501
Configuration correlation diagrams, 235–236
  and bonding, 244–246
Configuration interaction, 192–194
  dynamical, 250–251
  essential, 249–250
Configuration interaction method, 425, 427–429, 431–433
Configuration state functions, 225–226, 240–244. *See also* Slater-Condon rules
  complete-active-space methods, 427
  couplings differing by two spin-orbitals and one spin-orbital, 236
  list choices, 427
  multiple, 423–424
Configurations, 402
Conversion factors, 602
Coordinates
  cartesian, 4, 5, 287
  cartesian and spherical, 508
  curvilinear, 5
  spherical, 508–510
Core orbitals, 128–129, 138
Coupled-cluster method, 425–426, 429–430, 434–435
CSFs. *See* Configuration state functions
Curvilinear coordinates, 5

Density functional methods, 435–438
Density matrices, 407–408
Diatomic molecules
  rotation, 21–22, 45–48
  vibration, 45–48
Diffuse functions, 416–417
Dipole moment, 441–442
  derivatives, 331
Dirac notation, 228. *See also* Bra-ket notation
Doppler broadening, 349–350
Doppler shift, 349–350
Double-zeta basis, 413
Dynamical electron correlation, 251

Eigenfunctions, 5, 7, 519–520
complete sets, 30–33
and experiments, 520–521
rotation of rigid molecules, 282–283, 284–285
Eigenstate
and measurement process, 27–28
Eigenvalues, 4, 5, 7, 519–520
and boundary conditions, 9–10
and experimental measurement, 27
of Hermitian matrices, 488–493
rotation of rigid molecules, 282–283, 284–285
of square matrices, 481–488
Eigenvectors
of Hermitian matrices, 488–493
of square matrices, 481–488
Einstein *A* and *B* rate coefficients, 323–324
Electron attachment, 125
Electron correlation, 402–403
Electronic configurations, 197–198
Electronic energies, 44
Electronic wavefunctions, 44
configuration interaction, 192–194
and fluctuation potential, 191
and mean-field model, 189–192
and SCF potential, 191
and symmetry and angular momentum operators, 194–195
Electronic-vibration-rotation transitions, 335–340
and time correlation functions, 346–347
Electrons
configuration in orbitals, 189
spatial correlations, 194
spin angular momentum, 204–206
Energy
conversion factors for units of, 602
corrections, 40
degeneracy, 126
expression, 405
Energy levels
for bound states, 10
and quantized action, 10–12
Equilibrium, 323–324
Equilibrium probability, 341–342
Ethylene
spin states, 200–201
Expansion coefficients, 28–30
Expectation value, 35

Fermi-Wentzel "golden rule," 314–317
First overtone transitions, 331
Fluctuation potential, 191
Fock equations, 406–407. *See also* Hartree-Fock equations
Fock matrices, 561–564
Fourier series, 501–508
Fourier transforms, 499–500, 504–508
Franck-Condon factors, 336–338
Free-particle motion in two dimensions
bound states and energy levels, 10
boundary conditions, 9–10
example illustrating use of basic rules, 34–36
quantized action and energy levels, 10–12
and Schrödinger equation, 8
Fundamental transitions, 331

Gaussian peaks, 350, 352
Gaussian-type orbitals, 411–412
basis set libraries, 412
core and valence basis, 413–415
Gradients, 444–445
GTOs. *See* Gaussian-type orbitals

Hamiltonian operator, 5, 25, 281–282
and atomic orbitals, 127
and atomic symmetry, 146–147
defined, 6
and molecular symmetry, 140–141, 144
and perturbation theory, 39, 312–313
in variational method, 37–39
Harmonic oscillator, 47
Harmonic potential, 47, 286
Harmonic vibrational motion, 22–24. *See also* Anharmonicity
frequencies, 446–447
normal mode, 287–288
polyatomic molecules, 286–291
Hartree-Fock equations
canonical, 408. *See also* Fock equations
unrestricted, 409–410
Heisenberg definition, 342
Heisenberg homogeneous broadening, 353–354
Hellmann-Feynman force operator, 442–443
Hermitian matrices
eigenvalues, 488–493
eigenvectors, 488–493
and turnover rule, 495–496
Hermitian operators, 27, 518–519
commutative, 521–526
in variational method, 37
Hessian matrix, 444–445, 446
Hückel model, 156–158, 558–559
extended, 158–159, 559–561
Hybrid orbitals, 133
interaction topology, 135–138
and symmetry, 145–146
Hydrogen atom orbitals, 513–515
Hydrogen orbitals model. *See* One-electron model

Inhomogeneous broadening, 354–356
  and hole burning, 355–356
Integral transformations, 426–427
Integrals, 599–601
Inversion symmetry, 211–212

Jacobians, 508–509

Koopmans' theorem, 129, 419–420

Laguerre polynomial, 20
LCAO-MO expansion, 410–411
  in modeling orbitals, 155–156
LCAO-MO process, 127, 128, 129
Legendre polynomials, 17
  Associated, 17
Levels, 321–322
Lewis-base donation, 125
Lifetime broadening, 353–354
Line broadening, 347–349. *See also* Line shape
  Doppler broadening, 349–350
  Heisenberg homogeneous broadening, 353–354
  lifetime broadening, 353–354
  pressure broadening, 350–352
  rotational diffusion broadening, 352–353
  site inhomogeneous broadening, 354–356
Line shape, 343. *See also* Line broadening
  Gaussian, 350, 352
  Lorentzian, 351, 352
  spectral, 343
  Voight shape, 352
Linear molecules, 216–217. *See also* Non-linear mole- cules
  configuration wavefunctions, 214–215
  equivalent orbital term symbols, 213–214
  inversion symmetry, 215
  non-equivalent orbital term symbols, 213
  reflection symmetry, 215–216
  rotation, 48, 281–283, 327–329, 333–335
  rotational selection rules for electronic transitions, 340
  symmetry, 144–146
Linear-combination-of-atomic-orbital-molecular-orbital process. *See* LCAO-MO process
Linear variational calculations, 38
Lorentzian peaks, 351, 352

Many-body perturbation method. *See* Møller-Plesset perturbation method
Matrices, 477–480
  finding inverses, square roots, etc., 494
  Hermitian, 488–493, 495–496
  as linear operators, 480
  projector, 495
  representations of functions and operators, 497–499
  square, 480–481
  and vectors, 477–479
Maxwell-Boltzmann velocity distribution, 349–350. *See also* Boltzmann population
MBPT method (many-body perturbation theory method). *See* Møller-Plesset perturbation method
MCSCF method. *See* Multiconfigurational self-consistent field method
Mean-field model, 189–190, 402
  corrections to describe instantaneous Coulombic interactions, 423–424
  lack of accuracy, 190–192
  and SCF potential, 191–192
Mean-field potentials, 401, 402
Microscopic reversibility, 321
Minimal basis, 413
Models. *See also* Hückel model, Mean-field model, One-electron model, Particle-in-a-box model
  qualitative and quantitative, 403
Molecular mechanics, 377
Molecular orbitals, 127–128. *See also* Atomic orbitals
  bonding, nonbonding, and antibonding, 127, 129–131, 135, 138
  core orbitals, 128–129, 138
  and electronic configurations, 197–198
  hybrid orbitals, 133
  interaction topology, 135–138
  multicenter orbitals, 132
  and multiconfiguration wavefunctions, 405–408
  Rydberg orbitals, 131–132
  symmetry, 138–140
  symmetry of linear molecules, 144–146
  symmetry of nonlinear molecules, 140–144
  valence orbitals, 129–131
Molecules. *See also* Linear molecules, Non-linear molecules
  calculating non-energy properties, 439–441
Morse oscillator, 47–48
Møller-Plesset perturbation method, 425, 429–430, 433–434
  and time-independent perturbation theory, 531–532
MPPT method. *See* Møller-Plesset perturbation method
Mulliken notation, 228–229
Multicenter orbitals, 132
  interaction topology, 137

Multichannel dynamics, 367–369
  chemical relevance, 371–374
  coupled channel equations, 369–370
  perturbative treatment, 370–371
Multiconfigurational self-consistent field method, 424–425, 427–429, 431
  and dipole moment, 441–442
  and geometrical forces, 442–444

Newton equations
  of motion for vibration, 286–287
Non-linear molecules. *See also* Linear molecules
  degenerate representations, 219–223
  rotation, 48–50, 283–285, 330
  rotational selection rules for electronic transitions, 340
  symmetry, 140–144
  term symbols for non-degenerate point group symmetries, 217–218
  wavefunctions, 218–219
Nuclear motion, 357–358
  chemical relevance of multichannel dynamics, 371–374
  classical trajectories, 375–377
  classical treatment of, 374–379
  coherent state wavepackets, 379–381
  coupled channel equations, 369–370
  final conditions, 378–379
  initial conditions, 377–378
  multichannel problems, 367–374
  one dimensional scattering, 359–367
  perturbative treatment, 370–371
  scattering, 358

OCDs. *See* Orbital correlation diagrams
One dimensional scattering, 359
  and bound states, 360–362
  scattering states, 362–363
  shape resonance states, 363–367
One-electron model, 14–15, 20–21
  phi equation, 15–16
  radial equation, 17–20
  theta equation, 16–17
Operators, 4–5
  commuting, 30–33
  converting from cartesian to spherical, 509–510
Orbital angular momentum, 195
Orbital correlation diagrams, 151–154
Orbital diagrams, 241–242
Orbital energies, 402, 418–419
  Koopmans' theorem, 419–420
  and total energy, 420
Orbitals, 402
Orthonormal functions, 496–497
Orthonormal vectors, 496–497

Pariser-Parr-Pople method, 564–565
Particle-in-a-box model, 10–12
  in one dimension, 13–14
  in three dimensions, 12–13
Pauli principle, 12, 199
Periodic table, 3
Perturbation theory, 39–41, 311. *See also* Time-dependent perturbation theory, Time-independent perturbation theory
  electric and magnetic fields, 312
  and Hamiltonian operator, 39, 312–313
  time-dependent vector potential, 311–312
  time-independent, 527–533
Photons
  absorption, 125, 321–324, 341–343
  emission, 321–324, 342–343
Physical constants, 603
π-electron energies, 13–14
Point group symmetry, 217–218, 289–291, 546–547, 550
  ammonia, 535–536
  and electronic transition dipole, 335–336
  and matrices, 537–539
  character tables, 550–556
  direct products in *N*-electron wavefunctions, 548–549
  direct products in selection rules, 549–550
  nitrogen, 543–545
  operators, 195
  projector operators, 545–546
  reducible and irreducible representations, 539–543
Polarization functions, 416
Polyatomic molecules
  rotation, 48–50, 281–285
Potential, 156
  extended Hückel method, 158–159, 559–561
  Hückel parameterization, 156–158
PPP method. *See* Pariser-Parr-Pople method
Pressure broadening, 350–352
Properties
  average values, 33–34

Quantum mechanics
  basic rules, 24–34
  relationship of rules to experimental measurements, 24–25, 34
Quantum Mechanics in Chemistry programs, 593–597

Radiationless transitions, 371
Rate coefficients, 323–324
Reaction paths, 447–448
  orbital correlation diagrams, 151–154
  and reduction in symmetry, 149–151
Rigid diatomic molecules
  rotational motion, 21–22
Rigid rotor, 47, 282
Roothaan matrix process, 417–418
Rotation, 44
  and Born-Oppenheimer approximation, 45
  of diatomic molecules, 45–48
  harmonic oscillator, 47
  of linear molecules, 48, 281–283, 327–329
  of non-linear molecules, 48–50, 283–285, 330
  of polyatomic molecules, 48–50, 281–285
  of rigid diatomic molecules, 21–22
  rigid rotor, 47, 282
Rotational diffusion broadening, 352–353
Rotational transitions, 325–330
  and time correlation functions, 343–345
Rydberg orbitals, 131–132
  defined, 131

Saturated transitions, 323
Scattering, 358
  one dimensional, 359–367
  states, 362–363
SCDs. *See* State correlation diagrams
SCF method, 427–429
SCF potential, 191
  Roothaan matrix process, 417–418
Schrödinger equation, 3, 281
  and approximations, 401
  basic, 6
  defined, 6
  and free-particle motion, 8
  and free-particle motion in two dimensions, 8
  and harmonic vibrational motion, 22–24
  and nuclear motion, 357–358
  and one-electron model, 14–21
  orbital-level, 155–156
  and quantized action, 10–12
  and rotational motion for a rigid diatomic molecule, 21–22
  and scattering, 359
  separation into electronic and nuclear motion aspects, 43–45, 50
  time-dependent, 5–6, 25
  in time-dependent perturbation theory, 313
  time-independent, 7–8, 25
Screening, 21

Secular equations, 406–407
Self-consistent field method. *See* SCF method
Self-Consistent Field potential. *See* SCF potential
Separation of variables, 510–512
Shape. *See also* Line shape
  atomic orbitals, 125–126
  resonance states, 363–367
Shell model of nuclei, 12
Site inhomogeneous broadening, 354–356
Slater-Condon rules, 226–229, 234
  example applications, 229–234
Slater determinants, 199, 240, 243
Slater-type orbitals, 411
  basis set libraries, 412
  core and valence basis, 413–415
SO method. *See* Symmetric orthonormalization method
Software (Quantum Mechanics in Chemistry programs), 593–597
Spectral line shape, 343
Spherical coordinates, 508
Spherical tops, 284
Spin angular momentum, 195
  of electrons, 204–206
Spin-orbitals, 402
  occupied, 408
  virtual, 408
Spontaneous emission, 322
Square matrices, 480–481
  eigenvalues, 481–488
  eigenvectors, 481–488
  special kinds, 481
State correlation diagrams, 236–238
  and bonding, 246–249
State-to-state rates, 321, 341
Stimulated emission, 322
STOs. *See* Slater-type orbitals
Symmetric orthonormalization method, 128
Symmetric tops, 284–285, 333
  oblate, 284
  prolate, 284
Symmetry. *See also* Antisymmetry
  angular momentum, 203–207
  atomic orbitals, 138–140, 146–147
  and Hamiltonian operator,140–141, 144, 146–147
  in harmonic vibrational motion, 288–291
  inversion symmetry, 211–212
  linear molecules, 144–146
  molecular orbitals, 138–140
  nonlinear molecules, 140–144
  operators, 127
  point group, 217–218, 289–291, 535–557

point-group operators, 195
and reaction potential energy, 153
reduction in, 149–151
symmetry-imposed barriers, 238

Taylor series expansion, 286
Time correlation functions, 342
and line shape, 343
Time-dependent perturbation theory
and *A* and *B* rate coefficients, 323–324
application to electromagnetic perturbations, 314–318
and black body radiators, 323–324
and dipole moment derivatives, 331
and electric dipole transitions, 318–319
and electric quadrupole transitions, 319–321
and electronic transition dipole, 335–336
and electronic transitions, 340
and electronic-vibration-rotation transitions, 335–340
and equilibrium, 323–324
and equilibrium probability, 341–342
and first-order Fermi-Wentzel "golden rule," 314–317
first overtone transitions, 331
and Franck-Condon factors, 336–338
fundamental transitions, 331
and higher-order results, 317–318
and line shape, 343
long-wavelength approximation, 318–321
and magnetic dipole transitions, 319–321
and microscopic reversibility, 321
perturbative solution, 313–314
and phenomenological rate laws, 321–322
photon absorption and emission, 321–324, 341–343
and point group symmetry, 335–336
and rotational transitions, 325–330, 343–345
and saturated transitions, 323
Schrödinger equation, 313
and spontaneous emission, 322
and state-to-state rates, 321, 341
and stimulated emission, 322
time correlation function expressions for transition rates, 341–356
and time correlation functions, 343–347
and transparency, 323
and vibration-rotation transitions, 330–335
and vibronic effects, 338–340
Time-independent perturbation theory, 527–528
conceptual use, 532–533
and Møller-Plesset perturbation series, 531–532
structure, 528–531

Trajectories
classical, 375–377
ensembles, 377–378
Transitions
electric dipole, 318–319
electric quadrupole, 319–321
electronic, 340
electronic-vibration-rotation, 335–340, 346–347
first overtone, 331
fundamental, 331
magnetic dipole, 319–321
radiationless, 371
rate coefficients, 445–446
rotational, 325–330, 343–345
saturated, 323
vibration-rotation, 330–335, 345–346
Translation, 44
Transparency, 323
Trial wavefunctions, 38
Triple-zeta basis, 413
Turnover rule, 495–496

Units, 601

V. *See* Potential
Valence electron methods, 565–567
Valences
atomic, 3
bond analysis, 201
orbitals, 129–131
Variational method, 37–39, 406
Vectors, 477–478
orthonormal, 496–497
Vibration, 44
and Born-Oppenheimer approximation, 45
of diatomic molecules, 45–48
Vibration-rotation transitions, 330–335
and time correlation functions, 345–346
Voight shape, 352

Wavefunctions. *See also* Electronic wavefunctions
antisymmetric 198–201
of atomic configuration, 210–211
calculating properties other than energy, 440–441, 444
configuration interaction method, 425, 431–433
configuration state functions, 225–226, 240–244. *See also* Slater-Condon rules
and configurations, 223–224
coupled-cluster method, 425–426, 429–430, 434–435
defined, 5
density functional methods, 435–438

Wavefunctions (*cont'd*)
electronic, 44
energy expression, 405
expansion coefficients, 28–30
first-order, 40
Fock equations, 406–407
gerade, 215
Hartree-Fock equations, 408–410
and LCAO-MO expansion, 410–411
of linear molecule configuration, 214–215
methods for determining "best," 424–426, 430
Møller-Plesset perturbation method, 425, 429–430, 433–434
multiconfigurational self-consistent field method, 424–425, 427–429, 431
for non-degenerate, non-linear point molecules, 218–219
one- and two-electron density matrices, 407–408
secular equations, 406–407
single-determinant, 408
time evolution, 25–27
trial, 38
ungerade, 215
using geometrical energy derivatives, 444–448
variational method, 406
X-alpha methods, 435–438
Wavepackets, 379–381

X-alpha methods, 435–438

ZDO approximation. *See* Zero-differential overlap approximation
Zero-differential overlap approximation, 561